COMPLEX ALGEBRAIC THREEFOLDS

The first book on the explicit birational geometry of complex algebraic threefolds arising from the minimal model program, this text is sure to become an essential reference in the field of birational geometry. Threefolds remain the interface between low- and high-dimensional settings, and a good understanding of them is necessary in this actively evolving area.

Intended for advanced graduate students as well as researchers working in birational geometry, the book is as self-contained as possible. Detailed proofs are given throughout, and more than 100 examples help to deepen understanding of birational geometry.

The first part of the book deals with threefold singularities, divisorial contractions and flips. After a thorough explanation of the Sarkisov program, the second part is devoted to the analysis of outputs, specifically minimal models and Mori fibre spaces. The latter are divided into conical fibrations, del Pezzo fibrations and Fano threefolds according to the relative dimension.

Masayuki Kawakita is Associate Professor at the Research Institute for Mathematical Sciences, Kyoto University. He has established a classification of threefold divisorial contractions and is a leading expert in algebraic threefolds.

Complex Algebraic Threefolds

MASAYUKI KAWAKITA

Kyoto University

CAMBRIDGE
UNIVERSITY PRESS

Shaftesbury Road, Cambridge CB2 8EA, United Kingdom

One Liberty Plaza, 20th Floor, New York, NY 10006, USA

477 Williamstown Road, Port Melbourne, VIC 3207, Australia

314–321, 3rd Floor, Plot 3, Splendor Forum, Jasola District Centre,
New Delhi – 110025, India

103 Penang Road, #05–06/07, Visioncrest Commercial, Singapore 238467

Cambridge University Press is part of Cambridge University Press & Assessment,
a department of the University of Cambridge.

We share the University's mission to contribute to society through the pursuit of
education, learning and research at the highest international levels of excellence.

www.cambridge.org
Information on this title: www.cambridge.org/9781108844239

DOI: 10.1017/9781108933988

First published 2024

Printed in the United Kingdom by CPI Group Ltd, Croydon CR0 4YY

A catalogue record for this publication is available from the British Library

*A Cataloging-in-Publication data record for this book is available from the Library
of Congress*

ISBN 978-1-108-84423-9 Hardback

Cambridge University Press & Assessment has no responsibility for the persistence
or accuracy of URLs for external or third-party internet websites referred to in this
publication and does not guarantee that any content on such websites is, or will
remain, accurate or appropriate.

For my family

Contents

Preface

The present book treats explicit aspects of the birational geometry of complex algebraic threefolds. It is a fundamental problem in birational geometry to find and analyse a good representative of each birational class of algebraic varieties. The minimal model program, or the MMP for short, conjecturally realises this by comparing the canonical divisors of varieties. According as the MMP has developed, the book has arisen from two perspectives.

Firstly, since the MMP in dimension three was established about a quarter century ago, it is desirable to understand individual threefolds explicitly by means of the MMP. The initial step is to describe birational transformations in the MMP. Now we have a practical classification of them in dimension three. The ensuing important subject is to analyse the threefolds output by the MMP. In this direction one can mention the Sarkisov program, which decomposes every birational map of Mori fibre spaces.

Secondly, the MMP in higher dimensions is still evolving actively after the existence of flips was proved, as typified in the settlement of the Borisov–Alexeev–Borisov conjecture. A good knowledge of threefolds is useful and will be necessary in further development of higher dimensional birational geometry. This is comparable to the nature that most results on threefolds are based upon the classical theory of surfaces.

The book concentrates on the explicit study of algebraic threefolds by the MMP. The author has tried to elucidate the proofs rigorously and to make the book as self-contained as possible. A number of examples will help to deepen the understanding of the reader. The reader is strongly encouraged to verify the computations in examples. Though it does not cover important topics such as affine geometry, derived categories and positive characteristic aspects, the book will supply enough knowledge of threefold birational geometry to enter the field of higher dimensional birational geometry.

The book is intended for advanced graduate students who are interested in the birational geometry of algebraic varieties. It can also be used by researchers as a reference for the classification results on threefolds. The reader should be familiar with basic algebraic geometry at the level of Hartshorne's textbook [178]. Some knowledge of the MMP is helpful but it is not a prerequisite. One can learn the general theory of the MMP from a standard book such as that by Kollár and Mori [277] or by Matsuki [307]. Whilst it roughly corresponds to Chapter 1 of the book, the main body starts from Chapter 2 and concentrates on threefolds.

The volume edited by Corti and Reid [97] is an outstanding collection from the same standpoint. It played a guiding role in the explicit study of algebraic threefolds when it was published. However, great progress has been made since then. The present book aims at an organised treatment of threefolds, including recent results. It seeks to be somewhat of a threefold version of the book on surfaces by Barth, Hulek, Peters and Van de Ven [30] or by Beauville [35].

Chapter 1 summarises the theory of the MMP in an arbitrary dimension but excludes detailed proofs. The first part, Chapters 2 to 5, of the main body deals with objects which appear in the course of the MMP. Chapter 2 classifies three-fold singularities in the MMP completely. Then Chapters 3, 4 and 5 describe threefold birational transformations in the MMP, that is, divisorial contractions and flips. They contain all the necessary arguments by omitting only the parts which repeat the preceding arguments.

The second part, Chapters 6 to 10, is devoted to the analysis of outputs of the MMP, which are Mori fibre spaces and minimal models. After Chapter 6 explains the general theory of the Sarkisov program, Chapters 7, 8 and 9 investigate the geometry of threefold Mori fibre spaces according to the relative dimension of the fibre structure. Finally Chapter 10 discusses minimal threefolds from the point of view of abundance.

The author would like to express thanks to all the colleagues with whom he held discussions. Yujiro Kawamata introduced him to the subject of birational geometry as his academic supervisor. Alessio Corti and Miles Reid communicated their profound knowledge of threefolds to him during his visit to the University of Cambridge. When he stayed at the Institute for Advanced Study, he received warm hospitality from János Kollár. He also learnt a great deal from the regular seminar organised by Shigefumi Mori, Shigeru Mukai and Noboru Nakayama. Finally he would like to thank Philip Meyler and John Linglei Meng at Cambridge University Press for their support for the publication.

1

The Minimal Model Program

This chapter outlines the general theory of the minimal model program. We shall study algebraic threefolds thoroughly in the subsequent chapters in alignment with the program. The reader who is not familiar with the program may grasp the basic notions at first and refer back later.

Blowing up a surface at a point is not an essential operation from the birational point of view. Its exceptional curve is characterised numerically as a (-1)-curve. As is the case in this observation, the intersection number is a basic linear tool in birational geometry. The minimal model program, or the MMP for short, outputs a representative of each birational class that is minimal with respect to the numerical class of the canonical divisor.

The MMP grew out of the surface theory with allowing mild singularities. For a given variety, it produces a minimal model or a Mori fibre space after finitely many birational transformations, which are divisorial contractions and flips. Now the program is formulated in the logarithmic framework where we treat a pair consisting of a variety and a divisor.

The MMP functions subject to the existence and termination of flips. Hacon and McKernan with Birkar and Cascini proved the existence of flips in an arbitrary dimension. Considering a flip to be the relative canonical model, they established the MMP with scaling in the birational setting. The termination of threefold flips follows from the decrease in the number of divisors with small log discrepancy. Shokurov reduced the termination in an arbitrary dimension to certain conjectural properties of the minimal log discrepancy.

It is also important to analyse the representative output by the MMP. The Sarkisov program decomposes a birational map of Mori fibre spaces into elementary ones. For a minimal model, we expect the abundance which claims the freedom of the linear system of a multiple of the canonical divisor. It defines a morphism to the projective variety associated with the canonical ring, which we know is finitely generated.

1.1 Preliminaries

We shall fix the notation and recall the fundamentals of algebraic geometry. The book [178] by Hartshorne is a standard reference.

The *natural numbers* begin with zero. The symbol $R_{\geq r}$ for $R = \mathbf{N}, \mathbf{Z}, \mathbf{Q}$ or $\mathbf{R}$ stands for the subset $\{x \in R \mid x \geq r\}$ and similarly $R_{>r} = \{x \in R \mid x > r\}$. For instance, $\mathbf{N} = \mathbf{Z}_{\geq 0}$. The quotient $\mathbf{Z}_r = \mathbf{Z}/r\mathbf{Z}$ is the *cyclic group* of order r. The *round-down* $\lfloor r \rfloor$ of a real number r is the greatest integer less than or equal to r, whilst the *round-up* $\lceil r \rceil$ is defined as $\lceil r \rceil = -\lfloor -r \rfloor$.

Schemes   A *scheme* is always assumed to be separated. It is said to be *integral* if it is irreducible and reduced.

We work over the field $\mathbf{C}$ of complex numbers unless otherwise mentioned. An *algebraic scheme* is a scheme of finite type over Spec k for the algebraically closed ground field k, which is tacitly assumed to be $\mathbf{C}$. We call it a *complex scheme* when we emphasise that it is defined over $\mathbf{C}$. An algebraic scheme is said to be *complete* if it is proper over Spec k. A *point* in an algebraic scheme usually means a closed point.

A *variety* is an integral algebraic scheme. A *complex variety* is a variety over $\mathbf{C}$. A *curve* is a variety of dimension one and a *surface* is a variety of dimension two. An *n-fold* is a variety of dimension n. The *affine space* $\mathbf{A}^n$ is Spec $k[x_1, \ldots, x_n]$ and the *projective space* $\mathbf{P}^n$ is Proj $k[x_0, \ldots, x_n]$. The origin of $\mathbf{A}^n$ is denoted by o.

The *germ* $x \in X$ of a scheme is considered at a closed point unless otherwise specified. It is an equivalence class of the pair (X, x) of a scheme X and a point x in X where (X, x) is equivalent to (X', x') if there exists an isomorphism $U \simeq U'$ of open neighbourhoods $x \in U \subset X$ and $x' \in U' \subset X'$ sending x to x'. By a *singularity*, we mean the germ at a singular point as a rule.

For a locally free coherent sheaf $\mathscr{E}$ on an algebraic scheme X, the *projective space bundle* $\mathbf{P}(\mathscr{E}) = \mathrm{Proj}_X S\mathscr{E}$ over X is defined by the symmetric $\mathscr{O}_X$-algebra $S\mathscr{E} = \bigoplus_{i \in \mathbf{N}} S^i \mathscr{E}$ of $\mathscr{E}$. It is a $\mathbf{P}^n$*-bundle* if $\mathscr{E}$ is of rank $n + 1$. In particular, the projective space $\mathbf{P}V = \mathrm{Proj} \, SV$ is defined for a finite dimensional vector space V. It is regarded as the quotient space $(V^\vee \setminus 0)/k^\times$ of the dual vector space $V^\vee$ minus zero by the action of the multiplicative group $k^\times = k \setminus \{0\}$ of the ground field k. As used above, the symbol $^\vee$ stands for the dual and $^\times$ for the group of units.

Morphisms   For a morphism $\pi \colon X \to Y$ of schemes, the *image* $\pi(A)$ of a closed subset A of X and the *inverse image* $\pi^{-1}(B)$ of a closed subset B of Y are considered set-theoretically. When π is proper and A is a closed subscheme, we regard $\pi(A)$ as a reduced scheme. We also regard $\pi^{-1}(B)$ for a closed subscheme B as a reduced scheme and distinguish it from the scheme-theoretic fibre $X \times_Y B$.

A *rational map* $f \colon X \dashrightarrow Y$ of algebraic schemes is an equivalence class of a morphism $U \to Y$ defined on a dense open subset U of X. The *image* $f(X)$ of f is the image $p(\Gamma)$ of the graph Γ of f as a closed subscheme of $X \times Y$ by the projection $p \colon X \times Y \to Y$. We say that a morphism or a rational map is *birational* if it has an inverse as a rational map. Two algebraic schemes are *birational* if there exists a birational map between them. By definition, two varieties are birational if and only if they have the same function field.

Let $\pi \colon X \to Y$ be a morphism of algebraic schemes. We say that π is *projective* if it is isomorphic to $\mathrm{Proj}_Y \, \mathscr{R} \to Y$ by a graded $\mathscr{O}_Y$-algebra $\mathscr{R} = \bigoplus_{i \in \mathbf{N}} \mathscr{R}_i$ generated by coherent $\mathscr{R}_1$, with $\mathscr{R}_0 = \mathscr{O}_Y$. When Y is quasi-projective, the projectivity of π means that it is realised as a closed subscheme of a relative projective space $\mathbf{P}^n \times Y \to Y$. An invertible sheaf $\mathscr{L}$ on X is *relatively very ample* (or *very ample over Y* or *π-very ample*) if it is isomorphic to $\mathscr{O}(1)$ by an expression $X \simeq \mathrm{Proj}_Y \, \mathscr{R}$ as above. We say that $\mathscr{L}$ is *relatively ample (π-ample)* if $\mathscr{L}^{\otimes a}$ is relatively very ample for some positive integer a.

Suppose that $\pi \colon X \to Y$ is proper. We say that *π has connected fibres* if the natural map $\mathscr{O}_Y \to \pi_* \mathscr{O}_X$ is an isomorphism. This implies that the fibre $X \times_Y y$ at every $y \in Y$ is connected and non-empty [160, III corollaire 4.3.2]. The proof for projective morphism is in [178, III corollary 11.3]. In general, π admits the *Stein factorisation* $\pi = g \circ f$ with $f \colon X \to Z$ and $g \colon Z \to Y$ defined by $Z = \mathrm{Spec}_Y \, \pi_* \mathscr{O}_X$, for which f is proper with connected fibres and g is finite. If π is a proper birational morphism from a variety to a normal variety, then the factor g in the Stein factorisation is an isomorphism and hence π has connected fibres. This is referred to as *Zariski's main theorem*.

Lemma 1.1.1 *Let $\pi \colon X \to Y$ and $\varphi \colon X \to Z$ be morphisms of algebraic schemes such that π is proper and has connected fibres. If every curve in X contracted to a point by π is also contracted by φ, then φ factors through π as $\varphi = f \circ \pi$ for a morphism $f \colon Y \to Z$.*

Proof Let Y^m and Z^m denote the sets of closed points in Y and Z respectively. For $y \in Y^m$, the inverse image $\pi^{-1}(y)$ is connected and $\varphi(\pi^{-1}(y))$ is one point. Define $f^m \colon Y^m \to Z^m$ by $f^m(y) = \varphi(\pi^{-1}(y))$. Since π is proper and surjective, for any closed subset B of Z, $\pi(\varphi^{-1}(B))$ is closed in Y and $(f^m)^{-1}(B|_{Z^m}) = \pi(\varphi^{-1}(B))|_{Y^m}$. Thus f^m extends to a continuous map $f \colon Y \to Z$, which is a morphism of schemes by the natural map $\mathscr{O}_Z \to \varphi_* \mathscr{O}_X = f_* \pi_* \mathscr{O}_X = f_* \mathscr{O}_Y$. $\quad\square$

Chow's lemma [160, II §5.6] replaces the proper morphism $\pi \colon X \to Y$ by a projective morphism. It asserts the existence of a projective birational morphism $\mu \colon X' \to X$ such that $\pi \circ \mu \colon X' \to Y$ is projective. The *projection formula* and the *Leray spectral sequence*, formulated for ringed spaces in [160, 0 §12.2], will be frequently used. The reference [198, section 3.6] explains spectral sequences from our perspective.

Theorem 1.1.2 (Projection formula) *Let* $\pi \colon X \to Y$ *be a morphism of ringed spaces. Let* $\mathscr{F}$ *be an* $\mathscr{O}_X$-*module and let* $\mathscr{E}$ *be a finite locally free* $\mathscr{O}_Y$-*module. Then there exists a natural isomorphism* $R^i\pi_*\mathscr{F} \otimes \mathscr{E} \simeq R^i\pi_*(\mathscr{F} \otimes \pi^*\mathscr{E})$.

Theorem 1.1.3 (Leray spectral sequence) *Let* $f \colon X \to Y$ *and* $g \colon Y \to Z$ *be morphisms of ringed spaces. Let* $\mathscr{F}$ *be an* $\mathscr{O}_X$-*module. Then there exists a spectral sequence*

$$E_2^{p,q} = R^p g_* R^q f_* \mathscr{F} \Rightarrow E^{p+q} = R^{p+q}(g \circ f)_* \mathscr{F}.$$

In practice for a spectral sequence $E_2^{p,q} \Rightarrow E^{p+q}$, we assume that $E_2^{p,q}$ is zero whenever p or q is negative. Then there exists an exact sequence

$$0 \to E_2^{1,0} \to E^1 \to E_2^{0,1} \to E_2^{2,0} \to E^2.$$

If further $E_2^{p,q} = 0$ for all $p \geq 0$ and $q \geq 1$, then $E_2^{p,0} \simeq E^p$. Likewise if $E_2^{p,q} = 0$ for all $p \geq 1$ and $q \geq 0$, then $E_2^{0,q} \simeq E^q$.

Cohomologies We write $H^i(\mathscr{F})$ for the cohomology $H^i(X, \mathscr{F})$ of a sheaf $\mathscr{F}$ of abelian groups on a topological space X when there is no confusion. If X is noetherian, then $H^i(\mathscr{F})$ vanishes for all i greater than the dimension of X.

Let $\mathscr{F}$ be a coherent sheaf on an algebraic scheme X. If X is affine, then $H^i(\mathscr{F}) = 0$ for all $i \geq 1$. If $\pi \colon X \to Y$ is a proper morphism, then the higher direct image $R^i\pi_*\mathscr{F}$ is coherent [160, III théorème 3.2.1]. In particular if X is complete, then $H^i(\mathscr{F})$ is a finite dimensional vector space. The dimension of $H^i(\mathscr{F})$ is denoted by $h^i(\mathscr{F})$. The alternating sum $\chi(\mathscr{F}) = \sum_{i \in \mathbf{N}}(-1)^i h^i(\mathscr{F})$ is called the *Euler characteristic* of $\mathscr{F}$.

Let X be a complete scheme of dimension n. For a coherent sheaf $\mathscr{F}$ and an invertible sheaf $\mathscr{L}$ on X, the *asymptotic Riemann–Roch theorem* defines the *intersection number* $(\mathscr{L}^n \cdot \mathscr{F}) \in \mathbf{Z}$ by the expression

$$\chi(\mathscr{L}^{\otimes l} \otimes \mathscr{F}) = \frac{(\mathscr{L}^n \cdot \mathscr{F})}{n!} l^n + O(l^{n-1}),$$

where by *Landau's symbol* O, $f(l) = O(g(l))$ means the existence of a constant c such that $|f(l)| \leq c|g(l)|$ for any large l. By this, Grothendieck's *dévissage* yields the estimate $h^i(\mathscr{F} \otimes \mathscr{L}^{\otimes l}) = O(l^n)$ for all i [266, section VI.2].

If X is projective with a very ample sheaf $\mathscr{O}_X(1)$, then the Euler characteristic $\chi(\mathscr{F} \otimes \mathscr{O}_X(l))$ is described as a polynomial in $\mathbf{Q}[l]$, called the *Hilbert polynomial* of $\mathscr{F}$. The vanishing of $H^i(\mathscr{F} \otimes \mathscr{O}_X(l))$ below is known as *Serre vanishing*.

Theorem 1.1.4 (Serre) *Let $\mathscr{F}$ be a coherent sheaf on a projective scheme X. Then for any sufficiently large integer l, the twisted sheaf $\mathscr{F} \otimes \mathscr{O}_X(l)$ is generated by global sections and satisfies $H^i(\mathscr{F} \otimes \mathscr{O}_X(l)) = 0$ for all $i \geq 1$.*

We have the *cohomology and base change theorem* for flat families of coherent sheaves [160, III §§7.6–7.9], [361, section 5]. See also [178, section III.12].

Theorem 1.1.5 (Cohomology and base change) *Let $\pi \colon X \to T$ be a proper morphism of algebraic schemes. Let $\mathscr{F}$ be a coherent sheaf on X flat over T. Take the restriction $\mathscr{F}_t$ of $\mathscr{F}$ to the fibre $X_t = X \times_T t$ at a closed point t in T and consider the natural map*

$$\alpha_t^i \colon R^i\pi_*\mathscr{F} \otimes k(t) \to H^i(X_t, \mathscr{F}_t),$$

where $k(t)$ is the skyscraper sheaf of the residue field at t.

  (i) *The dimension $h^i(\mathscr{F}_t)$ is upper semi-continuous on T and the Euler characteristic $\chi(\mathscr{F}_t)$ is locally constant on T.*
 (ii) *Fix i and t and suppose that α_t^i is surjective. Then $\alpha_{t'}^i$ is an isomorphism for all t' in a neighbourhood at t. Further, $R^i\pi_*\mathscr{F}$ is locally free at t if and only if α_t^{i-1} is surjective.*
(iii) (Grauert) *Suppose that T is reduced. Fix i. If $h^i(\mathscr{F}_t)$ is locally constant, then $R^i\pi_*\mathscr{F}$ is locally free and α_t^i is an isomorphism.*

Divisors Let X be an algebraic scheme. We write $\mathscr{K}_X$ for the sheaf of total quotient rings of $\mathscr{O}_X$. If X is a variety, then it is the constant sheaf of the function field $K(X)$ of X. A *Cartier divisor* D on X is a global section of the quotient sheaf $\mathscr{K}_X^\times / \mathscr{O}_X^\times$ of multiplicative groups of units. It is associated with an invertible subsheaf $\mathscr{O}_X(D)$ of $\mathscr{K}_X$. If D is represented by local sections $f_i \in \mathscr{K}_{U_i}^\times$ with $f_i f_j^{-1} \in \mathscr{O}_{U_i \cap U_j}^\times$, then $\mathscr{O}_X(D)|_{U_i} = f_i^{-1}\mathscr{O}_{U_i}$. We say that D is *principal* if it is defined by a global section of $\mathscr{K}_X^\times$ or equivalently $\mathscr{O}_X(D) \simeq \mathscr{O}_X$. The principal divisor given by $f \in \Gamma(X, \mathscr{K}_X^\times)$ is denoted by $(f)_X$. If f_i belongs to $\mathscr{O}_{U_i} \cap \mathscr{K}_{U_i}^\times$ for all i, then D defines a closed subscheme of X and we say that D is *effective*.

The *Picard group* Pic X of X is the group of isomorphism classes of invertible sheaves on X. It has an isomorphism

$$\text{Pic } X \simeq H^1(\mathscr{O}_X^\times).$$

In fact this holds for any ringed space via Čech cohomology. The proof is found in [440, section 5.4]. The isomorphism for a variety X is derived at once from the vanishing of $H^1(\mathscr{K}_X^\times)$ for the flasque sheaf $\mathscr{K}_X^\times$.

By *Serre's criterion*, an algebraic scheme X is normal if and only if it satisfies the conditions R_1 and S_2 defined as

(R_i) for any $\eta \in X$, $\mathscr{O}_{X,\eta}$ is regular if $\mathscr{O}_{X,\eta}$ is of dimension at most i and
(S_i) for any $\eta \in X$, $\mathscr{O}_{X,\eta}$ is Cohen–Macaulay if $\mathscr{O}_{X,\eta}$ is of depth less than i,

in which we consider scheme-theoretic points $\eta \in X$. Let X be a normal variety. A closed subvariety of codimension one in X is called a *prime divisor*. A *Weil divisor* D on X, or simply called a *divisor*, is an element in the free abelian group $Z^1(X)$ generated by prime divisors on X. A Cartier divisor on a normal variety is a Weil divisor. Every Weil divisor on a smooth variety is Cartier. The divisor D is expressed as a finite sum $D = \sum_i d_i D_i$ of prime divisors D_i with non-zero integers d_i. The *support* of D is the union of D_i. The divisor D is *effective* if all d_i are positive, and it is *reduced* if all d_i equal one. We write $D \le D'$ if $D' - D$ is effective. The *linear equivalence* $D \sim D'$ of divisors means that $D' - D$ is principal.

The divisor D is associated with a divisorial sheaf $\mathscr{O}_X(D)$ on X. A *divisorial sheaf* is a reflexive sheaf of rank one, where a coherent sheaf $\mathscr{F}$ is said to be *reflexive* if the natural map $\mathscr{F} \to \mathscr{F}^{\vee\vee}$ to the double dual is an isomorphism. The sheaf $\mathscr{O}_X(D)$ is the subsheaf of $\mathscr{K}_X$ defined by

$$\Gamma(U, \mathscr{O}_X(D)) = \{ f \in K(X) \mid (f)_U + D|_U \ge 0 \},$$

in which zero is contained in the set on the right by convention. The *divisor class group* $\operatorname{Cl} X$ is the quotient of the group $Z^1(X)$ of Weil divisors divided by the subgroup of principal divisors. It is regarded as the group of isomorphism classes of divisorial sheaves on X and has an injection $\operatorname{Pic} X \hookrightarrow \operatorname{Cl} X$.

Linear systems Let X be a normal complete variety. Let D be a Weil divisor on X and let V be a vector subspace of global sections in $H^0(\mathscr{O}_X(D))$. The projective space $\Lambda = \mathbf{P}V^\vee = (V \setminus 0)/k^\times$ where k is the ground field is called a *linear system* on X. It defines a rational map $X \dashrightarrow \mathbf{P}V$. When $V = H^0(\mathscr{O}_X(D))$, we write $|D| = \mathbf{P}H^0(\mathscr{O}_X(D))^\vee$ and call it a *complete* linear system. By the inclusion $\mathscr{O}_X(D) \subset \mathscr{K}_X$, the linear system $|D|$ is regarded as the set of effective divisors D' linearly equivalent to D, and Λ is a subset of $|D|$. That is,

$$\Lambda \subset |D| = \{ D' \ge 0 \mid D' \sim D \}.$$

The *base locus* of Λ means the scheme-theoretic intersection $B = \bigcap_{D' \in \Lambda} D'$ in X. We say that the linear system Λ is *free* if B is empty. We say that Λ is *mobile* if B is of codimension at least two. The divisor D is said to be *free* (resp.

mobile) if $|D|$ is free (resp. mobile). By definition, a free Weil divisor is Cartier. When $\emptyset \neq \Lambda \subset |D|$, Λ is decomposed as $\Lambda = \Lambda' + F$ with a mobile linear system $\Lambda' \subset |D - F|$ and the maximal effective divisor F such that $F \leq D_1$ for all $D_1 \in \Lambda$. The constituents Λ' and F are called the *mobile part* and the *fixed part* of Λ respectively. The rational map defined by Λ' is isomorphic to $X \dashrightarrow \mathbf{P}V$. The linear system Λ is mobile if and only if F is zero.

Even if X is not complete, the linear system $\Lambda = \mathbf{P}V^\vee$ is defined for a finite dimensional vector subspace V of $H^0(\mathscr{O}_X(D))$. We consider $|D|$ to be the direct limit $\varinjlim_V \Lambda$ of linear systems.

A *general* point in a variety Z means a point in a dense open subset U of Z. A *very general* point in Z means a point in the intersection $\bigcap_{i \in \mathbf{N}} U_i$ of countably many dense open subsets U_i. Thus by the general member of the linear system Λ, we mean a general point in Λ as a projective space. Bertini's theorem asserts that a free linear system on a smooth complex variety has a smooth member. The statement for the hyperplane section holds even in positive characteristic.

Theorem 1.1.6 (Bertini's theorem) *Let $\Lambda = \mathbf{P}V^\vee$ be a free linear system on a smooth variety X and let $\varphi \colon X \to \mathbf{P}V$ be the induced morphism. Suppose that φ is a closed embedding or the ground field is of characteristic zero. Then the general member H of Λ is a smooth divisor on X, and if the image $\varphi(X)$ is of dimension at least two, then H is a smooth prime divisor.*

The canonical divisor It is the canonical divisor that plays the most important role in birational geometry. The *sheaf of differentials* on an algebraic scheme X is denoted by Ω_X. When X is smooth, Ω_X^i denotes the i-th exterior power $\bigwedge^i \Omega_X$.

Definition 1.1.7 The *canonical divisor* K_X on a normal variety X is the divisor defined up to linear equivalence by the isomorphism $\mathscr{O}_X(K_X)|_U \simeq \Omega_U^n$ on the smooth locus U in X, where n is the dimension of X.

Example 1.1.8 The projective space $\mathbf{P}^n$ has the canonical divisor $K_{\mathbf{P}^n} \sim -(n+1)H$ for a hyperplane H. This follows from the *Euler sequence*

$$0 \to \Omega_{\mathbf{P}^n} \to \mathscr{O}_{\mathbf{P}^n}(-1)^{\oplus(n+1)} \to \mathscr{O}_{\mathbf{P}^n} \to 0.$$

One can describe $K_{\mathbf{P}^n}$ in an explicit way. Take homogeneous coordinates $x_0, \ldots, x_n$ of $\mathbf{P}^n$. Let $U_i \simeq \mathbf{A}^n$ denote the complement of the hyperplane H_i defined by x_i. The chart U_0 admits a nowhere vanishing n-form $dy_1 \wedge \cdots \wedge dy_n$ with coordinates $y_1, \ldots, y_n$ for $y_i = x_i x_0^{-1}$. It is expressed on the chart U_1 having coordinates $z_0, z_2, \ldots, z_n$ for $z_i = x_i x_1^{-1}$ as the rational n-form $dz_0^{-1} \wedge d(z_2 z_0^{-1}) \wedge \cdots \wedge d(z_n z_0^{-1}) = -z_0^{-(n+1)} dz_0 \wedge dz_2 \wedge \cdots \wedge dz_n$, which has pole of order $n+1$ along H_0. Thus $K_{\mathbf{P}^n} \sim -(n+1)H_0$.

In spite of the ambiguity concerning linear equivalence, it is standard to treat the canonical divisor as if it were a specified divisor.

For a closed subscheme D of an algebraic scheme X, there exists an exact sequence $\mathscr{I}/\mathscr{I}^2 \to \Omega_X \otimes \mathcal{O}_D \to \Omega_D \to 0$, where $\mathscr{I}$ is the ideal sheaf in $\mathcal{O}_X$ defining D. This induces the *adjunction formula*, which connects the canonical divisor to that on a Cartier divisor.

Theorem 1.1.9 (Adjunction formula)   *Let X be a normal variety and let D be a reduced Cartier divisor on X which is normal. Then $K_D = (K_X + D)|_D$ in the sense that $\mathcal{O}_D(K_D) \simeq \mathcal{O}_X(K_X + D) \otimes \mathcal{O}_D$.*

Duality   Albeit Grothendieck's duality theory works in the derived category for proper morphisms [177], it is extremely hard to obtain the dualising complex and a trace map in a compatible manner. The theory becomes efficient if it is restricted to the Cohen–Macaulay projective case as explained in [178, section III.7] and [277, section 5.5]. For example, the dualising complex on a Cohen–Macaulay projective scheme X of pure dimension n is the shift $\omega_X[n]$ of the dualising sheaf ω_X.

Definition 1.1.10   Let X be a complete scheme of dimension n over an algebraically closed field k. The *dualising sheaf* ω_X for X is a coherent sheaf on X endowed with a *trace map* $t\colon H^n(\omega_X) \to k$ such that for any coherent sheaf $\mathscr{F}$ on X, the natural pairing

$$\mathrm{Hom}(\mathscr{F}, \omega_X) \times H^n(\mathscr{F}) \to H^n(\omega_X) \overset{t}{\to} k$$

induces an isomorphism $\mathrm{Hom}(\mathscr{F}, \omega_X) \simeq H^n(\mathscr{F})^{\vee}$.

The dualising sheaf is unique up to isomorphism if it exists. The projective space $\mathbf{P}^n$ has the dualising sheaf $\omega_{\mathbf{P}^n} \simeq \bigwedge^n \Omega_{\mathbf{P}^n}$. This with Lemma 1.1.11 yields the existence of ω_X for every projective scheme X by taking a finite morphism $X \to \mathbf{P}^n$ known as projective *Noether normalisation*. If X is embedded into a projective space P with codimension r, then $\omega_X \simeq \mathscr{E}xt^r_P(\mathcal{O}_X, \omega_P)$ [178, III proposition 7.5]. If X is a normal projective variety, then ω_X coincides with the sheaf $\mathcal{O}_X(K_X)$ associated with the canonical divisor.

For a finite morphism $\pi\colon X \to Y$ of algebraic schemes, the push-forward π_* defines an equivalence of categories from the category of coherent $\mathcal{O}_X$-modules to that of coherent $\pi_*\mathcal{O}_X$-modules. This associates every coherent sheaf $\mathscr{G}$ on Y functorially with a coherent sheaf $\pi^!\mathscr{G}$ on X satisfying $\pi_* \mathscr{H}om_X(\mathscr{F}, \pi^!\mathscr{G}) \simeq \mathscr{H}om_Y(\pi_*\mathscr{F}, \mathscr{G})$ for any coherent sheaf $\mathscr{F}$ on X.

Lemma 1.1.11 *Let $\pi\colon X \to Y$ be a finite morphism of complete schemes of the same dimension. If the dualising sheaf ω_Y for Y exists, then $\omega_X = \pi^!\omega_Y$ is the dualising sheaf for X.*

Proof Let n denote the common dimension of X and Y. For a coherent sheaf $\mathscr{F}$ on X, $\mathrm{Hom}_X(\mathscr{F}, \pi^!\omega_Y) = \mathrm{Hom}_Y(\pi_*\mathscr{F}, \omega_Y)$ is dual to $H^n(\mathscr{F}) = H^n(\pi_*\mathscr{F})$ by the property of ω_Y, where the latter equality follows from the Leray spectral sequence $H^p(R^q\pi_*\mathscr{F}) \Rightarrow H^{p+q}(\mathscr{F})$. □

The duality for Cohen–Macaulay sheaves on a projective scheme is derived from that on the projective space via projective Noether normalisation. See [277, theorem 5.71].

Theorem 1.1.12 (Serre duality) *Let X be a projective scheme of dimension n. Let $\mathscr{F}$ be a Cohen–Macaulay coherent sheaf on X with support of pure dimension n. Then $H^i(\mathscr{H}om_X(\mathscr{F}, \omega_X))$ is dual to $H^{n-i}(\mathscr{F})$ for all i.*

The *adjunction formula* $\omega_D \simeq \omega_X \otimes \mathscr{O}_X(D) \otimes \mathscr{O}_D$ holds for a Cohen–Macaulay projective scheme X of pure dimension and an effective Cartier divisor D on X. Compare it with Theorem 1.1.9.

Resolution of singularities A projective birational morphism is described as a blow-up. The *blow-up* of an algebraic scheme X along a coherent ideal sheaf $\mathscr{I}$ in $\mathscr{O}_X$, or along the closed subscheme defined by $\mathscr{I}$, is the projective morphism $\pi\colon B = \mathrm{Proj}_X \bigoplus_{i\in\mathbf{N}} \mathscr{I}^i \to X$. The pull-back $\mathscr{I}\mathscr{O}_B = \pi^{-1}\mathscr{I} \cdot \mathscr{O}_B$ in $\mathscr{O}_B$ is an invertible ideal sheaf. Notice that $\mathscr{I}\mathscr{O}_B$ is different from $\pi^*\mathscr{I}$. The blow-up π has the universal property that every morphism $\varphi\colon Y \to X$ that makes $\mathscr{I}\mathscr{O}_Y$ invertible factors through π as $\varphi = \pi \circ f$ for a morphism $f\colon Y \to B$.

Let $f\colon X \dashrightarrow Y$ be a birational map of varieties. The *exceptional locus* of f is the locus in X where f is not biregular. Let Z be a closed subvariety of X not contained in the exceptional locus of f. The *strict transform* f_*Z in Y of Z is the closure of the image of $Z \dashrightarrow Y$. When X and Y are normal, the *strict transform* f_*P in Y of an arbitrary prime divisor P on X is defined as a divisor in such a manner that f_*P is zero if P is in the exceptional locus of f. By linear extension, we define the strict transform f_*D in Y for any divisor D on X.

Resolution of singularities is a fundamental tool in complex birational geometry. We say that a reduced divisor D on a smooth variety X is *simple normal crossing*, or *snc* for short, if D is defined at every point x in X by the product $x_1 \cdots x_m$ of a part of a regular system $x_1, \ldots, x_n$ of parameters in $\mathscr{O}_{X,x}$.

Definition 1.1.13 A *resolution* of a variety X is a projective birational morphism $\mu\colon X' \to X$ from a smooth variety. The resolution μ is said to be *strong* if it is isomorphic on the smooth locus in X.

Definition 1.1.14   Let X be a normal variety, let Δ be a divisor on X and let $\mathscr{I}$ be a coherent ideal sheaf in $\mathscr{O}_X$. A *log resolution* of $(X, \Delta, \mathscr{I})$ is a resolution $\mu \colon X' \to X$ such that

- the exceptional locus E of μ is a divisor on X',
- the pull-back $\mathscr{I}\mathscr{O}_{X'}$ is invertible and hence defines a divisor D and
- $E + D + \mu_*^{-1}S$ has snc support for the support S of Δ.

The log resolution μ is said to be *strong* if it is isomorphic on the maximal locus U in X such that U is smooth, $\mathscr{I}|_U$ defines a divisor D_U and $D_U + S|_U$ has snc support. A (strong) log resolution of X means that of $(X, 0, \mathscr{O}_X)$, and those of (X, Δ) and $(X, \mathscr{I})$ are likewise defined.

The existence of these resolutions for complex varieties is due to Hironaka. The items (i) and (ii) below are derived from the main theorems I and II in [187] respectively.

Theorem 1.1.15 (Hironaka [187])   (i) *A strong resolution exists for every complex variety.*

(ii) *A strong log resolution exists for every pair $(X, \mathscr{I})$ of a smooth complex variety X and a coherent ideal sheaf $\mathscr{I}$ in $\mathscr{O}_X$.*

Hironaka's construction includes the existence of a strong log resolution $X' \to X$ equipped with an effective exceptional divisor E on X' such that $\mathscr{O}_{X'}(-E)$ is relatively ample.

Analytic spaces   We shall occasionally consider a complex scheme to be an analytic space in the Euclidean topology. Whilst an algebraic scheme is obtained by gluing affine schemes in $\mathbf{A}^n$, an analytic space is constructed by gluing analytic models in a domain in $\mathbf{C}^n$. A reference is [151]. The ring of convergent complex power series is denoted by $\mathbf{C}\{x_1, \ldots, x_n\}$.

Let D be a domain in the complex manifold $\mathbf{C}^n$. Let $\mathscr{O}_D$ denote the sheaf of holomorphic functions on D. Let $\mathscr{I}$ be an ideal sheaf in $\mathscr{O}_D$ generated by a finite number of global sections. The locally $\mathbf{C}$-ringed space $(V, (\mathscr{O}_D/\mathscr{I})|_V)$ for the support V of the quotient sheaf $\mathscr{O}_D/\mathscr{I}$ is called an *analytic model*, where being $\mathbf{C}$-*ringed* means having the structure sheaf of $\mathbf{C}$-algebras. An *analytic space* is a locally $\mathbf{C}$-ringed Hausdorff space such that every point has an open neighbourhood isomorphic to an analytic model.

Every complex scheme X is associated with an analytic space X_h. This defines a functor h from the category of complex schemes to the category of analytic spaces. There exists a natural morphism $X_h \to X$ of locally $\mathbf{C}$-ringed spaces which maps X_h bijectively to the set of closed points in X. It pulls back a coherent sheaf $\mathscr{F}$ on X to a coherent sheaf $\mathscr{F}_h$ on X_h. When X is complete, it induces an equivalence of categories. This is known as the *GAGA principle*, which takes the acronym from the title of Serre's paper [414].

Theorem 1.1.16 (GAGA principle [163, exposé XII], [414]) *Let X be a complete complex scheme and let X_h be the analytic space associated with X. Then the functor h induces an equivalence of categories from the category of coherent sheaves on X to the category of coherent sheaves on X_h.*

For an analytic space V, the exponential function $\exp(2\pi\sqrt{-1}t)$ defines a group homomorphism $\mathcal{O}_V \to \mathcal{O}_V^\times$. The induced exact sequence

$$0 \to \mathbf{Z} \to \mathcal{O}_V \to \mathcal{O}_V^\times \to 0$$

is called the *exponential sequence*.

In principle, one can deal with analytic spaces analogously to complex schemes as in [29]. For an analytic space V, the *Oka–Cartan theorem* asserts the coherence of every ideal sheaf in $\mathcal{O}_V$ that defines an analytic subspace of V. For a proper map $\pi\colon V \to W$ of analytic spaces, the higher direct image $R^i\pi_*\mathscr{F}$ of a coherent sheaf $\mathscr{F}$ on V is coherent on W. In particular, the image $\pi(V)$ is the support of the analytic subspace of W defined by the kernel of the map $\mathcal{O}_W \to \pi_*\mathcal{O}_V$, which is referred to as the *proper mapping theorem*.

The canonical divisor on a normal analytic space may not be defined as a finite sum of prime divisors. Some notions such as projectivity of resolution of singularities only make sense on a small neighbourhood about a fixed compact subset of an analytic space. These will pose no obstacles as we mainly work on the germ at a point in the analytic category.

Notation 1.1.17 The symbol $\mathfrak{D}^n$ denotes a domain in the complex space $\mathbf{C}^n$ which contains the origin o. For example, we write $o \in \mathfrak{D}^n$ for a germ of a complex manifold.

1.2 Numerical Geometry

The intersection number is a basic linear tool in birational geometry. We shall define it in the relative setting of a proper morphism $X \to S$. This section works over an algebraically closed field k of any characteristic.

One encounters divisors with rational coefficients naturally. For example for a finite surjective morphism $X \to Y$ of smooth varieties tamely ramified along a smooth prime divisor D on Y, the ramification formula which will be proved in Theorem 2.2.20 expresses K_X as the pull-back of $K_Y + (1 - 1/m)D$ with the ramification index m along D. One also has divisors with real coefficients taking limits. We begin with formulation of these notions.

Let X be a normal variety. Let $Z^1(X)$ denote the group of Weil divisors on X. A **Q**-*divisor* is an element in the rational vector space $Z^1(X) \otimes \mathbf{Q}$. In like manner, an **R**-*divisor* is an element in $Z^1(X) \otimes \mathbf{R}$. An **R**-divisor D is expressed as a finite sum $D = \sum_i d_i D_i$ of prime divisors D_i with real coefficients d_i, and D is a **Q**-divisor if d_i are rational. It is *effective* if $d_i \geq 0$ for all i and $D \leq D'$ means that $D' - D$ is effective. The *round-down* $\lfloor D \rfloor$ and the *round-up* $\lceil D \rceil$ are defined as $\lfloor D \rfloor = \sum_i \lfloor d_i \rfloor D_i$ and $\lceil D \rceil = \sum_i \lceil d_i \rceil D_i$. We sometimes say that a usual divisor is *integral* to distinguish it from a **Q**-divisor and an **R**-divisor.

Let $C^1(X)$ denote the subgroup of $Z^1(X)$ generated by Cartier divisors on X. A **Q**-*Cartier* **Q**-divisor is an element in the rational vector space $C^1(X) \otimes \mathbf{Q}$. In other words, a **Q**-divisor D is **Q**-Cartier if and only if there exists a non-zero integer r such that rD is integral and Cartier. Likewise an **R**-*Cartier* **R**-divisor is an element in $C^1(X) \otimes \mathbf{R}$. An **R**-Cartier **Q**-divisor is always **Q**-Cartier but a **Q**-Cartier integral divisor is not necessarily Cartier.

Example 1.2.1   Consider the prime divisor $D = (x_1 = x_2 = 0)$ on the surface $X = (x_1^2 - x_2 x_3 = 0) \subset \mathbf{A}^3$ with coordinates x_1, x_2, x_3. Then $2D$ is the Cartier divisor defined by x_2 and the scheme-theoretic intersection $2D \cap l$ with the line $l = (x_1 = x_2 = x_3)$ in X is of length one. It follows that D is not Cartier.

Let $\pi \colon Y \to X$ be a morphism of normal varieties. The *pull-back* $\pi^* D$ of an **R**-Cartier **R**-divisor D on X is defined as an **R**-Cartier **R**-divisor on Y by the natural map $\pi^* \colon C^1(X) \otimes \mathbf{R} \to C^1(Y) \otimes \mathbf{R}$. If D is a **Q**-divisor, then so is $\pi^* D$.

Definition 1.2.2   Let X be a normal variety. We say that X is **Q**-*Gorenstein* if the canonical divisor K_X is **Q**-Cartier. We say that X is **Q**-*factorial* if all divisors on X are **Q**-Cartier, that is, $\mathrm{Cl}\, X / \mathrm{Pic}\, X$ is torsion. It is said to be *factorial* if all divisors are Cartier, that is, $\mathrm{Pic}\, X = \mathrm{Cl}\, X$.

The **Q**-factoriality is not an analytically local property.

Example 1.2.3   The algebraic germ $o \in X = (x_1 x_2 + x_3 x_4 = 0) \subset \mathbf{A}^4$ is not **Q**-factorial. The prime divisor $D = (x_1 = x_4 = 0)$ on X is not **Q**-Cartier and the divisor class group $\mathrm{Cl}\, X$ is $\mathbf{Z}[D] \simeq \mathbf{Z}$. Indeed, the blow-up B of X at o resolves the projection $X \dashrightarrow \mathbf{P}^3$ from o as a morphism $B \to \mathbf{P}^3$ and it yields a line bundle $B \to S$ over the surface $S = (x_1 x_2 + x_3 x_4 = 0) \simeq \mathbf{P}^1 \times \mathbf{P}^1 \subset \mathbf{P}^3$. By this structure, $\mathrm{Pic}\, B$ is generated by the strict transforms D_B and E_B of D and $E = (x_2 = x_4 = 0)$. They satisfy the relation $D_B + E_B + F \sim 0$ for the exceptional divisor F of $B \to X$. Thus $\mathrm{Cl}\, X \simeq \mathrm{Pic}(B \setminus F) = \mathbf{Z}[D_B \setminus F]$.

On the other hand, the algebraic germ $o \in Y = (x_1 x_2 + x_3 x_4 + f = 0) \subset \mathbf{A}^4$ is factorial for a general cubic form f in $x_1, \ldots, x_4$. To see this, we compactify Y to $\bar{Y} = (x_0(x_1 x_2 + x_3 x_4) + f = 0) \subset \mathbf{P}^4$. The blow-up $\bar{B}$ of $\bar{Y}$ at o resolves the projection from o as $\bar{B} \to \mathbf{P}^3$, and this is the blow-up of $\mathbf{P}^3$ along the sextic curve $(x_1 x_2 + x_3 x_4 = f = 0)$. By this structure, Pic $\bar{B}$ is generated by the exceptional divisor $\bar{F}$ of $\bar{B} \to \bar{Y}$ and the strict transform $\bar{H}_B$ of $\bar{H} = \bar{Y} \setminus Y$. Thus Cl $Y \simeq \mathrm{Pic}(\bar{B} \setminus (\bar{F} + \bar{H}_B)) = 0$.

The two germs $o \in X$ and $o \in Y$ become isomorphic in the analytic category as will be seen in Proposition 2.3.3. See Remark 3.1.11 for further discussion.

We shall fix the base scheme S and work *relatively* on a proper morphism $\pi \colon X \to S$ of algebraic schemes, which is frequently denoted by X/S. Every terminology is accompanied by the reference to the relative setting. The reference is omitted when we consider a complete scheme X with the structure morphism $X \to S = \mathrm{Spec}\, k$.

A *relative subvariety* Z of X/S means a closed subvariety of X such that $\pi(Z)$ is a point in S. A *relative m-cycle* on X/S is an element in the free abelian group $Z_m(X/S)$ generated by relative subvarieties of dimension m in X/S. For invertible sheaves $\mathscr{L}_1, \dots, \mathscr{L}_m$ and a relative m-cycle Z on X, the *intersection number* $(\mathscr{L}_1 \cdots \mathscr{L}_m \cdot Z)$ is defined by the multilinear map

$$(\mathrm{Pic}\, X)^{\oplus m} \times Z_m(X/S) \to \mathbf{Z}$$

such that $(\mathscr{L}^m \cdot Z)$ for a relative subvariety Z coincides with $(\mathscr{L}^m \cdot \mathscr{O}_Z)$ in the asymptotic Riemann–Roch theorem $\chi(\mathscr{L}^{\otimes l} \otimes \mathscr{O}_Z) = (\mathscr{L}^m \cdot \mathscr{O}_Z) l^m / m! + O(l^{m-1})$. The *intersection number* $(D_1 \cdots D_m \cdot Z)$ with Cartier divisors D_i on X is defined as $(\mathscr{O}_X(D_1) \cdots \mathscr{O}_X(D_m) \cdot Z)$. If D_i are effective and intersect properly on a relative subvariety Z, then $(D_1 \cdots D_m \cdot Z)$ equals the length of the structure sheaf $\mathscr{O}_A$ of the artinian scheme $A = D_1 \cap \cdots \cap D_m \cap Z$. The length of $\mathscr{O}_{A,x}$ for $x \in A$ is referred to as the *local intersection number* at x and denoted by $(D_1 \cdots D_m \cdot Z)_x$. When X is a complete variety of dimension n with the structure morphism $X \to S = \mathrm{Spec}\, k$, we write $(\mathscr{L}_1 \cdots \mathscr{L}_n) = (\mathscr{L}_1 \cdots \mathscr{L}_n \cdot X)$ and $D_1 \cdots D_n = (D_1 \cdots D_n)_X = (D_1 \cdots D_n \cdot X)$.

By the extension $(\mathrm{Pic}\, X \otimes \mathbf{R}) \times (Z_1(X/S) \otimes \mathbf{R}) \to \mathbf{R}$, the *relative numerical equivalence* $\equiv_S$ is defined in both the real vector spaces Pic $X \otimes \mathbf{R}$ and $Z_1(X/S) \otimes \mathbf{R}$ in such a way that it induces a perfect pairing

$$N^1(X/S) \times N_1(X/S) \to \mathbf{R}$$

of vector spaces on the quotients $N^1(X/S) = (\mathrm{Pic}\, X \otimes \mathbf{R})/\equiv_S$ and $N_1(X/S) = (Z_1(X/S) \otimes \mathbf{R})/\equiv_S$. When $S = \mathrm{Spec}\, k$, we just write $\equiv$, $N^1(X)$ and $N_1(X)$ without reference to S as remarked above.

Definition 1.2.4 The spaces $N^1(X/S)$ and $N_1(X/S)$ are finite dimensional [254, IV§4, proposition 3]. The equal dimension of $N^1(X/S)$ and $N_1(X/S)$ is called the *relative Picard number* of X/S and denoted by $\rho(X/S)$. When

$S = \operatorname{Spec} k$, this number is called the *Picard number* of the complete scheme X and denoted by $\rho(X)$.

Let $\varphi \colon Y \to X$ be a proper morphism. It induces the *pull-back* $\varphi^* \colon \operatorname{Pic} X \to \operatorname{Pic} Y$ and the *push-forward* $\varphi_* \colon Z_m(Y/S) \to Z_m(X/S)$ as group homomorphisms. The push-forward $\varphi_* Z$ of a relative subvariety Z of Y/S is $d\varphi(Z)$ if the morphism $Z \to \varphi(Z)$ is generically finite of degree d, and $\varphi_* Z$ is zero if $\varphi(Z)$ is of dimension less than that of Z. These satisfy the *projection formula*

$$(\varphi^* \mathscr{L}_1 \cdots \varphi^* \mathscr{L}_m \cdot Z) = (\mathscr{L}_1 \cdots \mathscr{L}_m \cdot \varphi_* Z)$$

for invertible sheaves $\mathscr{L}_i$ on X and a relative m-cycle Z on Y. They yield $\varphi^* \colon N^1(X/S) \to N^1(Y/S)$ and dually $\varphi_* \colon N_1(Y/S) \to N_1(X/S)$. One also has the natural surjection $N^1(Y/S) \twoheadrightarrow N^1(Y/X)$ and injection $N_1(Y/X) \hookrightarrow N_1(Y/S)$. If φ is surjective, then $\varphi^* \colon N^1(X/S) \to N^1(Y/S)$ is injective and $\varphi_* \colon N_1(Y/S) \to N_1(X/S)$ is surjective.

Henceforth we fix a proper morphism $\pi \colon X \to S$ from a normal variety to a variety and make basic definitions for an **R**-Cartier **R**-divisor D on X. We say that integral divisors D and D' on X are *relatively linearly equivalent* and write $D \sim_S D'$ if the difference $D - D'$ is linearly equivalent to the pull-back $\pi^* B$ of some Cartier divisor B on S. Namely, $D - D'$ is zero in the quotient $\operatorname{Cl} X / \pi^* \operatorname{Pic} S$. For **R**-divisors D and D' on X, the *relative* **R**-*linear equivalence* $D \sim_{\mathbf{R},S} D'$ means that $D - D'$ is zero in $(\operatorname{Cl} X / \pi^* \operatorname{Pic} S) \otimes \mathbf{R}$. When D and D' are **Q**-divisors, this is referred to as the *relative* **Q**-*linear equivalence* and denoted by $D \sim_{\mathbf{Q},S} D'$. The space $\operatorname{Pic} X \otimes \mathbf{R}$ is regarded as that of **R**-linear equivalence classes of **R**-Cartier **R**-divisors on X. The intersection number $(D \cdot C)$ is defined for a pair of an **R**-Cartier **R**-divisor D on X and a relative one-cycle C on X/S. This makes the notion of *relative numerical equivalence* $D \equiv_S D'$ for **R**-Cartier **R**-divisors D and D' on X.

Definition 1.2.5 An **R**-Cartier **R**-divisor D on X/S is said to be *relatively nef* (or *nef over S* or *π-nef*) if $(D \cdot C) \geq 0$ for any relative curve C in X/S. When $S = \operatorname{Spec} k$, we just say that D is *nef* as usual.

A Cartier divisor D on X is said to be *relatively ample* (*π-ample*) if $\mathscr{O}_X(D)$ is a relatively ample invertible sheaf. It is said to be *relatively very ample* (*π-very ample*) if $\mathscr{O}_X(D)$ is a relatively very ample invertible sheaf. In spite of the geometric definition, the ampleness is characterised numerically.

Theorem 1.2.6 (Nakai's criterion) *Let $\pi \colon X \to S$ be a proper morphism of algebraic schemes. An invertible sheaf $\mathscr{L}$ on X is relatively ample if and only if $(\mathscr{L}^{\dim Z} \cdot Z) > 0$ for any relative subvariety Z of X/S.*

Kleiman's ampleness criterion rephrases Nakai's criterion in terms of the cones of divisors and curves. A *convex cone C*, or simply called a *cone*, in a finite dimensional real vector space V is a subset of V such that if $v, w \in C$ and $c \in \mathbf{R}_{>0}$, then $v + w \in C$ and $cv \in C$.

Definition 1.2.7 The *ample cone* $A(X/S)$ is the convex cone in $N^1(X/S)$ spanned by the classes of relatively ample Cartier divisors on X. The *closed cone* $\overline{NE}(X/S)$ *of curves* is the closure of the convex cone in $N_1(X/S)$ spanned by the classes of relative curves in X/S.

Notice that $A(X/S)$ is an open cone since for a relatively ample divisor A and a Cartier divisor D, the sum $D + lA$ is relatively ample for large l.

Theorem 1.2.8 (Kleiman's ampleness criterion [254]) *Let $\pi \colon X \to S$ be a proper morphism of algebraic schemes. Then a Cartier divisor D on X is relatively ample if and only if the class of D belongs to the ample cone $A(X/S)$. If π is projective, then $A(X/S)$ and $\overline{NE}(X/S)$ are dual with respect to the intersection pairing $N^1(X/S) \times N_1(X/S) \to \mathbf{R}$ in the sense that*

$$A(X/S) = \{y \in N^1(X/S) \mid (y, z) > 0 \text{ for all } z \in \overline{NE}(X/S) \setminus 0\},$$

$$\overline{NE}(X/S) \setminus 0 = \{z \in N_1(X/S) \mid (y, z) > 0 \text{ for all } y \in A(X/S)\}.$$

The theorem shows that if π is projective, then the closure of the ample cone $A(X/S)$ coincides with the *nef cone* $\mathrm{Nef}(X/S)$ in $N^1(X/S)$ spanned by relatively nef **R**-Cartier **R**-divisors. The duality of $A(X/S)$ and $\overline{NE}(X/S)$ still holds for a **Q**-factorial complete variety $X/S = \mathrm{Spec}\, k$ as studied in [254], but it fails for a proper morphism in general.

Example 1.2.9 Fujino [128] constructed an example of a non-projective complete toric threefold X with $\rho(X) = 1$ such that $\overline{NE}(X)$ is a half-line $\mathbf{R}_{\geq 0}$. The book [140] by Fulton is a standard introduction to toric varieties. Let $v_1 = (1, 0, 1)$, $v_2 = (0, 1, 1)$, $v_3 = (-1, -1, 1)$ and $w_1 = (1, 0, -1)$, $w_2 = (0, 1, -1)$, $w_3 = (-1, -1, -1)$ in $N = \mathbf{Z}^3$. Take the fan Δ which consists of faces of the cones $\langle v_1, v_2, v_3 \rangle$, $\langle w_1, w_2, w_3 \rangle$, $\langle v_1, v_2, w_1 \rangle$, $\langle v_2, w_1, w_2 \rangle$, $\langle v_2, v_3, w_2, w_3 \rangle$, $\langle v_3, v_1, w_3, w_1 \rangle$. The toric variety X associated with (N, Δ) is the example.

The numerical nature extends the notion of ampleness to **R**-divisors.

Definition 1.2.10 An **R**-Cartier **R**-divisor D on X/S is said to be *relatively ample* (π-*ample*) if the class of D belongs to the ample cone $A(X/S)$. In other words, D is expressed as a finite sum $D = \sum_i a_i A_i$ of relatively ample Cartier divisors A_i with $a_i \in \mathbf{R}_{>0}$.

We keep $\pi\colon X \to S$ being a proper morphism from a normal variety to a variety. For a Cartier divisor D on X, the natural map $\pi^*\pi_*\mathscr{O}_X(D) \to \mathscr{O}_X(D)$ defines a rational map

$$X \dashrightarrow \mathrm{Proj}_S \, S\pi_*\mathscr{O}_X(D)$$

over S for the symmetric $\mathscr{O}_S$-algebra $S\pi_*\mathscr{O}_X(D)$ of $\pi_*\mathscr{O}_X(D)$. The *relative base locus* of D is the closed subscheme B of X given by the ideal sheaf $\mathscr{I}_B$ in $\mathscr{O}_X$ such that the above map induces the surjection $\pi^*\pi_*\mathscr{O}_X(D) \twoheadrightarrow \mathscr{I}_B\mathscr{O}_X(D)$. We say that D is *relatively free* (*π-free*) if B is empty. We say that D is *relatively mobile* (*π-mobile*) if B is of codimension at least two. The definitions coincide with those on a normal complete variety. Unless $B = X$, there exists a maximal effective divisor F such that $\mathscr{I}_B \subset \mathscr{O}_X(-F)$. The divisors $D - F$ and F are called the *relative mobile part* (*π-mobile part*) and the *relative fixed part* (*π-fixed part*) of D respectively.

Definition 1.2.11 A Cartier divisor D on X/S is said to be *relatively semi-ample* (*π-semi-ample*) if aD is relatively free for some positive integer a. An **R**-Cartier **R**-divisor D on X is said to be *relatively semi-ample* (*π-semi-ample*) if it is expressed as a finite sum $D = \sum_i a_i A_i$ of relatively semi-ample Cartier divisors A_i with $a_i \in \mathbf{R}_{\geq 0}$. The definition is consistent by the next lemma.

Lemma 1.2.12 *Let $\pi\colon X \to S$ be a proper morphism from a normal variety to a variety. Let D and $A_1, \ldots, A_n$ be Cartier divisors on X such that $D = \sum_i a_i A_i$ with $a_i \in \mathbf{R}_{\geq 0}$. If all A_i are relatively free, then aD is relatively free for some positive integer a.*

Proof Let $Z^1(X)_{\mathbf{Q}}$ denote the rational vector space of **Q**-divisors on X. Let V be the vector subspace of $Z^1(X)_{\mathbf{Q}}$ spanned by $A_1, \ldots, A_n$. Then D belongs to $(V \otimes_{\mathbf{Q}} \mathbf{R}) \cap Z^1(X)_{\mathbf{Q}} = V$ and hence we may assume that $a_i \in \mathbf{Q}$ and further $a_i \in \mathbf{Z}$ by multiplying D. The assertion in this case follows from the existence of the natural map $\bigoplus_i (\pi^*\pi_*\mathscr{O}_X(A_i))^{\oplus a_i} \to \pi^*\pi_*\mathscr{O}_X(D)$. $\qquad\square$

We provide an alternative characterisation of semi-ampleness.

Lemma 1.2.13 *Let $\pi\colon X \to S$ be a proper morphism from a normal variety to a variety. An **R**-divisor D on X is relatively semi-ample if and only if there exists a projective morphism $\pi_Y\colon Y \to S$ from a normal variety through which π factors as $\pi = \pi_Y \circ \varphi$ for $\varphi\colon X \to Y$ such that $D \sim_{\mathbf{R}} \varphi^*A$ by a relatively ample **R**-divisor A on Y/S.*

Proof The if part is obvious. We shall prove the only-if part for a relatively semi-ample **R**-divisor D. Write D as a finite sum $D = \sum_i a_i B_i$ of relatively free divisors B_i with $a_i \in \mathbf{R}_{>0}$. The morphism $\varphi_i \colon X \to Y_i = \mathrm{Proj}_S \, S\pi_* \mathscr{O}_X(B_i)$ provides a relation $B_i \sim \varphi_i^* A_i$ by a relatively ample divisor A_i on Y_i/S.

Let $\varphi \colon X \to Y$ be the Stein factorisation of $X \to \mathrm{Proj}_S \, S\pi_* \mathscr{O}_X(\sum_i B_i)$. A relative curve C in X/S is contracted to a point by φ if and only if $(\sum_i B_i \cdot C) = 0$. This is equivalent to $(B_i \cdot C) = 0$ for all i since B_i are relatively nef. By Lemma 1.1.1, every φ_i factors through φ as $\varphi_i = \psi_i \circ \varphi$ for $\psi_i \colon Y \to Y_i$ and $D \sim_{\mathbf{R}} \varphi^* \sum_i a_i \psi_i^* A_i$. Then C is contracted by φ if and only if $(\varphi^*(\sum_i a_i \psi_i^* A_i) \cdot C) = 0$. This shows the relative ampleness of $\sum_i a_i \psi_i^* A_i$ on Y/S. $\qquad\square$

Definition 1.2.14 A Cartier divisor D on X/S is said to be *relatively big* (π-*big*) if there exists a positive integer a such that the rational map $X \dashrightarrow \mathrm{Proj}_S \, S\pi_* \mathscr{O}_X(aD)$ is birational to the image.

Assuming that π is projective, *Kodaira's lemma* characterises the bigness numerically.

Theorem 1.2.15 (Kodaira's lemma) *Let $\pi \colon X \to S$ be a projective morphism from a normal variety to a quasi-projective variety. A Cartier divisor D on X is relatively big if and only if there exist a relatively ample $\mathbf{Q}$-divisor A and an effective $\mathbf{Q}$-divisor B such that $D = A + B$.*

Proof The if part is obvious. We shall prove the only-if part for a relatively big divisor D. By Stein factorisation, we may assume that π has connected fibres. Multiplying D, we may assume that $X \dashrightarrow \mathrm{Proj}_S \, S\pi_* \mathscr{O}_X(D)$ is birational to the image Y. We write $\pi_Y \colon Y \to S$. Take an open subset U of X such that $\varphi \colon U \to Y$ is a morphism and such that the complement $X \setminus U$ is of codimension at least two. Then $\mathscr{O}_X(D)|_U \simeq \varphi^* \mathscr{O}_Y(1)$ and

$$\pi_{Y*} \mathscr{O}_Y(l) \subset \pi_{Y*} \varphi_* \varphi^* \mathscr{O}_Y(l) \simeq \pi_*(\mathscr{O}_X(lD)|_U) = \pi_* \mathscr{O}_X(lD)$$

for any $l \in \mathbf{Z}$. Hence there exists a positive rational constant c such that the rank of the $\mathscr{O}_S$-module $\pi_* \mathscr{O}_X(lD)$ is greater than cl^n for sufficiently large l, where $n = \dim X - \dim S$ is the dimension of the general fibre of π_Y.

Take a general very ample effective divisor H on X. Since the rank of $\pi_* \mathscr{O}_H(lD|_H)$ is estimated as $O(l^{n-1})$, the exact sequence

$$0 \to \pi_* \mathscr{O}_X(lD - H) \to \pi_* \mathscr{O}_X(lD) \to \pi_* \mathscr{O}_H(lD|_H)$$

yields the non-vanishing $\pi_* \mathscr{O}_X(lD - H) \neq 0$ for large l. Hence $H^0(\mathscr{O}_X(lD - H + \pi^* G)) = H^0(\pi_* \mathscr{O}_X(lD - H) \otimes \mathscr{O}_S(G)) \neq 0$ by a sufficiently very ample divisor G on S. Thus one can write $lD - H + \pi^* G = B_1 + (f)_X$ with an effective divisor B_1 and a principal divisor $(f)_X$ on X. Then $D = A + B$ with $A = l^{-1}(H - \pi^* G + (f)_X)$ and $B = l^{-1} B_1$. $\qquad\square$

Remark 1.2.16 Without the quasi-projectivity of the base variety, Kodaira's lemma gives the decomposition $D = A + B$ into a relatively ample **Q**-divisor A and a **Q**-divisor B such that $\pi_* \mathcal{O}_X(lB) \neq 0$ for some positive integer l. In some literature, a Cartier divisor D on X/S with $\pi_* \mathcal{O}_X(D) \neq 0$ is said to be *relatively effective*. Provided that S is quasi-projective, this means that D is relatively linearly equivalent to an effective divisor. We do not use this terminology for the reason that a relatively effective divisor over $S = \operatorname{Spec} \mathbf{C}$ is not necessarily effective but effective up to linear equivalence.

By definition, a Cartier divisor D on X is relatively big if and only if so is the restriction $D|_{\pi^{-1}(U)}$ over some open subset U of S containing the generic point of $\pi(X)$. Thus Kodaira's lemma with Kleiman's criterion implies that bigness on a projective morphism is a numerical condition. This provides grounds for considering the cone of big divisors.

Definition 1.2.17 Assume that $\pi\colon X \to S$ is projective. The *big cone* $B(X/S)$ is the convex cone in $N^1(X/S)$ spanned by the classes of relatively big Cartier divisors on X. This is an open cone containing the ample cone $A(X/S)$. An **R**-Cartier **R**-divisor D on X is said to be *relatively big* (π-*big*) if the class of D belongs to the big cone $B(X/S)$. Namely, D is expressed as a finite sum $D = \sum_i b_i B_i$ of relatively big Cartier divisors B_i with $b_i \in \mathbf{R}_{>0}$.

One can formulate Kodaira's lemma for **R**-divisors.

Corollary 1.2.18 *Let $\pi\colon X \to S$ be a projective morphism from a normal variety to a quasi-projective variety. An **R**-Cartier **R**-divisor D on X is relatively big if and only if there exist a relatively ample **Q**-divisor A and an effective **R**-divisor B such that $D = A + B$.*

Finally we introduce the notion of numerical limit of effective **R**-divisors.

Definition 1.2.19 The *pseudo-effective cone* $P(X/S)$ is the closure of the convex cone in $N^1(X/S)$ spanned by the classes of Cartier divisors D on X with $\pi_* \mathcal{O}_X(D) \neq 0$. An **R**-Cartier **R**-divisor D on X is said to be *relatively pseudo-effective* (π-*pseudo-effective*) if the class of D belongs to the pseudo-effective cone $P(X/S)$.

By Kodaira's lemma, if π is projective, then the pseudo-effective cone $P(X/S)$ coincides with the closure of the big cone $B(X/S)$. If S is quasi-projective, then a relatively pseudo-effective **R**-divisor on X is realised in the space $N^1(X/S)$ as a limit of a sequence of effective **R**-divisors.

Example 1.2.20 Let C be a smooth projective curve of genus g. The *Jacobian* $J(C)$ of C represents the subgroup $\operatorname{Pic}^0 C$ of $\operatorname{Pic} C$ which consists of invertible sheaves of degree zero. Refer to [318] for example. It is an abelian variety of dimension g. Thus as far as $g \geq 1$, C has an invertible sheaf $\mathscr{L}$ of degree zero which is not a torsion in $\mathrm{Pic}^0 C$. The divisor $D \equiv 0$ such that $\mathscr{L} \simeq \mathscr{O}_C(D)$ is pseudo-effective but not $\mathbf{Q}$-linearly equivalent to zero.

1.3 The Program

The classification theory in birational geometry seeks to find a good representative of each birational class of varieties and to analyse it. We shall explain how the surface theory has matured into the minimal model program in higher dimensions. The program will be generalised logarithmically and relatively in the next section. We shall tacitly work over $\mathbf{C}$ as mentioned in the first section. We define a contraction in the following manner.

Definition 1.3.1 A *contraction* $\pi\colon X \to Y$ is a projective morphism of normal varieties with connected fibres, namely $\mathscr{O}_Y = \pi_*\mathscr{O}_X$. It is said to be *of fibre type* if $\dim Y < \dim X$. Thus a contraction is either birational or of fibre type. We say that the contraction π is *extremal* if $\rho(X/Y) = 1$.

We recall the surface theory in brief. Standard books are [28], [30] and [35]. The treatise [403] is also excellent. Let S be a smooth projective surface. Though one can construct a new surface from S by blowing it up at an arbitrary point, this operation is not essential in the birational study of S. Its exceptional curve is superfluous. We want to obtain a simple birational model of S by contracting superfluous curves. *Castelnuovo's criterion* for contraction enables us to do so.

Definition 1.3.2 A curve C in a smooth surface is called a (-1)-*curve* if it is isomorphic to $\mathbf{P}^1$ with self-intersection number $(C^2) = -1$.

Theorem 1.3.3 (Castelnuovo's contraction theorem) *Let C be a (-1)-curve in a smooth projective surface S. Then C is the exceptional curve of the blow-up $S \to T$ of a smooth projective surface T at a point.*

The contraction of C decreases the Picard number by one as $\rho(T) = \rho(S) - 1$. Hence one eventually attains a surface without (-1)-curves by contracting them successively. This surface is characterised as follows.

Theorem 1.3.4 *If a smooth projective surface S has no (-1)-curves, then either*

(i) *the canonical divisor K_S is nef or*
(ii) *there exists a contraction $S \to B$ of fibre type which is a $\mathbf{P}^1$-bundle over a smooth curve or isomorphic to $\mathbf{P}^2 \to \mathrm{Spec}\,\mathbf{C}$.*

Notice that a surface S_1 with the property (i) is never birational to a surface S_2 with (ii). If there were a birational map $S_1 \dashrightarrow S_2$, then a general relative curve C_2 in the fibration $S_2 \to B$ would be mapped regularly to a curve C_1 in S_1 with $(K_{S_1} \cdot C_1) \leq (K_{S_2} \cdot C_2)$. This contradicts the intersection numbers $(K_{S_1} \cdot C_1) \geq 0$ and $(K_{S_2} \cdot C_2) < 0$.

In the case (ii), if B is a curve, then S/B is described as $\mathbf{P}(\mathcal{E})/B$ by a locally free sheaf $\mathcal{E}$ of rank two on B. When $B \simeq \mathbf{P}^1$, it is completely classified as below. The reader may refer to [377, 1 theorem 2.1.1] for the proof.

Theorem 1.3.5 (Grothendieck [159]) *Every locally free sheaf of rank r on $\mathbf{P}^1$ is isomorphic to a direct sum $\bigoplus_{i=1}^{r} \mathcal{O}_{\mathbf{P}^1}(a_i)$ with $a_i \in \mathbf{Z}$.*

Hence every $\mathbf{P}^1$-bundle over $\mathbf{P}^1$ is isomorphic to a Hirzebruch surface.

Definition 1.3.6 The *Hirzebruch surface* Σ_n for $n \in \mathbf{N}$ is the $\mathbf{P}^1$-bundle $\Sigma_n = \mathbf{P}(\mathcal{O}_{\mathbf{P}^1} \oplus \mathcal{O}_{\mathbf{P}^1}(-n))$ over $\mathbf{P}^1$.

The surface S in the case (i) is classified with showing the semi-ampleness of K_S, that is, lK_S is free for some positive integer l. Then the complete linear system $|lK_S|$ defines a morphism $S \to \mathbf{P}H^0(\mathcal{O}_S(lK_S))$. When l is sufficiently large and divisible, the induced surjection to the image in $\mathbf{P}H^0(\mathcal{O}_S(lK_S))$ is independent of l. One can refer to [30, chapter VI] for the classification below.

Theorem 1.3.7 *Let S be a smooth projective surface such that K_S is nef. Then K_S is semi-ample and S is one of the following in terms of $q = h^1(\mathcal{O}_S)$, $\chi = \chi(\mathcal{O}_S)$ and the dimension κ of the image of $S \to \mathbf{P}H^0(\mathcal{O}_S(lK_S))$ for sufficiently large and divisible l.*

 (i) *$\kappa = 0$. S is a K3 surface defined by $K_S \sim 0$ and $q = 0$. $\chi = 2$.*
 (ii) *$\kappa = 0$. S is an Enriques surface defined by $K_S \nsim 0$, $2K_S \sim 0$ and $q = 0$. $\chi = 1$.*
 (iii) *$\kappa = 0$. S is an abelian surface. $K_S \sim 0$, $q = 2$ and $\chi = 0$.*
 (iv) *$\kappa = 0$. S is a hyperelliptic surface defined as the quotient $(E \times F)/G$ of the product of elliptic curves E and F by a finite subgroup G of the translations of E which acts on F so that $F/G \simeq \mathbf{P}^1$. The least positive integer r such that $rK_S \sim 0$ is 2, 3, 4 or 6. $q = 1$ and $\chi = 0$.*
 (v) *$\kappa = 1$. $\chi \geq 0$.*
 (vi) *$\kappa = 2$. $\chi \geq 1$.*

Both the contraction $S \to T$ of a (-1)-curve and the fibration $S \to B$ in Theorem 1.3.4(ii) are extremal contractions in Definition 1.3.1 with respect to which the *anti-canonical divisor* $-K_S$ is relatively ample. Hence given a smooth projective surface S, the program for finding a model in Theorem 1.3.4 by Theorem 1.3.3 is described as the following algorithm.

1 If K_S is nef, then output S, which belongs to the case (i) in Theorem 1.3.4.
2 If K_S is not nef, then there exists an extremal contraction $\pi\colon S \to T$ to a smooth projective variety such that $-K_S$ is relatively ample.
3 If π is of fibre type, then output S, which belongs to the case (ii) in Theorem 1.3.4.
4 If π is birational, then replace S by T and go back to 1.

The minimal model program is a higher dimensional extension of this program. However, a naive extension is confronted with several obstacles as will be seen. Let X be a smooth projective variety and suppose the existence of an extremal contraction $X \to Y$ such that $-K_X$ is relatively ample. The first obstacle is that Y may be singular.

Example 1.3.8 Let $\mathbf{A}^3 = \operatorname{Spec} \mathbf{C}[x_1, x_2, x_3]$. Consider the action of $\mathbf{Z}_2$ on $\mathbf{A}^3$ given by the involution which sends (x_1, x_2, x_3) to $(-x_1, -x_2, -x_3)$. The quotient $\mu\colon \mathbf{A}^3 \to Y = \mathbf{A}^3/\mathbf{Z}_2 = \operatorname{Spec} R$ is defined by the invariant ring $R = \mathbf{C}[x_1^2, x_2^2, x_3^2, x_1 x_2, x_2 x_3, x_3 x_1]$. The germ $o \in Y$ at the image o of the origin of $\mathbf{A}^3$ is an isolated singularity known as the cyclic quotient singularity of type $\frac{1}{2}(1, 1, 1)$ in Definition 2.2.10. The canonical sheaf ω_Y is the invariant part of $\mu_* \omega_{\mathbf{A}^3}$, generated by $x_1 \theta, x_2 \theta, x_3 \theta$ for $\theta = dx_1 \wedge dx_2 \wedge dx_3$. This description shows that K_Y is not Cartier at o but $2K_Y$ is Cartier.

Let $\pi\colon X \to Y$ be the blow-up of Y at o, or to be precise, along the maximal ideal in $\mathcal{O}_Y$ defining o. Then X is smooth and the exceptional locus E in X is isomorphic to $\mathbf{P}^2$ with $\mathcal{O}_X(-E) \otimes \mathcal{O}_E \simeq \mathcal{O}_{\mathbf{P}^2}(2)$. It is an extremal contraction with $K_X = \pi^* K_Y + (1/2)E$ and $-K_X$ is π-ample.

The next example by Ueno reveals that we cannot avoid singularities. It exhibits a threefold which has no smooth birational models as in Theorem 1.3.4. We need the *negativity lemma*, which will be used at several places.

Theorem 1.3.9 (Negativity lemma) *Let $\pi\colon X \to Y$ be a proper birational morphism of normal varieties. Let E be a π-exceptional $\mathbf{R}$-divisor expressed as $E = M + F$ with a π-nef $\mathbf{R}$-divisor M and an effective $\mathbf{R}$-divisor F such that no π-exceptional prime divisor appears in F. Then $E \le 0$.*

Proof It suffices to work about the generic point η_i of the image $\pi(E_i)$ of each prime component E_i of E. Replacing π by the base change to the intersection of general hyperplane sections of Y about η_i, we may assume that $\pi(E_i)$ is a point in Y. By Chow's lemma, we may assume that π is a contraction. Then cutting it with general hyperplane sections of X, we may assume that $\pi\colon X \to Y$ is a birational morphism of surfaces. By resolution, we may also assume that X is smooth.

After this reduction, the exceptional **R**-divisor $E = M + F$ is π-nef on the smooth surface X. Since π is projective, there exists an effective divisor A on X supported on the exceptional locus of π such that $-A$ is π-ample. If $E \nleq 0$, then there would exist a positive real number r and an exceptional curve C such that $E - rA \leq 0$ and such that C has coefficient zero in $E - rA$. Then $((E - rA) \cdot C) \leq 0$, which contradicts the π-ampleness of $E - rA$. □

Example 1.3.10 (Ueno [458, section 16]) Let A be an abelian threefold. It is described as the quotient $\mathbf{C}^3/\Gamma$ of the complex threefold $\mathbf{C}^3$ by a lattice $\Gamma \simeq \mathbf{Z}^6$ spanning $\mathbf{C}^3$ as a real vector space. The involution of A which sends x to $-x$ has $2^6 = 64$ fixed points. The associated quotient $X = A/\mathbf{Z}_2$ has at each fixed point a cyclic quotient singularity of type $\frac{1}{2}(1, 1, 1)$ appeared in Example 1.3.8. The sheaf $\mathscr{O}_X(2K_X)$ is globally generated by $(dx_1 \wedge dx_2 \wedge dx_3)^{\otimes 2}$ for the coordinates x_1, x_2, x_3 of $\mathbf{C}^3$. Hence $2K_X \sim 0$. For the same reason as explained after Theorem 1.3.4, X is never birational to a threefold X' equipped with a contraction of fibre type with respect to which $-K_{X'}$ is relatively ample. Let Y be an arbitrary smooth projective variety birational to X. Using the negativity lemma, we shall prove that K_Y is never nef.

The blow-up B of X at all the 64 singular points is smooth and $2K_B \sim E$ for the sum $E = \sum_{i=1}^{64} E_i$ of the exceptional divisors $E_i \simeq \mathbf{P}^2$. Take a common log resolution W of (B, E) and Y with $p_1\colon W \to X$ and $p_2\colon W \to Y$. Let T_i denote the sum of p_i-exceptional prime divisors on W and let T denote the common part of T_1 and T_2. We write $2K_W = p_1^*(2K_X) + F_1 + G_1$ by effective divisors F_1 and G_1 with support T and $T_1 - T$ respectively. Similarly $K_W = p_2^* K_Y + F_2 + G_2$ by effective divisors F_2 and G_2 with support T and $T_2 - T$. Then

$$F_1 - 2F_2 + G_1 = -p_1^*(2K_X) + p_2^*(2K_Y) + 2G_2.$$

If K_Y were nef, then by the negativity lemma for p_1, the p_1-exceptional divisor $F_1 - 2F_2 + G_1$ would be negative in the sense that $F_1 - 2F_2 + G_1 \leq 0$. Hence $G_1 = 0$, by which $T = T_1$. It follows that the rational map $X \dashrightarrow Y$ produces no new divisors on Y. In particular, K_Y is the strict transform of K_X and hence $2K_Y \sim 0$. As a consequence, one obtains the relation $F_1 \sim 2F_2 + 2G_2$ of p_2-exceptional divisors. This is the equality $F_1 = 2F_2 + 2G_2$ by the negativity lemma for p_2. However, the strict transform of E_i appears in F_1 with coefficient one, which is absurd.

As a solution to these obstacles, the minimal model program admits mild singularities as in the above examples. We define *terminal* singularities by making the class of these singularities as small as possible.

Definition 1.3.11 Let $\pi\colon Y \to X$ be a proper birational morphism of normal varieties. Provided that X is $\mathbf{Q}$-Gorenstein, there exists a unique exceptional $\mathbf{Q}$-divisor $K_{Y/X}$ such that $K_Y = \pi^* K_X + K_{Y/X}$. We call $K_{Y/X}$ the *relative canonical divisor*.

Definition 1.3.12 A normal variety X is said to be *terminal*, or to have *terminal* singularities, if X is $\mathbf{Q}$-Gorenstein and for any resolution $X' \to X$, every exceptional prime divisor appears in $K_{X'/X}$ with positive coefficient.

The definition uses resolution of singularities. It is equivalent to the existence of some resolution $X' \to X$ satisfying the required property. See also Definition 1.4.3. Terminal threefold singularities will be classified in the next chapter.

Theorem 1.3.13 *A surface is terminal if and only if it is smooth.*

Proof Every surface S has the *minimal* resolution $S' \to S$, namely a unique resolution such that $K_{S'/S}$ is relatively nef. Since $K_{S'/S} \leq 0$ by the negativity lemma, if S is terminal, then S' has no exceptional curves, that is, S is smooth. The converse is obvious. □

Let X be a terminal projective variety. Unless K_X is nef, the *cone theorem* with the *contraction theorem* produces an extremal contraction $X \to Y$ such that $-K_X$ is relatively ample. Mori [332] first established the cone theorem for smooth varieties and constructed a contraction from a smooth threefold. Whereas he used the deformation theory of curves in positive characteristic, Kawamata [232] and others developed the extension to singular varieties by a cohomological method. Below we extract the part needed for the program. The precise statements will be provided in Theorems 1.4.7 and 1.4.9.

Theorem 1.3.14 *Let X be a terminal projective variety. If K_X is not nef, then there exists an extremal contraction $\pi\colon X \to Y$ to a normal projective variety such that $-K_X$ is relatively ample. It always satisfies $\mathrm{Pic}\,X/\pi^*\,\mathrm{Pic}\,Y \simeq \mathbf{Z}$.*

The most serious obstacle is that $\pi\colon X \to Y$ may be isomorphic in codimension one. We say that such π is *small* as defined below. In this case, the canonical divisor K_Y is never $\mathbf{Q}$-Cartier and it does not make sense to ask if K_Y is nef. If K_Y were $\mathbf{Q}$-Cartier, then the pull-back $K_X = \pi^* K_Y$ would contradict the π-ampleness of $-K_X$. Hence it is necessary to reconstruct a reasonable variety X^+ from Y, which is called the *flip* of X/Y. In the book, by a flip X is assumed to be terminal, and a generalised notion will be referred to as a log flip as defined in the next section.

Definition 1.3.15 Let $f\colon X \dashrightarrow Y$ be a birational map of normal varieties factorised as $f = q \circ p^{-1}$ with contractions $p\colon W \to X$ and $q\colon W \to Y$. We call f a *birational contraction map* if all p-exceptional prime divisors are q-exceptional. In other words, the strict transform defines a surjection $f_*\colon Z^1(X) \twoheadrightarrow Z^1(Y)$ of the groups of Weil divisors. We say that f is *small* if it is isomorphic in codimension one, that is, f_* is an isomorphism $Z^1(X) \simeq Z^1(Y)$.

Definition 1.3.16 A *flipping contraction* $\pi\colon X \to Y$ is a small contraction from a terminal variety such that $-K_X$ is π-ample. The *flip* of π is a small contraction $\pi^+\colon X^+ \to Y$ such that K_{X^+} is π^+-ample. The transformation $X \dashrightarrow X^+$ is also called the *flip* by abuse of language. We say that a flipping contraction $\pi\colon X \to Y$ and the flip of π are *elementary* if X is $\mathbf{Q}$-factorial and π is extremal.

Notation 1.3.17 For a Weil divisor D on a normal variety X, we write the graded $\mathscr{O}_X$-algebra $\mathscr{R}(X, D) = \bigoplus_{i \in \mathbf{N}} \mathscr{O}_X(iD)$.

The flip is described as $X^+ = \operatorname{Proj}_Y \mathscr{R}(Y, lK_Y)$ with a positive integer l such that lK_{X^+} is π^+-very ample. Note that $\mathscr{R}(Y, lK_Y) = \pi^+_* \mathscr{R}(X^+, lK_{X^+})$ as π^+ is small. Hence the flip is unique if it exists. Lemma 1.5.20 further provides the description $X^+ = \operatorname{Proj}_Y \mathscr{R}(Y, K_Y)$. One can consider the flip $X \dashrightarrow X^+$ to be the operation of replacing curves with negative intersection number with K_X by curves with positive intersection number with K_{X^+}. We shall explain the first example of a flip by Francia as a quotient of the Atiyah flop.

Example 1.3.18 (Atiyah [24]) Consider the germ $o \in Y = (x_1x_2 + x_3x_4 = 0) \subset \mathbf{A}^4$ discussed in Example 1.2.3. The canonical divisor K_Y is Cartier by the adjunction $K_Y = (K_{\mathbf{A}^4} + Y)|_Y$. Let $\pi\colon X \to Y$ be the blow-up along the ideal $(x_2, x_4)\mathscr{O}_Y$ in $\mathscr{O}_Y$. Then X is smooth and has exceptional locus $C = \pi^{-1}(o) \simeq \mathbf{P}^1$. The blow-up $\varphi\colon B \to Y$ at o factors through π as $\varphi = \pi \circ \mu$ for the contraction $\mu\colon B \to X$ of the exceptional locus $F \simeq \mathbf{P}^1 \times \mathbf{P}^1$ of φ to $C \simeq \mathbf{P}^1$.

Corresponding to the other contraction of F to $\mathbf{P}^1$, the morphism φ also factors through the blow-up $\pi^+\colon X^+ \to Y$ along the ideal $(x_1, x_4)\mathscr{O}_Y$ in $\mathscr{O}_Y$. It has exceptional locus $C^+ = (\pi^+)^{-1}(o) \simeq \mathbf{P}^1$. The transformation $X \to Y \leftarrow X^+$ is small with $K_X = \pi^* K_Y$ and $K_{X^+} = (\pi^+)^* K_Y$. This is known as the *Atiyah flop*, which is a classical example of a flop in Definition 5.1.6.

Example 1.3.19 (Francia [125]) We constructed the Atiyah flop $X \to Y \leftarrow X^+$ for $o \in Y = (x_1x_2 + x_3x_4 = 0) \subset \mathbf{A}^4$. Act $\mathbf{Z}_2$ on Y by the involution which sends (x_1, x_2, x_3, x_4) to $(-x_1, x_2, x_3, -x_4)$. The fixed locus in Y is the divisor

$D = (x_1 = x_4 = 0)$. This action extends to X and X^+ and one can consider the quotients $\pi': X' = X/\mathbf{Z}_2 \to Y' = Y/\mathbf{Z}_2$ and $\pi'^+: X'^+ = X^+/\mathbf{Z}_2 \to Y'$. Then X' has a cyclic quotient singularity of type $\frac{1}{2}(1,1,1)$ appeared in Example 1.3.8, whilst X'^+ remains smooth.

Let $C' = C/\mathbf{Z}_2 \simeq \mathbf{P}^1$ and $C'^+ = C^+/\mathbf{Z}_2 \simeq \mathbf{P}^1$ be the exceptional curves in X' and X'^+ respectively. We shall compute the intersection number $(K_{X'} \cdot C')$. One has $p_*C = 2C'$ as a cycle by $p: X \to X'$. On the other hand, the ramification formula in Theorem 2.2.20 gives $p^*K_{X'} = K_X - D_X$ with the strict transform D_X of D. Hence $(K_{X'} \cdot C') = ((K_X - D_X) \cdot C/2) = -1/2$ by the projection formula. In like manner, one has $(K_{X'^+} \cdot C'^+) = 1$ using $p^+: X^+ \to X'^+$ with $p_*^+C^+ = C'^+$. Thus π'^+ is the flip of π', which is known as the *Francia flip*.

The flip $X \dashrightarrow X^+$ retains the property of being terminal. If it is elementary, then it also keeps the Picard number unchanged.

Lemma 1.3.20 *Let $\pi^+: X^+ \to Y$ be the flip of a flipping contraction $\pi: X \to Y$. Then X^+ as well as X is terminal. If X and Y are projective and the flip $X \dashrightarrow X^+$ is elementary, then X^+ is $\mathbf{Q}$-factorial and projective and $\rho(X) = \rho(X^+)$.*

Proof Take a common resolution W of X and Y with $\mu: W \to X$ and $\mu^+: W \to X^+$. Then $K_{W/X} - K_{W/X^+} = (\mu^+)^*K_{X^+} - \mu^*K_X$ is μ-exceptional and μ-nef. Hence $K_{W/X} \leq K_{W/X^+}$ by the negativity lemma, showing that X^+ is terminal as is X.

Suppose that X and Y are projective and the flip is elementary. Then X^+ is projective. By Theorem 1.3.14, $(\text{Pic } X/\pi^* \text{Pic } Y) \otimes \mathbf{Q} \simeq \mathbf{Q}$ and it is generated by the π-ample divisor $-K_X$. Hence for any divisor D on X, there exists a rational number c such that $D_Y + cK_Y = \pi_*(D + cK_X)$ is $\mathbf{Q}$-Cartier where $D_Y = \pi_*D$. Then the strict transform D^+ in X^+ of D is expressed as $D^+ = (\pi^+)^*(D_Y + cK_Y) - cK_{X^+}$, which is $\mathbf{Q}$-Cartier. Thus X^+ is $\mathbf{Q}$-factorial and K_{X^+} generates $(\text{Pic } X^+/(\pi^+)^* \text{Pic } Y) \otimes \mathbf{Q}$. In particular, $\rho(X^+) = \rho(Y) + 1 = \rho(X)$. □

We are now in the position of stating the minimal model program. Given a $\mathbf{Q}$-factorial terminal projective variety X, it finds a birational contraction map $X \dashrightarrow Y$ such that Y is a minimal model or admits a structure of a Mori fibre space.

Definition 1.3.21 Let X be a $\mathbf{Q}$-factorial terminal projective variety. It is called a *minimal model* if K_X is nef. A *Mori fibre space* $X \to S$ is an extremal contraction of fibre type to a normal projective variety such that $-K_X$ is relatively ample. The base S will be proved to be $\mathbf{Q}$-factorial in Lemma 1.4.13.

If a $\mathbf{Q}$-factorial terminal projective variety X is not a minimal model, then Theorem 1.3.14 provides an extremal contraction $X \to Y$ to a normal projective variety such that $-K_X$ is relatively ample. By Lemma 1.3.23, it is a Mori fibre space, a flipping contraction or a divisorial contraction below.

Definition 1.3.22 A *divisorial contraction* is a birational contraction $\pi\colon X \to Y$ between terminal varieties such that $-K_X$ is π-ample and such that the exceptional locus E is a prime divisor on X. One can write $K_X = \pi^* K_Y + dE$ with a positive rational number d since Y is terminal. In particular, $-E$ is π-ample. We say that a divisorial contraction $\pi\colon X \to Y$ is *elementary* if X is **Q**-factorial and π is extremal.

Lemma 1.3.23 *Let $\pi\colon X \to Y$ be an extremal contraction from a **Q**-factorial terminal projective variety to a normal projective variety such that $-K_X$ is π-ample. If π is birational but not small, then π is a divisorial contraction, Y is **Q**-factorial and $\rho(X) = \rho(Y) + 1$.*

Proof We shall prove that the exceptional locus is a prime divisor. The remaining assertions are derived from this in the same way as for Lemma 1.3.20.

Take a hyperplane section H_Y of Y such that $\pi^* H_Y$ contains an exceptional prime divisor. Write $\pi^* H_Y = H + E$ with the strict transform H of H_Y and an exceptional divisor E. By $\rho(X/Y) = 1$, the divisor $-E \equiv_Y H$ is π-ample and the support of E contains a prime divisor F such that $-F$ is π-ample. If a curve C not in F were contracted to a point by π, then the intersection number $(F \cdot C) \geq 0$ would contradict the relative ampleness of $-F$. Thus F must coincide with the exceptional locus. □

Definition 1.3.24 The *minimal model program*, or the *MMP* for short, in the category $\mathscr{C}$ of **Q**-factorial terminal projective varieties is the algorithm for $X \in \mathscr{C}$ which outputs $Y \in \mathscr{C}$ with a birational contraction map $X \dashrightarrow Y$ in the following manner.

1. If K_X is nef, then output $X \in \mathscr{C}$ as a minimal model.
2. If K_X is not nef, then there exists an extremal contraction $\pi\colon X \to Y$ as in Theorem 1.3.14.
3. If π is a Mori fibre space, then output $X \in \mathscr{C}$.
4. If π is a divisorial contraction, then $Y \in \mathscr{C}$ and $\rho(Y) = \rho(X) - 1$ by Lemma 1.3.23. Replace X by Y and go back to 1.
5. If π is a flipping contraction, then construct the flip $\pi^+\colon X^+ \to Y$ of π, for which $X^+ \in \mathscr{C}$ and $\rho(X^+) = \rho(X)$ by Lemma 1.3.20. Replace X by X^+ and go back to 1.

In order for the MMP to function, the existence and termination of flips are necessary. The termination means the non-existence of an infinite loop by the step 5 of the algorithm as stated in Conjecture 1.3.26. Assuming them, the MMP ends with a minimal model or a Mori fibre space by induction of the Picard number. The *existence of flips* was first proved by Mori [335] for threefold flips and fully settled by Birkar, Cascini, Hacon and McKernan [48] as in Theorem 1.4.11. However, the full form of Conjecture 1.3.26 is still open in dimension greater than four in spite of the termination of flips with scaling in the setting of Corollary 1.5.13. The termination in dimension three will be demonstrated in Theorem 1.6.3. The result in dimension four is in [249, theorem 5.1.15].

Theorem 1.3.25 *The flip in Definition* 1.3.16 *exists.*

Conjecture 1.3.26 (Termination of flips) *Let X be a $\mathbf{Q}$-factorial terminal projective variety. Then there exists no infinite sequence $X = X_0 \dashrightarrow X_1 \dashrightarrow \cdots$ of elementary flips $X_i \to Y_i \leftarrow X_{i+1}$ with X_i and Y_i projective.*

As is the case with the program for surfaces, which of a minimal model or a Mori fibre space the MMP outputs is determined by the input variety.

Proposition 1.3.27 *Let X be a $\mathbf{Q}$-factorial terminal projective variety. If K_X is pseudo-effective, then the output by the MMP from X is always a minimal model. If K_X is not pseudo-effective, then the output is always a Mori fibre space.*

Proof Let $X = X_0 \dashrightarrow \cdots \dashrightarrow X_n = Y$ be an output of the MMP from X, where each $f_i \colon X_i \dashrightarrow X_{i+1}$ is a divisorial contraction or a flip. It suffices to prove the equivalence of the pseudo-effectivity of K_X and that of K_Y.

Fix an ample divisor A on X. Its strict transform A_Y in Y is big. For a small positive rational number ε, the strict transform $-(K_{X_i} + \varepsilon A_i)$ in X_i of $-(K_X + \varepsilon A)$ is f_i-ample when f_i is a divisorial contraction. Then for a sufficiently large and divisible integer l, the multiple $l(K_X + \varepsilon A)$ is integral and $H^0(\mathscr{O}_{X_i}(l(K_{X_i} + \varepsilon A_i))) \simeq H^0(\mathscr{O}_{X_{i+1}}(l(K_{X_{i+1}} + \varepsilon A_{i+1})))$. It follows that

$$H^0(\mathscr{O}_X(l(K_X + \varepsilon A))) \simeq H^0(\mathscr{O}_Y(l(K_Y + \varepsilon A_Y))).$$

Thus the limit K_X of $K_X + \varepsilon A$ with ε approaching zero is pseudo-effective if and only if so is the limit K_Y of $K_Y + \varepsilon A_Y$. □

The problem subsequent to the MMP is to study Mori fibre spaces and minimal models. The Sarkisov program, which will be introduced in Chapter 6, is a standard tool for analysing birational maps of Mori fibre spaces. In the study of minimal models, the rational map defined by a multiple of the canonical divisor plays an important role as in Theorem 1.3.7. We expect the following *abundance* conjecture. It is only known up to dimension three as will be demonstrated in Chapter 10. The designation of abundance is by the assertion in Theorem 1.7.12 that the canonical divisor is semi-ample if and only if it is nef and *abundant* in the sense of Definition 1.7.10.

Conjecture 1.3.28 (Abundance)   *If X is a minimal model, then K_X is semi-ample.*

1.4 Logarithmic and Relative Extensions

We formulated the MMP for projective varieties. However, it has turned out that the program becomes much more powerful by logarithmic and relative extensions. This perspective proves its worth even in the study of the original MMP. For example, the flip X^+/Y of a flipping contraction X/Y is considered to be the canonical model of X/Y. The purpose of this section is to outline the generalisations of the MMP. The books [277] and [307] elucidate the abstract side of the program. The treatise [249] is standard until now.

Let X be an algebraic scheme. Let Z be a closed subvariety of X and let $\mathfrak{p}$ denote the ideal sheaf in $\mathscr{O}_X$ defining Z. The *order* $\mathrm{ord}_Z \mathscr{I}$ along Z of a coherent ideal sheaf $\mathscr{I}$ in $\mathscr{O}_X$ is the maximal $l \in \mathbf{N} \cup \{\infty\}$ such that $\mathscr{I}\mathscr{O}_{X,\eta} \subset \mathfrak{p}^l \mathscr{O}_{X,\eta}$ at the generic point η of Z, where we define $\mathfrak{p}^\infty$ to be zero. We write $\mathrm{ord}_Z f = \mathrm{ord}_Z f\mathscr{O}_X$ for $f \in \mathscr{O}_X$. When X is normal, the *order* $\mathrm{ord}_Z D$ of a $\mathbf{Q}$-Cartier divisor D on X is defined as $r^{-1} \mathrm{ord}_Z \mathscr{O}_X(-rD)$ by a positive integer r such that rD is Cartier, which is independent of the choice of r. Beware of the difference between $\mathrm{ord}_Z D$ and $\mathrm{ord}_Z \mathscr{O}_X(-D)$. The notion of $\mathrm{ord}_Z D$ is extended linearly to $\mathbf{R}$-Cartier $\mathbf{R}$-divisors.

Let X be a variety. A divisor *over* X means the equivalence class of a prime divisor E on a normal variety Y equipped with a birational morphism $Y \to X$, where divisors E on Y/X and E' on Y'/X are equivalent if E is the strict transform of E' and vice versa, that is, E and E' define the same valuation on the function field of X. A divisor E over X is said to be *exceptional* if it is not realised as a prime divisor on X. The *order* $\mathrm{ord}_E \mathscr{I}$ along E of a coherent ideal sheaf $\mathscr{I}$ in $\mathscr{O}_X$ is defined as $\mathrm{ord}_E \mathscr{I}\mathscr{O}_Y$, which is independent of the realisation $E \subset Y/X$. The orders $\mathrm{ord}_E f$ and $\mathrm{ord}_E D$ are defined in like manner for a function f in $\mathscr{O}_X$ and an $\mathbf{R}$-Cartier $\mathbf{R}$-divisor D on X.

Definition 1.4.1   The *centre* $c_X(E)$ of a divisor E over X is the closure of the image $\pi(E)$ of E by $\pi\colon Y \to X$ on which E is realised as a prime divisor.

The logarithmic setting treats pairs (X, Δ) using $K_X + \Delta$ instead of K_X. A *pair* (X, Δ) consists of a normal variety X and an effective $\mathbf{R}$-divisor Δ on X such that $K_X + \Delta$ is $\mathbf{R}$-Cartier, in which Δ is referred to as the *boundary*.

Definition 1.4.2   Let (X, Δ) be a pair and let E be a divisor over X realised on $\pi \colon Y \to X$. One can write $K_Y = \pi^*(K_X + \Delta) + A$ uniquely with an **R**-divisor A such that $\pi_* A = -\Delta$. The *log discrepancy* $a_E(X, \Delta)$ of E with respect to (X, Δ) is defined as $a_E(X, \Delta) = 1 + \operatorname{ord}_E A$. When Δ is zero, we write $a_E(X)$ for $a_E(X, 0)$ and occasionally prefer the *discrepancy* of E which is defined as $a_E(X) - 1 = \operatorname{ord}_E K_{Y/X}$.

Definition 1.4.3   We say that a pair (X, Δ) is *terminal, canonical, purely log terminal* (*plt* for short) respectively if $a_E(X, \Delta) > 1, \geq 1, > 0$ respectively for all divisors E exceptional over X. We say that (X, Δ) is *Kawamata log terminal* (*klt*), *log canonical* (*lc*) respectively if $a_E(X, \Delta) > 0, \geq 0$ respectively for all divisors E over X. We say that (X, Δ) is *divisorially log terminal* (*dlt*) if it is lc and there exists a log resolution Y of (X, Δ) such that $a_E(X, \Delta) > 0$ for every prime divisor E on Y that is exceptional over X. When Δ is zero, in which X is **Q**-Gorenstein, we simply say that X is terminal, canonical and so forth. This coincides with Definition 1.3.12. The notions of klt, plt and dlt singularities for $\Delta = 0$ are the same and we just say that X is *log terminal* (*lt*).

The vanishing of higher cohomologies enables us to compute the dimension of global sections of a sheaf numerically. The *Kodaira vanishing* and its generalisations are indispensable in complex birational geometry. The classical vanishing by Kodaira was proved in the theory of harmonic analysis. It does not hold in positive characteristic.

Theorem 1.4.4 (Kodaira vanishing [257])   *Let X be a smooth projective complex variety and let $\mathscr{L}$ be an ample invertible sheaf on X. Then $H^i(\omega_X \otimes \mathscr{L}) = 0$ for all $i \geq 1$.*

The following generalisation is one of the most fundamental tools in the minimal model theory. It is obtained by the covering trick associated with a Kummer extension. It includes the *Grauert–Riemenschneider vanishing* $R^i \pi_* \omega_X = 0$ for $i \geq 1$ for a generically finite proper morphism π from a smooth variety X [152].

Theorem 1.4.5 (Kawamata–Viehweg vanishing [230], [463])   *Let (X, Δ) be a klt pair and let $\pi \colon X \to Y$ be a proper morphism to a variety. Let D be a* **Q**-*Cartier integral divisor on X such that $D - (K_X + \Delta)$ is π-nef and π-big. Then $R^i \pi_* \mathscr{O}_X(D) = 0$ for all $i \geq 1$.*

We shall formulate the minimal model program for an lc pair (X, Δ) with a projective morphism $\pi \colon X \to S$ to a fixed variety. The *contraction* and *cone theorems* provide extremal contractions in the program.

Definition 1.4.6 Let C be a convex cone in a finite dimensional real vector space V. An *extremal face* F of C means a convex subcone of C such that if $v, w \in C$ and $v + w \in F$, then $v, w \in F$. An extremal face R is called an *extremal ray* if R is a half-line $\mathbf{R}_{\geq 0} v \cap C$ with one generator $v \in R \setminus 0$. For $\lambda \in V^\vee = \mathrm{Hom}(V, \mathbf{R})$, we say that an extremal face F is *negative* with respect to λ, or λ-*negative*, if $\lambda(v) < 0$ for all $v \in F \setminus 0$.

We consider the closed cone $\overline{\mathrm{NE}}(X/S)$ of curves in $N_1(X/S)$ and regard $N^1(X/S)$ as the dual of $N_1(X/S)$ via the intersection pairing.

Theorem 1.4.7 (Contraction theorem) *Let (X, Δ) be an lc pair with a projective morphism $\pi \colon X \to S$ to a variety. Let F be a $(K_X + \Delta)$-negative extremal face of $\overline{\mathrm{NE}}(X/S)$. Then there exists a unique contraction $\varphi \colon X \to Y$ through which π factors as $\pi = \pi_Y \circ \varphi$ for a projective morphism $\pi_Y \colon Y \to S$ such that a relative curve C in X/S is contracted to a point in Y if and only if $[C] \in F$. Further the natural sequence*

$$0 \to \mathrm{Pic}\, Y \xrightarrow{\varphi^*} \mathrm{Pic}\, X \to N^1(X/Y)$$

is exact and in particular $\rho(X/S) = \rho(Y/S) + \rho(X/Y)$.

We say that the contraction $\varphi \colon X \to Y$ above is *associated with* the face F. The dimension of F equals the relative Picard number $\rho(X/Y)$. By Kleiman's criterion, $K_X + \Delta$ is relatively ample with respect to φ.

The contraction theorem is a corollary to the *base-point free theorem*. Kawamata [232] completed it for klt pairs using the vanishing theorem and Shokurov's *non-vanishing theorem* [423].

Theorem 1.4.8 (Base-point free theorem) *Let (X, Δ) be a klt pair with a proper morphism $\pi \colon X \to S$ to a variety. Let D be a π-nef Cartier divisor on X such that $aD - (K_X + \Delta)$ is π-nef and π-big for some positive integer a. Then there exists a positive integer l_0 such that lD is π-free for all $l \geq l_0$.*

Thanks to the cone theorem, every extremal face F treated by the contraction theorem is spanned by the classes of rational curves. The cone theorem originated in the work of Mori [331] on the Hartshorne conjecture on a characterisation of the projective space. Kawamata [232] supplemented by [259] established the theorem for klt pairs by a cohomological method. It was extended to lc pairs by Fujino [129].

Theorem 1.4.9 (Cone theorem) *Let (X, Δ) be an lc pair with a projective morphism $\pi \colon X \to S$ to a variety. Let H be a π-ample $\mathbf{R}$-divisor on X. Then*

$$\overline{\mathrm{NE}}(X/S) = \overline{\mathrm{NE}}(X/S)_{K_X + \Delta + H \geq 0} + \sum_i \mathbf{R}_{\geq 0}[C_i],$$

where $\overline{\mathrm{NE}}(X/S)_{K_X+\Delta+H\geq 0} = \{z \in \overline{\mathrm{NE}}(X/S) \mid ((K_X + \Delta + H) \cdot z) \geq 0\}$ *and* $\sum_i \mathbf{R}_{\geq 0}[C_i]$ *is a finite sum of extremal rays generated by the class of a rational relative curve C_i in X/S.*

It is remarkable that the degree of the generator C_i of an extremal ray is bounded. This is essentially due to Kawamata [237] and extended in [129].

Theorem 1.4.10 *Let (X, Δ) be an lc pair with a projective morphism $X \to S$ to a variety. Then every $(K_X + \Delta)$-negative extremal ray of $\overline{\mathrm{NE}}(X/S)$ is generated by the class of a rational relative curve C in X/S such that $0 < (-(K_X + \Delta) \cdot C) \leq 2 \dim X$.*

Fix the base variety S and let $\mathscr{C}^l$ denote the category of $\mathbf{Q}$-factorial lc pairs (X, Δ) projective over S, meaning that (X, Δ) is lc with X being $\mathbf{Q}$-factorial and equipped with a projective morphism to S.

Let $(X/S, \Delta) \in \mathscr{C}^l$. We call $(X/S, \Delta)$ a *log minimal model* over S if $K_X + \Delta$ is relatively nef. Let $\pi\colon X \to Y/S$ be an extremal contraction associated with a $(K_X + \Delta)$-negative extremal ray of $\overline{\mathrm{NE}}(X/S)$. The contraction π is a *log Mori fibre space* if it is of fibre type, a *log divisorial contraction* if it is birational but not small and a *log flipping contraction* if it is small. If π is birational but not small, then the exceptional locus is a prime divisor similarly to Lemma 1.3.23. When X is terminal and Δ is zero, a log minimal model X/S and a log Mori fibre space $X \to Y/S$ are usually called a *relative minimal model* and a *relative Mori fibre space*. These generalise Definition 1.3.21 by relaxing the abstract projectivity of X and Y.

Let (X, Δ) be an lc pair. As in Definitions 1.3.16 and 1.3.22, we use the terminology of a log divisorial or flipping contraction $\pi\colon X \to Y$ without assuming that it is elementary. We say that π is *elementary* if X is $\mathbf{Q}$-factorial and π is extremal. A *log flipping contraction* $\pi\colon X \to Y$ with respect to (X, Δ) is a small contraction such that $-(K_X + \Delta)$ is π-ample. The *log flip* of π is a small contraction $\pi^+\colon X^+ \to Y$ such that $K_{X^+} + \Delta^+$ is π^+-ample for the strict transform Δ^+ of Δ. The transformation $X \dashrightarrow X^+$ is also called the *log flip*. After Shokurov's establishment of threefold log flips [424] and his attempt at higher dimensional generalisation [426], Birkar, Cascini, Hacon and McKernan [48] proved the existence of log flips for klt pairs. This was extended to lc pairs substantially by Birkar [45] and Hacon–Xu [175].

Theorem 1.4.11 (Existence of log flips) *The log flip exists for an lc pair.*

Log divisorial contractions and log flips improve singularities as shown by the next lemma, which is proved essentially in the same manner as Lemmata 1.3.20 and 1.3.23 are proved.

Lemma 1.4.12   *Let* $(X/S, \Delta) \in \mathscr{C}^l$ *and let* $f : X \dashrightarrow Y/S$ *be a log divisorial contraction or a log flip associated with a* $(K_X + \Delta)$*-negative extremal ray of* $\overline{NE}(X/S)$. *Then* $(Y, f_*\Delta) \in \mathscr{C}^l$ *and the inequality* $a_E(X, \Delta) \leq a_E(Y, f_*\Delta)$ *of log discrepancies holds for any divisor* E *over* X. *Further* $a_E(X, \Delta) < a_E(Y, f_*\Delta)$ *if* f *is not isomorphic at the generic point of the centre* $c_X(E)$ *in* X. *In particular if* (X, Δ) *is klt, plt, dlt or lc, then so is* $(Y, f_*\Delta)$. *If* (X, Δ) *is terminal or canonical, and in addition* $\Delta = f_*^{-1}f_*\Delta$, *then so is* $(Y, f_*\Delta)$.

We remark that the base of a log Mori fibre space is **Q**-factorial.

Lemma 1.4.13   *Let* $\pi : X \to Y$ *be a log Mori fibre space. Then* Y *is* **Q**-*factorial. If* $\operatorname{Cl} X / \operatorname{Pic} X$ *is* r-*torsion for* $r \in \mathbf{Z}_{>0}$, *then so is* $\operatorname{Cl} Y / \operatorname{Pic} Y$.

Proof   Let U be the smooth locus in Y and let $U_X = \pi^{-1}(U)$. For an arbitrary divisor D on Y, there exists a divisor D_X on X such that the restriction $D_X|_{U_X}$ equals the pull-back of $D|_U$. Suppose that rD_X is Cartier. Note that $rD_X \equiv_Y 0$ by $\rho(X/Y) = 1$. By the exact sequence in Theorem 1.4.7 for $S = Y$, the invertible sheaf $\mathscr{O}_X(rD_X)$ is isomorphic to the pull-back of some invertible sheaf $\mathscr{L}$ on Y. By the projection formula, $\mathscr{L}|_U \simeq (\pi|_{U_X})_*\mathscr{O}_X(rD_X)|_{U_X} = \mathscr{O}_Y(rD)|_U$. This implies that $\mathscr{L} \simeq \mathscr{O}_Y(rD)$ since $Y \setminus U$ is of codimension at least two. Thus rD is Cartier. $\square$

The $(K_X + \Delta)$-*minimal model program* over S, or the $(K_X + \Delta)/S$-*MMP*, for $(X/S, \Delta) \in \mathscr{C}^l$ is the algorithm which outputs $(Y/S, \Gamma) \in \mathscr{C}^l$ with a birational contraction map $X \dashrightarrow Y/S$ in the following manner.

1 If $K_X + \Delta$ is relatively nef, then output $(X, \Delta) \in \mathscr{C}^l$ as a log minimal model.
2 If $K_X + \Delta$ is not relatively nef, then there exists a $(K_X + \Delta)$-negative extremal ray of $\overline{NE}(X/S)$ and it defines an extremal contraction $\pi : X \to Y/S$ by the cone and contraction theorems.
3 If π is a log Mori fibre space, then output $(X, \Delta) \in \mathscr{C}^l$.
4 If π is a log divisorial contraction, then $(Y, \pi_*\Delta) \in \mathscr{C}^l$ and $\rho(Y/S) = \rho(X/S) - 1$. Replace (X, Δ) by $(Y, \pi_*\Delta)$ and go back to 1.
5 If π is a log flipping contraction, then construct the log flip $f : X \dashrightarrow X^+$ of π, for which $(X^+, f_*\Delta) \in \mathscr{C}^l$ and $\rho(X^+/S) = \rho(X/S)$. Replace (X, Δ) by $(X^+, f_*\Delta)$ and go back to 1.

Remark 1.4.14   Nakayama [368] formulated the minimal model program for klt pairs in the analytic category, including the Kawamata–Viehweg vanishing, the base-point free theorem and the cone and contraction theorems.

The full form of termination of log flips is known up to dimension three [425]. The log abundance conjecture will be discussed in the last section.

Conjecture 1.4.15 (Termination of log flips)   *Let $(X/S, \Delta)$ be a* **Q***-factorial lc pair projective over a variety. Then there exists no infinite sequence $X = X_0 \dashrightarrow X_1 \dashrightarrow \cdots$ of elementary log flips $X_i \to Y_i \leftarrow X_{i+1}$ associated with an $(K_{X_i} + \Delta_i)$-negative extremal ray of $\overline{NE}(X_i/S)$ for the strict transform Δ_i of Δ.*

The following is the same as Proposition 1.3.27.

Proposition 1.4.16   *Let $(X/S, \Delta)$ be a* **Q***-factorial lc pair projective over a variety. If $K_X + \Delta$ is relatively pseudo-effective, then the output by the $(K_X + \Delta)/S$-MMP is always a log minimal model. If $K_X + \Delta$ is not relatively pseudo-effective, then the output is always a log Mori fibre space.*

We shall supplement the notions of plt, dlt and lc singularities. Whereas lc pairs form the maximal class in which the MMP works, the variety of an lc pair is not even Cohen–Macaulay in general.

Example 1.4.17   Let S be an abelian surface and let $\mathscr{L}$ be an ample invertible sheaf on S. Take the line bundle $B = \mathrm{Spec}_S \bigoplus_{i \in \mathbf{N}} \mathscr{L}^{\otimes i} \to S$. The natural projection $\pi \colon B \to X$ to the *affine cone* $X = \mathrm{Spec} \bigoplus_{i \in \mathbf{N}} H^0(\mathscr{L}^{\otimes i})$ of S is the contraction which contracts a section $E \simeq S$ of B/S to the vertex o. One has $K_B + E \sim 0$ since $(K_B + E)|_E = K_E \sim 0$ by adjunction. Hence $K_X = \pi_*(K_B + E) \sim 0$ and $K_B + E = \pi^* K_X$. In particular, X is lc. The spectral sequence gives an isomorphism $R^1 \pi_* \mathscr{O}_B \simeq H^1(\mathscr{O}_B) = H^1(\mathscr{O}_S) \simeq \mathbf{C}^2$.

We shall prove that X is not Cohen–Macaulay. Embed X into a projective variety $\bar{X}$ which is smooth outside o and extend π to the contraction $\bar{\pi} \colon \bar{B} \to \bar{X}$ isomorphic outside o. If X were Cohen–Macaulay, then by the duality and Serre vanishing, $H^2(\mathscr{O}_{\bar{X}}(-\bar{A})) = H^1(\omega_{\bar{X}} \otimes \mathscr{O}_{\bar{X}}(\bar{A}))^\vee = 0$ for a sufficiently ample Cartier divisor $\bar{A}$ on $\bar{X}$. On the other hand, $H^1(\mathscr{O}_{\bar{B}}(-\bar{\pi}^* \bar{A})) = H^2(\mathscr{O}_{\bar{B}}(K_{\bar{B}} + \bar{\pi}^* \bar{A}))^\vee = 0$ by Kawamata–Viehweg vanishing. Hence the spectral sequence $H^p(R^q \bar{\pi}_* \mathscr{O}_{\bar{B}}(-\bar{\pi}^* \bar{A})) \Rightarrow H^{p+q}(\mathscr{O}_{\bar{B}}(-\bar{\pi}^* \bar{A}))$ would give $R^1 \pi_* \mathscr{O}_B \simeq R^1 \bar{\pi}_* \mathscr{O}_{\bar{B}}(-\bar{\pi}^* \bar{A}) = 0$, which is a contradiction.

Dlt pairs form a class suited for inductive arguments on dimension. The variety of a dlt pair has *rational singularities* and it is Cohen–Macaulay as below. On the other hand, the definition of a dlt singularity is subtle. It is not an analytically local property as evidenced by Example 1.4.21.

Definition 1.4.18   Let X be a normal variety in characteristic zero. We say that X has *rational singularities* if $R^i \mu_* \mathscr{O}_{X'} = 0$ for all $i \geq 1$ for every resolution $\mu \colon X' \to X$. By Theorem 1.4.19, it is equivalent to the existence of some resolution μ having the required property.

Theorem 1.4.19 ([251, p.50 proposition]) *A resolution $\mu\colon X' \to X$ of a normal variety X satisfies $R^i\mu_*\mathscr{O}_{X'} = 0$ for all $i \geq 1$ if and only if X is Cohen–Macaulay with $\omega_X = \mu_*\omega_{X'}$.*

Theorem 1.4.20 ([249, theorem 1.3.6]) *If (X, Δ) is a dlt pair, then X has rational singularities and in particular it is Cohen–Macaulay.*

Example 1.4.21 The pair $(\mathbf{A}^2, D)$ with the sum D of the two axes given by $x_1 x_2$ is dlt but $(\mathbf{A}^2, C)$ with the nodal curve C given by $x_1 x_2 + x_1^3 + x_2^3$ is only lc. They are analytically isomorphic at origin. See also Example 2.1.2.

Szabó's result characterises a dlt singularity in an alternative way.

Theorem 1.4.22 (Szabó [437]) *A pair (X, Δ) is dlt if and only if there exists a smooth dense open subset U of X such that $\Delta|_U$ is reduced and snc and such that $a_E(X, \Delta) > 0$ for any divisor E over X with $c_X(E) \subset X \setminus U$. When X is quasi-projective, this implies the existence of an effective $\mathbf{R}$-divisor Δ' such that $(X, (1 - \varepsilon)\Delta + \varepsilon\Delta')$ is klt for any sufficiently small positive real number ε.*

By the *connectedness lemma* of Shokurov and Kollár, a dlt pair (X, Δ) is plt if and only if all the connected components of $\lfloor \Delta \rfloor$ are irreducible.

Theorem 1.4.23 (Connectedness lemma) *Let (X, Δ) be an algebraic or analytic pair and let $\pi\colon X \to Y$ be a proper morphism to a variety with connected fibres such that $-(K_X + \Delta)$ is π-nef and π-big. Let $\mu\colon X' \to X$ be a log resolution of (X, Δ) and write $K_{X'} + N = \mu^*(K_X + \Delta) + P$ with effective $\mathbf{R}$-divisors P and N without common components such that $\mu_*P = 0$ and $\mu_*N = \Delta$. Then the natural map $\mathscr{O}_Y \to (\pi \circ \mu)_*\mathscr{O}_{\lfloor N \rfloor}$ is surjective.*

Proof This is an application of Kawamata–Viehweg vanishing to the klt pair (X', B) for $B = (N - P) - \lfloor N - P \rfloor$ with $\pi' = \pi \circ \mu\colon X' \to Y$. The vanishing also holds in the analytic category as noted in Remark 1.4.14. Since $\lceil P \rceil - \lfloor N \rfloor - (K_{X'} + B) = -\mu^*(K_X + \Delta)$ is π'-nef and π'-big, it follows from Theorem 1.4.5 that $R^1\pi'_*\mathscr{O}_{X'}(\lceil P \rceil - \lfloor N \rfloor) = 0$. Thus one has the surjection

$$\pi'_*\mathscr{O}_{X'}(\lceil P \rceil) \twoheadrightarrow \pi'_*\mathscr{O}_{\lfloor N \rfloor}(\lceil P \rceil|_{\lfloor N \rfloor}).$$

The left-hand side equals $\mathscr{O}_Y$ since $\mu_*\mathscr{O}_{X'}(\lceil P \rceil) = \mathscr{O}_X$. Hence the above map factors through the natural map $\mathscr{O}_Y \to \pi'_*\mathscr{O}_{\lfloor N \rfloor}$, which must be surjective. $\square$

Finally we discuss formulae of adjunction type. Let $(X, S + B)$ be a pair such that S is reduced and has no common components with B. Take a log resolution $\mu\colon X' \to X$ of $(X, S + B)$ and write $K_{X'} + S' + B' = \mu^*(K_X + S + B)$ with the strict transform S' of S and an $\mathbf{R}$-divisor B' such that $\mu_*B' = B$. Choose it so that $S' \to S$ factors through the normalisation $\nu\colon S^\nu \to S$ with a contraction $\mu_S \colon S' \to S^\nu$. We define the **R**-divisor $B_{S^\nu} = \mu_{S*}(B'|_{S'})$ on S^ν. Then $K_{S^\nu} + B_{S^\nu}$ is an **R**-Cartier **R**-divisor such that $K_{S'} + B'|_{S'} = \mu_S^*(K_{S^\nu} + B_{S^\nu})$. If S is Cartier and normal and B is zero, then B_{S^ν} is zero by adjunction.

Definition 1.4.24 ([424, section 3]) The **R**-divisor $B_{S^\nu} = \mu_{S*}(B'|_{S'})$ is called the *different* on S^ν of the pair $(X, S + B)$. The different B_{S^ν} is effective and independent of the choice of the log resolution μ. It satisfies the *adjunction* $\nu^*((K_X + S + B)|_S) = K_{S^\nu} + B_{S^\nu}$ via the pull-back $\operatorname{Pic} X \otimes \mathbf{R} \to \operatorname{Pic} S^\nu \otimes \mathbf{R}$.

Example 1.4.25 Consider the surface X in $\mathbf{A}^3$ given by $x_1^2 - x_2 x_3$ and the x_3-axis D in Example 1.2.1. The blow-up $\mu \colon X' \to X$ at origin o is a log resolution of (X, D) with $K_{X'} + D' + (1/2)E = \mu^*(K_X + D)$ for the strict transform D' of D and the exceptional curve E. Hence $K_D + (1/2)o = (K_X + D)|_D$, that is, $(1/2)o$ is the different on D of the pair (X, D). One can regard X as the quotient $\mathbf{A}^2/\mathbf{Z}_2$ by the morphism $\mathbf{A}^2 \to X$ given by $(x_1, x_2, x_3) = (y_1 y_2, y_1^2, y_2^2)$ for the coordinates y_1, y_2 of $\mathbf{A}^2$. In general for the cyclic quotient $X = \mathbf{A}^2/\mathbf{Z}_r(1, w)$ in Definition 2.2.10 with w coprime to r, each axis l of X satisfies the formula $K_l + (1 - r^{-1})o = (K_X + l)|_l$.

The *inversion of adjunction* compares the singularities on X with those on S^ν. The only-if part of each item in the next theorem is evident. The if part of the first item follows from the connectedness lemma.

Theorem 1.4.26 (Inversion of adjunction) *Let $(X, S + B)$ be an algebraic or analytic pair such that S is reduced and has no common components with B. Let S^ν be the normalisation of S and let B_{S^ν} denote the different on S^ν of $(X, S + B)$.*

(i) *The pair $(X, S + B)$ is plt about S if and only if (S^ν, B_{S^ν}) is klt. In this case, S is normal.*

(ii) ([225]) *The pair $(X, S + B)$ is lc about S if and only if (S^ν, B_{S^ν}) is lc.*

There is a generalisation of the adjunction. Let (X, Δ) be a pair. A *non-klt centre* of (X, Δ) means the centre $c_X(E)$ in X of a divisor E over X such that $a_E(X, \Delta) \leq 0$. The union of all non-klt centres is called the *non-klt locus* of (X, Δ). It is the complement of the maximal open subset U of X where $(U, \Delta|_U)$ is klt. If (X, Δ) is lc, then a non-klt centre of (X, Δ) is often called an *lc centre*. An lc centre that is minimal with respect to inclusion is called a *minimal lc centre*. A minimal lc centre exists and it is normal [129, theorem 9.1].

Theorem 1.4.27 (Subadjunction formula [133, theorem 4.1], [244]) *Let (X, Δ) be an lc pair such that X is projective and let Z be a minimal lc centre of (X, Δ). Then Z admits a klt pair (Z, Δ_Z) which satisfies the adjunction $(K_X + \Delta)|_Z \sim_\mathbf{R} K_Z + \Delta_Z$.*

For fibrations, we state Ambro's adjunction formula. The extension to the case of $\mathbf{R}$-divisors is in [133, theorem 3.1].

Theorem 1.4.28 (Ambro [13, theorem 0.2]) *Let* (X, Δ) *be a klt pair such that* X *is projective and let* $\pi \colon X \to S$ *be a contraction to a normal projective variety with* $K_X + \Delta \sim_{\mathbf{R},S} 0$. *Then* S *admits a klt pair* (S, Δ_S) *which satisfies the adjunction* $K_X + \Delta \sim_{\mathbf{R}} \pi^*(K_S + \Delta_S)$ *via* $\pi^* \colon \operatorname{Pic} S \otimes \mathbf{R} \to \operatorname{Pic} X \otimes \mathbf{R}$.

1.5 Existence of Flips

The existence of flips [48], [172] by Hacon and McKernan with Birkar and Cascini is a landmark in the minimal model theory. This section is an introduction to their work. The books [94] and [169] treat it.

Definition 1.5.1 Let $f \colon X \dashrightarrow Y$ be a birational contraction map defined in Definition 1.3.15. Let D be an $\mathbf{R}$-Cartier $\mathbf{R}$-divisor on X such that $D_Y = f_* D$ is also $\mathbf{R}$-Cartier. We say that f is *non-positive* with respect to D (or D-*non-positive*) if $\operatorname{ord}_E D \geq \operatorname{ord}_E D_Y$ for all divisors E over X. We say that f is *negative* with respect to D (or D-*negative*) if it is D-non-positive and further $\operatorname{ord}_E D > \operatorname{ord}_E D_Y$ for every prime divisor E on X exceptional over Y.

We say that f is *crepant* with respect to D if $\operatorname{ord}_E D = \operatorname{ord}_E D_Y$ for all divisors E over X. This is equivalent to the equality $p^* D = q^* D_Y$ on a common resolution W with $p \colon W \to X$ and $q \colon W \to Y$. We simply say that f is *crepant* if it is crepant with respect to the canonical divisor K_X.

We define models of a pair over a fixed variety S.

Definition 1.5.2 Let X be a normal variety which is projective over a variety S. Let D be an $\mathbf{R}$-Cartier $\mathbf{R}$-divisor on X. Let $g \colon X \dashrightarrow Z/S$ be a rational map to a normal variety projective over S which is resolved as $g = q \circ p^{-1}$ by a resolution $p \colon W \to X$ and a contraction $q \colon W \to Z$. We call g the *ample model* of D over S if $p^* D \sim_{\mathbf{R},S} q^* A + E$ with a relatively ample $\mathbf{R}$-divisor A on Z and an effective $\mathbf{R}$-divisor E on W such that $E|_{U_W} \leq B$ for any open subset U of S and any effective $\mathbf{R}$-divisor $B \sim_{\mathbf{R},U} p^* D|_{U_W}$ on $U_W = W \times_S U$. The definition is independent of the choice of the resolution W.

Lemma 1.5.3 *The ample model is unique up to isomorphism if it exists.*

Proof Keep the notation in Definition 1.5.2. Suppose that another ample model $X \dashrightarrow Z'$ of D is realised with $q' \colon W \to Z'$ and $p^* D \sim_{\mathbf{R},S} (q')^* A' + E'$. If $E = E'$, then $q^* A \sim_{\mathbf{R},S} (q')^* A'$ and a relative curve in W/S is contracted by q if and only if it is contracted by q'. This means the isomorphism $q \simeq q'$

by Lemma 1.1.1. To see the equality $E = E'$, we may assume that S is quasi-projective. Since A' is relatively ample, there exists an effective **R**-divisor $B' \sim_{\mathbf{R},S} (q')^* A'$ which has no common components with E. It follows from the property of E that $E \le B' + E'$ and hence $E \le E'$. By symmetry, $E' \le E$ and thus $E = E'$. □

We have the following characterisation of the ample model when it is birational. In particular, the log flip $X \to Y \leftarrow X^+$ with respect to (X, Δ) is the ample model of $K_X + \Delta$ over Y.

Proposition 1.5.4 *Notation as in Definition 1.5.2. Suppose that g is birational. Then g is the ample model of D if and only if g is a D-non-positive birational contraction map such that $g_* D$ is relatively ample.*

Proof The if part is easy. By assumption, one can write $p^* D = q^*(g_* D) + E$ with relatively ample $g_* D$ and q-exceptional $E \ge 0$. Suppose that $B \sim_{\mathbf{R},S} p^* D$ is effective. Let $B' = B - q_*^{-1} q_* B$ be the q-exceptional part of B, for which $E - B' \sim_{\mathbf{R},Z} B - B' \ge 0$. Then $E \le B' \le B$ by the negativity lemma, showing that g is the ample model.

Conversely suppose that g is the ample model with $p^* D \sim_{\mathbf{R},S} q^* A + E$ as in Definition 1.5.2. Taking W suitably, we may assume that the q-exceptional locus Q equals the support of an effective **Q**-divisor F such that $q^* A - F$ is ample over S. Then $H = q^* A - F + \varepsilon E$ is still ample over S for small positive ε. Since $p^* D \sim_{\mathbf{R},S} H + F + (1 - \varepsilon)E$, the property of the ample model shows that $E \le F + (1 - \varepsilon)E$. Hence E is supported in Q.

Let P be a p-exceptional prime divisor. For a general curve C in P contracted by p, one has $(q^* A \cdot C) + (E \cdot C) = 0$ and $(q^* A \cdot C) \ge 0$. If $(E \cdot C) \ge 0$, then $(q^* A \cdot C) = 0$ and q contracts C. If $(E \cdot C) < 0$, then P appears in E. In both cases, P is q-exceptional. It follows that g is a birational contraction map and $g_* D \sim_{\mathbf{R},S} q_*(q^* A + E) = A$ is relatively ample. Then $p^* D = q^*(g_* D) + E$ and in particular g is D-non-positive. □

Definition 1.5.5 Let (X, Δ) be an lc pair projective over a variety S. Let $f \colon X \dashrightarrow Y/S$ be a birational contraction map over S to a normal variety Y projective over S. We call f (or Y) a *log minimal model* of $(X/S, \Delta)$ if f is $(K_X + \Delta)$-negative, $K_Y + f_* \Delta$ is relatively nef and Y is **Q**-factorial. If the identity $X \to X/S$ is a log minimal model, then $(X/S, \Delta)$ is a log minimal model in the sense in the preceding section. We call f (or Y) the *log canonical model* of $(X/S, \Delta)$ if f is the ample model of $K_X + \Delta$ over S. When X is terminal and Δ is zero, a log minimal model and the log canonical model of X/S are called a *minimal model* and the *canonical model* of X/S.

Log minimal models are isomorphic in codimension one and crepant with respect to the pairs.

Proposition 1.5.6 *Let $(X/S, \Delta)$ be an lc pair projective over a variety. Let $f: X \dashrightarrow X_1/S$ be a $(K_X + \Delta)$-negative birational contraction map over S to a normal variety projective over S, which makes an lc pair $(X_1/S, \Delta_1)$ with $\Delta_1 = f_* \Delta$. Then every log minimal model of $(X/S, \Delta)$ is a log minimal model of $(X_1/S, \Delta_1)$ and vice versa.*

Proof The vice-versa part is obvious. We shall prove the former part. Let $(X_2/S, \Delta_2)$ be a log minimal model of $(X/S, \Delta)$. It suffices to show that $g: X_1 \dashrightarrow X_2$ is a $(K_{X_1} + \Delta_1)$-negative birational contraction map.

The idea has appeared in Example 1.3.10. Take a common log resolution Y of (X, Δ), (X_1, Δ_1) and (X_2, Δ_2) with $\mu: Y \to X$ and $\mu_i: Y \to X_i$. Let T_i denote the sum of μ_i-exceptional prime divisors and let T denote the common part of T_1 and T_2. Write $\mu^*(K_X + \Delta) = \mu_i^*(K_{X_i} + \Delta_i) + E_i + F_i$ with **R**-divisors E_i and F_i supported in T and $T_i - T$ respectively. Since $X \dashrightarrow X_i$ is $(K_X + \Delta)$-negative, E_i and F_i are effective and the support of F_i equals $T_i - T$.

By the negativity lemma, the μ_1-exceptional **R**-divisor

$$E_1 - E_2 + F_1 = -\mu_1^*(K_{X_1} + \Delta_1) + \mu_2^*(K_{X_2} + \Delta_2) + F_2$$

is negative, that is, $E_1 \leq E_2$ and $F_1 \leq 0$. In particular $F_1 = 0$ and thus $T_1 = T \leq T_2$, which means that g is a birational contraction map. The $(K_{X_1} + \Delta_1)$-negativity of g follows from the expression $\mu_1^*(K_{X_1} + \Delta_1) = \mu_2^*(K_{X_2} + \Delta_2) + (E_2 - E_1) + F_2$ where $E_2 - E_1 \geq 0$.   □

Corollary 1.5.7 *Let $(X/S, \Delta)$ be an lc pair projective over a variety. Let $(X_i/S, \Delta_i)$ for $i = 1, 2$ be log minimal models of $(X/S, \Delta)$. Then $X_1 \dashrightarrow X_2$ is small and crepant with respect to $K_{X_1} + \Delta_1$.*

Proof By Proposition 1.5.6, $(X_2/S, \Delta_2)$ is a log minimal model of $(X_1/S, \Delta_1)$ and vice versa.   □

Let $(X/S, \Delta)$ be a **Q**-factorial lc pair projective over a variety. Assume that $K_X + \Delta$ is relatively pseudo-effective. Then every output $(Y/S, \Gamma)$ of the $(K_X + \Delta)/S$-MMP is a log minimal model of $(X/S, \Delta)$. If the log abundance holds for $(Y/S, \Gamma)$, that is, $K_Y + \Gamma$ is relatively semi-ample, then by Lemma 1.2.13, Y admits a contraction $\varphi: Y \to Z/S$ for which $K_Y + \Gamma \sim_\mathbf{R} \varphi^* A$ with a relatively ample **R**-divisor A on Z. The rational map $X \dashrightarrow Z$ is the ample model of $K_X + \Delta$ over S.

Birkar, Cascini, Hacon and McKernan proved the finiteness of models simultaneously with the existence of log flips for klt pairs. In Definition 1.5.5, we call f a *weak log canonical model* of $(X/S, \Delta)$ if f is $(K_X + \Delta)$-non-positive and $K_Y + f_*\Delta$ is relatively nef. Let $V_{\mathbf{Q}}$ be a finite dimensional rational vector space and let $V = V_{\mathbf{Q}} \otimes_{\mathbf{Q}} \mathbf{R}$. A *rational polytope* in V is the convex hull of a finite number of points in $V_{\mathbf{Q}}$.

Theorem 1.5.8 ([48, corollary 1.1.5, theorem E]) *Let X be a normal variety projective over a quasi-projective variety S. Let $V = V_{\mathbf{Q}} \otimes_{\mathbf{Q}} \mathbf{R}$ be the extension of a finite dimensional vector subspace $V_{\mathbf{Q}}$ of the rational vector space $Z^1(X) \otimes \mathbf{Q}$ of $\mathbf{Q}$-divisors on X. Let C be a rational polytope in V such that for all $\Delta \in C$, (X, Δ) is klt and $A_\Delta \leq \Delta$ for some relatively ample $\mathbf{Q}$-divisor A_Δ. Define the set $\mathscr{E}$ of $\Delta \in C$ such that $K_X + \Delta$ is relatively pseudo-effective.*

(i) *There exist finitely many rational maps $g_i : X \dashrightarrow Y_i/S$ such that*

$$\mathscr{E} = \bigsqcup_i \mathscr{A}_i, \qquad \mathscr{A}_i = \{\Delta \in \mathscr{E} \mid g_i \text{ the ample model of } K_X + \Delta \text{ over } S\}.$$

Every $\mathscr{A}_i$ has the closure $\bar{\mathscr{A}}_i$ which is a finite union of rational polytopes. Moreover if $\bar{\mathscr{A}}_i \cap \mathscr{A}_j \neq \emptyset$, then there exists a contraction $g_{ji} : Y_i \to Y_j$ such that $g_j = g_{ji} \circ g_i$.

(ii) *There exist finitely many birational maps $f_i : X \dashrightarrow W_i/S$ such that*

$$\mathscr{E} = \bigcup_i \mathscr{B}_i, \qquad \mathscr{B}_i = \{\Delta \in \mathscr{E} \mid f_i \text{ a weak lc model of } (X/S, \Delta)\}$$

and such that every weak lc model of $(X/S, \Delta)$ with $\Delta \in \mathscr{E}$ is isomorphic to some f_i. Every $\mathscr{B}_i$ is a rational polytope. For any $\mathscr{B}_i$, there exist some $\mathscr{A}_j$ and a contraction $h_{ji} : W_i \to Y_j$ such that $\mathscr{B}_i \subset \bar{\mathscr{A}}_j$ and $g_j = h_{ji} \circ f_i$.

Remark 1.5.9 In the theorem with the quasi-projective condition relaxed,

$$\mathscr{P} = \{\Delta \in V \mid (X, \Delta) \text{ lc}, \ K_X + \Delta \text{ relatively nef}\}$$

is also a rational polytope and called *Shokurov's polytope* [425]. The proof in [44, proposition 3.2], [131, theorem 4.7.2] uses the boundedness of length of an extremal ray in Theorem 1.4.10.

By the finiteness of models, one can run the MMP *with scaling* practically.

Definition 1.5.10 Let (X, Δ) be a $\mathbf{Q}$-factorial klt pair projective over a variety S. Fix a relatively big $\mathbf{R}$-divisor A such that $(X, \Delta + A)$ is klt and such that $K_X + \Delta + A$ is relatively nef. The $(K_X + \Delta)$-*minimal model program* over S with *scaling* of A is a $(K_X + \Delta)/S$-MMP $(X, \Delta) = (X_0, \Delta_0) \dashrightarrow (X_1, \Delta_1) \dashrightarrow \cdots$ running in the following manner, where $A_0 = A$ and $t_{-1} = 1$.

1 If $K_{X_i} + \Delta_i$ is relatively nef, then output $(X_i/S, \Delta_i)$.

2 If $K_{X_i} + \Delta_i$ is not relatively nef, then define the *nef threshold* $t_i \in \mathbf{R}_{>0}$ as the least real number such that $K_{X_i} + \Delta_i + t_i A_i$ is relatively nef. It satisfies $t_i \leq t_{i-1}$. By Lemma 1.5.11, there exists a $(K_{X_i} + \Delta_i)$-negative extremal ray $\mathbf{R}_{\geq 0}[C_i]$ of $\overline{\mathrm{NE}}(X_i/S)$ such that $((K_{X_i} + \Delta_i + t_i A_i) \cdot C_i) = 0$. Take the extremal contraction $\pi_i \colon X_i \to Y_i/S$ associated with this ray.

3 If π_i is a log Mori fibre space, then output $(X_i/S, \Delta_i)$.

4 If π_i is a log divisorial contraction, then take $X_{i+1} = Y_i$. If π_i is a log flipping contraction, then construct the log flip $X_{i+1} \to Y_i$ of π_i by Theorem 1.4.11. In each case, set $f_i \colon X_i \dashrightarrow X_{i+1}$, $\Delta_{i+1} = f_{i*}\Delta_i$ and $A_{i+1} = f_{i*}A_i$. Then f_i is crepant with respect to $K_{X_i} + \Delta_i + t_i A_i$ and hence $K_{X_{i+1}} + \Delta_{i+1} + t_i A_{i+1}$ is relatively nef. Replace (X_i, Δ_i) by (X_{i+1}, Δ_{i+1}) and go back to 1.

Lemma 1.5.11 *Let $(X/S, \Delta)$ be a klt pair projective over a variety. Let A be a relatively big $\mathbf{R}$-divisor such that $(X, \Delta + A)$ is klt and such that $K_X + \Delta + A$ is relatively nef but $K_X + \Delta + tA$ is not relatively nef for any $t < 1$. Then there exists a $(K_X + \Delta)$-negative extremal ray $\mathbf{R}_{\geq 0}[C]$ of $\overline{\mathrm{NE}}(X/S)$ such that $((K_X + \Delta + A) \cdot C) = 0$.*

Proof   We shall verify the finiteness of the number of $(K_X + \Delta + A/2)$-negative extremal rays of $\overline{\mathrm{NE}}(X/S)$, which implies the lemma. We may assume that S is quasi-projective. By Kodaira's lemma, one can write $A = H + E$ with a relatively ample $\mathbf{Q}$-divisor H and an effective $\mathbf{R}$-divisor E. Take $D = \Delta + (1-\varepsilon)A/2 + \varepsilon E/2$ with small positive ε. Then (X, D) is klt and $K_X + \Delta + A/2 = K_X + D + \varepsilon H/2$. Thus the finiteness follows from the cone theorem.   □

Remark 1.5.12   The lemma actually holds for an lc pair $(X/S, \Delta)$ and an $\mathbf{R}$-divisor A such that $(X, \Delta + A)$ is lc. This is an elementary consequence [43, lemma 3.1] of Theorem 1.4.10 and Remark 1.5.9. Hence the MMP with scaling may be formulated for lc pairs without relative bigness of A.

The advantage of the MMP in Definition 1.5.10 is that each step X_i is a weak lc model of $(X/S, \Delta + t_i A)$. This yields the following termination. As a special case, the MMP with scaling functions when $X \to S$ is generically finite because all divisors are relatively big.

Corollary 1.5.13   *Let $(X/S, \Delta)$ be a $\mathbf{Q}$-factorial klt pair projective over a variety. If Δ is relatively big or $K_X + \Delta$ is not relatively pseudo-effective, then the $(K_X + \Delta)$-MMP over S with scaling terminates.*

Proof   Keep the notation in Definition 1.5.10. When Δ is relatively big, let $\varepsilon = 0$. When $K_X + \Delta$ is not relatively pseudo-effective, choose a rational number $0 < \varepsilon \leq 1$ such that $K_X + \Delta + \varepsilon A$ is not relatively pseudo-effective. Take the closed interval $I = [\varepsilon, 1]$. Then $\Delta + tA$ is relatively big for all $t \in I$, and locally on S, each step X_i of the $(K_X + \Delta)/S$-MMP with scaling of A is a weak lc model of $(X/S, \Delta + t_i A)$ with $t_i \in I$. By Theorem 1.5.8(ii), the number of weak lc models of $(X/S, \Delta + tA)$ with $t \in I$ is finite and thus so is the number of models X_i. Hence the MMP must terminate. Note that $X_i \dashrightarrow X_j$ with $i < j$ is never an isomorphism by Lemma 1.4.12. □

Corollary 1.5.14 *Let $(X/S, \Delta)$ be a $\mathbf{Q}$-factorial klt pair projective over a variety. If $K_X + \Delta$ is relatively big, then the $(K_X + \Delta)$-MMP over S with scaling terminates with a log minimal model $(Y/S, \Gamma)$ such that $K_Y + \Gamma$ is relatively semi-ample.*

Proof We may assume that S is quasi-projective. Then $(X, \Delta + B)$ is klt for a general effective $\mathbf{R}$-divisor $B \sim_{\mathbf{R},S} \varepsilon(K_X + \Delta)$ with small positive ε. Every $(K_X + \Delta)/S$-MMP with scaling of A is a $(K_X + \Delta + B)/S$-MMP with scaling of $(1 + \varepsilon)A$. By Corollary 1.5.13, it terminates with a log minimal model $(Y/S, \Gamma)$ with relatively big $K_Y + \Gamma$, which is relatively semi-ample by the base-point free theorem. □

Combining this with the work of Fujino and Mori [134], one obtains the finite generation of the log canonical ring.

Definition 1.5.15 The *canonical ring* of a normal variety X is the graded ring $\bigoplus_{i \in \mathbf{N}} H^0(\mathscr{O}_X(iK_X))$. The *log canonical ring* of a pair (X, Δ) is the graded ring $\bigoplus_{i \in \mathbf{N}} H^0(\mathscr{O}_X(\lfloor i(K_X + \Delta) \rfloor))$.

Theorem 1.5.16 *Let (X, Δ) be a klt pair such that X is complete and Δ is a $\mathbf{Q}$-divisor. Then the log canonical ring of (X, Δ) is finitely generated.*

The next proposition with $I = \emptyset$ shows the existence of a $\mathbf{Q}$-factorialisation of X which admits a klt pair (X, Δ). A $\mathbf{Q}$-*factorialisation* of a normal variety means a small contraction from a $\mathbf{Q}$-factorial normal variety to it.

Proposition 1.5.17 ([48, corollary 1.4.3]) *Let (X, Δ) be a klt pair and let I be an arbitrary subset of divisors E exceptional over X with $a_E(X, \Delta) \leq 1$. Then there exists a birational contraction $\pi \colon Y \to X$ from a $\mathbf{Q}$-factorial normal variety such that I coincides with the set of π-exceptional prime divisors on Y.*

Proof Take a log resolution $\mu \colon X' \to X$ of (X, Δ) on which every $E \in I$ is realised as a divisor. Write $K_{X'} + N = \mu^*(K_X + \Delta) + P$ with effective $\mathbf{R}$-divisors P and N without common components such that $\mu_* P = 0$ and $\mu_* N = \Delta$. Take the sum $S = \sum_i S_i$ of all the μ-exceptional prime divisors S_i not in I and make a klt pair $(X', N + \varepsilon S)$ with small positive ε.

We run the $(K_{X'} + N + \varepsilon S)/X$-MMP with scaling. It ends with a log minimal model (Y, N_Y) over X for the strict transform N_Y of $N + \varepsilon S$. Since $K_{X'} + N +$

$\varepsilon S \equiv_X P + \varepsilon S$, the negativity lemma shows that $f : X' \dashrightarrow Y$ contracts all prime divisors that appear in $P + \varepsilon S$. On the other hand, the $(P + \varepsilon S)$-negative map f does not contract any divisors in I. Hence f exactly contracts μ-exceptional prime divisors not in I and the result $Y \to X$ is a desired contraction. □

We quote a clever application which transforms an lc pair to a dlt pair.

Theorem 1.5.18 (Hacon [270, theorem 3.1]) *Let (X, Δ) be an lc pair such that X is quasi-projective. Then there exists a birational contraction $\pi : Y \to X$ from a $\mathbf{Q}$-factorial normal variety such that (Y, Δ_Y) is dlt for the $\mathbf{R}$-divisor Δ_Y defined by $K_Y + \Delta_Y = \pi^*(K_X + \Delta)$ with $\pi_* \Delta_Y = \Delta$.*

In the remainder of the section, we shall explain how the existence of flips is derived from the lower dimensional MMP. We follow the exposition [248]. The argument requires familiarity with the abstract theory of the MMP. The reader may treat this part only for reference.

Shokurov reduced the existence to that of pre-limiting flips. A *pre-limiting flip* (*pl flip*) is the log flip of a log flipping contraction from a $\mathbf{Q}$-factorial dlt pair (X, Δ) such that $\lfloor \Delta \rfloor$ contains a prime divisor S with $-S$ relatively ample. The following assertion is sufficient after perturbation of Δ.

Theorem 1.5.19 *Let (X, Δ) be a $\mathbf{Q}$-factorial plt pair such that Δ is a $\mathbf{Q}$-divisor and $S = \lfloor \Delta \rfloor$ is a prime divisor. Let $\pi : X \to Z$ be an elementary log flipping contraction with respect to (X, Δ) such that $-S$ is π-ample. Then the log flip of π exists.*

Fix a positive integer l such that $l(K_X + \Delta)$ is integral and Cartier. The log flip of π is, if it exists, described as $\mathrm{Proj}_Z \mathscr{R}(Z, l(K_Z + \pi_* \Delta)) \to Z$ by the graded $\mathscr{O}_Z$-algebra $\mathscr{R}(Z, l(K_Z + \pi_* \Delta))$ in Notation 1.3.17. As will be seen in Lemma 5.1.2, the existence of the log flip is equivalent to the finite generation of $\mathscr{R}(Z, l(K_Z + \pi_* \Delta))$. Since $\rho(X/Z) = 1$ and $-S$ is π-ample, it is also equivalent to the finite generation of $\mathscr{R}(Z, S_Z)$ for $S_Z = \pi_* S$. The lemma below is used implicitly.

Lemma 1.5.20 *Let X be a variety and let $\mathscr{R} = \bigoplus_{i \in \mathbf{N}} \mathscr{R}_i$ be a graded $\mathscr{O}_X$-algebra which is an integral domain. Fix a positive integer l. Then $\mathscr{R}$ is a finitely generated $\mathscr{O}_X$-algebra if and only if so is the truncation $\mathscr{S} = \bigoplus_{i \in \mathbf{N}} \mathscr{R}_{il}$.*

Proof The only-if part is obvious. To see the converse, suppose that $\mathscr{S}$ is finitely generated, by which $\mathscr{S}$ is a noetherian domain. It suffices to prove that $\mathscr{M}_j = \bigoplus_{i \in \mathbf{N}} \mathscr{R}_{il+j}$ is a finite $\mathscr{S}$-module for each $0 \le j < l$. We may assume that $\mathscr{M}_j \ne 0$ and locally take a non-zero member m_j of $\mathscr{M}_j$. The multiplication by m_j^{l-1} defines an injection $\mathscr{M}_j \hookrightarrow \mathscr{S}$ of $\mathscr{S}$-modules. Since $\mathscr{S}$ is noetherian, the finiteness of $\mathscr{M}_j$ follows. □

By Theorem 1.4.26(i), S is normal. By Proposition 2.2.23, the quotient $\mathscr{Q}_i = \mathscr{O}_X(iS)/\mathscr{O}_X((i-1)S)$ is a divisorial sheaf on S. Let $\mathscr{S}_i$ denote the image of the induced map $\mathscr{O}_Z(iS_Z) = \pi_*\mathscr{O}_X(iS) \to \pi_*\mathscr{Q}_i$. It fits into the exact sequence

$$0 \to \mathscr{O}_Z((i-1)S_Z) \to \mathscr{O}_Z(iS_Z) \to \mathscr{S}_i \to 0.$$

Then $\mathscr{R}(Z, S_Z)$ is finitely generated if and only if so is the graded $\mathscr{O}_{S_Z}$-algebra $\bigoplus_{i\in\mathbf{N}} \mathscr{S}_i$. This is also equivalent to the finite generation of $\bigoplus_{i\in\mathbf{N}} \mathscr{R}_i$ for the image $\mathscr{R}_i$ of the restriction map

$$\pi_*\mathscr{O}_X(il(K_X + \Delta)) \to \pi_*\mathscr{O}_S(il(K_X + \Delta)|_S).$$

We shall demonstrate this finite generation applying the *extension theorem* due to Hacon–McKernan [170] and Takayama [438]. What will be used is the following variant.

Theorem 1.5.21 (Extension theorem [171, theorem 5.4.21]) *Let $\pi\colon X \to Z$ be a projective morphism from a smooth variety to a variety. Let (X, Δ) be a plt pair such that Δ is a $\mathbf{Q}$-divisor with snc support and $S = \lfloor\Delta\rfloor$ is a prime divisor. Take a positive integer l such that $L = l(K_X + \Delta)$ is integral. Suppose that $\Delta = S + A + C$ with a π-ample $\mathbf{Q}$-divisor A and an effective $\mathbf{Q}$-divisor C in which S does not appear. Suppose that the relative base locus of aL contains no lc centres of $(X, \lceil\Delta\rceil)$ for some positive integer a. Then the natural map $\pi_*\mathscr{O}_X(L) \to \pi_*\mathscr{O}_S(L|_S)$ is surjective.*

We fix a log resolution $\mu\colon Y \to X$ of (X, Δ) with $\pi_Y = \pi \circ \mu\colon Y \to Z$. Let T denote the strict transform in Y of S and let $\pi_T = \pi_Y|_T\colon T \to S_Z$, which is birational. We write $K_Y + \Delta_Y = \mu^*(K_X + \Delta) + P$ with effective $\mathbf{Q}$-divisors P and Δ_Y without common components such that $\mu_*P = 0$ and $\mu_*\Delta_Y = \Delta$. We may choose μ so that the support of $\Delta_Y - T$ is a disjoint union of prime divisors. Indeed for the expression $\Delta_Y = T + \sum_i e_i E_i$ with prime divisors E_i and $e_i \in \mathbf{Q}_{>0}$, let a be the maximum of $e_i + e_j$ such that $E_i \cap E_j \neq \emptyset$ and let b be the number of pairs (E_i, E_j) such that $E_i \cap E_j \neq \emptyset$ and $e_i + e_j = a$. Then the blow-up of Y along $E_i \cap E_j$ decreases (a, b) with respect to lexicographic order. We write $K_T + \Delta_T = (K_Y + \Delta_Y)|_T$ by $\Delta_T = (\Delta_Y - T)|_T$. Then (T, Δ_T) is terminal.

We build a tower $\cdots \to Y_2 \to Y_1 \to Y_0 = Y$ of log resolutions inductively.

Lemma 1.5.22 *There exists a sequence $\{(\mu_i, D_i)\}_{i\geq 1}$ of pairs of a log resolution $\mu_i\colon Y_i \to X$ of (X, Δ) factoring through Y_{i-1} as $Y_i \to Y_{i-1} \to X$ and a $\mathbf{Q}$-divisor D_i on Y_i which satisfies the following. Let $\pi_i = \pi \circ \mu_i\colon Y_i \to Z$ and $T_i = \mu_{i*}^{-1}S$ and write $K_{Y_i} + \Delta_i = \mu_i^*(K_X + \Delta) + P_i$ with $\mu_{i*}P_i = 0$ and $\mu_{i*}\Delta_i = \Delta$ in the same manner as above. Then for all i and j,*

- *the induced morphism $T_i \to T$ is an isomorphism,*
- $T_i \leq D_i \leq \Delta_i$ *and* ilD_i *is integral,*
- $il(K_{Y_i} + D_i)$ *and* $il(K_{Y_i} + \Delta_i)$ *have the same* π_i-*mobile part* M_i,
- $\pi_{i*}\mathcal{O}_{Y_i}(ail(K_{Y_i}+D_i)) \to \pi_{T*}\mathcal{O}_T(ail(K_T+C_i))$ *is surjective for any positive integer* a, *where* $C_i = (D_i - T_i)|_T$ *via* $T_i \simeq T$ *and*
- $iC_i + jC_j \leq (i+j)C_{i+j}$.

Proof For a log resolution $\mu_i \colon Y_i \to X$ which factors through Y_{i-1} and induces an isomorphism $T_i \simeq T$, we take the decomposition $il(K_{Y_i} + \Delta_i) = M_i + F_i$ into the π_i-mobile part M_i and the π_i-fixed part F_i. Note that F_i is π_i-exceptional. Let Σ_i denote the support of $\Delta_i - T_i$. We take Y_{i-1} as Y_i initially and blow up Y_i along $G \cap T_i$ for a prime divisor $G \subset \Sigma_i$ successively until the relative base locus of M_i contains no components of $\Sigma_i \cap T_i$. This blow-up keeps T_i isomorphic to T. We further blow up Y_i along $G \cap G'$ for prime divisors $G, G' \subset \Sigma_i$ until Σ_i is a disjoint union of prime divisors.

Let E_i be the maximal divisor on Y_i such that $E_i \leq il(\Delta_i - T_i)$ and $E_i \leq F_i$. In other words, E_i is the componentwise minimum of $il(\Delta_i - T_i)$ and F_i. We take $D_i = \Delta_i - E_i/il$. Then $T_i \leq D_i \leq \Delta_i$ and

$$il(K_{Y_i} + D_i) = M_i + (F_i - E_i)$$

with the effective divisor $F_i - E_i$ which has no common components with D_i. In particular, $il(K_{Y_i} + D_i)$ still has the π_i-mobile part M_i. One can apply Theorem 1.5.21 to the plt pair $(Y_i/Z, D_i)$, which asserts the surjection $\pi_{i*}\mathcal{O}_{Y_i}(ail(K_{Y_i} + D_i)) \twoheadrightarrow \pi_{T*}\mathcal{O}_T(ail(K_T + C_i))$ for any $a \in \mathbf{Z}_{>0}$ with $C_i = (D_i - T_i)|_T$ via $T_i \simeq T$.

The inequality $K_{Y_j} + \Delta_j \geq \mu_{ij}^*(K_{Y_i} + \Delta_i)$ by $\mu_{ij} \colon Y_j \to Y_i$ for $i \leq j$ is an equality about T_j, and $\Delta_j - T_j$ equals $\mu_{ij}^*(\Delta_i - T_i)$ about T_j. Hence about T_{i+j}, one has $F_{i+j} \leq \mu_{i,i+j}^*F_i + \mu_{j,i+j}^*F_j$ and $i\mu_{i,i+j}^*(D_i - T_i) + j\mu_{j,i+j}^*(D_j - T_j) \leq (i+j)(D_{i+j} - T_{i+j})$. This yields $iC_i + jC_j \leq (i+j)C_{i+j}$. $\square$

By the convexity $iC_i + jC_j \leq (i+j)C_{i+j}$ and the boundedness $C_i \leq \Delta_T$, the sequence $\{C_i\}_i$ converges to some effective **R**-divisor $C \leq \Delta_T$. The pair (T, C) is terminal as so is (T, Δ_T).

Take a log minimal model $f_i \colon T \dashrightarrow \bar{T}_i/S_Z$ of the terminal pair $(T, C_i)/S_Z$ by the lower dimensional MMP. We apply the finiteness of log minimal models in Theorem 1.5.8(ii) in the neighbourhood at C in the real vector space generated by prime divisors appearing in Δ_T. It shows that there exist only finitely many models $\bar{T}_i$ up to isomorphism. Then replacing μ by a higher log resolution, we can assume that every f_i is a contraction $T \to \bar{T}_i/S_Z$.

By the base-point free theorem, the relatively nef and big divisor $il(K_{\bar{T}_i} + f_{i*}C_i)$ is relatively semi-ample. Hence there exists a positive integer a_i such that $a_i i l (K_{\bar{T}_i} + f_{i*} C_i)$ is relatively free. The pull-back $Q_i = a_i i l f_i^* (K_{\bar{T}_i} + f_{i*} C_i)$ coincides with the π_T-mobile part of $a_i i l (K_T + C_i)$ and it is relatively free. By the surjection in Lemma 1.5.22, the π_i-mobile part R_i of $a_i i l (K_{Y_i} + D_i)$ becomes relatively free with $Q_i = R_i |_T$ after blowing up Y_i away from T_i if necessary.

We keep the notation M_i in Lemma 1.5.22 and define $L_i = M_i |_T$. Write $K_T = (\mu |_T)^* ((K_X + \Delta)|_S) + F$ with $F = P|_T - \Delta_T$. Then $\lceil F \rceil \geq 0$.

Lemma 1.5.23 *The sequence $\{L_i / i\}_i$ converges to an **R**-divisor A on T. The limit A is π_T-semi-ample and the π_T-mobile part of $\lceil iA + F \rceil$ is at most L_i.*

Proof Keep the notation $\mu_{ij} : Y_j \to Y_i$. Observe that $\mu_{i,i+j}^* M_i + \mu_{j,i+j}^* M_j \leq M_{i+j}$ and hence $L_i + L_j \leq L_{i+j} \leq (i + j) l (K_T + \Delta_T)$. Thus the limit A of $\{L_i / i\}_i$ exists with $A \leq l (K_T + \Delta_T)$. From $a_i M_i \leq R_i$ and $\mu_{i,a_i i}^* R_i \leq M_{a_i i}$, one has $a_i L_i \leq Q_i \leq L_{a_i i}$, that is, $L_i / i \leq Q_i / a_i i \leq L_{a_i i} / a_i i$. Hence A is also the limit of the sequence of $Q_i / a_i i = l f_i^* (K_{\bar{T}_i} + f_{i*} C_i)$. Choosing one of the finitely many models $\bar{T}_i$ up to isomorphism, one can describe A as $l f_i^* (K_{\bar{T}_i} + f_{i*} C)$. This is the pull-back of the relatively nef and big **R**-divisor $l (K_{\bar{T}_i} + f_{i*} C)$, which is relatively semi-ample by the base-point free theorem.

For a divisor B on Y_j or T, we write $\mathrm{Mb}(B)$ for the relative mobile part of B over Z or S_Z. For all $i, j \geq 1$, since

$$\left(\frac{iR_j}{a_{jj}} + P_j - \Delta_j \right) - K_{Y_j} = \frac{iR_j}{a_{jj}} - \mu_j^* (K_X + \Delta)$$

is relatively nef and big, $R^1 \pi_{j*} \mathcal{O}_{Y_j} (\lceil iR_j / a_{jj} + P_j - \Delta_j \rceil) = 0$ by Kawamata–Viehweg vanishing. This provides the surjection $\pi_{j*} \mathcal{O}_{Y_j} (\lceil iR_j / a_{jj} + \Gamma_j \rceil) \twoheadrightarrow \pi_{T*} \mathcal{O}_T (\lceil iQ_j / a_{jj} + F \rceil)$ for $\Gamma_j = P_j + T_j - \Delta_j$. Hence

$$\mathrm{Mb}\left(\left\lceil \frac{iQ_j}{a_{jj}} + F \right\rceil \right) \leq \mathrm{Mb}\left(\left\lceil \frac{iR_j}{a_{jj}} + \Gamma_j \right\rceil \right)\bigg|_T .$$

From $R_j / a_{jj} \leq l (K_{Y_j} + \Delta_j) = l \mu_j^* (K_X + \Delta) + l P_j$ and $\lceil \Gamma_j \rceil = \lceil P_j \rceil$, one has

$$\mathrm{Mb}\left(\left\lceil \frac{iR_j}{a_{jj}} + \Gamma_j \right\rceil \right)\bigg|_T \leq \mathrm{Mb}(i l \mu_j^* (K_X + \Delta) + i l P_j + \lceil P_j \rceil)|_T$$

$$= \mathrm{Mb}(i l \mu_j^* (K_X + \Delta))|_T = \mathrm{Mb}(i l \mu_i^* (K_X + \Delta))|_T = L_i .$$

Join these two inequalities into $\mathrm{Mb}(\lceil iQ_j / a_{jj} + F \rceil) \leq L_i$. This tends to the inequality $\mathrm{Mb}(\lceil iA + F \rceil) \leq L_i$ when j goes to infinity. $\square$

Proof of Theorem 1.5.19 We have observed that the theorem is equivalent to the finite generation of $\mathscr{R} = \bigoplus_{i \in \mathbf{N}} \mathscr{R}_i$ for the image $\mathscr{R}_i$ of $\pi_* \mathcal{O}_X (i l (K_X + \Delta)) \to \pi_* \mathcal{O}_S (i l (K_X + \Delta)|_S)$. By Lemma 1.5.22, $\mathscr{R}_i = \pi_{T*} \mathcal{O}_T (L_i)$. Note that $L_i \leq iA$

for the π_T-semi-ample **R**-divisor A in Lemma 1.5.23 since $\{L_{2^p i}/2^p i\}_{p \in \mathbf{N}}$ is non-decreasing and converges to A.

If A is a **Q**-divisor, then there exists a positive integer n such that nA is integral, Cartier and π_T-free. Then $inA \leq L_{in}$ by Lemma 1.5.23 and hence $L_{in} = inA$. Now $\mathscr{R}_{in} = \pi_{T*}\mathscr{O}_T(inA)$ and nA is π_T-free. Thus $\bigoplus_{i \in \mathbf{N}} \mathscr{R}_{in}$ is finitely generated and so is $\mathscr{R}$ by Lemma 1.5.20.

If A were not a **Q**-divisor, then by the Diophantine approximation in [48, lemma 3.7.6], for any $\varepsilon > 0$, there would exist $i \geq 1$ and a π_T-free Cartier divisor G such that $G \nleq iA$ and such that every prime divisor has coefficient at least $-\varepsilon$ in $iA - G$. The latter property implies that $\lceil F + iA - G \rceil$ is effective if ε is sufficiently small. Consequently $G \leq G + \lceil F + iA - G \rceil = \lceil iA + F \rceil$ and hence $G \leq L_i \leq iA$ by Lemma 1.5.23, which contradicts $G \nleq iA$.   □

1.6 Termination of Flips

In spite of the establishment of the termination of the MMP with scaling in the setting of Corollary 1.5.13, we still do not know whether an arbitrary MMP terminates or not. The full termination is only known up to dimension three.

Theorem 1.6.1 ([240], [424, theorem 4.1])   *The termination of log flips in Conjecture* 1.4.15 *holds in dimension three.*

The proof relies on the decreasing property of an invariant named the *difficulty*, which grew out of the work of Shokurov [423, definition 2.15]. We shall demonstrate the termination for terminal threefold pairs. Before this, we see that a terminal threefold singularity is isolated.

Lemma 1.6.2   *If* (X, Δ) *is a terminal threefold pair, then* X *has isolated singularities.*

Proof   We may assume that X is affine. Take a log resolution $\mu \colon X' \to X$ of (X, Δ) and write $K_{X'} + \Delta' = \mu^*(K_X + \Delta) + P$ with the strict transform Δ' of Δ and an exceptional **R**-divisor P. Every exceptional prime divisor has positive coefficient in P. Let C be an arbitrary curve in X. Take the general hyperplane section S of X, which intersects C properly at a general point x in C. It suffices to prove that S is smooth at x, from which we deduce that X is smooth at x.

By the general choice of S, μ is also a log resolution of $(X, \Delta + S)$ and $K_{S'} + \Delta'|_{S'} \equiv_S P|_{S'}$ with $S' = \mu_*^{-1}S = \mu^*S$. Unless $P|_{S'}$ is zero, $K_{S'}$ is not μ-nef by the negativity lemma and one can find and contract a relative (-1)-curve in S'/S. One eventually attains a smooth surface T/S in which the strict transform of $P|_{S'}$ becomes zero. Then $\mu_T \colon T \to S$ is finite. It is an isomorphism because S is Cohen–Macaulay by Theorem 1.4.20, or from the following direct argument. The natural map $\mathscr{O}_X(-S) \otimes R^1 \mu_* \mathscr{O}_{X'} = R^1 \mu_* \mathscr{O}_{X'}(-S') \to R^1 \mu_* \mathscr{O}_{X'}$ is surjective by $R^1 \mu_* \mathscr{O}_{S'} \simeq R^1 \mu_{T*} \mathscr{O}_T = 0$. Hence $R^1 \mu_* \mathscr{O}_{X'}$ is zero about S and $\mathscr{O}_X \to \mu_* \mathscr{O}_{S'} = \mu_{T*} \mathscr{O}_T$ is surjective. Then $T \simeq S$ and S is smooth. $\qquad\square$

We do not assume relative setting in the following theorem. We do not need **Q**-factoriality if the boundary is zero.

Theorem 1.6.3 *Let* (X, Δ) *be a* **Q***-factorial canonical threefold pair such that* $\lfloor \Delta \rfloor = 0$. *Then there exists no infinite sequence* $X = X_0 \dashrightarrow X_1 \dashrightarrow \cdots$ *of elementary log flips with respect to* (X_i, Δ_i) *for the strict transform* Δ_i *of* Δ.

Proof Write $\Delta = \sum_{j=1}^n d_j D_j$ with prime divisors D_j and $d_j \in \mathbf{R}_{>0}$. We shall prove the theorem by induction on the number n of components of Δ. We write the log flip as $X_i \to Y_i \leftarrow X_{i+1}$.

Let d be the maximum of d_j, where we define $d = 0$ if $n = 0$, that is, $\Delta = 0$. Discussing on a log resolution of (X, Δ), we have the finiteness of the number of divisors E over X with log discrepancy $a_E(X, \Delta) < 2 - d$. Define the finite set $S = \{s \in d + \sum_j \mathbf{N}d_j \mid s < 1\}$, where $S = \{0\}$ if $n = 0$. For $s \in S$, let $N(i, s)$ denote the number of divisors E over X_i such that $a_E(X_i, \Delta_i) < 2 - s$. We define the *difficulty*

$$D(i) = \sum_{s \in S} N(i, s).$$

The sequence $\{D(i)\}_i$ is non-increasing since the inequality $a_E(X_i, \Delta_i) \leq a_E(X_{i+1}, \Delta_{i+1})$ holds in our situation similarly to Lemma 1.4.12.

Assume that $n = 0$, in which $D(i) = N(i, 0)$. Take a curve C^+ in X_{i+1} contracted to a point in Y_i. In the same manner as for Lemma 1.4.12, every divisor E over X_{i+1} with centre $c_{X_{i+1}}(E) = C^+$ has $1 \leq a_E(X_i) < a_E(X_{i+1})$. Hence X_{i+1} is terminal at the generic point η of C^+ and thus smooth at η by Lemma 1.6.2. By blowing up X_{i+1} along C^+ about η, one obtains the exceptional divisor F with $a_F(X_{i+1}) = 2$. Since $a_F(X_i) < a_F(X_{i+1}) = 2$, F is counted in $N(i, 0)$ but not in $N(i + 1, 0)$. Thus the strict inequality $D(i + 1) < D(i)$ holds and the sequence must terminate.

When $n \geq 1$, we assume without loss of generality that $d_n = d$. Let D_{ji} be the strict transform in X_i of D_j. If there exists a curve C^+ in $D_{n,i+1}$ contracted to a point in Y_i, then the same argument as above shows that X_{i+1} is smooth at the generic point η of C^+. The divisor F obtained at η by the blow-up of X_{i+1} along C^+ has $a_F(X_i, \Delta_i) < a_F(X_{i+1}, \Delta_{i+1}) = 2 - s$ for $s = \sum_j d_j \operatorname{ord}_F D_{j,i+1} \in S$. It is counted in $N(i, s)$ but not in $N(i + 1, s)$. Hence for the normalisation T_i of D_{ni}, the induced map $T_i \dashrightarrow T_{i+1}$ is a morphism for all i after truncation of the sequence. Take a compactification $\bar{T}_0$ of T_0 which is smooth about $\bar{T}_0 \setminus T_0$, and compactify $T_i \to T_{i+1}$ to $\bar{T}_i \to \bar{T}_{i+1}$ naturally.

If D_{ni} contains a curve C contracted to a point in Y_i, then $\bar{T}_i \to \bar{T}_{i+1}$ is not isomorphic above C and hence $\rho(\bar{T}_{i+1}) < \rho(\bar{T}_i)$ as $(-K_{X_i} \cdot C) \neq 0$. Truncating the sequence again, we attain the situation where $T_i \simeq T_{i+1}$ for all i. Then D_{ni} is nef over Y_i and $X_i \dashrightarrow X_{i+1}$ is a log flip with respect to $(X_i, \sum_{j=1}^{n-1} d_j D_{ji})$. By the assumption of induction on n, the sequence terminates. $\qquad\square$

Shokurov reduced the termination in an arbitrary dimension to the two properties in Conjecture 1.6.5 of an invariant of singularity called the minimal log discrepancy. We include the proof for the sake of completion.

Definition 1.6.4 Let (X, Δ) be a pair. Let η be a scheme-theoretic point in X with closure $Z = \overline{\{\eta\}}$ in X. The *minimal log discrepancy* $\mathrm{mld}_\eta(X, \Delta)$ of the pair (X, Δ) at η is defined as

$$\mathrm{mld}_\eta(X, \Delta) = \inf\{a_E(X, \Delta) \mid E \text{ divisor over } X, \ c_X(E) = Z\}.$$

Taking a log resolution of $(X, \Delta, \mathfrak{p})$ for the ideal sheaf $\mathfrak{p}$ in $\mathcal{O}_X$ defining Z, one sees that $\mathrm{mld}_\eta(X, \Delta)$ is either a non-negative real number or minus infinity and that it is actually the minimum unless $\mathrm{mld}_\eta(X, \Delta) = -\infty$. It satisfies the relation $\mathrm{mld}_z(X, \Delta) = \mathrm{mld}_\eta(X, \Delta) + \dim Z$ for a general closed point z in Z. It is sometimes convenient to use the *minimal log discrepancy* of (X, Δ) *in a closed subset W of X* which is defined as

$$\mathrm{mld}_W(X, \Delta) = \inf\{a_E(X, \Delta) \mid E \text{ divisor over } X, \ c_X(E) \subset W\}.$$

We say that a subset I of the real numbers satisfies the *ascending chain condition*, or the *ACC* for short, if there exists no strictly increasing infinite sequence in I. We say that I satisfies the *descending chain condition*, or the *DCC*, if there exists no strictly decreasing infinite sequence in I.

Conjecture 1.6.5   (i) (Lower semi-continuity) *For a pair* (X, Δ), *the function* $X^m \to \mathbf{R}_{\geq 0} \cup \{-\infty\}$ *from the set X^m of closed points in X which sends x to* $\mathrm{mld}_x(X, \Delta)$ *is lower semi-continuous.*

(ii) (ACC) *Fix a positive integer n and a subset I of the positive real numbers which satisfies the DCC. Then the set*

$$\{\mathrm{mld}_x(X, \Delta) \mid (X, \Delta) \text{ pair}, \ \dim X = n, \ \Delta \in I\}$$

satisfies the ACC, where $\Delta \in I$ means that all coefficients in Δ belong to I.

The lower semi-continuity was verified in dimension three by Ambro [12] as an easy consequence of the MMP. On the other hand, the ACC was only proved in dimension two by Alexeev [7]. The proof requires deep numerical analysis of surface singularities.

We assume the projectivity of the variety in the following theorem.

Theorem 1.6.6 (Shokurov [427]) *Let (X, Δ) be an lc pair such that X is projective. Assume Conjecture 1.6.5 in the dimension of X. Then there exists no infinite sequence $X = X_0 \dashrightarrow X_1 \dashrightarrow \cdots$ of log flips $X_i \dashrightarrow X_{i+1}$ associated with an $(K_{X_i} + \Delta_i)$-negative extremal face of $\overline{\mathrm{NE}}(X_i)$ for the strict transform Δ_i of Δ.*

Proof We write the log flip $f_i \colon X_i \dashrightarrow X_{i+1}$ as $X_i \to Y_i \leftarrow X_{i+1}$ where X_i and Y_i are projective. Let Z_i and Z_i^+ denote the exceptional loci of $X_i \to Y_i$ and $X_{i+1} \to Y_i$ respectively.

Step 1 We use the ACC in this step. Let $a_i = \mathrm{mld}_{Z_i}(X_i, \Delta_i)$. We claim the existence of a real number a such that $a \le a_i$ for all i after truncation and such that $a = a_i$ for infinitely many i.

Consider the non-decreasing sequence $\alpha_0 \le \alpha_1 \le \cdots$ of $\alpha_i = \inf\{a_j \mid i \le j\}$. It suffices to show that for any i, there exists $j \ge i$ such that $\alpha_i = a_j$. Indeed, then $\alpha_i = \mathrm{mld}_\eta(X_j, \Delta_j) = \mathrm{mld}_z(X_j, \Delta_j) - \dim \overline{\{\eta\}}$ with some scheme-theoretic point η in Z_j and a general point z in $\overline{\{\eta\}}$. The ACC provides a real number a such that $a = \alpha_i$ for all sufficiently large i, which satisfies the property in the claim.

Let $\alpha_{il} = \min\{a_j \mid i \le j \le l\}$, which equals $a_j = a_E(X_j, \Delta_j)$ for some $i \le j \le l$ and a divisor E over X_j with $c_{X_j}(E) \subset Z_j$. Similarly to Lemma 1.4.12,

$$a_E(X_i, \Delta_i) \le a_E(X_{i+1}, \Delta_{i+1}) \le \cdots \le a_E(X_j, \Delta_j).$$

If $c_{X_p}(E)$ were contained in Z_p for some $i \le p < j$, then the strict inequality $a_E(X_p, \Delta_p) < a_E(X_{p+1}, \Delta_{p+1})$ would hold as in the same lemma and hence $a_p \le a_E(X_p, \Delta_p) < a_j$, which contradicts $a_j = \alpha_{il} \le a_p$. Thus $c_{X_p}(E) \not\subset Z_p$ for all $i \le p < j$. It follows that $X_i \dashrightarrow X_j$ is isomorphic at the generic point η_E of $c_{X_i}(E)$ and $\alpha_{il} = \mathrm{mld}_{\eta_E}(X_i, \Delta_i)$. This number belongs to a finite set given by the fixed pair (X_i, Δ_i). Hence the non-increasing sequence $\{\alpha_{il}\}_{l \ge i}$ stabilises at the infimum α_i.

Step 2 Take the truncation as in Step 1. For infinitely many i, Z_i has a scheme-theoretic point η_i with $\mathrm{mld}_{\eta_i}(X_i, \Delta_i) = \mathrm{mld}_{Z_i}(X_i, \Delta_i) = a$. Let d be the maximum number such that for infinitely many i, there exists $\eta_i \in Z_i$ with $\mathrm{mld}_{\eta_i}(X_i, \Delta_i) = a$ and $d = \dim \overline{\{\eta_i\}}$. We take the Zariski closure W_i in X_i of the subset of scheme-theoretic points

$$\{\eta \in X_i \mid \mathrm{mld}_\eta(X_i, \Delta_i) = a, \ d = \dim \overline{\{\eta\}}\}.$$

Then $W_{i+1} = \overline{W_{i+1} \setminus Z_i^+}$ because every divisor E over X_{i+1} with $c_{X_{i+1}}(E) \subset Z_i^+$ has $a \le a_i \le a_E(X_i, \Delta_i) < a_E(X_{i+1}, \Delta_{i+1})$. Thus f_i induces a birational map $W_i' \dashrightarrow W_{i+1}$ from the closure W_i' of $f_i^{-1}(W_{i+1} \setminus Z_i^+) \subset W_i$. By noetherian induction, every f_i induces a birational map $W_i \dashrightarrow W_{i+1}$ after truncation.

Step 3 We use the lower semi-continuity in this step. By the definition of W_i, closed points z in W_i with $\mathrm{mld}_z(X_i, \Delta_i) = a + d$ form a Zariski dense subset of W_i. By the lower semi-continuity, every closed point z in W_i has $\mathrm{mld}_z(X_i, \Delta_i) \le a + d$.

We shall prove that $W_i \cap Z_i$ is of dimension at most d and $W_{i+1} \cap Z_i^+$ is of dimension less than d. Let C be an irreducible component of $W_i \cap Z_i$ with generic point η. Then for a general point z in C,

$$a \le \mathrm{mld}_\eta(X_i, \Delta_i) = \mathrm{mld}_z(X_i, \Delta_i) - \dim C \le a + d - \dim C,$$

by which $\dim C \le d$. In like manner for an irreducible component C^+ of $W_{i+1} \cap Z_i^+$ with generic point η^+,

$$a \le \mathrm{mld}_{Z_i}(X_i, \Delta_i) < \mathrm{mld}_{\eta^+}(X_{i+1}, \Delta_{i+1}) \le a + d - \dim C^+$$

where the middle inequality follows as in Lemma 1.4.12. Thus $\dim C^+ < d$.

Therefore $W_i \dashrightarrow W_{i+1}$ is isomorphic outside loci in W_i and W_{i+1} of dimension at most d and it generates no new d-cycles on W_{i+1}. Further by the definition of d, it really contracts a d-cycle for infinitely many i. One can find an irreducible component of W_0 such that for the normalisation V_i of its strict transform in X_i, the induced map $V_i \dashrightarrow V_{i+1}$ has the same properties as $W_i \dashrightarrow W_{i+1}$ has, namely the properties mentioned above. Write $V_i \to U_i \leftarrow V_{i+1}$ with the normalisation U_i of the same image in Y_i of V_i and V_{i+1}.

Step 4 This step follows [127]. We consider the *Borel–Moore homology* $H_{2d}^{\mathrm{BM}}(V_i)$ of V_i as an analytic space [141, example 19.1.1]. Let $\Lambda_{2d}(V_i)$ be the subgroup of $H_{2d}^{\mathrm{BM}}(V_i)$ generated by algebraic d-cycles on V_i. We also define $\Lambda_{2d}(U_i) \subset H_{2d}^{\mathrm{BM}}(U_i)$ in the same manner.

Take the induced map

$$\Lambda_{2d}(V_i) \twoheadrightarrow \Lambda_{2d}(U_i) \simeq \Lambda_{2d}(V_{i+1}),$$

where the isomorphism $\Lambda_{2d}(V_{i+1}) \simeq \Lambda_{2d}(U_i)$ follows from that of $V_{i+1} \to U_i$ outside a locus in V_{i+1} of dimension less than d. By the projectivity of V_i, whenever $V_i \to U_i$ contracts a d-cycle, which occurs for infinitely many i, this cycle generates a non-trivial sublattice $\mathbf{Z}$ in the kernel of $\Lambda_{2d}(V_i) \twoheadrightarrow \Lambda_{2d}(U_i)$. Consequently the rank of $\Lambda_{2d}(V_i)$ would drop infinitely many times. This is a contradiction.   □

The minimal log discrepancy can be described in terms of the arc space. We quickly explain this when the variety is smooth.

For any natural number i, the functor on the category of algebraic schemes over an algebraically closed field k which sends X to $X \times \operatorname{Spec} k[t]/(t^{i+1})$ has the right adjoint functor J_i. A closed point in $J_i X$ corresponds to a morphism $\operatorname{Spec} k[t]/(t^{i+1}) \to X$. There exists a natural morphism $J_{i+1} X \to J_i X$. Note that $J_0 X = X$. We define the *arc space* of X as the inverse limit $J_\infty X = \varprojlim_i J_i X$. It is a noetherian scheme but not of finite type over $\operatorname{Spec} k$ in general. A closed point in $J_\infty X$ corresponds to a morphism $\gamma\colon \operatorname{Spec} k[[t]] \to X$ from the spectrum of the ring of formal power series. For a function f in $\mathcal{O}_X$, the order function $J_\infty X \to \mathbf{N} \cup \{\infty\}$ is defined by sending γ to the order of $\gamma^* f$ in the discrete valuation ring $k[[t]]$.

Let X be a smooth variety of dimension n. Then $J_{i+1} X \to J_i X$ is a vector bundle of rank n. A subset S of $J_\infty X$ is called a *cylinder* if it is the inverse image $\pi_i^{-1}(S_i)$ by $\pi_i\colon J_\infty X \to J_i X$ of a constructible subset S_i of $J_i X$ for some i, for which S_i equals $\pi_i(S)$. The *dimension* of S is defined as

$$\dim S = \dim \pi_i(S) - (i+1)n \in \mathbf{Z}$$

for any such i. An effective $\mathbf{R}$-divisor Δ on X defines the order function $\operatorname{ord}_\Delta\colon J_\infty X \to \mathbf{R}_{\geq 0} \cup \{\infty\}$, and the inverse image $(\operatorname{ord}_\Delta)^{-1}(r)$ is a cylinder in $J_\infty X$ for all $r \in \mathbf{R}_{\geq 0}$.

Theorem 1.6.7 (Ein–Mustață–Yasuda [117]) *Let (X, Δ) be a klt pair such that X is smooth and let W be a closed subset of X. Then for a real number a, $\operatorname{mld}_W(X, \Delta) < a$ if and only if there exists a non-negative real number r such that the cylinder $(\operatorname{ord}_\Delta)^{-1}(r) \cap \pi^{-1}(W)$ in $J_\infty X$ is of dimension greater than $-r - a$, where π denotes the projection $J_\infty X \to X$.*

This was used to prove the lower semi-continuity of minimal log discrepancies and the *precise inversion of adjunction*, a more precise version of Theorem 1.4.26, on local complete intersection (lci) varieties.

Conjecture 1.6.8 (Precise inversion of adjunction) *Let (X, Δ) be a pair such that $S = \lfloor \Delta \rfloor$ is reduced and normal. Let Δ_S denote the different on S of (X, Δ). Then $\operatorname{mld}_x(X, \Delta) = \operatorname{mld}_x(S, \Delta_S)$ for any $x \in S$.*

Theorem 1.6.9 *Conjecture 1.6.8 holds in the following cases.*

(i) ([48]) *(X, B) is klt for some B, S is $\mathbf{Q}$-Cartier and $\operatorname{mld}_x(X, \Delta) \leq 1$.*
(ii) ([116], [117]) *X and S are lci.*

Whilst the ACC for minimal log discrepancies remains open, that for log canonical thresholds has been established completely.

Definition 1.6.10　Let (X, Δ) be an lc pair and let A be a non-zero effective **R**-Cartier **R**-divisor on X. The *log canonical threshold* of A on (X, Δ) is the greatest real number t such that $(X, \Delta + tA)$ is lc.

Theorem 1.6.11 (Hacon–McKernan–Xu [174])　*Fix a positive integer n and a subset I of the positive real numbers which satisfies the DCC. Then the set of log canonical thresholds of A on (X, Δ) where X is of dimension n and the coefficients in Δ and A belong to I satisfies the ACC.*

The study of $\mathrm{mld}_x(X, \Delta)$ of an lc pair pertains to the analysis of a divisor E over X which *computes* $\mathrm{mld}_x(X, \Delta)$ in the sense that $a_E(X, \Delta) = \mathrm{mld}_x(X, \Delta)$ with $c_X(E) = x$. The perspective of the above theorem tempts us to want A such that $a_E(X, \Delta + A) = \mathrm{mld}_x(X, \Delta + A) = 0$, but it is not always possible.

Example 1.6.12 ([229, example 2])　Consider $o \in X = \mathbf{A}^2$ with coordinates x_1, x_2. Let C be the cuspidal curve in X defined by $x_1^2 + x_2^3$ and take the pair (X, Δ) with $\Delta = (2/3)C$. Let X_1 be the blow-up of X at o and let E_1 be the exceptional curve. For $i = 2$ and 3, let X_i be the blow-up of X_{i-1} at the point at which the strict transform of C meets E_{i-1} and let E_i be the exceptional curve in X_i. Then X_3 is a log resolution of (X, Δ) and $a_{E_1}(X, \Delta) = 2/3$, $a_{E_2}(X, \Delta) = 1, a_{E_3}(X, \Delta) = 1$. Thus E_1 is a unique divisor over X that computes $\mathrm{mld}_o(X, \Delta) = 2/3$. If an effective **R**-divisor A satisfies $a_{E_1}(X, \Delta + A) = 0$, then $\mathrm{ord}_{E_1} A = a_{E_1}(X, \Delta) = 2/3$. Since $\mathrm{ord}_{E_3} A \geq \mathrm{ord}_{E_1} A \cdot \mathrm{ord}_{E_3} E_1 = (2/3) \cdot 2 = 4/3$, one has $\mathrm{mld}_o(X, \Delta + A) \leq a_{E_3}(X, \Delta + A) \leq 1 - 4/3 < 0$.

1.7 Abundance

Recall that a *log minimal model* is a **Q**-factorial lc pair $(X/S, \Delta)$ projective over a variety such that $K_X + \Delta$ is relatively nef. The *log abundance* is the following generalisation of Conjecture 1.3.28. It is known up to dimension three. We remark that the boundary Δ may be assumed to be a **Q**-divisor by the rationality of Shokurov's polytope in Remark 1.5.9.

Conjecture 1.7.1 (Log abundance)　*If $(X/S, \Delta)$ is a log minimal model, then $K_X + \Delta$ is relatively semi-ample.*

Theorem 1.7.2 (Keel–Matsuki–McKernan [250])　*The log abundance in Conjecture* 1.7.1 *holds in dimension three.*

We shall explain the reduction of the abundance to the equality of Kodaira dimension κ and numerical Kodaira dimension ν.

Definition 1.7.3 Let X be a normal complete variety and let D be a **Q**-Cartier **Q**-divisor on X. The *Iitaka dimension* $\kappa(X, D)$ of D is defined as follows. We define $\kappa(X, D) = -\infty$ if $H^0(\mathscr{O}_X(\lfloor lD \rfloor)) = 0$ for all positive integers l. Otherwise we take the rational map $\varphi_l : X \dashrightarrow \mathbf{P}H^0(\mathscr{O}_X(\lfloor lD \rfloor))$ for each $l \in \mathbf{Z}_{>0}$ such that $H^0(\mathscr{O}_X(\lfloor lD \rfloor)) \neq 0$. There exists a contraction $\varphi' : X' \to Y'$ with a birational contraction $\mu : X' \to X$ such that for any sufficiently large and divisible l, the rational map $X \dashrightarrow \varphi_l(X)$ to the image of φ_l is birational to φ'. Namely, $\varphi_l \circ \mu = f_l \circ \varphi'$ for a birational map $f_l : Y' \dashrightarrow \varphi_l(X)$. We define $\kappa(X, D)$ to be the dimension of Y'. Such φ' is called the *Iitaka fibration* associated with D. Note that $\kappa(F', \mu^* D|_{F'}) = 0$ for the fibre F' of φ' at a *very* general point in Y'.

If X is terminal, then we write $\kappa(X)$ for $\kappa(X, K_X)$ and call $\kappa(X)$ the *Kodaira dimension* of X. The Kodaira dimension is a birational invariant because every proper birational morphism $\mu : X' \to X$ of terminal varieties satisfies $\mu_* \mathscr{O}_{X'}(lK_{X'}) = \mathscr{O}_X(lK_X)$ for all $l \geq 0$. The variety X is said to be of *general type* if the Kodaira dimension $\kappa(X)$ equals the dimension of X.

Standard references for Iitaka fibrations are [196, chapter 10], [334] and [458]. The Iitaka dimension $\kappa(X, D)$ is minus infinity or a natural number up to the dimension of X. It attains the dimension of X if and only if D is big. One has $\kappa(X, D) = \kappa(X, qD)$ for any positive rational number q.

Theorem 1.7.4 (Easy addition) *Let $X \to Y$ be a contraction between normal projective varieties and let F be the general fibre. Let D be a Cartier divisor on X. Then $\kappa(X, D) \leq \kappa(F, D|_F) + \dim Y$.*

Proof We write $\kappa = \kappa(X, D)$. The inequality is trivial if $\kappa = -\infty$. If $\kappa \geq 0$, then $\kappa(F, D|_F) \geq 0$. In particular, the inequality is also evident when $\kappa = 0$. We shall assume that $\kappa \geq 1$. Take a positive integer l such that $\varphi_l : X \dashrightarrow \varphi_l(X) \subset P = \mathbf{P}H^0(\mathscr{O}_X(lD))$ is birational to the Iitaka fibration associated with D.

Let Z be the image of the rational map $X \dashrightarrow P \times Y$ defined by φ_l and $X \to Y$. The first projection $Z \to \varphi_l(X)$ is surjective, by which the dimension κ of $\varphi_l(X)$ satisfies

$$\kappa \leq \dim Z = \dim Z_y + \dim Y$$

for the general fibre $Z_y = Z \times_Y y$ at the image y in Y of F. Let V be the image of the restriction map $H^0(\mathscr{O}_X(lD)) \to H^0(\mathscr{O}_F(lD|_F))$. Then $Z_y \subset \mathbf{P}V \subset P$, and for the rational map $\varphi_{Fl} : F \dashrightarrow P_F = \mathbf{P}H^0(\mathscr{O}_F(lD|_F))$ given by $lD|_F$, the projection $P_F \dashrightarrow \mathbf{P}V$ induces a dominant map $\varphi_{Fl}(F) \dashrightarrow Z_y$. Hence

$$\dim Z_y \leq \dim \varphi_{Fl}(F) \leq \kappa(F, D|_F).$$

These two inequalities are combined into the one in the theorem. □

Iitaka originally defined $\kappa(X, D)$ as the order of growth of $h^0(\mathscr{O}_X(\lfloor lD \rfloor))$ as below. The proof is found in [196, theorem 10.2] and [288, corollary 2.1.38].

Theorem 1.7.5 *Let D be a Cartier divisor on a normal complete variety X and let $\kappa = \kappa(X, D)$. Then there exist a positive integer l_0 and positive real numbers a and b such that $al^\kappa \leq h^0(\mathscr{O}_X(ll_0D)) \leq bl^\kappa$ for all $l \in \mathbf{Z}_{>0}$.*

Definition 1.7.6 Let X be a normal complete variety and let D be a nef $\mathbf{Q}$-Cartier $\mathbf{Q}$-divisor on X. The *numerical Iitaka dimension* $\nu(X, D)$ of D is the maximal number ν such that $D^\nu \not\equiv 0$, that is, $(D^\nu \cdot Z) \neq 0$ for some ν-cycle Z. We define $\nu(X, D) = 0$ if $D \equiv 0$. If X is terminal and K_X is nef, then we write $\nu(X)$ for $\nu(X, K_X)$ and call $\nu(X)$ the *numerical Kodaira dimension* of X.

The numerical Iitaka dimension $\nu(X, D)$ is a natural number up to the dimension of X. The intersection number (D^n) of the nef $\mathbf{Q}$-divisor D, where n is the dimension of X, is the *volume* of D.

Definition 1.7.7 Let X be a normal complete variety of dimension n. The *volume* of a Cartier divisor D on X is the non-negative real number

$$\mathrm{vol}(D) = \limsup_{l \to \infty} \frac{h^0(\mathscr{O}_X(lD))}{l^n/n!}.$$

The volume of a $\mathbf{Q}$-Cartier $\mathbf{Q}$-divisor D is defined as $a^{-n} \mathrm{vol}(aD)$ by a positive integer a such that aD is integral and Cartier. By Theorem 1.7.5, D has positive volume if and only if D is big. The volume $\mathrm{vol}(D)$ is actually defined by the numerical equivalence class of D and it extends to a continuous function $\mathrm{vol} \colon N^1(X) \to \mathbf{R}_{\geq 0}$. Refer to [288, subsection 2.2.C] for the proof.

Example 1.7.8 Cutkosky and Srinivas [99, example 4] constructed an explicit example of a divisor with irrational volume. Let $S = C \times C$ be the product of an elliptic curve C. Let Δ be the diagonal on S and let $A = o \times C$ and $B = C \times o$ by a point o in C. Take the $\mathbf{P}^1$-bundle $\pi \colon X = \mathbf{P}(\mathscr{O}_S(-A - B - \Delta) \oplus \mathscr{O}_S) \to S$. The projection $\mathscr{O}_S(-A - B - \Delta) \oplus \mathscr{O}_S \twoheadrightarrow \mathscr{O}_S(-A - B - \Delta)$ gives a section S in X with $\mathscr{O}_X(S) \simeq \mathscr{O}_X(1)$. They computed the volume of the divisor $D = 2S + \pi^*(A + 2B + 3\Delta)$ as $\mathrm{vol}(D) = 36 + 4/\sqrt{3}$.

Lemma 1.7.9 *Let X be a normal complete variety of dimension n and let D be a nef $\mathbf{Q}$-Cartier $\mathbf{Q}$-divisor on X. Then $\kappa(X, D) \leq \nu(X, D)$. Moreover, $\kappa(X, D) = n$ if and only if $\nu(X, D) = n$, and $\mathrm{vol}(D) = (D^n)$ in this case.*

Proof We write $\kappa = \kappa(X, D)$ and $\nu = \nu(X, D)$. By Chow's lemma and Hironaka's resolution, we have a resolution $\mu \colon X' \to X$ such that X' is projective. Replacing X by X' and D by $l\mu^*D$ for suitable l, we may and shall assume that X is a smooth projective variety and D is a Cartier divisor such that $\varphi \colon X \dashrightarrow \varphi(X) \subset P = \mathbf{P}H^0(\mathscr{O}_X(D))$ is birational to the Iitaka fibration associated with D. Resolving φ, we may assume that it is a morphism $X \to P$.

One can write $D = \varphi^* H + E$ with a hyperplane H in P and an effective divisor E on X. Take an ample divisor A on X. Since D is nef, one has

$$0 < (\varphi^* H^\kappa \cdot A^{n-\kappa}) \le (\varphi^* H^{\kappa-1} \cdot DA^{n-\kappa}) \le \cdots \le (D^\kappa A^{n-\kappa}),$$

which implies that $D^\kappa \not\equiv 0$. Thus $\kappa \le \nu$.

Suppose that $\nu = n$. We take as A the general hyperplane section of X such that $A - K_X$ is ample. By Kleiman's criterion, $lD + A - K_X$ is ample for any $l \ge 0$ and hence $H^i(\mathscr{O}_X(lD + A)) = 0$ for all $i \ge 1$ by Kodaira vanishing. Thus $h^0(\mathscr{O}_X(lD + A)) = \chi(\mathscr{O}_X(lD + A))$, which is the sum of $\chi(\mathscr{O}_X(lD))$ and $\chi(\mathscr{O}_A((lD + A)|_A))$. The first summand $\chi(\mathscr{O}_X(lD))$ is expressed as $cl^n + O(l^{n-1})$ with $c = (D^n)/n! > 0$ by the asymptotic Riemann–Roch theorem. The second is $O(l^{n-1})$ by an application of Grothendieck's dévissage. Hence $h^0(\mathscr{O}_X(lD + A)) = cl^n + O(l^{n-1})$. With $h^0(\mathscr{O}_A((lD + A)|_A)) = O(l^{n-1})$, the exact sequence

$$0 \to H^0(\mathscr{O}_X(lD)) \to H^0(\mathscr{O}_X(lD + A)) \to H^0(\mathscr{O}_A((lD + A)|_A))$$

yields $h^0(\mathscr{O}_X(lD)) = cl^n + O(l^{n-1})$. Then $\kappa = n$ by Theorem 1.7.5, and $\mathrm{vol}(D) = n! \cdot c = (D^n)$. □

The inequality $\kappa(X, D) \le \nu(X, D)$ is not an equality in general. For instance, the divisor $D \equiv 0$ on a curve C of positive genus in Example 1.2.20 has $\kappa(C, D) = -\infty$ but $\nu(C, D) = 0$. On the other hand if D is semi-ample, then $\kappa(X, D) = \nu(X, D)$ since a multiple of D is the pull-back of an ample divisor. Thus the abundance includes the following conjecture.

Definition 1.7.10 Let X be a normal complete variety. We say that a nef **Q**-Cartier **Q**-divisor D on X is *abundant* if $\kappa(X, D) = \nu(X, D)$.

Conjecture 1.7.11 *Let* (X, Δ) *be a log minimal model such that* Δ *is a* **Q***-divisor. Then* $K_X + \Delta$ *is abundant.*

Kawamata [234] derived the abundance from the corresponding statement of this conjecture. This was generalised logarithmically and relatively by Nakayama [367, theorem 5] and reproved in [130].

Theorem 1.7.12 (Kawamata) *Let* $(X/S, \Delta)$ *be a klt pair such that* $X \to S$ *has connected fibres and* Δ *is a* **Q***-divisor. If* $K_X + \Delta$ *is relatively nef and the restriction* $(K_X + \Delta)|_F$ *to the general fibre* F *of* $X \to S$ *is abundant, then* $K_X + \Delta$ *is relatively semi-ample.*

The base-point free theorem provides the log abundance for a klt log minimal model $(X/S, \Delta)$ such that $K_X + \Delta$ is relatively big. In view of Lemma 1.7.9, this amounts to the case when the numerical Iitaka dimension ν is maximal. We also have the log abundance at the opposite extreme when ν is zero as stated below, proved by Nakayama [369, V theorem 4.8] for klt pairs and extended to lc pairs in [65], [144], [247]. This is formulated without the minimality of the pair.

Theorem 1.7.13 *Let (X, Δ) be an lc projective pair such that Δ is a $\mathbf{Q}$-divisor. Suppose that $K_X + \Delta$ is pseudo-effective and for any ample Cartier divisor A on X, there exists a constant c such that $h^0(\mathscr{O}_X(\lfloor l(K_X + \Delta) \rfloor + A)) \leq c$ for all positive l. Then $\kappa(X, K_X + \Delta) = 0$. In particular if (X, Δ) is a log minimal model with $\nu(X, K_X + \Delta) = 0$, then $\kappa(X, K_X + \Delta) = 0$ and $K_X + \Delta \sim_{\mathbf{Q}} 0$.*

Definition 1.7.14 Let $(X/S, \Delta)$ be an lc pair projective over a variety such that $K_X + \Delta$ is relatively pseudo-effective. We say that a log minimal model $(Y/S, \Gamma)$ of $(X/S, \Delta)$ is *good* if $K_Y + \Gamma$ is relatively semi-ample. If $(X/S, \Delta)$ has a good log minimal model, then by Corollary 1.5.7, every log minimal model of $(X/S, \Delta)$ is good.

We shall explain Lai's inductive approach to the abundance in the case when the Iitaka dimension is non-negative. Let (X, Δ) be a klt projective pair such that Δ is a $\mathbf{Q}$-divisor. Suppose that $\kappa(X, K_X + \Delta) \geq 0$. Take a log resolution $\mu\colon X' \to X$ which admits the Iitaka fibration $\varphi'\colon X' \to Y'$ associated with $K_X + \Delta$. One has a klt pair (X', Δ') with a $\mathbf{Q}$-divisor Δ' such that $\mu_* \Delta' = \Delta$ and such that μ is $(K_{X'} + \Delta')$-negative. Namely, $K_{X'} + \Delta' = \mu^*(K_X + \Delta) + P$ with an effective $\mathbf{Q}$-divisor P the support of which equals the μ-exceptional locus. Then by Proposition 1.5.6, every log minimal model of (X, Δ) is a log minimal model of (X', Δ') and vice versa. By the following theorem, (X, Δ) has a good log minimal model if so does the restriction $(F, \Delta'|_F)$ to the very general fibre of φ', where $\Delta'|_F$ can be defined since F meets the singular locus of X' in codimension at least two in F.

Theorem 1.7.15 (Lai [282]) *Let (X, Δ) be a klt projective pair such that Δ is a $\mathbf{Q}$-divisor. Suppose that $\kappa(X, K_X + \Delta) \geq 0$ and X admits the Iitaka fibration $\varphi\colon X \to Y$ associated with $K_X + \Delta$. Let F be the very general fibre of φ. If the klt pair $(F, \Delta|_F)$ has a good log minimal model, then so does (X, Δ).*

Thus by induction on dimension, the abundance for a klt projective pair (X, Δ) with $K_X + \Delta$ nef is reduced to the two statements which are

- the *non-vanishing* $\kappa(X, K_X + \Delta) \geq 0$ and
- the implication from $\nu(X, K_X + \Delta) \geq 1$ to $\kappa(X, K_X + \Delta) \geq 1$.

The proof of Theorem 1.7.15 uses the next application [282, proposition 2.7] of the work of Birkar, Cascini, Hacon and McKernan.

Proposition 1.7.16 *Let $(X/S, \Delta)$ be a $\mathbf{Q}$-factorial klt pair with a contraction $X \to S$. Suppose that the restriction $(F, \Delta|_F)$ to the very general fibre F of $X \to S$ has a good log minimal model. Then after finitely many steps in the $(K_X + \Delta)/S$-MMP with scaling, one attains a $\mathbf{Q}$-factorial klt pair $(X'/S, \Delta')$ such that the restriction $(F', \Delta'|_{F'})$ to the very general fibre F' of $X' \to S$ has semi-ample $K_{F'} + \Delta'|_{F'}$.*

Proof of Theorem 1.7.15   The Iitaka dimension $\kappa = \kappa(X, K_X + \Delta)$ equals the dimension of Y. The assertion is trivial if $\kappa = 0$ in which $F = X$. If κ equals the dimension of X, then the theorem follows from Corollary 1.5.14. We shall assume that $1 \leq \kappa < \dim X$. By $\mathbf{Q}$-factorialisation in Proposition 1.5.17, X may be assumed to be $\mathbf{Q}$-factorial.

Step 1   There exists a positive integer l such that $l(K_X + \Delta)$ is integral and such that $l(K_X + \Delta) = \varphi^* A + G$ with a hyperplane section A of Y and the fix part G of $l(K_X + \Delta)$. Applying Proposition 1.7.16 to φ and replacing X, we may assume that $K_F + \Delta|_F$ is semi-ample. Then $G|_F \sim l(K_F + \Delta|_F) \sim_{\mathbf{Q}} 0$ by $\kappa(F, K_F + \Delta|_F) = 0$. Hence $G|_F = 0$. In other words, G does not dominate Y. It suffices to derive the equality $G = 0$.

The reduction above is made by running the $(K_X + \Delta)/Y \equiv_Y l^{-1}G$-MMP with scaling of some big H. After further replacement of X, we may assume that this MMP no longer contracts any divisors on X. Then for any $\varepsilon \in \mathbf{Q}_{>0}$, a multiple of $G + \varepsilon H$ is mobile over Y. Indeed thanks to Corollary 1.5.13, the $(K_X + \Delta)/Y$-MMP with scaling of H produces a log minimal model $(X'/Y, \Delta' + l^{-1}\varepsilon H')$ of $(X/Y, \Delta + l^{-1}\varepsilon H)$. By the base-point free theorem, the big $\mathbf{Q}$-divisor $K_{X'} + \Delta' + l^{-1}\varepsilon H'$ is semi-ample over Y. This implies the relative mobility of a multiple of $K_X + \Delta + l^{-1}\varepsilon H \sim_{\mathbf{Q},Y} l^{-1}(G + \varepsilon H)$ because the rational map $X \dashrightarrow X'$ is small by assumption.

Step 2   Let E be a prime component of G. If the image $P = \varphi(E)$ were of codimenison at least two in Y, then as in the proof of Theorem 1.3.9, we would construct a birational contraction $\varphi|_S \colon S \to T$ from a normal surface by taking the base change to the intersection T of general hyperplane sections of Y and cutting it with general hyperplane sections of X. Then $G|_S$ is a non-zero exceptional divisor on S/T. By the negativity lemma, $G|_S$ is not $\varphi|_S$-nef and hence contains a curve C with $(G \cdot C) < 0$. Curves C realised in this manner cover some divisor E' on X with $\varphi(E') = P$. They have $((G + \varepsilon H) \cdot C) < 0$ for small positive ε, contradicting the relative mobility in Step 1.

Thus $P = \varphi(E)$ is a prime divisor on Y. Using Step 1 in like manner, one can also verify that $G|_{\varphi^{-1}(U)}$ is proportional to the pull-back $\varphi^* P|_U$ restricted to some smooth neighbourhood U in Y at the generic point of P.

Step 3 By Theorem 1.5.16, we may choose l so that jG is the fix part of $j\varphi^* A + jG$ for all $j \geq 0$. Then for any $0 \leq i \leq j$,

$$H^0(\mathscr{O}_X(j\varphi^* A)) = H^0(\mathscr{O}_X(j\varphi^* A + iG)) = H^0(\mathscr{O}_Y(jA) \otimes \varphi_* \mathscr{O}_X(iG)).$$

Thus for $0 \leq i < j$, the quotient $\mathscr{Q}_i = \varphi_* \mathscr{O}_X((i+1)G)/\varphi_* \mathscr{O}_X(iG)$ satisfies the inclusion

$$H^0(\mathscr{O}_Y(jA) \otimes \mathscr{Q}_i) \subset H^1(\mathscr{O}_Y(jA) \otimes \varphi_* \mathscr{O}_X(iG)).$$

Given $i \geq 0$, whenever j is sufficiently large, $\mathscr{O}_Y(jA) \otimes \mathscr{Q}_i$ is generated by global sections and $H^1(\mathscr{O}_Y(jA) \otimes \varphi_* \mathscr{O}_X(iG)) = 0$. Hence $\mathscr{Q}_i = 0$ from the above inclusion. It follows that $\mathscr{O}_Y = \varphi_* \mathscr{O}_X(iG)$ for all $i \geq 0$.

If $G \neq 0$ and has a component E, then by Step 2, $P = \varphi(E)$ is a divisor and there exists a smooth open subset $\iota : U \hookrightarrow Y$ such that the complement $Y \setminus U$ is of codimension at least two and such that $\varphi^* P|_U \leq iG|_{\varphi^{-1}(U)}$ for some $i \geq 1$. But then

$$\mathscr{O}_Y(P) = \iota_* \mathscr{O}_U(P|_U) \subset \iota_*(\varphi_* \mathscr{O}_X(iG))|_U = \iota_* \mathscr{O}_U = \mathscr{O}_Y,$$

which is absurd. Thus G must be zero.   □

The existence of good minimal models includes the *Iitaka conjecture*. Below we chronicle several known results. The case (iv) contains (i) and now also contains (v) by Corollary 1.5.14.

Conjecture 1.7.17 (Iitaka conjecture $C_{n,m}$) *Let $X \rightarrow Y$ be a contraction between smooth projective varieties, where X and Y are of dimension n and m respectively, and let F be the very general fibre. Then $\kappa(X) \geq \kappa(F) + \kappa(Y)$.*

Theorem 1.7.18 *Conjecture 1.7.17 holds in the following cases.*

  (i) ($C_{n,n-1}$, Viehweg [460]) *F is a curve.*
 (ii) ($C_{n,1}$, Kawamata [231]) *Y is a curve.*
(iii) (Viehweg [462], [464, corollary IV]) *Y is of general type.*
(iv) (Kawamata [233]) *F has a good minimal model.*
 (v) (Kollár [261]) *F is of general type.*

2

Singularities

In the study of birational geometry using the minimal model program, one inevitably encounters singularities. The singularity is an obstacle to the treatment of algebraic varieties but at the same time enriches the geometry by allowing more birational transformations such as flips.

Since a terminal threefold singularity is isolated, it is often more flexible to treat it in the analytic category. By Artin's algebraisation theorem, every isolated analytic singularity is algebraically realised. Tougeron's implicit function theorem and the Weierstrass preparation theorem are fundamental analytic tools for finding a simple function which defines a given germ.

Taking quotient produces singularities. We clarify the notion of quotient and define the weighted blow-up in the context of which cyclic quotient singularities appear. The converse of cyclic quotient is a cyclic cover and it is constructed by a divisor. The singularities upstairs and downstairs are compared by the ramification formula.

We furnish a complete classification of terminal threefold singularities due to Reid and Mori, which is a foundation of the explicit study of threefolds. The classification is divided into two stages. First we deal with singularities of index one by a method not relying on the theory of elliptic singularities. Next we describe singularities of higher index by taking the index-one cover. It turns out that the general member of the anti-canonical system of a terminal threefold singularity is always Du Val. This insight is known as the general elephant conjecture and plays a leading role in the analysis of threefold contractions.

Every terminal threefold singularity is deformed to a finite number of cyclic quotient singularities. Applying this, Reid established an explicit formula of Riemann–Roch type on a terminal projective threefold. It is a principal tool for the numerical classification of divisorial contractions in the next chapter. We also discuss canonical threefold singularities and bound the index by means of the singular Riemann–Roch formula.

2.1 Analytic Germs

Although we study a complex variety in the Zariski topology, it is often adapt-
able to treat a germ of it as an analytic space. The isomorphism class of the
germ of an analytic space is determined by the *completion* of the local ring.

Theorem 2.1.1 (Artin [18], Hironaka–Rossi [188])   *Two germs $v \in V$ and*
$w \in W$ of analytic spaces are isomorphic if and only if the completions $\hat{\mathcal{O}}_{V,v}$
and $\hat{\mathcal{O}}_{W,w}$ of the local rings are isomorphic.

Example 2.1.2   The germs $o \in D = (x_1 x_2)_{\mathbf{A}^2}$ and $o \in C = (x_1 x_2 + x_1^3 + x_2^3)_{\mathbf{A}^2}$
at origin in Example 1.4.21 are analytically but not algebraically isomorphic.
Indeed, $\hat{\mathcal{O}}_{D,o} \simeq \hat{\mathcal{O}}_{C,o} \simeq \mathbf{C}[[x_1,x_2]]/(x_1 x_2)$.

The essential step in the proof is to approximate a formal power series
solution to a system of analytic equations by a convergent one. This is known
as *Artin approximation*, the analytic form of which is as follows.

Theorem 2.1.3 (Artin approximation [18], [19])   *Let $f(x,y)$ be a system*
$(f_1(x,y),\ldots,f_k(x,y))$ of a finite number of convergent power series in $x =$
$(x_1,\ldots,x_n)$ and $y = (y_1,\ldots,y_m)$. Let $\hat{\mathfrak{m}}$ denote the maximal ideal in the ring
of formal power series in x. Let $\hat{y}(x) = (\hat{y}_1(x),\ldots,\hat{y}_m(x))$ with $\hat{y}_i(x) \in \hat{\mathfrak{m}}$ be
a formal power series solution to $f = 0$, that is, $f(x,\hat{y}(x)) = 0$. Then for any
positive integer l, there exists $y(x) = (y_1(x),\ldots,y_m(x))$ with convergent power
series $y_i(x)$ in x such that $f(x,y(x)) = 0$ and such that $y_i(x) - \hat{y}_i(x) \in \hat{\mathfrak{m}}^l$ for
all i.

A technical ingredient is *Tougeron's implicit function theorem* recalled in [19,
lemma 5.11]. We state a variant for a single equation as in [277, lemma 4.64].
Recall the symbol $o \in \mathfrak{D}^n$ in Notation 1.1.17.

Theorem 2.1.4 (Tougeron's implicit function theorem [449])   *Let $\mathbf{C}\{x\}$ denote*
the ring of convergent power series in $x = (x_1,\ldots,x_n)$. Let I be a proper ideal
in $\mathbf{C}\{x\}$. Let $f(x), g(x) \in \mathbf{C}\{x\}$ and put $J = (f, \partial f/\partial x_1,\ldots,\partial f/\partial x_n)\mathbf{C}\{x\}$. If
$f - g$ belongs to IJ^2, then there exist $d(x) \in I$ and $y(x) = (y_1(x),\ldots,y_n(x))$
with $y_i \in \mathbf{C}\{x\}$ such that $(1+d(x))f(x) = g(y(x))$ and such that $y_i(x) - x_i \in I$
for all i.

Corollary 2.1.5   *Let $o \in \mathfrak{D}^n$ be the analytic germ of a complex manifold and*
let $\mathfrak{m}$ denote the maximal ideal in $\mathcal{O}_{\mathfrak{D}^n}$ defining o. Let f be a function in $\mathcal{O}_{\mathfrak{D}^n}$
which defines an isolated hypersurface singularity. Then for any positive integer
l, there exists a positive integer L such that if $g \in \mathcal{O}_{\mathfrak{D}^n}$ satisfies $f - g \in \mathfrak{m}^L$,
then there exists an automorphism σ of $\mathcal{O}_{\mathfrak{D}^n,o}$ which sends f to ug with some
unit u and which induces the identity of $\mathcal{O}_{\mathfrak{D}^n}/\mathfrak{m}^l$.

Proof Take coordinates $x = (x_1, \ldots, x_n)$ of $o \in \mathfrak{D}^n$, by which $\mathscr{O}_{\mathfrak{D}^n, o} = \mathbf{C}\{x\}$. The ideal $J = (f, \partial f / \partial x_1, \ldots, \partial f / \partial x_n) \mathscr{O}_{\mathfrak{D}^n, o}$ is $\mathfrak{m}$-primary because f defines an isolated singularity. Hence $\mathfrak{m}^i \subset J$ for some i.

For any integer l at least two, we shall verify that $L = l + 2i$ satisfies the condition. Suppose that $f - g \in \mathfrak{m}^{l+2i}$. Then $f - g \in \mathfrak{m}^l J^2$ and hence there exist d and $y = (y_1, \ldots, y_n)$ as in Theorem 2.1.4 for $I = \mathfrak{m}^l$, which satisfies $(1 + d(x)) f(x) = g(y(x))$ and $y_i(x) - x_i \in \mathfrak{m}^l \subset \mathfrak{m}^2$. In particular, $y_i(x)$ form a regular system of parameters in $\mathscr{O}_{\mathfrak{D}^n, o}$. Define the automorphism σ of $\mathscr{O}_{\mathfrak{D}^n, o}$ by $\sigma(y_i(x)) = x_i$. It induces the identity of $\mathscr{O}_{\mathfrak{D}^n} / \mathfrak{m}^l$ and satisfies $(1 + d) f = \sigma^{-1}(g)$, that is, $\sigma(f) = ug$ with the unit $u = (1 + \sigma(d))^{-1}$. □

Remark 2.1.6 In the corollary if the cyclic group $\mathbf{Z}_r$ acts on $o \in \mathfrak{D}^n$ by which f is semi-invariant in Definition 2.2.9, then one can choose as σ an equivariant automorphism. One has only to replace σ by $r^{-1} \sum_{i=0}^{r-1} \tau^{-i} \sigma \tau^i$ assuming that $l \geq 2$, where τ is the automorphism of $\mathscr{O}_{\mathfrak{D}^n, o}$ given by a generator of $\mathbf{Z}_r$.

Example 2.1.7 Tougeron's theorem finds a simple polynomial which defines a given singularity. For example we consider a non-smooth isolated hypersurface singularity $o \in H \subset \mathfrak{D}^n$ defined by a convergent power series $f \in \mathbf{C}\{x_1, \ldots, x_n\}$. Suppose that f has quadratic part $x_1 x_2 + g_2(x_3, \ldots, x_n)$. Then one can choose as f a polynomial of form $x_1 x_2 + g(x_3, \ldots, x_n)$ after coordinate change which induces the identity of $\mathfrak{m} / \mathfrak{m}^2$ for the maximal ideal $\mathfrak{m}$ in $\mathscr{O}_{\mathfrak{D}^n}$ defining o.

To see this, let f_l be the truncation of f which consists of the terms in f of degree at most l. We claim that f_l becomes of form $x_1 x_2 + g_l(x_3, \ldots, x_n)$ after coordinate change. This holds for $l = 2$ by assumption. By induction, it suffices to derive the claim for $l + 1$ from the expression $f_l = x_1 x_2 + g_l$. Write

$$f_{l+1} = x_1 x_2 + g_l + a(x_1, \ldots, x_n) x_1 + b(x_2, \ldots, x_n) x_2 + c(x_3, \ldots, x_n)$$
$$= (x_1 + b)(x_2 + a) + g_l + c - ab$$

with homogeneous polynomials a, b, c of degree l, l, $l + 1$ unless zero. By replacements of x_1 by $x_1 + b$ and x_2 by $x_2 + a$, the right-hand side becomes $x_1 x_2 + g_l + c - ab$ with $ab \in \mathfrak{m}^{2l} \subset \mathfrak{m}^{l+2}$. Thus the truncation f_{l+1} with respect to these new coordinates is $x_1 x_2 + g_l + c$, which shows the claim for $l + 1$.

Apply Corollary 2.1.5 for $l = 2$ and take the integer L. One can assume that $f_{L-1} = x_1 x_2 + g_{L-1}(x_3, \ldots, x_n)$. Since $f - f_{L-1}$ belongs to $\mathfrak{m}^L$, there exists an automorphism σ of $\mathscr{O}_{\mathfrak{D}^n, o}$ which sends f to $u f_{L-1}$ with a unit u and which induces the identity on $\mathfrak{m} / \mathfrak{m}^2$. Then H is defined by $f_{L-1}(y_1, \ldots, y_n)$ with the coordinates $y_i = \sigma^{-1}(x_i)$, which implies the assertion.

In like manner, every isolated hypersurface singularity is defined by a polynomial for suitable coordinates. This is actually the case with an arbitrary isolated singularity.

Theorem 2.1.8 (Artin's algebraisation theorem [19, theorem 3.8]) *Let $v \in V$ be the germ of an analytic space which is an isolated singularity. Then there exists an algebraic germ of a complex scheme analytically isomorphic to $v \in V$.*

The *embedding dimension* of the germ $x \in X$ of an algebraic scheme or an analytic space is the dimension of the vector space $\mathfrak{m}/\mathfrak{m}^2$ for the maximal ideal $\mathfrak{m}$ in $\mathcal{O}_X$ defining x. Plainly an analytic germ of embedding dimension e is embedded into the domain $\mathfrak{D}^e$. This is the case even in the étale topology.

Lemma 2.1.9 *Let $x \in X$ be the germ of an algebraic scheme of embedding dimension e. Then there exists an étale morphism from $x \in X$ to the germ $o \in Y$ of a subscheme of the affine space $\mathbf{A}^e$.*

Proof We work over an algebraically closed field k. Take elements $x_1, \ldots, x_e$ in $\mathfrak{m}$ which form a basis of $\mathfrak{m}/\mathfrak{m}^2$ for the maximal ideal $\mathfrak{m}$ in $\mathcal{O}_X$ defining x. Define by these x_i a morphism from $x \in X$ to $\mathbf{A}^e = \operatorname{Spec} k[x_1, \ldots, x_e]$. The kernel of the induced map $k[x_1, \ldots, x_e] \to \mathcal{O}_{X,x}$ defines a closed subscheme Y of $\mathbf{A}^e$. Take the germ $o \in Y \subset \mathbf{A}^e$ at origin. The map $\mathcal{O}_{Y,o} \to \mathcal{O}_{X,x}$ is injective. By the faithful flatness of the completion of a noetherian local ring [308, theorem 8.14], it extends to an injection $\hat{\mathcal{O}}_{Y,o} \hookrightarrow \hat{\mathcal{O}}_{X,x}$ of the completions, which is also surjective by the choice of x_i. Hence $\hat{\mathcal{O}}_{Y,o} \simeq \hat{\mathcal{O}}_{X,x}$, via which $\mathcal{O}_{Y,o} \subset \mathcal{O}_{X,x} \subset \hat{\mathcal{O}}_{Y,o}$ and $X \to Y$ is étale at x.   □

The following basic theorem makes the defining function simple.

Theorem 2.1.10 (Weierstrass preparation theorem) *Let f be a convergent power series in $x_1, \ldots, x_n$. Suppose that f is not identically zero on the x_1-axis. Then f is uniquely expressed as $f = u \cdot (x_1^d + a_1 x_1^{d-1} + \cdots + a_d)$ with a unit u and convergent power series a_i in $x_2, \ldots, x_n$ vanishing at origin.*

Remark 2.1.11 Elimination of the second highest term is also useful for finding a simpler function. For a function $x_1^d + a_1 x_1^{d-1} + \cdots + a_d$ such that a_i are functions in $x_2, \ldots, x_n$ vanishing at origin, one can eliminate the term $a_1 x_1^{d-1}$ by replacing x_1 by $x_1 + d^{-1} a_1$.

As an application, we shall describe canonical hypersurface singularities of dimension two as *Du Val* singularities. Later in Theorem 2.3.1, we shall prove that every canonical surface singularity is Du Val.

Definition 2.1.12 A germ of an algebraic or analytic surface is called a *Du Val* singularity if it is analytically isomorphic to the hypersurface singularity given by one of the functions

A_n $x_1^2 + x_2^2 + x_3^{n+1}$ for $n \geq 1$,

D_n $x_1^2 + x_2^2 x_3 + x_3^{n-1}$ for $n \geq 4$,

E_6 $x_1^2 + x_2^3 + x_3^4$,

E_7 $x_1^2 + x_2^3 + x_2 x_3^3$,

E_8 $x_1^2 + x_2^3 + x_3^5$.

It is said to be of *type* A_n, D_n, E_6, E_7 or E_8 correspondingly. We occasionally say that the germ of a smooth surface is Du Val of type A_0.

For a polynomial or a convergent power series f in $x_1, \ldots, x_n$, the *order* of f means the minimum of the degrees of monomials that appear in f. It is denoted by ord f. The order of zero is infinity by convention. As a function in $\mathscr{O}_{\mathfrak{D}^n}$, the order of f equals the order $\mathrm{ord}_E f$ along the divisor E obtained by the blow-up of $\mathfrak{D}^n$ at origin.

Proposition 2.1.13 *Let $o \in S \subset \mathfrak{D}^3$ be the germ of an analytic surface in $\mathfrak{D}^3$. If $o \in S$ is canonical, then it is Du Val.*

Proof In the course of the proof, we shall describe a monomial valuation in terms of weighted blow-up in Definition 2.2.11. The reader may return after learning it. By inversion of adjunction in Theorem 1.4.26, the pair $(\mathfrak{D}^3, S)$ is plt and it is further canonical since $K_{\mathfrak{D}^3} + S$ is Cartier.

Let f be the function in $\mathscr{O}_{\mathfrak{D}^3}$ defining S. Consider the divisor E obtained by the blow-up of $\mathfrak{D}^3$ at o. Since $(\mathfrak{D}^3, S)$ is canonical, the log discrepancy $a_E(\mathfrak{D}^3, S) = 3 - \mathrm{ord}\, f$ is at least one. That is, ord $f = 1$ or 2. If ord $f = 1$, then S is smooth. We shall assume that ord $f = 2$. Applying Theorem 2.1.10 and then Remark 2.1.11, one can write $f = x_1^2 + g(x_2, x_3)$ with suitable coordinates x_1, x_2, x_3 of $\mathfrak{D}^3$.

Next we consider the divisor F obtained by the weighted blow-up of $\mathfrak{D}^3$ with $\mathrm{wt}(x_1, x_2, x_3) = (2, 1, 1)$. Then $a_F(\mathfrak{D}^3, S) = 4 - \min\{4, \mathrm{ord}\, g\}$, which must also be at least one. Hence ord $g = 2$ or 3. If ord $g = 2$, then by Theorem 2.1.10, $f = x_1^2 + x_2^2 + x_3^{n+1}$ after coordinate change, which is of type A_n. We shall assume that ord $g = 3$. Take the germ $c \in C = (g = 0) \subset \mathfrak{D}^2$ at origin c in the domain $\mathfrak{D}^2$ with coordinates x_2, x_3. Let B be the blow-up of $\mathfrak{D}^2$ at c and let C_B be the strict transform in B of C.

We look at the cubic part g_3 of g. Suppose that g_3 is not a cube. Then after coordinate change, $g_3 = x_2 h_2(x_2, x_3)$ in which x_3^2 appears in h_2. The divisor C_B on B is decomposed into at least two components in accordance with the factorisation of g_3. By the proper mapping theorem, the image in $\mathfrak{D}^2$ of each component is a component of C and g is factorised accordingly. Thus changing coordinates, one attains the form $g = x_2 h(x_2, x_3)$ in which x_3^2 appears in h, and further $g = x_2(x_3^2 + x_2^{n-2})$ by Theorem 2.1.10 for h and Remark 2.1.11. Then $f = x_1^2 + x_2 x_3^2 + x_2^{n-1}$, which is of type D_n.

Now we may assume that $g_3 = x_2^3$. By Theorem 2.1.10 for g with Remark 2.1.11, one can write $g = x_2^3 + ux_2 x_3^a + vx_3^b$ after coordinate change, in which u and v are a unit or zero. We set $a = \infty$ when $u = 0$ and $b = \infty$ when $v = 0$ for convenience. Consider the divisor G obtained by the weighted blow-up of $\mathfrak{D}^3$ with $\mathrm{wt}(x_1, x_2, x_3) = (3, 2, 1)$. Then $a_G(\mathfrak{D}^3, S) = 6 - \min\{6, 2+a, b\}$, which must be at least one. Hence

$$a \geq 3, \ b = 4 \qquad \text{or} \qquad a = 3, \ b \geq 5 \qquad \text{or} \qquad a \geq 4, \ b = 5.$$

Suppose the first case $b = 4$. If $a \geq 4$, then $g = x_2^3 + wx_3^4$ with a unit $w = ux_2 x_3^{a-4} + v$ and f is of type E_6. If $a = 3$, then one can write $g = px_2^3 + qx_2^2 x_3^2 + vx_3^4$ with units p and q by Remark 2.1.11 for x_3^4, and further $g = px_2^3 + rx_3^4$ with a unit r by the same remark for x_2^3. Thus f is of type E_6. By a similar argument, the third case corresponds to a Du Val singularity of type E_8.

Suppose the second case $a = 3$ with $b \geq 5$. At the origin of the chart of B with coordinates $y_2 = x_2 x_3^{-1}$ and x_3, C_B is given by $y_2^3 + uy_2 x_3 + vx_3^{b-3}$ and it is a union of two smooth curves intersecting transversally. Applying the proper mapping theorem as above with Remark 2.1.11, one has $g = x_2(x_2^2 + x_3^3)$ by coordinate change. Then $f = x_1^2 + x_2^3 + x_2 x_3^3$, which is of type E_7. □

2.2 Quotients and Coverings

Taking quotient is a basic operation to produce singularities. We begin with the definition of quotient. We shall formulate it over an algebraically closed field k in the framework where an algebraic group G acts on an algebraic scheme X from left by a morphism $\sigma \colon G \times X \to X$. By an *algebraic group*, we mean a group scheme smooth over Spec k. For a closed point x in X, the *orbit* of x by G is the image of the restriction $\sigma_x \colon G \simeq G \times x \to X$ of σ. The scheme-theoretic fibre $G \times_X x$ of σ_x at x is the *stabiliser* of x and denoted by G_x.

For an element g in G and a function f in $\mathscr{O}_X$, let $g(f)$ denote the pull-back $\sigma_g^* f$ by the automorphism $\sigma_g \colon X \simeq g \times X \to X$. The group of closed points of G acts on $\mathscr{O}_X$ from right in which $(hg)(f) = g(h(f))$ for $g, h \in G$. A function f in $\mathscr{O}_X$ is said to be *invariant* by the action of G if $\sigma^* f = p^* f$ in $\mathscr{O}_{G \times X}$ where $p \colon G \times X \to X$ is the projection. In other words, $g(f) = f$ for all $g \in G$. We write $\mathscr{O}_X^G$ for the subalgebra of invariant functions in $\mathscr{O}_X$. In general when G acts on a space M, we write M^G for the subspace of invariant elements in M.

Definition 2.2.1 Let X be an algebraic scheme on which an algebraic group G acts by $\sigma \colon G \times X \to X$. The *geometric quotient* of X by G is a surjective morphism $\pi \colon X \to Y$ of algebraic schemes such that

- $\pi \circ \sigma = \pi \circ p$ where $p \colon G \times X \to X$ is the projection,
- two closed points in X belong to the same orbit by G if and only if they are mapped to the same point by π,
- $U \subset Y$ is open if and only if $\pi^{-1}(U) \subset X$ is open and
- $\mathcal{O}_Y = \pi_* \mathcal{O}_X^G$.

The scheme Y is called the *quotient* of X by G and denoted by X/G.

The geometric quotient may not exist. The existence requires every orbit to be closed because it should be realised as a fibre. The geometric quotient is also the *categorical quotient* defined by a universal property [364, proposition 0.1]. In particular, the geometric quotient $X \to X/G$ is unique if it exists, and if X is reduced or normal, then so is X/G.

Example 2.2.2 The multiplicative group $k^\times$ admits the structure of the algebraic group $\mathbf{G}_m = \operatorname{Spec} k[t, t^{-1}]$. It acts on $\mathbf{A}^1$ in such a way that $t \in \mathbf{G}_m$ sends the coordinate x to tx. The line $\mathbf{A}^1$ consists of two orbits o and $\mathbf{A}^1 \setminus o$. The geometric quotient of $\mathbf{A}^1$ by $\mathbf{G}_m$ does not exist.

We shall consider the case when $G = \operatorname{Spec} A$ is an affine algebraic group. A *representation* of G means a linear map $\rho \colon V \to V \otimes A$ with a vector space V over k, which may be of infinite dimension, such that for the symmetric algebra SV of V, the map $SV \to SV \otimes A$ of k-algebras induced by ρ defines a left action of G on $\operatorname{Spec} SV$. Here we admit the scheme $\operatorname{Spec} SV$ not necessarily of finite type over k. An element v in V is *invariant* by G if $\rho(v) = v \otimes 1$. The vector subspace of invariant elements in V is denoted by V^G. When G acts on an algebraic scheme X by $G \times X \to X$, the associated map $\mathcal{O}_X \to \mathcal{O}_X \otimes A$ is a representation of G.

We say that an affine algebraic group G is *linearly reductive* if for every surjective map $V \twoheadrightarrow W$ of representations of G, the induced map $V^G \to W^G$ is also surjective. For example, a finite group in characteristic zero, the *general linear group* $\mathrm{GL}(n, \mathbf{C})$ and the *special linear group* $\mathrm{SL}(n, \mathbf{C})$ are linearly reductive. The reader can refer to [357, subsection 4.3(a)].

Theorem 2.2.3 ([364, section 1.2]) *Let X be an affine algebraic scheme on which a linearly reductive affine algebraic group G acts. If the orbit by G of every closed point in X is closed, then the geometric quotient $X \to X/G$ exists and X/G is an affine algebraic scheme.*

Proof We write $G = \operatorname{Spec} A$ and $X = \operatorname{Spec} R$ and take $Y = \operatorname{Spec} R^G$ for the ring R^G of invariants in R. We shall prove that $\pi \colon X \to Y$ is the geometric quotient.

Step 1 It is a prerequisite to show that R^G is a finitely generated k-algebra. We first claim the existence of a finite dimensional k-vector subspace V of R invariant by G which induces a surjection $SV \twoheadrightarrow R$ from the symmetric algebra of V. Take generators $r_1, \ldots, r_n$ of the k-algebra R and write $\sigma^* r_i = \sum_j r_{ij} \otimes t_j \in R \otimes A$ with linearly independent elements t_j in A, where $\sigma^* \colon R \to R \otimes A$ is the representation associated with the action $\sigma \colon G \times X \to X$. Let V be the vector subspace of R spanned by all r_{ij}. The identity $\varepsilon \colon \operatorname{Spec} k \to G$ gives $r_i = \sum_j \varepsilon^* t_j \cdot r_{ij} \in V$. The product $\mu \colon G \times G \to G$ gives $\sum_j \sigma^* r_{ij} \otimes t_j = \sum_j r_{ij} \otimes \mu^* t_j \in V \otimes A \otimes A$. Hence $\sigma^* r_{ij} \in V \otimes A$ from the linear independence of t_j. Thus V fulfils the claim.

We consider $S = SV$ to be a polynomial ring over k. The group G acts on $\operatorname{Spec} S$ preserving the grading of S so that $S \twoheadrightarrow R$ is equivariant. Since G is linearly reductive, the induced map $S^G \to R^G$ is surjective. Thus the finite generation of R^G is reduced to that of S^G.

Let I be the homogeneous ideal in S generated by invariant polynomials of positive degree. Choose homogeneous generators $f_1, \ldots, f_m$ of I and take the subalgebra $S' = k[f_1, \ldots, f_m]$ of S generated by them. For any homogeneous element f in S^G of positive degree, we shall prove that $f \in S'$ by induction on the degree of f. Because the surjection $S^{\oplus m} \twoheadrightarrow I$ given by $f_1, \ldots, f_m$ yields the surjection $(S^G)^{\oplus m} \twoheadrightarrow I^G$ by the linearly reductive property of G, one can write $f = \sum_i g_i f_i$ with homogeneous $g_i \in S^G$. Then $g_i \in S'$ by the inductive hypothesis, which implies that $f \in S'$. It follows that $S' = S^G$ and this completes the finite generation of R^G.

Step 2 We shall verify the conditions in Definition 2.2.1. For an ideal $\mathfrak{b}$ in R^G, take a surjection $R^{\oplus n} \to \mathfrak{b} R$ by fixing generators $b_1, \ldots, b_n$ of $\mathfrak{b}$. This yields the surjection $(R^G)^{\oplus n} \twoheadrightarrow (\mathfrak{b} R)^G$, that is, $\mathfrak{b} = \sum_i R^G b_i = (\mathfrak{b} R)^G$. In particular, $\mathfrak{b} R \neq R$ when $\mathfrak{b} \neq R^G$. Hence π is surjective.

The first condition in Definition 2.2.1 is evident. The fourth will follow from the equality $(R^G)_f = (R_f)^G$ on localisation by $f \in R^G$. To see this, we regard $(R^G)_f$ and $(R_f)^G$ as R^G-submodules of R_f. Consider the R-submodule $R f^{-l}$ of R_f for each $l \in \mathbf{N}$. The surjection $R \twoheadrightarrow R f^{-l}$ induces the surjection $R^G \twoheadrightarrow (R f^{-l})^G$. Hence $R^G f^{-l} = (R f^{-l})^G$ and $(R^G)_f = \bigcup_l R^G f^{-l} = \bigcup_l (R f^{-l})^G = (R_f)^G$.

For the third condition, let U be a subset of Y such that $\pi^{-1}(U)$ is open. Let $\mathfrak{a}$ be the radical ideal in R which defines the support of the complement $X \setminus \pi^{-1}(U)$ as a reduced closed subscheme of X. The ideal $\mathfrak{a}$ is invariant by G and the surjection $R \twoheadrightarrow R/\mathfrak{a}$ induces the surjection $R^G \twoheadrightarrow (R/\mathfrak{a})^G$, by which $(R/\mathfrak{a})^G \simeq R^G/\mathfrak{a}^G$. Hence $Y \setminus U = \pi(X \setminus \pi^{-1}(U))$ is the support of the closed subscheme $\operatorname{Spec} R^G/\mathfrak{a}^G$ of Y via $\operatorname{Spec}(R/\mathfrak{a}) \to \operatorname{Spec}(R/\mathfrak{a})^G$, which has been seen to be surjective. Thus U is open.

Finally for the second, we let O_x denote the orbit by G of a closed point x in X. It is closed by assumption. We regard O_x as a reduced subscheme of X and write $\mathfrak{a}_x$ for the ideal in R defining O_x. The image $\pi(O_x)$ is the closed point $\pi(x)$ since π is equivariant by the trivial action on Y. If $O_x \cap O_y = \emptyset$ for $x, y \in X$, then $\mathfrak{a}_x + \mathfrak{a}_y = R$. The surjection $\mathfrak{a}_x \twoheadrightarrow R/\mathfrak{a}_y$ induces the surjection $\mathfrak{a}_x^G \twoheadrightarrow (R/\mathfrak{a}_y)^G \simeq R^G/\mathfrak{a}_y^G$, by which $\mathfrak{a}_x^G + \mathfrak{a}_y^G = R^G$. Thus $\pi(x) \neq \pi(y)$ because they are defined by the ideals $\mathfrak{a}_x^G$ and $\mathfrak{a}_y^G$ in R^G. $\qquad\square$

Henceforth the ground field is assumed to be the field $\mathbf{C}$ of complex numbers. For the purpose of this section, the case when G is a finite group is sufficient. The geometric quotient is also defined in the analytic category [66], [190], [191] and it exists when the action is by a finite group.

Theorem 2.2.4 *Let $x \in X$ be the germ of the quotient of a smooth complex variety by a finite group. Then there exist étale morphisms $x' \in X' \to x \in X$ and $x' \in X' \to o \in \mathbf{A}^n/G$ of germs of complex varieties such that $o \in \mathbf{A}^n/G$ is the germ at origin of the quotient of the affine space $\mathbf{A}^n$ by a finite subgroup G of $\mathrm{GL}(n, \mathbf{C})$ with the natural action.*

Proof We write $\pi \colon \bar{X} \to X = \bar{X}/G$ with a smooth variety $\bar{X}$ and a finite group G. Let $\bar{x}$ be a closed point in $\pi^{-1}(x)$. For the stabiliser $G_{\bar{x}}$ of $\bar{x}$, the induced morphism $\bar{X}/G_{\bar{x}} \to X$ is étale at x. Replacing X by $\bar{X}/G_{\bar{x}}$, we may and shall assume that $\pi^{-1}(x)$ consists of one point $\bar{x}$.

Let $\bar{\mathfrak{m}}$ denote the maximal ideal in $\mathscr{O}_{\bar{x}}$ defining $\bar{x}$. The action of G defines a representation $\rho \colon G \to \mathrm{GL}(\bar{\mathfrak{m}}/\bar{\mathfrak{m}}^2)$. Take a regular system $z_1, \dots, z_n$ of parameters in $\mathscr{O}_{\bar{X},\bar{x}}$. Their images $\bar{z}_1, \dots, \bar{z}_n$ form a basis of the complex vector space $\bar{\mathfrak{m}}/\bar{\mathfrak{m}}^2$. With respect to this basis, $\rho(g)$ for $g \in G$ is a matrix in $\mathrm{GL}(n, \mathbf{C})$ satisfying ${}^t(g(\bar{z}_1), \dots, g(\bar{z}_n)) = \rho(g)\,{}^t(\bar{z}_1, \dots, \bar{z}_n)$ for the image $g(\bar{z}_i)$ in $\bar{\mathfrak{m}}/\bar{\mathfrak{m}}^2$ of $g(z_i)$, where t stands for the transpose. Define $y_i \in \mathscr{O}_{\bar{X}}$ by

$$
{}^t(y_1, \dots, y_n) = \frac{1}{N} \sum_{g \in G} \rho(g^{-1})\,{}^t(g(z_1), \dots, g(z_n))
$$

with the order N of G. Then $y_i - z_i \in \bar{\mathfrak{m}}^2$ and hence $y_1, \dots, y_n$ form a regular system of parameters in $\mathscr{O}_{\bar{X},\bar{x}}$. Writing $z = {}^t(z_1, \dots, z_n)$ and $y = {}^t(y_1, \dots, y_n)$ for brevity, one obtains for all $h \in G$,

$$
Nh(y) = \sum_{g \in G} \rho(g^{-1})h(g(z)) = \sum_{g \in G} \rho(h(gh)^{-1})(gh)(z)
$$
$$
= \sum_{g \in G} \rho(hg^{-1})g(z) = N\rho(h)y.
$$

Thus the action of G on X is linear with respect to $y_1, \dots, y_n$.

We define by y_i an étale morphism from $\bar{x} \in \bar{X}$ to $\mathbf{A}^n = \text{Spec } \mathbf{C}[y_1, \ldots, y_n]$. The group G acts on the subalgebra $\mathbf{C}[y_1, \ldots, y_n]$ of $\mathcal{O}_{X,x}$, by which $\mathbf{A}^n$ admits an action of G such that $\bar{X} \to \mathbf{A}^n$ is equivariant. It induces a morphism from the germ $x \in X = \bar{X}/G$ to the germ $o \in \mathbf{A}^n/G$ at origin, which is étale since the induced map $\hat{\mathcal{O}}_{\mathbf{A}^n/G,o} \to \hat{\mathcal{O}}_{X,x}$ of the complete local rings is an isomorphism. The quotient $\mathbf{A}^n/G$ is the same as $\mathbf{A}^n/\rho(G)$ by the linear group $\rho(G)$.   □

Remark 2.2.5   The proof above shows that if $x \in \bar{X}/G$ is the germ of the quotient of a smooth variety $\bar{X}$ by a finite group G which fixes a point $\bar{x}$ above x, then there exists a regular system of parameters in $\mathcal{O}_{\bar{X},\bar{x}}$ with respect to which G acts linearly on $\bar{x} \in \bar{X}$. Even if $\bar{x} \in \bar{X}$ is singular with embedding dimension e, there exists an e-tuple of elements in $\bar{\mathfrak{m}}$ forming a basis of $\bar{\mathfrak{m}}/\bar{\mathfrak{m}}^2$ with respect to which G acts linearly. In particular by Lemma 2.1.9, the germ $x \in \bar{X}/G$ admits an étale morphism to a closed subscheme of $\mathbf{A}^e/\rho(G)$ with $\rho(G) \subset \text{GL}(e, \mathbf{C})$.

Definition 2.2.6   A germ of a normal variety is called a *quotient singularity* if it is analytically isomorphic to the germ at origin of the quotient $\mathbf{A}^n/G$ by a finite subgroup G of $\text{GL}(n, \mathbf{C})$.

Remark 2.2.7   An element in $\text{GL}(n, \mathbf{C})$ of finite order but identity is called a *pseudo-reflection* if it fixes a hyperplane in $\mathbf{A}^n$ pointwise. A *reflection* is a pseudo-reflection of order two. We say that a finite subgroup of $\text{GL}(n, \mathbf{C})$ is *small* if it contains no pseudo-reflections. The quotient of $\mathbf{A}^n$ by a finite group generated by pseudo-reflections remains smooth [86], [415]. Hence in the study of a quotient singularity $\mathbf{A}^n/G$, it is harmless to assume that G is a small linear group by regarding it as $(\mathbf{A}^n/H)/(G/H)$ for the normal subgroup H of G generated by pseudo-reflections.

Note that a quotient singularity is $\mathbf{Q}$-factorial. For a prime divisor D on the analytic germ $o \in \mathfrak{D}^n/G$, the norm $\prod_{g \in G} g(f)$ in $\mathcal{O}_{\mathfrak{D}^n/G}$ of the function f in $\mathcal{O}_{\mathfrak{D}^n}$ defining the inverse image of D defines a multiple of D.

Example 2.2.8   Every Du Val singularity in Definition 2.1.12 is realised as the quotient of $\mathbf{A}^2$ by a finite subgroup of $\text{SL}(2, \mathbf{C})$. A finite subgroup of $\text{SL}(2, \mathbf{C})$ is conjugate to one of the following groups in accordance with the type of the quotient singularity. See [431, section 4.4].

A_n the *cyclic group* C_{n+1} of order $n + 1$ generated by $\begin{pmatrix} \zeta & 0 \\ 0 & \zeta^{-1} \end{pmatrix}$ for a primitive $(n + 1)$-th root ζ of unity

D_n the *binary dihedral group* $2D_{n-2}$ of order $4(n-2)$ generated by $C_{2(n-2)}$ and $\begin{pmatrix} 0 & \sqrt{-1} \\ \sqrt{-1} & 0 \end{pmatrix}$

E_6 the *binary tetrahedral group* $2T$ of order 24 generated by $2D_2$ and the matrix $\dfrac{1}{\sqrt{2}}\begin{pmatrix} \varepsilon & \varepsilon^3 \\ \varepsilon & \varepsilon^7 \end{pmatrix}$ for a primitive eighth root ε of unity

E_7 the *binary octahedral group* $2O$ of order 48 generated by C_8 and $2T$

E_8 the *binary icosahedral group* $2I$ of order 120 generated by C_{10} and the matrix $\dfrac{1}{\sqrt{5}}\begin{pmatrix} -\eta+\eta^4 & \eta^2-\eta^3 \\ \eta^2-\eta^3 & \eta-\eta^4 \end{pmatrix}$ for a primitive fifth root η of unity

For instance, the invariant ring $\mathbf{C}[x_1,x_2]^{2D_2}$ by $2D_2$ is generated by

$$y_1 = (x_1^4 + x_2^4)/\sqrt{-3}, \qquad y_2 = 2x_1x_2(x_1^4 - x_2^4), \qquad y_3 = 2x_1^2x_2^2$$

with a relation $y_2^2 + 6y_1^2y_3 + 2y_3^3 = 0$. Hence $\mathbf{A}^2/2D_2$ is of type D_4. Further, the invariant ring $\mathbf{C}[x_1,x_2]^{2T}$ by $2T$ is generated by y_2 and

$$z_1 = (y_1 + y_3)^3, \qquad z_2 = (y_1 - y_3)^3, \qquad z_3 = y_1^2 - y_3^2$$

with relations $z_1 - z_2 + y_2^2 = z_1z_2 - z_3^3 = 0$. By elimination of z_2, $\mathbf{C}[x_1,x_2]^{2T}$ is generated by y_2, z_1, z_3 with a relation $z_1(z_1 + y_2^2) - z_3^3 = 0$. Hence $\mathbf{A}^2/2T$ is of type E_6.

We shall discuss the most typical case when G is a cyclic group. Let $x \in \bar{X}/\mathbf{Z}_r$ be the quotient of the germ $\bar{x} \in \bar{X}$ of a smooth variety by the cyclic group $\mathbf{Z}_r$ of order r which fixes $\bar{x}$. We fix a *primitive character* $\chi\colon \mathbf{Z}_r \to \mathbf{C}^\times$, that is, a group monomorphism to the multiplicative group $\mathbf{C}^\times$.

Definition 2.2.9 We say that a function f in $\mathscr{O}_{\bar{X}}$ is *semi-invariant* if there exists an integer a modulo r such that $g(f) = \chi(g)^a f$ for all $g \in \mathbf{Z}_r$. We call a the *weight* of f with respect to χ. It is denoted by wt f when there is no ambiguity. The weight wt f may be regarded as an element in $\mathbf{Z}_r$.

Because every element in $\mathrm{GL}(n, \mathbf{C})$ of finite order is diagonalisable, there exists a regular system of semi-invariant parameters in $\mathscr{O}_{\bar{X},\bar{x}}$ by Remark 2.2.5. Hence one can formulate a cyclic quotient singularity as follows.

Definition 2.2.10 Let $\mathbf{A}^n$ be the affine space with coordinates $x_1, \ldots, x_n$. We write $\mathbf{A}^n/\mathbf{Z}_r(a_1, \ldots, a_n)$ for the quotient of $\mathbf{A}^n$ by $\mathbf{Z}_r$ in which every x_i is semi-invariant with wt $x_i = a_i$ with respect to χ. We call these $x_1, \ldots, x_n$ the *orbifold coordinates* of the quotient $\mathbf{A}^n/\mathbf{Z}_r(a_1, \ldots, a_n)$. We use the notation $o \in \mathfrak{D}^n/\mathbf{Z}_r(a_1, \ldots, a_n)$ for the analytic germ at origin of $\mathbf{A}^n/\mathbf{Z}_r(a_1, \ldots, a_n)$.

A *cyclic quotient singularity* is a singularity analytically isomorphic to some $o \in \mathfrak{D}^n / \mathbf{Z}_r(a_1, \ldots, a_n)$, and in this case, it is said to be of *type* $\frac{1}{r}(a_1, \ldots, a_n)$. By taking χ^l instead of χ for an integer l coprime to r, one can replace $a = (a_1, \ldots, a_n)$ by $a' = (a'_1, \ldots, a'_n)$ such that $a \equiv la' \bmod r$.

Cyclic quotient singularities appear in the context of weighted blow-up, which can be described explicitly in terms of toric geometry. Whilst the reader can learn *toric varieties* from [140] or [198, section 4.4] for example, the description below of the weighted blow-up of $\mathbf{A}^n / \mathbf{Z}_r(w_1, \ldots, w_n)$ with weights $\frac{1}{r}(w_1, \ldots, w_n)$ is sufficient for the purpose of the book.

Consider the cyclic quotient $X = \mathbf{A}^n / \mathbf{Z}_r(a_1, \ldots, a_n)$ with orbifold coordinates $x_1, \ldots, x_n$. We allow the case $r = 1$. Define the lattice $N = \mathbf{Z}^n + \mathbf{Z} \cdot \frac{1}{r}(a_1, \ldots, a_n)$. Then X is the toric variety associated with (N, Δ) for the fan Δ which consists of faces of the cone $\langle e_1, \ldots, e_n \rangle$ spanned by the standard basis of $\mathbf{Z}^n \otimes \mathbf{R} = N \otimes \mathbf{R}$.

Definition 2.2.11   A non-zero element e in the lattice N is said to be *primitive* if $\mathbf{Z}e = \mathbf{R}e \cap N$ in $N \otimes \mathbf{R}$. Fix a primitive element $e = \frac{1}{r}(w_1, \ldots, w_n)$ in N which belongs to the cone $\langle e_1, \ldots, e_n \rangle$. Note that r may be one. Let Δ_e be the fan which consists of faces of the cones $\langle e_1, \ldots, e_{i-1}, e, e_{i+1}, \ldots, e_n \rangle$ for all $1 \leq i \leq n$. The *weighted blow-up* of $X = \mathbf{A}^n / \mathbf{Z}_r(a_1, \ldots, a_n)$ with *weights* $\mathrm{wt}(x_1, \ldots, x_n) = \frac{1}{r}(w_1, \ldots, w_n)$ is the natural projective morphism $\pi \colon B \to X$ from the toric variety B associated with (N, Δ_e). It is isomorphic outside the locus $Z = \bigcap_{w_i \neq 0}(x_i = 0)$ in X.

Let $o \in Y$ be a closed subvariety of the germ $o \in X$ at origin which contains Z properly. The induced morphism $Y_B \to Y$ from the strict transform Y_B in B is the *weighted blow-up* of $o \in Y \subset X$ with the weights above. When an algebraic or analytic germ $v \in V$ admits an étale morphism to $o \in Y$, the *weighted blow-up* of V is defined as the base change $Y_B \times_Y V \to V$.

Let f be a semi-invariant function in $\mathscr{O}_{\mathbf{A}^n}$. The *weighted order* of f with respect to weights $\mathrm{wt}(x_1, \ldots, x_n) = \frac{1}{r}(w_1, \ldots, w_n)$ is the order of f in which the *weighted degree* w_i / r is assigned to x_i for each i. We write w-ord f for the weighted order of f when the assigned weights are clear in the context.

The semi-invariant function f defines an effective $\mathbf{Q}$-Cartier divisor $(f)_X$ on the quotient $X = \mathbf{A}^n / \mathbf{Z}_r(a_1, \ldots, a_n)$ unless it is zero. Let $\pi \colon B \to X$ be the weighted blow-up with $\mathrm{wt}(x_1, \ldots, x_n) = \frac{1}{r}(w_1, \ldots, w_n)$ as above. Let E denote the exceptional divisor. The order of $(f)_X$ along E equals the weighted order w-ord f. In other words,

$$\pi^*(f)_X = F + (\text{w-ord } f)E$$

for the strict transform F of $(f)_X$ and one can write $\mathrm{ord}_E f = \text{w-ord } f$. The next formula follows from the description of the canonical divisor [251, p.28 theorem 9], [373, p.118 remark].

Lemma 2.2.12 *Suppose that* $\mathbf{Z}_r \subset \mathrm{GL}(n, \mathbf{C})$ *is small. Then* $a_E(X) = \sum_i w_i/r$. *In other words,*

$$K_B = \pi^* K_X + \left(\frac{1}{r} \sum_{i=1}^n w_i - 1\right) E.$$

We focus on the case when $(a_1, \ldots, a_n) \equiv (w_1, \ldots, w_n) \bmod r$. Hence B is the weighted blow-up of $X = \mathbf{A}^n/\mathbf{Z}_r(w_1, \ldots, w_n)$ with $\mathrm{wt}(x_1, \ldots, x_n) = \frac{1}{r}(w_1, \ldots, w_n)$. Then B is covered by the affine charts

$$U_i \simeq \mathbf{A}^n/\mathbf{Z}_{w_i}(w_1, \ldots, -r, \ldots, w_n)$$

for $1 \leq i \leq n$, where $-r$ is taken as the i-th entry instead of w_i. The chart U_i is endowed with orbifold coordinates

$$(y_1, \ldots, y_i, \ldots, y_n) = (x_1 x_i^{-w_1/w_i}, \ldots, x_i^{r/w_i}, \ldots, x_n x_i^{-w_n/w_i}),$$

in which $(x_1, \ldots, x_i, \ldots, x_n) = (y_1 y_i^{w_1/r}, \ldots, y_i^{w_i/r}, \ldots, y_n y_i^{w_n/r})$. In fact, U_i is the toric variety which corresponds to the cone $\langle e_1, \ldots, e, \ldots, e_n \rangle$. If all w_i are positive, then the exceptional divisor E of $B \to X$ is isomorphic to the weighted projective space $\mathbf{P}(w_1, \ldots, w_n)$ below.

Definition 2.2.13 Let $a_0, \ldots, a_n$ be positive integers. Equip the polynomial ring $k[x_0, \ldots, x_n]$ over the ground field k with the structure of a graded ring by assigning degree a_i to x_i. The *weighted projective space* $\mathbf{P}(a_0, \ldots, a_n)$ is $\mathrm{Proj}\, k[x_0, \ldots, x_n]$ associated with this graded ring. It is the quotient of $\mathbf{A}^{n+1} \setminus o$ by the action of $\mathbf{G}_m$ such that $t(x_i) = t^{a_i} x_i$ for $t \in \mathbf{G}_m = k^\times$. In particular, the space $P = \mathbf{P}(a_0, \ldots, a_n)$ is isomorphic to $\mathbf{P}(l a_0, \ldots, l a_n)$ and to $\mathbf{P}(l a_0, \ldots, a_i, \ldots, l a_n)$, where a_i lies instead of $l a_i$, for any positive integer l. It has $\omega_P \simeq \mathscr{O}_P(-\sum_i a_i)$ and $H^i(\mathscr{O}_P(l)) = 0$ for all $0 < i < n$ and $l \in \mathbf{Z}$. The reader can consult the excellent treatise [106].

There is an alternative description of the weighted blow-up. For simplicity, we consider the germ $x \in X$ of a smooth variety. Let $x_1, \ldots, x_m$ be a part of a regular system of parameters in $\mathscr{O}_{X,x}$ and let $w_1, \ldots, w_m$ be positive integers. For a natural number l, let $\mathscr{I}_l$ denote the ideal in $\mathscr{O}_X$ generated by all monomials $x_1^{s_1} \cdots x_m^{s_m}$ such that $\sum_{i=1}^m s_i w_i \geq l$. The *weighted blow-up* of X with $\mathrm{wt}(x_1, \ldots, x_m) = (w_1, \ldots, w_m)$ is $B = \mathrm{Proj}_X \bigoplus_{l \in \mathbf{N}} \mathscr{I}_l \to X$. If $y_1, \ldots, y_m$ is a part of another regular system of parameters with $y_i \in \mathscr{I}_{w_i} \setminus \mathscr{I}_{w_i+1}$, then the weighted blow-up of X with $\mathrm{wt}(y_1, \ldots, y_m) = (w_1, \ldots, w_m)$ is the same as B.

We provide a combinatoric criterion for a cyclic quotient singularity to be terminal or canonical.

Notation 2.2.14 Fix a positive integer r. For an integer l, we let $\bar{l}$ denote the residue of l modulo r, that is, $\bar{l} = l - \lfloor l/r \rfloor r$.

Proposition 2.2.15 *Let $o \in X = \mathbf{A}^n / \mathbf{Z}_r(a_1, \ldots, a_n)$ be the germ at origin of a cyclic quotient singularity with orbifold coordinates $x_1, \ldots, x_n$, where r is at least two. Suppose that $\mathbf{Z}_r$ is small or equivalently there exists no common factor of r and any $n - 1$ integers out of a_i. Then $o \in X$ is terminal (resp. canonical) if and only if $\sum_{i=1}^n \overline{la_i} > r$ (resp. $\geq r$) for all $0 < l < r$.*

Proof We use the description of X in terms of toric geometry. Let $N = \mathbf{Z}^n + \mathbf{Z} \cdot \frac{1}{r}(a_1, \ldots, a_n)$ and let Δ be the fan of faces of the cone spanned by the standard basis of $N \otimes \mathbf{R}$. Then X is the toric variety associated with (N, Δ).

For $0 < l < r$, we take a primitive element $\frac{1}{r}(b_1, \ldots, b_n)$ in N from the ray $\mathbf{R}_{\geq 0} \cdot \frac{1}{r}(\overline{la_1}, \ldots, \overline{la_n})$. Let E_l be the divisor obtained by the weighted blow-up of X with $\mathrm{wt}(x_1, \ldots, x_n) = \frac{1}{r}(b_1, \ldots, b_n)$, which is exceptional since at least two of b_i are positive. By Lemma 2.2.12, it has log discrepancy $a_{E_l}(X) = \sum_i b_i/r$. If X is terminal (resp. canonical), then it is greater than one (resp. at least one). Hence $\sum_i \overline{la_i} \geq \sum_i b_i = r a_{E_l}(X) > r$ (resp. $\geq r$).

To see the converse, we take a unimodular subdivision of (N, Δ) which provides a log resolution of X. Every exceptional prime divisor E corresponds to a ray in $N \otimes \mathbf{R}$ generated by a primitive element $\frac{1}{r}(c_1, \ldots, c_n)$ in N in which at least two of c_i are positive. For each E, there exists an integer $0 \leq l < r$ such that $(c_1, \ldots, c_n) \equiv (la_1, \ldots, la_n) \bmod r$. Then $c_i \geq \overline{la_i}$ and hence

$$a_E(X) = \frac{1}{r} \sum_{i=1}^n c_i \geq \begin{cases} 2 & \text{if } l = 0, \\ \sum_i \overline{la_i}/r & \text{if } 0 < l < r. \end{cases}$$

Thus if $\sum_i \overline{la_i} > r$ (resp. $\geq r$) for all $0 < l < r$, then $a_E(X) > 1$ (resp. ≥ 1) for any E, that is, X is terminal (resp. canonical). $\qquad\Box$

We shall construct a cyclic cover by a Weil divisor. This amounts to the converse of cyclic quotient. Let X be a normal variety and let D be a divisor on X. Let s be a non-zero global section in $\Gamma(X, \mathscr{O}_X(rD))$ with a positive integer r. The section s is regarded as a map $s \colon \mathscr{O}_X \to \mathscr{O}_X(rD)$ and it defines an effective divisor $S \sim rD$ on X. Define the finite $\mathscr{O}_X$-algebra $\mathscr{R} = \bigoplus_{i=0}^{r-1} \mathscr{R}_i$ with $\mathscr{R}_i = \mathscr{O}_X(-iD)$ where the multiplication for $i + j \geq r$ is given by

$$\mathscr{R}_i \otimes \mathscr{R}_j \to \mathscr{O}_X(-(i+j)D) = \mathscr{O}_X(-(i+j)D) \otimes \mathscr{O}_X$$

$$\xrightarrow{\mathrm{id} \otimes s} \mathscr{O}_X(-(i+j)D) \otimes \mathscr{O}_X(rD) \to \mathscr{R}_{i+j-r}.$$

The finite morphism $\pi \colon \bar{X} = \mathrm{Spec}_X \mathscr{R} \to X$ is the *cyclic cover* ramified along S. Notice that $\bar{X}$ may be reducible.

Lemma 2.2.16 *The cyclic cover* $\pi\colon \bar{X} \to X$ *ramified along* S *is the geometric quotient of* $\bar{X}$ *by an action of the cyclic group* $\mathbf{Z}_r$. *If* S *is reduced, then* $\bar{X}$ *is normal. If* S *is zero in which* $rD \sim 0$, *then* π *is étale on the locus where* D *is Cartier, and the pull-back* π^*D *is principal on* $\bar{X}$.

Proof We define the action of $\mathbf{Z}_r$ on $\bar{X}$ so that $g \in \mathbf{Z}_r$ sends an element f in $\mathscr{R}_i$ to $\chi(g)^i f$ for a fixed primitive character $\chi\colon \mathbf{Z}_r \to \mathbf{C}^\times$. Then $\mathscr{O}_X = \pi_*\mathscr{O}_{\bar{X}}^{\mathbf{Z}_r}$ and π is the geometric quotient by Theorem 2.2.3.

Let U be the locus in X where D is Cartier. It is covered by open subsets U_i on which $\mathscr{R}|_{U_i} \simeq \mathscr{O}_{U_i}[t]/(t^r - f_i)$ with a function f_i in $\mathscr{O}_{U_i}$ defining S. Hence if S is reduced, then $\pi^{-1}(U)$ is normal by Serre's criterion and $\bar{X}$ is normal by the reflexive property $\mathscr{R} = \iota_*(\mathscr{R}|_U)$ for $\iota\colon U \hookrightarrow X$.

Suppose that S is zero. Then $f_i \in \mathscr{O}_{U_i}^\times$ and hence π is étale on U. Since the divisorial sheaf $\mathscr{O}_{\bar{X}}(-\pi^*D)$ is given by the double dual of $\mathscr{O}_X(-D) \otimes \mathscr{R}$, one has $\pi_*\mathscr{O}_{\bar{X}}(-\pi^*D) = \bigoplus_{i=1}^r \mathscr{M}_i$ with $\mathscr{M}_i = \mathscr{O}_X(-iD)$ where the multiplication $\mathscr{R}_i \otimes \mathscr{M}_j \to \mathscr{M}_{i+j-r}$ for $i + j > r$ is given by $s\colon \mathscr{O}_X \simeq \mathscr{O}_X(rD)$. Thus $\mathscr{O}_{\bar{X}}(-\pi^*D) = \mathscr{O}_{\bar{X}}m$ by a generator m of $\mathscr{M}_r \simeq \mathscr{O}_X$, which shows that π^*D is principal. $\qquad\qquad\square$

The following definition is applicable to **Q**-Cartier divisors on a germ.

Definition 2.2.17 Let X be a normal variety and let D be a divisor on X which is a torsion in $\mathrm{Cl}\, X$. Take the least positive integer r such that rD is principal and fix an isomorphism $s\colon \mathscr{O}_X \simeq \mathscr{O}_X(rD)$. The *cyclic cover* of X associated with D is $\pi\colon \bar{X} = \mathrm{Spec}_X \bigoplus_{i=0}^{r-1} \mathscr{O}_X(-iD) \to X$ constructed as above from s. Note that the divisor S defined by s is zero.

Lemma 2.2.18 *Let* $\pi\colon \bar{X} \to X$ *be the cyclic cover associated with* D *in Definition 2.2.17. Then* $\bar{X}$ *is a normal variety and* π^*D *is principal. The inverse image* $\pi^{-1}(x)$ *of a point* x *in* X *consists of* r/r' *points where* r' *is the least positive integer such that* $r'D$ *is Cartier at* x.

Proof We have already seen in Lemma 2.2.16 that $X = \bar{X}/\mathbf{Z}_r$, $\bar{X}$ is normal and π^*D is principal. Let n be the number of irreducible components of $\bar{X}$. Elements in $\mathbf{Z}_r$ fixing $\bar{X}$ componentwise form a subgroup $\mathbf{Z}_{r/n}$. In particular, n is a divisor of r. The quotient X' of $\bar{X}$ by $\mathbf{Z}_{r/n}$ is the cyclic cover of X associated with $(r/n)D$. Each irreducible component X_i' of X' is normal and the finite morphism $X_i' \to X$ is birational. Hence $X_i' \simeq X$ by Zariski's main theorem and $(r/n)D$ is principal as so is the pull-back to X_i'. From the minimality of r, it follows that $n = 1$.

To see the assertion on $\pi^{-1}(x)$, we take the quotient X'' of $\bar{X}$ by $\mathbf{Z}_{r'}$ as the cyclic cover of X associated with $r'D$. By Lemma 2.2.16, $X'' \to X$ is étale over x.

Replacing $x \in X$ by the germ $x'' \in X''$ at a point x'' above x, we may assume that $r' = r$. Write $\bar{X} = \mathrm{Spec}_X \mathscr{R}$ by $\mathscr{R} = \bigoplus_{i=0}^{r-1} \mathscr{R}_i$ with $\mathscr{R}_i = \mathcal{O}_X(-iD)$. Then the fibre of π at x is the spectrum of $\mathscr{R}/\mathfrak{m}\mathscr{R} = \bigoplus_{i=0}^{r-1} \mathscr{R}_i/\mathfrak{m}\mathscr{R}_i$ for the maximal ideal $\mathfrak{m}$ in $\mathcal{O}_X$ defining x. Since $r' = r$, the map $\mathscr{R}_i^{\otimes r} \to \mathcal{O}_X(-riD)$ for $0 < i < r$ is not surjective at x, that is, the image is contained in $\mathfrak{m}\mathcal{O}_X(-riD)$. Thus $\bigoplus_{i=1}^{r-1} \mathscr{R}_i/\mathfrak{m}\mathscr{R}_i$ is a nilpotent ideal in $\mathscr{R}/\mathfrak{m}\mathscr{R}$, which turns out to be the only maximal ideal. This shows that $\pi^{-1}(x)$ consists of one point.   □

The cyclic cover associated with the canonical divisor is most important and has a name of its own.

Definition 2.2.19   The *index* of a **Q**-Gorenstein normal variety X is the least positive integer r such that rK_X is Cartier. The *index-one cover* of the germ $x \in X$ of a **Q**-Gorenstein normal variety is the cyclic cover $\pi \colon \bar{X} \to X$ associated with K_X. Then $K_{\bar{X}} = \pi^* K_X$ is Cartier and $\bar{X}$ is of index one.

We shall investigate how a finite morphism affects singularities.

Theorem 2.2.20 (Ramification formula)   *Let $\pi \colon Y \to X$ be a generically finite surjective morphism of smooth varieties. Let D and E be snc divisors on X and Y respectively such that $\pi^{-1}(D) = E$. Suppose that π is étale outside E. Then $K_Y + E = \pi^*(K_X + D) + A$ with an effective divisor A on Y such that $\pi_* A = 0$.*

Proof   It suffices to verify the formula about the generic point of each prime component F of E. Take a general point y in F and let $x = \pi(y)$. Let f be the function in $\mathcal{O}_{Y,y}$ which defines F.

First suppose that $\pi_* F$ is zero. There exists a regular system $x_1, \dots, x_n$ of parameters in $\mathcal{O}_{X,x}$ such that D is defined at x by the product $x_1 \cdots x_d$. Then $\mathcal{O}_X(K_X + D)$ is generated at x by $(x_1 \cdots x_d)^{-1} dx_1 \wedge \cdots \wedge dx_n$. Write $\pi^* x_i = u_i f^{l_i}$ with a unit u_i and a positive integer l_i for $1 \le i \le d$, where the positivity of l_i follows from the general choice of y. Writing $u = u_1 \cdots u_d$ and $\theta = d\pi^* x_{d+1} \wedge \cdots \wedge d\pi^* x_n$ for brevity, one computes the pull-back as

$$\pi^* \left(\frac{dx_1 \wedge \cdots \wedge dx_n}{x_1 \cdots x_d} \right) = \frac{d(u_1 f^{l_1}) \wedge \cdots \wedge d(u_d f^{l_d}) \wedge \theta}{u f^{\sum_{i=1}^d l_i}}$$

$$= \frac{1}{uf} \left(f \, du_1 \wedge \cdots \wedge du_d + \sum_{i=1}^d l_i u_i du_1 \wedge \cdots \wedge df \wedge \cdots \wedge du_d \right) \wedge \theta,$$

where df in $du_1 \wedge \cdots \wedge df \wedge \cdots \wedge du_d$ lies in the i-th position. This belongs to $\mathcal{O}_Y(K_Y + E)$ at y, which means that F has non-negative coefficient in A.

Second suppose that $\pi(F)$ is a divisor. Take a regular system $x_1, \dots, x_n$ of parameters in $\mathcal{O}_{X,x}$ such that D is defined at x by x_1. Write $\pi^* x_1 = u f^l$ with a unit u and a positive integer l. By the general choice of y, $E \to D$ is étale at y. Thus $f, \pi^* x_2, \ldots, \pi^* x_n$ form a regular system of parameters in $\mathcal{O}_{Y,y}$. The generator $x_1^{-1} dx_1 \wedge \cdots \wedge dx_n$ at x of $\mathcal{O}_X(K_X + D)$ is transformed to

$$\pi^*(x_1^{-1} dx_1 \wedge \cdots \wedge dx_n) = (uf^l)^{-1} d(uf^l) \wedge \theta$$
$$= (uf)^{-1}(f\,du + lu\,df) \wedge \theta = vf^{-1} df \wedge \theta$$

with $\theta = d\pi^* x_2 \wedge \cdots \wedge d\pi^* x_n$ and some unit v, which generates $\mathcal{O}_X(K_Y + E)$ at y. Thus F has coefficient zero in A. □

Corollary 2.2.21 *Let (X, Δ) and $(\bar{X}, \bar{\Delta})$ be pairs. Let $\pi \colon \bar{X} \to X$ be a finite surjective morphism such that $K_{\bar{X}} + \bar{\Delta} = \pi^*(K_X + \Delta)$. Then (X, Δ) is klt (or lc) if and only if so is $(\bar{X}, \bar{\Delta})$. If (X, Δ) is terminal (or canonical), then so is $(\bar{X}, \bar{\Delta})$.*

Proof Take resolutions $\mu \colon Y \to X$ and $\bar{\mu} \colon \bar{Y} \to \bar{X}$ inducing a proper morphism $\pi_Y \colon \bar{Y} \to Y$ such that $\mu \circ \pi_Y = \pi \circ \bar{\mu}$. Possibly replacing them by higher models, we may assume the existence of snc divisors E on Y and $\bar{E}$ on $\bar{Y}$ with $\pi_Y^{-1}(E) = \bar{E}$ such that μ and $\bar{\mu}$ are isomorphic outside E and $\bar{E}$ respectively and such that the strict transforms of Δ and $\bar{\Delta}$ are supported in E and $\bar{E}$ respectively. We may further assume that π_Y is étale outside $\bar{E}$. Let E_e be the sum of μ-exceptional prime divisors on Y. In like manner, $\bar{E}_e$ is the sum of $\bar{\mu}$-exceptional prime divisors on $\bar{Y}$. We may assume that $\pi_Y^{-1}(E_e) = \bar{E}_e$ and that $E - E_e$ and $\bar{E} - \bar{E}_e$ are disjoint unions of prime divisors.

By Theorem 2.2.20, one can write $K_{\bar{Y}} + \bar{E} = \pi_Y^*(K_Y + E) + A$ with an effective divisor A such that $\pi_{Y*} A = 0$. Pushing it forward downstairs, one obtains the equality $K_{\bar{X}} + \bar{D} = \pi^*(K_X + D)$ for $D = \mu_* E$ and $\bar{D} = \bar{\mu}_* \bar{E}$, where $\bar{\mu}_* A = 0$ by the finiteness of π. Then $\bar{D} - \bar{\Delta} = \pi^*(D - \Delta)$. We may assume that $\Delta \leq D$ and hence $\bar{\Delta} \leq \bar{D}$. We write $K_Y + E = \mu^*(K_X + \Delta) + R$ with an **R**-divisor R such that $\mu_* R = D - \Delta$. Then

$$K_{\bar{Y}} + \bar{E} = \pi_Y^*(\mu^*(K_X + \Delta) + R) + A = \bar{\mu}^*(K_{\bar{X}} + \bar{\Delta}) + \pi_Y^* R + A,$$

in which $\bar{\mu}_* \pi_Y^* R = \bar{D} - \bar{\Delta}$.

Since A is effective with $\pi_{Y*} A$ zero, one sees that R is effective if and only if so is $\pi_Y^* R + A$. This is the equivalence of the log canonicity of (X, Δ) and $(\bar{X}, \bar{\Delta})$. When this is the case, R has support E if and only if $\pi_Y^* R + A$ has support $\bar{E}$, which is the equivalence of the klt property. If in addition $R - E_e$ is effective (and has support E), then so is $\pi_Y^* R + A - \bar{E}_e$ (and has support $\bar{E}$). This means the assertion on the terminal and canonical properties. □

Corollary 2.2.22 *A quotient singularity is log terminal.*

Proof Let $x \in X = \bar{X}/G$ be the germ of the quotient of a smooth variety by a finite group. By Remark 2.2.7, we may assume that G acts on $\bar{X}$ freely in codimension one. Then $K_{\bar{X}} = \pi^* K_X$ for $\pi\colon \bar{X} \to X$ and the assertion follows from Corollary 2.2.21. □

The converse in dimension two will be seen in Remark 4.4.15.

The next result taken from [277, corollary 5.25, proposition 5.26] is useful in the study of threefolds.

Proposition 2.2.23 *Let (X, Δ) be a dlt pair and let D be a **Q**-Cartier divisor on X. Then $\mathscr{O}_X(D)$ is Cohen–Macaulay. If S is an effective **Q**-Cartier divisor on X, then S is a Cohen–Macaulay scheme and the quotient $\mathscr{O}_X(D)/\mathscr{O}_X(D-S)$ is a Cohen–Macaulay $\mathscr{O}_S$-module. One can write $\mathscr{O}_S(D|_S) = \mathscr{O}_X(D)/\mathscr{O}_X(D-S)$ if D is Cartier at every scheme-theoretic point of codimension one in S.*

Proof Work locally on the germ at a closed point of X. Thanks to Theorem 1.4.22, we may assume that (X, Δ) is klt. Take the cyclic cover $\pi\colon \bar{X} \to X$ associated with $-D$. Then $(\bar{X}, \pi^*\Delta)$ is klt by Corollary 2.2.21 and $\bar{X}$ is Cohen–Macaulay by Theorem 1.4.20. Hence the direct summand $\mathscr{O}_X(D)$ of $\pi_* \mathscr{O}_{\bar{X}} = \bigoplus_{i=0}^{r-1} \mathscr{O}_X(iD)$ is also Cohen–Macaulay, where r is the least positive integer such that rD is Cartier. If D is effective, then the Cartier divisor $\bar{D} = \pi^* D$ is Cohen–Macaulay and so is D since $\mathscr{O}_D$ is a direct summand of $\pi_* \mathscr{O}_{\bar{D}}$.

Consider an effective **Q**-Cartier divisor S on X. By the above observation, S and $\bar{S} = \pi^* S$ are Cohen–Macaulay. Take the exact sequence

$$0 \to \pi_* \mathscr{O}_{\bar{X}}(\bar{D} - \bar{S}) \to \pi_* \mathscr{O}_{\bar{X}}(\bar{D}) \to \pi_* \mathscr{O}_{\bar{S}}(\bar{D}|_{\bar{S}}) \to 0$$

of Cohen–Macaulay sheaves. In the direct summand $\mathscr{O}_X(D)$ of $\pi_* \mathscr{O}_{\bar{X}}(\bar{D}) = \bigoplus_{i=1}^{r} \mathscr{O}_X(iD)$, this becomes an exact sequence $0 \to \mathscr{F} \to \mathscr{O}_X(D) \to \mathscr{Q} \to 0$ of Cohen–Macaulay sheaves where $\mathscr{F} = \pi_* \mathscr{O}_{\bar{X}}(\bar{D} - \bar{S}) \cap \mathscr{O}_X(D)$. The equality $\mathscr{F} = \mathscr{O}_X(D - S)$ of Cohen–Macaulay subsheaves of $\mathscr{O}_X(D)$ follows from the restriction to the locus where D is Cartier. Thus $\mathscr{Q} = \mathscr{O}_X(D)/\mathscr{O}_X(D - S)$, which is a Cohen–Macaulay $\mathscr{O}_S$-module. □

2.3 Terminal Singularities of Index One

In this and the next section, we shall describe terminal threefold singularities completely. This section deals with those of index one. As seen in Theorem 1.3.13 and Lemma 1.6.2, a terminal surface is smooth and a terminal threefold singularity is isolated. These also hold in the analytic category by the proofs. Whilst we treat an isolated singularity in the algebraic category, the results still hold in the analytic category by Artin's algebraisation in Theorem 2.1.8.

Theorem 2.3.1 *A surface singularity is canonical if and only if it is Du Val.*

This is one of the numerous characterisations of a Du Val singularity such as listed in [113]. It was originally proved by analysing the configuration of exceptional curves in the minimal resolution. In Proposition 2.1.13 with Lemma 2.1.9, we have already seen that a canonical surface singularity of embedding dimension three is Du Val. We shall provide a new proof of the theorem simultaneously with a characterisation of a terminal threefold singularity of index one.

Definition 2.3.2 The germ $x \in X$ of a threefold is called a *compound Du Val* or *cDV* singularity if the general hyperplane section $x \in S$ is Du Val. We say that $x \in X$ is of *type cA_n, cD_n or cE_n* correspondingly to the type A_n, D_n or E_n of $x \in S$.

By definition, a cDV singularity is analytically isomorphic to a hypersurface singularity given by $f(x_1, x_2, x_3) + x_4 g(x_1, x_2, x_3, x_4)$ in which f is one of the polynomials in Definition 2.1.12. Below is a more explicit expression.

Proposition 2.3.3 *Every cDV singularity is analytically given by one of the convergent power series*

$$cA_{\leq 1} \ \ x_1^2 + x_2^2 + x_3^2 + g(x_4),$$
$$cA_{\geq 2} \ \ x_1^2 + x_2^2 + g(x_3, x_4) \ \textit{with ord } g \geq 3,$$
$$cD_4 \ \ x_1^2 + g(x_2, x_3, x_4) \ \textit{with ord } g = 3 \ \textit{and with the cubic part of } g \textit{ having no}$$
$$\textit{multiple factor,}$$
$$cD_{\geq 5} \ \ x_1^2 + x_2^2 x_3 + g(x_2, x_3, x_4) \ \textit{with ord } g \geq 4,$$
$$cE \ \ x_1^2 + x_2^3 + x_2 g(x_3, x_4) + h(x_3, x_4) \ \textit{with ord } g \geq 3 \ \textit{and ord } h \geq 4$$

in accordance with the type. If a cD singularity is isolated, then it is given by a refined function $x_1^2 + x_2^2 x_3 + \lambda x_2 x_4^k + g(x_3, x_4)$.

Proof Let f be a function in $\mathscr{O}_{\mathbb{D}^4}$ which defines a cDV singularity at origin. Applying Theorem 2.1.10 and Remark 2.1.11 repeatedly, one attains f either of form in $cA_{\leq 1}$ or $cA_{\geq 2}$ in the statement or of form $x_1^2 + g(x_2, x_3, x_4)$ with ord $g = 3$ for suitable coordinates. We shall assume the latter case, in which f is of type cD or cE.

If the cubic part g_3 of g has no multiple factor, then it defines a cubic in $\mathbf{P}^2$ which does not contain a double line. Choosing x_4 generally, one has $g_3(x_2, x_3, 0) = x_2^2 x_3 + x_3^3$ after coordinate change, in which f is of type cD_4. If g_3 has a multiple factor, then g_3 is $x_2^2 x_3$ or x_2^3 after coordinate change. The former case $g_3 = x_2^2 x_3$ corresponds to $cD_{\geq 5}$ and the latter to cE by Theorem 2.1.10 and Remark 2.1.11.

Now suppose that $x_1^2 + g(x_2, x_3, x_4)$ defines an isolated cD singularity. The cubic part of g is $x_2^2 x_3 + \mu x_2 x_4^2 + g_3(x_3, x_4)$ with $\mu \in \mathbf{C}$ by coordinate change. We write

$$g = x_2^2(x_3 + p(x_2, x_3, x_4)) + x_2 q(x_3, x_4) + r(x_3, x_4).$$

Consider the weighted order with respect to $\mathrm{wt}(x_2, x_3, x_4) = (2, 3, 3)$. Then $w = \text{w-ord } p \geq 4$ and $v = \text{w-ord}(x_2 q + r) \geq 8$. Set $y_3 = x_3 + p$. The part in $x_2 q(y_3 - p, x_4) + r(y_3 - p, x_4)$ divisible by x_2^2 is of weighted order at least $v + (w - 3) \geq w + 5$ with respect to $\mathrm{wt}(x_2, y_3, x_4) = (2, 3, 3)$. Hence if one expresses g by the new coordinates x_2, y_3, x_4 as above, then the term corresponding to p is of weighted order at least $(w + 5) - 4 = w + 1$. Thus one can attain the above expression of g with p of arbitrarily high weighted order.

Hence by Corollary 2.1.5, we can take $p = 0$ and $g = x_2^2 x_3 + x_2 q + r$. Write $q = 2a(x_3, x_4) x_3 + b(x_4)$ and $g = (x_2 + a)^2 x_3 + (x_2 + a)b + r - a^2 x_3 - ab$. Replacing x_2 by $x_2 + a$, g becomes of form $x_2^2 x_3 + x_2 b(x_4) + s(x_3, x_4)$, from which the last assertion is derived. □

The main theorem in this section is the following.

Theorem 2.3.4 (Reid [399]) *A threefold singularity is terminal of index one if and only if it is an isolated cDV singularity.*

The original proof by Reid relies on the classification of elliptic surface singularities as will be explained later. Our proof, following [226], uses the Riemann–Roch theorem. By Theorem 1.4.20, a canonical singularity of index one is Gorenstein. We make a framework in which the machinery is applied to both Theorems 2.3.1 and 2.3.4.

Framework 2.3.5 Let $x \in X$ be the germ of an isolated Gorenstein singularity of dimension n at least two. Let $\mathfrak{m}$ denote the maximal ideal in $\mathscr{O}_X$ defining x. We fix a strong log resolution $\mu \colon \bar{X} \to X$ of $(X, \mathfrak{m})$, which is isomorphic outside x. We define the exceptional divisors E and K by $\mathfrak{m}\mathscr{O}_{\bar{X}} = \mathscr{O}_{\bar{X}}(-E)$ and $K = K_{\bar{X}/X}$ respectively. Then $\mu^* H = \bar{H} + E$ for the general hyperplane section $x \in H$ and the strict transform $\bar{H}$. Take n general hyperplane sections $H_1, \ldots, H_n$ and let $S = H_1 \cap \cdots \cap H_{n-2}$, $C = S \cap H_{n-1}$ and $\operatorname{Spec} R = C \cap H_n$. By Bertini's theorem, S is a normal surface, C is a curve and R is an artinian Gorenstein local ring. For the strict transform $\bar{H}_i$ in $\bar{X}$ of H_i, the intersections $\bar{S} = \bar{H}_1 \cap \cdots \cap \bar{H}_{n-2}$ and $\bar{C} = \bar{S} \cap \bar{H}_{n-1}$ are smooth and admit strong resolutions $\bar{S} \to S$ and $\bar{C} \to C$.

We write $\operatorname{mult}_x X$ for the *multiplicity* of X at x. When X is Cohen–Macaulay as in the present framework, the multiplicity coincides with the *intersection multiplicity* [308, section 14]. Hence $\mathrm{mult}_x X = E\bar{H}^{n-1}$. For an integer l, we define the dimension

$$d_l = \dim \mu_* \mathscr{O}_{\bar{S}}((K - lE)|_{\bar{S}})/\mu_* \mathscr{O}_{\bar{S}}((K - (l+1)E)|_{\bar{S}})$$

of the vector space.

Lemma 2.3.6   (i) $E\bar{H}^{n-1} = \mathrm{mult}_x X = d_{n-1} - d_{n-2}$.
(ii) $K\bar{H}^{n-1} = (n-1)\,\mathrm{mult}_x X - 2d_{n-2}$.

Proof   We compute the Euler characteristic $p_l = \chi(\mathscr{O}_{E|_S}((K - lE)|_{E|_S}))$. Compactifying the germ $x \in X$, we may assume that X is projective. Then $p_l = q_l - q_{l+1}$ for $q_l = \chi(\mathscr{O}_{\bar{S}}((K - lE)|_{\bar{S}}))$ from the exact sequence

$$0 \to \mathscr{O}_{\bar{S}}((K - (l+1)E)|_{\bar{S}}) \to \mathscr{O}_{\bar{S}}((K - lE)|_{\bar{S}}) \to \mathscr{O}_{E|_S}((K - lE)|_{E|_S}) \to 0.$$

The Riemann–Roch formula on $\bar{S}$ with $K_{\bar{S}/S} = (K - (n-2)E)|_{\bar{S}}$ provides

$$q_l = \frac{1}{2}(K - lE)((K - lE) - (K - (n-2)E))\bar{H}^{n-2} + \chi(\mathscr{O}_{\bar{S}})$$

$$= \frac{1}{2}(l + 2 - n)(K - lE)\bar{H}^{n-1} + \chi(\mathscr{O}_{\bar{S}}).$$

Hence $p_l = q_l - q_{l+1}$ is computed as

$$p_l = \frac{1}{2}(2l + 3 - n)E\bar{H}^{n-1} - \frac{1}{2}K\bar{H}^{n-1},$$

from which $E\bar{H}^{n-1} = p_{n-1} - p_{n-2}$ and $K\bar{H}^{n-1} = (n-1)p_{n-1} - (n+1)p_{n-2}$. It suffices to prove the equality $p_l = d_l$ for $l \geq n - 2$.

Suppose that $l \geq n - 2$. Since $-E$ is relatively free and $(K - lE)|_{\bar{S}} = K_{\bar{S}/S} - (l + 2 - n)E|_{\bar{S}}$, the Kawamata–Viehweg vanishing $R^i\mu_*\mathscr{O}_{\bar{S}}((K - lE)|_{\bar{S}}) = 0$ holds for $i \geq 1$. Hence the above sequence gives

$$H^0(\mathscr{O}_{E|_S}((K - lE)|_{E|_S})) \simeq \mu_*\mathscr{O}_{\bar{S}}((K - lE)|_{\bar{S}})/\mu_*\mathscr{O}_{\bar{S}}((K - (l+1)E)|_{\bar{S}})$$

and $H^i(\mathscr{O}_{E|_S}((K - lE)|_{E|_S})) = 0$ for $i \geq 1$. These imply that $p_l = d_l$.   □

We shall translate the lemma into the language of C and R. For a moment, we work in the general setting where $x \in X$ is the germ of an isolated singularity, which is not necessarily normal. Let $\mu\colon \bar{X} \to X$ be a strong resolution. For an arbitrary divisor D on $\bar{X}$ supported in $\mu^{-1}(x)$, we define the ideal $\mathscr{O}_X(D)$ in $\mathscr{O}_X$ by $\mathscr{O}_X(D) = \mu_*\mathscr{O}_{\bar{X}}(D) \cap \mathscr{O}_X$, which is expressed as

$$\mathscr{O}_X(D) = \{f \in \mathscr{O}_X \mid (f)_{\bar{X}} + D \geq 0\}.$$

Let H be an effective Cartier divisor on X and write $\mu^*H = \bar{H} + E$ with the strict transform $\bar{H}$ of H and an exceptional divisor E. Note that $\mathscr{O}_X(D) \subset \mathscr{O}_X(D+E)$.

Lemma 2.3.7 $\dim \mathscr{O}_X(D + E)/\mathscr{O}_X(D) = \dim \mathscr{O}_H/\mathscr{O}_X(D)\mathscr{O}_H$.

Proof   We write $I_l = \mathscr{O}_X(D + lE) \subset \mathscr{O}_X$ for $l \in \mathbf{Z}$ and claim the equality

$$\dim I_{l+1}/I_l = \dim I_{l+1}\mathscr{O}_H/I_l\mathscr{O}_H + \dim I_{l+2}/I_{l+1}.$$

Take the function h in $\mathscr{O}_X$ defining H. Then $\dim I_{l+1}/I_l$ equals the sum of the dimensions of the two vector spaces

$$I_{l+1}/(I_l + I_{l+1} \cap h\mathscr{O}_X) \simeq (I_{l+1} + h\mathscr{O}_X)/(I_l + h\mathscr{O}_X) \simeq I_{l+1}\mathscr{O}_H/I_l\mathscr{O}_H,$$
$$(I_l + I_{l+1} \cap h\mathscr{O}_X)/I_l \simeq (I_{l+1} \cap h\mathscr{O}_X)/(I_l \cap h\mathscr{O}_X).$$

For a function f in $\mathscr{O}_X$, the function hf is contained in I_l if and only if $(f)_{\bar{X}} + D + (l + 1)E \geq 0$, that is, $f \in \mathscr{O}_X(D + (l + 1)E) = I_{l+1}$. Hence $I_l \cap h\mathscr{O}_X = hI_{l+1}$. The equality for $l + 1$ is $I_{l+1} \cap h\mathscr{O}_X = hI_{l+2}$. Thus $(I_{l+1} \cap h\mathscr{O}_X)/(I_l \cap h\mathscr{O}_X) \simeq I_{l+2}/I_{l+1}$ and the claim follows.

Fix a positive integer i such that $D + iE$ is effective. Summing up the equalities above for all $0 \leq l < i$, one obtains $\dim I_1/I_0 = \dim I_i\mathscr{O}_H/I_0\mathscr{O}_H + \dim I_{i+1}/I_i$. This shows the lemma because $I_i = I_{i+1} = \mathscr{O}_X$ by $D + iE \geq 0$. □

Returning to Framework 2.3.5 where $x \in X$ is an isolated Gorenstein singularity, we apply Lemma 2.3.7 to $\bar{S} \to S$ and $\bar{C} \to C$. To avoid intricacy, we use the notation $\mathscr{O}_S(D) = \mu_*\mathscr{O}_{\bar{S}}(D|_{\bar{S}})$ and $\mathscr{O}_C(D) = \mu_*\mathscr{O}_{\bar{C}}(D|_{\bar{C}}) \cap \mathscr{O}_C$ for an exceptional divisor D on $\bar{X}$. The restriction $\mathscr{O}_C(D)R = \mathscr{O}_C(D) \cdot R$ is an ideal in R.

Lemma 2.3.8   *The ideal $\mathscr{O}_C(K - nE)R$ in R is zero. If X is canonical, then $K\bar{H}^{n-1} = (n - 1)\dim R - 2\sum_{l=1}^{n-1} \dim R/\mathscr{O}_C(K - lE)R$.*

Proof   Apply Lemma 2.3.7 to $\bar{S} \to S$, $H = C$ and $D = (K - (l + 1)E)|_{\bar{S}}$ and obtain the equality $d_l = \dim \mathscr{O}_C/\mathscr{O}_S(K - (l + 1)E)\mathscr{O}_C$. For $l \geq n - 1$, the Kawamata–Viehweg vanishing $R^1\mu_*\mathscr{O}_{\bar{S}}((K - lE)|_{\bar{S}} - \bar{C}) = 0$ holds by $(K - lE)|_{\bar{S}} - \bar{C} \equiv_S K_{\bar{S}} - (l + 1 - n)E|_{\bar{S}}$ and the natural map $\mathscr{O}_S(K - lE) = \mu_*\mathscr{O}_{\bar{S}}((K-lE)|_{\bar{S}}) \to \mu_*\mathscr{O}_{\bar{C}}((K-lE)|_{\bar{C}})$ is surjective. Hence $\mathscr{O}_S(K-lE)\mathscr{O}_C = \mathscr{O}_C(K - lE)$. Consequently $d_l = \dim \mathscr{O}_C/\mathscr{O}_C(K - (l + 1)E)$ for $l \geq n - 2$.

By this expression, we rewrite Lemma 2.3.6(i) as

$$\dim R = \dim \mathscr{O}_C/\mathscr{O}_C(K - nE) - \dim \mathscr{O}_C/\mathscr{O}_C(K - (n - 1)E)$$
$$= \dim \mathscr{O}_C(K - (n - 1)E)/\mathscr{O}_C(K - nE).$$

Applying Lemma 2.3.7 to $\bar{C} \to C$, $H = \operatorname{Spec} R$ and $D = (K - nE)|_{\bar{C}}$, we see that the right-hand side equals $\dim R/\mathscr{O}_C(K-nE)R$. Thus $\mathscr{O}_C(K-nE)R = 0$.

We also rewrite Lemma 2.3.6(ii) as

$$K\bar{H}^{n-1} = (n - 1)\dim R - 2\dim \mathscr{O}_C/\mathscr{O}_C(K - (n - 1)E).$$

If X is canonical, then K is effective. In this case, $\mathscr{O}_C(K) = \mathscr{O}_C$ and the dimension of $\mathscr{O}_C/\mathscr{O}_C(K - (n-1)E)$ equals the sum $\sum_{l=1}^{n-1} \dim \mathscr{O}_C(K - (l-1)E)/\mathscr{O}_C(K - lE)$. The formula on $K\bar{H}^{n-1}$ in the lemma is thus reduced to the equality $\dim \mathscr{O}_C(K - (l-1)E)/\mathscr{O}_C(K - lE) = \dim R/\mathscr{O}_C(K - lE)R$, which is derived from Lemma 2.3.7 as in the preceding paragraph.   □

Corollary 2.3.9 $\dim \mathscr{O}_C(K - (n-1)E)R \le 1$.

Proof By definition, $\mathscr{O}_C(K - (n-1)E) \cdot \mathfrak{m}\mathscr{O}_C$ is contained in $\mathscr{O}_C(K - nE)$. We restrict this to R. From $\mathscr{O}_C(K - nE)R = 0$ in Lemma 2.3.8, the ideal $\mathscr{O}_C(K - (n-1)E)R$ in R is contained in the *socle* $(0 : \mathfrak{m}R)$ of R. In general for an artinian Gorenstein local ring R with maximal ideal $\mathfrak{m}$, the *socle* $(0 : \mathfrak{m})$ is isomorphic to $R/\mathfrak{m}$ [308, theorem 18.1]. The corollary follows from this property.   □

Applying these results, we establish Theorems 2.3.1 and 2.3.4.

Proof of Theorem 2.3.1   One can directly verify the if part by resolving each Du Val singularity explicitly. For the only-if part, by Proposition 2.1.13 and Lemma 2.1.9, it suffices to show that a canonical surface singularity $x \in S$ is of embedding dimension at most three.

We shall prove that S is Gorenstein by working in the analytic category. Take the minimal resolution $\pi\colon S' \to S$. Since S is canonical, the relative canonical divisor $K_{S'/S}$ is numerically trivial. Hence $\omega_{S'}$ has degree zero along every exceptional curve. By Theorem 1.4.20, S has rational singularities, that is, $R^1\pi_*\mathscr{O}_{S'} = 0$. Thus $R^1\pi_*\mathscr{O}_{S'}^\times \simeq R^2\pi_*\mathbf{Z}$ from the exponential sequence $0 \to \mathbf{Z} \to \mathscr{O}_{S'} \to \mathscr{O}_{S'}^\times \to 0$. On the analytic germ $x \in S$, one has

$$\text{Pic } S' = H^1(\mathscr{O}_{S'}^\times) \simeq H^2(S', \mathbf{Z}) \simeq H^2(\pi^{-1}(x), \mathbf{Z}),$$

where the last isomorphism is deduced from Theorem 2.3.10. It follows that $\omega_{S'} \simeq \mathscr{O}_{S'}$ and $\pi_*\omega_{S'} \simeq \mathscr{O}_S$. Then the double dual ω_S of $\pi_*\omega_{S'}$ is isomorphic to $\mathscr{O}_S$, that is, S is Gorenstein.

Work in Framework 2.3.5 for $X = S$ with $n = 2$. Since $K = K_{\bar{S}/S}$ is effective, the inclusion $\mathfrak{m}\mathscr{O}_C = \mathscr{O}_C(-E) \subset \mathscr{O}_C(K - E)$ holds. Hence $\dim \mathfrak{m}R \le 1$ by Corollary 2.3.9 and therefore $\dim \mathfrak{m}/\mathfrak{m}^2 = 2 + \dim \mathfrak{m}R/\mathfrak{m}^2R \le 3$.   □

We used a result of *triangulation* found in [30, I theorem 8.8] and [198, theorem 2.4.11]. See [297] for the proof.

Theorem 2.3.10 *Let X be an analytic space in a countable complex manifold and let Z be a closed analytic subspace of X. Then there exists a triangulation of X such that Z is the support of a subcomplex. In particular, Z has an open neighbourhood U in X such that Z is a deformation retract of U.*

Proof of Theorem 2.3.4　　The if part is a consequence of inversion of adjunction as will be proved in Proposition 2.3.11. To see the only-if part, we let $x \in X$ be a terminal threefold singularity of index one. It is isolated and Gorenstein by Lemma 1.6.2 and Theorem 1.4.20. We work in Framework 2.3.5 for $n = 3$.

The formula in Lemma 2.3.8 for $n = 3$ is

$$K\bar{H}^2 = 2 \dim \mathscr{O}_C(K - 2E)R - 2 \dim R/\mathscr{O}_C(K - E)R.$$

Since X is terminal, K is effective and has support exactly $\pi^{-1}(x)$. In particular $K\bar{H}^2$ is positive. By Corollary 2.3.9, $\dim \mathscr{O}_C(K - 2E)R \leq 1$. Thus the above expression of $K\bar{H}^2$ implies that $K\bar{H}^2 = 2$, $\mathscr{O}_C(K - E) = \mathscr{O}_C$ and $\dim \mathscr{O}_C(K - 2E)R = 1$. Then $\mathfrak{m}R = \mathscr{O}_C(K - E) \cdot \mathscr{O}_C(-E)R \subset \mathscr{O}_C(K - 2E)R$ and hence $\text{mult}_x X = \dim R = 1 + \dim \mathfrak{m}R \leq 2$. If $\text{mult}_x X = 1$, then $x \in X$ is smooth. We shall assume that $\text{mult}_x X = 2$. By $\mathscr{O}_C(K - E) = \mathscr{O}_C$, the restriction $(K - E)|_{\bar{C}}$ is effective, that is, $K|_{\bar{C}} \geq E|_{\bar{C}}$. On the other hand, $\deg E|_{\bar{C}} = \text{mult}_x X = 2$ and $\deg K|_{\bar{C}} = K\bar{H}^2 = 2$. Hence $K|_{\bar{C}} = E|_{\bar{C}}$ and thus $K_{\bar{S}/S}|_{\bar{C}}$ is zero.

Starting from $\bar{S}/S$, we contract a relative (-1)-curve away from $\bar{C}$ successively as far as it exists. If we attain the minimal resolution T of S, then the relatively nef divisor $K_{T/S}$ is zero as so is $K_{T/S}|_{\bar{C}}$. Thus S is canonical and X is cDV from Theorem 2.3.1. Suppose that we would attain a smooth surface T/S on which $\bar{C}$ intersects a relative (-1)-curve l_0. Write $E = \sum_i m_i E_i$ and $K = \sum_i k_i E_i$ with prime divisors E_i and positive integers m_i, k_i. We set the strict transform l_i in T of $E_i|_{\bar{S}}$, possibly reducible if $E_i\bar{H}^2 = 0$. The curve l_0, not appearing in $K_{T/S}$, intersects some l_1 with $k_1 < m_1$ because $(K_{T/S} \cdot l_0)_T < 0$. Then $(\bar{C} \cdot l_0)_T = (-\sum m_i l_i \cdot l_0)_T \leq m_0 - m_1 \leq E\bar{H}^2 - 2 = 0$, which is absurd. □

Proposition 2.3.11　*An algebraic or analytic cDV singularity is canonical and Gorenstein. If it is isolated, then it is terminal.*

Proof　Let $x \in X$ be a cDV singularity. Take the general hyperplane section $x \in S$, which is Du Val. Since S is normal and Gorenstein, so is X. Since S is canonical by Theorem 2.3.1, the pair (X, S) is plt by inversion of adjunction in Theorem 1.4.26. It is further canonical as $K_X + S$ is Cartier. In particular, X itself is canonical. The minimal log discrepancy is estimated as $\text{mld}_x X > \text{mld}_x(X, S) \geq 1$ from the canonicity of (X, S). This shows that $x \in X$ is terminal if it is an isolated singularity. □

We shall explain the original proof of Theorem 2.3.4 which requires the notion of elliptic singularity.

Definition 2.3.12　The germ $x \in S$ of a normal Gorenstein surface is called an *elliptic singularity* if $\dim R^1\mu_*\mathscr{O}_{S'} = 1$ for every resolution $\mu: S' \to S$. This is equivalent to the existence of some resolution μ having this property because the Leray spectral sequence gives the isomorphism $R^1\mu_*\mathcal{O}_{S'} \simeq R^1(\mu \circ \pi)_*\mathcal{O}_T$ for any higher resolution $\pi\colon T \to S'$.

Lemma 2.3.13 *Let* $\mu\colon S' \to S$ *be a resolution of a normal surface. Then* $\dim R^1\mu_*\mathcal{O}_{S'} = \dim \omega_S/\mu_*\omega_{S'}$.

Proof The statement is local on S. By compactification, we may assume that S is projective. Fix an ample invertible sheaf $\mathcal{L}$ on S such that $H^p(R^q\mu_*\mathcal{O}_{S'} \otimes \mathcal{L}) = 0$ for all $p \geq 1$ and $q \geq 0$. The spectral sequence $H^p(R^q\mu_*\mathcal{O}_{S'} \otimes \mathcal{L}) \Rightarrow H^{p+q}(\mu^*\mathcal{L})$ yields

$$H^0(R^1\mu_*\mathcal{O}_{S'} \otimes \mathcal{L}) \simeq H^1(\mu^*\mathcal{L}).$$

By Serre duality, the space $H^i(\omega_S \otimes \mathcal{L}^{-1})$ is dual to $H^{2-i}(\mathcal{L})$, which is zero for $i = 0$ and 1. From the exact sequence

$$0 \to \mu_*\omega_{S'} \otimes \mathcal{L}^{-1} \to \omega_S \otimes \mathcal{L}^{-1} \to (\omega_S/\mu_*\omega_{S'}) \otimes \mathcal{L}^{-1} \to 0,$$

one obtains the isomorphism $H^0((\omega_S/\mu_*\omega_{S'}) \otimes \mathcal{L}^{-1}) \simeq H^1(\mu_*\omega_{S'} \otimes \mathcal{L}^{-1})$. By Grauert–Riemenschneider vanishing, $R^q\mu_*\omega_{S'} = 0$ for $q \geq 1$. Hence the spectral sequence $H^p(R^q\mu_*\omega_{S'} \otimes \mathcal{L}^{-1}) \Rightarrow H^{p+q}(\omega_{S'} \otimes \mu^*\mathcal{L}^{-1})$ yields $H^1(\mu_*\omega_{S'} \otimes \mathcal{L}^{-1}) \simeq H^1(\omega_{S'} \otimes \mu^*\mathcal{L}^{-1})$. Thus

$$H^0((\omega_S/\mu_*\omega_{S'}) \otimes \mathcal{L}^{-1}) \simeq H^1(\omega_{S'} \otimes \mu^*\mathcal{L}^{-1}).$$

The two spaces $H^1(\mu^*\mathcal{L})$ and $H^1(\omega_{S'} \otimes \mu^*\mathcal{L}^{-1})$ are dual by Serre duality. The artinian sheaves $R^1\mu_*\mathcal{O}_{S'}$ and $\omega_S/\mu_*\omega_{S'}$ therefore have the same dimension. □

The results on elliptic singularities by Laufer and Reid are stated in terms of the fundamental cycle, which Artin introduced in the study of rational surface singularities.

Definition 2.3.14 Let $x \in S$ be a normal surface singularity. The *fundamental cycle* on a resolution S' of $x \in S$ is the least non-zero effective exceptional divisor Z on S' such that $-Z$ is relatively nef. It always exists.

Theorem 2.3.15 (Artin [17]) *Let* $x \in S$ *be a rational surface singularity and let* $\mathfrak{m}$ *denote the maximal ideal in* $\mathcal{O}_S$ *defining* x. *Let* Z *be the fundamental cycle on a resolution* S' *of* $x \in S$. *Then for any positive integer* l, $\mathfrak{m}^l\mathcal{O}_{S'} = \mathcal{O}_{S'}(-lZ)$, $H^0(\mathcal{O}_{lZ}) = \mathcal{O}_S/\mathfrak{m}^l$ *and* $\dim \mathfrak{m}^l/\mathfrak{m}^{l+1} = -lZ^2 + 1$. *In particular,* $x \in S$ *is of multiplicity* $-Z^2$ *and of embedding dimension* $-Z^2 + 1$.

Theorem 2.3.16 (Laufer [287], Reid [396]) *Let* $x \in S$ *be an elliptic Gorenstein surface singularity of embedding dimension* e *and let* $\mathfrak{m}$ *denote the maximal ideal in* $\mathcal{O}_S$ *defining* x. *Let* Z *be the fundamental cycle on the minimal resolution* S' *of* $x \in S$. *Then* $K_{S'/S} = -Z$. *If* $-Z^2 \geq 2$, *then* $\mathcal{O}_{S'}(-Z) = \mathfrak{m}\mathcal{O}_{S'}$. *If* $-Z^2 = 1$, *then* $\dim \mathscr{O}_{S'}(-Z)/\mathfrak{m}\mathscr{O}_{S'} = 1$. *The following hold more explicitly and the weighted blow-up B of S with* $\text{wt}(x_1, \ldots, x_e) = w$ *prescribed there is Du Val. Notice that $B \to S$ is defined in the algebraic category.*

  (i) *If* $-Z^2 \geq 3$, *then* $e = -Z^2$. *Set* $w = (1, \ldots, 1)$.
 (ii) *If* $-Z^2 = 2$, *then* $e = 3$ *and* $x \in S$ *is analytically given by* $x_1^2 + f(x_2, x_3)$
      with $\text{ord } f \geq 4$. *Set* $w = (2, 1, 1)$.
(iii) *If* $-Z^2 = 1$, *then* $e = 3$ *and* $x \in S$ *is analytically given by* $x_1^2 + x_2^3 + x_2 f(x_3) + g(x_3)$ *with* $\text{ord } f \geq 4$ *and* $\text{ord } g \geq 6$. *Set* $w = (3, 2, 1)$.

Proposition 2.3.17   *The general hyperplane section of a canonical threefold singularity $x \in X$ of index one is either Du Val or elliptic. It is Du Val if and only if* $\text{mld}_x X$ *is greater than one.*

Proof   The general hyperplane section S of $x \in X$ is normal and Gorenstein as so is X by Theorem 1.4.20. Take a log resolution $\mu \colon X' \to X$ of (X, S) and write S' for the strict transform in X' of S. We define the exceptional divisors E and K on X' by $\mu^* S = S' + E$ and $K = K_{X'/X}$ respectively. Then $K_{X'} + S' = \mu^*(K_X + S) + K - E$ and $\omega_{S'} = \mu^*\omega_S \otimes \mathscr{O}_{S'}((K - E)|_{S'})$. Note that K is effective since X is canonical.

Let $\mathfrak{m} = \mu_* \mathscr{O}_{S'}(-E|_{S'})$ be the maximal ideal in $\mathscr{O}_S$ defining x, which is contained in $\mu_* \mathscr{O}_{S'}((K - E)|_{S'})$. Then $\mathfrak{m}\omega_S \subset \omega_S \otimes \mu_* \mathscr{O}_{S'}((K - E)|_{S'}) = \mu_*\omega_{S'} \subset \omega_S$. If $\mu_*\omega_{S'} = \omega_S$, then $\mu_* \mathscr{O}_{S'}((K - E)|_{S'}) = \mathscr{O}_S$, which means the effectivity of $K_{S'/S} = (K - E)|_{S'}$. Hence S is canonical. If $\mu_*\omega_{S'} = \mathfrak{m}\omega_S$, then S is elliptic by Lemma 2.3.13. This is the first assertion.

If S is Du Val, then (X, S) is canonical by inversion of adjunction. Hence $\text{mld}_x X > \text{mld}_x(X, S) \geq 1$. To see the converse, suppose that S is elliptic. The morphism $S' \to S$ factors through the minimal resolution T of S with a contraction $\pi \colon S' \to T$. By Theorem 2.3.16, $K_{T/S} = \pi_*((K - E)|_{S'})$ equals $-\pi_*(E|_{S'})$, the support of which is the exceptional locus of $T \to S$. Thus there exists an exceptional prime divisor F on X' such that $\text{ord}_F(K - E) = \text{ord}_F(-E) < 0$ and such that $\pi_*(F|_{S'}) \neq 0$. It has log discrepancy $a_F(X) = 1 + \text{ord}_F K = 1$. The positivity of $\text{ord}_F E$ means that F has centre $c_X(F) \subset \mu(E) = x$. Hence $\text{mld}_x X \leq a_F(X) = 1$ and the second assertion is completed. $\square$

Original proof of Theorem 2.3.4   The theorem follows from Lemma 1.6.2 and Propositions 2.3.11 and 2.3.17. $\square$

Finally we provide two important properties of terminal threefold singularities of index one. One pertains to the divisor class group.

Theorem 2.3.18   *Every* **Q**-*Cartier integral divisor on a terminal threefold of index one is Cartier.*

Thanks to Theorem 2.3.4, this is a special case of the following theorem.

Theorem 2.3.19 *Let $x \in X$ be an isolated hypersurface singularity of dimension at least three, which is normal. Then there exists an analytic neighbourhood U in X at x such that $U \setminus x$ is simply connected. In particular, every $\mathbf{Q}$-Cartier integral divisor on $x \in X$ is Cartier.*

Proof The proof is due to Reid and Ue. The singularity $x \in X$ is normal by Serre's criterion. By Lemmata 2.2.16 and 2.2.18, every $\mathbf{Q}$-Cartier divisor on X produces a cyclic cover of $x \in X$ as a normal variety étale outside x. Thus the second assertion is reduced to the first one.

Consider the analytic germ $x \in X$. By Corollary 2.1.5 or Theorem 2.1.8, we may assume that it is defined at origin in the domain $\mathfrak{D}^{n+1}$ by a polynomial, where n is the dimension of X. For a sufficiently small positive real number ε, we define the closed disc $D_{\varepsilon} = \{z \in \mathfrak{D}^{n+1} \mid \|z\| \leq \varepsilon\}$ and its boundary $S_{\varepsilon} = \{z \in \mathfrak{D}^{n+1} \mid \|z\| = \varepsilon\}$ for the Euclidean norm $\| \ \|$. Then by [320, theorem 2.10], $X \cap D_{\varepsilon}$ is homeomorphic to the cone over $X \cap S_{\varepsilon}$. Further by [320, theorem 5.2], $X \cap S_{\varepsilon}$ is simply connected for $n \geq 3$. Thus $U \setminus x$ is simply connected for the interior U of $X \cap D_{\varepsilon}$. □

Corollary 2.3.20 *A terminal quotient threefold singularity is cyclic.*

Proof By Theorem 2.2.4 and Remark 2.2.7, it is enough to consider the quotient $X = \mathbf{A}^3/G$ by a small subgroup G of $\mathrm{GL}(3, \mathbf{C})$. Suppose that $o \in X$ is terminal. It is an isolated singularity and $\mathbf{A}^3 \to X$ is étale outside o. The group homomorphism $\det \colon G \to \mathbf{C}^{\times}$ which takes the determinant has kernel $H = G \cap \mathrm{SL}(3, \mathbf{C})$. The quotient $\mathbf{A}^3/H$ is a cyclic cover of X, which is also terminal by Corollary 2.2.21 and of index one. Then $H = 1$ by Theorem 2.3.19 and G is isomorphic to a finite subgroup of $\mathbf{C}^{\times}$, which is cyclic. □

The other is on the minimal log discrepancy. The original proof of the next theorem is by finding a divisor with log discrepancy two from the explicit expression in Proposition 2.3.3.

Theorem 2.3.21 (Markushevich [303]) *Let $x \in X$ be a terminal threefold singularity of index one. Then $\mathrm{mld}_x X = 2$ unless it is smooth.*

Proof Since X is terminal of index one, $\mathrm{mld}_x X$ is an integer greater than one. We need to prove the opposite inequality $\mathrm{mld}_x X \leq 2$ when it is singular. By Theorem 2.3.4 and Lemma 2.1.9, the germ $x \in X$ admits an étale morphism to a hypersurface in $\mathbf{A}^4$, by which we may assume that X is a hypersurface in $\mathbf{A}^4$. Then $\mathrm{mld}_x X = \mathrm{mld}_x(\mathbf{A}^4, X)$ by precise inversion of adjunction in Theorem 1.6.9. It is at most the log discrepancy $a_E(\mathbf{A}^4, X) = 4 - \mathrm{mult}_x X = 2$ of the divisor E obtained by the blow-up of $\mathbf{A}^4$ at x. □

Example 2.3.22   Let $x \in X$ be the hypersurface singularity given by $x_1^2 + x_2^3 + x_3^5 + x_4^5$, which is terminal and cE_8. For each integer $0 \le i \le 4$, the divisor E_i obtained by the weighted blow-up of X with $\mathrm{wt}(x_1, x_2, x_3 + \zeta^i x_4, x_4) = (3, 2, 2, 1)$, where ζ is a primitive fifth root of unity, has log discrepancy $a_{E_i}(X) = 2$. Let Y be the weighted blow-up of X with $\mathrm{wt}(x_1, x_2, x_3, x_4) = (3, 2, 1, 1)$. It is normal and every E_i appears on it as a divisor. The exceptional locus in Y is given in $\mathbf{P}(3, 2, 1, 1)$ by $x_3^5 + x_4^5 = \prod_{i=0}^{4}(x_3 + \zeta^i x_4)$.

2.4 Terminal Singularities of Higher Index

This section is devoted to the analysis of terminal threefold singularities of higher index. Let $x \in X$ be a terminal threefold singularity of index r. Let $\bar{x} \in \bar{X} \to x \in X$ be the index-one cover. By Lemma 2.2.16, $\bar{X} \to X$ is the quotient by the cyclic group $\mathbf{Z}_r$. The action of $\mathbf{Z}_r$ on $\bar{X}$ is free outside $\bar{x}$ because the singularity $x \in X$ is isolated by Lemma 1.6.2. The germ $\bar{x} \in \bar{X}$ is terminal of index one by Corollary 2.2.21, and hence it is an isolated cDV singularity by Theorem 2.3.4. By Theorem 2.2.4 with Remark 2.2.5, the germ $x \in X$ is analytically isomorphic to either a cyclic quotient singularity $\mathfrak{D}^3/\mathbf{Z}_r(a_1, a_2, a_3)$ or a singularity at origin defined in $\mathfrak{D}^4/\mathbf{Z}_r(a_1, a_2, a_3, a_4)$ by a semi-invariant function. We shall classify $x \in X$ in the analytic category. We write $\bar{l} = l - \lfloor l/r \rfloor r$ as in Notation 2.2.14.

In dimension three, the condition in Proposition 2.2.15 is well understood.

Theorem 2.4.1 (White [472])   *Let r be a positive integer and let a_1, a_2, a_3 be integers. If $\overline{la_1} + \overline{la_2} + \overline{la_3} > r$ for all $0 < l < r$, then $a_i + a_j$ is divisible by r for some distinct i and j.*

We explain an abstract approach due to Morrison and Stevens whilst the proof by White is elementary. Take the multiplicative group $\mathbf{Z}_r^\times$ of units in $\mathbf{Z}_r$. A *character* of $\mathbf{Z}_r^\times$ is a group homomorphism $\chi \colon \mathbf{Z}_r^\times \to \mathbf{C}^\times$. The *conductor* of χ is the greatest divisor s of r such that χ factors through $\mathbf{Z}_r^\times \to \mathbf{Z}_s^\times$. The character χ is said to be *even* if $\chi(-1) = 1$ and *odd* if $\chi(-1) = -1$. For a rational number q, we let

$$B_1(q) = \begin{cases} q - \lfloor q \rfloor - 1/2 & \text{for } q \notin \mathbf{Z}, \\ 0 & \text{for } q \in \mathbf{Z}. \end{cases}$$

Note that $B_1(-q) = -B_1(q)$ for any q. We define the *generalised Bernoulli number*

$$B_{1,\chi} = \sum_{a \in \mathbf{Z}_s^\times} \chi(a) B_1\left(\frac{a}{s}\right) \in \mathbf{C}$$

by regarding each $a \in \mathbf{Z}_s = \mathbf{Z}/s\mathbf{Z}$ as an integer $0 \leq a < s$. The following classical theorem is essentially due to Dirichlet. See [215, §2 theorem 2] for the proof.

Theorem 2.4.2 *If χ is odd, then $B_{1,\chi}$ is not zero.*

Let $V = \mathbf{C}[\mathbf{Z}_r^\times]$ be the group algebra of $\mathbf{Z}_r^\times$ over $\mathbf{C}$. It is generated by elements σ_a for $a \in \mathbf{Z}_r^\times$. For $q \in r^{-1}\mathbf{Z}/\mathbf{Z}$, we define an element $S(q)$ in V as

$$S(q) = \sum_{a \in \mathbf{Z}_r^\times} B_1(aq)\sigma_a$$

by regarding B_1 as a function $\mathbf{Q}/\mathbf{Z} \to \mathbf{Q}$. Let W be the linear subspace of V spanned by the image of $S\colon r^{-1}\mathbf{Z}/\mathbf{Z} \to V$. The dual vector space $V^\vee$ of V has the dual basis $\{\sigma_a^\vee\}$ of $\{\sigma_a\}$. Let $W^\perp$ be the orthogonal complement in $V^\vee$ of W defined as the kernel of $V^\vee \to \operatorname{Hom}(W, \mathbf{C})$. Then for any $a \in \mathbf{Z}_r^\times$, $\lambda_a = \sigma_a^\vee + \sigma_{-a}^\vee$ belongs to $W^\perp$ since $B_1(-aq) = -B_1(aq)$.

Theorem 2.4.3 (Morrison–Stevens [349]) *The space $W^\perp$ is spanned by all λ_a for $a \in \mathbf{Z}_r^\times$.*

Proof The dimension of V equals the *Euler function* $\varphi(r)$ and the subspace of $V^\vee$ spanned by all λ_a is of dimension $\lceil \varphi(r)/2 \rceil$. Thus it suffices to prove the estimate $\dim W \geq \lfloor \varphi(r)/2 \rfloor$.

Let $\chi\colon \mathbf{Z}_r^\times \to \mathbf{C}^\times$ be an arbitrary character with conductor s. Using the map $\mathbf{Z}_s = \mathbf{Z}/s\mathbf{Z} \simeq s^{-1}\mathbf{Z}/\mathbf{Z} \hookrightarrow \mathbf{Q}/\mathbf{Z}$ which sends $a \in \mathbf{Z}_s$ to $a/s \in \mathbf{Q}/\mathbf{Z}$, we define an element w_χ in W as

$$w_\chi = \sum_{a \in \mathbf{Z}_s^\times} \chi(a) S\left(\frac{a}{s}\right) = \sum_{b \in \mathbf{Z}_r^\times} \sum_{a \in \mathbf{Z}_s^\times} \chi(a) B_1\left(\frac{ab}{s}\right) \sigma_b.$$

By the action of $\mathbf{Z}_r^\times$ on V, an element $c \in \mathbf{Z}_r^\times$ sends w_χ to

$$\sum_{b \in \mathbf{Z}_r^\times} \sum_{a \in \mathbf{Z}_s^\times} \chi(a) B_1\left(\frac{ab}{s}\right) \sigma_{cb} = \sum_{b \in \mathbf{Z}_r^\times} \sum_{a \in \mathbf{Z}_s^\times} \chi(a) B_1\left(\frac{ac^{-1}b}{s}\right) \sigma_b$$

$$= \sum_{b \in \mathbf{Z}_r^\times} \sum_{a \in \mathbf{Z}_s^\times} \chi(ac) B_1\left(\frac{ab}{s}\right) \sigma_b = \chi(c) w_\chi.$$

Thus w_χ belongs to the χ-eigenspace of V. In particular, all non-zero w_χ are linearly independent in W.

Since there exist exactly $\lfloor \varphi(r)/2 \rfloor$ odd characters, one has only to show that $w_\chi \neq 0$ for all odd χ. This follows from Theorem 2.4.2 because the coefficient of σ_1 in w_χ is the generalised Bernoulli number $B_{1,\chi}$. $\qquad \square$

Proof of Theorem 2.4.1   We write $a(l) = \overline{la_1} + \overline{la_2} + \overline{la_3}$ for brevity. Note that $a(l) + a(r - l)$ equals r times the number of i such that $\overline{la_i} \neq 0$.

For $0 < l < r$, one has $a(l) + a(r - l) = 3r$ from the assumptions $a(l) > r$ and $a(r - l) > r$. Hence all a_i are coprime to r. Further $r < a(l) < 2r$ again by $a(l) > r$ and $a(r - l) > r$. Thus $a(l) = \overline{l(a_1 + a_2 + a_3)} + r$, and $a_1 + a_2 + a_3$ is also coprime to r. This expression of $a(l)$ is equivalent to the equality

$$\left(\overline{la_1} - \frac{r}{2}\right) + \left(\overline{la_2} - \frac{r}{2}\right) + \left(\overline{la_3} - \frac{r}{2}\right) = \left(\overline{l(a_1 + a_2 + a_3)} - \frac{r}{2}\right).$$

Hence $B_1(a_1q) + B_1(a_2q) + B_1(a_3q) = B_1((a_1 + a_2 + a_3)q)$ for any $q \in r^{-1}\mathbf{Z}/\mathbf{Z}$.

Consider $\mu = \sigma_{a_1}^\vee + \sigma_{a_2}^\vee + \sigma_{a_3}^\vee - \sigma_{a_1+a_2+a_3}^\vee \in V^\vee$. For any $q \in r^{-1}\mathbf{Z}/\mathbf{Z}$, μ sends $S(q)$ to $B_1(a_1q) + B_1(a_2q) + B_1(a_3q) - B_1((a_1+a_2+a_3)q) = 0$, meaning that $\mu \in W^\perp$. Thus by Theorem 2.4.3, $a_1 + a_2 + a_3$ must be congruent to some a_k modulo r. That is, $a_i + a_j \equiv 0 \bmod r$ for distinct i, j and k. $\square$

Theorem 2.4.4 (Terminal lemma [100], [349])   *A cyclic quotient threefold singularity is terminal if and only if it is of type $\frac{1}{r}(1, -1, b)$ for some coprime integers r and b.*

Proof   By Lemma 1.6.2, we consider a cyclic quotient singularity of type $\frac{1}{r}(a_1, a_2, a_3)$ such that all a_i are coprime to r. If it is terminal, then by Proposition 2.2.15 and Theorem 2.4.1, $a_1 + a_2 = r$ after permutation. For an integer a such that $\overline{a_1a} = 1$, it becomes of type $\frac{1}{r}(1, -1, a_3a)$ by a suitable primitive character $\chi \colon \mathbf{Z}_r \to \mathbf{C}^\times$, in which a_3a is coprime to r. Conversely by Proposition 2.2.15, the singularity of type $\frac{1}{r}(1, -1, b)$ is terminal if b is coprime to r. $\square$

We proceed to the study of terminal threefold singularities with singular index-one cover. We extend Proposition 2.2.15 to the cyclic quotient of a hypersurface singularity.

Proposition 2.4.5   *Let $o \in A = \mathfrak{D}^n/\mathbf{Z}_r(a_1, \ldots, a_n)$ be the germ at origin of a cyclic quotient analytic singularity with orbifold coordinates $x_1, \ldots, x_n$. Let $o \in X$ be the analytic subspace of A defined by a semi-invariant function ξ in $\mathcal{O}_{\mathfrak{D}^n}$. Suppose that $\mathbf{Z}_r$ acts on the hypersurface in $\mathfrak{D}^n$ defined by ξ freely in codimension one. If $o \in X$ is terminal (resp. canonical), then for any $\frac{1}{r}(b_1, \ldots, b_n) \in N \cap (\mathbf{R}_{\geq 0})^n$ where $N = \mathbf{Z}^n + \mathbf{Z} \cdot \frac{1}{r}(a_1, \ldots, a_n)$ such that at least three of b_i are positive, the weighted order of ξ with respect to $\mathrm{wt}(x_1, \ldots, x_n) = (b_1, \ldots, b_n)$ satisfies the inequality $\sum_{i=1}^n b_i > \mathrm{w\text{-}ord}\,\xi + r$ (resp. $\geq \mathrm{w\text{-}ord}\,\xi + r$).*

Proof   We may assume that $\frac{1}{r}(b_1, \ldots, b_n)$ is primitive in N. Take the weighted blow-up $\pi \colon B \to A$ with $\mathrm{wt}(x_1, \ldots, x_n) = \frac{1}{r}(b_1, \ldots, b_n)$. Let E be the exceptional divisor and let X_B be the strict transform in B of X. Then $K_B + X_B =$

$\pi^*(K_A + X) + aE$ with $a = (\sum_i b_i - \text{w-ord}\,\xi - r)/r$. Because at least three of b_i are positive, $\pi(X_B \cap E)$ is of codimension at least two in X.

There exists a relative minimal model $\mu \colon B' \to B$ in the toric category [135], [398]. Let X' be the strict transform in B' of X_B. Since $K_{B'}$ is μ-nef, by the negativity lemma, one can write $K_{B'} + X' = \mu^*(K_B + X_B) - N$ with an effective μ-exceptional $\mathbf{Q}$-divisor N. Then

$$K_{B'} + X' = (\pi \circ \mu)^*(K_A + X) + a\mu^* E - N.$$

One can treat π and μ in the algebraic category. Since the terminal variety B' is smooth in codimension two by Lemma 1.6.2, there exists a Zariski open subset U_B of B meeting $X_B \cap E$ such that $U = \mu^{-1}(U_B)$ is smooth. Then $V = X'|_U$ is Gorenstein with $\omega_V \simeq \mathcal{O}_U(K_U + V) \otimes \mathcal{O}_V$. For a sufficiently divisible integer s, the invertible sheaf $\omega_V^{\otimes s}$ is associated with the Cartier divisor $s(\varphi^* K_X + (a\mu^* E - N)|_V)$ for $\varphi \colon V \to X$.

Let $\nu \colon V^\nu \to V$ be the normalisation and let C denote the *conductor* in V^ν defined by the ideal sheaf $\nu^! \mathcal{O}_V$ in $\mathcal{O}_{V^\nu}$ as in Definition 2.4.6. Then by Lemma 1.1.11, $\omega_{V^\nu} = \nu^* \omega_V \otimes \mathcal{O}_{V^\nu}(-C)$ and hence

$$K_{V^\nu} = (\varphi \circ \nu)^* K_X + \nu^*((a\mu^* E - N)|_V) - C.$$

By construction, the $\mathbf{Q}$-divisor $K_{V^\nu/X} = \nu^*((a\mu^* E - N)|_V) - C$ is $(\varphi \circ \nu)$-exceptional and regarded as the relative canonical divisor. Choose a prime divisor P on V^ν which dominates an irreducible component of $X_B \cap E$. It is $(\varphi \circ \nu)$-exceptional. If the coefficient of P in $K_{V^\nu/X}$ is positive (or non-negative), then so is $a = (\sum_i b_i - \text{w-ord}\,\xi - r)/r$. This implies the proposition. □

Definition 2.4.6 Let X be a reduced algebraic scheme with normalisation $\nu \colon X^\nu \to X$. The ideal sheaf $\mathcal{H}om_X(\nu_* \mathcal{O}_{X^\nu}, \mathcal{O}_X)$ in $\mathcal{H}om_X(\mathcal{O}_X, \mathcal{O}_X) = \mathcal{O}_X$ is called the *conductor ideal* of X. It is expressed as the push-forward $\nu_* \nu^! \mathcal{O}_X$ of the ideal sheaf $\nu^! \mathcal{O}_X$ in $\mathcal{O}_{X^\nu}$.

Consider the analytic germ $o \in A = \mathfrak{D}^4/\mathbf{Z}_r(a_1, a_2, a_3, a_4)$ with orbifold coordinates x_1, x_2, x_3, x_4 where a primitive character $\chi \colon \mathbf{Z}_r \to \mathbf{C}^\times$ is fixed. Let $o \in X$ be the analytic subspace of A defined by a semi-invariant function ξ in $\mathcal{O}_{\mathfrak{D}^4}$. The germ $o \in X \subset A$ is the quotient of $\bar{o} \in \bar{X} \subset \mathfrak{D}^4$ in which $\bar{X}$ is defined by ξ. Suppose that $o \in X$ is a terminal threefold singularity of index r and that $\bar{o} \in \bar{X}$ is its index-one cover. At the beginning of the section, we have seen that every such $o \in X$ is realised in this manner. The germ $\bar{o} \in \bar{X}$ is an isolated cDV singularity and the action of $\mathbf{Z}_r$ on $\bar{X}$ is free outside $\bar{o}$. In particular, the following hold.

Lemma 2.4.7 (i) *There exists no common factor of r, a_i, a_j for any distinct i and j.*

(ii) *If a_i is not coprime to r, then some power of x_i appears in ξ.*

The following with Theorem 2.4.16 is a complete classification of $o \in X$.

Theorem 2.4.8 (Mori [333]) *Let $o \in X$ be the germ of a terminal threefold analytic singularity of index r greater than one. It is realised as above so that $o \in X = (\xi = 0) \subset A = \mathfrak{D}^4/\mathbf{Z}_r(a_1, a_2, a_3, a_4)$ with a semi-invariant function ξ in $\mathcal{O}_{\mathfrak{D}^4}$. Then one of the following holds after changing expression of type $\frac{1}{r}(a_1, a_2, a_3, a_4)$ and orbifold coordinates x_1, x_2, x_3, x_4, where $\mathfrak{m}$ is the maximal ideal in $\mathcal{O}_{\mathfrak{D}^4}$ defining origin. Notice that terminal cyclic quotient singularities are included in the case cA/r.*

cA/r $\xi = x_1 x_2 + f(x_3, x_4^r)$ and A is of type $\frac{1}{r}(1, -1, 0, b)$ with b coprime to r.

$cAx/4$ $\xi = x_1^2 + x_2^2 + f(x_3, x_4^2)$, $f \in \mathfrak{m}^2$ and A is of type $\frac{1}{4}(1, 3, 2, 1)$.

$cAx/2$ $\xi = x_1^2 + x_2^2 + f(x_3, x_4)$, $f \in \mathfrak{m}^4$ and A is of type $\frac{1}{2}(1, 0, 1, 1)$.

$cD/3$ $\xi = x_1^2 + f(x_2, x_3, x_4)$, $f \in \mathfrak{m}^3$, the cubic part of f is $x_2^3 + x_3^3 + x_4^3$, $x_2^3 + x_3 x_4^2$ or $x_2^3 + x_3^3$ and A is of type $\frac{1}{3}(0, 1, 2, 2)$.

$cD/2$ $\xi = x_1^2 + f(x_2, x_3, x_4)$, $f \in \mathfrak{m}^3$, $x_2 x_3 x_4$ or $x_2^2 x_3$ appears in f and A is of type $\frac{1}{2}(1, 1, 0, 1)$.

$cE/2$ $\xi = x_1^2 + x_2^3 + f(x_3, x_4)x_2 + g(x_3, x_4)$, $f \in \mathfrak{m}^4$, $g \in \mathfrak{m}^4 \setminus \mathfrak{m}^5$ and A is of type $\frac{1}{2}(1, 0, 1, 1)$.

Definition 2.4.9   A terminal threefold singularity of index greater than one is said to be of *type* $cA/r, \ldots, cE/2$ in accordance with the case in Theorem 2.4.8.

The proof depends on the type of the cDV singularity $\bar{o} \in \bar{X} \subset \mathfrak{D}^4$. We shall demonstrate the theorem in the case when $\bar{X}$ is cA, in which all the essential ideas of the proof appear. Suppose that $\bar{o} \in \bar{X}$ is an isolated cA singularity. The case when $\bar{X}$ is smooth has been settled by Theorem 2.4.4. We shall assume that $\bar{X}$ is singular. In other words, $\xi \in \mathfrak{m}^2$ and the quadratic part ξ_2 has rank at least two.

Lemma 2.4.10   *If $\xi \in \mathfrak{m}^2$ and the quadratic part ξ_2 has rank at least two, then after orbifold coordinate change, either*

(i) $\xi = x_1 x_2 + f(x_3, x_4)$ or

(ii) $\xi = x_1^2 + x_2^2 + f(x_3, x_4)$ and the monomial $x_3 x_4$ does not appear in ξ_2.

Proof   By orbifold coordinate change, we shall see that ξ_2 is of form either $x_1 x_2 + f_2(x_3, x_4)$ or $x_1^2 + x_2^2 + g_2(x_3, x_4)$ without $x_3 x_4$ appearing in g_2. From this one can derive a desired expression of ξ by an equivariant version of the argument in Example 2.1.7, which functions by Remark 2.1.6. This argument does not affect ξ_2.

If x_i^2 appears in ξ_2, then one can eliminate the linear term in x_i as in Remark 2.1.11. Thus after orbifold coordinate change, one can write $\xi_2 = p_2(x_1, \ldots, x_i) + x_{i+1}^2 + \cdots + x_4^2$ for some $0 \le i \le 4$ so that p_2 does not contain any x_l^2. If $p_2 = 0$, then $\xi_2 = x_1^2 + x_2^2 + x_3^2 + x_4^2$ is of the second form. If $p_2 \ne 0$, then we may assume that $x_1 x_2$ appears in p_2 with $i \ge 2$. Express ξ_2 up to unit as $x_1 x_2 + a(x_3, x_4)x_1 + b(x_3, x_4)x_2 + c(x_3, x_4)$, that is, $\xi_2 = (x_1 + b)(x_2 + a) + c - ab$. Replacing x_1 by $x_1 + b$ and x_2 by $x_2 + a$, one attains the first form $\xi_2 = x_1 x_2 + f_2$. □

Proof of Theorem 2.4.8 in the case (i) *of Lemma 2.4.10* We shall derive the case cA/r or $cAx/4$.

For $0 < l < r$, we let w-ord$_l$ h denote the weighted order of a semi-invariant function h in $\mathscr{O}_{\mathfrak{D}^4}$ with respect to $\text{wt}(x_1, x_2, x_3, x_4) = (\overline{la_1}, \overline{la_2}, \overline{la_3}, \overline{la_4})$. By Lemma 2.4.7(i), at least three of $\overline{la_i}$ are positive. Then by Proposition 2.4.5,

$$\text{w-ord}_l\, \xi + r < \overline{la_1} + \overline{la_2} + \overline{la_3} + \overline{la_4}.$$

Step 1 We take the greatest common divisor $p = \gcd(a_1 + a_2, r)$. By Lemma 2.4.7(ii), a_1 and a_2 are coprime to r. We shall see that a_3 or a_4 is divisible by p. Indeed, otherwise $p \ge 2$ and all $\overline{sa_i}$ would be positive for $s = r/p$. Possibly replacing each a_i by $-a_i$ by the change of the character to χ^{-1}, one may assume that $\overline{sa_3} + \overline{sa_4} \le r$. On the other hand, $\overline{sa_1} + \overline{sa_2} = r$ since p divides $a_1 + a_2$. Then w-ord$_s$ $\xi + r < \sum_i \overline{sa_i} \le 2r$, that is, w-ord$_s$ $\xi < r$. This contradicts the relation w-ord$_s$ $\xi \equiv$ w-ord$_s$ $x_1 x_2 \equiv s(a_1 + a_2) \equiv 0 \bmod r$.

Thus possibly permuting x_3 and x_4, we may and shall assume that a_3 is divisible by p. Then by Lemma 2.4.7(i), a_4 is coprime to p. We shall prove that a_4 is coprime to r. If $q = \gcd(a_4, r) > 1$, then by Lemma 2.4.7(ii), some multiple of x_4 would appear in ξ. In particular, $a_1 + a_2 = \text{wt}\,x_1 x_2$ is divisible by q. Then q is a factor of p by the definition of p, which contradicts the coprimeness of a_4 and p.

We have seen that a_1, a_2, a_4 are coprime to r and a_3 is divisible by $p = \gcd(a_1 + a_2, r)$. Hence one can write

$$\xi = x_1 x_2 + f(x_3, x_4^p).$$

We write $\gcd(a_3, r) = pg$. If $g \ge 2$, then by Lemma 2.4.7(ii), some multiple of x_3 would appear in f and hence pg must divide $p = \gcd(a_1 + a_2, r)$, which is absurd. Thus $g = 1$, that is, $p = \gcd(a_1 + a_2, r) = \gcd(a_3, r)$.

If $p = r$, then $(a_1, a_2, a_3, a_4) \equiv (a_1, -a_1, 0, a_4) \bmod r$ where a_1 and a_4 are coprime to r, and the case cA/r holds. Henceforth we shall assume that $p < r$.

Step 2 We consider integers $0 < l < r$ such that

$$l(a_1 + a_2) \equiv p \text{ or } -p \bmod r, \qquad \overline{la_3} + \overline{la_4} \le r.$$

Note that such l is coprime to r/p, by which $\overline{l a_3}$ as well as other $\overline{l a_i}$ is positive. The number N of l satisfies

$$N \geq \begin{cases} p & \text{if } r > 2p, \\ p/2 & \text{if } r = 2p. \end{cases}$$

Indeed, the number of l satisfying the first condition is $2p$ (resp. p) if $r > 2p$ (resp. $r = 2p$). For any such l, either l or $r - l$ satisfies the second.

One has the estimate w-ord$_l \xi + r < \sum_i \overline{l a_i} \leq \overline{l a_1} + \overline{l a_2} + r$, that is, w-ord$_l \xi < \overline{l a_1} + \overline{l a_2}$. Hence $\overline{l a_1} + \overline{l a_2} =$ w-ord$_l \xi + r$ and exactly one of the following holds.

(i) w-ord$_l \xi = p \geq 2$ and $\overline{l a_1} + \overline{l a_2} = r + p$.
(ii) $r > 2p$, w-ord$_l \xi = r - p$ and $\overline{l a_1} + \overline{l a_2} = 2r - p$.

Let N_1 (resp. N_2) be the number of l in the case (i) (resp. (ii)). Then

$$N_1 \leq 1, \qquad N_2 \leq p - 1.$$

Indeed, the condition w-ord$_l \xi = p$ implies that the monomial x_4^p appears in f with w-ord$_l x_4 = 1$. Hence $N_1 \leq 1$. The condition $\overline{l a_1} + \overline{l a_2} = 2r - p$ implies that $r - p < \overline{l a_1} < r$ and hence $N_2 \leq p - 1$.

Step 3 If $r = 2p$, then $p/2 \leq N = N_1 \leq 1$. Hence $p = 2$ and $r = 4$, and one can set $(a_1, a_2, a_3, a_4) = (1, 1, 2, a_4)$ with $a_4 = 1$ or 3. The inequality w-ord$_1 \xi + 4 < \sum_i a_i$ gives $a_4 = 3$. It follows from the existence of l in the case (i) that x_4^2 appears in f. Thus after orbifold coordinate change, one can write $\xi = 4 x_1 x_2 + x_3^n + x_4^2 = (x_1 + x_2)^2 + x_4^2 + x_3^n - (x_1 - x_2)^2$ and the case $cAx/4$ holds for the orbifold coordinates $x_1 + x_2, x_4, x_3, \sqrt{-1}(x_1 - x_2)$.

We shall assume that $r > 2p$ and derive a contradiction. If $r > 2p$, then $p \leq N = N_1 + N_2 \leq p$, that is, $N_1 = 1$ and $N_2 = p - 1$. For $i = 1$ and 2, we take the integer $0 < l_i < r$ such that $\overline{l_i a_1} = r - i$. Then $l_2 = 2l_1$. Unless $p = 2$, these l_i must satisfy the case (ii) by $N_2 = p - 1$, in which $\overline{l_i a_2} = r - p + i$. But then $2(r - p + 1) \equiv 2 l_1 a_2 \equiv l_2 a_2 \equiv r - p + 2 \bmod r$, that is, p is divisible by r. This contradicts the assumption $p < r$.

We have $p = 2$ and $N_1 = N_2 = 1$. Let l_1 (resp. l_2) be the integer l in the case (i) (resp. (ii)). Then $\overline{l_2 a_1} = \overline{l_2 a_2} = r - 1$. Hence $a_1 = a_2$ and $\overline{l_1 a_1} = \overline{l_1 a_2} = r/2 + 1$. By w-ord$_{l_1} \xi = 2$, the monomial x_4^2 appears in f with $\overline{l_1 a_4} = 1$. Thus $(\overline{2 l_1 a_1}, \overline{2 l_1 a_2}, \overline{2 l_1 a_4}) = (2, 2, 2)$ and w-ord$_{\overline{2 l_1}} \xi = 4$ by $r > 2p = 4$. However w-ord$_{\overline{2 l_1}} \xi + r < \sum_i \overline{2 l_1 a_i} = 6 + \overline{2 l_1 a_3} \leq 4 + r$, which is absurd. □

Remark 2.4.11 By the above proof, we have seen that in the case (i) in Lemma 2.4.10, $o \in X$ is cA/r or $cAx/4$, and x_4^2 appears in ξ if it is $cAx/4$.

Proof of Theorem 2.4.8 *in the case* (ii) *of Lemma* 2.4.10   We shall derive the case $cA/2$, $cAx/4$ or $cAx/2$.

First we claim that r is a power of two. Otherwise r would have an odd factor s. By Corollary 2.2.21, the quotient $\bar{X}/\mathbf{Z}_s$ is also terminal. Taking $\bar{X}/\mathbf{Z}_s$ instead of $X = \bar{X}/\mathbf{Z}_r$, we assume that r is odd. Then $a_1 = a_2$ because $x_1^2 + x_2^2$ is semi-invariant. Replacing x_1 by $x_1 + \sqrt{-1}x_2$ and x_2 by $x_1 - \sqrt{-1}x_2$, one can make f as in the case (i) in Lemma 2.4.10, in which Theorem 2.4.8 has been proved. Since r is odd, the case cA/r would hold by Remark 2.4.11, in which a_1 is coprime to r and $a_2 \equiv -a_1 \bmod r$. By $a_1 = a_2$, this can occur only when $r = 2$, which contradicts the assumption that r is odd.

Next we claim that $r = 2$ or 4. Otherwise taking $\bar{X}/\mathbf{Z}_8$, we assume that $r = 8$. Then $a_1 \equiv a_2 \bmod 4$ since $x_1^2 + x_2^2$ is semi-invariant. One can apply to $\bar{X}/\mathbf{Z}_4$ the result of the case (i) in Lemma 2.4.10 after replacing x_1 by $x_1 + \sqrt{-1}x_2$ and x_2 by $x_1 - \sqrt{-1}x_2$ as in the preceding paragraph. The case $cA/4$ never occurs since $a_1 \equiv a_2 \bmod 4$. Hence by Remark 2.4.11, the case $cAx/4$ would occur and the rank of ξ_2 is at least three. We may assume that x_1^2, x_2^2, x_3^2 appear in ξ_2 and $a_1 \equiv a_2 \equiv a_3 \bmod 4$. Then $a_1 = a_2$ after permutation, which contradicts Remark 2.4.11 applied to $\bar{X}/\mathbf{Z}_8$.

Therefore $r = 2$ or 4. Suppose that $r = 4$. If $a_1 = a_2$, then one can apply Remark 2.4.11 as above to derive the case $cAx/4$. The condition $a_1 = a_2$ is satisfied after permutation whenever the rank of ξ_2 is at least three. The case when $\xi_2 = x_1^2 + x_2^2$ with $a_1 \neq a_2$ remains. In this case, we may assume that $(a_1, a_2) = (1, 3)$. Apply to $\bar{X}/\mathbf{Z}_2$ the result of the case (i) in Lemma 2.4.10 and see that exactly one of a_3 and a_4 is even. Without loss of generality, we assume that a_3 is even. Then by Lemma 2.4.7(ii), some multiple of x_3 appears in ξ besides $x_1^2 + x_2^2$. Hence a_3 must be two. Possibly permuting x_1 and x_2 and changing a_i with $4 - a_i$, one can take $\xi_2 = x_1^2 + x_2^2$ and $(a_1, a_2, a_3, a_4) = (1, 3, 2, 1)$, which is $cAx/4$.

Suppose that $r = 2$. If $a_1 = a_2$, then the case $cA/2$ holds by Remark 2.4.11. In the remaining case when $\xi_2 = x_1^2 + x_2^2$ with $a_1 \neq a_2$, we may assume that $(a_1, a_2) = (1, 0)$. Then $a_3 = a_4 = 1$ by Lemma 2.4.7(i), which is $cAx/2$. $\square$

Reid pointed out that the general member of the anti-canonical system $|-K_X|$ on the germ $o \in X$ in Theorem 2.4.8 is Du Val as in the case of index one. We state it as Theorem 2.4.15 below. He named it the *general elephant* and proposed a remarkable question of when the general elephant is Du Val.

Definition 2.4.12 (Reid [401])   Let X be a normal variety. The general member of the anti-canonical system $|-K_X|$ is called the *general elephant* of X.

Question 2.4.13 (General elephant conjecture) *Let X be a terminal three-fold such that $-K_X$ is ample. Does the general elephant of X have Du Val singularities?*

This question plays a guiding role in the study of threefold contractions. Fundamental results in dimension three such as the existence of flips and the classification of divisorial contractions are obtained through the affirmation of the general elephant conjecture as will be seen in the following chapters.

Remark 2.4.14 Consider the germ $o \in X \subset A = \mathfrak{D}^4/\mathbf{Z}_r(a_1, a_2, a_3, a_4)$ in Theorem 2.4.8. By adjunction, $\omega_{\bar{X}}$ is generated by $\theta = (dx_1 \wedge dx_2 \wedge dx_3 \wedge dx_4)/\xi$. Recall that an element $g \in \mathbf{Z}_r$ sends x_i to $\chi(g)^{a_i} x_i$ for the fixed primitive character χ. Hence g sends θ to $\chi(g)^{a_4}\theta$ as $a_1 + a_2 + a_3 + a_4 - \mathrm{wt}\,\xi \equiv a_4 \bmod r$ in every case in Theorem 2.4.8. Thus the general elephant $S \in |-K_X|$ is defined by a general semi-invariant function of weight a_4.

Theorem 2.4.15 *Let $o \in X$ be the algebraic or analytic germ of a terminal threefold singularity. Then the general elephant $o \in S$ of X is Du Val. Suppose that X is of index greater than one and take the inverse image $\bar{S}$ of S by the index-one cover $\bar{o} \in \bar{X} \to o \in X$. Then $\bar{S}$ as well as S is Du Val and the type of $\bar{o} \in \bar{S} \to o \in S$ is as follows in accordance with the list in Theorem 2.4.8.*

cA/r $A_{n-1} \to A_{rn-1}$ with $n = \mathrm{ord}\,f(x_3, 0)$
$cAx/4$ $A_{2n-2} \to D_{2n+1}$ with $2n - 1 = \mathrm{ord}\,f(x_3, 0)$
$cAx/2$ $A_{2n-1} \to D_{n+2}$ with $2n = \mathrm{ord}\,f(x_3, x_4)$
$cD/3$ $D_4 \to E_6$
$cD/2$ $D_{n+1} \to D_{2n}$ with $n = \mathrm{ord}\,f(0, x_3, 0)$
$cE/2$ $E_6 \to E_7$

Proof We work in the analytic category. The algebraic version also holds by the proof. The case of index one is Theorem 2.3.4. We assume that $o \in X$ is of index greater than one. We shall demonstrate the assertion in the $cAx/4$ case. The assertion in other cases is verified by the same argument.

Consider $o \in X \subset A = \mathfrak{D}^4/\mathbf{Z}_4(1, 3, 2, 1)$ given by $\xi = x_1^2 + x_2^2 + f(x_3, x_4^2)$ in the $cAx/4$ case of Theorem 2.4.8. By Remark 2.4.14, the general elephant S is defined by $x_4 + ux_1 + vx_2x_3 + wx_3^2$ with invariant units u, v, w. Thus S is given in $\mathfrak{D}^3/\mathbf{Z}_4(1, 3, 2)$ by $px_1^2 + ax_1x_2x_3 + qx_2^2 + rx_3^{2n-1}$ with units p, q, r and a function a which are invariant. It becomes $x_1^2 + x_2^2 + x_3^{2n-1}$ after elimination of the part $ax_1x_2x_3$ as in Remark 2.1.11 with orbifold coordinate change. In particular, $\bar{o} \in \bar{S}$ is of type A_{2n-2}.

Whilst S is the quotient of $\bar{S}$ by $\mathbf{Z}_4$, we first consider the quotient $\bar{S}/\mathbf{Z}_2$ by the subgroup $\mathbf{Z}_2$. It is embedded in $D = \mathfrak{D}^3/\mathbf{Z}_2$. The quotient D is regarded as the hypersurface in $\mathfrak{D}^4$ with coordinates $(y_1, y_2, y_3, x_3) = (x_1^2, x_1x_2, x_2^2, x_3)$

defined by $y_1 y_3 - y_2^2$, and $\bar{S}/\mathbf{Z}_2$ is given in D by $y_1 + y_3 + x_3^{2n-1}$. By elimination of y_3, $\bar{S}/\mathbf{Z}_2$ is given in $\mathfrak{D}^3$ by $y_1(y_1 + x_3^{2n-1}) + y_2^2$. It becomes $y_1^2 + y_2^2 + x_3^{4n-2}$ after replacements of y_1 by $y_1 + x_3^{2n-1}/2$ and x_3 by $x_3/(-4)^{1/(4n-2)}$.

We have $S = (\bar{S}/\mathbf{Z}_2)/\mathbf{Z}_2 \subset D' = \mathfrak{D}^3/\mathbf{Z}_2(1, 0, 1)$ with orbifold coordinates y_1, y_2, x_3. The quotient D' is regarded as the hypersurface in $\mathfrak{D}^4$ with coordinates $(z_1, z_2, z_3, y_2) = (y_1^2, y_1 x_3, x_3^2, y_2)$ defined by $z_1 z_3 - z_2^2$, and S is given in D' by $z_1 + y_2^2 + z_3^{2n-1}$. By elimination of z_1, S is given in $\mathfrak{D}^3$ by $z_2^2 + (y_2^2 + z_3^{2n-1})z_3$, which is of type D_{2n+1}. □

We complete the classification by a sufficient condition for X to be terminal.

Theorem 2.4.16 (Kollár–Shepherd-Barron [278]) *Suppose that an isolated cDV singularity $\bar{o} \in \bar{X}$ in $\bar{A} = \mathfrak{D}^4$ is given by one of ξ in Theorem 2.4.8 with the action of $\mathbf{Z}_r$. If $\mathbf{Z}_r$ acts on $\bar{X}$ freely outside $\bar{o}$, then the quotient $o \in X = \bar{X}/\mathbf{Z}_r$ is terminal.*

Proof The general elephant S of $o \in X$ is canonical by the argument for Theorem 2.4.15 which does not use the terminal property of X. Thus (X, S) is canonical by inversion of adjunction, where $K_X + S$ is Cartier. Hence the isolated singularity $o \in X$ must be terminal. □

The following two properties have already been proved in the case of index one.

Theorem 2.4.17 *Let $x \in X$ be a terminal threefold singularity of index r. Then $\pi_1(U \setminus x) \simeq \mathbf{Z}_r$ for an analytic neighbourhood U in X at x. Further every $\mathbf{Q}$-Cartier integral divisor on the germ $x \in X$ is linearly equivalent to lK_X for some integer l.*

Proof The case when $r = 1$ is treated by Theorems 2.3.18 and 2.3.19. One has only to prove the latter assertion for the analytic germ $o \in X \subset A = \mathfrak{D}^4/\mathbf{Z}_r(a_1, a_2, a_3, a_4)$ in Theorem 2.4.8 which is the quotient of $\bar{o} \in \bar{X} \in \mathfrak{D}^4$. Let D be an arbitrary $\mathbf{Q}$-Cartier divisor on X. By Theorem 2.3.18, the pullback $\pi^* D$ by $\pi\colon \bar{X} \to X$ is Cartier and it is defined by a semi-invariant rational function of some weight w. There exists an integer l such that $w \equiv -la_4 \bmod r$ since a_4 is coprime to r. Then by Remark 2.4.14, the divisor $D - lK_X$ is defined by an invariant rational function, which means that it is Cartier. □

Theorem 2.4.18 (Kawamata [424, appendix]) *Let $x \in X$ be a terminal threefold singularity of index r. Then $\mathrm{mld}_x X = 1 + 1/r$ unless it is smooth.*

Proof The assertion for $r = 1$ is Theorem 2.3.21. Assuming that $r \geq 2$, we shall find a divisor with log discrepancy $1 + 1/r$ explicitly. We may work in the analytic category. Every terminal threefold analytic singularity is realised as $o \in X \subset A = \mathfrak{D}^4/\mathbf{Z}_r(a_1, a_2, a_3, a_4)$ with orbifold coordinates x_1, x_2, x_3, x_4 in which X is defined by one of ξ in Theorem 2.4.8. We take the weighted blow-up Y of X with $\mathrm{wt}(x_1, x_2, x_3, x_4) = w$ as follows.

cA/r Let $k = $ w-ord f with respect to $\mathrm{wt}(x_3, x_4) = (1, 1/r)$ and take $0 < a < r$ such that A is of type $\frac{1}{r}(a, -a, 0, 1)$. Set $w = \frac{1}{r}(a+lr, (k-l)r-a, r, 1)$ for any integer $0 \leq l < k$.

$cAx/4$ Let $k = $ w-ord f with respect to $\mathrm{wt}(x_3, x_4) = (1, 1/2)$, which is an odd integer. Set $w = \frac{1}{4}(k, k+2, 2, 1)$ if $k \equiv 1 \bmod 4$ and $w = \frac{1}{4}(k+2, k, 2, 1)$ if $k \equiv 3 \bmod 4$.

$cAx/2$ Let $k = $ w-ord f with respect to $\mathrm{wt}(x_3, x_4) = (1/2, 1/2)$. Set $w = \frac{1}{2}(k, k+1, 1, 1)$ if k is odd and $w = \frac{1}{2}(k+1, k, 1, 1)$ if k is even.

$cD/3$ Let f_3 be the cubic part of f. Set $w = \frac{1}{3}(3, 2, 1, 1)$ if $f_3 = x_2^3 + x_3^3 + x_4^3$ and $w = \frac{1}{3}(3, 2, 4, 1)$ if $f_3 = x_2^3 + x_3 x_4^2$ or $x_2^3 + x_3^3$.

$cD/2$ Set $w = \frac{1}{2}(3, 1, 2, 1)$.

$cE/2$ Let g_4 be the quartic part of g. If g_4 is a square, then one may assume that g_4 is $x_3^2 x_4^2$ or x_3^4. Set $w = \frac{1}{2}(3, 2, 1, 1)$ if g_4 is not a square and $w = \frac{1}{2}(3, 2, 3, 1)$ if $g_4 = x_3^2 x_4^2$ or x_3^4.

It suffices to check the existence of a normal open subset U of Y and an exceptional prime divisor F of $U \to X$ with log discrepancy $a_F(X) = 1 + 1/r$. One can perform this explicitly by taking some chart. For example, the demonstration in the $cD/2$ case is as follows.

Consider $o \in X \subset A = \mathfrak{D}^4/\mathbf{Z}_2(1, 1, 0, 1)$ given by $\xi = x_1^2 + f(x_2, x_3, x_4)$ in the $cD/2$ case of Theorem 2.4.8. Then Y is the weighted blow-up with weights $w = \frac{1}{2}(3, 1, 2, 1)$. Take the chart $Y_1 = Y|_{U_1} \subset U_1$ for $U_1 \simeq \mathfrak{D}^4/\mathbf{Z}_3(1, 1, 2, 1)$ with orbifold coordinates $(y_1, y_2, y_3, y_4) = (x_1^{2/3}, x_2 x_1^{-1/3}, x_3 x_1^{-2/3}, x_4 x_1^{-1/3})$ as in the paragraph after Lemma 2.2.12. Then Y_1 is defined in U_1 by $y_1 + y_1^{-2} g(y_2 y_1^{1/2}, y_3 y_1, y_4 y_1^{1/2})$ and the exceptional locus E of $U_1 \to A$ is given by y_1. The chart Y_1 is normal and $Y_1 \cap E$ is given in $E \simeq \mathfrak{D}^3/\mathbf{Z}_3(1, 2, 1)$ by a linear combination of $y_2 y_3 y_4, y_2^2 y_3, y_3 y_4^2, y_2^4, y_2^3 y_4, y_2^2 y_4^2, y_2 y_4^3, y_4^4$ in which $y_2 y_3 y_4$ or $y_2^2 y_3$ appears. Thus $Y_1 \cap E$ has a reduced component F. It has $a_F(X) = 3/2$. $\quad \square$

2.5 Singular Riemann–Roch Formula

This section introduces a formula for the Euler characteristic on a projective variety with isolated singularities. It is a basic tool in the explicit study of terminal threefolds. We begin with recalling the *Hirzebruch–Riemann–Roch theorem*. Let X be a projective complex variety of dimension n. We write $H^\bullet(X, R) = \bigoplus_{i=0}^{2n} H^i(X, R)$ for the *cohomology ring* with coefficients in the ring R. The highest cohomology $H^{2n}(X, R)$ is naturally isomorphic to R by the cap product with the *fundamental class* $[X]$ in $H_{2n}(X, R)$.

Theorem 2.5.1 (Chern [85]) *With every locally free coherent sheaf $\mathscr{E}$ on a smooth projective variety X, one can uniquely associate a class $c(\mathscr{E}) = \sum_{i=0}^{r} c_i(\mathscr{E})$ in $H^\bullet(X, \mathbf{Z})$, where $c_i(\mathscr{E}) \in H^{2i}(X, \mathbf{Z})$ and r is the rank of $\mathscr{E}$, by the axioms*

- $c_0(\mathscr{E}) = 1$ *for any $\mathscr{E}$,*
- *for an invertible sheaf $\mathscr{L}$, $c_1(\mathscr{L})$ is the image of $\mathscr{L}$ by the natural map* $\mathrm{Pic}\, X \simeq H^1(\mathscr{O}_X^\times) \to H^2(X, \mathbf{Z})$,
- $c(\pi^*\mathscr{E}) = \pi^* c(\mathscr{E})$ *for a morphism π of smooth projective varieties and*
- $c(\mathscr{E}) = c(\mathscr{E}')c(\mathscr{E}'')$ *for an exact sequence $0 \to \mathscr{E}' \to \mathscr{E} \to \mathscr{E}'' \to 0$ of locally free sheaves.*

Definition 2.5.2 We call $c(\mathscr{E}) = \sum_i c_i(\mathscr{E}) \in H^\bullet(X, \mathbf{Z})$ in the theorem the *total Chern class* of $\mathscr{E}$ and $c_i(\mathscr{E})$ the i-th *Chern class* of $\mathscr{E}$. We define $c(X) = c(\mathscr{T}_X)$ and $c_i(X) = c_i(\mathscr{T}_X)$ by the *tangent sheaf* $\mathscr{T}_X = \Omega_X^\vee$.

Example 2.5.3 The projective space $\mathbf{P}^n$ has cohomology ring $H^\bullet(\mathbf{P}^n, \mathbf{Z}) = \mathbf{Z}[\xi]/(\xi^{n+1})$ for $\xi = c_1(\mathscr{O}_{\mathbf{P}^n}(1))$ [466, theorem 7.14]. By the dual $0 \to \mathscr{O}_{\mathbf{P}^n} \to \mathscr{O}_{\mathbf{P}^n}(1)^{\oplus(n+1)} \to \mathscr{T}_{\mathbf{P}^n} \to 0$ of the Euler sequence appeared in Example 1.1.8, one computes $c(\mathbf{P}^n)$ to be $c(\mathscr{O}_{\mathbf{P}^n}(1))^{n+1} = (1 + \xi)^{n+1}$ in which $\xi^{n+1} = 0$.

One can compute the Chern class of a locally free sheaf $\mathscr{E}$ on X by means of the splitting principle below. Let $\pi\colon Y \to X$ be a birational morphism of smooth projective varieties. By the *projection formula* $\pi_*(\pi^*a \cap [Y]) = a \cap [X]$ for $a \in H^\bullet(X, \mathbf{Z})$ and the *Poincaré duality* $\cap[X]\colon H^i(X, \mathbf{Z}) \simeq H_{2n-i}(X, \mathbf{Z})$ where n is the dimension of X, the induced map π^* satisfies the following.

Lemma 2.5.4 *The map $\pi^*\colon H^\bullet(X, \mathbf{Z}) \to H^\bullet(Y, \mathbf{Z})$ is injective in such a way that $H^\bullet(X, \mathbf{Z})$ is a direct summand of $H^\bullet(Y, \mathbf{Z})$ by the Gysin map $H^\bullet(Y, \mathbf{Z}) \to H^\bullet(X, \mathbf{Z})$.*

The *splitting principle* means the existence of π by which $\pi^*\mathscr{E}$ admits a filtration $0 = \mathscr{E}_0 \subset \mathscr{E}_1 \subset \cdots \subset \mathscr{E}_r = \pi^*\mathscr{E}$ such that every quotient $\mathscr{E}_i/\mathscr{E}_{i-1}$ is invertible. This is a consequence of Hironaka's resolution. The decomposition $\pi^*c(\mathscr{E}) = \prod_i c(\mathscr{E}_i/\mathscr{E}_{i-1})$ holds inside $H^\bullet(X, \mathbf{Z}) \hookrightarrow H^\bullet(Y, \mathbf{Z})$. For instance, one obtains the expression $c_1(\mathscr{E}) = c_1(\det \mathscr{E})$ using the splitting principle.

Remark 2.5.5 The top Chern class $c_r(\mathscr{E})$ of a locally free sheaf $\mathscr{E}$ on X of rank r equals the *Euler class* of $\mathscr{E}$ [189, 4.11]. For a global section $s \in \Gamma(X, \mathscr{E})$ or equivalently a map $s\colon \mathscr{O}_X \to \mathscr{E}$, if the closed subscheme Z of X defined by the image $s^\vee(\mathscr{E}^\vee)$ of the dual map $s^\vee\colon \mathscr{E}^\vee \to \mathscr{O}_X$ is of codimension r, then

$[Z] = c_r(\mathcal{E}) \cap [X]$ in $H_{2n-2r}(X, \mathbf{Z})$ with the dimension n of X. As a special case, the top Chern class $c_n(X) \in H^{2n}(X, \mathbf{Z}) \simeq \mathbf{Z}$ equals the *topological Euler characteristic* $\chi(X)$ defined as the alternating sum $\chi(X) = \sum_{i \in \mathbf{N}} (-1)^i b_i(X)$ of the Betti numbers, where the i-th *Betti number* $b_i(X)$ is the rank of $H_i(X, \mathbf{Z})$.

For a locally free sheaf $\mathcal{E}$ of rank r, we take the *Chern polynomial* $c_t(\mathcal{E}) = \sum_{i=0}^{r} c_i(\mathcal{E}) t^i$ with an indeterminate t and write $c_t(\mathcal{E}) = \prod_{j=1}^{r}(1 + a_j t)$ with formal symbols a_j. Then $c_i(\mathcal{E})$ is considered to be the elementary symmetric polynomial s_i in $a_1, \ldots, a_r$ of degree i. We define the *Chern character* $\mathrm{ch}(\mathcal{E})$ and the *Todd class* $\mathrm{td}(\mathcal{E})$ of $\mathcal{E}$ in $H^{\bullet}(X, \mathbf{Q})$ as

$$\mathrm{ch}(\mathcal{E}) = \sum_{j=1}^{r} e^{a_j}, \qquad \mathrm{td}(\mathcal{E}) = \prod_{j=1}^{r} \frac{a_j}{1 - e^{-a_j}}$$

by expanding the right-hand side in $s_i = c_i(\mathcal{E})$ with $1 \leq i \leq r$. Explicitly

$$\mathrm{ch}(\mathcal{E}) = r + c_1 + \frac{1}{2}(c_1^2 - 2c_2) + \frac{1}{6}(c_1^3 - 3c_1 c_2 + 3c_3) + \cdots,$$

$$\mathrm{td}(\mathcal{E}) = 1 + \frac{1}{2}c_1 + \frac{1}{12}(c_1^2 + c_2) + \frac{1}{24}c_1 c_2 + \cdots$$

with $c_i = c_i(\mathcal{E})$ for $1 \leq i \leq r$ and $c_i = 0$ for $i > r$.

Theorem 2.5.6 (Hirzebruch–Riemann–Roch [189]) *Let X be a smooth projective variety of dimension n and let $\mathcal{T}_X$ denote the tangent sheaf of X. Let $\mathcal{E}$ be a locally free coherent sheaf on X. Then*

$$\chi(\mathcal{E}) = (\mathrm{ch}(\mathcal{E}) \cdot \mathrm{td}(\mathcal{T}_X))_{2n},$$

where $_{2n}$ stands for the entry in $H^{2n}(X, \mathbf{Q}) = \mathbf{Q}$.

Grothendieck [53] formulated this over an arbitrary algebraically closed field by taking Chern classes in the *Chow ring* $\bigoplus_i A^i(X)$, where $A^i(X)$ is the group of cycles of codimension i on X modulo rational equivalence. It has been generalised as the *Atiyah–Singer index theorem* [26], [27]. The formula for an invertible sheaf $\mathcal{L}_n$ on X of lower dimension n is expressed with $\xi = c_1(\mathcal{L}_n)$ and $c_i = c_i(X)$ as

$$\chi(\mathcal{L}_1) = \xi + \frac{1}{2}c_1,$$

$$\chi(\mathcal{L}_2) = \frac{1}{2}\xi^2 + \frac{1}{2}c_1 \xi + \frac{1}{12}(c_1^2 + c_2),$$

$$\chi(\mathcal{L}_3) = \frac{1}{6}\xi^3 + \frac{1}{4}c_1 \xi^2 + \frac{1}{12}(c_1^2 + c_2)\xi + \frac{1}{24}c_1 c_2.$$

We shall extend the formula to normal varieties with isolated singularities, following Reid [401, chapter III]. Let D be a $\mathbf{Q}$-divisor on a smooth projective variety X. Take a non-zero integer r such that rD is integral. The Euler characteristic $\chi_{rD}(l) = \chi(\mathscr{O}_X(lrD))$ for $l \in \mathbf{Z}$ is a polynomial in l. We take the rational number $\chi_{rD}(l)$ for $l \in \mathbf{Q}$ by this polynomial and define

$$\chi_{\mathbf{Q}}(D) = \chi_{rD}(1/r),$$

which is independent of the choice of r.

Lemma 2.5.7 *Let D be a $\mathbf{Q}$-Cartier $\mathbf{Q}$-divisor on a normal projective variety X. Then $\chi_{\mathbf{Q}}(\mu^*D)$ for a resolution $\mu\colon X' \to X$ is independent of the choice of μ.*

Proof By the definition of $\chi_{\mathbf{Q}}$, we may assume that D is a Cartier divisor, for which $\chi_{\mathbf{Q}}(\mu^*D) = \chi(\mathscr{O}_{X'}(\mu^*D))$. Hence it suffices to see the coincidence $\chi(\mathscr{O}_Y(\pi^*D)) = \chi(\mathscr{O}_X(D))$ for a resolution $\pi\colon Y \to X$ of a smooth projective variety and a divisor D on X. This is derived from the Leray spectral sequence since $R^i\pi_*\mathscr{O}_Y(\pi^*D) = 0$ for $i \geq 1$. □

Theorem 2.5.8 (Singular Riemann–Roch formula) *With every germ $x \in X$ of an isolated lt singularity and a $\mathbf{Q}$-Cartier integral divisor D on $x \in X$, one can uniquely associate a rational number $c_x(D)$ by the axioms*

- $c_x(D)$ *is zero if $x \in X$ is smooth and*
- *for a resolution $\mu\colon X' \to X$ of a projective variety X with only isolated lt singularities and a $\mathbf{Q}$-Cartier integral divisor D on X,*

$$\chi(\mathscr{O}_X(D)) = \chi_{\mathbf{Q}}(\mu^*D) + \sum_{x \in X} c_x(D).$$

Proof For each germ $x \in X$ of a non-smooth isolated lt singularity, we fix a strong log resolution $\mu_x\colon Y_x \to X$. Compactifying $x \in X$ to a projective variety with a unique singular point x, we define $c_x(D) = \chi(\mathscr{O}_X(D)) - \chi_{\mathbf{Q}}(\mu_x^*D)$ for a $\mathbf{Q}$-Cartier integral divisor D on X. This is a unique candidate for $c_x(D)$ because of the equality in the second axiom for μ_x.

We shall express $c_x(D)$ more explicitly. Write $\lceil K_{Y_x/X} + \mu_x^*D \rceil = \mu_x^*D + E_{xD}$ with an exceptional $\mathbf{Q}$-divisor E_{xD}. Since X is lt, $\lceil E_{xD} \rceil$ is effective and $\mu_{x*}\mathscr{O}_{Y_x}(\mu_x^*D + E_{xD}) = \mathscr{O}_X(D)$ by Proposition 2.5.9 below. On the other hand, $R^i\mu_{x*}\mathscr{O}_{Y_x}(\mu_x^*D + E_{xD}) = 0$ for $i \geq 1$ by Kawamata–Viehweg vanishing. Hence the Leray spectral sequence yields the equality $\chi(\mathscr{O}_X(D)) = \chi(\mathscr{O}_{Y_x}(\mu_x^*D + E_{xD})) = \chi_{\mathbf{Q}}(\mu_x^*D + E_{xD})$. Thus

$$c_x(D) = \chi_{\mathbf{Q}}(\mu_x^*D + E_{xD}) - \chi_{\mathbf{Q}}(\mu_x^*D) = \chi_{\mathbf{Q}}(E_{xD}) - \chi_{\mathbf{Q}}(0),$$

where the latter equality follows from $\mu_x^*D \cdot E_{xD} = 0$ in $H^4(Y_x, \mathbf{Q})$ since E_{xD} is supported in $\mu_x^{-1}(x)$. The right-hand side is determined by the values

$\chi(\mathcal{O}_{Y_x}(lE_{xD})) - \chi(\mathcal{O}_{Y_x}) = \chi(\mathcal{O}_{lP}(lP|_{lP})) - \chi(\mathcal{O}_{lN}(lP|_{lN}))$ at several divisible integers l, where we write $E_{xD} = P - N$ with effective **Q**-divisors P and N without common components. It follows that $c_x(D)$ is independent of the compactification of $x \in X$.

We shall verify the second axiom. Let X and D be as in the axiom. By Lemma 2.5.7, it suffices to check the equality in the axiom for some $\mu: X' \to X$. Let I denote the set of singular points of X. We construct μ by gluing $\mu_x: Y_x \to X$ for all $x \in I$. Then $\lceil K_{X'/X} + \mu^* D \rceil = \mu^* D + \sum_{x \in I} E_{xD}$. As in the preceding paragraph, one has $\chi(\mathcal{O}_X(D)) = \chi_{\mathbf{Q}}(\mu^* D + \sum_{x \in I} E_{xD})$ and

$$
\begin{aligned}
\chi(\mathcal{O}_X(D)) - \chi_{\mathbf{Q}}(\mu^* D) &= \chi_{\mathbf{Q}}(\mu^* D + \textstyle\sum_{x \in I} E_{xD}) - \chi_{\mathbf{Q}}(\mu^* D) \\
&= \chi_{\mathbf{Q}}(\textstyle\sum_{x \in I} E_{xD}) - \chi_{\mathbf{Q}}(0) \\
&= \sum_{x \in I} (\chi_{\mathbf{Q}}(E_{xD}) - \chi_{\mathbf{Q}}(0)) = \sum_{x \in I} c_x(D). \qquad \square
\end{aligned}
$$

Proposition 2.5.9 *Let $\pi: Y \to X$ be a proper birational morphism of normal varieties. Let D be a divisor on Y such that $D_X = \pi_* D$ is **Q**-Cartier. Write $D = \pi^* D_X + \sum_{i \in I} e_i E_i$ with exceptional prime divisors E_i and $e_i \in \mathbf{Q}$. Suppose that $e_i \geq -1$ for all $i \in I$. Let $Z = \bigcup_{j \in J} \pi(E_j)$ for $J = \{i \in I \mid e_i = -1\}$. If D_X is Cartier, then $\pi_* \mathcal{O}_Y(D) = \mathcal{I} \otimes \mathcal{O}_X(D_X)$ for the ideal sheaf $\mathcal{I}$ in $\mathcal{O}_X$ defining Z. If D_X is not Cartier at any point in Z, then $\pi_* \mathcal{O}_Y(D) = \mathcal{O}_X(D_X)$.*

Proof We write $E = \sum_{i \in I} e_i E_i$ for brevity. If D_X is Cartier, then E is integral and $\pi_* \mathcal{O}_Y(E) = \mathcal{I}$. Hence the first assertion follows from the projection formula. We shall see the second.

Let $\mathcal{K}_X$ denote the constant sheaf of the function field of X. The sheaves $\mathcal{O}_X(D_X)$ and $\pi_* \mathcal{O}_Y(D)$ are considered to be subsheaves of $\mathcal{K}_X$ as

$$
\begin{aligned}
\mathcal{O}_X(D_X) &= \{f \in \mathcal{K}_X \mid (f)_X + D_X \geq 0\}, \\
\pi_* \mathcal{O}_Y(D) &= \{f \in \mathcal{K}_X \mid (f)_Y + \pi^* D_X + E \geq 0\}.
\end{aligned}
$$

In particular, $\pi_* \mathcal{O}_Y(D) \subset \mathcal{O}_X(D_X)$. Take an arbitrary element $f \in \mathcal{O}_X(D_X)$, which satisfies $(f)_Y + \pi^* D_X \geq 0$. If D_X is not Cartier at any point in Z, then the support of the effective divisor $S = (f)_X + D_X$ necessarily contains Z. The round-up $\lceil \pi^* S + E \rceil$ is effective because every E_j for $j \in J$ has positive coefficient in $\pi^* S$. Hence the integral divisor $(f)_Y + D = (f)_Y + \pi^* D_X + E = \pi^* S + E$ is effective, meaning that f belongs to $\pi_* \mathcal{O}_Y(D)$. Thus $\pi_* \mathcal{O}_Y(D) = \mathcal{O}_X(D_X)$. $\qquad \square$

Definition 2.5.10 The rational number $c_x(D)$ in Theorem 2.5.8 is called the *contribution* of $x \in X$ with respect to D. It depends only on the divisorial sheaf $\mathcal{O}_X(D)$ by the equality $c_x(D) = \chi(\mathcal{O}_X(D)) - \chi_{\mathbf{Q}}(\mu_x^* D)$. For a divisorial sheaf $\mathcal{L} = \mathcal{O}_X(D)$, one can write $c_x(\mathcal{L})$ for $c_x(D)$.

Remark 2.5.11 The contribution $c_x(D)$ is determined by the analytic germ. Indeed, it is determined by the values $\chi(\mathcal{O}_{Y_x}(lE_{xD})) - \chi(\mathcal{O}_{Y_x})$ as in the proof of Theorem 2.5.8. Take as μ_x the blow-up along an ideal in $\mathcal{O}_X$ which is trivial outside x. Then its restriction to the analytic germ extends to a resolution of every algebraic germ that has the same analytic germ as $x \in X$.

The contribution of a Cartier divisor is zero as expected.

Lemma 2.5.12 $c_x(0) = 0$.

Proof For $\mu_x \colon Y_x \to X$ in the proof of Theorem 2.5.8, $c_x(0) = \chi(\mathcal{O}_X) - \chi_\mathbf{Q}(\mu_x^* 0) = \chi(\mathcal{O}_X) - \chi(\mathcal{O}_{Y_x})$, which is zero by Theorem 1.4.20. □

The next lemma asserts deformation invariance of contributions.

Lemma 2.5.13 *Let $X \to T$ be a contraction to a smooth curve. Let D be a $\mathbf{Q}$-Cartier divisor on X such that $\mathcal{O}_X(D)$ satisfies S_3 and is flat over T. For every $t \in T$, suppose that the fibre $X_t = X \times_T t$ has only isolated lt singularities and that X_t is not contained in the support of D, by which the restriction $D_t = D|_{X_t}$ on X_t is defined as a $\mathbf{Q}$-Cartier divisor. Then the contribution $\sum_{x_t \in X_t} c_{x_t}(D_t)$ is independent of $t \in T$.*

Proof By the definition of contribution, it suffices to verify that $\chi(\mathcal{O}_{X_t}(D_t))$ and $\chi_\mathbf{Q}(\mu_t^* D_t)$ are both independent of $t \in T$, where $\mu_t \colon Y_t \to X_t$ is a resolution. Since $\mathcal{O}_X(D)$ satisfies S_3, the divisorial sheaf $\mathcal{O}_{X_t}(D_t)$ on X_t is nothing but $\mathcal{O}_X(D) \otimes \mathcal{O}_{X_t}$. Hence $\chi(\mathcal{O}_{X_t}(D_t)) = \chi(\mathcal{O}_X(D) \otimes \mathcal{O}_{X_t})$, which is constant since $\mathcal{O}_X(D)$ is flat over T. Fix a non-zero integer r such that rD is Cartier. The constancy of $\chi_\mathbf{Q}(\mu_t^* D_t)$ is then reduced to that of the polynomial $\chi(\mathcal{O}_{Y_t}(\mu_t^*(lrD_t)))$ in l. By Theorem 1.4.20, $\chi(\mathcal{O}_{Y_t}(\mu_t^*(lrD_t))) = \chi(\mathcal{O}_{X_t}(lrD_t)) = \chi(\mathcal{O}_X(lrD) \otimes \mathcal{O}_{X_t})$. This is constant since X is flat over T. □

We shall compute the contribution of a cyclic quotient singularity, which is lt by Corollary 2.2.22. Fixing a primitive character $\chi \colon \mathbf{Z}_r \to \mathbf{C}^\times$, we take the cyclic quotient singularity $o \in X = \mathbf{A}^n/\mathbf{Z}_r(a_1, \ldots, a_n)$ in Definition 2.2.10. We assume that all a_i are coprime to r because we are interested in isolated singularities. Let $\mathscr{L}_l$ denote the subsheaf of the direct image $\pi_* \mathcal{O}_{\mathbf{A}^n}$ by $\pi \colon \mathbf{A}^n \to X$ which consists of semi-invariant functions of weight l with respect to χ. It is a divisorial sheaf on X.

Take a primitive r-th root ζ of unity and define

$$\sigma_l(\tfrac{1}{r}(a_1, \ldots, a_n)) = \sum_{j=1}^{r-1} \frac{\zeta^{-jl}}{\prod_{i=1}^n (1 - \zeta^{ja_i})}$$

for an integer l. It is independent of the choice of ζ. In the setting where $g \in \mathbf{Z}_r$ sends x_i to $\chi(g)^{a_i} x_i$, the automorphism σ_g of $\mathbf{A}^n$ given by g sends the tangent vector $\partial/\partial x_i$ at origin to $(\sigma_g)_* \partial/\partial x_i = \chi(g)^{a_i} \partial/\partial x_i$. One should correct [401, p.405 remark] and the subsequent arguments accordingly, for which the definition of σ_l above is different from that in [401, theorem 8.5]. One has $\mathscr{O}_X(K_X) \simeq \mathscr{L}_{-\sum_i a_i}$ because the generator $dx_1 \wedge \cdots \wedge dx_n$ of $\mathscr{O}_{\mathbf{A}^n}(K_{\mathbf{A}^n})$ has weight $\sum_i a_i$. The type of K_X in [401, theorem 10.2] should be minus one.

Theorem 2.5.14 ([401, p.407 corollary])   *The contribution of the germ $o \in X = \mathbf{A}^n/\mathbf{Z}_r(a_1, \ldots, a_n)$ with respect to the sheaf $\mathscr{L}_l$ of semi-invariant functions of weight l is*

$$c_o(\mathscr{L}_l) = \frac{1}{r}(\sigma_l(\tfrac{1}{r}(a_1, \ldots, a_n)) - \sigma_0(\tfrac{1}{r}(a_1, \ldots, a_n))).$$

The proof needs the *holomorphic Lefschetz fixed-point formula* of form [25, theorem 4.12]. See also [154, pp.422–426].

Theorem 2.5.15 (Holomorphic Lefschetz fixed-point formula)   *Let σ be an automorphism of a compact complex manifold X. Define the holomorphic Lefschetz number $L(\sigma, \mathscr{O}_X) = \sum_{i \in \mathbf{N}}(-1)^i \operatorname{tr} \sigma^*|_{H^i(\mathscr{O}_X)} \in \mathbf{C}$ where $\operatorname{tr} \sigma^*|_{H^i(\mathscr{O}_X)}$ stands for the trace of the induced automorphism of $H^i(\mathscr{O}_X)$. Suppose that the set I of points in X fixed by σ is finite and that $\det(1 - d\sigma_x) \neq 0$ for all $x \in I$ where $d\sigma_x$ denotes the induced automorphism of the tangent space $T_{X,x}$ at x. Then*

$$L(\sigma, \mathscr{O}_X) = \sum_{x \in I} \frac{1}{\det(1 - d\sigma_x)}.$$

Proof of Theorem 2.5.14   We realise a projective variety X with only cyclic quotient singularities of the same type $\frac{1}{r}(a_1, \ldots, a_n)$ in the following way. Consider the action of $\mathbf{Z}_r$ on $\mathbf{P}^{2n}$ with homogeneous coordinates $x_0, \ldots, x_{2n}$ such that $g \in \mathbf{Z}_r$ sends x_i to x_i for $0 \le i \le n$ and x_{n+i} to $\chi(g)^{a_i} x_{n+i}$ for $1 \le i \le n$. The group $\mathbf{Z}_r$ fixes the subspace $\Sigma = (x_{n+1} = \cdots = x_{2n} = 0) \simeq \mathbf{P}^n$ of $\mathbf{P}^{2n}$ pointwise, outside which $\mathbf{Z}_r$ acts freely in codimension n. Let $\bar{X}$ be a complete intersection of n general hypersurfaces in $\mathbf{P}^{2n}$ defined by invariant homogeneous polynomials. The group $\mathbf{Z}_r$ acts on $\bar{X}$ freely outside $\bar{X} \cap \Sigma$ which consists of d closed points, where d is the degree of $\bar{X}$. Then $X = \bar{X}/\mathbf{Z}_r$ has d singular points, all of which are of type $\frac{1}{r}(a_1, \ldots, a_n)$.

We write $\pi \colon \bar{X} \to X$. Let $\mathscr{L}_l$ denote the subsheaf of $\pi_* \mathscr{O}_{\bar{X}}$ which consists of semi-invariant functions of weight l with respect to χ. It is a divisorial sheaf and the double dual of $\mathscr{L}_l^{\otimes r}$ is isomorphic to $\mathscr{O}_X$. Hence $\mathscr{L}_l \simeq \mathscr{O}_X(D_l)$ with some integral divisor $D_l \sim_{\mathbf{Q}} 0$. Take a resolution $\mu \colon X' \to X$. The germ at every singular point together with $\mathscr{L}_l$ is analytically isomorphic to the germ

$o \in X$ with $\mathscr{L}_l$ in the statement. Hence $\chi(\mathscr{L}_l) = \chi_{\mathbf{Q}}(\mu^* D_l) + dc_o(\mathscr{L}_l)$ by the definition of contribution. It follows from $D_l \sim_{\mathbf{Q}} 0$ that $\chi_{\mathbf{Q}}(\mu^* D_l) = \chi(\mathscr{O}_{X'})$, which equals $\chi(\mathscr{O}_X)$ since X has rational singularities by Theorem 1.4.20. Thus

$$dc_o(\mathscr{L}_l) = \chi(\mathscr{L}_l) - \chi(\mathscr{O}_X).$$

We shall compute the right-hand side by applying the holomorphic Lefschetz fixed-point formula. Take a generator g of $\mathbf{Z}_r$ and let $\zeta = \chi(g)$. For $0 < j < r$, the element g^j provides an automorphism of $\bar{X}$. The decomposition $\pi_* \mathscr{O}_{\bar{X}} = \bigoplus_{l=0}^{r-1} \mathscr{L}_l$ yields $H^i(\mathscr{O}_{\bar{X}}) = \bigoplus_{l=0}^{r-1} H^i(\mathscr{L}_l)$ for all i. Hence the holomorphic Lefschetz number $L(g^j, \mathscr{O}_{\bar{X}})$ is

$$L(g^j, \mathscr{O}_{\bar{X}}) = \sum_{i=0}^{n} (-1)^i \sum_{l=0}^{r-1} h^i(\mathscr{L}_l) \zeta^{jl} = \sum_{l=0}^{r-1} \chi(\mathscr{L}_l) \zeta^{jl}.$$

On the other hand for $\bar{o} \in \bar{X} \cap \Sigma$, the automorphism $(dg^j)_{\bar{o}}$ of $T_{\bar{X},\bar{o}}$ induced by g^j is diagonalised to $(\zeta^{ja_1}, \ldots, \zeta^{ja_n})$. Hence $\det(1 - (dg^j)_{\bar{o}}) = \prod_i (1 - \zeta^{ja_i})$. Thus Theorem 2.5.15 shows the equality

$$\sum_{l=0}^{r-1} \chi(\mathscr{L}_l) \zeta^{jl} = \frac{d}{\prod_i (1 - \zeta^{ja_i})}.$$

This and the relation $\sum_l \chi(\mathscr{L}_l) = \chi(\mathscr{O}_{\bar{X}})$ are expressed by the matrix $M = (\zeta^{(j-1)(l-1)})_{jl}$ as

$$M \begin{pmatrix} \chi(\mathscr{L}_0) \\ \chi(\mathscr{L}_1) \\ \vdots \\ \chi(\mathscr{L}_{r-1}) \end{pmatrix} = \begin{pmatrix} \chi(\mathscr{O}_{\bar{X}}) \\ d/\prod_i (1 - \zeta^{a_i}) \\ \vdots \\ d/\prod_i (1 - \zeta^{(r-1)a_i}) \end{pmatrix}.$$

Multiplying this by $rM^{-1} = (\zeta^{-(j-1)(l-1)})_{jl}$, one obtains

$$r\chi(\mathscr{L}_l) = \chi(\mathscr{O}_{\bar{X}}) + \sum_{j=1}^{r-1} \frac{d\zeta^{-jl}}{\prod_i (1 - \zeta^{ja_i})} = \chi(\mathscr{O}_{\bar{X}}) + d\sigma_l(\tfrac{1}{r}(a_1, \ldots, a_n)).$$

Taking the difference between $\chi(\mathscr{L}_l)$ and $\chi(\mathscr{L}_0)$,

$$r(\chi(\mathscr{L}_l) - \chi(\mathscr{O}_X)) = d(\sigma_l(\tfrac{1}{r}(a_1, \ldots, a_n)) - \sigma_0(\tfrac{1}{r}(a_1, \ldots, a_n))).$$

We have proved that the left-hand side equals $rdc_o(\mathscr{L}_l)$. The theorem is thus completed. $\square$

We shall compute $\sigma_l(\tfrac{1}{r}(a_1, \ldots, a_n))$ explicitly. The next lemma follows at once from the definition of $\sigma_l(\tfrac{1}{r}(a_1, \ldots, a_n))$.

Lemma 2.5.16   (i) $\sigma_l(\frac{1}{r}(a_1,\ldots,a_n)) = \sigma_{lb}(\frac{1}{r}(a_1 b,\ldots,a_n b))$ *for any integer b coprime to r.*

(ii) $\sum_{l=0}^{r-1} \sigma_l(\frac{1}{r}(a_1,\ldots,a_n)) = 0.$

(iii) $\sigma_l(\frac{1}{r}(a_1,\ldots,a_n)) - \sigma_{l-a_n}(\frac{1}{r}(a_1,\ldots,a_n)) = \sigma_l(\frac{1}{r}(a_1,\ldots,a_{n-1}))$. *This is valid even if n = 1 by defining* $\sigma_l(\frac{1}{r}(\emptyset)) = r - 1$ *for* $l \equiv 0 \bmod r$ *and* $\sigma_l(\frac{1}{r}(\emptyset)) = -1$ *for* $l \not\equiv 0 \bmod r.$

Lemma 2.5.17   *For any natural number l, let* $\sigma_l = \sigma_l(\frac{1}{r}(a_1,\ldots,a_{n-1},-1))$ *and let* $\tau_l = \sigma_l(\frac{1}{r}(a_1,\ldots,a_{n-1}))$. *Then*

(i) $\sigma_l - \sigma_0 = -\sum_{j=0}^{l-1} \tau_j$ *and*

(ii) $\sigma_l = -r^{-1} \sum_{j=1}^{r-1} j\tau_j - \sum_{j=0}^{l-1} \tau_j.$

Proof By Lemma 2.5.16(iii), $\sigma_{j+1} - \sigma_j = -\tau_j$. Summing up these for $0 \leq j < l$, one obtains (i). Moreover, $\sum_{j=1}^{r-1} j(\sigma_{j+1} - \sigma_j) = -\sum_{j=1}^{r-1} j\tau_j$. The left-hand side equals $(r-1)\sigma_r - \sum_{j=1}^{r-1}\sigma_j = r\sigma_0$ by Lemma 2.5.16(ii). Hence $\sigma_0 = -r^{-1}\sum_{j=1}^{r-1} j\tau_j$. Substituting this into (i), one obtains (ii). $\square$

Proposition 2.5.18   *For $0 \leq l < r$ with Notation 2.2.14, one has*

(i) $\sigma_l(\frac{1}{r}(1,-1)) - \sigma_0(\frac{1}{r}(1,-1)) = -\dfrac{l(r-l)}{2}$ *and*

(ii) $\sigma_{-la}(\frac{1}{r}(1,-1,a)) - \sigma_0(\frac{1}{r}(1,-1,a)) = -l\dfrac{r^2-1}{12} + \sum_{j=0}^{l-1}\dfrac{\overline{ja}(r-\overline{ja})}{2}.$

Proof By Lemma 2.5.17(ii), $\sigma_l(\frac{1}{r}(-1)) = r^{-1}\sum_{j=1}^{r-1} j - ((r-1)-(l-1)) = l - (r+1)/2$ for $0 < l \leq r$. With Lemma 2.5.16(i), $\sigma_l(\frac{1}{r}(1)) = \sigma_{r-l}(\frac{1}{r}(-1)) = (r-1)/2 - l$ for $0 \leq l < r$. Using this and Lemma 2.5.17(i), one obtains for $0 \leq l < r$,

$$\sigma_l(\tfrac{1}{r}(1,-1)) - \sigma_0(\tfrac{1}{r}(1,-1)) = -\sum_{j=0}^{l-1}\left(\frac{r-1}{2} - j\right),$$

from which (i) follows. By Lemma 2.5.17(ii), for $0 \leq l < r$,

$$\sigma_l(\tfrac{1}{r}(1,-1)) = -\frac{1}{r}\sum_{j=1}^{r-1} j\left(\frac{r-1}{2}-j\right) - \sum_{j=0}^{l-1}\left(\frac{r-1}{2}-j\right) = \frac{r^2-1}{12} - \frac{l(r-l)}{2}.$$

Take an integer b such that $ab \equiv 1 \bmod r$. Then

$$\sigma_{-la}(\tfrac{1}{r}(1,-1,a)) - \sigma_0(\tfrac{1}{r}(1,-1,a)) = \sigma_l(\tfrac{1}{r}(-b,b,-1)) - \sigma_0(\tfrac{1}{r}(-b,b,-1))$$
$$= -\sum_{j=0}^{l-1}\sigma_j(\tfrac{1}{r}(-b,b)) = -\sum_{j=0}^{l-1}\sigma_{-\overline{ja}}(\tfrac{1}{r}(1,-1)),$$

where the first and third equalities follow from Lemma 2.5.16(i) and the second from Lemma 2.5.17(i). These two equalities provide (ii). □

We write down the singular Riemann–Roch formula in lower dimensions explicitly.

Theorem 2.5.19 ([401, theorem 9.1]) *Let S be an lt projective surface and let D be a $\mathbf{Q}$-Cartier divisor on S. Then*

$$\chi(\mathscr{O}_S(D)) = \frac{1}{2}D(D - K_S) + \chi(\mathscr{O}_S) + \sum_{x \in S} c_x(D)$$

with contribution $c_x(D)$.

If $x \in S$ is a cyclic quotient singularity of type $\frac{1}{r}(1, -1)$ and D is analytically defined at x by a semi-invariant function of weight l where $0 \le l < r$, then

$$c_x(D) = -\frac{l(r - l)}{2r}.$$

Proof Take a resolution $\mu \colon S' \to S$. By Theorem 2.5.8, the formula on $\chi(\mathscr{O}_S(D))$ is reduced to the equality $\chi_{\mathbf{Q}}(\mu^*D) = D(D - K_S)/2 + \chi(\mathscr{O}_S)$. This follows from the Riemann–Roch theorem on S' and the coincidence $\chi(\mathscr{O}_{S'}) = \chi(\mathscr{O}_S)$ by the property of rational singularities. The description of $c_x(D)$ for a cyclic quotient singularity follows from Theorem 2.5.14 and Proposition 2.5.18(i) with $c_x(D) = c_o(\mathscr{L}_l)$ by the notation in that theorem. □

For the description in dimension three, we introduce the notion of *fictitious singularities*.

Definition 2.5.20 Let $o \in X$ be a terminal threefold singularity of index r greater than one. It is analytically expressed as the cyclic quotient $o \in X = \bar{X}/\mathbf{Z}_r \subset \mathfrak{D}^4/\mathbf{Z}_r$ of a hypersurface singularity $\bar{o} \in \bar{X} = (\xi = 0) \subset \mathfrak{D}^4$ as in Theorem 2.4.8. Let $\bar{l} \simeq \mathfrak{D}^1$ denote the axis in $\mathfrak{D}^4$ where the action of $\mathbf{Z}_r$ is not free. The *axial multiplicity* of $o \in X$ is the local intersection number $(\bar{X} \cdot \bar{l})_{\bar{o}}$ on the germ $\bar{o} \in \mathfrak{D}^4$. It is independent of the expression of X and computed explicitly as below with the notation in Theorem 2.4.8.

$o \in X$	cA/r	cAx/4	cAx/2	cD/3	cD/2	cE/2
axial mult.	ord $f(x_3, 0)$	ord $f(x_3, 0)$	2	2	ord $f(0, x_3, 0)$	3

Let $o \in X$ be the germ of a terminal threefold analytic singularity of index r. It admits a $\mathbf{Q}$-smoothing as will be constructed below. A $\mathbf{Q}$-*smoothing* means a flat morphism $Y \to T$ from a normal variety to the germ $0 \in T$ of a smooth curve such that the central fibre $Y_0 = Y \times_T 0$ has a unique singular point y_0, such that $y_0 \in Y_0$ is analytically isomorphic to $o \in X$ and such that other fibres $Y_t = Y \times_T t$ with $t \ne 0$ have only terminal quotient singularities.

By Theorems 2.3.4 and 2.4.8, $o \in X$ is realised as an isolated singularity in $\mathfrak{D}^4/\mathbf{Z}_r(a_1, a_2, a_3, a_4)$ defined by a semi-invariant function ξ. By Corollary 2.1.5 and Remark 2.1.6, one may assume that ξ is a polynomial and it is taken as in Theorem 2.4.8 if $r \geq 2$. Then $o \in X$ is algebraically defined by a semi-invariant polynomial ξ in $A = \mathbf{A}^4/\mathbf{Z}_r(a_1, a_2, a_3, a_4)$ with orbifold coordinates x_1, x_2, x_3, x_4. Take the algebraic germ $0 \in T = \mathbf{A}^1$ at origin with coordinate t and the hypersurface Y in $A \times T$ defined by the semi-invariant function $\xi + tz$, where z is a general linear form in x_1, x_2, x_3, x_4 if $r = 1$ and is the unique coordinate x_i that has the same weight as ξ if $r \geq 2$. Shrinking Y, the morphism $Y \to T$ becomes a $\mathbf{Q}$-smoothing of $o \in X$. We can make it a projective morphism by compactification and desingularisation away from $o \times T$.

The types of singularities of Y_t for $t \neq 0$ are as in the following lemma, where $n \times \frac{1}{s}(b_1, b_2, b_3)$ means that n cyclic quotient singularities of type $\frac{1}{s}(b_1, b_2, b_3)$ appear in Y_t. Taking the index-one cover of the germ $y_0 \in Y$, we see that the types are independent of the $\mathbf{Q}$-smoothing.

Lemma 2.5.21   (i) *If $o \in X$ is of index one, then Y_t is smooth.*

(ii) *If $o \in X$ is of index r greater than one and of axial multiplicity n but it is not $cAx/4$, then Y_t has $n \times \frac{1}{r}(1, -1, b)$.*

(iii) *If $o \in X$ is $cAx/4$ and of axial multiplicity $2k+1$, then Y_t has $1 \times \frac{1}{4}(1, 3, 1)$ and $k \times \frac{1}{2}(1, 1, 1)$.*

Definition 2.5.22   The quotient singularities on Y_t for $t \neq 0$ are called the *fictitious singularities* from $o \in X$. The *basket* of a terminal threefold X is the set of fictitious singularities arising from the singularities of X.

Theorem 2.5.23 ([401, theorem 10.2])   *Let X be a terminal projective threefold and let D be a $\mathbf{Q}$-Cartier divisor on X. Then*

$$\chi(\mathscr{O}_X(D)) = \frac{1}{12}D(D - K_X)(2D - K_X) + \frac{1}{12}Dc_2(X) + \chi(\mathscr{O}_X) + \sum_{x \in X} c_x(D)$$

with contribution $c_x(D)$, where $Dc_2(X)$ is defined as $\mu^* D \cdot c_2(X')$ for a strong resolution $\mu \colon X' \to X$ and it is independent of the choice of μ.

On the germ $x \in X$ of a terminal threefold singularity, a $\mathbf{Q}$-Cartier divisor D is linearly equivalent to lK_X for some integer l and $c_x(D) = c_x(lK_X) = \sum_{x_i \in I} c_{x_i}(lK_{X_i})$ for the set I of fictitious singularities $x_i \in X_i$ from $x \in X$. If $x \in X$ is a terminal quotient singularity of type $\frac{1}{r}(1, -1, b)$, then for $0 \leq l < r$,

$$c_x(lK_X) = -l\frac{r^2 - 1}{12r} + \sum_{j=0}^{l-1} \frac{\overline{jb}(r - \overline{jb})}{2r}.$$

Proof Take a strong resolution $\mu\colon X' \to X$. The formula on $\chi(\mathscr{O}_X(D))$ is obtained from the Riemann–Roch theorem on X' in the same manner as that on $\chi(\mathscr{O}_S(D))$ in Theorem 2.5.19 is obtained. This includes the independence of $\mu^*D \cdot c_2(X')$ from the choice of μ.

The local linear equivalence $D \sim lK_X$ for some l is from Theorem 2.4.17. Recall that $c_x(lK_X)$ is determined by the analytic germ as in Remark 2.5.11. We have observed that as an analytic germ, $x \in X$ admits a projective **Q**-smoothing $X' \to T$ over a smooth curve. Hence the relation $c_x(lK_X) = \sum c_{x_i}(lK_{X_i})$ follows from Lemma 2.5.13. Finally the description of $c_x(lK_X)$ for a terminal quotient singularity follows from Theorem 2.5.14 and Proposition 2.5.18(ii) with $c_x(lK_X) = c_o(\mathscr{L}_{-lb})$ by the notation in that theorem. □

2.6 Canonical Singularities

We discuss strictly canonical threefold singularities in this section.

Definition 2.6.1 A canonical singularity which is not terminal is said to be *strictly canonical*.

The argument for terminal singularities in the previous sections can be applied to cyclic quotients of hypersurface singularities. We have a satisfactory classification of cyclic quotient threefold singularities, which is based on the criterion in Proposition 2.2.15.

Theorem 2.6.2 (Morrison [346], Ishida–Iwashita [197]) *A cyclic quotient threefold singularity is canonical if and only if it is of type* $\frac{1}{r}(1,-1,b)$, $\frac{1}{r}(b_1, b_2, -b_1 - b_2)$, $\frac{1}{4s}(1, 2s+1, -2)$, $\frac{1}{14}(1, 9, 11)$ *or* $\frac{1}{9}(1, 4, 7)$.

However, as studied by Hayakawa and Takeuchi, there are already a large number of types for a canonical threefold singularity which is a quotient of a hypersurface singularity. We abstract their result by focusing on the index.

Theorem 2.6.3 (Hayakawa–Takeuchi [184]) *Let* $o \in X = \bar{X}/\mathbf{Z}_s$ *be a canonical threefold analytic singularity of index r which is the quotient of a hypersurface singularity $\bar{o} \in \bar{X}$ in $\mathfrak{D}^4$ by the cyclic group $\mathbf{Z}_s$. Suppose that $\mathbf{Z}_s$ acts on $\bar{X}$ freely outside $\bar{o}$. Then either $\bar{o} \in \bar{X}$ is cDV with $s = r$ or the index r is at most four.*

Remark 2.6.4 The precise result of Hayakawa and Takeuchi is explicit and provides a list of the semi-invariant function defining $\bar{X}$ and the action on $\mathfrak{D}^4$. If the singularity $o \in X = \bar{X}/\mathbf{Z}_s$ in the theorem is isolated and strictly canonical, then the index r is at most four. If in addition r equals four, then it is analytically isomorphic to $o \in (x_1 x_2 + x_3^2 + x_4^2 = 0) \subset \mathfrak{D}^4/\mathbf{Z}_8(1,5,3,7)$ with orbifold coordinates x_1, x_2, x_3, x_4.

It is natural to ask if the index of a strictly canonical threefold singularity $x \in X$ is bounded. This was conjectured by Shokurov and supported by the above theorems. By an easy observation as below, it is necessary to assume the condition $\mathrm{mld}_x X = 1$. Then we settle the conjecture effectively.

Example 2.6.5 Let $o \in X = (x_1 x_2 + x_3^2 = 0) \subset \mathbf{A}^4/\mathbf{Z}_r(1, -1, 0, 1)$ with orbifold coordinates x_1, x_2, x_3, x_4 where $r \geq 2$. It is of index r and singular along the x_4-axis l. Let $\pi \colon Y \to X$ be the weighted blow-up with $\mathrm{wt}(x_1, x_2, x_3, x_4) = \frac{1}{r}(1, r-1, r, 1)$. The exceptional locus E is the sum of two prime divisors. By Lemma 2.2.12 and the subsequent description, $K_Y = \pi^* K_X + (1/r)E$ and Y has two terminal quotient singularities of types $\frac{1}{r-1}(-1, 1, 1)$ and $\frac{1}{r}(1, -1, 1)$ outside the strict transform l_Y of l. The blow-up $W \to Y$ along l_Y is crepant and W is smooth about the exceptional divisor. Hence X is strictly canonical but $\mathrm{mld}_x X = 1 + 1/r$.

Theorem 2.6.6 (Kawakita [228]) *The algebraic germ $x \in X$ of a canonical threefold singularity such that $\mathrm{mld}_x X = 1$ is of index at most six.*

Examples of index up to four appear in Theorems 2.6.2 and 2.6.3. Prokhorov constructed those of index five and six, which will be explained later.

The proof uses the singular Riemann–Roch formula. The information on the index r of $x \in X$ is deduced from the periodic property that the natural map $\mathcal{O}_X(iK_X) \otimes \mathcal{O}_X(-iK_X) \to \mathcal{O}_X$ is surjective for each i divisible by r. By the Euler characteristic of a certain sheaf, we shall realise the periodic function $\delta(i)$ for $i \in \mathbf{Z}$ defined as

$$\delta(i) = \left\lfloor \frac{i}{r} \right\rfloor - \left\lfloor \frac{i-1}{r} \right\rfloor = \begin{cases} 1 & \text{if } i \text{ is divisible by } r, \\ 0 & \text{otherwise.} \end{cases}$$

Proposition 2.6.7 *Let $x \in X$ be a canonical singularity of index r such that $\mathrm{mld}_x X = 1$. Let $\mathfrak{m}$ be the maximal ideal in $\mathcal{O}_X$ defining x. Then there exist a crepant birational contraction $\pi \colon Y \to X$ from a $\mathbf{Q}$-factorial terminal variety and an effective divisor E on Y supported in $\pi^{-1}(x)$ such that for any integer i,*

$$\pi_* \mathcal{O}_Y(iK_Y - E) = \begin{cases} \mathfrak{m} \mathcal{O}_X(iK_X) & \text{if } i \text{ is divisible by } r, \\ \mathcal{O}_X(iK_X) & \text{otherwise} \end{cases}$$

and $R^p \pi_* \mathcal{O}_Y(iK_Y - E) = 0$ for all $p \geq 1$.

Proof The number $e = e(X)$ of divisors F exceptional over X such that $a_F(X) = 1$ is finite, which will be formulated in Notation 5.1.7. Set $X_0 = X$.

One can inductively construct a sequence of crepant birational contractions $\pi_l \colon X_l \to X_{l-1}$ for $1 \le l \le e$ such that for every $l \ge 1$,

- X_l is **Q**-factorial and canonical and further X_e is terminal and
- π_l has exactly one exceptional prime divisor F_l and $-F_l$ is π_l-nef

and such that $\pi_1(F_1) = x$. Indeed, the existence of $\pi_1 \colon X_1 \to X_0 = X$ without the condition for $-F_1$ to be π_1-nef is a consequence of Proposition 1.5.17 for an arbitrary divisor F_1 over X with $a_{F_1}(X) = 1$ and $c_X(F_1) = x$. For the strict transform H_1 of the general hyperplane section of $x \in X$ and a small positive real number ε, the $(K_{X_1} + \varepsilon H_1)/X$-MMP by Corollary 1.5.13 ends with a required contraction to X. Every π_l is constructed in like manner from a divisor F_l exceptional over X_{l-1} with $a_{F_l}(X_{l-1}) = a_{F_l}(X) = 1$.

We take $\pi \colon Y = X_e \to X$ and shall find a divisor E. The divisor $E_1 = F_1$ on X_1 satisfies for any i,

$$\pi_{1*}\mathcal{O}_{X_1}(iK_{X_1} - E_1) = \begin{cases} m\mathcal{O}_X(iK_X) & \text{if } i \text{ is divisible by } r, \\ \mathcal{O}_X(iK_X) & \text{otherwise,} \end{cases}$$

$$R^p\pi_{1*}\mathcal{O}_{X_1}(iK_{X_1} - E_1) = 0 \quad \text{for } p \ge 1.$$

The former follows from Proposition 2.5.9 and the latter is the Kawamata–Viehweg vanishing. Starting with $E_1 = F_1$, we define the divisor $E_l = \lceil \pi_l^* E_{l-1} \rceil$ on X_l for $l > 1$ inductively. Then for $l > 1$ and $i \in \mathbf{Z}$, it satisfies $\pi_{l*}\mathcal{O}_{X_l}(iK_{X_l} - E_l) = \mathcal{O}_{X_{l-1}}(iK_{X_{l-1}} - E_{l-1})$ and $R^p\pi_{l*}\mathcal{O}_{X_l}(iK_{X_l} - E_l) = 0$ for $p \ge 1$ for the same reason. Notice that $-E_l$ is π_l-nef. One can see that $E = E_e$ is a desired divisor by applying the spectral sequence $R^p\pi_{l*}R^q\varphi_{l*}\mathcal{O}_Y(iK_Y - E) \Rightarrow R^{p+q}\varphi_{l-1*}\mathcal{O}_Y(iK_Y - E)$ for $1 \le l < e$ inductively, where φ_l denotes the contraction $Y \to X_l$. □

We discuss the algebraic germ $x \in X$ of a canonical threefold singularity of index r such that $\mathrm{mld}_x X = 1$. Fix $\pi \colon Y \to X$ and E as in Proposition 2.6.7, in which π is crepant. Consider the exact sequence

$$0 \to \mathcal{O}_Y(iK_Y - E) \to \mathcal{O}_Y(iK_Y) \to \mathcal{O}_E(iK_Y|_E) \to 0,$$

which makes sense by Proposition 2.2.23. By the description of $\pi_*\mathcal{O}_Y(iK_Y - E)$ in Proposition 2.6.7, the periodic function $\delta(i) = \lfloor i/r \rfloor - \lfloor (i-1)/r \rfloor$ is realised as $\delta(i) = \dim \mathcal{O}_X(iK_X)/\pi_*\mathcal{O}_Y(iK_Y - E)$. One has the Kawamata–Viehweg vanishing $R^p\pi_*\mathcal{O}_Y(iK_Y) = 0$ and the vanishing $R^p\pi_*\mathcal{O}(iK_Y - E) = 0$ in Proposition 2.6.7 for $p \ge 1$. It follows that $\delta(i) = h^0(\mathcal{O}_E(iK_Y|_E)) = \chi(\mathcal{O}_E(iK_Y|_E))$ and thus

$$\delta(i) = \chi(\mathcal{O}_Y(iK_Y)) - \chi(\mathcal{O}_Y(iK_Y - E))$$

after a compactification of Y to a terminal projective threefold.

We compute the right-hand side by the singular Riemann–Roch formula. Let $\{q_\iota$ of type $\frac{1}{r_\iota}(1, -1, b_\iota)\}_{\iota \in I_0}$ be the basket of fictitious singularities in Definition 2.5.22 from the singularities of Y. Note that b_ι is coprime to r_ι. For $\iota \in I_0$, we take an integer e_ι such that $E \sim e_\iota K_Y$ at q_ι which exists by Theorem 2.4.17. By replacing b_ι by $r_\iota - b_\iota$ if necessary, we may assume that $v_\iota = \overline{e_\iota b_\iota} \le r_\iota/2$, where $\bar{l}$ denotes the residue of l modulo r_ι as in Notation 2.2.14. Let $I = \{\iota \in I_0 \mid \bar{e}_\iota \neq 0\}$.

Definition 2.6.8 Define $A_\iota(l)$ and $B_\iota(l)$ as

$$A_\iota(l) = -\bar{l}\frac{r_\iota^2 - 1}{12r_\iota} + \sum_{i=1}^{\bar{l}-1} \frac{\overline{ib_\iota}(r_\iota - \overline{ib_\iota})}{2r_\iota}, \qquad B_\iota(l) = \frac{\bar{l}(r_\iota - \bar{l})}{2r_\iota}.$$

One sees the following relations by direct computation.

Lemma 2.6.9 (i) $A_\iota(l+1) - A_\iota(l) = -(r_\iota^2 - 1)/12r_\iota + B_\iota(lb_\iota)$.
(ii) $A_\iota(l) + A_\iota(-l) = -B_\iota(lb_\iota)$.

With this notation, we express $\delta(i) = \chi(\mathcal{O}_Y(iK_Y)) - \chi(\mathcal{O}_Y(iK_Y - E))$ by Theorem 2.5.23 as

$$\delta(i) = \frac{1}{12}(i(i-1)(2i-1)K_Y^3$$

$$- (iK_Y - E)((i-1)K_Y - E)((2i-1)K_Y - 2E))$$

$$+ \frac{1}{12}(iK_Y - (iK_Y - E))c_2(Y) + \sum_{\iota \in I}(A_\iota(i) - A_\iota(i - e_\iota)).$$

Since $K_Y E = \pi^* K_X \cdot E = 0$ in $H^4(Y, \mathbf{Q})$, this is computed as

$$\delta(i) = \frac{1}{6}E^3 + \frac{1}{12}Ec_2(Y) + \sum_{\iota \in I}(A_\iota(i) - A_\iota(i - e_\iota)).$$

Using Lemma 2.6.9(i), one obtains the following description.

Lemma 2.6.10 $\delta(i+1) - \delta(i) = \sum_{\iota \in I}(B_\iota(ib_\iota) - B_\iota(ib_\iota - v_\iota))$.

Lemma 2.6.11 *The index r of $x \in X$ equals the least common multiple (lcm) of r_ι for all $\iota \in I$, where we define the lcm to be one if I is empty.*

Proof Since $rK_Y = r\pi^* K_X$ is Cartier, the index r_ι of $q_\iota \in Y_\iota$ is a factor of r for all $\iota \in I$. On the other hand, the function $\sum_{\iota \in I}(B_\iota(ib_\iota) - B_\iota(ib_\iota - v_\iota))$ for $i \in \mathbf{Z}$ is periodic and the period divides the lcm of r_ι. Hence by Lemma 2.6.10, the period r of $\delta(i+1) - \delta(i)$ divides the lcm of r_ι. $\square$

Proof of Theorem 2.6.6   We have $\pi\colon Y \to X$ and E in Proposition 2.6.7 and Y is compactified. Let r denote the index of $x \in X$. Lemma 2.6.10 for $i = 0$ is

$$\sum_{\iota \in I} B_\iota(v_\iota) = 1$$

unless $r = 1$. Since $v_\iota/4 \le B_\iota(v_\iota) < v_\iota/2$ by $v_\iota \le r_\iota/2$, $J' = \{v_\iota\}_{\iota \in I}$ is $\{1,1,1,1\}$, $\{1,1,2\}$, $\{1,1,1\}$, $\{2,2\}$, $\{1,3\}$, $\{1,2\}$, $\{4\}$, $\{3\}$ or $\emptyset$.

For each of these candidates for J', one can solve the equation $\sum_\iota B_\iota(v_\iota) = 1$ for r_ι explicitly and obtain the list of the solutions $J = \{(r_\iota, v_\iota)\}_{\iota \in I}$ in Table 2.1, in which r is determined by Lemma 2.6.11. For example, suppose the case $J' = \{1,2\}$ and write $J = \{(r_1, 1), (r_2, 2)\}$. Then the equation becomes $1/r_1 + 4/r_2 = 1$ and $(r_1, r_2) = (2,8), (3,6)$ or $(5,5)$ as in the table.

Table 2.1 *Candidates for J*

J	r	J	r	J	r
$(2,1), (2,1), (2,1), (2,1)$	2	$(4,2), (4,2)$	4	$(8,4)$	8
$(2,1), (2,1), (4,2)$	4	$(2,1), (6,3)$	6	$(9,3)$	9
$(2,1), (3,1), (6,1)$	6	$(2,1), (8,2)$	8	$\emptyset$	1
$(2,1), (4,1), (4,1)$	4	$(3,1), (6,2)$	6		
$(3,1), (3,1), (3,1)$	3	$(5,1), (5,2)$	5		

For each candidate for J in Table 2.1, one can compute the right-hand side of the equality in Lemma 2.6.10 explicitly. It must coincide with $\delta(i+1) - \delta(i)$, which occurs only if $\tilde{J} = \{(r_\iota, v_\iota, b_\iota)\}_{\iota \in I}$ belongs to Table 2.2. For example when $J = \{(2,1), (3,1), (6,1)\}$ and $\tilde{J} = \{(2,1,1), (3,1,b_2), (6,1,b_3)\}$ with $b_2 = 1$ or 2 and $b_3 = 1$ or 5, Lemma 2.6.10 for $i = 1$ becomes $\delta(2) - \delta(1) = 1$, $1/3, 2/3, 0$ if $(b_2, b_3) = (1,1), (1,5), (2,1), (2,5)$ respectively. Hence (b_2, b_3) must be $(2,5)$, for which the equality in the lemma certainly holds for all i.

Table 2.2 *Classification of $\tilde{J}$*

$\tilde{J}$	r	$\tilde{J}$	r
$(2,1,1), (2,1,1), (2,1,1), (2,1,1)$	2	$(3,1,2), (3,1,2), (3,1,2)$	3
$(2,1,1), (3,1,2), (6,1,5)$	6	$(5,1,4), (5,2,3)$	5
$(2,1,1), (4,1,3), (4,1,3)$	4	$\emptyset$	1

Table 2.2 completes the theorem. Every case in the table occurs as remarked right after the statement of the theorem. □

We shall explain the examples for $r = 5$ and 6 due to Prokhorov [390, section 5]. Let S be a *del Pezzo surface*, namely a smooth projective surface with ample anti-canonical divisor $-K_S$. We assume that the *degree* $d = K_S^2$ is at least three. Then by the description in Theorem 8.1.3, $-K_S$ is very ample and it defines an embedding $S = S_d \subset \mathbf{P}^d$ as a projectively normal variety. Take the affine cone $Z = \operatorname{Spec} \bigoplus_{i \in \mathbf{N}} H^0(\mathscr{O}_S(-iK_S))$ as in Example 1.4.17, which is normal by the projective normality of $S \subset \mathbf{P}^d$. The line bundle $\varphi \colon B = \operatorname{Spec}_S \bigoplus_{i \in \mathbf{N}} \mathscr{O}_S(-iK_S) \to S$ is obtained as the blow-up $\pi \colon B \to Z$ at the vertex $z \in Z$. The exceptional divisor E of π is a section of φ, which is isomorphic to S.

Lemma 2.6.12   *The cone Z is canonical of index one and π is crepant.*

Proof   The idea has appeared in Example 1.4.17. The restriction $\mathscr{O}_B(-E)|_E$ is isomorphic to $\mathscr{O}_S(1) = \mathscr{O}_S(-K_S)$ via $E \simeq S$, by which $\mathscr{O}_E(K_E) \simeq \mathscr{O}_B(E)|_E$. Then $K_B|_E \sim 0$ by the adjunction $(K_B + E)|_E = K_E$, from which $K_B \sim 0$. Hence $K_Z = \pi_* K_B \sim 0$ and $K_B = \pi^* K_Z$. These imply the lemma.   □

From now on, we assume that the cyclic group $\mathbf{Z}_r$ acts on S. This induces an action on $H^0(\mathscr{O}_S(-K_S))$, by which $\mathbf{Z}_r$ acts on Z and B so that π and φ are $\mathbf{Z}_r$-equivariant. Since the action of $\mathbf{Z}_r$ on $H^0(\mathscr{O}_S(-K_S))$ is diagonalisable, B is covered by invariant affine charts. Thus the geometric quotient $B/\mathbf{Z}_r$ as well as $Z/\mathbf{Z}_r$ exists from Theorem 2.2.3. This implies the existence of $E/\mathbf{Z}_r$ and hence that of $S/\mathbf{Z}_r$.

The action defines a group homomorphism $\alpha \colon \mathbf{Z}_r \to \operatorname{Aut} B$ to the automorphism group of B. On the other hand, the multiplicative group $\mathbf{G}_m = \mathbf{C}^\times$ acts on $H^0(\mathscr{O}_S(-K_S))$ naturally by multiplication, which induces an action σ_β of $\mathbf{G}_m$ on B. Fixing an embedding $\mathbf{Z}_r \hookrightarrow \mathbf{G}_m$, we restrict σ_β to an action of $\mathbf{Z}_r$ on B and express it by a group homomorphism $\beta \colon \mathbf{Z}_r \to \operatorname{Aut} B$.

The two actions α and β are commutative by construction. Hence one can define a new action by their product $\gamma = \alpha \cdot \beta \colon \mathbf{Z}_r \to \operatorname{Aut} B$, that is,

$$\gamma(g) = \alpha(g)\beta(g) = \beta(g)\alpha(g) \in \operatorname{Aut} B$$

for $g \in \mathbf{Z}_r$. In order to distinguish it from the original action, we say that $\gamma(\mathbf{Z}_r)$ acts on B. Then $\gamma(\mathbf{Z}_r)$ also acts on Z and S by which π and φ are $\gamma(\mathbf{Z}_r)$-equivariant. Note that the new actions on E and S are the same as the original ones. Let $Y = B/\gamma(\mathbf{Z}_r)$, $X = Z/\gamma(\mathbf{Z}_r)$ and $x = z/\mathbf{Z}_r$ as in the diagram

$$
\begin{array}{ccccc}
E/\mathbf{Z}_r & \subset & Y = B/\gamma(\mathbf{Z}_r) & \longrightarrow & S/\mathbf{Z}_r \\
\downarrow & & \downarrow & & \\
x = z/\mathbf{Z}_r & \in & X = Z/\gamma(\mathbf{Z}_r) & &
\end{array}
$$

Theorem 2.6.13 ([390, proposition 5.5]) *Suppose that $\mathbf{Z}_r$ acts on S freely outside finitely many points and that the quotient $S/\mathbf{Z}_r$ is Du Val. Then $x \in X$ is an isolated strictly canonical singularity of index r.*

Proof Let $p \in S$ be a point such that the stabiliser G_p in $\mathbf{Z}_r$ is not trivial. Then $G_p \simeq \mathbf{Z}_s$ for some factor s of r. By Theorem 2.2.4 and Remark 2.2.5, there exists an analytic neighbourhood $p \in \mathfrak{D}^2$ in S on which the action of G_p is diagonalised with respect to coordinates x_1, x_2. To be precise, there exists a primitive character $\chi \colon G_p \to \mathbf{C}^\times$ such that $g \in G_p$ sends x_i to $\chi(g)^{a_i} x_i$ for some integer a_i. Since $\mathbf{Z}_r$ acts freely on a punctured neighbourhood at p, a_1 and a_2 are coprime to s. It follows that $S/\mathbf{Z}_r$ has a cyclic quotient singularity of type $\frac{1}{s}(a_1, a_2)$ at the image of p. By assumption, it is Du Val and hence $a_2 \equiv -a_1 \bmod s$.

Take the expression

$$\varphi^{-1}(p) = \mathrm{Spec}(\bigoplus_{i \in \mathbf{N}} \mathscr{O}_S(-iK_S) \otimes \mathscr{O}_S/\mathfrak{m}) \simeq \mathrm{Spec}\, \mathbf{C}[t] = \mathbf{A}^1$$

with the maximal ideal $\mathfrak{m}$ in $\mathscr{O}_S$ defining p, in which the analytic generator $t = (dx_1 \wedge dx_2)^{-1}$ at p of $\mathscr{O}_S(-K_S)$ is regarded as the indeterminate. By the original action of G_p induced by α, an element $g \in G_p$ sends t to $(d\chi(g)^{a_1} x_1 \wedge d\chi(g)^{-a_1} x_2)^{-1} = t$. This means that G_p acts on $\varphi^{-1}(p)$ trivially by α.

Hence the action of G_p on $\varphi^{-1}(p)$ defined by γ is the same as that by β. It is free on $\varphi^{-1}(p)$ outside the closed point $q = \varphi^{-1}(p) \cap E$. Thus the germ $q \in \varphi^{-1}(p) \simeq \mathbf{A}^1$ has a coordinate x_3 such that $g \in G_p$ sends x_3 to $\chi(g)^{a_3} x_3$ for some integer a_3 coprime to s. The action of $\gamma(G_p)$ on the tangent space $T_{B,q}$ is therefore diagonalised as $(\chi^{a_1}, \chi^{-a_1}, \chi^{a_3})$. It follows that $Y = B/\gamma(\mathbf{Z}_r)$ has a cyclic quotient singularity of type $\frac{1}{s}(a_1, -a_1, a_3)$ at the image of q, which is terminal.

By the above analysis of the germ $p \in S$, we have seen that $\gamma(\mathbf{Z}_r)$ acts on B freely outside finitely manly points and that Y is terminal with index being the lcm of the orders of the stabilisers G_p for all $p \in S$. We claim that Y is of index r. It suffices to show the existence of a point in S which is fixed by the automorphism $\sigma \colon S \to S$ defined by a generator of $\mathbf{Z}_r$. This follows from the holomorphic Lefschetz fixed-point formula in Theorem 2.5.15 because $L(\sigma, \mathscr{O}_S) = \mathrm{tr}\,\sigma^*|_{H^0(\mathscr{O}_S)} = 1$ by the Kodaira vanishing $H^1(\mathscr{O}_S) = H^2(\mathscr{O}_S) = 0$.

The image X of the terminal threefold Y has an isolated singularity at x. The relation $K_B = \pi^* K_Z$ in Lemma 2.6.12 descends to $K_Y = \pi_X^* K_X$ for $\pi_X \colon Y \to X$. Thus $x \in X$ is strictly canonical. Since K_Z is Cartier by Lemma 2.6.12, rK_X is Cartier on $X = Z/\gamma(\mathbf{Z}_r)$. By $K_Y = \pi_X^* K_X$, the index r of Y divides that of X. Hence X is also of index r. $\square$

Example 2.6.14 Consider $\mathbf{P}^1 \times \mathbf{P}^1 \times \mathbf{P}^1$ with homogeneous coordinates (x_0, x_1), (y_0, y_1), (z_0, z_1). Let S be the surface in $\mathbf{P}^1 \times \mathbf{P}^1 \times \mathbf{P}^1$ defined by $x_0 y_0 z_0 - x_1 y_1 z_1$. It is a del Pezzo surface of degree six. Define an automorphism σ of S by

$$\sigma([x_0, x_1], [y_0, y_1], [z_0, z_1]) = ([y_1, y_0], [z_1, z_0], [x_1, x_0]).$$

It has order six in Aut S, defining an action of $\mathbf{Z}_6$ on S. Every point in S with non-trivial stabiliser belongs to one of the orbits of $p_1 = ([1, 1], [1, 1], [1, 1])$, $p_2 = ([1, \omega], [1, \omega], [1, \omega])$ and $p_3 = ([1, 1], [1, -1], [1, -1])$ with stabilisers $\mathbf{Z}_6$, $\mathbf{Z}_3$ and $\mathbf{Z}_2$ respectively, where ω is a primitive cubic root of unity.

The quotient $S/\mathbf{Z}_6$ is Du Val of type A_5, A_2, A_1 at the images of p_1, p_2, p_3 respectively. To see this, we take coordinates $(u_1, u_2) = (x_1 x_0^{-1}, y_1 y_0^{-1})$ on the chart $(x_0 y_0 z_0 \neq 0)$ in S, for which $\sigma^*(u_1, u_2) = (u_2^{-1}, u_1 u_2)$. The two-form $\theta = du_1 \wedge du_2$ is pulled back as $\sigma^*\theta = d(u_2^{-1}) \wedge d(u_1 u_2) = u_2^{-1}\theta$. Let p be one of p_1, p_2, p_3 in $(x_0 y_0 z_0 \neq 0)$, which has stabiliser $\mathbf{Z}_s$ with $s = 6, 3, 2$ respectively. For example, p_1 is the point $(u_1, u_2) = (1, 1)$. Let θ_p be the image in $\omega_S \otimes k(p)$ of θ for the residue field $k(p) \simeq \mathbf{C}$ at p regarded as a skyscraper sheaf. Then $(\sigma^{6/s})^*\theta_p = \theta_p$ by direct computation using $\sigma^*\theta = u_2^{-1}\theta$. Hence the automorphism on the tangent space $T_{S,p}$ induced by $\sigma^{6/s}$ is diagonalisable to (ζ, ζ^{-1}) with a primitive s-th root ζ of unity. Thus $S/\mathbf{Z}_6$ has a cyclic quotient singularity of type $\frac{1}{s}(1, -1)$ at the image of p, which is of type A_{s-1}. By Theorem 2.6.13, $x \in X = Z/\gamma(\mathbf{Z}_6)$ is strictly canonical of index six.

Example 2.6.15 Consider $\mathbf{P}^2$ with homogeneous coordinates x_0, x_1, x_2. Let $\pi \colon S \to \mathbf{P}^2$ be the blow-up at the four points $p_1 = [1, 0, 0]$, $p_2 = [0, 1, 0]$, $p_3 = [0, 0, 1]$ and $p_4 = [1, 1, 1]$. Then S is a del Pezzo surface of degree five. Define a birational self-map $\sigma_0 \colon \mathbf{P}^2 \dashrightarrow \mathbf{P}^2$ by

$$\sigma_0[x_0, x_1, x_2] = [x_0(x_2 - x_1), x_2(x_0 - x_1), x_0 x_2].$$

It induces an automorphism σ of S. Indeed, the indeterminacy points of σ_0 are p_1, p_2, p_3 and eliminated by π, that is, $\sigma_0 \circ \pi$ is a morphism. The contraction $\sigma_0 \circ \pi \colon S \to \mathbf{P}^2$ is not isomorphic over the points p_1, p_2, p_3, p_4 in $\mathbf{P}^2$. By [35, proposition II.8], it factors through the blow-up π at these four points, that is, $\sigma_0 \circ \pi = \pi \circ \sigma$ for a morphism σ.

The automorphism σ has order five in Aut S, defining an action of $\mathbf{Z}_5$ on S. More precisely, let e_i be the curve in S mapped to p_i and let l_{ij} be the strict transform in S of the line in $\mathbf{P}^2$ joining p_i and p_j. Then σ sends these curves isomorphically as $e_1 \to l_{14} \to l_{23} \to e_2 \to l_{12} \to e_1$ and $e_3 \to l_{24} \to l_{34} \to l_{13} \to e_4 \to e_3$. Using this, one sees that σ fixes exactly the two points $q_+ = [-1 + \sqrt{5}, 3 - \sqrt{5}, 2]$ and $q_- = [-1 - \sqrt{5}, 3 + \sqrt{5}, 2]$.

Take coordinates $(u_1, u_2) = (x_0 x_2^{-1}, x_1 x_2^{-1})$ on the complement in S of $\bigcup_{i,j} e_i \cup l_{ij}$, for which $\sigma_0^*(u_1, u_2) = (1 - u_2, 1 - u_1^{-1} u_2)$. The two-form $\theta = du_1 \wedge du_2$ is pulled back as $\sigma_0^* \theta = d(1 - u_2) \wedge d(1 - u_1^{-1} u_2) = u_1^{-2} u_2 \theta$. Since $q_\pm$ is the point $(u_1, u_2) = ((-1 \pm \sqrt{5})/2, (3 \mp \sqrt{5})/2)$ having the relation $u_1^{-2} u_2 = 1$, the image $\theta_{q_\pm}$ in $\omega_S \otimes k(q_\pm)$ of θ, where $k(q_\pm) \simeq \mathbf{C}$ is the residue field, satisfies $\sigma^* \theta_{q_\pm} = \theta_{q_\pm}$. As in the preceding example, this implies that $S/\mathbf{Z}_5$ is Du Val of type A_4 at the images of q_+ and q_-. By Theorem 2.6.13, $x \in X = Z/\gamma(\mathbf{Z}_5)$ is strictly canonical of index five.

3

Divisorial Contractions to Points

A divisorial contraction is a contraction in the MMP which has exceptional locus of codimension one. In dimension three, it contracts a prime divisor to a point or a curve. In this chapter, we furnish a systematic classification of threefold divisorial contractions which contract the divisor to a point, mainly due to the author. We regard the contraction as a divisorial extraction from a terminal threefold singularity. This perspective has turned out to be natural with the development of the Sarkisov program.

The classification is founded on a numerical one obtained by the singular Riemann–Roch formula, which makes a list of all the possibilities for the basket of fictitious singularities along the exceptional divisor. The list consists of a series of ordinary types and several exceptional types. The discrepancy a/n in the case of exceptional type is very small, where n is the index of the target singularity. In fact it equals the minimum $1/n$ if n is at least two and coprime to a. When a and n have a common factor, it produces a cyclic cover of the contraction which imposes a very strict condition upon the pair of a and n.

The general elephant conjecture plays a central role in the geometric classification. We establish it for the divisorial contraction by a delicate analysis of a tree of rational curves realised as the intersection of a certain surface with the exceptional divisor. The analysis further describes the general elephant as a partial resolution of the Du Val singularity. This narrows down the candidates for the target singularity.

The singular Riemann–Roch formula computes the dimensions of parts in lower degrees of the graded ring for the contraction restricted to the exceptional divisor. We recover the graded ring from these numerical data by finding orbifold coordinates compatible with the structure of the ring. As a result, we nearly conclude that the divisorial contraction is a certain weighted blow-up of the cyclic quotient of a complete intersection inside a smooth fivefold. Examples are collected in accordance with the classification.

3.1 Identification of the Divisor

Recall that a *divisorial contraction* $\pi: Y \to X$ is by Definition 1.3.22 a birational contraction between terminal varieties such that $-K_Y$ is relatively ample and such that the exceptional locus E is a prime divisor. The divisor $-E$ is relatively ample by the expression $K_Y = \pi^* K_X + dE$ with positive discrepancy d.

Threefold divisorial contractions were originally studied as contractions from a given threefold. Mori classified them when the threefold is smooth and Cutkosky extended the result to the case when it is Gorenstein. Below is the result in the case when the divisor is contracted to a point. The other case will be treated in Theorem 4.1.9.

Theorem 3.1.1 (Mori [332], Cutkosky [98]) *Let $\pi: E \subset Y \to x \in X$ be an elementary divisorial contraction from a Gorenstein threefold which contracts the divisor E to a point x. Then $x \in X$ is smooth, cA_1 or a quotient singularity of type $\frac{1}{2}(1, 1, 1)$, and π is the blow-up of X at x.*

We regard the contraction π in the theorem as a *divisorial extraction* from the fixed threefold singularity $x \in X$. This point of view is more natural in the context of the Sarkisov program and enables us to work on the analytic germ at x. For this reason, we relax the condition for π to be elementary. Remark that Definition 1.3.22 is not consistent once one shifts to the analytic category, because the exceptional divisor E may become reducible on the analytic germ. This does not matter as far as E is contracted to a point.

A divisorial extraction from a fixed variety is determined by the valuation given by the exceptional divisor.

Lemma 3.1.2 *Let X_1 and X_2 be normal varieties projective over a variety S. Let $f: X_1 \dashrightarrow X_2$ be a small birational map over S. If there exists a relatively ample $\mathbf{R}$-divisor A_1 on X_1 transformed to a relatively ample $\mathbf{R}$-divisor $A_2 = f_* A_1$ on X_2, then f is an isomorphism.*

Proof Take a common resolution $\mu_i: Y \to X_i$ with $\mu_2 = f \circ \mu_1$. The difference between $\mu_1^* A_1$ and $\mu_2^* A_2$ is μ_i-exceptional for $i = 1, 2$ and hence they coincide by the negativity lemma in Theorem 1.3.9. Further, μ_1 and μ_2 are both the ample models of $\mu_1^* A_1 = \mu_2^* A_2$ over S. Thus the assertion follows from the uniqueness of the ample model in Lemma 1.5.3. $\square$

In this section, we determine a divisorial extraction from a certain threefold singularity $x \in X$. One case is when $x \in X$ is a quotient singularity by Kawamata, and the other is when it is an ordinary double point by Corti. This will demonstrate how to identify the divisor with the aid of the next lemma.

Lemma 3.1.3 *Let* $\pi \colon Y \to X$ *be a divisorial contraction to the germ* $x \in X$. *Then* $x \in X$ *admits a canonical pair* (X, Δ) *with a* **Q**-*divisor* Δ *such that* (Y, Δ_Y) *is terminal for the strict transform* Δ_Y *of* Δ *and such that* π *is crepant with respect to* $K_Y + \Delta_Y$.

Proof Since $-K_Y$ is π-ample, the pair $(Y, l^{-1}H)$ is terminal for the general member H of $|-lK_Y|$ with a sufficiently large and divisible integer l. Then $K_Y + l^{-1}H \sim_{\mathbf{Q}} 0$ and it is pushed forward to X as $K_X + \Delta \sim_{\mathbf{Q}} 0$ for $\Delta = l^{-1}\pi_* H$. The pair (X, Δ) has the required property by $K_Y + l^{-1}H = \pi^*(K_X + \Delta)$. □

Let $x \in X$ be a terminal quotient threefold singularity of index r. By Corollary 2.3.20 and Theorem 2.4.4, it is analytically isomorphic to the germ $o \in \mathfrak{D}^3/\mathbf{Z}_r(a, -a, 1)$ with orbifold coordinates x_1, x_2, x_3 in which $0 < a < r$ and a is coprime to r.

Definition 3.1.4 The weighted blow-up of X with $\mathrm{wt}(x_1, x_2, x_3) = \frac{1}{r}(a, r - a, 1)$ is called the *Kawamata blow-up* of X at x. It exists in the algebraic category and is independent of the analytic isomorphism $x \in X \simeq \mathfrak{D}^3/\mathbf{Z}_r(a, -a, 1)$.

Let $\pi \colon B \to X$ be the Kawamata blow-up and let F be the exceptional divisor. Then $K_B = \pi^* K_X + (1/r)F$ and B is covered by the three charts $U_i = (x_i \neq 0)$ for $i = 1, 2, 3$, for which $U_1 \simeq \mathfrak{D}^3/\mathbf{Z}_a(-r, r, 1)$ and $U_2 \simeq \mathfrak{D}^3/\mathbf{Z}_{r-a}(r, -r, 1)$ analytically. Thus B has at most two terminal quotient singularities, which are of index less than r, and π is a divisorial contraction.

Kawamata proved that π is the only divisorial contraction to the germ $x \in X$. The result includes the non-existence of a threefold divisorial contraction which contracts the divisor to a curve through the quotient singularity.

Theorem 3.1.5 (Kawamata [241]) *Let* $\pi \colon Y \to X$ *be a threefold divisorial contraction to the germ* $x \in X$ *of a terminal quotient singularity. Then* π *is the Kawamata blow-up.*

We shall realise the exceptional divisor by a finite sequence of blow-ups. The blow-up does not increase the order.

Theorem 3.1.6 ([187, III lemmata 7, 8]) *Let* X *be a smooth variety and let* Z *be a smooth closed subvariety of* X. *Let* $\mathscr{I}$ *be a coherent ideal sheaf in* $\mathcal{O}_X$. *Then* $\mathrm{ord}_Z \mathscr{I} \leq \mathrm{ord}_W \mathscr{I}$ *for any closed subvariety* W *of* Z.

Consider the blow-up X' *of* X *along* Z *and let* E *denote the exceptional divisor. Define the weak transform* $\mathscr{I}' = \mathscr{I}\mathcal{O}_{X'}((\mathrm{ord}_Z \mathscr{I})E)$ *in* $\mathcal{O}_{X'}$ *of* $\mathscr{I}$. *Then* $\mathrm{ord}_{Z'} \mathscr{I}' \leq \mathrm{ord}_Z \mathscr{I}$ *for any closed subvariety* Z' *of* X' *mapped onto* Z.

Lemma 3.1.7 *Let* $x \in X$ *be a terminal quotient threefold singularity of index* r. *Let* B *be the Kawamata blow-up of* X *at* x *and let* E *be the exceptional* *divisor. Let* Δ *be an effective* **Q**-*divisor on* X, *which is* **Q**-*Cartier, and take the strict transform* Δ_B *in* B. *Let* Z *be a closed subvariety of* E.

(i) *If* Z *is a curve or a smooth point of* B, *then* $\mathrm{ord}_Z \Delta_B \le r \, \mathrm{ord}_E \Delta$.

(ii) *If* Z *is a singular point of* B *of index* s, *then* $s \, \mathrm{ord}_F \Delta_B \le r \, \mathrm{ord}_E \Delta$ *where* F *is the divisor obtained by the Kawamata blow-up of* B *at* Z.

Proof Multiplying Δ, we may assume that Δ and Δ_B are integral and Cartier. Passing through the analytic category, we can work on the germ $x \in X = \mathbb{A}^3/\mathbb{Z}_r(a, -a, 1)$ with orbifold coordinates x_1, x_2, x_3 where $0 < a < r$. If Z is a curve, then by Theorem 3.1.6, $\mathrm{ord}_Z \Delta_B \le \mathrm{ord}_y \Delta_B$ for a general closed point y in Z. Replacing Z by y, we reduce the statement (i) to the case when Z is a smooth point of B. In this case,

$$\mathrm{ord}_Z \Delta_B = \mathrm{mult}_Z \Delta_B \le \mathrm{mult}_Z \Delta_B|_E \le \text{w-deg}\,\Delta_B|_E = r \, \mathrm{ord}_E \Delta,$$

where w-deg stands for the weighted degree in $E \simeq \mathbf{P}(a, r - a, 1)$.

In order to see (ii), by symmetry, we may assume that Z is the origin of the chart $U_1 \simeq \mathfrak{D}^3/\mathbb{Z}_a(-r, r, 1)$ with orbifold coordinates $(y_1, y_2, y_3) = (x_1^{r/a}, x_2 x_1^{-(r-a)/a}, x_3 x_1^{-1/a})$. Then $a \, \mathrm{ord}_F \Delta_B$ equals the weighted order w of Δ_B with respect to $\mathrm{wt}(y_1, y_2, y_3) = (a - \bar{r}, \bar{r}, 1)$ for the residue $\bar{r}$ of r modulo a. Since $(a - \bar{r}, \bar{r}, 1) \le (\infty, r - a, 1)$ componentwise, one has $a \, \mathrm{ord}_F \Delta_B = w \le w_E$ for the weighted order w_E of $\Delta_B|_E$ with respect to $\mathrm{wt}(y_2, y_3) = (r - a, 1)$. Note that E is defined at Z by y_1. It is estimated as $w_E \le \text{w-deg}\,\Delta_B|_E = r \, \mathrm{ord}_E \Delta$. $\square$

Let X be a threefold with only terminal quotient singularities and let E be a divisor exceptional over X. By a *tower* on X with respect to E, we mean a finite sequence of contractions $X_{i+1} \to X_i$ for $0 \le i < n$ between threefolds with only terminal quotient singularities, where $X_0 = X$, such that

- if the centre $Z_i = c_{X_i}(E)$ is a curve or a smooth point of X_i, then X_{i+1} is about the generic point of Z_i the blow-up of X_i along Z_i,
- if Z_i is a singular point of X_i, then X_{i+1} is the Kawamata blow-up of X_i at Z_i,
- E_{i+1} is the unique prime divisor on X_{i+1} mapped onto Z_i and
- $E_n = E$ as a divisor over X.

The inductive construction of the tower terminates after finitely many steps. Indeed, we may start from a smooth threefold X because the Kawamata blow-up generates only singularities of lower index. When X is smooth, the log discrepancy $a_{E_i}(X)$ is greater than i and the process terminates after at most $a_E(X) - 1$ steps.

Corollary 3.1.8 *Notation as above. Consider an effective* **Q**-*divisor* Δ *on* X. *Let* r_i *be the index of* X_i *at the generic point of* Z_i *and let* Δ_i *be the strict* transform in X_i of Δ. *Then the numbers* $c_i = r_i \operatorname{ord}_{E_{i+1}} \Delta_i$ *for* $0 \le i < n$ *form a non-increasing sequence.*

Proof Notice that c_i equals $\operatorname{ord}_{Z_i} \Delta_i$ unless Z_i is a singular point. If Z_i is a curve or a smooth point, then $c_{i+1} = \operatorname{ord}_{Z_{i+1}} \Delta_{i+1} \le \operatorname{ord}_{Z_i} \Delta_i = c_i$ by Theorem 3.1.6. If Z_i is a singular point, then $c_{i+1} \le c_i$ by Lemma 3.1.7.   □

Proof of Theorem 3.1.5   We build a tower $X_n \to \cdots \to X_0 = X$ with respect to the exceptional divisor E of π and use the notation Z_i and E_i above. We construct a canonical pair (X, Δ) from π by Lemma 3.1.3. Then E is the only divisor exceptional over X such that $a_E(X, \Delta) = 1$. Following Corollary 3.1.8, we write r_i for the index of X_i at the generic point of Z_i and write Δ_i for the strict transform in X_i of Δ. By Lemma 3.1.2, it suffices to prove that $E = E_1$ as a divisor over X or equivalently $a_{E_1}(X, \Delta) = 1$.

First consider the case $\pi(E) = x$. Then $X_1 \to X$ is the Kawamata blow-up. It follows from the canonicity of (X, Δ) that $a_{E_1}(X, \Delta) = 1 + 1/r_0 - \operatorname{ord}_{E_1} \Delta_0 \ge 1$, that is, $r_0 \operatorname{ord}_{E_1} \Delta_0 \le 1$. Then by Corollary 3.1.8, $r_i \operatorname{ord}_{E_{i+1}} \Delta_i \le 1$ for all i. In particular, $a_{E_{i+1}}(X_i, \Delta_i) \ge 1 + 1/r_i - \operatorname{ord}_{E_{i+1}} \Delta_i \ge 1$. Let $a_i = a_{E_i}(X, \Delta) \ge 1$. Since $X_i \to X$ is crepant with respect to $K_{X_i} + \Delta_i - \sum_{j=1}^{i}(a_j - 1)E_j^i$ for the strict transform E_j^i of E_j, one has

$$a_{i+1} - 1 = a_{E_{i+1}}(X_i, \Delta_i) - 1 + \sum_{j=1}^{i}(a_j - 1)\operatorname{ord}_{E_{i+1}} E_j^i \ge \frac{a_i - 1}{r_i}$$

and hence $0 \le a_1 - 1 \le r_1 \cdots r_{n-1}(a_n - 1) = 0$ by $a_n = a_E(X, \Delta) = 1$. This shows that $a_{E_1}(X, \Delta) = a_1 = 1$.

One can easily modify this argument for the case when $Z = \pi(E)$ is a curve, once one verifies the inequality $\operatorname{ord}_Z \Delta \le r \operatorname{ord}_F \Delta$ for the index r of X and the divisor F obtained by the Kawamata blow-up of X at x. To verify this, we take the index-one cover $\bar{x} \in \bar{X} \to x \in X$, the pull-back $\bar{\Delta}$ of Δ and the inverse image $\bar{Z}$ of Z. By successive blow-ups at a point, one obtains a strong resolution $Y \to \bar{X}$ in which the strict transform Z_Y of $\bar{Z}$ becomes smooth. Then for a point $y \in Z_Y$ above $\bar{x}$ and the strict transform Δ_Y in Y of $\bar{\Delta}$, Theorem 3.1.6 gives $\operatorname{ord}_Z \Delta = \operatorname{ord}_{Z_Y} \Delta_Y \le \operatorname{ord}_y \Delta_Y \le \operatorname{ord}_{\bar{x}} \bar{\Delta}$. By the analytic identification $x \in X = \mathfrak{D}^3/\mathbf{Z}_r(a, -a, 1)$ where $0 < a < r$, the order $\operatorname{ord}_{\bar{x}}$ is defined by the weights $(1, 1, 1)$ whilst $r \operatorname{ord}_F$ is by $(a, r - a, 1)$. Hence $\operatorname{ord}_{\bar{x}} \bar{\Delta} \le r \operatorname{ord}_F \Delta$. These are combined into the inequality $\operatorname{ord}_Z \Delta \le r \operatorname{ord}_F \Delta$.   □

A threefold ordinary double point has already been discussed in Examples 1.2.3 and 1.3.18. Corti applied the *connectedness lemma*.

Definition 3.1.9 An *ordinary double point* is a singularity analytically iso-morphic to the hypersurface singularity defined in $\mathfrak{D}^{n+1}$ by $x_1^2 + \cdots + x_{n+1}^2$. It is a terminal singularity if $n \geq 3$.

Theorem 3.1.10 (Corti [93, theorem 3.10]) *Let $\pi \colon E \subset Y \to x \in X$ be a threefold divisorial contraction which contracts the divisor E to an ordinary double point x. Then π is the blow-up of X at x.*

Proof Passing through the analytic category, we may assume that $x \in X$ is the hypersurface in $\mathbf{A}^4$ given by $x_1 x_2 + x_3 x_4$. Take the blow-up $\varphi \colon B \to X$ at x and let F be the exceptional divisor. Then $F = (x_1 x_2 + x_3 x_4 = 0) \subset \mathbf{P}^3$. The reducible hypersurface in X given by x_4 consists of two prime divisors $S_i = (x_i = x_4 = 0)$ for $i = 1, 2$. The blow-up φ factors through the blow-up X_i of X along S_i with a morphism $\varphi_i \colon B \to X_i$ which contracts $F \simeq \mathbf{P}^1 \times \mathbf{P}^1$ to $\mathbf{P}^1$ by one of the rulings. In fact, $X_1 \to X \leftarrow X_2$ is the Atiyah flop in Example 1.3.18. We write S_{iB} for the strict transform in B of S_i. The restriction $f_i = S_{iB}|_F$ is a section of $\varphi_i|_F \colon \mathbf{P}^1 \times \mathbf{P}^1 \to \mathbf{P}^1$. Possibly permuting x_1, x_2, x_3, x_4, we may and shall assume that $f_1 \cup f_2$ does not contain the centre $Z = c_B(E)$.

We construct a canonical pair (X, Δ) from π by Lemma 3.1.3, for which $a_E(X, \Delta) = 1$. The log discrepancy $a = a_F(X, \Delta)$ satisfies $1 \leq a \leq a_F(X) = 2$. By Lemma 3.1.2, it suffices to show that $a = 1$.

Let $D_B = \Delta_B + S_{1B} + S_{2B} + (2 - a)F$ with the strict transform Δ_B of Δ. Then $K_B + D_B = \varphi^*(K_X + \Delta + S_1 + S_2)$. We shall analyse the non-klt locus N of the pair (B, D_B). It contains Z since $a_E(X, \Delta + S_1 + S_2) = 1 - \operatorname{ord}_E x_4 \leq 0$. Hence $Z \cup S_{1B} \cup S_{2B} \subset N$. By the connectedness lemma in Theorem 1.4.23, the locus $N|_F$ is connected along every fibre of $\varphi_i|_F$. Because $Z \cup f_1 \cup f_2 \subset N|_F$ with $Z \not\subset f_1 \cup f_2$, the only candidate for $N|_F$ is the whole F. This means that F has coefficient at least one in the boundary D_B, that is, $a \leq 1$. Thus $a = 1$. □

Remark 3.1.11 If the algebraic germ $x \in X$ above is not $\mathbf{Q}$-factorial, then the relative Picard number $\rho(B/X)$ remains two in the algebraic category. However if it is $\mathbf{Q}$-factorial, then $\rho(B/X)$ becomes one and S_i cannot extend to a divisor on the algebraic germ. In fact, the blow-up $X_i \to X$ along S_i does not extend to a morphism in the algebraic category.

3.2 Numerical Classification

We proceed to the systematic classification [223], [224] of threefold divisorial contractions which contract the divisor to a point. We set up the framework.

Framework 3.2.1 Let $\pi: E \subset Y \to x \in X$ be the germ of a threefold divisorial contraction which contracts the divisor E to a point x. We let n denote the index of $x \in X$ and write a/n for the *discrepancy* of E so that

$$K_Y = \pi^* K_X + \frac{a}{n} E.$$

Let $\{q_\iota$ of type $\frac{1}{r_\iota}(1, -1, b_\iota)\}_{\iota \in I_0}$ be the basket of *fictitious singularities* in Definition 2.5.22 from the singularities of Y. For $\iota \in I_0$, we take an integer e_ι such that $E \sim e_\iota K_Y$ at q_ι by Theorem 2.4.17. Note that $n \equiv a e_\iota$ mod r_ι by the global relation $n K_Y \sim a E$. We shall assume that $v_\iota = \overline{e_\iota b_\iota} \le r_\iota/2$ by replacing b_ι by $r_\iota - b_\iota$ if necessary, where $\overline{l}$ is the residue of l modulo r_ι as in Notation 2.2.14. Let $I = \{\iota \in I_0 \mid \overline{e}_\iota \ne 0\}$ and $J = \{(r_\iota, v_\iota)\}_{\iota \in I}$. By abuse of notation, we identify I with the subset $\{q_\iota\}_{\iota \in I}$ of the basket and hence $q_\iota \in I$ means $\iota \in I$.

We shall study the divisorial contraction π in Framework 3.2.1 throughout the chapter. This section features the numerical data of π. We begin with a numerical classification of the singularities on Y.

Theorem 3.2.2 ([224, theorem 1.1]) *Notation as in Framework 3.2.1. Every contraction π is of one of the types in Tables 3.1 and 3.2. The case when J is of form $\{(r_1, 1), (r_2, 1)\}$ is divided so that π is of type o3 if J comes from two non-Gorenstein points of Y and of type e2 or e3 if J comes from one non-Gorenstein point of Y. Type e4 does not occur. Any other type has an example.*

Table 3.1 *Ordinary type*

type	J	$(a/n)E^3$
o1	$\emptyset$	2
o2	$(r, 1)$	$1 + 1/r$
o3	$(r_1, 1), (r_2, 1), r_1 \le r_2$	$1/r_1 + 1/r_2$

Table 3.2 *Exceptional type*

type	J	$(a/n)E^3$	type	J	$(a/n)E^3$
e1	$(r, 2)$	$4/r$	e9	$(5, 2), (3, 1)$	$2/15$
e2	$(r, 1), (r, 1)$	$2/r$	e10	$(5, 2), (4, 1)$	$1/20$
e3	$(2, 1), (4, 1)$	$3/4$	e11	$(6, 2), (2, 1)$	$1/6$
e4	$(6, 3)$	$1/2$	e12	$(7, 2), (2, 1)$	$1/14$
e5	$(7, 3)$	$2/7$	e13	$(2, 1), (2, 1), (r, 1)$	$1/r$
e6	$(8, 3)$	$1/8$	e14	$(2, 1), (3, 1), (3, 1)$	$1/6$
e7	$(4, 2), (r, 1)$	$1/r$	e15	$(2, 1), (3, 1), (4, 1)$	$1/12$
e8	$(5, 2), (2, 1)$	$3/10$	e16	$(2, 1), (3, 1), (5, 1)$	$1/30$

Definition 3.2.3 The divisorial contraction π is said to be of *ordinary type* if it belongs to Table 3.1 and of *exceptional type* if it belongs to Table 3.2.

Type e4 will be excluded by Theorems 3.2.8 and 3.2.9. Examples for other types will be given in the last section of the chapter. The theorem encodes no information on the singularities of Y at which E is Cartier for the reason that one can only compute the difference along E for Euler characteristics.

Definition 3.2.4 A non-Gorenstein point q of Y is called a *hidden non-Gorenstein point* if E is Cartier at q. From $nK_Y \sim aE$, the local index at a hidden non-Gorenstein point is a factor of n. In particular if $x \in X$ is of index one, then no hidden non-Gorenstein point of Y exists.

We prove the theorem by the singular Riemann–Roch formula.

Notation 3.2.5 For integers i and j, we define the divisor $D_{ij} = iK_Y + jE$ and take the natural exact sequence $0 \to \mathcal{O}_Y(D_{i,j-1}) \to \mathcal{O}_Y(D_{ij}) \to \mathcal{Q}_{ij} \to 0$, in which the quotient $\mathcal{Q}_{ij} = \mathcal{O}_E(D_{ij}|_E)$ is a Cohen–Macaulay $\mathcal{O}_E$-module by Proposition 2.2.23. Note that $\mathcal{Q}_{00} = \mathcal{O}_E$ and $\mathcal{Q}_{11} = \omega_E$. Define

$$d(i, j) = \chi(\mathcal{Q}_{ij}) = \chi(\mathcal{O}_Y(D_{ij})) - \chi(\mathcal{O}_Y(D_{i,j-1})),$$

where the latter expression makes sense after a compactification of $x \in X$. The number $d(i, j)$ is determined by the class of (i, j) in $\mathbf{Z}^2/\mathbf{Z}(n, -a)$.

Lemma 3.2.6 *If $ia/n + j \le a/n$, then $d(i, j) = \dim \pi_* \mathcal{O}_Y(D_{ij})/\pi_* \mathcal{O}_Y(D_{ij} - E)$ and in particular*

$$d(i, j) = \begin{cases} 0 & \text{if } 0 \le ia/n + j \le a/n \text{ with } (i, j) \notin \mathbf{Z}(n, -a), \\ 1 & \text{if } (i, j) \in \mathbf{Z}(n, -a). \end{cases}$$

Proof The first expression is a consequence of the Kawamata–Viehweg vanishing $R^p \pi_* \mathcal{O}_Y(D_{ij}) = 0$ for $ia/n + j \le a/n$ and $p \ge 1$. To see the second, we apply Proposition 2.5.9 to $\pi_* \mathcal{O}_Y(D_{ij})$. If $ia/n + j > -1$ or if $ia/n + j = -1$ and iK_X is not Cartier, then $\pi_* \mathcal{O}_Y(D_{ij}) = \mathcal{O}_X(iK_X)$. If $ia/n + j = -1$ and iK_X is Cartier, then $\pi_* \mathcal{O}_Y(D_{ij}) = \mathfrak{m}\mathcal{O}_X(iK_X)$ for the maximal ideal $\mathfrak{m}$ in $\mathcal{O}_X$ defining x. From these the second expression follows. $\square$

We compute $d(i, j)$ using

$$A_\iota(l) = -\bar{l}\frac{r_\iota^2 - 1}{12r_\iota} + \sum_{i=1}^{\bar{l}-1} \frac{\overline{ib}_\iota(r_\iota - \overline{ib}_\iota)}{2r_\iota}, \qquad B_\iota(l) = \frac{\bar{l}(r_\iota - \bar{l})}{2r_\iota}$$

in Definition 2.6.8 as in the proof of Theorem 2.6.6.

Lemma 3.2.7 *One has the formulae*

(i) $d(i,j) = \dfrac{1}{12}\Big(6\big(\tfrac{a}{n}i+j\big)^2 - 6\big(\tfrac{a}{n}+1\big)\big(\tfrac{a}{n}i+j\big) + \big(\tfrac{a}{n}+1\big)\big(\tfrac{a}{n}+2\big)\Big)E^3$

$\qquad\qquad + \dfrac{1}{12}Ec_2(Y) + \displaystyle\sum_{\iota\in I}(A_\iota(i+je_\iota) - A_\iota(i+(j-1)e_\iota)),$

(ii) $d(i+1,j) - d(i,j) = \big(\tfrac{a}{n}i+j-\tfrac{1}{2}\big)\dfrac{a}{n}E^3$

$\qquad\qquad + \displaystyle\sum_{\iota\in I}(B_\iota(ib_\iota + jv_\iota) - B_\iota(ib_\iota + (j-1)v_\iota)).$

Proof The formula (i) is a direct consequence of Theorem 2.5.23, whilst (ii) follows from (i) and Lemma 2.6.9(i). □

Proof of Theorem 3.2.2 By Lemma 3.2.6, $d(1,0) = 0$ and $d(0,0) = 1$. Hence Lemma 3.2.7(ii) for $(i,j) = (0,0)$ is

$$-1 = -\frac{1}{2}\Big(\frac{a}{n}E^3\Big) - \sum_{\iota\in I}\frac{v_\iota(r_\iota - v_\iota)}{2r_\iota}.$$

All the possibilities for J to satisfy this equality with positive E^3 are listed in Tables 3.1 and 3.2. Notice that $\sum_{\iota\in I} v_\iota \le 3$ by $v_\iota/4 \le v_\iota(r_\iota - v_\iota)/2r_\iota$. □

A further numerical analysis will reveal the discrepancy a/n for exceptional type to be small. The following are the main results in this section. Compare them with the geometric results in Theorems 3.5.5 to 3.5.7.

Theorem 3.2.8 *Suppose that $n = 1$.*

(i) *If $a = 1$, then π is of type other than e4, e7 or e11.*
(ii) *If $a \ge 2$, then the pair (type, a) is (o1, 2), (o2, a), (o3, a), (e1, 4), (e1, 2), (e2, 2), (e3, 3), (e5, 2) or (e9, 2).*

Theorem 3.2.9 *Suppose that $n \ge 2$.*

(i) *If $a = 1$, then the pair (type, n) is (o1, 2), (o2, n), (o3, n), (e1, 4), (e1, 2), (e2, 2), (e3, 3), (e5, 2), (e7, 2), (e8, 3) or (e9, 2).*
(ii) *If $a \ge 2$, then the triple (type, a, n) is (o2, a, n), (o3, a, n), (e1, 4, 2), (e1, 2, 2), (e2, 2, 2) or (e11, 2, 2).*

Lemma 3.2.10 *Let R be the lcm of r_ι for all $\iota \in I$ and let R' be the lcm of $r_\iota/\gcd(r_\iota, v_\iota)$ for all $\iota \in I$. Then $(a/n)^2 RE^3$ and $R'E^3$ are integers.*

Proof The contribution part in the expression of $d(0,j)$ in Lemma 3.2.7(i) is determined by j modulo R'. Hence $d(0, R' + j) - d(0, j) = (j + (R' - a/n - 1)/2)R'E^3$, showing that $R'E^3$ is an integer. By Lemma 3.2.7(ii), $(d(R+i+1, j) - d(R+i, j)) - (d(i+1, j) - d(i, j)) = (a/n)^2 RE^3$, which must be an integer. □

Proof of Theorem 3.2.8   By $n = 1$, $K_Y \sim aE \sim ae_\iota K_Y$ and thus e_ι is coprime to r_ι. This excludes types e4, e7 and e11. By Lemma 3.2.10, RE^3 is an integer for the lcm R of r_ι for $\iota \in I$. Using this, one finds all the possibilities of $a \geq 2$ in Tables 3.1 and 3.2 except for exclusion of type e8 with $a = 3$. Note that type e13 with $a = 2$ does not occur since a is coprime to all r_ι.

If π were of type e8 with $a = 3$, then $E^3 = 1/10$ and $\{(r_\iota, v_\iota, b_\iota)\}_{\iota \in I} = \{(5, 2, 1), (2, 1, 1)\}$. By Lemma 3.2.7(i), $d(0, 1) = 1/60 + Ec_2(Y)/12 - 21/40$ and $d(0, 2) = -1/30 + Ec_2(Y)/12 + 21/40$. These contradict the values $d(0, 1) = d(0, 2) = 0$ in Lemma 3.2.6.   □

Proof of Theorem 3.2.9(i)   Since RE^3/n^2 is an integer for the lcm R of r_ι for $\iota \in I$ by Lemma 3.2.10, one has only to exclude the cases (type, n) = (e4, 3), (e7, 4) and (e13, 2). The case (e4, 3) is excluded by the computation $d(0, 0) - d(-1, 0) = -2/3$ from Lemma 3.2.7(ii), whilst the cases (e7, 4) and (e13, 2) are done by the relation $v_\iota \equiv nb_\iota \bmod r_\iota$ for $\iota \in I$ from $E \sim nK_Y$.   □

Lemma 3.2.10 admits an important corollary which almost reduces Theorem 3.2.9(ii) to the coprimeness of a and n.

Corollary 3.2.11   *If a and n are at least two and coprime, then π is of type* o2 *or* o3.

Proof   One can exclude each type other than o2 or o3 using Lemma 3.2.10. For example if π were of type o1, then $R'E^3 = 2n/a$ and $(a/n)^2 RE^3 = 2a/n$, both of which must be integers. This is impossible if a and n are at least two and coprime.   □

By virtue of the corollary, in order to prove Theorem 3.2.9(ii), we may and shall assume that a and n have a common factor p. We write $a = pa'$ and $n = pn'$.

Take the cyclic cover $\mu \colon x' \in X' \to x \in X$ associated with $n'K_X$ as in Definition 2.2.17 and lift it to the cyclic cover $\mu_Y \colon Y' \to Y$ about E associated with $n'\pi^*K_X = n'K_Y - a'E$. We have the diagram

$$
\begin{array}{ccc}
E' \subset Y' & \xrightarrow{\mu_Y} & E \subset Y \\
{\scriptstyle \pi'} \downarrow & & \downarrow {\scriptstyle \pi} \\
x' \in X' & \xrightarrow{\mu} & x \in X
\end{array}
$$

in which E' is the reduced divisor $\mu_Y^* E$. Then $K_{Y'} = \mu_Y^* K_Y$ and $-E'$ is π'-ample. Further Y' as well as X' is terminal by Corollary 2.2.21. Hence π' satisfies the conditions of a divisorial contraction except the irreducibility of E'. Theorem 3.2.2 still holds for π' without irreducibility of E' by the proof.

We examine how this covering changes the numerical data. It is obvious that $(a'/n')E'^3 = p \cdot (a/n)E^3$. In order to analyse the behaviour of the basket, we assume p to be prime, because the covering μ is decomposed into those of prime degrees. Let q_ι be a fictitious singularity of type $\frac{1}{r_\iota}(1, -1, b_\iota)$ from Y. If $n'K_Y - a'E$ is Cartier at q_ι, then μ_Y is étale at q_ι and q_ι splits into p fictitious singularities of type $\frac{1}{r_\iota}(1, -1, b_\iota)$ from Y'. If $n'K_Y - a'E$ is not Cartier at q_ι, then p is a factor of r_ι and μ_Y is ramified at q_ι. The point q_ι changes into a fictitious singularity of type $\frac{1}{r_\iota/p}(1, -1, b_\iota)$ from Y'. On the other hand, the value e_ι is always preserved modulo index since $K_{Y'} = \mu_Y^* K_Y$ and $E' = \mu_Y^* E$. Therefore an element (r_ι, v_ι) in J is altered to either p elements (r_ι, v_ι) or one element $(r_\iota/p, \pm v_\iota)$ in the data J' of Y' constructed in the same manner as J, where $\pm v_\iota$ is the minimum of $\overline{v_\iota}$ and $\overline{-v_\iota}$. Note that fictitious singularities q_ι from one non-Gorenstein point q of Y change in a common way unless q is of type $cAx/4$. If q is of type $cAx/4$ and $n'K_Y - a'E$ is not Cartier at q, then $p = 2$ and the basket $\{1 \times \frac{1}{4}(1, -1, \pm 1), k \times \frac{1}{2}(1, 1, 1)\}$ from q changes into $\{(2k + 1) \times \frac{1}{2}(1, 1, 1)\}$.

Lemma 3.2.12 *If p is a common prime factor of a and n, then μ_Y belongs to one of the cases in Table 3.3.*

Table 3.3 The cyclic cover

type of π	p	J'	type of π	p	J'
o3	p	$(r_1/p, 1), (r_2/p, 1)$	e11	2	$(2, 1), (2, 1), (3, 1)$
e1	p	$(r/p, 2), r/p \geq 2$		3	$(2, 1), (2, 1), (2, 1)$
e2	p	$(r/p, 1), (r/p, 1)$	e13	2	$(2, 1), (2, 1), (r/2, 1)$
e4	3	$(2, 1)$		2	$(r, 1), (r, 1)$
e5	7	$\emptyset$	e14	3	$(2, 1), (2, 1), (2, 1)$
e7	2	$(r, 1), (r, 1)$	e15	2	$(2, 1), (3, 1), (3, 1)$

Proof This is proved by the same argument for each type of π. For example, suppose that π is of type e4. Then $(a/n)E^3 = 1/2$. Let q_ι be the fictitious singularity from Y with $(r_\iota, v_\iota) = (6, 3)$. Because J' can contain at most one $(6, 3)$ by Tables 3.1 and 3.2, μ_Y must be ramified at q_ι. Hence $p = 2$ or 3. If $p = 2$, then $J' = \emptyset$ and $(a'/n')E'^3 = 1$, which is impossible by the same tables applied to π'. If $p = 3$, then $J' = \{(2, 1)\}$ and $(a'/n')E'^3 = 3/2$, in which π' is of type o2. $\qquad\square$

We use a numerical application of Theorem 2.4.18.

Lemma 3.2.13 *Suppose that $x \in X$ is singular. Take a divisor F over X with centre $c_X(F) = x$ and log discrepancy $a_F(X) = 1 + 1/n$, which exists by Theorem 2.4.18. Then $a \operatorname{ord}_F E + n(a_F(Y) - 1) = 1$. If $a/n \neq 1/n$, then the*

centre $c_Y(F)$ *is a non-Gonrestein point q of Y which is not hidden. Let r be the
index of q $\in Y$ and write E $\sim eK_Y$ at q. Then a* $\gcd(e, r) + n \leq r$.

Proof The first assertion follows from the expression $a_F(X) = a_F(Y) + (a/n) \operatorname{ord}_F E$. If $a \neq 1$, then $E \neq F$ and $a_F(Y) > 1$ since Y is terminal. Then $\operatorname{ord}_F E$ must be less than one, which occurs only if $c_Y(F)$ is a non-Gorenstein point q of Y at which E is not Cartier. Since $a_F(Y) - 1 \in r^{-1}\mathbf{Z}$ and $\operatorname{ord}_F E \in r^{-1}\gcd(e, r)\mathbf{Z}$, the first assertion implies that $a \gcd(e, r)/r + n/r \leq 1$. □

Proof of Theorem 3.2.9(ii)   The exclusion of type e2 with $(a, n) = (4, 2)$ requires a geometric argument which will be elucidated in the following sections. Here we shall derive the list with the extra entry (type, a, n) $= (e2, 4, 2)$. Thanks to Corollary 3.2.11, we assume that a and n have a common factor p throughout the proof. We write $a = pa'$ and $n = pn'$ and construct the covering $\mu_Y : Y' \to Y$ associated with $n'\pi^*K_X$ as above. Let $d_\iota = \overline{n' - a'e_\iota}$ denote the residue of $n' - a'e_\iota$ modulo r_ι.

Step 1   First we claim that π is not of type e4, e5 or e14. We also claim that $(a, n) = (2, 2)$ if π is of type e11 or e15.

We discuss type e11. The proof is similar for other types. We use the divisor F in Lemma 3.2.13. When π is of type e11, Y has two quotient singularities q_1 and q_2 equipped with $(6, 2) \in J$ and $(2, 1) \in J$ respectively. The estimate $a \gcd(e, r) + n \leq r$ in Lemma 3.2.13 shows that $c_Y(F) = q_1$, for which $2a + n \leq 6$. Thus (a, n) must be $(2, 2)$.

Step 2   Next we claim that π is of type neither e7, e13 nor e15. To see this, we may assume that p is prime. Then $pd_\iota \equiv 0 \bmod r_\iota$, and d_ι is zero if and only if μ_Y is étale at q_ι. By Table 3.3, the set $\{(r_\iota, v_\iota, d_\iota)\}_{\iota \in I}$ is

$$\{(4, 2, 2), (r, 1, 0)\}, \ \{(2, 1, 1), (2, 1, 1), (r, 1, 0)\}, \ \{(2, 1, 1), (3, 1, 0), (4, 1, 2)\}$$

if π is of type e7, e13, e15 respectively. Note that all d_ι with $r_\iota = 2$ are the same by $e_\iota = 1$.

Let $c_\iota = -A_\iota(-e_\iota) - A_\iota(d_\iota) + A_\iota(d_\iota - e_\iota)$. By Lemma 3.2.7(i), $d(0, 0) - d(n', -a') = \sum_{\iota \in I} c_\iota$. Since $d(0, 0) = 1$ and $d(n', -a') = 0$ by Lemma 3.2.6, one obtains $\sum_{\iota \in I} c_\iota = 1$. This contradicts the direct computation of $\{c_\iota\}_{\iota \in I}$ as

$$\{1/2, 0\}, \qquad \{1/4, 1/4, 0\}, \qquad \{1/4, 0, 0 \text{ or } 1/2\}$$

for type e7, e13, e15 respectively, in which the order of elements in $\{c_\iota\}_\iota$ is compatible with that in $\{(r_\iota, v_\iota, d_\iota)\}_\iota$.

Step 3   By Corollary 3.2.11, Lemma 3.2.12 and Steps 1 and 2, it remains to show that $(a, n) = (4, 2)$ or $(2, 2)$ if π is of type e1 or e2. Assume that π is of type e1 or e2. The data $r_\iota, b_\iota, e_\iota, v_\iota, d_\iota, B_\iota$ are independent of $\iota \in I$. We write r, b, e, v, d, B for them for simplicity. The cardinality of I is then $2/v$.

We shall see that $\gcd(a, n) = 2$. Indeed by Lemma 3.2.7(ii), $(d(n'+1, -a') - d(n', -a')) - (d(1,0) - d(0,0)) = (2/v)c_B$ for $c_B = B(db) - B(db-v) + B(-v)$. From this and Lemma 3.2.6, one obtains $c_B = v/2$. Hence $d = r/2$ since

$$c_B = \frac{\overline{db}(r - \overline{db})}{2r} - \frac{(\overline{db} - v)(r - \overline{db} + v)}{2r} + \frac{v(r - v)}{2r} = \frac{v(r - \overline{db})}{r}.$$

Thus if p is prime, then it must be two. In other words, $\gcd(a, n)$ is a power of two. On the other hand by Lemma 3.2.12, μ_Y is totally ramified at the point q of Y at which E is not Cartier. In particular if p were four, then $(nK_Y - aE)/2$ would not be Cartier at q, that is, $2d \not\equiv 0 \bmod r$. This contradicts $d = r/2$. Thus one cannot take $p = 4$, meaning that $\gcd(a, n) = 2$.

Therefore $a = 2a'$ and $n = 2n'$ with a' and n' coprime. By Lemma 3.2.10, if π is of type e1, then $R'E^3 = 2n'/a'$ and $(a/n)^2 RE^3 = 4a'/n'$ are integral. Likewise if type e2, then $R'E^3 = 2n'/a'$ and $(a/n)^2 RE^3 = 2a'/n'$ are integral. Thus $(a, n) = (4, 2)$, $(2, 2)$, $(2, 4)$ or $(2, 8)$ for type e1 and $(a, n) = (4, 2)$, $(2, 2)$ or $(2, 4)$ for type e2.

In order to exclude the case $n > 2$, we compute $d(2, 0) - d(1, 0)$ modulo **Z**. Notice that r is even by $d = r/2$. If (type, a, n) = (e1, 2, 4), then by Lemma 3.2.7(ii), $d(2, 0) - d(1, 0) = B(b) - B(b-2) \equiv (2 - 2b)/r \bmod$ **Z**. It is integral only if $b = 1$ or $r/2+1$. Since $v = \overline{eb} = 2$ and $e = r/2+2$ by $d = \overline{2 - e} = r/2$, one has $b = r/2+1$ with $r/2$ odd. This contradicts the coprimeness of b and r. If (type, a, n) = (e1, 2, 8) or (e2, 2, 4), then $d(2, 0) - d(1, 0) \equiv (1 - 2b)/r \bmod$ **Z** by the same lemma, which is never integral. $\qquad\square$

3.3 General Elephants for Exceptional Type

The general elephant conjecture in Question 2.4.13 plays a central role in the geometric classification of the divisorial contraction $\pi\colon E \subset Y \to x \in X$ in Framework 3.2.1. In this and the next section, we shall establish the conjecture for π.

Theorem 3.3.1   *Let $E \subset Y \to x \in X$ be a threefold divisorial contraction which contracts the divisor E to a point x. Then the general elephant of Y over the algebraic germ $x \in X$ is Du Val.*

The theorem will be proved as Theorems 3.3.5 and 3.4.1. If the discrepancy a/n is small, then the conjecture is easily derived from the local version for the germ $x \in X$.

Theorem 3.3.2   *Let $x \in S_X$ be the general elephant of $x \in X$ and let S be the strict transform in Y. If $d(-1, j) = 0$ for all $1 \le j < a/n$, then S is the general elephant of Y and it is Du Val.*

Proof   By Theorem 2.4.15, S_X is a Du Val surface. From $S_X \sim -K_X$, one can write $\pi^* S_X = S + (a/n + m)E$ with an integer m. The assumption means the equality $\pi_* \mathscr{O}_Y(-K_Y) = \pi_* \mathscr{O}_Y(-K_Y + (\lceil a/n \rceil - 1)E)$. By Proposition 2.5.9 for $-K_Y + (\lceil a/n \rceil - 1)E = -\pi^* K_X + (\lceil a/n \rceil - a/n - 1)E$, it becomes

$$\pi_* \mathscr{O}_Y(-K_Y) = \begin{cases} \mathfrak{m}\mathscr{O}_X(-K_X) & \text{if } n = 1, \\ \mathscr{O}_X(-K_X) & \text{if } n \ge 2, \end{cases}$$

where $\mathfrak{m}$ is the maximal ideal in $\mathscr{O}_X$ defining x. This shows that $S + mE \sim -K_Y$ is the general elephant of Y. In particular, $m \ge 0$.

By Proposition 2.2.23, S is Cohen–Macaulay and has the adjunction $\omega_S = \mathscr{O}_S((K_Y + S)|_S) = \pi_S^* \omega_{S_X} \otimes \mathscr{O}_S(-mE|_S)$ for $\pi_S \colon S \to S_X$. Hence $K_{S^\nu} = \nu^* \pi_S^* K_{S_X} - m\nu^* E|_S - C$ on the normalisation $\nu \colon S^\nu \to S$ with the conductor C defined by the ideal sheaf $\nu^! \mathscr{O}_S$ in $\mathscr{O}_{S^\nu}$. Since S_X is canonical, $m = 0$ and $C = 0$. By $C = 0$, S satisfies R_1 and hence it is normal. By $m = 0$, S is the general elephant of Y and $K_S = \pi_S^* K_{S_X}$. It is canonical as so is S_X.   □

Corollary 3.3.3   *If π is of exceptional type and (type, a, n) is not (e1, 4, 1), (e1, 2, 1), (e2, 2, 1) or (e3, 3, 1), then it satisfies the assumption in Theorem 3.3.2 and in particular the general elephant conjecture holds for π.*

Proof   The assumption in Theorem 3.3.2 is automatically satisfied if $a/n \le 1$. Hence by Theorems 3.2.8 and 3.2.9, one has only to check the equality $d(-1, 1) = 0$ in the cases (type, a, n) = (e1, 4, 2), (e5, 2, 1) and (e9, 2, 1). In these cases, $d(0, 1) = 0$ by Lemma 3.2.6. Thus Lemma 3.2.7(ii) for $(i, j) = (-1, 1)$ provides $d(-1, 1) = 3E^3 + \sum_{\iota \in I}(B_\iota(b_\iota) - B_\iota(b_\iota - v_\iota))$. One can compute $d(-1, 1)$ to be zero by this formula with Table 3.2 in the following way.

Note that $b_\iota = \overline{av_\iota}$ if $n = 1$ and that if (type, a, n) = (e1, 4, 2), then $K_Y - 2E$ is not Cartier at q_ι by Table 3.3 and hence $b_\iota - 2v_\iota \equiv r_\iota/2 \bmod r_\iota$. For instance if (type, a, n) = (e1, 4, 2), then $r \ge 10$ by Lemma 3.2.13, and further $\{(r_\iota, v_\iota, b_\iota)\}_{\iota \in I} = \{(r, 2, r/2 + 4)\}$ and $E^3 = 2/r$. Thus

$$d(-1, 1) = \frac{6}{r} + \frac{(r/2 + 4)(r/2 - 4)}{2r} - \frac{(r/2 + 2)(r/2 - 2)}{2r} = 0. □$$

Let $\pi \colon E \subset Y \to x \in X$ be a divisorial contraction in Framework 3.2.1. The general elephant conjecture remains to be proved in the excluded cases in Corollary 3.3.3 and in the case of ordinary type. Let S be the general elephant of Y. It is defined by the general element in the submodule $\pi_* \mathscr{O}_Y(-K_Y)$ of $\mathscr{O}_X(-K_X)$. The push-forward $S_X = \pi_* S$ is normal by Serre's criterion and

Gorenstein by the adjunction $K_{S_X} = (K_X + S_X)|_{S_X} \sim 0$. One has $K_Y + S = \pi^*(K_X + S_X)$. A priori, S may be a reducible divisor containing a multiple of E.

Assume the general elephant conjecture for π. Then S is irreducible and Du Val with $K_S = \pi_S^* K_{S_X}$ for the induced morphism $\pi_S \colon S \to S_X$. Hence $x \in S_X$ is also Du Val and the minimal resolution of S_X factors through π_S. We shall make the statement in the remaining cases by describing the dual graph of the partial resolution π_S. In the *dual graph* on a smooth surface, every *vertex* stands for a curve in the surface, and two vertices are joined by the same number of *edges* as the intersection number of the corresponding two curves.

Notation 3.3.4   We fix the notation of the *dual graph* of the minimal resolution of a Du Val singularity as below. The configuration of exceptional curves is one of the *Dynkin diagrams*. The vertex ○ stands for an exceptional curve and ● for the strict transform of the general hyperplane section. The dual graph for A_n has two vertices ● corresponding to the analytic branches of the section. Each exceptional curve F_i is marked with the coefficient in the fundamental cycle on the minimal resolution.

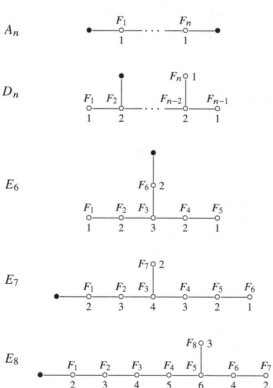

We address the general elephant conjecture for exceptional type in this section. It is beneficial for the reader to learn the proof in the simpler case of exceptional type ahead. The theorem is the following. As in Notation 3.3.4, ∘ stands for an exceptional curve and • for the strict transform of the general hyperplane section.

Theorem 3.3.5 *Suppose that π is of exceptional type. Then the general elephant S of Y is Du Val. Further if (type, a, n) is (e1, 4, 1), (e1, 2, 1), (e2, 2, 1) or (e3, 3, 1) as left in Corollary 3.3.3, then the dual graph of the partial resolution $\pi_S \colon S \to S_X = \pi_* S$ is as follows.*

The attached number at ∘ is the coefficient in the fundamental cycle. The graph means that all curves ∘ and • intersect at q. The point q is the unique non-Gorenstein point of Y. The surface S is at q of type A_{r-1} if π is of type e1, A_{2r-1} or E_6 if type e2 and D_5 if type e3, and it is smooth outside q. The number of curves ∘ is at most four if type e1 and at most a if type e2 or e3.

We have only to study the excluded cases in Corollary 3.3.3, by which we may assume that $n = 1$ in the rest of the section. By virtue of this assumption, we simplify the notation $\mathcal{Q}_{ij}$ and $d(i, j)$ in Notation 3.2.5.

Notation 3.3.6 Provided that $n = 1$, in which $K_Y \sim aE$, we define $\mathcal{Q}_i = \mathcal{Q}_{0i}$ and $d(i) = d(0, i) = \chi(\mathcal{Q}_i)$. The quotient $\mathcal{Q}_i = \mathcal{O}_E(iE|_E)$ fits into the exact sequence $0 \to \mathcal{O}_Y((i-1)E) \to \mathcal{O}_Y(iE) \to \mathcal{Q}_i \to 0$.

Lemmata 3.2.6 and 3.2.7(ii) become as follows.

Lemma 3.3.7 (i) $d(i) = \dim \pi_* \mathcal{O}_Y(iE)/\pi_* \mathcal{O}_Y((i-1)E)$ *for $i \leq a$.*

(ii) $d(i+a) - d(i) = \left(i - \dfrac{1}{2}\right) aE^3 + \sum_{\iota \in I}(B_\iota(iv_\iota) - B_\iota((i-1)v_\iota))$.

Lemma 3.3.8 $H^0(\mathcal{Q}_i) \simeq \pi_* \mathcal{O}_Y(iE)/\pi_* \mathcal{O}_Y((i-1)E)$ *and $H^1(\mathcal{Q}_i) = 0$ for all i. $H^2(\mathcal{Q}_i) = 0$ for $i \leq a$.*

Proof The assertion for $i \leq a$ follows from the Kawamata–Viehweg vanishing $R^p \pi_* \mathcal{O}_Y(iE) = 0$ for $i \leq a$ and $p \geq 1$. Then the assertion for $i > a$ follows from the duality of $H^p(\mathcal{Q}_i)$ and $H^{2-p}(\mathcal{Q}_{a+1-i})$ in Theorem 1.1.12. □

Let j be a positive integer. Let $L_j \sim -jE$ be an effective divisor on Y in which E does not appear. Such L_j exists if and only if $\pi_* \mathcal{O}_Y(-(j+1)E) \neq \pi_* \mathcal{O}_Y(-jE)$. The inclusion $\mathcal{O}_Y \subset \mathcal{O}_Y(L_j) \simeq \mathcal{O}_Y(-jE)$ defines a map

$$\mathcal{O}_Y((i+j)E) \to \mathcal{O}_Y((i+j)E) \otimes \mathcal{O}_Y(-jE) \to \mathcal{O}_Y(iE).$$

The restriction to E gives a map $\mathcal{Q}_{i+j} \to \mathcal{Q}_i$. It has kernel of finite length and hence it is injective because the $\mathcal{O}_E$-module $\mathcal{Q}_{i+j}$ satisfies S_2. If $i = 0$, then the inclusion $\mathcal{Q}_j \subset \mathcal{Q}_0 = \mathcal{O}_E$ is nothing but the ideal sheaf in $\mathcal{O}_E$ defining $L_j \cap E$. This follows from the commutative diagram

$$\begin{array}{ccccccc}
\mathcal{O}_Y(-L_j) \otimes \mathcal{O}_Y(-E) & \longrightarrow & \mathcal{O}_Y(-L_j) & \longrightarrow & \mathcal{O}_Y(-L_j) \otimes \mathcal{O}_E & \longrightarrow & 0 \\
\downarrow & & \| & & \downarrow & & \\
0 \longrightarrow \mathcal{O}_Y(-L_j - E) & \longrightarrow & \mathcal{O}_Y(-L_j) & \longrightarrow & \mathcal{Q}_j & \longrightarrow & 0 .
\end{array}$$

Notation 3.3.9   We write $\mathcal{R}_{ij}$ for the cokernel of the injection $\mathcal{Q}_{i+j} \hookrightarrow \mathcal{Q}_i$. It fits into the exact sequence $0 \to \mathcal{Q}_{i+j} \to \mathcal{Q}_i \to \mathcal{R}_{ij} \to 0$. By construction, it induces a surjection $\mathcal{Q}_i \otimes \mathcal{O}_{L_j \cap E} \twoheadrightarrow \mathcal{R}_{ij}$ which has kernel of finite length. It is an isomorphism $\mathcal{O}_{L_j \cap E} \simeq \mathcal{R}_{0j}$ when i is zero.

Lemma 3.3.10   *The map $H^0(\mathcal{Q}_i) \to H^0(\mathcal{R}_{ij})$ is always surjective. The cohomology $H^1(\mathcal{R}_{ij})$ vanishes if $i + j \le a$, and $H^1(\mathcal{R}_{a1}) \simeq \mathbf{C}$.*

Proof   This is a consequence of Lemma 3.3.8. Remark that $H^1(\mathcal{R}_{a1}) \simeq H^2(\mathcal{Q}_{a+1})$ is dual to $H^0(\mathcal{O}_E) = \mathbf{C}$.   □

Corollary 3.3.11   *For all j, $H^0(\mathcal{O}_{L_j \cap E}) = \mathbf{C}$ and in particular $L_j \cap E$ has no embedded points. If $j \le a$, then $H^1(\mathcal{O}_{L_j \cap E}) = 0$ and in particular the support of $L_j \cap E$ is a tree of $\mathbf{P}^1$ defined below.*

Definition 3.3.12   A connected reduced complete scheme C of dimension one is called a *tree* of $\mathbf{P}^1$ if $H^1(\mathcal{O}_C) = 0$. It is a union of smooth rational curves without circle in the sense that it contains no distinct curves $C_1, \ldots, C_n$ and distinct points $x_1, \ldots, x_n$ such that $x_i \in C_i \cap C_{i+1}$ for all i, where we set $C_{n+1} = C_1$.

Example 3.3.13   The union of three lines in $\mathbf{P}^3$ meeting at one point is a tree if and only if it is not contained in any plane.

We shall analyse a smooth rational curve C realised as a component of $L_j \cap E$ in Corollary 3.3.11. For this purpose, we introduce the integer $s_C(i)$ using the notation consistent with that in Chapter 5.

Definition 3.3.14   Let C be a smooth rational curve in E. For a divisorial sheaf $\mathcal{M}$ on Y, let $\mathrm{gr}^0_C \mathcal{M}$ denote the double dual of the $\mathcal{O}_C$-module $\mathcal{M} \otimes \mathcal{O}_C$.

It is the quotient of $\mathcal{M}$ by the maximal submodule $F^1\mathcal{M}$ containing $\mathcal{I}\mathcal{M}$ such that $F^1\mathcal{M}/\mathcal{I}\mathcal{M}$ is of finite length, where $\mathcal{I}$ is the ideal sheaf in $\mathcal{O}_Y$ defining C. We define the integer $s_C(i)$ for $i \in \mathbf{Z}$ as the degree of the invertible sheaf $\mathrm{gr}_C^0\,\mathcal{O}_Y(iE)$ on C so that

$$\mathrm{gr}_C^0\,\mathcal{O}_Y(iE) \simeq \mathcal{O}_{\mathbf{P}^1}(s_C(i)).$$

We write $s(i)$ for $s_C(i)$ simply when the curve C is clear in the context. Notice that $s(i) = i(E \cdot C)$ if iE is Cartier along C.

Next we define the integer $o(w)$ in the general setting where $q \in C \subset Y$ is the germ of a terminal *analytic* threefold Y endowed with a smooth curve C. This will be discussed in a more systematic manner in Section 5.2. We only abstract necessary notions here.

Let r be the index of $q \in Y$ and let $\mu\colon \bar{q} \in \bar{Y} \to q \in Y$ be the index-one cover. Let $\bar{C} = \mu^{-1}(C)$ and let s be the number of irreducible analytic components of $\bar{C}$, which is a divisor of r. Take the normalisation $\bar{q}^\nu \in \bar{C}^\nu$ of one of the irreducible components of $\bar{q} \in \bar{C}$ and fix a *uniformising parameter* $t^{s/r}$ in $\mathcal{O}_{\bar{C}^\nu, \bar{q}^\nu}$, that is, a generator of the maximal ideal in $\mathcal{O}_{\bar{C}^\nu}$ defining $\bar{q}^\nu$. Then $t = (t^{s/r})^{r/s}$ is a uniformising parameter in $\mathcal{O}_{C,q}$. By Theorem 2.4.8, there exists an identification $\bar{q} \in \bar{Y} = \mathfrak{D}^3$ or $\bar{q} \in \bar{Y} \subset \mathfrak{D}^4$ by which $Y = \mathfrak{D}^3/\mathbf{Z}_r(w_1, w_2, w_3)$ or $Y = \bar{Y}/\mathbf{Z}_r \subset \mathfrak{D}^4/\mathbf{Z}_r(w_1, w_2, w_3, w_4)$ with orbifold coordinates x_1, x_2, x_3 (and x_4) with respect to a fixed primitive character $\chi\colon \mathbf{Z}_r \to \mathbf{C}^\times$. Consider a semi-invariant function f in $\mathcal{O}_{\bar{Y}}$ of weight w_i vanishing at $\bar{q}$ and write a_f/r for the order of $f|_{\bar{C}^\nu}$ with respect to t. Namely, $a_f = s\,\mathrm{ord}_{\bar{q}^\nu}\,f|_{\bar{C}^\nu} > 0$. We define a_i to be the minimum of a_f for all f of weight w_i. Note that $(a_1, \dots, a_d) \in \mathbf{Z} \cdot s(w_1, \dots, w_d) + r\mathbf{Z}^d$ for the embedding dimension $d = 3$ or 4 of $\bar{q} \in \bar{Y}$. One can take $x_1, \dots, x_d$ so that they make a normal form below.

Definition 3.3.15 The description either $q \in Y = \mathfrak{D}^3/\mathbf{Z}_r(w_1, w_2, w_3)$ or $q \in Y \subset \mathfrak{D}^4/\mathbf{Z}_r(w_1, w_2, w_3, w_4)$ with orbifold coordinates x_1, x_2, x_3 (and x_4) is called a *normal form* of $q \in C \subset Y$ if $x_i|_{\bar{C}^\nu} = t^{a_i/r}$ for all i. Define the submonoid

$$\mathrm{ow}\,C = \sum_{i=1}^d \mathbf{N} \cdot (a_i, w_i) \subset \mathbf{Z} \times \mathbf{Z}_r$$

where $d = 3$ or 4 accordingly. For $w \in \mathbf{Z}_r$, we define $o_q(w)$ as the least integer such that $(o_q(w), w) \in \mathrm{ow}\,C$. We write $o(w)$ for $o_q(w)$ when there is no ambiguity. Note that $o(0) = 0$ and $o(w_i) = a_i$ unless w_i is zero. An element (l, w) belongs to $\mathrm{ow}\,C$ if and only if there exists a semi-invariant function f of weight w such that $f|_{\bar{C}^\nu} = t^{l/r}$. In particular, $(r, 0) \in \mathrm{ow}\,C$.

By definition, $o(w)$ has the following basic property.

Lemma 3.3.16 *Let $\mathscr{L}_w$ denote the divisorial subsheaf of $\mu_*\mathcal{O}_{\tilde{Y}}$ which consists of semi-invariant functions of weight w. If $w_1 + \cdots + w_l = 0$, then the natural map $\mathscr{L}_{w_1} \otimes \cdots \otimes \mathscr{L}_{w_l} \otimes \mathcal{O}_C \to \mathcal{O}_C$ has cokernel of length $\sum_{i=1}^{l} o(w_i)/r$.*

We deal with the case $(\text{type}, a, n) = (\text{e3}, 3, 1)$ and give the proof of Theorem 3.3.5. The other three cases can be handled in a similar way.

Lemma 3.3.17 *Suppose that $n = 1$ and that π is of ordinary type or of type e2 or e3. Define the numbers r_1 and r_2 by $r_1 = r_2 = 1$ if π is type o1, by $r_1 = 1$ and $J = \{(r_2, 1)\}$ if type o2 and by $J = \{(r_1, 1), (r_2, 1)\}$ if type e2, e3 or o3. Then*

$$\dim \pi_*\mathcal{O}_Y(-iE)/\pi_*\mathcal{O}_Y(-(i+1)E) = \begin{cases} 1 + \lfloor i/r_1 \rfloor + \lfloor i/r_2 \rfloor & \text{for } 0 \le i < a, \\ 2 + \lfloor a/r_1 \rfloor + \lfloor a/r_2 \rfloor & \text{for } i = a. \end{cases}$$

Proof By Lemma 3.3.7(i), the left-hand side for $i \ge 0$ equals $d(-i)$. By Tables 3.1 and 3.2, $aE^3 = 1/r_1 + 1/r_2$ and $v_\iota = 1$ for $\iota \in I$. By Lemma 3.3.7(ii),

$$d(-i) = d(a-i) + \left(i + \frac{1}{2}\right)\left(\frac{1}{r_1} + \frac{1}{r_2}\right) + \sum_{j=1}^{2}\left(\frac{(\bar{i}+1)(r_j - \bar{i} - 1)}{2r_j} - \frac{\bar{i}(r_j - \bar{i})}{2r_j}\right)$$

$$= d(a-i) + 1 + \left\lfloor\frac{i}{r_1}\right\rfloor + \left\lfloor\frac{i}{r_2}\right\rfloor.$$

Since $d(0) = 1$ and $d(i) = 0$ for $1 \le i \le a$ by Lemma 3.3.7(i), this shows the lemma. $\square$

In what follows, we assume that π is of type e3 with $(a, n) = (3, 1)$. Since $n = 1$, Y has no hidden non-Gorenstein points in Definition 3.2.4. Thus Y has a unique non-Gorenstein point q, from which two fictitious singularities emerge with $J = \{(2, 1), (4, 1)\}$. By Lemma 2.5.21, $q \in Y$ is of type $cAx/4$ and analytically described as

$$q \in Y = (y_1^2 + y_2^2 + f(y_3, y_4^2) = 0) \subset \mathfrak{D}^4/\mathbf{Z}_4(1, 3, 2, 1)$$

with ord $f(y_3, 0) = 3$. It has $E^3 = 1/4$ from Table 3.2.

By $\pi_*\mathcal{O}_Y(-E)/\pi_*\mathcal{O}_Y(-2E) \simeq \mathbf{C}$ in Lemma 3.3.17, the general hyperplane section H_X of $x \in X$ has order one along E, that is, $\pi^*H_X = H + E$ with the strict transform H. Then $H \sim -E$. By Corollary 3.3.11 for $L_1 = H$, the scheme $H \cap E$ has no embedded points and its support is a tree of $\mathbf{P}^1$. The intersection number $(H \cdot H \cdot E) = E^3 = 1/4$ shows that $H \cap E$ is integral. Let $C = H \cap E \simeq \mathbf{P}^1$. Since $(H \cdot C) = 1/4$, C passes through the unique non-Gorenstein point q. Recall the notation $s(i) = s_C(i)$ in Definition 3.3.14. Note that $\mathrm{gr}_C^0 \mathcal{O}_Y(iE) \simeq \mathcal{O}_{\mathbf{P}^1}(s(i))$ is the double dual $\mathscr{R}_{i1}^{\vee\vee}$ of the $\mathcal{O}_C$-module $\mathscr{R}_{i1}$ in Notation 3.3.9.

Lemma 3.3.18 $s(1) = s(2) = -1$, $s(3) = -2$ *and* $s(i) \geq -1$ *for* $i \leq -1$.

Proof It follows from $\mathcal{O}_{\mathbf{P}^1}(s(i)) \simeq \mathcal{R}_{i1}^{\vee\vee}$ that $H^1(\mathcal{O}_{\mathbf{P}^1}(s(i))) \simeq H^1(\mathcal{R}_{i1})$. By Lemma 3.3.10, it is zero for $i \leq 2$ and isomorphic to $\mathbf{C}$ for $i = 3$. Hence $s(i) \geq -1$ for $i \leq 2$ and $s(3) = -2$. The existence of the natural map $\mathcal{O}_Y(iE)^{\otimes 4} \otimes \mathcal{O}_C \to \mathcal{O}_Y(4iE) \otimes \mathcal{O}_C$ provides the inequality $4s(i) \leq 4i(E \cdot C) = -i$. In particular $s(i) \leq -1$ for $i \geq 1$, which completes the lemma. $\qquad\square$

Following Definition 3.3.15, we take a normal form of $q \in C \subset Y$ and set $(w_1, w_2, w_3, w_4) = (1, 3, 2, 1)$ and ow $C = \sum_{i=1}^{4} \mathbf{N} \cdot (a_i, w_i)$ with $a_i = o(w_i)$.

Lemma 3.3.19 $(a_1, a_2, a_3, a_4) = (3, 5, 2, 3)$ *and* $s(-3) = 0$.

Proof First we shall see that $s(-1) = -1$. By Lemma 3.3.18, it suffices to show the estimate $s(-1) \leq -1$. By Lemmata 3.3.8 and 3.3.10, the composite $\pi_* \mathcal{O}_Y(-E) \to H^0(\mathcal{Q}_{-1}) \to H^0(\mathcal{R}_{-1,1})$ is surjective. From this and $\mathcal{R}_{-1,1}^{\vee\vee} \simeq \mathcal{O}_{\mathbf{P}^1}(s(-1))$, one has the surjection

$$\pi_* \mathcal{O}_Y(-E)/\pi_* \mathcal{O}_Y(-2E) \twoheadrightarrow H^0(\mathcal{R}_{-1,1}) \twoheadrightarrow H^0(\mathcal{O}_{\mathbf{P}^1}(s(-1))).$$

The generator of $\pi_* \mathcal{O}_Y(-E)/\pi_* \mathcal{O}_Y(-2E) \simeq \mathbf{C}$ vanishes at the generic point of $H \cap E = C$. Hence $H^0(\mathcal{O}_{\mathbf{P}^1}(s(-1)))$ must be zero, that is, $s(-1) \leq -1$.

The isomorphism $\mathcal{O}_Y(-E) \simeq \mathcal{O}_Y(K_Y) \simeq \mathcal{L}_3$ holds on the analytic germ $q \in Y$ for the sheaf $\mathcal{L}_3$ in Lemma 3.3.16. Applying Lemma 3.3.16 to the map $\mathcal{O}_Y(-E)^{\otimes 4} \otimes \mathcal{O}_Y(4E) \otimes \mathcal{O}_C \to \mathcal{O}_C$, one has the equality $4s(-1) + 4(E \cdot C) = -o(3)$. Thus $o(3) = 5$ by $s(-1) = -1$ and $(E \cdot C) = -1/4$. Applying Lemma 3.3.16 to the map $\mathcal{O}_Y(E)^{\otimes 4} \otimes \mathcal{O}_Y(-4E) \otimes \mathcal{O}_C \to \mathcal{O}_C$, one has $4s(1) - 4(E \cdot C) = -o(1)$. Thus $o(1) = 3$ by $s(1) = -1$ in Lemma 3.3.18.

By $o(1) = 3$ and $o(3) = 5$, one can write $(a_1, a_2, a_3, a_4) = (3, 5, 2, 3)$, where $a_3 = 2$ follows from $(4, 0) \in$ ow C. Finally we apply Lemma 3.3.16 to the map $\mathcal{O}_Y(-3E)^{\otimes 4} \otimes \mathcal{O}_Y(12E) \otimes \mathcal{O}_C \to \mathcal{O}_C$. Then $4s(-3) + 12(E \cdot C) = -o(1)$. Hence $s(-3) = 0$. $\qquad\square$

Proof of Theorem 3.3.5 for $(\text{type}, a, n) = (e3, 3, 1)$ One has the surjection

$$\pi_* \mathcal{O}_Y(-3E)/\pi_* \mathcal{O}_Y(-4E) \twoheadrightarrow H^0(\mathcal{R}_{-3,1}) \twoheadrightarrow H^0(\mathcal{O}_{\mathbf{P}^1})$$

by Lemmata 3.3.8 and 3.3.10 with $s(-3) = 0$ in Lemma 3.3.19. Hence the general elephant $S \sim -3E$ on Y does not contain $C = H \cap E$ and it is irreducible. By $(S \cdot C) = 3/4$, the surface S intersects C only at q with local intersection number $(S \cdot C)_q = 3/4$. In particular, $|-K_Y|$ is free outside the support q of $S \cap 3H \cap E$ for the members S and $3H$ of $|-K_Y|$. By Bertini's theorem, S is smooth outside q and $S \cap E$ is generically reduced. Then by Corollary 3.3.11, the scheme $S \cap E$ is a tree of $\mathbf{P}^1$. By $(H \cdot S \cdot E) = 3/4$, the number of irreducible components of $S \cap E$ is at most three and every component D must pass through q with $(H \cdot D)_q < 1$. Hence for Theorem 3.3.5, it remains to prove that S is Du Val of type D_5 at q.

Take a normal form of $q \in C \in Y \subset \mathfrak{D}^4/\mathbf{Z}_4(1, 3, 2, 1)$ with orbifold coordinates x_1, x_2, x_3, x_4. By Lemma 3.3.19, $(x_1, x_2, x_3, x_4)|_C = (t^{3/4}, t^{5/4}, t^{1/2}, t^{3/4})$ with a uniformising parameter t in $\mathcal{O}_{C,q}$. Let $g = g(x_1, x_2, x_3, x_4)$ be the semi-invariant function of weight one which defines S in $q \in Y$. Its linear part $g_1(x_1, x_4)$ satisfies $g_1(t^{3/4}, t^{3/4}) \neq 0$ because the order of $g(t^{3/4}, t^{5/4}, t^{1/2}, t^{3/4})$ with respect to t equals $(S \cdot C)_q = 3/4$. If g_1 is a general linear form in x_1 and x_4, then one can observe that S is Du Val of type D_5 at q, just as in the proof of Theorem 2.4.15, from the local description $q \in Y = (y_1^2 + y_2^2 + f(y_3, y_4^2) = 0) \subset \mathfrak{D}^4/\mathbf{Z}_4(1, 3, 2, 1)$ with ord $f(y_3, 0) = 3$.

We analyse the semi-invariant function $h = h(x_1, x_2, x_3, x_4)$ of weight one which defines E in Y. Without using the general elephant conjecture, we shall have a complete classification of π in Theorem 3.5.1 when $x \in X$ is smooth, in which π is never of type e3. Hence we may assume that $x \in X$ is singular. Take a divisor F as in Lemma 3.2.13, which has $c_Y(F) = q$ and $3 \operatorname{ord}_F E + (a_F(Y) - 1) = 1$. This equality holds only if $a_F(Y) = 5/4$ and $\operatorname{ord}_F E = 1/4$. The linear part $h_1(x_1, x_4)$ of h is not zero by $\operatorname{ord}_F E = 1/4$ but satisfies $h_1(t^{3/4}, t^{3/4}) = 0$ since $h(t^{3/4}, t^{5/4}, t^{1/2}, t^{3/4}) = 0$ by $C \subset E$.

By $g_1(t^{3/4}, t^{3/4}) \neq 0$ and $h_1(t^{3/4}, t^{3/4}) = 0$, the non-zero g_1 and h_1 form a basis of the vector space $\mathbf{C}x_1 + \mathbf{C}x_4$. Take another general elephant $T \in |-K_Y|$, which is linearly equivalent to S and $E + 4H$. The linear system $|4H|$ is free along C from the surjection

$$\pi_* \mathcal{O}_Y(4H)/\pi_* \mathcal{O}_Y(4H - E) \twoheadrightarrow H^0(\mathcal{R}_{-4,1}) \twoheadrightarrow H^0(\mathcal{O}_{\mathbf{P}^1}(1))$$

by Lemmata 3.3.8 and 3.3.10 with $s(-4) = -4(E \cdot C) = 1$. It follows that the semi-invariant function in x_1, x_2, x_3, x_4 defining T has a general combination of g_1 and h_1 as the linear part. Thus one concludes that $q \in T$ is Du Val of type D_5 and so is $q \in S$ by the general choice of S. $\square$

3.4 General Elephants for Ordinary Type

This section is devoted to the general elephant conjecture for ordinary type. We keep Framework 3.2.1. In order to state the theorem uniformly, we use the following notation.

- If π is of type o1, then we set $r_1 = r_2 = 1$.
- If π is of type o2, then there exists a quotient singularity $q \in Y$ of type $\frac{1}{r}(1, -1, b)$ with $(r, 1) \in J$. We set $r_1 = 1$ and $r_2 = r$.

- If π is of type o3, then there exist two quotient singularities $q_i \in Y$ of type $\frac{1}{r_i}(1, -1, b_i)$ with $(r_i, 1) \in J$ for $i = 1, 2$. We arrange them so that $r_1 \le r_2$.

Theorem 3.4.1 *Suppose that π is of ordinary type. Then the general elephant S of Y is Du Val. Further unless $a = 1$, the partial resolution $\pi_S : S \to S_X = \pi_* S$ satisfies one of the following.*

(i) *The dual graph of π_S is one of the following.*

The attached number at $\circ$ is the coefficient in the fundamental cycle. The points q_1 and q_2 are two smooth points of Y if π is of type o1, a smooth point of Y and the point q in I if type o2 and the points q_1 and q_2 in I if type o3. The surface S is at q_i of type A_{r_i-1} for each i and at z of type A, possibly smooth, and it is smooth outside these points. Consequently $x \in S_X$ is of type A and $x \in X$ is cA or cA/n.

(ii) *The dual graph of π_S is one of the following.*

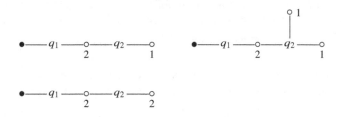

In this case, π is of type o3 and $n = 1$ or 2. The points q_1 and q_2 are those in I. The surface S is at q_i of type A_{r_i-1} for each i. If S has the first or second graph, then S is smooth outside q_1 and q_2. If S has the third graph, then S may have extra singularities on the rightmost curve $\circ$ with coefficient two. If $n = 2$, then S_X is the general elephant of $x \in X$.

The assertion for $a = 1$ is Theorem 3.3.2. If π is of type o1 with $a \ge 2$, then $(a, n) = (2, 1)$ by Theorems 3.2.8 and 3.2.9. This is a very special case and the assertion is also a corollary to Theorem 3.1.1. We shall discuss the case when π is of type o2 in which all the arguments necessary for type o3 are contained. At the end of the section, we shall explain how the case (ii) in Theorem 3.4.1 occurs when π is of type o3.

In the case when X is smooth, a complete description of π will be proved independently of Theorem 3.4.1. We thus assume that X is singular. In order to make the ideas clear, we shall assume that $n = 1$, by which the simplified notation $\mathscr{Q}_i = \mathscr{Q}_{0,i}$ and $d(i) = d(0,i)$ in Notation 3.3.6 can be applied.

Henceforth we assume that π is of type o2 with $a \geq 2$ and $n = 1$ and that $x \in X$ is singular. Then Y has a unique non-Gorenstein point q as remarked in Definition 3.2.4 and $q \in Y$ is a quotient singularity of type $\frac{1}{r}(1, -1, a)$. By Table 3.1 and Lemma 3.2.13 respectively, π satisfies

$$aE^3 = 1 + \frac{1}{r}, \qquad a < r.$$

By $\pi_*\mathcal{O}_Y(-E)/\pi_*\mathcal{O}_Y(-2E) \simeq \mathbf{C}^2$ in Lemma 3.3.17, the general hyperplane section H_X of $x \in X$ satisfies $\pi^*H_X = H + E$ with the strict transform H. By Corollary 3.3.11, the scheme $H \cap E$ has no embedded points and its support is a tree of $\mathbf{P}^1$. By $(H \cdot H \cdot E) = E^3 = a^{-1}(1+1/r) < 1$, every irreducible component of $H \cap E$ passes through the unique non-Gorenstein point q. By Lemma 3.3.8 and the S_2 property of $\mathscr{Q}_{-1}$ in Proposition 2.2.23, the isomorphism

$$\pi_*\mathcal{O}_Y(-E)/\pi_*\mathcal{O}_Y(-2E) \simeq H^0(\mathscr{Q}_{-1}) = H^0(E^\circ, \mathcal{O}_E(-E|_E)|_{E^\circ})$$

holds for $E^\circ = E \setminus q$. The space $H^0(E^\circ, \mathcal{O}_E(-E|_E)|_{E^\circ}) \simeq \mathbf{C}^2$ defines a linear system on E° containing non-zero mobile part, which makes sense even if E° is not normal. Hence there exists an irreducible component $C \simeq \mathbf{P}^1$ of $H \cap E$ not contained in the strict transform H' of another general hyperplane section of $x \in X$.

The next lemma on $s(l) = s_C(l)$ is derived from Lemmata 3.3.10 and 3.4.3 in the same manner as Lemma 3.3.18 is derived. We omit the proof.

Lemma 3.4.2 $(s(i), s(-i)) = (-1, 0)$ for $1 \leq i \leq a - 1$ and $(s(a), s(-a)) = (-2, 1)$, $(-2, 0)$ or $(-1, 0)$.

Lemma 3.4.3 Let $q \in Y = \mathfrak{D}^3/\mathbf{Z}_r(1, -1, a)$ be a terminal quotient threefold singularity with $\mu : \mathfrak{D}^3 \to Y$. Let $\mathscr{L}_w$ denote the subsheaf of $\mu_*\mathcal{O}_{\mathfrak{D}^3}$ which consists of semi-invariant functions of weight w. Then the image of the natural map $\mathscr{L}_1 \otimes \mathscr{L}_{-1} \to \mathcal{O}_Y$ is the maximal ideal defining q.

Proof This follows from the property that every invariant monomial but 1 in the orbifold coordinates is the product of two monomials of weights 1 and -1. □

Following Definition 3.3.15, we take a normal form of $q \in C \subset Y$ and set $(w_1, w_2, w_3) = (1, -1, a)$ and ow $C = \sum_{i=1}^3 \mathbf{N} \cdot (a_i, w_i)$ with $a_i = o(w_i)$.

Lemma 3.4.4 If $s(-a) = 1$, then the case (i) in Theorem 3.4.1 holds.

Proof Apply Lemma 3.3.16 to the map $\mathscr{O}_Y(-aE)^{\otimes r} \otimes \mathscr{O}_Y(arE) \otimes \mathscr{O}_C \to \mathscr{O}_C$. Then $rs(-a) + ar(E \cdot C) = -o(a)$. From the assumption $s(-a) = 1$, one has

$$1 + \frac{o(a)}{r} = a(H \cdot C) \le a(H \cdot H \cdot E) = 1 + \frac{1}{r}.$$

Hence $C = H \cap E$ and $o(a) = 1$. Lemmata 3.3.8 and 3.3.10 yield the surjection

$$\pi_* \mathscr{O}_Y(-aE) \twoheadrightarrow H^0(\mathscr{R}_{-a,1}) \twoheadrightarrow H^0(\mathscr{O}_{\mathbf{P}^1}(1)).$$

Thus the general elephant $S \sim -aE$ on Y intersects C properly at one general point q' in C besides q. Since $(S \cdot C) = (S \cdot H \cdot E) = 1 + 1/r$, S intersects C exactly at q and q' with local intersection numbers $(S \cdot C)_q = 1/r$ and $(S \cdot C)_{q'} = 1$. In particular, one can take orbifold coordinates x_1, x_2, x_3 at $q \in Y \simeq \mathfrak{D}^3/\mathbf{Z}_r(1, -1, a)$ so that x_3 defines S. It follows that S is Du Val of type A_{r-1} at q.

On the other hand, $|-K_Y|$ is free outside the support $q \cup q'$ of $S \cap aH \cap E$ for the members S and aH of $|-K_Y|$. Since q' is a general point in C, the linear system $|-K_Y|$ turns out to be free outside q. By Bertini's Theorem, S is smooth outside q and $S \cap E$ is generically reduced. Then by Corollary 3.3.11, $S \cap E$ is a tree of $\mathbf{P}^1$. Since H is π-ample, every irreducible component of $S \cap E$ must pass through q or q'. The component of $S \cap E$ through q is unique by $(H \cdot S \cdot E)_q = 1/r$ and so is the component through q' by $(H \cdot S \cdot E)_{q'} = 1$. Thus $S \cap E$ consists of at most two curves, and the case (i) in Theorem 3.4.1 holds. □

For a moment, we consider the general setting where $q \in C \subset Y$ is the germ of a terminal analytic threefold Y endowed with a smooth curve C. Let $\mathscr{I}$ denote the ideal sheaf in $\mathscr{O}_Y$ defining C and let $\mathscr{I}^{(2)}$ denote the second symbolic power of $\mathscr{I}$ defined below.

Definition 3.4.5 Let I be an ideal in a noetherian ring R. The n-th *symbolic power* of I is the ideal $I^{(n)} = \bigcap_{\mathfrak{p} \in \mathrm{Ass}(R/I)} I^n R_{\mathfrak{p}} \cap R$ in R for the set $\mathrm{Ass}(R/I)$ of associated primes of R/I. Here by abuse of notation, $I^n R_{\mathfrak{p}} \cap R$ denotes the inverse image of $I^n R_{\mathfrak{p}}$ by $R \to R_{\mathfrak{p}}$.

Recall the notion of $\mathrm{gr}_C^0 \, \omega_Y$ in Definition 3.3.14. We analyse the cokernel of the map

$$\alpha : \left(\bigwedge^2 \mathscr{I}/\mathscr{I}^{(2)} \right) \otimes \omega_C \to \mathrm{gr}_C^0 \, \omega_Y$$

induced by the exact sequence $\mathscr{I}/\mathscr{I}^2 \to \Omega_Y \otimes \mathscr{O}_C \to \omega_C \to 0$. The next lemma demonstrates how to compute it in the simplest case. A systematic treatment will be provided in Section 5.2.

Lemma 3.4.6 *Consider the analytic germ* $q \in C \subset Y = \mathfrak{D}^3/\mathbf{Z}_r(1, -1, a)$ *with orbifold coordinates* x_1, x_2, x_3 *such that* $(x_1, x_2, x_3)|_C = (t^{c_1/r}, t^{c_2/r}, 0)$ *with* $c_1 + c_2 = r$ *for a uniformising parameter* t *in* $\mathcal{O}_{C,q}$. *Then* α *has cokernel of length* $\min\{c_1, c_2\}$.

Proof The map $\omega_Y \otimes \mathcal{O}_C \to \mathrm{gr}^0_C \, \omega_Y$ is locally identified with

$$\gamma \colon \omega_Y \otimes \mathcal{O}_C \to \mathcal{O}_C, \quad f(x_1, x_2, x_3) dx_1 \wedge dx_2 \wedge dx_3 \mapsto \frac{f(t^{c_1/t}, t^{c_2/t}, 0)}{t^{o(-a)/r}}.$$

Since ω_C is generated by $d(x_1 x_2|_C)$, the cokernel of α has the same length as the cokernel of the composite $\gamma \circ \delta$, where δ is defined as

$$\delta \colon \mathcal{I}/\mathcal{I}^{(2)} \times \mathcal{I}/\mathcal{I}^{(2)} \to \omega_Y \otimes \mathcal{O}_C, \quad (f_1, f_2) \mapsto df_1 \wedge df_2 \wedge d(x_1 x_2).$$

The $\mathcal{O}_C$-module $\mathcal{I}/\mathcal{I}^{(2)}$ is generated by functions in $\mathcal{O}_Y$ of form $(x_1^{c_2} - x_2^{c_1}) f(x_1, x_2)$ or $x_3 g(x_1, x_2, x_3)$ in which f and g are semi-invariant of weight c_1 and $-a$ respectively. Hence the length of the cokernel of $\gamma \circ \delta$ equals the minimum of the order with respect to t of the image by γ of

$$fg \cdot d(x_1^{c_2} - x_2^{c_1}) \wedge dx_3 \wedge d(x_1 x_2) = -fg(c_2 x_1^{c_2} + c_1 x_2^{c_1}) dx_1 \wedge dx_2 \wedge dx_3.$$

The order of $f|_C$ attains the minimum when f is either $x_1^{c_1}$ or $x_2^{c_2}$, whilst the minimum of the order of $g|_C$ is $o(-a)/r$. Further $(c_2 x_1^{c_2} + c_1 x_2^{c_1})|_C = rt^{c_1 c_2/r}$. Thus the minimal length is $\min\{c_1^2/r, c_2^2/r\} + c_1 c_2/r = \min\{c_1, c_2\}$. $\square$

We return to our situation and proceed with the proof of Theorem 3.4.1 in the case of type o2 with $a \geq 2$, $n = 1$ and X singular. Recall that $C \simeq \mathbf{P}^1$ is a component of $H \cap E$ not contained in the strict transform H' of another general hyperplane section of $x \in X$. We consider the global map $\alpha \colon (\wedge^2 \mathcal{I}/\mathcal{I}^{(2)}) \otimes \omega_C \to \mathrm{gr}^0_C \, \omega_Y$ for the ideal sheaf $\mathcal{I}$ in $\mathcal{O}_Y$ defining C.

Lemma 3.4.7 *The map* α *has cokernel of length at most* $4 + s(a)$.

Proof The *degree* of a locally free coherent sheaf on a smooth curve means the degree of its determinant. The length of the cokernel of α equals

$$\deg \mathrm{gr}^0_C \, \omega_Y - \deg \omega_C - \deg \mathcal{I}/\mathcal{I}^{(2)} = s(a) + 2 - \deg \mathcal{I}/\mathcal{I}^{(2)}.$$

By Theorem 1.3.5, the locally free sheaf $\mathcal{I}/\mathcal{I}^{(2)}$ on C is isomorphic to a direct sum of invertible sheaves on $\mathbf{P}^1$. Thus it suffices to prove the vanishing $H^1(\mathcal{I}/\mathcal{I}^{(2)}) = 0$.

One has $H^2(\mathcal{O}_Y(-2H) \otimes \mathcal{O}_{2E}) = 0$ from the surjection $\mathcal{O}_Y(-2H) \twoheadrightarrow \mathcal{O}_Y(-2H) \otimes \mathcal{O}_{2E}$ with the Kawamata–Viehweg vanishing $R^2 \pi_* \mathcal{O}_Y(-2H) = 0$. One also has $H^1(\mathcal{O}_{2E}) = 0$ from the exact sequence $0 \to \mathcal{O}_Y(-2E) \to \mathcal{O}_Y \to \mathcal{O}_{2E} \to 0$ with Kawamata–Viehweg vanishing. Hence $H^1(\mathcal{O}_{2H \cap 2E}) = 0$ by the exact sequence $\mathcal{O}_Y(-2H) \otimes \mathcal{O}_{2E} \to \mathcal{O}_{2E} \to \mathcal{O}_{2H \cap 2E} \to 0$. The inclusion $\mathcal{O}_Y(-2H) + \mathcal{O}_Y(-2E) \subset \mathcal{I}^{(2)}$ yields the surjection $\mathcal{O}_{2H \cap 2E} \twoheadrightarrow \mathcal{O}_Y/\mathcal{I}^{(2)}$, from which $H^1(\mathcal{O}_Y/\mathcal{I}^{(2)}) = 0$. Now the required vanishing $H^1(\mathcal{I}/\mathcal{I}^{(2)}) = 0$ follows from the exact sequence $0 \to \mathcal{I}/\mathcal{I}^{(2)} \to \mathcal{O}_Y/\mathcal{I}^{(2)} \to \mathcal{O}_C \to 0$. $\square$

Recall that $(w_1, w_2, w_3) = (1, -1, a)$ and $\mathrm{ow}\,C = \sum_{i=1}^3 \mathbf{N} \cdot (a_i, w_i)$ with $a_i = o(w_i)$.

Lemma 3.4.8 *The monoid* $\mathrm{ow}\,C$ *is generated by at most two of* $(a_1, 1)$, $(a_2, -1)$ *and* (a_3, a). *More precisely, either*

(i) $\mathrm{ow}\,C$ *is generated by* $(a_1, 1)$ *and* (a_3, a) *or by* $(a_2, -1)$ *and* (a_3, a) *or*
(ii) $a = 2$, $r \geq 5$, $s(2) = -2$ *and* $\mathrm{ow}\,C$ *is generated by* $(a_1, 1)$ *and* $(a_2, -1)$.

Proof We express $(r, 0) \in \mathrm{ow}\,C$ as

$$(r, 0) = n_1(a_1, 1) + n_2(a_2, -1) + n_3(a_3, a)$$

in $\mathrm{ow}\,C$ with $n_i \in \mathbf{N}$. If $n_1 = n_2 = 0$, then $n_3 = r$ and $a_3 = 1$, in which $\mathrm{ow}\,C$ is generated by (a_3, a). If $n_1 \geq 1$ and $n_2 = 0$, then $(r - a_1, -1) = (n_1 - 1)(a_1, 1) + n_3(a_3, a) \in \mathrm{ow}\,C$ and hence $a_2 = r - a_1$, in which $\mathrm{ow}\,C$ is generated by $(a_1, 1)$ and (a_3, a). For the same reason, if $n_1 = 0$ and $n_2 \geq 1$, then $\mathrm{ow}\,C$ is generated by $(a_2, -1)$ and (a_3, a). Thus one has only to deal with the case when $n_1 \geq 1$ and $n_2 \geq 1$. Then $(r - a_1 - a_2, 0) \in \mathrm{ow}\,C$, by which $a_1 + a_2 = r$ and $(n_1, n_2, n_3) = (1, 1, 0)$. From now on, we assume that $(n_1, n_2, n_3) = (1, 1, 0)$ is the only solution to the equation above. This implies that $o(w) + o(-w) \geq 2r$ unless $w \equiv 0$ or $\pm 1 \bmod r$.

By permutation of x_i, we may assume that $r \geq 5$. Then $o(2) + o(-2) \geq 2r$. By Lemma 3.3.16, the natural map $\mathcal{O}_Y(-2E) \otimes \mathcal{O}_Y(2E) \otimes \mathcal{O}_C \to \mathcal{O}_C$ has image in $\mathcal{O}_C(-2q)$. Hence $s(-2) + s(2) \leq -2$. By Lemma 3.4.2, this occurs only if $a = 2$, $s(2) = -2$ and $s(-2) = 0$. In this case, $o(1) + o(-1) = r$ and $o(2) + o(-2) = 2r$, from which $a_3 = o(2) = 2o(1) = 2a_1$ and $o(-2) = 2o(-1)$. Thus $\mathrm{ow}\,C$ is generated by $(a_1, 1)$ and $(a_2, -1)$. $\square$

Lemma 3.4.9 *If* $s(-a) \neq 1$, *then* $\mathrm{ow}\,C$ *is generated either by* $(1, 1)$, *by* $(2, 1)$ *and* $(r - 2, -1)$ *or by* $(r - 2, 1)$ *and* $(2, -1)$. *The first case occurs if in addition* $r \leq 4$.

Proof The lemma will be proved in accordance with Lemma 3.4.8. Applying Lemma 3.3.16 to the map $\mathcal{O}_Y(-E)^{\otimes r} \otimes \mathcal{O}_Y(rE) \otimes \mathcal{O}_C \to \mathcal{O}_C$, one has $rs(-1) + r(E \cdot C) = -o(1)$. Hence $a_1 = o(1) = r(H \cdot C)$ by $s(-1) = 0$ in Lemma 3.4.2. In like manner from the map $\mathcal{O}_Y(-aE)^{\otimes r} \otimes \mathcal{O}_Y(arE) \otimes \mathcal{O}_C \to \mathcal{O}_C$, one has $rs(-a) + ar(E \cdot C) = -o(a)$. Hence $a_3 = o(a) = -rs(-a) + ar(H \cdot C)$. Suppose that $s(-a) \neq 1$. Then $s(-a) = 0$ by Lemma 3.4.2 and $a_3 = ar(H \cdot C) = aa_1$.

In this case if ow C is generated by $(a_1, 1)$ and (a_3, a), then it is generated by $(a_1, 1)$ and hence $a_1 = 1$.

If ow C is generated by $(a_2, -1)$ and (a_3, a), then $(r, 0) \in$ ow C is expressed as $(r, 0) = n_2(a_2, -1) + n_3(a_3, a)$ with $n_i \in \mathbf{N}$. One can choose n_2 and n_3 so that both are positive. Then $(r - a_2 - a_3, 1 - a) \in$ ow C, that is, $o(1 - a) = r - a_2 - a_3$. Apply Lemma 3.3.16 to the map $\mathcal{O}_Y(-aE) \otimes \mathcal{O}_Y(E) \otimes \mathcal{O}_Y((a - 1)E) \otimes \mathcal{O}_C \to \mathcal{O}_C$. Then $s(-a) + s(1) + s(a - 1) = -r^{-1}(o(a) + o(-1) + o(1 - a)) = -1$. Thus $s(-a) = 1$ by Lemma 3.4.2.

Now it suffices to discuss the second case in Lemma 3.4.8. By the preceding results, we may assume that a_1 and a_2 are at least two. In this case, $a_1 + a_2 = r$ by $(r, 0) \in$ ow C and there exist orbifold coordinates x_1, x_2, x_3 at $q \in Y \simeq \mathfrak{D}^3/\mathbf{Z}_r(1, -1, a)$ such that $(x_1, x_2, x_3)|_C = (t^{a_1/r}, t^{a_2/r}, 0)$. By Lemma 3.4.7, the map $\alpha \colon (\bigwedge^2 \mathscr{I}/\mathscr{I}^{(2)}) \otimes \omega_C \to \mathrm{gr}_C^0 \omega_Y$ has cokernel of length at most $4 + s(2) = 2$. Hence a_1 or a_2 is two by Lemma 3.4.6.   □

We shall complete the proof of the theorem in our case.

Proof of Theorem 3.4.1 *for type* o2 *with $n = 1$*   We can assume that X is singular. Thanks to Lemma 3.4.4, we have only to exclude the three cases in Lemma 3.4.9. Suppose one of these cases. Then $a_1 = 1, 2$ or $r - 2$, and $a_1 = 1$ if $r \le 4$. We have seen that $(H \cdot C) = a_1/r$ at the beginning of the proof of Lemma 3.4.9. We take orbifold coordinates x_1, x_2, x_3 at $q \in Y \simeq \mathfrak{D}^3/\mathbf{Z}_r(1, -1, a)$ such that $(x_1, x_2, x_3)|_C = (t^{a_1/r}, t^{1 - a_1/r}, 0)$ for a uniformising parameter t in $\mathcal{O}_{C,q}$.

Let H' be the strict transform in Y of another general hyperplane section of $x \in X$. It follows from the choice of C that C intersects H' properly. Since $(H' \cdot C) = a_1/r < 1$, C intersects H' only at q with $(H' \cdot C)_q = a_1/r$. Thus the order of $h'(t^{a_1/r}, t^{1 - a_1/r}, 0)$ equals a_1/r for the semi-invariant function $h' = h'(x_1, x_2, x_3)$ defining H' in $q \in Y$. This occurs only if the monomial x_1 appears in h', which means that h' can be an orbifold coordinate at $q \in Y \simeq \mathfrak{D}^3/\mathbf{Z}_r(1, -1, a)$.

This is also the case with H, that is, the semi-invariant function $h = h(x_1, x_2, x_3)$ defining H can be an orbifold coordinate. Then the monomial x_1 must appear in h since $1 \not\equiv -1$ or $a \bmod r$ by $2 \le a < r$. However, x_1 is the only monomial in x_1, x_2, x_3 of weight one the restriction to C of which is $t^{a_1/r}$. Thus $h(t^{a_1/r}, t^{1 - a_1/r}, 0)$ is never zero, which contradicts the inclusion $C \subset H$.   □

The strategy for the proof of Theorem 3.4.1 for type o3 is similar. In this case, provided that $n = 1$, Y has exactly two non-Gorenstein singularities q_1

and q_2, where $q_i \in Y$ is of type $\frac{1}{r_i}(1, -1, a)$ and $r_1 \leq r_2$. Assuming that $a \geq 2$ and X is singular, we take $C \simeq \mathbf{P}^1$ as an irreducible component of $H \cap E$ through both q_1 and q_2. The curve C exists by $(H \cdot H \cdot E) < 1/r_1 + 1/r_2$ and it is unique because the support of $H \cap E$ is a tree. One can show that $s(-a) = 0$ or -1 like Lemma 3.4.2 and that the condition $s(-a) = 0$ implies the case (i) in Theorem 3.4.1 like Lemma 3.4.4. However, the case $s(-a) = -1$ can really occur for type o3 whereas the proof for type o2 was done with exclusion of the case $s(-a) = 0$. When $s(-a) = -1$, the following case remains left.

Studying a normal form of $q_i \in C \subset Y \simeq \mathfrak{D}^3/\mathbf{Z}_{r_i}(1, -1, a_i)$ similarly to Lemmata 3.4.8 and 3.4.9, when $s(-a) = -1$, one can find orbifold coordinates x_{i1}, x_{i2}, x_{i3} at $q_i \in Y$ such that $(x_{11}, x_{12}, x_{13})|_C = (t^{1/r_1}, 0, 0)$ and $(x_{21}, x_{22}, x_{23})|_C = (0, t^{1/r_2}, 0)$, in the course of which one also obtains the estimate $a < r_1 < r_2$. Applying Lemma 3.3.16 to the map $\mathcal{O}_Y(E)^{\otimes r_1 r_2} \otimes \mathcal{O}_Y(-r_1 r_2 E) \otimes \mathcal{O}_C \to \mathcal{O}_C$, one has $s(1) + (H \cdot C) = -(r_1 - 1)/r_1 - 1/r_2$ and hence $(H \cdot C) = 1/r_1 - 1/r_2$ by $s(1) = -1$. Let S be the general elephant of Y. The linear system $|-K_Y|$ contains non-zero mobile part since $H^0(\mathcal{Q}_{-a}) \simeq \mathbf{C}^2$ by Lemma 3.3.17. Hence $S \cap E$ has an irreducible component other than C. The case left is when $[S \cap E] = [C] + [D]$ cycle-theoretically in which D is a subscheme of $S \cap E$ not containing C, with $H^1(\mathcal{O}_D) = 0$ and $(H \cdot D) = 2/r_2$.

In what follows, we shall prove the implication from the above remaining case to the case (ii) in Theorem 3.4.1. The situation is summarised as

- $(x_{11}, x_{12}, x_{13})|_C = (t^{1/r_1}, 0, 0)$ and $(x_{21}, x_{22}, x_{23})|_C = (0, t^{1/r_2}, 0)$,
- $2 \leq a < r_1 < r_2$,
- $[S \cap E] = [C] + [D]$ with $C \not\subset D$ and $H^1(\mathcal{O}_D) = 0$ and
- $(H \cdot C) = 1/r_1 - 1/r_2$ and $(H \cdot D) = 2/r_2$.

Since $(H \cdot D) = 2/r_2 < 1/r_1 + 1/r_2$, the next lemma and corollary imply that D intersects H properly only at q_2 with $(H \cdot D)_{q_2} = 2/r_2$.

Lemma 3.4.10 *One has $[H \cap E] = u[C]$ with $u \geq 2$ cycle-theoretically.*

Proof Write $[H \cap E] = u[C] + [F]$ cycle-theoretically with $u \geq 1$ and a subscheme F of $H \cap E$ not containing C. Then $(H \cdot F) = a^{-1}(1/r_1 + 1/r_2) - u(1/r_1 - 1/r_2) < 1/r_1 - (1/r_1 - 1/r_2) = 1/r_2$. Hence $F = 0$ because no irreducible component of F contains both q_1 and q_2. If $u = 1$, that is, $H \cap E = C$, then the semi-invariant function in x_{21}, x_{22}, x_{23} defining E in $q_2 \in Y$ would contain a linear form. It must contain the monomial x_{21} or x_{23} by $(x_{21}, x_{22}, x_{23})|_C = (0, t^{1/r_2}, 0)$, in which $-1 \equiv 1$ or $a \mod r_2$. This contradicts $a < r_1 < r_2$. $\square$

Corollary 3.4.11　*The number r_1 does not divide $2r_2$.*

Proof　If $lr_1 = 2r_2$ for an integer $l \geq 3$, then $1/r_1 + 2/lr_1 = (S \cdot H \cdot E) = au(H \cdot C) = au(1/r_1 - 2/lr_1)$, that is, $l + 2 = au(l - 2)$. This contradicts the conditions $a \geq 2$ and $u \geq 2$.　　　　　　　　　　　　　　　　□

Lemma 3.4.12　*One can choose the orbifold coordinates x_{21}, x_{22}, x_{23} so that H is defined in $q_2 \in Y$ by x_{21}.*

Proof　Let $h = h(x_{21}, x_{22}, x_{23})$ be the semi-invariant function of weight one which defines H in $q_2 \in Y$. It suffices to show that the linear part of h is not zero. From $(H \cdot D)_{q_2} = 2/r_2$, this is the case unless D is integral. When $D \simeq \mathbf{P}^1$, we take a normal form x_1', x_2', x_3' of $q_2 \in Y \simeq \mathfrak{D}^3/\mathbf{Z}_{r_2}(1, -1, a)$ for a uniformising parameter t' in $\mathscr{O}_{D, q_2}$. Then $h|_D$ has order $2/r_2$ with respect to t'. If the linear part of h is zero, then $x_2'|_D$ or $x_3'|_D$ must be $(t')^{1/r_2}$, in which $1 \equiv -2$ or $2a \bmod r_2$. From this and $2 \leq a < r_1 < r_2$, one has $a = (r_2 + 1)/2$. In this case, the equality between $(S \cdot H \cdot E) = 1/r_1 + 1/r_2$ and $au(H \cdot C) = au(1/r_1 - 1/r_2)$ in Lemma 3.4.10 becomes $2(r_1 + r_2) = u(r_2 + 1)(r_2 - r_1)$. Then $(r_2 + 1)(r_2 - r_1) \leq r_1 + r_2$, from which $r_2 = r_1 + 1$ and thus $(a, u, r_1, r_2) = (3, 3, 4, 5)$.

Suppose that $(a, u, r_1, r_2) = (3, 3, 4, 5)$. Let L be an effective divisor on the analytic germ $q_2 \in Y \simeq \mathfrak{D}^3/\mathbf{Z}_5(1, 4, 3)$ which is defined by a general semi-invariant function of weight four. Then $(L \cdot H \cdot E)_{q_2} = 3(L \cdot C) = 3/5$ by $(x_{21}, x_{22}, x_{23})|_C = (0, t^{1/5}, 0)$, and $(\bar{L} \cdot \bar{H} \cdot \bar{E})_{\bar{q}_2} = 3$ on the index-one cover $\bar{q}_2 \in \bar{Y} \simeq \mathfrak{D}^3$ of $q_2 \in Y$ for the pull-backs $\bar{L}, \bar{H}, \bar{E}$ of L, H, E. It follows from $C \subset E$ that $\mathrm{ord}_{\bar{q}_2} \bar{E} \neq 1$. Thus $\mathrm{ord}_{\bar{q}_2} \bar{H} = 1$ as $\mathrm{ord}_{\bar{q}_2} \bar{L} \cdot \mathrm{ord}_{\bar{q}_2} \bar{H} \cdot \mathrm{ord}_{\bar{q}_2} \bar{E} \leq (\bar{L} \cdot \bar{H} \cdot \bar{E})_{\bar{q}_2} = 3$, showing that h has non-zero linear part.　　□

We take the notation $H_X = \pi_* H$ and $S_X = \pi_* S$. Let B denote the curve in Y which is the strict transform of $H_X \cap S_X$.

Lemma 3.4.13　*One has $[H \cap S] = [C] + [B]$ cycle-theoretically. The curve B intersects E only at q_1.*

Proof　By $[H \cap E] = u[C]$ in Lemma 3.4.10, one can write $[H \cap S] = v[C] + [B]$ with $v \geq 1$ cycle-theoretically. By Lemma 3.4.12, H is smooth at the generic point η_C of C. Then about η_C,

$$E \cap S \cap H = E|_H \cap S|_H = uC \cap vC = \min\{u, v\}C$$

as a divisor on $\eta_C \in H$. This equals C since $[S \cap E] = [C] + [D]$. Hence $v = 1$ as $u \geq 2$, that is, $[H \cap S] = [C] + [B]$.

Since $(E \cdot B) = (E \cdot H \cdot S) - (E \cdot C) = 2/r_1 < 1$, B does not intersect E outside q_1 and q_2. By Corollary 3.4.11, the intersection number $2/r_1$ cannot be expressed as $l_1/r_1 + l_2/r_2$ with $l_1 \in \mathbf{N}$ and $l_2 \in \mathbf{N}_{>0}$. Thus B does not intersect E at q_2, by which B intersects E only at q_1. □

Proof of Theorem 3.4.1 *for type* o3 *with* $n = 1$ We discuss the remaining case above. By $S \cap E = C$ at q_1 and $(x_{11}, x_{12}, x_{13})|_C = (t^{1/r_1}, 0, 0)$, S is Du Val of type A_{r_1-1} at q_1. In like manner by $H \cap S = C$ at q_2 in Lemma 3.4.13 and $(x_{21}, x_{22}, x_{23})|_C = (0, t^{1/r_2}, 0)$, S is Du Val of type A_{r_2-1} at q_2. Further S is smooth about $C \setminus (q_1 \cup q_2)$ because $C \simeq \mathbf{P}^1$ is the intersection of the divisors S and E outside D, which are Cartier outside q_1 and q_2 by Theorem 2.3.18. By Lemma 3.4.13, C is the support of $H \cap S \cap E$ with $S \sim aH$. Hence $|-K_Y|$ is free outside C. By Bertini's theorem, the general elephant S is smooth outside C, and D is reduced. To sum up, S is Du Val of type A_{r_i-1} at q_i for each i and smooth outside q_1 and q_2. In particular, the general elephant conjecture holds for π.

By $[S \cap E] = [C] + [D]$ with $(H \cdot D) = (H \cdot D)_{q_2} = 2/r_2$, the dual graph of the partial resolution $\pi_S \colon S \to S_X$ is the first or second one in the case (ii) in Theorem 3.4.1. The coefficient in the fundamental cycle is determined by the expression $\pi_S^*(H_X|_{S_X}) = H|_S + E|_S = (C + B) + (C + D) = B + 2C + D$. □

3.5 Geometric Classification

The purpose of this section is to describe the divisorial contraction $\pi \colon E \subset Y \to x \in X$ in Framework 3.2.1 in terms of weighted blow-up. We begin with the following theorem which is the first classification result by means of the singular Riemann–Roch formula.

Theorem 3.5.1 (Kawakita [221]) *Let $\pi \colon E \subset Y \to x \in X$ be a threefold divisorial contraction which contracts the divisor E to a smooth point x. Then there exists a regular system x_1, x_2, x_3 of parameters in $\mathcal{O}_{X,x}$ such that π is the weighted blow-up with $\mathrm{wt}(x_1, x_2, x_3) = (1, r_1, r_2)$ with coprime positive integers r_1 and r_2.*

We write $K_Y = \pi^* K_X + aE$. If the discrepancy a is two, then π is the blow-up of X at x by Lemma 3.1.2 because E is the unique divisor with centre x and discrepancy two. In this case, it is of type o1 and the theorem holds with $r_1 = r_2 = 1$. We shall assume that a is at least three. Then by Theorem 3.2.8, either

(o) π is of type o2 or o3 and $a \geq 3$,
(e1) π is of type e1 and $a = 4$ or
(e3) π is of type e3 and $a = 3$.

In order to treat the cases (o) and (e3) uniformly, we define the numbers r_1 and r_2 by $r_1 = 1$ and $J = \{(r_2, 1)\}$ if π is of type o2, and by $J = \{(r_1, 1), (r_2, 1)\}$ and $r_1 \leq r_2$ if type o3 or e3. With this notation, we let

$$(s_1, s_2) = \begin{cases} (\min\{r_1, a\}, \min\{r_2, a\}) & \text{in the cases (o) and (e3),} \\ (2, 2) & \text{in the case (e1).} \end{cases}$$

Note that $2 \leq s_2$ and $a \leq s_1 + s_2$. The latter inequality is obvious unless $r_2 < a$ in the case (o). If $r_2 < a$ in the case (o), then $a \leq a r_1 r_2 E^3 = r_1 + r_2 = s_1 + s_2$ from Table 3.1 with Lemma 3.2.10.

Lemma 3.5.2 *There exists a regular system* x_1, x_2, x_3 *of parameters in* $\mathscr{O}_{X,x}$ *such that* $\mathrm{ord}_E \, x_1 = 1$, $\mathrm{ord}_E \, x_2 \geq s_1$ *and* $\mathrm{ord}_E \, x_3 \geq s_2$.

Proof Let $\mathfrak{m} = \pi_* \mathscr{O}_Y(-E)$ denote the maximal ideal in $\mathscr{O}_X$ defining x. First suppose the case (e1). Then $E^3 = 1/r$ and $J = \{(r, 2)\}$ with $r \geq 4$ from Table 3.2. By Lemma 3.3.7(i), $\dim \mathfrak{m}/\pi_* \mathscr{O}_Y(-2E) = d(-1) = -(d(3) - d(-1))$, which equals $6/r - 2(r-2)/2r + 4(r-4)/2r = 1$ by Lemma 3.3.7(ii). Take a regular system x_1, x_2, x_3 of parameters in $\mathscr{O}_{X,x}$ such that x_1 forms a basis of $\mathfrak{m}/\pi_* \mathscr{O}_Y(-2E) \simeq \mathbf{C}$. Then $\mathrm{ord}_E \, x_1 = 1$. For $j = 2, 3$, the coordinate x_j is congruent to $c_j x_1$ with some $c_j \in \mathbf{C}$ modulo $\pi_* \mathscr{O}_Y(-2E)$. Replacing x_j by $x_j - c_j x_1$, one has $x_j \in \pi_* \mathscr{O}_Y(-2E)$, that is, $\mathrm{ord}_E \, x_j \geq 2 = s_{j-1}$.

Next suppose the case (o) or (e3). For brevity, we write R_i for the vector space $\pi_* \mathscr{O}_Y(-iE)/\pi_* \mathscr{O}_Y(-(i+1)E)$. Lemma 3.3.17 includes

$$\dim R_i = \begin{cases} 1 & \text{for } 0 \leq i < s_1, \\ 1 + \lfloor i/r_1 \rfloor & \text{for } s_1 \leq i < s_2. \end{cases}$$

This formula for $i = 1$ computes the dimension of $R_1 = \mathfrak{m}/\pi_* \mathscr{O}_Y(-2E)$ to be two if π is of type o2 and to be one if type o3 or e3. When π is of type o3 or e3, there exists a regular system x_1, x_2, x_3 of parameters in $\mathscr{O}_{X,x}$ such that x_1 forms a basis of $R_1 \simeq \mathbf{C}$. By $\mathrm{ord}_E \, x_1^i = i$, the monomial x_1^i forms a basis of $R_i \simeq \mathbf{C}$ for $1 \leq i < s_1$. Hence for $j = 2, 3$, x_j is congruent to some polynomial $p_j(x_1)$ modulo $\pi_* \mathscr{O}_Y(-s_1 E)$. Replacing x_j by $x_j - p_j$, one has x_2 and x_3 belonging to $\pi_* \mathscr{O}_Y(-s_1 E)$. In particular, the assertion holds if $s_1 = s_2$. If $s_1 < s_2$, then $s_1 = r_1$ and the monomials $x_1^{r_1}, x_2, x_3$ span the vector space $R_{r_1} \simeq \mathbf{C}^2$. Up to permutation, we may assume that $x_1^{r_1}$ and x_2 form a basis of R_{r_1}. This also holds when π is of type o2 in which $R_1 \simeq \mathbf{C}^2$.

We are in the case when $\mathrm{ord}_E \, x_1 = 1$, $\mathrm{ord}_E \, x_2 = s_1 = r_1$, $\mathrm{ord}_E \, x_3 \geq s_1$ and $x_1^{r_1}, x_2$ form a basis of R_{r_1}. We claim that for any i, the monomials in x_1, x_2 of weighted degree i with respect to $\mathrm{wt}(x_1, x_2) = (1, r_1)$ are linearly independent in the vector space R_i. Indeed, every weighted homogeneous polynomial $f = f(x_1, x_2)$ of weighted degree i is factorised as

$$x_1^{i-dr_1} \prod_{l=1}^{d} (c_l x_1^{r_1} + x_2)$$

with $c_l \in \mathbf{C}$ up to constant, where $d \le i/r_1$ is the degree of f in x_2. Here $\mathrm{ord}_E (c_l x_1^{r_1} + x_2)$ equals r_1 by the linear independence of $x_1^{r_1}$ and x_2 in the space R_{r_1}. Thus $\mathrm{ord}_E f = i$, that is, f is non-zero in R_i, which shows the claim.

The number of monomials in x_1, x_2 of weighted degree i is $1 + \lfloor i/r_1 \rfloor$, which coincides with the dimension of R_i as far as $i < s_2$. Hence for $i < s_2$, these monomials form a basis of R_i. Consequently x_3 is congruent to some polynomial $q(x_1, x_2)$ modulo $\pi_* \mathscr{O}_Y (-s_2 E)$. Replacing x_3 by $x_3 - q$, one has $x_3 \in \pi_* \mathscr{O}_Y (-s_2 E)$ and this completes the lemma. □

Proof of Theorem 3.5.1   We can assume that the discrepancy a is at least three. Take a regular system x_1, x_2, x_3 of parameters in $\mathscr{O}_{X,x}$ as in Lemma 3.5.2. For $j = 2, 3$, adding $c x_1^{s_{j-1}}$ to x_j for general $c \in \mathbf{C}$, we may assume that $\mathrm{ord}_E x_j = s_{j-1}$. Let $\varphi \colon B \to X$ be the weighted blow-up with $\mathrm{wt}(x_1, x_2, x_3) = (1, s_1, s_2)$ and let F be the exceptional divisor. Then $a_E (X) = 1 + a \le 1 + s_1 + s_2 = a_F (X)$. It follows from Proposition 3.5.3 below that $E = F$ as a divisor over X. Hence π is isomorphic to φ over X by Lemma 3.1.2. The threefold B has cyclic quotient singularities of types $\frac{1}{s_1}(1, -1, s_2)$ and $\frac{1}{s_2}(1, s_1, -1)$. Because $B \simeq Y$ is terminal, s_1 and s_2 must be coprime by Theorem 2.4.4. □

Proposition 3.5.3   *Let $x \in X$ be the germ of a smooth variety with a regular system $x_1, \dots, x_n$ of parameters in $\mathscr{O}_{X,x}$. Let B be the weighted blow-up of X with $\mathrm{wt}(x_1, \dots, x_n) = (w_1, \dots, w_n)$ and let F be the exceptional divisor. Let E be a divisor over X such that $c_X(E) \subset c_X(F)$. If the vector $(w_1, \dots, w_n)$ is parallel to $(\mathrm{ord}_E x_1, \dots, \mathrm{ord}_E x_n)$, then the centre $c_B(E)$ is not contained in the strict transform of the snc divisor on X defined by the product $x_1 \cdots x_n$. In particular $a_F (X) \le a_E (X)$ and the equality holds if and only if $E = F$ as a divisor over X.*

Proof   We may assume that $w_1, \dots, w_n$ have no common factors. Let H_i denote the strict transform in B of the divisor $(x_i)_X$. Then

$$\mathrm{ord}_E x_i = \mathrm{ord}_E H_i + \mathrm{ord}_F x_i \cdot \mathrm{ord}_E F = \mathrm{ord}_E H_i + w_i \, \mathrm{ord}_E F,$$

where $\mathrm{ord}_E F$ is positive by the inclusion $c_X(E) \subset c_X(F)$. The order $\mathrm{ord}_E H_i$ is positive if and only if the centre $c_B(E)$ is inside H_i. Since the intersection $\bigcap_i H_i$ is empty, at least one of $\mathrm{ord}_E H_i$ is zero. Because $(w_1, \dots, w_n)$ is parallel to $(\mathrm{ord}_E x_1, \dots, \mathrm{ord}_E x_n)$, all $\mathrm{ord}_E H_i$ must be zero, which shows the first assertion. Since B is smooth outside $\sum_i H_i$, the second follows from the expression $a_E (X) = a_E (Y) + (a_F (X) - 1) \, \mathrm{ord}_E F$. □

The description in Theorem 3.5.1 no longer holds when Y admits canonical singularities.

Example 3.5.4   Consider the germ $x \in X = (x_1 x_2 + x_3^2 + x_4 = 0) \subset \mathbf{A}^4$ at origin of the smooth threefold and let $Y \to X$ be the weighted blow-up with $\mathrm{wt}(x_1, x_2, x_3, x_4) = (5, 1, 3, 8)$. Then Y has two isolated singularities. One is a quotient singularity of type $\frac{1}{5}(4, 3, 3)$ and the other is $o \in (y_1 y_2 + y_3^2 + y_4^2 = 0) \subset \mathbf{A}^4/\mathbf{Z}_8(5, 1, 3, 7)$. These are canonical by Proposition 2.2.15 and Remark 2.6.4 respectively.

The general elephant of a divisorial contraction was first studied in the classification of contractions to cA_1 points [222]. Applying the general elephant conjecture in a systematic way, we establish the following geometric classification.

Theorem 3.5.5 ([223], [224, theorem 1.2(i)])   *In Framework 3.2.1, suppose that π is of ordinary type and that $x \in X$ is cA or cA/n. Then there exists an analytic identification*

$$x \in X \simeq o \in (x_1 x_2 + f(x_3^n, x_4) = 0) \subset \mathfrak{D}^4/\mathbf{Z}_n(1, -1, b, 0)$$

such that π is the weighted blow-up with $\mathrm{wt}(x_1, x_2, x_3, x_4) = \frac{1}{n}(r_1, r_2, a, n)$, where n divides $a - br_1$, an divides $r_1 + r_2$ and $(a - br_1)/n$ is coprime to r_1. Further, f is of weighted order $(r_1 + r_2)/n$ with respect to $\mathrm{wt}(x_3, x_4) = (a/n, 1)$ and the monomial $x_3^{(r_1 + r_2)/a}$ appears in f. Conversely, every such weighted blow-up is a divisorial contraction.

Theorem 3.5.6 ([223], [224, theorem 1.2(ii)])   *In Framework 3.2.1, suppose that π is of ordinary type and that $x \in X$ is neither cA nor cA/n. If the discrepancy a/n is not $1/n$, then π is of type $o3$, $x \in X$ is cD or $cD/2$ and one of the following holds.*

(i) *There exists an analytic identification of $x \in X$ with either*

$$o \in (x_1^2 + x_1 f(x_3, x_4) + x_2^2 x_4 + \lambda x_2 x_3^2 + \mu x_3^3 + g(x_2, x_3, x_4) = 0) \subset \mathfrak{D}^4$$

with $\mathrm{ord}\, g \geq 4$ for $n = 1$ or

$$o \in (x_1^2 + x_1 x_3 f(x_3^2, x_4) + x_2^2 x_4 + \lambda x_2 x_3^{2k-1} + g(x_3^2, x_4) = 0)$$
$$\subset \mathfrak{D}^4/\mathbf{Z}_2(1, 1, 1, 0)$$

for $n = 2$, such that π is the weighted blow-up with $\mathrm{wt}(x_1, x_2, x_3, x_4) = \frac{1}{n}(r + n, r, a, n)$. The number a is an odd factor of $2r + n$ and $a \neq 2r + n$. If $n = 2$, then r is odd.

(ii) *There exists an analytic identification of $x \in X$ with either*

$$o \in \left(\begin{array}{c} x_1^2 + x_2 x_5 + f(x_2, x_3, x_4) = 0 \\ x_2 x_4 + x_3^{(r+1)/a} + g(x_3, x_4)x_4 + x_5 = 0 \end{array} \right) \subset \mathfrak{D}^5$$

with ord $f \geq 4$ for $n = 1$ or

$$o \in \left(\begin{array}{c} x_1^2 + x_2 x_5 + f(x_3^2, x_4) = 0 \\ x_2 x_4 + x_3^{(r+2)/a} + g(x_3^2, x_4)x_3 x_4 + x_5 = 0 \end{array} \right) \subset \mathfrak{D}^5/\mathbf{Z}_2(1,1,1,0,1)$$

for $n = 2$, such that π is the weighted blow-up with $\mathrm{wt}(x_1, x_2, x_3, x_4, x_5) = \frac{1}{n}(r+n, r, a, n, r+2n)$. The number a is a factor of $r + n$ and $a \neq r + n$. If $n = 2$, then $(r+2)/a$ is odd.

Theorem 3.5.7 ([224, theorem 1.3], [474]) *In Framework 3.2.1, suppose that π is of exceptional type. If the discrepancy a/n is $1/n$, then $x \in X$ is neither cA nor cA/n. If a/n is not $1/n$, then π belongs to one of the cases in Table 3.4. Every case in the table has an example.*

Table 3.4 *Exceptional type with $a/n \neq 1/n$*

type	x	a/n	type	x	a/n	type	x	a/n
e1	cA_1, cD	4	e2	cD, cE_6	2	e5	cE_7	2
	cD	2		$cD/2$	2/2	e9	$cE_{7,8}$	2
	$cD/2$	4/2	e3	cA_2, cD	3	e11	$cE/2$	2/2
	$cD/2$	2/2						

In the case of type e1 with $(a, n) = (4, 1)$, the claim $r = 5$ in [223, theorem 3.5(iii)] is not true as pointed out by Yamamoto. One can only say that $r \equiv 3$ or 5 mod 8 for $J = \{(r, 2)\}$. See Example 3.6.4. The case when x is cD should be added to [223, corollary 1.15] and [224, theorem 1.3].

Yamamoto [474] claims that every π belonging to Table 3.4 is a weighted blow-up in $\mathfrak{D}^5/\mathbf{Z}_n(a_1, \ldots, a_5)$. However, the argument seems insufficient in the following points. He uses a certain linear independence of monomials unconditionally, but for example in [474, lemma 4.4], some linear combination of x_3^2 and x_4^3 may have order seven along E and $\hat{x}_1$ may have order six. He also deduces the isomorphism $\pi \simeq \varphi$ from Proposition 3.5.11 or its extension without examining singularities of B along $c_B(E)$ or the coincidence $a_F(X) = a_E(X)$. See Remark 3.5.17 for our improvement.

There exist only finitely many divisorial contractions with discrepancy at most one to the fixed germ $x \in X$. Hayakawa studied divisors over X with discrepancy at most one explicitly and proved that if X is not Gorenstein and the discrepancy a/n is at most one, then π is described as a weighted blow-up in $\mathfrak{D}^5/\mathbf{Z}_n(a_1,\ldots,a_5)$ [181], [182], [183]. This also holds when π is of type e1 with $(a,n) = (4,2)$ [227]. In conclusion, one obtains the following ultimate description. We remark that Hayakawa asserts this even when $(a,n) = (1,1)$ as found in [474, p.533].

Theorem 3.5.8 *Let* $\pi\colon E \subset Y \to x \in X$ *be a threefold divisorial contraction which contracts the divisor E to a non-Gorenstein point x. Then there exists an analytic identification*

$$x \in X \simeq o \in \left(\begin{array}{c} f(x_1,x_2,x_3,x_4,x_5) = 0 \\ g(x_1,x_2,x_3,x_4) + x_5 = 0 \end{array} \right) \subset \mathfrak{D}^5/\mathbf{Z}_n(a_1,a_2,a_3,a_4,a_5)$$

such that π is a suitable weighted blow-up with respect to the orbifold coordinates x_1,x_2,x_3,x_4,x_5.

Theorems 3.5.5 to 3.5.7 are proved for each type by the same procedure. First we narrow down the candidates for the singularity $x \in X$ by the general elephant conjecture. Next we recover the structure of the graded ring

$$R = \bigoplus_{(i,j)\in S} R_{ij}, \quad R_{ij} = \pi_*\mathcal{O}_Y(iK_Y - jE)/\pi_*\mathcal{O}_Y(iK_Y - (j+1)E)$$

for $S = \mathbf{Z}^2/\mathbf{Z}(n,a)$ from the numerical data dim R_{ij}. Finally we take a weighted blow-up compatible with the structure of R and confirm an identification of it with π by means of Proposition 3.5.3. In the rest of the section, we shall prove the theorems in the most typical non-Gorenstein case when X is cA/n, and then we shall pick the case of type e5 and demonstrate a complete classification of π, for which an embedding of $x \in X$ into $\mathfrak{D}^5$ is necessary. Examples for Table 3.4 will be provided in the next section.

Assuming that $x \in X$ is cA/n with $n \geq 2$, we shall prove Theorems 3.5.5 and 3.5.7. The case of quotient singularity is settled by Theorem 3.1.5 and the case of discrepancy $1/n$ is by Hayakawa's work [181, theorem 6.4]. Thus we shall assume that $x \in X$ is not a quotient singularity and that the discrepancy a/n is greater than $1/n$. Analysing the dual graph of the partial resolution $\pi_S\colon S \to S_X$, one can exclude the types other than o2 or o3 in Theorem 3.2.9. Hence π is of type o2 or o3 and it belongs to the case (i) in Theorem 3.4.1.

As before, we let H_X be the general hyperplane section of $x \in X$ and let H be the strict transform in Y. Let S be the general elephant of Y and let S_X be the strict transform in X. Then $\pi^*H_X = H + E$ and $\pi^*S_X = S + (a/n)E$. By Theorem 3.4.1, the surfaces H, E and S intersect properly at two points q_1 and q_2 in Y. These points are a smooth point and the point q in I if π is of type o2, and the points q_1 and q_2 in I if type o3. We use the common notation that $q_i \in Y$ is of type $\frac{1}{r_i}(1, -1, b_i)$, where $r_1 = 1$ if type o2.

Theorem 2.4.8 provides an analytic isomorphism

$$x \in X \simeq o \in (x_1 x_2 + f(x_3^n, x_4) = 0) \subset \mathfrak{D}^4/\mathbf{Z}_n(1, -1, b, 0)$$

with orbifold coordinates x_1, x_2, x_3, x_4. Let $h = h(x_1, x_2, x_3, x_4)$ be the invariant function which defines H_X. Let $s = s(x_1, x_2, x_3, x_4)$ be the semi-invariant function of weight b which defines S_X.

Lemma 3.5.9 *One can choose the isomorphism above so that $h - x_4$ belongs to $\pi_* \mathcal{O}_Y(-2E)$ and $s - x_3$ belongs to $\pi_* \mathcal{O}_Y(-K_Y - E)$.*

Proof The proof in [224, lemma 6.3] does not work when x_3^2 appears in f. In this case, we need the existence of orbifold coordinates y_1, y_2, y_3, y_4 of $\mathfrak{D}^4/\mathbf{Z}_2(1, 1, 1, 0)$ either with $\mathrm{ord}_E y_i \geq a/2$ for all $1 \leq i \leq 3$ or with $\mathrm{ord}_E y_1 > \mathrm{ord}_E y_2 = a/2 > \mathrm{ord}_E y_3 = r_1/2$, in the latter case of which they have no relations in $R_i = \pi_* \mathcal{O}_Y(-iE)/\pi_* \mathcal{O}_Y(-(i+1)E)$ for $i \leq (r_1+a)/2$. This is obtained by such a computation as in the proof of Lemma 3.5.2 or 3.5.16 or in [224, section 7]. Henceforth we assume that x_3^2 does not appear in f.

Since S_X is Du Val of type A by Theorem 3.4.1, the monomial x_3 appears in s. Write $s \equiv u x_3 + x_1 p(x_1, x_4) + q(x_2, x_4) \mod \pi_* \mathcal{O}_Y(-K_Y - E)$ with a unit u. Replacing x_3 by $u x_3 + x_1 p$ and changing x_2 accordingly, one has a new expression $s \equiv v x_3 + x_1^{n-1} p^n g(x_1, x_4) + r(x_2, x_4)$ with a unit v. In other words, p is replaced by $x_1^{n-2} p^n g$ where $x_1^{n-2} p^{n-1} g$ is not a unit on our assumption. Iterating this procedure finitely many times, one can eventually eliminate the part $x_1 p$. The same argument also eliminates q, by which s is congruent to x_3 modulo $\pi_* \mathcal{O}_Y(-K_Y - E)$ up to unit. Writing $h \equiv u x_4 + x_1 p(x_1, x_3) + q(x_2, x_3)$ mod $\pi_* \mathcal{O}_Y(-2E)$, one can eliminate $x_1 p$ and q in like manner, in the course of which the orbifold coordinate x_3 is preserved. $\square$

We take an expression of $x \in X$ as in Lemma 3.5.9. Then $\mathrm{ord}_E x_3 = a/n$ and $\mathrm{ord}_E x_4 = 1$. Define $w = \mathrm{wt}(x_1, x_2, x_3, x_4)$ by $\mathrm{wt}\, x_i = \mathrm{ord}_E x_i$. For a semi-invariant function g, let w-ord g denote the weighted order of g with respect to w. Notice that w-ord $g \leq \mathrm{ord}_E g$ for all g and the equality holds if g is a monomial in x_1, x_2, x_3, x_4. The following lemma is crucial.

Lemma 3.5.10 w-ord $f = $ w-ord $x_1 x_2$.

Proof Let $f_w = f_w(x_3^n, x_4)$ denote the weighted homogeneous part in f of weighted degree w-ord f with respect to w. Then f_w is always factorised as

$$x_4^{\text{w-ord} f - da} \prod_{l=1}^{d} (x_3^n + c_l x_4^a)$$

with $c_l \in \mathbf{C}$ up to constant, where $d \le (\text{w-ord } f)/a$ is the degree of f_w in x_3^n. If $\text{ord}_E (x_3^n + c_l x_4^a) = a$ for all l, then $\text{ord}_E f_w = \text{w-ord } f$. In this case,

$$\text{ord}_E f_w = \text{w-ord } f < \text{w-ord}(f - f_w) \le \text{ord}_E (f - f_w).$$

Since $f_w + x_1 x_2 = f_w - f$ on X, this strict inequality implies that $\text{ord}_E f_w = \text{ord}_E x_1 x_2$, which equals w-ord $x_1 x_2$.

Thus it suffices to see that $\text{ord}_E (x_3^n + c_l x_4^a) = a$. By Lemma 3.5.9, this is equivalent to the equality $\text{ord}_E g = a$ for $g = s^n + c_l h^a$. By $(H \cdot E \cdot S) = (a/n)E^3 = 1/r_1 + 1/r_2$ in Table 3.1, the local intersection number $(H \cdot E \cdot S)_{q_i}$ is $1/r_i$ for each i. Hence one can take orbifold coordinates x_{i1}, x_{i2}, x_{i3} at $q_i \in Y \simeq \mathfrak{D}^3/\mathbf{Z}_{r_i}(1, -1, b_i)$ so that H, E, S are defined by x_{i1}, x_{i2}, x_{i3} respectively. In particular on the analytic germ $q_i \in Y$, the functions s^n and h^a are expressed as $s^n = u x_{i3}^n x_{i2}^a$ and $h^a = v x_{i1}^a x_{i2}^a$ with units u and v. Then $g = (u x_{i3}^n + c_l v x_{i1}^a) x_{i2}^a$, which has order a along the divisor E defined by x_{i2}.   □

Proof of Theorems 3.5.5 *and* 3.5.7 *for type* cA/n We apply an immediate extension of Proposition 3.5.3, which is stated as Proposition 3.5.11 without proof. We write f_w for the weighted homogeneous part in f of weighted degree w-ord f, which equals w-ord $x_1 x_2$ by Lemma 3.5.10.

We take $w = \frac{1}{n}(w_1, w_2, a, n)$ by $w_i = n \, \text{ord}_E x_i$ for $i = 1, 2$. Let $\varphi \colon B \to X$ be the weighted blow-up with respect to weights w. The exceptional locus F is given in $\mathbf{P}(w_1, w_2, a, n)$ by $x_1 x_2 + f_w(x_3^n, x_4)$, which is integral. Hence B is normal by Serre's criterion. The exceptional divisor F has log discrepancy $a_F(X) = (w_1 + w_2 + a + n)/n - \text{w-ord } x_1 x_2 = a/n + 1 = a_E(X)$.

We shall describe B explicitly. It is embedded in the weighted blow-up $\bar{B}$ of $\mathfrak{D}^4/\mathbf{Z}_n(1, -1, b, 0)$ with weights w. The ambient space $\bar{B}$ is covered by the four charts $\bar{U}_i = (x_i \ne 0)$ for $1 \le i \le 4$. We write $U_i = \bar{U}_i|_B$. Let $N = \mathbf{Z}^4 + \mathbf{Z} \cdot \frac{1}{n}(1, -1, b, 0)$ and let e_1, e_2, e_3, e_4 denote the standard basis of $N \otimes \mathbf{R}$. Note that $w_1 + w_2$, $a - bw_1$ and $a + bw_2$ are all divisible by n.

(i) The lattice N is generated by w, e_2, e_3, e_4 and

$$\frac{1}{n}(-1, 1, -b, 0) = \frac{1}{w_1}\left(-w + \frac{w_1 + w_2}{n}e_2 + \frac{a - bw_1}{n}e_3 + e_4\right).$$

Hence $\bar{U}_1 \simeq \mathfrak{D}^4/\mathbf{Z}_{w_1}(-1, (w_1 + w_2)/n, (a - bw_1)/n, 1)$ with orbifold coordinates $(y_1, y_2, y_3, y_4) = (x_1^{n/w_1}, x_2 x_1^{-w_2/w_1}, x_3 x_1^{-a/w_1}, x_4 x_1^{-n/w_1})$. The locus U_1 is given in $\bar{U}_1$ by $y_2 + f_w(y_3^n, y_4)$ modulo y_1. It has a quotient singularity of type $\frac{1}{w_1}(-1, (a - bw_1)/n, 1)$ at origin.

(ii) $\bar{U}_2 \simeq \mathfrak{D}^4/\mathbf{Z}_{w_2}((w_1 + w_2)/n, -1, (a + bw_2)/n, 1)$ and U_2 has a quotient singularity of type $\frac{1}{w_2}(-1, (a + bw_2)/n, 1)$.

(iii) The lattice N is generated by e_1, e_2, w, e_4 and

$$\frac{1}{n}(-b', b', -1, 0) = \frac{1}{a}\left(\frac{w_1 - b'a}{n}e_1 + \frac{w_2 + b'a}{n}e_2 - w + e_4\right)$$

for an integer b' such that $bb' \equiv 1 \bmod n$. Hence $\bar{U}_3 \simeq \mathfrak{D}^4/\mathbf{Z}_a((w_1 - b'a)/n, (w_2 + b'a)/n, -1, 1)$ and U_3 is given by $y_1 y_2 + f_w(1, y_4)$ modulo y_3.

(iv) $\bar{U}_4 \simeq \mathfrak{D}^4/\mathbf{Z}_n(1, -1, b, 0)$ and U_4 is given by $y_1 y_2 + f_w(y_3^n, 1)$ modulo y_4.

By (i), B is smooth outside the strict transform of the divisor $(x_1 x_2)_X$. Thus π is isomorphic to φ by Proposition 3.5.11. We have to confirm the conditions in Theorem 3.5.5. Since $B \simeq Y$ is terminal, w_1 is coprime to $(a - bw_1)/n$ by (i) and w_2 is coprime to $(a + bw_2)/n$ by (ii). Then by (iii), U_3 is terminal only if it does not contain the origin of $\bar{U}_3$, that is, $f_w(1, 0) \neq 0$. Hence the monomial x_3 to the power of (n/a) w-ord $f = (w_1 + w_2)/a$ appears in f. This implies that $(w_1 + w_2)/a$ is divisible by n, that is, an divides $w_1 + w_2$. Using [224, lemma 6.6], one can directly check that any weighted blow-up in Theorem 3.5.5 is a divisorial contraction. $\square$

Proposition 3.5.11 *Set $o \in A = \mathfrak{D}^n/\mathbf{Z}_r(a_1, \ldots, a_n)$ with orbifold coordinates $x_1, \ldots, x_n$. Let $o \in X = (\xi = 0) \subset A$ be a normal analytic subspace defined by a semi-invariant function ξ. Suppose that $\mathbf{Z}_r$ acts on the hypersurface in $\mathfrak{D}^n$ defined by ξ freely in codimension one. Let $\bar{B}$ be the weighted blow-up of A with $\mathrm{wt}(x_1, \ldots, x_n) = \frac{1}{r}(w_1, \ldots, w_n)$ and let $\bar{F}$ be the exceptional divisor. Let B be the strict transform in $\bar{B}$ of X. Suppose that the support F of $\bar{F} \cap B$ is irreducible and $\bar{F} \cap B$ is generically reduced. Thus B is normal and has the exceptional divisor F. Let E be a divisor over X such that $c_X(E) \subset c_X(F)$. If $(w_1, \ldots, w_n)$ is parallel to $(\mathrm{ord}_E x_1, \ldots, \mathrm{ord}_E x_n)$, then the centre $c_B(E)$ is not contained in the strict transform H of the divisor on X defined by the product $x_1 \cdots x_n$. If in addition $B \setminus H$ is terminal, then $a_F(X) \leq a_E(X)$ and the equality holds if and only if $E = F$ as a divisor over X.*

We proceed to the demonstration of the following theorem.

Theorem 3.5.12 *Let $\pi\colon E \subset Y \to x \in X$ be a threefold divisorial contraction of type e5. If the discrepancy a/n is not $1/n$, then $x \in X$ is cE_7, $a = 2$ and there exists an analytic identification*

$$x \in X \simeq o \in \left(\begin{array}{c} x_1^2 + x_2 x_5 + f(x_3, x_4) = 0 \\ x_2^2 + g(x_3, x_4) + x_5 = 0 \end{array}\right) \subset \mathfrak{D}^5$$

such that π is the weighted blow-up with $\mathrm{wt}(x_1, x_2, x_3, x_4, x_5) = (5, 3, 2, 2, 7)$, where $\mathrm{ord}\, f = 5$, $\mathrm{ord}\, g = 3$ and the homogeneous part of lowest degree in f and that in g have no common factors.

From Theorems 3.2.8 and 3.2.9, $(a, n) = (2, 1)$. From Table 3.2, $J = \{(7, 3)\}$ and $E^3 = 1/7$. Thus Y has a unique non-Gorenstein point q and $q \in Y$ is a quotient singularity of type $\frac{1}{7}(1, 6, 6)$. Using Lemma 3.3.7, one can compute $d(i)$ inductively from $d(1) = 0$ and $d(0) = 1$ as below.

Lemma 3.5.13 $d(-1) = 0$, $d(-2) = 2$, $d(-3) = 1$, $d(-4) = 3$, $d(-5) = 3$, $d(-6) = 4$, $d(-7) = 6$.

Let $x \in S_X$ be the general hyperplane section of $x \in X$ and let S be the strict transform in Y. By Corollary 3.3.3, S is the general elephant of Y and $\pi^* S_X = S + 2E$. We study the dual graph of the partial resolution $\pi_S \colon S \to S_X$.

Lemma 3.5.14 *The dual graph of $\pi_S \colon S \to S_X$ is one of the following.*

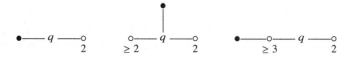

The attached number at $\circ$ is the coefficient in the fundamental cycle. The point q is that in I. The surface S is at q of type A_6 or worse and smooth elsewhere.

Proof By Corollary 3.3.11, $S \cap E$ has no embedded points and its support is a tree of $\mathbf{P}^1$. By $(-E \cdot S \cdot E) = 2E^3 = 2/7$, the one-cycle $[S \cap E]$ consists of at most two $\mathbf{P}^1$ and every $\mathbf{P}^1$ passes through q. From Lemma 3.3.8 and $d(-2) = 2$ in Lemma 3.5.13, one has the isomorphism $\mathbf{C}^2 \simeq \pi_* \mathcal{O}_Y(-2E)/\pi_* \mathcal{O}_Y(-3E) \simeq H^0(\mathcal{Q}_{-2})$. As in the argument before Lemma 3.4.2, the corresponding linear system Λ on $E \setminus q$ contains non-zero mobile part. Hence by the general choice of S_X, $S \cap E$ is a reduced scheme consisting of one or two $\mathbf{P}^1$. In particular, S is smooth outside q by Theorem 2.3.18. On the other hand by the intersection number $E^2 S = -2/7$, the Du Val singularity $q \in S$ is of type A_6 or worse.

Let $x \in H_X$ be another general hyperplane section of $x \in X$. Then $\pi^* H_X = H + 2E$ with the strict transform H of H_X. Let B be the strict transform in S of the curve $B_X = H_X \cap S_X$. If Λ has no fixed component, then $[H \cap S] = [B]$ cycle-theoretically and $\pi_S^* B_X = H|_S + 2E|_S = B + 2E|_S$. Further B intersects E only at q by $(E \cdot B) = (E \cdot H \cdot S) = 4/7$. Hence π_S has the first or second dual graph in the statement. If Λ has a fixed component, then $[S \cap E] = [C] + [D]$ and $[H \cap E] = [C] + [D']$ with different curves C, D and D'. In this case, $[H \cap S] = c[C] + [B]$ with some positive c and $\pi_S^* B_X = H|_S + 2E|_S = (cC + B) + 2(C + D)$. Further H intersects D only at q by $(H \cdot D) < (H \cdot S \cdot E) = 4/7$. Hence π_S has the second or third dual graph in the statement. $\qquad \square$

Corollary 3.5.15 *The germ $x \in X$ is cE_7.*

Proof We use the labels F_i in Notation 3.3.4 for the exceptional curves on the minimal resolution of a Du Val singularity. Let Z be the push-forward to S of the fundamental cycle on the minimal resolution of S_X. In order for $\pi_S \colon S \to S_X$ to satisfy Lemma 3.5.14, either

(i) $x \in S_X$ is of type E_7 and $Z = 2F_7$ or
(ii) $x \in S_X$ is of type E_8 and $Z = 2F_7$ or $2F_7 + 3F_8$.

Suppose that S_X were of type E_8. If $Z = 2F_7$, then $(-E|_S \cdot E|_S) = -(F_7 \cdot F_7)_S = 1/4$, contradicting $(-E|_S \cdot E|_S) = 2E^3 = 2/7$. If $Z = 2F_7 + 3F_8$, then $(-E|_S \cdot F_8) = -((F_7 + F_8) \cdot F_8)_S = -1/7$, contradicting the relative ampleness of $-E|_S$. Thus the general hyperplane section S_X is of type E_7 and X is cE_7. □

The next lemma indicates how to recover the graded ring $R = \bigoplus R_{ij}$ with $R_{ij} = \pi_* \mathscr{O}_Y(iK_Y - jE)/\pi_* \mathscr{O}_Y(iK_Y - (j+1)E)$ from the numerical data.

Lemma 3.5.16 *Consider an analytic embedding of $x \in X$ into $o \in \mathfrak{D}^4$ with coordinates x_1, x_2, x_3, x_4.*

(i) *Up to permutation of x_1, x_2, x_3, x_4, the monomials x_3 and x_4 form a basis of $\pi_* \mathscr{O}_Y(-2E)/\pi_* \mathscr{O}_Y(-3E) \simeq \mathbf{C}^2$. In particular for $i = 1, 2$, there exists a unique linear form $l_i(x_3, x_4)$ such that $y_i = x_i + l_i(x_3, x_4)$ has $\mathrm{ord}_E\, y_i \geq 3$.*
(ii) *Up to permutation of y_1 and y_2, $\mathrm{ord}_E\, y_2 = 3$.*
(iii) *There exists a unique element $p(y_2, x_3, x_4)$ in $\mathbf{C}y_2 + \mathbf{C}x_3^2 + \mathbf{C}x_3x_4 + \mathbf{C}x_4^2$ such that $z_1 = y_1 + p(y_2, x_3, x_4)$ has $\mathrm{ord}_E\, z_1 \geq 5$. Further $\mathrm{ord}_E\, z_1 = 5$.*
(iv) *There exists a unique non-zero cubic form $q(x_3, x_4)$ such that $h = y_2^2 + q(x_3, x_4)$ has $\mathrm{ord}_E\, h \geq 7$. Further $\mathrm{ord}_E\, h = 7$ and there exists a natural exact sequence*

$$0 \to \mathbf{C}h \to \mathbf{C}y_2^2 \oplus \bigoplus_{i=0}^{3} \mathbf{C}x_3^{3-i}x_4^i \to \pi_* \mathscr{O}_Y(-6E)/\pi_* \mathscr{O}_Y(-7E) \to 0.$$

Proof By Lemma 3.3.7(i), the vector space $R_i = \pi_* \mathscr{O}_Y(-iE)/\pi_* \mathscr{O}_Y(-(i+1)E)$ for $i \in \mathbf{N}$ is of dimension $d(-i)$, which is computed in Lemma 3.5.13 for $i \leq 7$. All $\mathrm{ord}_E\, x_j$ are at least two by $d(-1) = 0$ and the assertion (i) follows from $d(-2) = 2$. Note that for the basis x_3, x_4 of R_2, the monomials in x_3, x_4 of degree i are linearly independent in R_{2i} because every homogeneous polynomial in x_3, x_4 is decomposed into linear factors.

The assertion (ii) follows from $d(-3) = 1$. The function y_2 with $\mathrm{ord}_E\, y_2 = 3$ is a basis of R_3. By $d(-4) = 3$, the monomials x_3^2, x_3x_4, x_4^2 form a basis of R_4. This shows the unique existence of p in (iii). Then R_5 is generated by z_1, y_2x_3, y_2x_4, which must form a basis of R_5 by $d(-5) = 3$. Hence $\mathrm{ord}_E\, z_1 = 5$, showing (iii).

The space R_6 is generated by y_2^2, x_3^3, $x_3^2 x_4$, $x_3 x_4^2$, x_4^3 but $d(-6) = 4$. Thus the surjection $\mathbf{C}y_2^2 \oplus \bigoplus_i \mathbf{C}x_3^{3-i}x_4^i \twoheadrightarrow R_6$ has kernel of dimension one. Since the monomials in x_3, x_4 are linearly independent, the kernel has a generator $h = y_2^2 + q(x_3, x_4)$ with a unique non-zero cubic form q. Then R_7 is generated by h, $z_1 x_3$, $z_1 x_4$, $y_2 x_3^2$, $y_2 x_3 x_4$, $y_2 x_4^2$, which must form a basis of R_7 by $d(-7) = 6$. Hence $\text{ord}_E h = 7$, which completes (iv). □

Proof of Theorem 3.5.12 By Corollary 3.5.15, $x \in X$ is cE_7. By Proposition 2.3.3, there exists an analytic isomorphism

$$x \in X \simeq o \in (x_1^2 + x_2^3 + x_2 g(x_3, x_4) + f(x_3, x_4) = 0) \subset \mathfrak{D}^4$$

in which $\text{ord}\, g = 3$ and $\text{ord}\, f \geq 5$. Let g_3 be the homogeneous part in g of degree three, which is non-zero, and let f_5 be that in f of degree five.

Note that $\text{ord}_E x_j \geq 2$ for all j. We shall see that $\text{ord}_E x_2 \geq 3$. If $\text{ord}_E x_2 = 2$ contrarily, then $\text{ord}_E x_1 = 3$ by $\text{ord}_E x_1^2 = \text{ord}_E(x_2^3 + x_2 g + f) = 6$. By Lemma 3.5.16(i), x_2 and x_4 form a basis of $\pi_* \mathcal{O}_Y(-2E)/\pi_* \mathcal{O}_Y(-3E)$ up to permutation of x_3 and x_4. Then the triple (x_1, x_2, x_4) plays the role of (y_2, x_3, x_4) in Lemma 3.5.16. By $\text{ord}_E(x_1^2 + x_2^3) = \text{ord}_E(x_2 g + f) \geq 8$, the polynomial $x_1^2 + x_2^3$ must play the role of h in Lemma 3.5.16. This h should have $\text{ord}_E h = 7$, which is absurd.

Thus $\text{ord}_E x_2 \geq 3$, and $\text{ord}_E x_1 \geq 5$ by $\text{ord}_E x_1^2 = \text{ord}_E(x_2^3 + x_2 g + f) \geq 9$. From Lemma 3.5.16, one derives $\text{ord}_E x_1 = 5$, $\text{ord}_E x_2 = 3$ and $\text{ord}_E x_3 = \text{ord}_E x_4 = 2$. Then $\text{ord}_E(x_2^2 + g) \geq 7$ by $\text{ord}_E x_2(x_2^2 + g) = \text{ord}_E(x_1^2 + f) \geq 10$. Hence $\text{ord}_E(x_2^2 + g) = 7$ by Lemma 3.5.16(iv).

Letting $x_5 = -(x_2^2 + g)$ and replacing x_2 by $-x_2$, one obtains the analytic identification

$$x \in X \simeq o \in \left(\begin{array}{c} x_1^2 + x_2 x_5 + f(x_3, x_4) = 0 \\ x_2^2 + g(x_3, x_4) + x_5 = 0 \end{array} \right) \subset \mathfrak{D}^5.$$

Let $\varphi \colon B \to X$ be the weighted blow-up with respect to $\text{wt}(x_1, x_2, x_3, x_4, x_5) = (5, 3, 2, 2, 7)$. The exceptional locus F is given in $\mathbf{P}(5, 3, 2, 2, 7)$ by $x_1^2 + x_2 x_5 + f_5(x_3, x_4)$ and $x_2^2 + g_3(x_3, x_4)$, which is integral as $g_3 \neq 0$. Hence B is normal. The exceptional divisor F has $a_F(X) = 3 = a_E(X)$.

The threefold B is covered by the five charts $U_i = (x_i \neq 0)$ for $1 \leq i \leq 5$. The chart U_5 is given in $\mathfrak{D}^5/\mathbf{Z}_7(5, 3, 2, 2, 6)$ by $y_1^2 + y_2 + f_5(y_3, y_4)$ and $y_2^2 + g_3(y_3, y_4) + y_5$ modulo y_5^2, which has a quotient singularity of type $\frac{1}{7}(5, 2, 2)$ at origin. In particular, B is smooth outside the strict transform of the divisor $(x_1 x_2 x_3 x_4 x_5)_X$. Thus one obtains the isomorphism $\pi \simeq \varphi$ by a straightforward extension of Proposition 3.5.3 or 3.5.11 to a complete intersection threefold in $\mathfrak{D}^5$.

We shall check that f_5 and g_3 have no common factors. The chart U_3 is given in $\mathfrak{D}^5/\mathbf{Z}_2(1,1,1,0,1)$ by $y_1^2 + y_2 y_5 + f_5(1, y_4)$ and $y_2^2 + g_3(1, y_4) + y_3 y_5$ modulo y_3^2. Since $B \simeq Y$ is terminal, U_3 must be away from the y_4-axis. Hence $f_5(1, y_4)$ and $g_3(1, y_4)$ have no common factors. In like manner on U_4, $f_5(y_3, 1)$ and $g_3(y_3, 1)$ have no common factors. This completes the theorem. □

Remark 3.5.17 Table 3.4 excludes the cases (type, x) = (e2, cE_7), (e3, cE_6) from [224, table 3]. It also excludes the case (type, x, a) = (e1, cA_2, 4) treated in [474]. We shall explain this briefly.

If (type, x) = (e3, cE_6), then $x \in X$ is given by $x_1^2 + x_2^3 + x_2 f(x_3, x_4) + g(x_3, x_4)$ with ord $f \geq 3$ and ord $g = 4$. Similarly to Lemma 3.5.16, there exist coordinates y_1, y_2, y_3, y_4 with $\mathrm{ord}_E(y_1, y_2, y_3, y_4) = (4, 3, 2, 1)$ [474, lemma 4.23], which have no relations of weighted degree less than six in the direct sum of $R_i = \pi_* \mathscr{O}_Y(-iE)/\pi_* \mathscr{O}_Y(-(i+1)E)$. One sees that $\mathrm{ord}_E x_1 \geq 3$ and $\mathrm{ord}_E x_2 \geq 2$ and may assume that $\mathrm{ord}_E x_3 \geq 2$ and $\mathrm{ord}_E x_4 = 1$. Then the linear part of the function defining the general elephant S_X of X is a combination of x_1, x_2, x_3 containing x_1. Since S_X is of type E_6 from Theorem 3.3.5, g must contain x_4^4. This would generate a relation of weighted degree four, which is a contradiction.

If (type, x) = (e2, cE_7), then $J = \{(3, 1), (3, 1)\}$ as in [224, table 3] and S_X is of type E_7 from Theorem 3.3.5. With [474, lemma 4.7], the same strategy yields the expression $x_1^2 + (x_2 + cx_4^2)^3 + x_2 f(x_3, x_4) + g(x_3, x_4)$ of X with $\mathrm{ord}_E(x_1, x_2, x_3, x_4) = (3, 3, 2, 1)$ where $c \in \mathbf{C}$, ord $f = 3$, ord $g \geq 5$ and f contains x_4^3. Then the weighted blow-up B of X with $\mathrm{wt}(x_1, x_2, x_3, x_4) = (3, 3, 2, 1)$ is smooth outside the strict transform of $(x_1 x_2 x_3 x_4)_X$. By Proposition 3.5.11, it would be the divisorial contraction. However, B is singular along a curve.

If (type, x, a) = (e1, cA_2, 4), then $J = \{(r, 2)\}$ with $r > 5$ [223, lemma 6.3] and S_X is of type D from Theorem 3.3.5. Then $r \geq 11$ and the same strategy works with [474, lemma 4.18].

3.6 Examples

We collect examples of divisorial contractions $\pi \colon E \subset Y \to x \in X$ in Framework 3.2.1. Those up to Example 3.6.12 are found in [223], [224], [474]. A system of weights in this section means $\mathrm{wt}(x_1, x_2, x_3, x_4)$ or $\mathrm{wt}(x_1, x_2, x_3, x_4, x_5)$. We begin with concrete examples for Theorem 3.5.6.

Example 3.6.1 (i) Let $x \in X$ be the cD point

$$o \in (x_1^2 + x_2^2 x_4 + x_3^{(2r+1)/a} + x_4^{2r+1} = 0) \subset \mathfrak{D}^4$$

where a divides $2r + 1$ and $a \neq 2r + 1$. If $r \neq 1$, then the weighted blow-up of X with weights $(r + 1, r, a, 1)$ is a divisorial contraction of type o3 with discrepancy a and $J = \{(r, 1), (r + 1, 1)\}$. If $r = 1$, then it is of type o2.

(ii) Let $x \in X$ be the $cD/2$ or $cAx/2$ point

$$o \in (x_1^2 + x_2^2 x_4 + x_3^{2(r+1)/a} + x_4^{r+1} = 0) \subset \mathfrak{D}^4/\mathbf{Z}_2(1, 1, 1, 0)$$

where a divides $r + 1$, and a and r are odd. If $r \neq 1$, then X is $cD/2$ and the weighted blow-up of X with weights $\frac{1}{2}(r + 2, r, a, 2)$ is a divisorial contraction of type o3 with discrepancy $a/2$ and $J = \{(r, 1), (r + 2, 1)\}$. If $r = 1$, then X is $cAx/2$ and the contraction is of type o2.

Example 3.6.2 (i) Let $x \in X$ be the cD point

$$o \in \left(\begin{array}{l} x_1^2 + x_2 x_5 + x_4^{2r+2} = 0 \\ x_2 x_4 + x_3^{(r+1)/a} + x_5 = 0 \end{array} \right) \subset \mathfrak{D}^5$$

where a divides $r + 1$ and $a \neq r + 1$. If $r \neq 1$, then the weighted blow-up of X with weights $(r + 1, r, a, 1, r + 2)$ is a divisorial contraction of type o3 with discrepancy a and $J = \{(r, 1), (r + 2, 1)\}$. If $r = 1$, then it is of type o2.

(ii) Let $x \in X$ be the $cD/2$ point

$$o \in \left(\begin{array}{l} x_1^2 + x_2 x_5 + x_4^{r+2} = 0 \\ x_2 x_4 + x_3^{(r+2)/a} + x_5 = 0 \end{array} \right) \subset \mathfrak{D}^5/\mathbf{Z}_2(1, 1, 1, 0, 1)$$

where a divides $r + 2$, $(r + 2)/a$ is odd and $a \neq r + 2$. If $r \neq 1$, then the weighted blow-up of X with weights $\frac{1}{2}(r + 2, r, a, 2, r + 4)$ is a divisorial contraction of type o3 with discrepancy $a/2$ and $J = \{(r, 1), (r + 4, 1)\}$. If $r = 1$, then it is of type o2.

We have a complete classification when X is cA_1.

Example 3.6.3 ([222]) Let $\pi \colon E \subset Y \to x \in X$ be a threefold divisorial contraction which contracts the divisor E to a cA_1 point x. The classification for ordinary type is included in Theorem 3.5.5. If π is of exceptional type, then there exists an analytic identification

$$x \in X \simeq o \in (x_1 x_2 + x_3^2 + x_4^3 = 0) \subset \mathfrak{D}^4$$

such that π is the weighted blow-up with weights $(5, 1, 3, 2)$. It is of type e1 with discrepancy four and $J = \{(5, 2)\}$.

Below are examples of other cases of type e1, in which X is cD or $cD/2$.

Example 3.6.4 Let $x \in X$ be the cD point either

$$o \in \left(\begin{array}{l} x_1^2 + x_4 x_5 + x_3^{(r+1)/4} = 0 \\ x_2^2 + x_1 x_3^{(r-3)/8} + x_4^{r-1} + x_5 = 0 \end{array} \right) \subset \mathfrak{D}^5$$

with $r \equiv 3 \bmod 8$ and $r \neq 3$ or

$$o \in \left(\begin{array}{c} x_1^2 + x_4 x_5 + x_2 x_3^{(r+3)/8} = 0 \\ x_2^2 + x_3^{(r-1)/4} + x_4^{r-1} + x_5 = 0 \end{array} \right) \subset \mathfrak{D}^5$$

with $r \equiv 5 \bmod 8$ and $r \neq 5$. The weighted blow-up of X with weights $((r+1)/2, (r-1)/2, 4, 1, r)$ is a divisorial contraction of type e1 with discrepancy four and $J = \{(r, 2)\}$.

Example 3.6.5 Let $x \in X$ be the $cD/2$ point either

$$o \in \left(\begin{array}{c} x_1^2 + x_4 x_5 + x_2 x_3^{(s+3)/4} = 0 \\ x_2^2 + x_3^{(s-1)/2} + x_4^{s-1} + x_5 = 0 \end{array} \right) \subset \mathfrak{D}^5 / \mathbf{Z}_2(1, 1, 1, 0, 0)$$

with $s \equiv 1 \bmod 8$ and $s \neq 1$ or

$$o \in \left(\begin{array}{c} x_1^2 + x_4 x_5 + x_3^{(s+1)/2} = 0 \\ x_2^2 + x_1 x_3^{(s-3)/4} + x_4^{s-1} + x_5 = 0 \end{array} \right) \subset \mathfrak{D}^5 / \mathbf{Z}_2(1, 1, 1, 0, 0)$$

with $s \equiv 7 \bmod 8$. The weighted blow-up of X with weights $((s+1)/2, (s-1)/2, 2, 1, s)$ is a divisorial contraction of type e1 with discrepancy $4/2$ and $J = \{(2s, 2)\}$. The double cover constructed as in the argument preceding Lemma 3.2.12 is also a divisorial contraction of type e1.

Example 3.6.6 Let $x \in X$ be the $cD/2$ point

$$o \in \left(\begin{array}{c} x_1^2 + x_4 x_5 + x_3^{s+1} = 0 \\ x_2^2 + x_3^{s-1} + x_4^{s-1} + x_5 = 0 \end{array} \right) \subset \mathfrak{D}^5 / \mathbf{Z}_2(1, 1, 1, 0, 0)$$

with s odd and $s \neq 1$. The weighted blow-up of X with weights $((s+1)/2, (s-1)/2, 1, 1, s)$ is a divisorial contraction of type e1 with discrepancy $2/2$ and $J = \{(2s, 2)\}$. Its double cover is also a divisorial contraction of type e1.

We shall exhibit examples for other types in Table 3.4.

Example 3.6.7 (i) Let $x \in X$ be the cD point

$$o \in (x_1^2 + x_2^2 x_4 + x_3^r + x_4^{2r} = 0) \subset \mathfrak{D}^4$$

with r odd and $r \neq 1$. The weighted blow-up of X with weights $(r, r, 2, 1)$ is a divisorial contraction of type e2 with discrepancy two in which $J = \{(r, 1), (r, 1)\}$ comes from a cA/r point.

(ii) Let $x \in X$ be the cE_6 point

$$o \in (x_1^2 + (x_2 - x_3)^3 + x_3^4 + x_2 x_4^3 = 0) \subset \mathfrak{D}^4.$$

The weighted blow-up X with weights $(3, 3, 2, 1)$ is a divisorial contraction of type e2 with discrepancy two in which $J = \{(3, 1), (3, 1)\}$ comes from a $cD/3$ point.

Example 3.6.8 Let $x \in X$ be the $cD/2$ point

$$o \in (x_1^2 + x_2^2 x_4 + x_3^{2s} + x_4^{2s} = 0) \subset \mathfrak{D}^4/\mathbf{Z}_2(1, 1, 1, 0)$$

with s even. The weighted blow-up of X with weights$(s, s, 1, 1)$ is a divisorial contraction of type e2 with discrepancy $2/2$ and $J = \{(2s, 1), (2s, 1)\}$ from a $cA/2s$ point. Its double cover is also a divisorial contraction of type e2.

Example 3.6.9 Let $x \in X$ be the cA_2 or cD_4 point

$$o \in (x_1^2 + x_2^k + x_3^3 + x_2 x_4^2 = 0) \subset \mathfrak{D}^4$$

with $k \geq 2$. It is cA_2 if $k = 2$ and cD_4 if $k \geq 3$. The weighted blow-up of X with weights $(3, 4, 2, 1)$ is a divisorial contraction of type e3 with discrepancy three.

Example 3.6.10 Let $x \in X$ be the cE_7 point

$$o \in \left(\begin{array}{c} x_1^2 + x_2 x_5 + x_3^5 = 0 \\ x_2^2 + x_4^3 + x_5 = 0 \end{array} \right) \subset \mathfrak{D}^5.$$

The weighted blow-up of X with weights $(5, 3, 2, 2, 7)$ is a divisorial contraction of type e5 with discrepancy two.

Example 3.6.11 Let $x \in X$ be the cE_7 point

$$o \in (x_1^2 + x_2^3 + x_2 x_3^3 + x_4^7 = 0) \subset \mathfrak{D}^4$$

or the cE_8 point

$$o \in (x_1^2 + x_2^3 + x_3^5 + x_4^7 = 0) \subset \mathfrak{D}^4.$$

The weighted blow-up of X with weights $(7, 5, 3, 2)$ is a divisorial contraction of type e9 with discrepancy two.

Example 3.6.12 Let $x \in X$ be the $cE/2$ point

$$o \in (x_1^2 + x_2^3 + x_3^4 + x_4^8 = 0) \subset \mathfrak{D}^4/\mathbf{Z}_2(1, 0, 1, 1).$$

The weighted blow-up of X with weights $(4, 3, 2, 1)$ is a divisorial contraction of type e11 with discrepancy $2/2$. Its double cover is a divisorial contraction of type e13 with $J = \{(2, 1), (2, 1), (3, 1)\}$.

We shall complete a list of examples for all types but e4 in Tables 3.1 and 3.2. Note that type e4 never occurs as stated in Theorem 3.2.2. The types for which we have not yet given any examples are e6, e7, e8, e10, e12, e14, e15 and e16, in which the discrepancy a/n must be $1/n$ by Theorem 3.5.7.

Example 3.6.13 Let $x \in X$ be the $cD/2$ or $cAx/2$ point

$$o \in (x_1^2 + x_2^2 x_3 + x_3^{(r+3)/2} + x_4^{2r+4} = 0) \subset \mathfrak{D}^4/\mathbf{Z}_2(1, 1, 0, 1)$$

with r odd. If $r \neq 1$, then X is $cD/2$ and the weighted blow-up of X with weights $\frac{1}{2}(r + 2, r, 4, 1)$ is a divisorial contraction of type e7 with discrepancy $1/2$ and $J = \{(4, 2), (r, 1)\}$. If $r = 1$, then X is $cAx/2$ and the contraction is of type e1.

Example 3.6.14 Let $x \in X$ be the cE_6 point

$$o \in (x_1^2 + x_2^3 + x_3^4 + x_4^{12} = 0) \subset \mathfrak{D}^4.$$

The weighted blow-up of X with weights $(6, 4, 3, 1)$ is a divisorial contraction of type e14 with discrepancy one.

Example 3.6.15 Let $x \in X$ be the cE_7 point

$$o \in (x_1^2 + x_2^3 + x_2 x_3^3 + x_4^k = 0) \subset \mathfrak{D}^4.$$

The weighted blow-up of X with weights w for $(k, w) = (9, (5, 3, 2, 1))$, $(18, (9, 6, 4, 1))$ is a divisorial contraction of type e8, e15 respectively, with discrepancy one.

Example 3.6.16 Let $x \in X$ be the cE_8 point

$$o \in (x_1^2 + x_2^3 + x_3^5 + x_4^k = 0) \subset \mathfrak{D}^4.$$

The weighted blow-up of X with weights w for $(k, w) = (15, (8, 5, 3, 1))$, $(24, (12, 8, 5, 1))$, $(20, (10, 7, 4, 1))$, $(30, (15, 10, 6, 1))$ is a divisorial contraction of type e6, e10, e12, e16 respectively, with discrepancy one.

4

Divisorial Contractions to Curves

Every threefold divisorial contraction that contracts the divisor to a curve is the usual blow-up about the generic point of the curve. Considered to be an extraction, it is uniquely described as the symbolic blow-up as far as it exists.

The contraction bears similarities to a threefold flipping contraction because the central fibre is of dimension one. The general elephant conjecture is settled by Kollár and Mori when the fibre is irreducible. On the assumption of this conjecture, the symbolic blow-up always exists as a contraction from a canonical threefold. We want to determine whether the threefold is further terminal.

Two cases have been analysed in detail. One is by Tziolas when the extraction is from a smooth curve in a Gorenstein terminal threefold, and the other is by Ducat when it is from a singular curve in a smooth threefold. They follow the same division into cases based upon the divisor class of the curve in the Du Val section. In principle, Tziolas describes the symbolic blow-up as a certain weighted blow-up, whilst Ducat realises it by serial unprojections. The unprojection is an operation to construct a new Gorenstein variety from a simpler one.

The contraction can be regarded as a one-parameter deformation of the birational morphism of surfaces cut out by a hyperplane section. In reverse, one can construct a threefold contraction by deforming an appropriate surface morphism. We obtain examples of threefold divisorial contractions in the analytic category in this manner.

4.1 Contractions from Gorenstein Threefolds

The object in this chapter is a threefold divisorial contraction $\pi \colon E \subset Y \to C \subset X$ which contracts the divisor E to a curve C.

Definition 4.1.1   Let $\mathscr{I}$ be a coherent ideal sheaf on an algebraic scheme X. The *symbolic power algebra* of $\mathscr{I}$ is the graded $\mathscr{O}_X$-algebra $\mathscr{S} = \bigoplus_{i \in \mathbf{N}} \mathscr{I}^{(i)}$ for the i-th symbolic power $\mathscr{I}^{(i)}$ of $\mathscr{I}$ in Definition 3.4.5. Provided that $\mathscr{S}$ is finitely generated, the projective morphism $\mathrm{Proj}_X \mathscr{S} \to X$ is called the *symbolic blow-up* of X along $\mathscr{I}$, or along the closed subscheme defined by $\mathscr{I}$.

Example 4.1.2   The symbolic power algebra is not necessarily finitely generated [395]. A simple example is that of the ideal I in $\mathbf{C}[x_1, x_2, x_3]$ which defines the monomial curve $(x_1, x_2, x_3) = (t^{7n-3}, t^{8n-3}, t^{(5n-2)n})$ in $\mathbf{A}^3$ parametrised by t, where $n \geq 4$ and $n \not\equiv 0 \bmod 3$ [146]. The ideal I is generated by the maximal minors of the matrix $\begin{pmatrix} x_1^{2n-1} & x_2^n & x_3 \\ x_2^{2n-1} & x_3^2 & x_1^n \end{pmatrix}$.

The contraction π is determined by the pair $C \subset X$.

Proposition 4.1.3   *Let C be a curve in a terminal threefold X. Then a divisorial contraction $\pi \colon Y \to X$ which contracts the divisor onto C is unique if it exists. In this case, π is the symbolic blow-up of X along C and $K_Y = \pi^* K_X + E$ with the exceptional divisor E.*

Proof   We may assume that X is affine. Suppose the existence of a divisorial contraction $\pi \colon E \subset Y \to C \subset X$ which contracts the prime divisor E onto C. Write $K_Y = \pi^* K_X + dE$ with discrepancy $d \in \mathbf{Q}_{>0}$. Let H be the general hyperplane section of X, which intersects C, and let H_Y be the strict transform in Y. Then H and H_Y are smooth and $H_Y = \pi^* H$.

The exceptional locus of $H_Y \to H$ is a disjoint union $\bigcup_i f_i$ of general fibres f_i of $E \to C$. Every fibre f_i is connected and $E|_{H_Y} = \sum_i f_i$. By the relative version of Theorem 1.3.4, each f_i contains a (-1)-curve l_i on H_Y. In fact $f_i = l_i$ because $((f_i - l_i) \cdot l_i)_{H_Y} = (E \cdot l_i)_Y + 1 \leq 0$ from the relative ampleness of $-E$. By Castelnuovo's contraction theorem, H_Y is the blow-up of H at the union $H \cap C$ of finitely many points. Hence $d = 1$ from the adjunction $K_{H_Y/H} = d \sum_i f_i$. This means that E is obtained at the generic point η of C by the blow-up of X along C. Then π is unique by Lemma 3.1.2.

The ideal sheaf $\pi_* \mathscr{O}_Y(-iE)$ consists of functions in $\mathscr{O}_X$ of order at least i along E. It is expressed as $(\pi_* \mathscr{O}_Y(-iE) \cdot \mathscr{O}_{X,\eta}) \cap \mathscr{O}_X$. Since $\pi_* \mathscr{O}_Y(-iE) \cdot \mathscr{O}_{X,\eta} = \mathscr{I}^i \mathscr{O}_{X,\eta}$ for the ideal sheaf $\mathscr{I}$ in $\mathscr{O}_X$ defining C, one has the equality $\pi_* \mathscr{O}_Y(-iE) = \mathscr{I}^i \mathscr{O}_{X,\eta} \cap \mathscr{O}_X = \mathscr{I}^{(i)}$. Thus $Y = \mathrm{Proj}_X \bigoplus_{i \in \mathbf{N}} \pi_* \mathscr{O}_Y(-iE)$ is the symbolic blow-up of X along $\mathscr{I}$. $\square$

Example 4.1.4   The proposition no longer holds when one allows canonical singularities. For example, the germ $o \in \mathbf{A}^2$ with coordinates x_1, x_2 admits the weighted blow-up B_i with $\mathrm{wt}(x_1, x_2) = (1, i)$ for any positive integer i. The total space Y_i of the base change $Y_i = B_i \times \mathbf{A}^1 \to X = \mathbf{A}^2 \times \mathbf{A}^1$ is canonical.

As in the preceding chapter, the general elephant is a fundamental object in the study of the threefold divisorial contraction $\pi \colon E \subset Y \to C \subset X$. Working locally on the germ $x \in X$ at a point in C, we take the linear system $\Lambda \subset |-K_X|$ on $x \in X$ defined by the subsheaf $\pi_* \mathcal{O}_Y(-\pi^* K_X - E)$ of $\mathcal{O}_X(-K_X)$. The general member S of Λ is an effective divisor on X linearly equivalent to $-K_X$ which contains $C = \pi(E)$.

The divisor S is a normal surface. Indeed, it is smooth outside the support C of the base locus of Λ. On the other hand for the general hyperplane section H of $x \in X$ containing C and the general elephant G of $x \in X$, the sum $H + G$ belongs to Λ and it is smooth at the generic point η of C. Hence the general member S is also smooth at η and it satisfies R_1. Since it also satisfies S_2 by Proposition 2.2.23, it is normal by Serre's criterion.

One can write $\pi^* S = S_Y + E$ with the strict transform S_Y of S. The surface S_Y is the general elephant of Y and $\omega_{S_Y} = \pi^* \omega_S$. Most of the classification results on a flipping contraction are applicable to the divisorial contraction π. Amongst them is the general elephant conjecture in the case when the central fibre is irreducible. The following theorem will be proved as Theorem 5.5.1 in the next chapter. Note that by Lemma 4.1.7, S_Y is Du Val if and only if so is S.

Theorem 4.1.5 (Kollár–Mori [276]) *Let $\pi \colon E \subset Y \to C \subset X$ be a threefold divisorial contraction which contracts the divisor E to a curve C. Work on the algebraic germ $x \in X$ at a point in C. Assume that $\pi^{-1}(x)$ is irreducible. Then the general elephant of Y is Du Val.*

Remark 4.1.6 The original result is analytic. On the analytic germ $x \in C_h \subset X_h$ of $C \subset X$, it asserts that the general member $S_h \in |-K_{X_h}|$ containing C_h is Du Val. From this one can derive the existence of a Du Val member $C \subset S \in |-K_X|$ on the *algebraic* germ $x \in X$ as follows. Fix a resolution $X' \to X$ which gives a strong log resolution of the analytic pair (X_h, S_h). The isolated singularity $x \in S_h$ is Du Val if and only if every exceptional prime divisor F on X' that intersects the strict transform of S_h has log discrepancy $a_F(X_h, S_h) \geq 1$. The latter condition is preserved by the replacement of S_h by an algebraic approximation S modulo $\mathscr{I} \cap \mathfrak{m}^n$ for sufficiently high n, where $\mathscr{I}$ and $\mathfrak{m}$ denote the ideals in $\mathcal{O}_X$ defining C and x respectively. Further X' is still a strong log resolution of (X, S). Thus $x \in S$ is Du Val.

Lemma 4.1.7 *Let $\pi \colon T \to S$ be a proper birational morphism from a Cohen–Macaulay surface to a normal Gorenstein surface such that $\omega_T = \pi^* \omega_S$. Then S is canonical if and only if so is T.*

Proof The idea has appeared in the proof of Theorem 3.3.2. On the normalisation $\nu \colon T^\nu \to T$, one has $K_{T^\nu} = \nu^* \pi^* K_S - C$ with the conductor C defined by the ideal sheaf $v^! \mathscr{O}_T$ in $\mathscr{O}_{T^v}$. If S is canonical, then C is zero, that is, T satisfies R_1 and hence it is normal. Thus in order to prove the lemma, we may assume that T is normal. Then the assertion follows from the relation $K_T = \pi^* K_S$. $\quad\square$

Supported by Theorem 4.1.5, we shall assume the existence of a Du Val member $S \in |-K_X|$ containing C in the following two sections. The assumption implies the existence and canonicity of the symbolic blow-up.

Proposition 4.1.8 *Let C be a curve in a $\mathbf{Q}$-factorial terminal threefold X. Suppose the existence of a Du Val member S of $|-K_X|$ containing C. Then the symbolic power algebra $\mathscr{S} = \bigoplus_{i \in \mathbf{N}} \mathscr{I}^{(i)}$ of the ideal sheaf $\mathscr{I}$ in $\mathscr{O}_X$ defining C is finitely generated, and the symbolic blow-up $Y = \mathrm{Proj}_X \mathscr{S}$ is $\mathbf{Q}$-factorial and canonical with relative Picard number $\rho(Y/X) = 1$.*

Proof The pair (X, S) is plt by inversion of adjunction in Theorem 1.4.26 and it is further canonical since $K_X + S$ is Cartier. Let E be the divisor over X obtained at the generic point of C by the blow-up of X along C. It has log discrepancy $a_E(X, S) = 1$. By Proposition 1.5.17 after slight perturbation of S, there exists a birational contraction $\pi \colon Y \to X$ from a $\mathbf{Q}$-factorial normal variety such that E appears on Y as the only exceptional prime divisor. Since X and Y are $\mathbf{Q}$-factorial, $\rho(Y/X)$ equals one and $-E$ is relatively ample. Thus Y is described as $\mathrm{Proj}_X \bigoplus_{i \in \mathbf{N}} \pi_* \mathscr{O}_Y(-iE)$, which equals $\mathrm{Proj}_X \mathscr{S}$ by $\pi_* \mathscr{O}_Y(-iE) = \mathscr{I}^{(i)}$. In particular, $\mathscr{S}$ is finitely generated. Plainly $K_Y + S_Y = \pi^*(K_X + S)$ with the strict transform S_Y of S. Hence (Y, S_Y) is canonical as so is (X, S), and Y itself is also canonical. $\quad\square$

In the remainder of the section, we shall prove the following classification due to Mori and Cutkosky in the case when Y is Gorenstein, which was obtained in the same framework as for Theorem 3.1.1. We follow Cutkosky's cohomological argument, whilst Mori obtained the classification for smooth Y by means of the deformation theory of curves.

Theorem 4.1.9 (Mori [332], Cutkosky [98]) *Let $\pi \colon E \subset Y \to C \subset X$ be a divisorial contraction from a Gorenstein threefold which contracts the divisor E to a curve C. Then X is smooth about C and C is locally of embedding dimension at most two. The contraction π is the blow-up of X along C and Y has only cA singularities about E.*

Lemma 4.1.10 *Let $Y \to X$ be a threefold divisorial contraction which contracts the divisor to a curve. Then the index of Y is divisible by that of X.*

Proof The reader should compare this with Lemma 1.4.13. We have to show that $r K_X$ is Cartier for the index r of Y. Write $\pi \colon Y \to X$ and let E denote the exceptional divisor. Then $K_Y - E = \pi^* K_X$ by Proposition 4.1.3 and $r(K_Y - E)$

is Cartier by Theorem 2.4.17. The contraction theorem stated as Theorem 1.4.7 shows that the invertible sheaf $\mathcal{O}_Y(r(K_Y - E))$ is the pull-back of some invertible sheaf on X, which should be $\mathcal{O}_X(rK_X)$ by the equality $r(K_Y - E) = r\pi^* K_X$. Thus rK_X is Cartier.
□

Lemma 4.1.11 *Let $Y \to X$ be a threefold divisorial contraction which contracts the divisor to a curve. If Y is Gorenstein, then $-K_Y$ is relatively free.*

Proof This is a special case of Lipman's result [295, theorem 12.1]. We write $\pi \colon E \subset Y \to C \subset X$ where the divisor E is contracted to the curve C, and work on the germ $x \in C \subset X$ at a point.

We claim that the inverse image $f = \pi^{-1}(x)$ regarded as a reduced scheme is a smooth rational curve. Since π is about the generic point of C the blow-up along C, the general fibre l of $E \to C$ is a smooth rational curve with intersection number $(-K_Y \cdot l) = 1$. For the induced morphism $E^\nu \to C^\nu$ of the normalisations of E and C, the scheme-theoretic fibre $l^\nu = E^\nu \times_{C^\nu} z$ at every point $z \in C^\nu$ has intersection number $(-K_Y|_{E^\nu} \cdot l^\nu) = 1$ with the pull-back of the Cartier divisor $-K_Y$. Hence l^ν must be integral by the relative ampleness of $-K_Y|_{E^\nu}$. In particular, the image f is irreducible. From the surjection $\mathcal{O}_Y \twoheadrightarrow \mathcal{O}_f$ with the Kawamata–Viehweg vanishing $R^1\pi_*\mathcal{O}_Y = 0$, one obtains $H^1(\mathcal{O}_f) = 0$. Hence the curve f is smooth and rational with $(-K_Y \cdot f) = 1$.

Fix an effective Cartier divisor A on the curve C with support x and take the pull-back $F = A \times_C E$ to E. We shall see the surjectivity of the restriction map $\pi_*\mathcal{O}_Y(-K_Y) \to H^0(\mathcal{O}_Y(-K_Y) \otimes \mathcal{O}_F)$. For this, we factorise it as the composite of the two maps $\pi_*\mathcal{O}_Y(-K_Y) \to \pi_*(\mathcal{O}_Y(-K_Y) \otimes \mathcal{O}_E)$ and $\pi_*(\mathcal{O}_Y(-K_Y) \otimes \mathcal{O}_E) \to H^0(\mathcal{O}_Y(-K_Y) \otimes \mathcal{O}_F)$. The first map is surjective by the Kawamata–Viehweg vanishing $R^1\pi_*\mathcal{O}_Y(-K_Y - E) = 0$. The second has cokernel inside $R^1\pi_*(\mathcal{O}_Y(-K_Y) \otimes \mathcal{O}_E(-F)) \simeq R^1\pi_*(\mathcal{O}_Y(-K_Y) \otimes \mathcal{O}_E) \otimes \mathcal{O}_C(-A)$, which is zero from $R^1\pi_*\mathcal{O}_Y(-K_Y) = 0$.

The lemma is therefore reduced to the freedom of the Cartier divisor $-K_Y|_F$ on the scheme F. Note that $H^1(\mathcal{O}_F) = 0$ by $R^1\pi_*\mathcal{O}_Y = 0$. By Lemma 4.1.12 below, one has the isomorphism $\operatorname{Pic} F \simeq \operatorname{Pic} f = \mathbf{Z}$ for the support f of F. Hence the isomorphism class of an invertible sheaf on F is determined by the degree of the restriction to $f \simeq \mathbf{P}^1$. In particular, the invertible sheaf $\mathcal{O}_Y(-K_Y) \otimes \mathcal{O}_F$ is isomorphic to $\mathcal{O}_F(y_F)$ with an effective Cartier divisor y_F on F such that $y = y_F \times_F f$ is a general closed point in f. Moving y, one obtains the freedom of $-K_Y|_F \sim y_F$.
□

Lemma 4.1.12 ([16, section 1]) *If a connected complete scheme C of dimension one satisfies $H^1(\mathcal{O}_C) = 0$, then the support C_0 of C is a tree of $\mathbf{P}^1$ and $\operatorname{Pic} C \simeq \operatorname{Pic} C_0 = \mathbf{Z}^{\oplus n}$ where n is the number of irreducible components of C_0.*

Proof By the surjection $\mathscr{O}_C \twoheadrightarrow \mathscr{O}_{C_0}$, the vanishing of $H^1(\mathscr{O}_C)$ implies that of $H^1(\mathscr{O}_{C_0})$. Hence C_0 is a tree of $\mathbf{P}^1$. Let $\mathscr{N}$ be the ideal sheaf of nilpotents in $\mathscr{O}_C$. It defines the reduced scheme C_0 and fits into the exact sequence $0 \to \mathscr{N} \to \mathscr{O}_C \to \mathscr{O}_{C_0} \to 0$. One has $H^1(\mathscr{N}) = 0$ from $H^1(\mathscr{O}_C) = 0$ with $H^0(\mathscr{O}_{C_0}) \simeq \mathbf{C}$. We consider the exact sequence

$$1 \to 1 + \mathscr{N} \to \mathscr{O}_C^\times \to \mathscr{O}_{C_0}^\times \to 1$$

of sheaves of multiplicative groups. It induces the exact sequence $H^1(1+\mathscr{N}) \to \operatorname{Pic} C \to \operatorname{Pic} C_0 = \mathbf{Z}^{\oplus n}$. Since $\operatorname{Pic} C \to \mathbf{Z}^{\oplus n}$ is surjective, it suffices to prove the vanishing $H^1(1 + \mathscr{N}) = 0$.

We shall see this by noetherian induction on C. The assertion is trivial when $C = C_0$. We shall assume that $\mathscr{N}^l \neq 0$ but $\mathscr{N}^{l+1} = 0$ for some positive integer l. Let C' be the subscheme of C defined by $\mathscr{N}^l$, which has the exact sequence $0 \to \mathscr{N}/\mathscr{N}^l \to \mathscr{O}_{C'} \to \mathscr{O}_{C_0} \to 0$. Note that $H^1(\mathscr{O}_{C'}) = 0$ by $H^1(\mathscr{O}_C) = 0$. By noetherian induction, we can assume the vanishing $H^1(1 + (\mathscr{N}/\mathscr{N}^l)) = 0$.

Consider the exact sequence

$$1 \to 1 + \mathscr{N}^l \to 1 + \mathscr{N} \to 1 + (\mathscr{N}/\mathscr{N}^l) \to 1$$

of sheaves of multiplicative groups. As a sequence of sheaves of sets, this is obtained from $0 \to \mathscr{N}^l \to \mathscr{N} \to \mathscr{N}/\mathscr{N}^l \to 0$ by adding 1 to every element. This operation does not preserve group structure in general. In spite of this, the induced map $\mathscr{N}^l \to 1 + \mathscr{N}^l$ is a group isomorphism from an additive group to a multiplicative group because $(\mathscr{N}^l)^2 \subset \mathscr{N}^{l+1} = 0$. Hence in the associated diagram below, the map β is an isomorphism of groups. Notice that α is merely a map of sets.

$$
\begin{array}{ccccc}
H^0(\mathscr{N}/\mathscr{N}^l) & \xrightarrow{\ \delta\ } & H^1(\mathscr{N}^l) & \longrightarrow & H^1(\mathscr{N}) = 0 \\
\downarrow{\scriptstyle \alpha} & & \downarrow{\scriptstyle \beta}{\wr} & & \\
H^0(1 + (\mathscr{N}/\mathscr{N}^l)) & \xrightarrow{\ \delta'\ } & H^1(1 + \mathscr{N}^l) & &
\end{array}
$$

Since δ is surjective by $H^1(\mathscr{N}) = 0$, the map $\delta' \colon H^0(1 + (\mathscr{N}/\mathscr{N}^l)) \to H^1(1 + \mathscr{N}^l)$ is also surjective, by which $H^1(1 + \mathscr{N}) \to H^1(1 + (\mathscr{N}/\mathscr{N}^l))$ is injective. Thus the vanishing of $H^1(1 + \mathscr{N})$ follows from the inductive assumption $H^1(1 + (\mathscr{N}/\mathscr{N}^l)) = 0$. $\qquad\square$

Remark 4.1.13 In Lemma 4.1.12, an invertible sheaf $\mathscr{L}$ on C is generated by global sections if and only if the degree of the restriction $\mathscr{L}|_D$ is non-negative for every irreducible component D of C_0. Lipman also showed that $\mathscr{L}$ is very ample if and only if the degree of $\mathscr{L}|_D$ is positive for every D.

Proof of Theorem 4.1.9 We work on the germ $x \in C \subset X$ at a point and take the general elephant S_Y of Y. By Lemma 4.1.11, S_Y is a smooth surface intersecting $\pi^{-1}(x)$ properly. The induced morphism $S_Y \to S = \pi_* S_Y$ is finite. Since $\pi^* S = S_Y + E$, the surface S is smooth at the generic point of C. Hence S satisfies R_1 and it is normal. Then $S_Y \to S$ is an isomorphism by Zariski's main theorem. Thus $x \in S$ is smooth and $x \in C$ is of embedding dimension at most two. In particular, C is lci. Since X is Gorenstein by Lemma 4.1.10, $S \sim -K_X$ is Cartier. Hence $x \in X$ is smooth as so is $x \in S$. The remaining assertions follow from Propositions 4.1.3 and 4.1.14. □

The following is a converse to Theorem 4.1.9. It supplies a criterion for the existence of a divisorial extraction from an lci curve in a smooth threefold.

Proposition 4.1.14 ([111, lemma 1.6]) *Let $x \in C \subset X$ be the germ of a smooth threefold X with an lci curve C. Then the blow-up Y of X along C is the symbolic blow-up of X along C. The blow-up Y is terminal if and only if the general hyperplane section of X containing C is smooth, and in this case Y has one singularity of type cA_{m-1} for the multiplicity m of C at x.*

Proof The lci curve C is defined by a regular sequence f_1, f_2 in $\mathcal{O}_{X,x}$ [308, theorem 21.2] and the blow-up $\pi \colon Y \to X$ along C is given in $X \times \mathbf{P}^1$ by $y_1 f_2 - y_2 f_1$ with the homogeneous coordinates y_1, y_2 of $\mathbf{P}^1$. This description shows that Y is normal and the exceptional locus E of π is a prime Cartier divisor. Then $-E$ is relatively ample and $Y = \operatorname{Proj} \bigoplus_{i \in \mathbf{N}} \pi_* \mathcal{O}_Y(-iE)$ is the symbolic blow-up of X along C.

Let $\mathfrak{m}$ be the maximal ideal in $\mathcal{O}_X$ defining x. If both f_1 and f_2 belong to $\mathfrak{m}^2$, then Y is singular along $\pi^{-1}(x)$ and it is not terminal. If f_1 or f_2 belongs to $\mathfrak{m} \setminus \mathfrak{m}^2$, then Y has one singularity of type cA_{m-1} for $m = \operatorname{mult}_x C$. Indeed when $f_1 \in \mathfrak{m} \setminus \mathfrak{m}^2$ for example, the germ $x \in X$ is analytically isomorphic to $o \in \mathfrak{D}^3$ with coordinates f_1, x_2, x_3 on which one may assume that f_2 is a function $f_2(x_2, x_3)$ in x_2 and x_3. The singularity of Y is given by $f_2(x_2, x_3) - z_2 f_1$ where $z_2 = y_2 y_1^{-1}$. From these observations, we derive the latter assertion. □

4.2 Contractions to Smooth Curves

We are regarding the divisorial contraction $\pi \colon E \subset Y \to C \subset X$ as an extraction from the germ $x \in C \subset X$ of a terminal threefold with a curve. Theorem 3.1.5 shows that no such π exists if $x \in X$ is a quotient singularity. In general by Proposition 4.1.3, the symbolic blow-up of X along C is a unique candidate for π. By the discussion preceding Theorem 4.1.5, we shall assume the existence of a Du Val member $x \in S \in |-K_X|$ containing C. Then by Proposition 4.1.8, the symbolic blow-up always exists as a canonical threefold. We are faced with the question of when this canonical threefold is terminal.

Following a series of works of Tziolas [453], [454], [457], we shall discuss the case when $x \in X$ is Gorenstein and terminal and C is smooth. By Theorem 2.3.18, a $\mathbf{Q}$-factorial Gorenstein terminal threefold is factorial.

Framework 4.2.1 Let $x \in C \subset X$ be the germ of a non-smooth factorial terminal threefold X with a smooth curve C. Suppose that the general hyperplane section $x \in S$ of X containing C is Du Val. Let $\mathscr{I}$ denote the ideal in $\mathscr{O}_X$ defining C. By Proposition 4.1.8, the symbolic blow-up $Y = \mathrm{Proj}_X \mathscr{S}$ for $\mathscr{S} = \bigoplus_{i \in \mathbf{N}} \mathscr{I}^{(i)}$ is $\mathbf{Q}$-factorial and canonical.

We begin with a simple example.

Example 4.2.2 Let $x \in X$ be the ordinary double point defined by $x_1 x_2 + x_3 x_4$ in $\mathfrak{D}^4$ and let C be the x_1-axis, which is contained in X. We do not care about factoriality. Take the weighted blow-up $\pi \colon Y \to X$ with $\mathrm{wt}(x_2, x_3, x_4) = (2, 1, 1)$. It is the blow-up along C outside x and the exceptional locus E is irreducible, from which Y is normal. It satisfies $K_Y = \pi^* K_X + E$ and $-E$ is relatively ample. It follows that π is the symbolic blow-up of X along C. The chart $U_2 = (x_2 \neq 0)$ of Y is given in $\mathfrak{D}^4/\mathbf{Z}_2(0, 1, 1, 1)$ by $x_1 + y_3 y_4$ for the orbifold coordinates $(x_1, y_2, y_3, y_4) = (x_1, x_2^{1/2}, x_3 x_2^{-1/2}, x_4 x_2^{-1/2})$. Hence $U_2 \simeq \mathfrak{D}^3/\mathbf{Z}_2(1, 1, 1)$. It has a quotient singularity at origin o and this is the only singularity on Y. Thus Y is terminal and π is a divisorial contraction.

The inverse image $\pi^{-1}(x)$ is reducible. It is given in $\mathbf{P}(2, 1, 1)$ by $x_3 x_4$ and consists of two lines intersecting at o. The ideal $\mathscr{I} = (x_2, x_3, x_4) \mathscr{O}_X$ defining C has the second symbolic power $\mathscr{I}^{(2)} = (x_2, x_3^2, x_3 x_4, x_4^2) \mathscr{O}_X$, and Y is the blow-up of X along $\mathscr{I}^{(2)}$.

Although Proposition 4.1.8 is a consequence of the MMP, we shall construct the symbolic blow-up in a more specific manner by a certain MMP starting from a $\mathbf{Q}$-factorialisation of the usual blow-up of X along C.

Lemma 4.2.3 *In Framework* 4.2.1, *let B be the blow-up of X along C. Then B is Gorenstein and canonical, and the exceptional locus of $B \to X$ consists of two prime divisors E and F which are mapped to C and x respectively. Further, F is isomorphic to the projective plane and the scheme-theoretic intersection $E \cap F$ is a line in $F \simeq \mathbf{P}^2$.*

Proof By Theorem 2.3.4, the germ $x \in X$ is of embedding dimension four. Working analytically, we embed $x \in X$ into $\mathfrak{D}^4$. By assumption, the general hypersurface H in $\mathfrak{D}^4$ containing C defines a Du Val surface $S = H \cap X$.

Let $\bar{\varphi}: \bar{B} \to \mathfrak{D}^4$ be the blow-up along C and let $\bar{E}$ be the exceptional divisor. Then B is the strict transform in $\bar{B}$ of X and it is Gorenstein. We write $\varphi: B \to X$. Surfaces in B exceptional over X are the component E of $\bar{E}|_B$ dominating C and possibly $F = \bar{\varphi}^{-1}(x) \simeq \mathbf{P}^2$. Let H_B be the strict transform in $\bar{B}$ of H. The exceptional locus $\bar{E}|_{H_B}$ of $H_B \to H$ is a $\mathbf{P}^1$-bundle over C and the scheme-theoretic intersection $S_B = H_B \cap B$ is a Gorenstein surface.

One has $K_{\bar{B}} = \bar{\varphi}^* K_{\mathfrak{D}^4} + 2\bar{E}$, $\bar{\varphi}^* H = H_B + \bar{E}$ and $\bar{\varphi}^* X = B + \bar{E}$. Hence $(K_{\bar{B}} + H_B + B)|_{S_B} = \bar{\varphi}^* (K_{\mathfrak{D}^4} + H + X)|_S$, from which $\omega_{S_B} = (\varphi|_{S_B})^* \omega_S$ and S_B is canonical by Lemma 4.1.7. In particular, B is normal about S_B. Because S_B intersects both E and F, the blow-up B satisfies R_1 and it is normal.

We write $\bar{E}|_B = E + dF$ with $d \in \mathbf{N}$, where d is positive if and only if B contains F. Then $K_B = \varphi^* K_X + E + dF$. Let $x \in T$ be the general hyperplane section of the germ $x \in X$, which is singular and Du Val, and let T_B be the strict transform in B. Then $T_B \to T$ is the blow-up at the closed point $x = C \cap T$ and hence $K_{T_B} = (\varphi|_{T_B})^* K_T$. If d were zero, then $T_B = \varphi^* T$ and $K_{T_B} = (\varphi|_{T_B})^* K_T + E|_{T_B}$ by adjunction. Hence $E|_{T_B} = 0$. It follows that $\bar{E} \cap T_B = E \cap T_B = \emptyset$ since $T_B = \varphi^* T$ is a Cartier divisor on B, which is absurd. Thus F lies on B.

We have divisors E and F on the total space $\bar{E}$ of the $\mathbf{P}^2$-bundle $\bar{E} \to C$. The divisor E is flat over C and the general fibre of $E \to C$ is a line in the fibre $\mathbf{P}^2$ of $\bar{E} \to C$. Hence the fibre $E \cap F$ at x is also a line in $F = \mathbf{P}^2$. It remains to show that B is canonical. We have seen that B is normal and Gorenstein. Since S is canonical, the pair (X, S) is plt by inversion of adjunction. It is further canonical since $K_X + S$ is Cartier. Then (B, S_B) is canonical by $K_B + S_B = \varphi^* (K_X + S)$, and B itself is also canonical. $\square$

The positive integer d is in fact an invariant of the germ $x \in C \subset S$.

Lemma 4.2.4 *In Framework* 4.2.1, *keep the notation in Lemma* 4.2.3 *and write* $\mathscr{I} \mathscr{O}_B = \mathscr{O}_B(-E - dF)$. *Let* S_B *be the strict transform in B of S, which is the blow-up of S along C. Then* S_B *is normal and the exceptional locus l of* $S_B \to S$ *is a prime divisor. Further* $\mathscr{O}_S(-C) \cdot \mathscr{O}_{S_B} = \mathscr{O}_{S_B}(-C_B - dl)$ *with the strict transform C_B of C.*

Proof We have already seen that S_B is normal in the course of the proof of Lemma 4.2.3. Since S is taken generally, the fibre $F \cap S_B$ of $S_B \to S$ at x is a general line l in $F \simeq \mathbf{P}^2$. On the other hand, the scheme $E \cap S_B$ equals C_B outside F and the restriction $(E \cap S_B)|_F = E \cap l$ is of dimension zero. Hence $[E \cap S_B] = [C_B]$ cycle-theoretically. Thus the invertible ideal sheaf $\mathscr{O}_S(-C) \cdot \mathscr{O}_{S_B} = \mathscr{I} \mathscr{O}_{S_B}$ in $\mathscr{O}_{S_B}$ defines the divisor $(E + dF)|_{S_B} = C_B + dl$ on S_B. $\square$

Following Lemma 4.2.3, we take the blow-up $\varphi\colon B \to X$ along C, which has two exceptional prime divisors E and F with $\varphi(E) = C$ and $\varphi(F) = x$. Proposition 1.5.17 provides a **Q**-*factorialisation* $B' \to B$, which is by definition a small contraction from a **Q**-factorial normal variety. Let S' and F' denote the strict transforms in B' of S and F. Then $K_{B'} + S'$ is the pull-back of $K_X + S$ and the pair (B', S') is canonical as so is (X, S) by inversion of adjunction.

Take the strict transform T' in B' of another general hyperplane section of X containing C. Then (B', Δ') is klt for $\Delta' = (1 - \varepsilon)S' + \varepsilon T' + \varepsilon F'$ with a small positive rational number ε. We run the $(K_{B'} + \Delta') \equiv_X \varepsilon F'$-MMP over X where $\rho(B'/X) = 2$. By the negativity lemma stated as Theorem 1.3.9, it ends with a **Q**-factorial lt variety Y over X such that $Y \to X$ is an elementary log divisorial contraction which contracts the strict transform of E. The contraction $Y \to X$ is the symbolic blow-up along C. This construction is known as the *two-ray game* below.

Definition 4.2.5 Let $X \to S$ be a projective morphism from a **Q**-factorial normal variety to a variety with relative Picard number two. For a pair (X, Δ), the $(K_X + \Delta)/S$-MMP is uniquely determined by the first extremal contraction from X as far as it runs, because the closed cone $\overline{\mathrm{NE}}(X/S)$ of curves possesses exactly two extremal rays whenever $\rho(X/S) = 2$. If X_i is an intermediate output by the MMP after a log flip $X_{i-1} \to Y_{i-1} \leftarrow X_i$, then the next extremal contraction must be the contraction associated with the unique extremal ray of $\overline{\mathrm{NE}}(X_i/S)$ other than that contracted by $X_i \to Y_{i-1}$. This machinery is referred to as the *two-ray game*.

We constructed Y by the $(K_{B'} + \Delta')/X$-MMP from a **Q**-factorialisation $\alpha\colon B' \to B$. If the first extremal contraction β equals α, then we can replace B' by its log flip to assume that $\beta \neq \alpha$. Now α and β correspond to the two extremal rays of $\overline{\mathrm{NE}}(B'/X)$. In this setting, the MMP is also a part of the $K_{B'}$-MMP. Indeed, $-K_{B'} = -\alpha^* K_B$ is relatively ample with respect to the first contraction β because $-K_B$ is ample over X. By the negativity lemma, the strict transform of $K_{B'} \equiv_X \alpha^*(E + dF)$ never becomes nef over X during our $(K_{B'} + \Delta')/X$-MMP. Then by the nature of the two-ray game, the anti-canonical divisor is relatively ample with respect to every new extremal contraction in the MMP.

By the above choice of the **Q**-factorialisation, one can answer the question of when the symbolic blow-up Y is terminal, in terms of the singularities of B'.

Proposition 4.2.6 *In Framework* 4.2.1, *keep the notation in Lemma* 4.2.3. *Take a* **Q**-*factorialisation* $\alpha\colon B' \to B$ *from which the symbolic blow-up* Y *is obtained by the* $K_{B'}$-MMP *over* X. *Then* Y *is terminal if and only if* B' *is* *terminal outside the strict transform of F. This is the case if B is cDV along* $E \cap F$.

Proof We have constructed Y from B' by the $(K_{B'} + \Delta')/X$-MMP. The rational map $B' \dashrightarrow Y$ is isomorphic exactly outside the strict transform F' of F because $K_{B'} + \Delta' \equiv_X \varepsilon F'$. Hence if Y is terminal, then so is $B' \setminus F'$. Conversely suppose that B' is terminal outside F'. Recall that B' is canonical and that Y is also an intermediate output of the $K_{B'}$-MMP. By Lemma 1.4.12, Y is terminal.

The blow-up B is smooth outside F and $\alpha \colon B' \to B$ is isomorphic outside finitely many points in B. Since $K_{B/X} = E + dF$ is Cartier, α is isomorphic outside $E \cap F$. If B is cDV at $b \in E \cap F$, then there exists a Du Val section $b \in T$ for which the pair (B, T) is plt and canonical by inversion of adjunction. Hence $\mathrm{mld}_{\alpha^{-1}(b)} B' = \mathrm{mld}_b B > \mathrm{mld}_b(B, T) \geq 1$, from which one deduces that $B' \setminus F'$ is terminal about $\alpha^{-1}(b) \setminus F'$. This shows the last assertion. $\square$

Whereas the original approach by Tziolas uses this proposition, we derive a stronger result by a direct construction of the symbolic blow-up. The complete classification of the germ $x \in C \subset S$ by Jaffe will be crucial.

Theorem 4.2.7 (Jaffe [216]) *Every germ $o \in C \subset S$ of a Du Val analytic surface S with a smooth curve C is isomorphic to one of the following pairs.*

$$
\begin{array}{lll}
A_n^i & (x_1 - x_3^i = x_2 + x_3^{n+1-i} = 0) & \subset (x_1 x_2 + x_3^{n+1} = 0) \quad \text{for } 1 \leq i \leq n \\
D_n^l & (x_1 = x_3 = 0) & \subset (x_1^2 + x_2^2 x_3 + x_3^{n-1} = 0) \\
D_{2m}^r & (x_1 = x_2 + \sqrt{-1} x_3^{m-1} = 0) & \subset (x_1^2 + x_2^2 x_3 + x_3^{2m-1} = 0) \\
D_{2m+1}^r & (x_2 = x_1 + \sqrt{-1} x_3^m = 0) & \subset (x_1^2 + x_2^2 x_3 + x_3^{2m} = 0) \\
E_6 & (x_2 = x_1 + \sqrt{-1} x_3^2 = 0) & \subset (x_1^2 + x_2^3 + x_3^4 = 0) \\
E_7 & (x_1 = x_2 = 0) & \subset (x_1^2 + x_2^3 + x_2 x_3^3 = 0)
\end{array}
$$

Remark 4.2.8 Realise each pair in the above list as the germ $o \in C \subset S = A^2/G$ with $G \subset \mathrm{SL}(2, \mathbf{C})$ as in Example 2.2.8. Let f be the semi-invariant function in $\mathcal{O}_{A^2}$ which defines the inverse image of C. It yields a one-dimensional representation $\rho \colon G \to \mathbf{C}^\times$ by $g(f) = \rho(g)f$ for $g \in G$. By the *McKay correspondence* [312], ρ corresponds to some vertex $\circ$ with attached number one in the Dynkin diagram of the Du Val singularity $o \in S$ given in Notation 3.3.4. Up to symmetry, the vertices in the cases A_n^i, D_n^l, D_n^r, E_6, E_7 are F_i, F_1, F_n, F_1, F_6 respectively. Note that A_n^i is isomorphic to A_n^{n+1-i} and D_4^l is to D_4^r.

Remark 4.2.9 The divisor class group $\mathrm{Cl}\, S$ of the analytic germ $o \in S$ of a Du Val singularity is as follows [295, section 24]. Up to symmetry, every non-trivial element in $\mathrm{Cl}\, S$ is represented by some curve listed in Theorem 4.2.7.

S	A_n	D_{2m}	D_{2m+1}	E_6	E_7	E_8
Cl S	$\mathbf{Z}_{n+1}$	$\mathbf{Z}_2 \times \mathbf{Z}_2$	$\mathbf{Z}_4$	$\mathbf{Z}_3$	$\mathbf{Z}_2$	1

The E_8 singularity is analytically factorial. In fact as shown by Brieskorn [58, Satz 3.3], a normal surface singularity is analytically factorial if and only if it is smooth or E_8. Note that every rational surface singularity is analytically **Q**-factorial [198, theorem 7.3.2].

We seek to describe the symbolic blow-up Y explicitly by finding a simple function for X. In fact we do not need the factoriality of X in Theorem 4.2.10, 4.2.12 or 4.2.14 below. Firstly we consider the case when S is of type A.

Theorem 4.2.10 ([453, theorem 5.1]) *In Framework* 4.2.1, *suppose that the germ $x \in C \subset S$ belongs to the case A_n^1 in Theorem* 4.2.7. *Then Y is the blow-up of X along $\mathscr{I}^{(n+1)}$. It has one quotient singularity of type $\frac{1}{n+1}(1, -1, 1)$, and possibly, isolated cA singularities. In particular, Y is terminal. Analytically the germ $x \in C \subset S \subset X$ is isomorphic to*

$$o \in (x_1\text{-axis}) \subset (f = x_4 = 0) \subset (f = 0) \subset \mathfrak{D}^4$$

where $f = x_1x_2 + x_3^{n+1} + x_4g(x_3, x_4) + x_4^{n+2}h(x_1, x_2, x_3, x_4)$ with ord $g \geq n$. By this description, Y is the weighted blow-up of X with $\mathrm{wt}(x_2, x_3, x_4) = (n + 1, 1, 1)$.

Proof By Theorem 4.2.7, the germ $x \in C \subset S$ is analytically given by $o \in (x_3\text{-axis}) \subset (g = 0)$ for $g = (x_1 + x_3)(x_2 - x_3^n) + x_3^{n+1}$. Replacing x_3 by $x_1 + x_3$, the function g becomes $x_3(x_2 - (x_3 - x_1)^n) + (x_3 - x_1)^{n+1}$, that is, $g = x_3(x_2 + x_1p(x_1, x_3)) + (-x_1)^{n+1}$ with some p. Replace (x_1, x_2, x_3) by $(x_3, x_2 + x_1p, -x_1)$. Then $x \in C \subset S$ is analytically isomorphic to $o \in (x_1\text{-axis}) \subset (x_1x_2 + x_3^{n+1} = 0)$ in $\mathfrak{D}^3$.

Thus one can choose the function f defining X as

$$f = x_1x_2 + x_3^{n+1} + x_4g(x_3, x_4) + x_4^l h(x_1, x_2, x_3, x_4)$$

with some g and h and with $l = 1$. This expression is rewritten as

$$f = (x_1 + x_4^l h_2(x_2, x_3, x_4))(x_2 + x_4^l h_1(x_1, x_2, x_3, x_4))$$
$$+ x_3^{n+1} + x_4g'(x_3, x_4) - x_4^{2l} h_1 h_2$$

with some g', h_1 and h_2. Replacing x_1 by $x_1 + x_4^l h_2$ and x_2 by $x_2 + x_4^l h_1$, one obtains a new expression of f in which $x_4^l h$ is replaced by $x_4^{2l} h$. Repeating this procedure, one may assume that $l = n + 2$. Then g has order at least n because the general hyperplane section of X containing the x_1-axis is Du Val of type A_n. This completes the description of $x \in C \subset S \subset X$ in the statement.

By this description, $\mathscr{I}^{(n+1)} = x_2\mathcal{O}_X + (x_3, x_4)^{n+1}\mathcal{O}_X$ and $\mathscr{I}^{((n+1)i)} = (\mathscr{I}^{(n+1)})^i$ for all i. Hence Y is the blow-up of X along $\mathscr{I}^{(n+1)}$ and it is the weighted blow-up with $\text{wt}(x_2, x_3, x_4) = (n + 1, 1, 1)$. The chart $U_2 = (x_2 \neq 0)$ of Y is given in $\mathfrak{D}^4/\mathbf{Z}_{n+1}(0, -1, 1, 1)$ by $x_1 + y_3^{n+1} + y_2^{-n} y_4 g(y_3 y_2, y_4 y_2) + y_2 y_4^{n+2} h(x_1, y_2^{n+1}, y_3 y_2, y_4 y_2)$ for the orbifold coordinates x_1, y_2, y_3, y_4. Thus $U_2 \simeq \mathfrak{D}^3/\mathbf{Z}_{n+1}(-1, 1, 1)$. Every singularity of Y outside U_2 is isolated cA. $\square$

Example 4.2.11 By Theorem 4.2.10, Y is always terminal if S is A_1 or A_2. On the other hand, Y may not be terminal if S is of type A_3 [453, example 5.3]. Let $x \in X$ be the cA_3 point defined by $x_1^2 + x_2^2 + x_3^4 + x_1 x_4^2$ in $\mathfrak{D}^4$ and let C be the x_4-axis. The general hyperplane section S containing C is Du Val of type A_3 at x and the germ $x \in C \subset S$ belongs to the case A_3^2 in Theorem 4.2.7. Let Y be the weighted blow-up of X with $\text{wt}(x_1, x_2, x_3) = (2, 1, 1)$. It is normal and is the symbolic blow-up of X along C. The chart $U_1 = (x_1 \neq 0)$ of Y is given in $\mathfrak{D}^4/\mathbf{Z}_2(1, 1, 1, 0)$ by $y_1^2 + y_2^2 + y_1^2 y_3^4 + x_4^2$ for the orbifold coordinates y_1, y_2, y_3, x_4, which is singular along the y_3-axis. In particular, Y is not terminal.

We explain the original argument by Tziolas for its own sake.

Original proof of terminal property of Y in Theorem 4.2.10 We apply Proposition 4.2.6. It suffices to show that B is cDV at any point b in $E \cap F$ with the notation in Lemma 4.2.3. The number d in Lemma 4.2.4 is one and hence $E + F$ is Cartier. By Lemma 4.2.3, E and F are both smooth. Analytically on $x \in X \subset \mathfrak{D}^4$, E and F are divisors on the $\mathbf{P}^2$-bundle $\bar{E} \to C$ from the exceptional divisor $\bar{E}$ of the blow-up $\bar{B} \to \mathfrak{D}^4$ along C. Since they intersect transversally in $\bar{E}$ by Lemma 4.2.3, there exists an analytic identification $b \in \bar{B} \simeq \mathfrak{D}^4$ with coordinates x_1, x_2, x_3, x_4 such that $\bar{E}, E, F$ are defined by x_4, (x_1, x_4), (x_2, x_4) respectively. Then from $B \cap \bar{E} = E + F$, the germ $b \in B$ is given by the function $x_1 x_2 + x_4 g$ for some $g = g(x_1, x_2, x_3, x_4)$. Thus $b \in B$ is a (not necessarily isolated) cA singularity. $\square$

Secondly we consider the case when S is of type D. In the cases D_n^l and D_{2m}^r, we describe Y as a weighted blow-up.

Theorem 4.2.12 ([453, theorem 6.1]) *In Framework* 4.2.1, *suppose that the germ* $x \in C \subset S$ *belongs to the case* D_n^l *or* D_{2m}^r *in Theorem* 4.2.7. *Then* Y *is the blow-up of* X *along* $\mathscr{I}^{(2)}$. *It is never terminal in the case* D_n^l *with* $n \geq 5$. *Analytically the germ* $x \in C \subset S \subset X$ *is isomorphic to*

$$o \in (x_2\text{-axis}) \subset (f = x_4 = 0) \subset (f = 0) \subset \mathfrak{D}^4,$$

D_n^l $f = x_1^2 + x_2^2 x_3 + x_3^{n-1} + x_4 g(x_2, x_3, x_4)$ *with* $g \in x_3\mathfrak{m} + x_4\mathfrak{m}$,
D_{2m}^r $f = x_1^2 + x_2 x_3^2 + x_2^m x_3 + x_4 g(x_2, x_3, x_4)$ *with* $g \in x_3\mathfrak{m} + x_4\mathfrak{m}$

for the maximal ideal $\mathfrak{m}$ *in* $\mathbf{C}\{x_2, x_3, x_4\}$. *By this description,* Y *is the weighted blow-up of* X *with* $\text{wt}(x_1, x_3, x_4) = (1, 2, 1)$.

Proof The strategy is the same as that for Theorem 4.2.10. We shall only discuss the case D_n^l. The proof in the other case D_{2m}^r is similar, where no monomial x_2^k with $k < m$ appears in g because otherwise one could take as S a Du Val section of type better than D_{2m}.

In the case D_n^l, by Theorem 4.2.7 one can choose f as

$$f = x_1^2 + x_2^2 x_3 + x_3^{n-1} + x_4 g(x_1, x_2, x_3, x_4)$$

with $g \in \mathbf{C}\{x_1, x_2, x_3, x_4\}$. Since the general hyperplane section of X containing the x_2-axis is Du Val of type D, one may assume that ord $g \geq 2$ possibly after replacing x_1 by $x_1 + cx_4$ with some $c \in \mathbf{C}$. By the Weierstrass preparation theorem stated as Theorem 2.1.10, f is written as $u \cdot (x_1^2 + h)$ for $h = h_1 x_1 + h_0$ with a unit u in $\mathbf{C}\{x_1, x_2, x_3, x_4\}$ and functions h_i in $\mathbf{C}\{x_2, x_3, x_4\}$ vanishing at origin. Then $x_1^2 + x_2^2 x_3 + x_3^{n-1} = u(x_1, x_2, x_3, 0) \cdot (x_1^2 + h(x_1, x_2, x_3, 0))$, from which $u(x_1, x_2, x_3, 0) = 1$ and $h(x_1, x_2, x_3, 0) = x_2^2 x_3 + x_3^{n-1}$. Replacing f by $x_1^2 + h$, one can take f so that $g \in \mathbf{C}\{x_2, x_3, x_4\} x_1 + \mathbf{C}\{x_2, x_3, x_4\}$. Eliminate x_1 from g by Remark 2.1.11 and express f as $x_1^2 + x_2^2(x_3 + x_4 p(x_2)) + x_3^{n-1} + x_4 h(x_2, x_3, x_4)$ with $h \in x_3 \mathfrak{m} + x_4 \mathfrak{m}$. Replacing x_3 by $x_3 + x_4 p$, one attains the expression in the statement.

By this description, $\mathscr{I}^{(2)} = (x_3, x_1^2, x_1 x_4, x_4^2) \mathscr{O}_X$ and $\mathscr{I}^{(2i)} = (\mathscr{I}^{(2)})^i$ for all i. Hence Y is the blow-up of X along $\mathscr{I}^{(2)}$ and it is the weighted blow-up with $\mathrm{wt}(x_1, x_3, x_4) = (1, 2, 1)$. In the case D_n^l, if $x_2 x_4$ or x_4^2 appears in g, then the hyperplane section in X given by $x_4 + \lambda x_3$ for general $\lambda \in \mathbf{C}$ is Du Val of type D_4. Hence when $n \geq 5$, neither $x_2 x_4$ nor x_4^2 appears in g. In this case, the chart $U_4 = (x_4 \neq 0)$ of Y, which is given in $\mathfrak{D}^4$ by $y_1^2 + x_2^2 y_3 + y_3^{n-1} x_4^{2n-4} + x_4^{-1} g(x_2, y_3 x_4^2, x_4)$ for the coordinates y_1, x_2, y_3, x_4, is singular along the y_3-axis. Thus Y is not terminal. □

One cannot apply the above argument to the case D_{2m+1}^r because a suitable weighted blow-up still has reducible exceptional locus. The following example suggests how to remove this obstacle by adding extra coordinates.

Example 4.2.13 ([454, example 3.7]) Let $x \in X$ be the cD_5 point defined by $x_1^2 + x_2^2 x_3 + x_1 x_3^2 + x_4^l$ in $\mathfrak{D}^4$ for $l \geq 4$ and let C be the x_3-axis. The general hyperplane section S containing C is Du Val of type D_5 at x and the germ $x \in C \subset S$ belongs to the case D_5^r. The germ $x \in X$ is expressed as

$$o \in \left(\begin{array}{c} x_1^2 + x_3 x_5 + x_4^l = 0 \\ x_2^2 + x_1 x_3 - x_5 = 0 \end{array} \right) \subset \mathfrak{D}^5.$$

The weighted blow-up Y of X with $\mathrm{wt}(x_1, x_2, x_4, x_5) = (2, 1, 1, 4)$ is the symbolic blow-up of X along C.

The chart $U_5 = (x_5 \neq 0)$ of Y is given in $\mathfrak{D}^5/\mathbf{Z}_4(2, 1, 0, 1, 3)$ by $y_1^2 + x_3 + y_4^l y_5^{l-4}$ and $y_2^2 + y_1 x_3 - y_5^2$ for the orbifold coordinates y_1, y_2, x_3, y_4, y_5. Eliminating x_3, U_5 is given in $\mathfrak{D}^4/\mathbf{Z}_4(2, 1, 1, 3)$ by $y_2^2 - y_1(y_1^2 + y_4^l y_5^{l-4}) - y_5^2$. If $l \geq 5$, then U_5 is singular along the y_4-axis and hence Y is not terminal. If $l = 4$, then U_5 has a terminal singularity of type $cAx/4$ at origin and Y is smooth elsewhere.

Finally we observe the theorem in the case when S is of type E. It is proved similarly to Theorems 4.2.10 and 4.2.12. See [457] for further discussion.

Theorem 4.2.14 *In Framework 4.2.1, suppose that the germ $x \in C \subset S$ belongs to the case E_6 or E_7 in Theorem 4.2.7. Then Y is the blow-up of X along $\mathscr{I}^{(3)}$ in E_6 and along $\mathscr{I}^{(2)}$ in E_7. Analytically the germ $x \in C \subset S \subset X$ is isomorphic to*

$$o \in (x_3\text{-axis}) \subset (f = x_4 = 0) \subset (f = 0) \subset \mathfrak{D}^4,$$

E_6 $f = x_1^2 + x_2^3 + x_1 x_3^2 + x_1 x_4 h(x_2, x_3, x_4) + x_4 g(x_2, x_3, x_4)$ *with* $g \in x_2^2\mathfrak{m} + x_2 x_4\mathfrak{m} + x_4^2\mathfrak{m}$ *and* $h \in \mathfrak{m}$,
E_7 $f = x_1^2 + x_2^3 + x_2 x_3^3 + x_4 g(x_2, x_3, x_4)$ *with* $g \in x_2\mathfrak{m} + x_4\mathfrak{m}$

for the maximal ideal $\mathfrak{m}$ in $\mathbf{C}\{x_2, x_3, x_4\}$. By this description, Y is the weighted blow-up of X with $\mathrm{wt}(x_1, x_2, x_4) = (3, 1, 1)$ in E_6 and $(1, 2, 1)$ in E_7.

4.3 Contractions to Singular Curves

Ducat [111] studied the divisorial extraction from a singular curve in a smooth threefold. Proposition 4.1.14 treats the case when the curve is lci. The essential case is when it is not lci.

Let $x \in C \subset X$ be the germ of a smooth threefold together with a curve. As in the preceding section, we assume the existence of a Du Val section S of X containing C. By Proposition 4.1.8, the germ admits the symbolic blow-up $\pi \colon E \subset Y \to C \subset X$ in which Y is canonical.

We begin with the analysis of the ideal $\mathscr{O}_S(-C)$ in $\mathscr{O}_S$ for the germ $x \in C \subset S$ of a Du Val singularity with a curve. By Remark 4.2.9, every such ideal is generated by at most two elements. See [214, theorem 10.5] for the perspective of the McKay correspondence mentioned in Remark 4.2.8.

Lemma 4.3.1 *Let $x \in S$ be an algebraic or analytic Du Val singularity. Then for any effective divisor C on S, the ideal $\mathscr{O}_S(-C)$ in $\mathscr{O}_S$ is generated by at most two elements.*

Thanks to Eisenbud, one can construct a minimal free resolution of $\mathscr{O}_S(-C)$ by taking matrix factorisation.

Definition 4.3.2 Let R be a noetherian ring and let M be a finite R-module. A *free resolution* of M is the exact sequence $F_\bullet$: $\cdots \to F_2 \to F_1 \to F_0$ of finite free R-modules such that M is the cokernel of $F_1 \to F_0$. Suppose that R is local and let $\mathfrak{m}$ denote the maximal ideal in R. The free resolution $F_\bullet$ is said to be *minimal* if every map in the complex $F_\bullet \otimes R/\mathfrak{m}$ is zero. In other words, the image of $F_i \to F_{i-1}$ is contained in $\mathfrak{m}F_{i-1}$ for all i. The minimal free resolution exists uniquely up to isomorphism of complexes.

Definition 4.3.3 A *matrix factorisation* of an element x in a noetherian ring R is a pair $(\varphi \colon F \to G, \psi \colon G \to F)$ of maps of finite free R-modules such that $\psi \circ \varphi = x \cdot \mathrm{id}_F$ and $\varphi \circ \psi = x \cdot \mathrm{id}_G$.

Theorem 4.3.4 (Eisenbud [118]) *Let x be an element in a regular local ring R and let M be a maximal Cohen–Macaulay R/xR-module without free summand. Then there exists a matrix factorisation $(\varphi \colon F \to G, \psi \colon G \to F)$ of $x \in R$ such that M admits a minimal free resolution*

$$\cdots \xrightarrow{\bar{\psi}} F/xF \xrightarrow{\bar{\varphi}} G/xG \xrightarrow{\bar{\psi}} F/xF \xrightarrow{\bar{\varphi}} G/xG$$

by the maps $\bar{\varphi}$ and $\bar{\psi}$ induced by φ and ψ.

Now we suppose an embedding $x \in C \subset S \subset X$ into a smooth threefold X and apply the theorem to $M = \mathcal{O}_S(-C)$. We are interested in a non-lci curve C, in which C is not a Cartier divisor on S. Then by Theorem 4.3.4 with Lemma 4.3.1, $\mathcal{O}_S(-C)$ admits a minimal free resolution

$$\cdots \xrightarrow{\bar{\psi}} \mathcal{O}_S^{\oplus 2} \xrightarrow{\bar{\varphi}} \mathcal{O}_S^{\oplus 2} \to \mathcal{O}_S(-C)$$

where $\bar{\varphi}$ and $\bar{\psi}$ arise from a matrix factorisation $(\varphi \colon \mathcal{O}_X^{\oplus 2} \to \mathcal{O}_X^{\oplus 2}, \psi \colon \mathcal{O}_X^{\oplus 2} \to \mathcal{O}_X^{\oplus 2})$ of the function f in $\mathcal{O}_X$ defining S. By $\det \varphi \cdot \det \psi = \det(f \cdot \mathrm{id}_{\mathcal{O}_X^{\oplus 2}}) = f^2$, one can take the matrix factorisation so that $\det \varphi = \det \psi = f$. Then as a 2×2 matrix, ψ is the adjugate matrix of φ. Below is an explicit form as in [218, section 5], which is consistent with the classification in Theorem 4.2.7.

Proposition 4.3.5 *Up to analytic isomorphism, the first matrix φ is one of the following.*

$$A_n^i \begin{pmatrix} x_1 & x_3^{n+1-i} \\ x_3^i & x_2 \end{pmatrix} \qquad D_n^l \begin{pmatrix} x_1 & x_2^2 + x_3^{n-2} \\ x_3 & x_1 \end{pmatrix} \qquad D_{2m}^r \begin{pmatrix} x_1 & x_2 x_3 + x_3^m \\ x_2 & x_1 \end{pmatrix}$$

$$D_{2m+1}^r \begin{pmatrix} x_1 & x_2 x_3 \\ x_2 & x_1 + x_3^m \end{pmatrix} \qquad E_6 \begin{pmatrix} x_1 & x_2^2 \\ x_2 & x_1 + x_3^2 \end{pmatrix} \qquad E_7 \begin{pmatrix} x_1 & x_2^2 + x_3^3 \\ x_2 & x_1 \end{pmatrix}$$

Composing the above resolution of $\mathcal{O}_S(-C)$ with the short exact sequence $0 \to \mathcal{O}_S(-C) \to \mathcal{O}_S \to \mathcal{O}_C \to 0$, one obtains the exact sequence

$$\mathcal{O}_S^{\oplus 2} \xrightarrow{\bar{\varphi}} \mathcal{O}_S^{\oplus 2} \xrightarrow{\bar{\alpha}} \mathcal{O}_S \to \mathcal{O}_C \to 0.$$

Fix a lift $\alpha \colon \mathcal{O}_X^{\oplus 2} \to \mathcal{O}_X$ of $\bar{\alpha}$. By $\bar{\varphi} \circ \bar{\alpha} = 0$, the composite $\varphi \circ \alpha \colon \mathcal{O}_X^{\oplus 2} \to \mathcal{O}_X$ is given by a row vector of form $-(fg, fh)$ with some $g, h \in \mathcal{O}_X$. Let $(\alpha, f) \colon \mathcal{O}_X^{\oplus 3} \to \mathcal{O}_X$ be the sum of α and the multiplication map $f \colon \mathcal{O}_X \to \mathcal{O}_X$ by f. Then one makes the exact sequence

$$0 \to \mathcal{O}_X^{\oplus 2} \xrightarrow{{}^t A} \mathcal{O}_X^{\oplus 3} \xrightarrow{(\alpha, f)} \mathcal{O}_X \to \mathcal{O}_C \to 0,$$

where the map $\mathcal{O}_X^{\oplus 2} \to \mathcal{O}_X^{\oplus 3}$ is defined by the transpose ${}^t A$ of the 2×3 matrix

$$A = \begin{pmatrix} {}^t \varphi & \begin{matrix} g \\ h \end{matrix} \end{pmatrix}.$$

The exactness is obvious except that the kernel of $(\alpha, f) \colon \mathcal{O}_X^{\oplus 3} \to \mathcal{O}_X$ is in the image of ${}^t A \colon \mathcal{O}_X^{\oplus 2} \to \mathcal{O}_X^{\oplus 3}$. To see this, we suppose that $(\alpha, f)(x, y) = 0$ for $x \in \mathcal{O}_X^{\oplus 2}$ and $y \in \mathcal{O}_X$. Then $\bar{\alpha}(\bar{x}) = 0$ for the image $\bar{x} \in \mathcal{O}_S$ of x and hence one can write $\bar{x} = \bar{\varphi}(\bar{z})$ with some $\bar{z} \in \mathcal{O}_S^{\oplus 2}$, that is, $x = \varphi(z) + fw = \varphi(z + \psi w)$ with some $z, w \in \mathcal{O}_X^{\oplus 2}$. It follows from the relation $(\alpha, f)(\varphi(z + \psi w), y) = 0$ that $y = (g, h)(z + \psi w)$. Thus $(x, y) = ({}^t A)(z + \psi w)$.

Let $\mathscr{I}_C$ denote the ideal in $\mathcal{O}_X$ defining C. We apply the Hilbert–Burch theorem, for which the reader may refer to [119, theorem 20.15].

Notation 4.3.6 For a matrix M of elements in a ring R, we write $I_r(M)$ for the ideal in R generated by all $r \times r$ minors of M. For instance, $I_1(M)$ is the ideal generated by all entries of M.

Theorem 4.3.7 (Hilbert–Burch [63]) *Let I be an ideal in a noetherian local ring R. If R/I admits a minimal free resolution $0 \to R^{\oplus m} \xrightarrow{\varphi} R^{\oplus n} \to R$, then $m = n - 1$ and $I = x I_{n-1}(\varphi)$ with some regular element x in R.*

Therefore the ideal $\mathscr{I}_C$ equals $I_2(A)$. Explicitly by writing

$${}^t \varphi = \begin{pmatrix} a & c \\ b & d \end{pmatrix}, \qquad \xi_1 = \det \begin{pmatrix} a & g \\ b & h \end{pmatrix}, \qquad \xi_2 = \det \begin{pmatrix} c & g \\ d & h \end{pmatrix},$$

we have seen the following structure of $\mathscr{I}_C$. Neither g nor h is a unit because C is not lci.

Proposition 4.3.8 *One has $\mathscr{I}_C = (f, \xi_1, \xi_2)\mathcal{O}_X$ with syzygy*

$$\begin{pmatrix} a & c & g \\ b & d & h \end{pmatrix} \begin{pmatrix} \xi_2 \\ -\xi_1 \\ f \end{pmatrix} = 0.$$

Henceforth we fix the germ $x \in C \subset X$ as follows.

Framework 4.3.9 Let $x \in C \subset X$ be the germ of a smooth threefold X with a non-lci curve C. Let S be the general hyperplane section of X containing C. We assume that $x \in S$ is Du Val. The preceding argument applied to $x \in C \subset S \subset X$ constructs $A = \begin{pmatrix} {}^t\varphi & g \\ & h \end{pmatrix}$. By Proposition 4.1.8, the symbolic blow-up Y of X along C exists and it is **Q**-factorial and canonical. Let S_Y denote the strict transform in Y of S, which is canonical by Lemma 4.1.7.

The next lemma stems from the generality of S.

Lemma 4.3.10 *In Framework* 4.3.9, *one has* $I_1(A) = I_1(\varphi)$.

Proof We shall demonstrate the equality in the case when φ is of form A_n^i in Proposition 4.3.5. The proof in other cases is similar. We may assume without loss of generality that $i \leq n/2$ and hence $I_1(\varphi) = (x_1, x_2, x_3^i)\mathscr{O}_X$. Keep the notation in Proposition 4.3.8 and consider the surface defined by $\det \varphi_{\lambda,\mu} = f - \lambda\xi_1 + \mu\xi_2$, where

$$\varphi_{\lambda,\mu} = \begin{pmatrix} x_1 - \mu g & x_3^{n+1-i} - \mu h \\ x_3^i - \lambda g & x_2 - \lambda h \end{pmatrix}$$

for general $\lambda, \mu \in \mathbf{C}$. Then $x_1' = x_1 - \lambda^{-1}\mu x_3^i$ and $x_2' = x_2 - \mu^{-1}\lambda x_3^{n+1-i}$ belong to $I_1(\varphi_{\lambda,\mu})$, by which one can write $I_1(\varphi_{\lambda,\mu}) = (x_1', x_2', x_3^{i'})\mathscr{O}_X$ with some $i' \in \mathbf{Z}_{>0}$. If g or h were not in $I_1(\varphi) = (x_1', x_2', x_3^i)\mathscr{O}_X$, then the exponent i' would be less than i, that is, $I_1(\varphi) \subsetneq I_1(\varphi_{\lambda,\mu})$. This contradicts the general choice of S. $\qquad\square$

We start with the blow-up $\beta\colon B \to X$ along C. By Proposition 4.3.8, B is realised as the closed subvariety of $X \times \mathbf{P}^2$ defined by

$$\begin{pmatrix} {}^t\varphi & g \\ & h \end{pmatrix}\begin{pmatrix} y_2 \\ -y_1 \\ y_0 \end{pmatrix} = 0$$

with the homogeneous coordinates y_0, y_1, y_2 of $\mathbf{P}^2$. This is never the symbolic blow-up of X along C since the central fibre $\beta^{-1}(x)$ is the whole plane $\Pi = \mathbf{P}^2$. Our strategy is to contract Π birationally by the unprojection explained below.

Inspired by the work of Kustin and Miller [281], Papadakis and Reid [380] developed the theory of *unprojection*, which constructs a new Gorenstein ring from a simpler one. Let $x \in X$ be the germ of a Gorenstein variety and let D be a Gorenstein closed subscheme of codimension one in X. Let $\mathscr{I}_D$ denote the ideal in $\mathscr{O}_X$ defining D. Applying the functor $\mathscr{H}om(\,\cdot\,, \omega_X)$ to the exact sequence $0 \to \mathscr{I}_D \to \mathcal{O}_X \to \mathcal{O}_D \to 0$, one obtains the *adjunction exact sequence*

$$0 \to \omega_X \to \mathcal{H}om(\mathscr{I}_D, \omega_X) \to \omega_D \to 0.$$

Here we used the adjunction $\omega_D = \mathcal{E}xt^1(\mathcal{O}_D, \omega_X)$, which can be found in [10, I proposition 2.3]. Embed the germ $x \in D \subset X$ into an affine space $A = \mathbf{A}^e$. Then $\omega_X = \mathcal{E}xt_A^c(\mathcal{O}_X, \omega_A)$ and $\omega_D = \mathcal{E}xt_A^{c+1}(\mathcal{O}_D, \omega_A)$ for the codimension c of X in A as explained after Definition 1.1.10. The adjunction follows from the spectral sequence $\mathcal{E}xt_X^p(\mathcal{O}_D, \mathcal{E}xt_A^q(\mathcal{O}_X, \omega_A)) \Rightarrow \mathcal{E}xt_A^{p+q}(\mathcal{O}_D, \omega_A)$.

Take an element s in $\mathcal{H}om(\mathscr{I}_D, \omega_X) \simeq \mathcal{H}om(\mathscr{I}_D, \mathcal{O}_X)$ defining a generator of $\omega_D \simeq \mathcal{O}_D$ via $\mathcal{H}om(\mathscr{I}_D, \omega_X) \twoheadrightarrow \omega_D$. It can be regarded as a rational function in the sheaf $\mathcal{K}_X$ of the function field of X. The subring $\mathcal{O}_Y = \mathcal{O}_X[s] \subset \mathcal{K}_X$ is independent of the choice and identification of $s \in \mathcal{K}_X$. The variety $Y = \operatorname{Spec} \mathcal{O}_Y$ is called the *(Kustin–Miller) unprojection* of $x \in D \subset X$.

Theorem 4.3.11 (Papadakis–Reid [380, theorem 1.5]) *The unprojection $Y = \operatorname{Spec} \mathcal{O}_Y$ is Gorenstein.*

The map $s\colon \mathscr{I}_D \to \omega_X \simeq \mathcal{O}_X$ induces an isomorphism $\mathscr{I}_D \simeq s(\mathscr{I}_D)$ of ideals in $\mathcal{O}_X$ as $\mathcal{O}_X$-modules. Let N be the closed subscheme of X defined by the ideal $s(\mathscr{I}_D)$. Then by the unprojection $X \dashrightarrow Y$, the locus $D \setminus N$ disappears to infinity whilst $N \setminus D$ becomes the Cartier divisor defined by s.

Example 4.3.12 For the surface X in $\mathbf{A}^3$ given by $x_1^2 - x_2 x_3$ and the x_3-axis D in Example 1.2.1, one can take $s = x_2^{-1} x_1$ and N is the x_2-axis. The unprojection $\operatorname{Spec} \mathcal{O}_X[s]$ is isomorphic to the chart $U_2 = (x_2 \neq 0)$ of the blow-up of X along D, in which the strict transform of D does not appear.

The construction is valid in the following projective case as discussed in [380, 2.4]. Let $x \in X$ be the germ of a variety and take $Y = \operatorname{Proj}_X \mathscr{R}$ for a finitely generated integral graded $\mathcal{O}_X$-algebra $\mathscr{R} = \bigoplus_{i \in \mathbf{N}} \mathscr{R}_i$ with $\mathscr{R}_0 = \mathcal{O}_X$. Let D be a closed subscheme of codimension one in Y, defined by a graded ideal $\mathscr{I}$ in $\mathscr{R}$. We assume that Y and D are *projectively Gorenstein*, that is, the associated graded algebras $\mathscr{R}$ and $\mathscr{R}/\mathscr{I}$ are Gorenstein. See [61, section 3.6] or [316, section 4.1] for the basic properties of projectively Gorenstein schemes. Writing $\omega_Y = \mathcal{O}_Y(k_Y)$ and $\omega_D = \mathcal{O}_D(k_D)$ with $k_Y, k_D \in \mathbf{Z}$, we impose the assumption $k_Y > k_D$. Then it makes sense to define the *unprojection* $Y' = \operatorname{Proj}_X \mathscr{R}'$ of $D \subset Y$ by constructing $\mathscr{R}' = \mathscr{R}[s]$ from the affine cone $\operatorname{Spec}_X \mathscr{R}$ in the same manner as above.

The new element s is of degree $l = k_Y - k_D$ in Y'. Take $C(Y) = \operatorname{Spec}_X \mathscr{R}$ and $C(D) = \operatorname{Spec}_X \mathscr{R}/\mathscr{I}$ and twist the adjunction sequence $0 \to \omega_{C(Y)} \to$

$\mathcal{H}om(\mathcal{I}, \omega_{C(Y)}) \to \omega_{C(D)} \to 0$ by $\mathcal{R}(-k_D)$. This yields the exact sequence $0 \to \mathcal{R}(l) \to \mathcal{H}om(\mathcal{I}, \mathcal{R}(l)) \to \mathcal{R}/\mathcal{I} \to 0$ which induces

$$0 \to \mathcal{O}_Y(l) \to \mathcal{H}om(\mathcal{I}_D, \mathcal{O}_Y)(l) \to \mathcal{O}_D \to 0$$

for the ideal sheaf $\mathcal{I}_D = \tilde{\mathcal{I}}$ in $\mathcal{O}_Y$ defining D.

In our setting, the blow-up B of X along C is projectively Gorenstein. We take the unprojection $\beta_1 \colon B_1 \to X$ of $\Pi = \beta^{-1}(x) \subset B$. Note that $\omega_B = \mathcal{O}_B(-1)$ and $\omega_\Pi = \mathcal{O}_\Pi(-3)$. If the fibre $\beta_1^{-1}(x)$ contains a locus Π_1 of codimension one again, then we seek to unproject Π_1. We attempt to continue this operation until a model $\beta_n \colon B_n \to X$ with one-dimensional central fibre $\beta_n^{-1}(x)$ is achieved, and then to verify that it is the symbolic blow-up.

We shall convey the explicit way of computation by describing the case of type A_1 in detail. The calculation first appeared in [391]. We shall prove the following theorem.

Theorem 4.3.13 ([111]) *In Framework* 4.3.9, *suppose that* $x \in S$ *is of type* A_1. *Then the symbolic blow-up is described as a model* $Y \subset X \times \mathbf{P}(1, 1, 1, 2)$. *The model* Y *is terminal if and only if the type of* C *in the sense of Definition* 4.3.14 *is as follows.*

$$\overset{\circ}{3}$$

Definition 4.3.14 In Framework 4.3.9, we say that C is of *type* Γ if Γ is the dual graph of the minimal resolution of $x \in S$ in which a vertex $\circ$ stands for a curve exceptional over S_Y and $\bullet$ for a curve not exceptional over S_Y but exceptional over S and in which the intersection number of the strict transform of C with the curve $\circ$ is attached to $\circ$ if they intersect. Notice that the minimal resolution of S factors through S_Y.

Suppose that S is of type A_1. We shall compute in the analytic category. By Proposition 4.3.8 and Lemma 4.3.10, C is defined by the ideal $I_2(A)$ in $\mathcal{O}_X$ for

$$A = \begin{pmatrix} x_1 & x_3 & g \\ x_3 & x_2 & h \end{pmatrix}$$

with $g, h \in (x_1, x_2, x_3)\mathcal{O}_X$. Here one can take $g = g(x_2, x_3)$ and $h = h(x_1, x_3)$. Indeed if we write $g = p(x_1, x_3)x_1 + q(x_2, x_3)$ and $h = r(x_2, x_3)x_2 + s(x_1, x_3)$ using the relation $x_1 x_2 = x_3^2$ on S, then the vector (g, h) equals $p \cdot (x_1, x_3) + r \cdot (x_3, x_2) + (q - rx_3, s - px_3)$. Replace (g, h) by $(q - rx_3, s - px_3)$.

We thus take $g = -(ax_2 + bx_3)$ and $h = cx_1 + dx_3$ with $a, b, c, d \in \mathcal{O}_X$. The blow-up $B \subset X \times \mathbf{P}^2$ of X along C is given by

$$\begin{pmatrix} x_1 & x_3 & -(ax_2 + bx_3) \\ x_3 & x_2 & cx_1 + dx_3 \end{pmatrix} \begin{pmatrix} y_2 \\ -y_1 \\ y_0 \end{pmatrix} = 0,$$

which is projectively Gorenstein over X. We rearrange this as

$$\begin{pmatrix} y_2 & ay_0 & -(by_0 + y_1) \\ cy_0 & y_1 & dy_0 + y_2 \end{pmatrix} \begin{pmatrix} x_1 \\ -x_2 \\ x_3 \end{pmatrix} = 0.$$

The maximal minors of the left 2×3 matrix has the syzygy

$$\begin{pmatrix} y_2 & ay_0 & -(by_0 + y_1) \\ cy_0 & y_1 & dy_0 + y_2 \end{pmatrix} \begin{pmatrix} ay_0(dy_0 + y_2) + y_1(by_0 + y_1) \\ -(y_2(dy_0 + y_2) + cy_0(by_0 + y_1)) \\ y_1 y_2 - acy_0^2 \end{pmatrix} = 0.$$

Define the ratio z of the two column vectors by

$$z \begin{pmatrix} x_1 \\ x_2 \\ x_3 \end{pmatrix} = \begin{pmatrix} ay_0(dy_0 + y_2) + y_1(by_0 + y_1) \\ y_2(dy_0 + y_2) + cy_0(by_0 + y_1) \\ y_1 y_2 - acy_0^2 \end{pmatrix}.$$

These three equations together with the two equations for $B \subset X \times \mathbf{P}^2$ define a new variety B_1 in $X \times \mathbf{P}(1, 1, 1, 2)$ with weighted homogeneous coordinates y_0, y_1, y_2, z, where z is a new coordinate of degree two. The threefold B_1 is the unprojection of $\Pi = \beta^{-1}(x) \subset B$. In fact, z gives a map $\mathscr{I}_\Pi \to \mathcal{O}_B(2)$ from the ideal sheaf $\mathscr{I}_\Pi$ in $\mathcal{O}_B$ defining Π. The centre of Π becomes on B_1 the closed point $q_z = (x, [0, 0, 0, 1]) \in x \times \mathbf{P}(1, 1, 1, 2)$, at which B_1 has a cyclic quotient singularity of type $\frac{1}{2}(1, 1, 1)$. This is the only singularity away from the locus $(z = 0)$. The inverse $B_1 \dashrightarrow B$ of the unprojection is regarded as the projection from q_z.

The five equations for B_1 are described as the maximal Pfaffians of the skew-symmetric 5×5 matrix

$$\begin{pmatrix} z & y_2 & by_0 + y_1 & -ay_0 \\ & -cy_0 & dy_0 + y_2 & y_1 \\ & & x_2 & x_3 \\ & & & x_1 \end{pmatrix},$$

where the lower triangle part as well as the diagonal of zeros is omitted by anti-symmetry. This matches the Buchsbaum–Eisenbud theorem below.

Let A be a skew-symmetric $n \times n$ matrix. If n is even, then $\det A$ is the square of a homogeneous polynomial of degree $n/2$ in the entries of A, and the *Pfaffian* Pf A of A is defined so that $\det A = (\text{Pf } A)^2$ [56, section 5.2]. For instance when $n = 4$, $A = (a_{ij})$ has Pf $A = a_{12}a_{34} - a_{13}a_{24} + a_{14}a_{23}$. If n is odd, then $\det A$ is zero. In this case, a *maximal Pfaffian* of A is the Pfaffian of an $(n-1) \times (n-1)$ submatrix of A obtained by deletion of the i-th row and column for some i.

Let I be a proper ideal in a noetherian local ring R. The *grade g* of I is the common length of a maximal regular sequence in I. We say that I is *perfect* if g equals the projective dimension of R/I. We say that I is *Gorenstein* if it is perfect and $\text{Ext}_R^g(R/I, R) \simeq R/I$. When R is Gorenstein and I is perfect, I is Gorenstein if and only if R/I is Gorenstein [61, theorem 3.3.7].

Theorem 4.3.15 (Buchsbaum–Eisenbud [62]) *Let I be an ideal of grade three in a noetherian local ring R. Then I is Gorenstein if and only if R/I admits a minimal free resolution*

$$0 \to R \to R^{\oplus(2m+1)} \xrightarrow{\varphi} R^{\oplus(2m+1)} \to R$$

such that φ is given by a skew-symmetric matrix and such that I is generated by the maximal Pfaffians of φ.

Proof of Theorem 4.3.13   The unprojection $B_1 \subset X \times \mathbf{P}(1, 1, 1, 2)$ is normal and the scheme-theoretic central fibre $B_1 \times_X x \subset \mathbf{P}(1, 1, 1, 2)$ of $\beta_1 : B_1 \to X$ is given by the maximal minors of

$$\begin{pmatrix} y_2 & \bar{a}y_0 & -(\bar{b}y_0 + y_1) \\ \bar{c}y_0 & y_1 & \bar{d}y_0 + y_2 \end{pmatrix}$$

with the values $\bar{a}, \bar{b}, \bar{c}, \bar{d} \in \mathbf{C}$ at x of a, b, c, d, which is of dimension one. The exceptional locus E_1 in B_1 is a prime divisor and $-E_1$ is relatively ample. Hence B_1 is the symbolic blow-up of X along C.

We shall investigate the singularities of B_1. The central fibre $\beta_1^{-1}(x)$ is the union of lines joining $q_z = c_{B_1}(\Pi)$ and a point q_l in the plane $(z = 0) \subset \mathbf{P}(1, 1, 1, 2)$. Since y_0 never vanishes on $\beta_1^{-1}(x) \setminus q_z$, it suffices to examine the locus $(y_0 \neq 0)$ where B_1 is embedded in $\mathfrak{D}^6$ with coordinates x_1, x_2, x_3 and $y_1' = y_1 y_0^{-1}, y_2' = y_2 y_0^{-1}, z' = z y_0^{-2}$ and defined by the maximal Pfaffians of

$$\begin{pmatrix} z' & y_2' & b + y_1' & -a \\ & -c & d + y_2' & y_1' \\ & & x_2 & x_3 \\ & & & x_1 \end{pmatrix}.$$

If neither of a, b, c, d is a unit in $\mathcal{O}_{X,x}$, then B_1 contains the origin o in $\mathfrak{D}^6$, at which all the derivations $\partial/\partial x_1, \ldots, \partial/\partial z'$ annihilate the maximal Pfaffians above. Hence the tangent space $T_{B_1,o}$ equals $T_{\mathfrak{D}^6,o}$. By Theorem 2.3.4, B_1 is never terminal at o. Notice that $o \in B_1$ is Gorenstein by Theorem 4.3.15.

On the other hand if one of a, b, c, d is a unit, then B_1 is terminal. One has only to check this at each point q_l on $B_1 \cap (z = 0)$. After coordinate change, a becomes a unit. Eliminating the variables x_2 and y_2' by the relations

$$y_2' x_1 - ax_2 - (b + y_1')x_3 = z'x_1 - a(d + y_2') - (b + y_1')y_1' = 0,$$

one sees that B_1 is defined by the equation

$$z'(y_1'x_1 - ax_3) = (y_1')^3 + b(y_1')^2 + ady_1' + a^2c$$

where a is a unit. Thus B_1 is smooth (resp. at worst cA_1, at worst cA_2) at q_l if q_l lies on a simple (resp. double, triple) root of the cubic equation $(y_1')^3 + \bar{b}(y_1')^2 + \bar{a}\bar{d}y_1' + \bar{a}^2\bar{c} = 0$.

Regard $S = (x_1x_2 - x_3^2 = 0)$ as the quotient $\mathfrak{D}^2/\mathbf{Z}_2$ by $(x_1, x_2, x_3) = (u_1^2, u_2^2, u_1u_2)$ for the orbifold coordinates u_1, u_2. The inverse image $\bar{C}$ in $\mathfrak{D}^2$ of C is given by $cu_1^3 + du_1^2u_2 + bu_1u_2^2 + au_2^3$. Thus B_1 is terminal if and only if $\bar{C}$ has multiplicity three at origin, that is, the type of C is as in the statement. This completes the theorem. □

Example 4.3.16 This example goes back to Hironaka [186]. See also [249, example 5.2.6] or [307, example 3.1.9]. For $x \in C \subset X$, suppose that $x \in C$ is analytically the union of the three axes in $X = \mathfrak{D}^3$ defined by $\mathscr{I} = (x_1x_2, x_2x_3, x_3x_1)\mathscr{O}_X$. The general hyperplane section S containing C is Du Val of type A_1. The unprojection B_1 from the blow-up B of X along C can be reconstructed as follows.

Let W be the blow-up of X at x and let Π_W be the exceptional divisor. Let V be the blow-up of W along the strict transform of C and let E be the exceptional divisor. The relative canonical divisor $K_{V/X} = 2\Pi + E$ is anti-nef over X and $\mathscr{I}\mathscr{O}_V = \mathscr{O}_V(-K_{V/X})$, where Π is the strict transform of Π_W. The divisor Π is the blow-up of $\Pi_W \simeq \mathbf{P}^2$ at three points. We write l_i for $i = 1, 2, 3$ for the strict transforms in Π of the three lines in Π_W joining two of these three points.

The two-ray game in Definition 4.2.5 from $V \to X$ generates a flop $V \to Z \leftarrow V_1$ in Definition 5.1.6 followed by a divisorial contraction $V_1 \to Z_1$. The curves l_i are contracted to points in Z and Z is the blow-up B of X along C. The threefold V_1 is smooth and the fibre of $V_1 \to X$ at x consists of the strict transform $\Pi_1 \simeq \mathbf{P}^2$ of Π and three new lines. The surface Π_1 is contracted to a quotient singularity of type $\frac{1}{2}(1, 1, 1)$ in Z_1 as in Theorem 3.1.1, and Z_1 is the unprojection B_1.

Ducat obtained the structure of the symbolic blow-up in several cases by the same method as for Theorem 4.3.13.

Theorem 4.3.17 ([111]) *Work in Framework 4.3.9.*

(i) *Suppose that $x \in S$ is of type A_2. Then the symbolic blow-up is described as a model $Y \subset X \times \mathbf{P}(1, 1, 1, 2, 3)$. If Y is terminal, then the type of C is one of the following.*

$$\underset{3}{\circ}\!\!-\!\!-\!\!-\!\!\underset{1}{\circ} \qquad \underset{4}{\circ}\!\!-\!\!-\!\!-\!\!\circ$$

(ii) *Suppose that $x \in C \subset S$ belongs to D_n^l, D_{2m}^r or E_7 in Proposition 4.3.5. Then the symbolic blow-up is described as a model $Y \subset X \times \mathbf{P}(1, 1, 1, 2)$. It is never terminal.*

(iii) *Suppose that $x \in S$ is of type E_6 and S_Y is isomorphic to S. Then the symbolic blow-up is described as a model $Y \subset X \times \mathbf{P}(1, 1, 1, 2, 3)$. If Y is terminal, then the type of C is as follows.*

(iv) *Suppose that $x \in S$ is of type E_6 and S_Y is not isomorphic to S. Then the symbolic blow-up is described as a model $Y \subset X \times \mathbf{P}(1, 1, 1, 2, 3, 4)$. If Y is terminal, then the type of C is as follows.*

Example 4.3.18 ([110, example 4.3])   This is an example of the last case (iv) in Theorem 4.3.17. Let $x \in C \subset X$ be the monomial curve $(x_1, x_2, x_3) = (t^{11}, t^7, t^5)$ defined by the maximal minors of $A = \begin{pmatrix} x_1 & x_2 & x_3^2 \\ x_3^3 & x_1 & x_2^2 \end{pmatrix}$. The general hyperplane section $x \in S$ containing C is Du Val of type E_6.

The blow-up B of X along C is defined in $X \times \mathbf{P}^2$ by $A\,{}^t(y_2, -y_1, y_0) = 0$. For the central fibre $\Pi \simeq \mathbf{P}^2$, the unprojection B_1 of $\Pi \subset B$ is given in $X \times \mathbf{P}(1, 1, 1, 2)$ by the maximal Pfaffians of

$$\begin{pmatrix} z & y_0 & y_1 & y_2 \\ & -x_3 y_2 & x_2 y_0 & y_1 \\ & & x_1 & x_2 \\ & & & x_3^2 \end{pmatrix}.$$

Unproject the locus Π_1 in B_1 defined by $(x_1, x_2, x_3, y_1)\mathcal{O}_{B_1}$ to obtain a variety B_2 with a new coordinate w of degree three. Then unproject Π_2 in B_2 defined by $(x_1, x_2, x_3, y_1, y_0 y_2) \mathscr{O}_{B_2}$ to obtain B_3 with new v of degree four. Here w and v are defined by

$$
w \begin{pmatrix} x_1 \\ x_2 \\ x_3 \\ y_1 \end{pmatrix} = \begin{pmatrix} x_3 y_0^3 + y_1^2 y_2 \\ x_3 y_0 z + y_1 y_2^2 \\ y_1 z - y_0^2 y_2 \\ x_3 z^2 + y_0 y_2^3 \end{pmatrix}, \qquad v \begin{pmatrix} x_1 \\ x_2 \\ x_3 \\ y_1 \\ y_0 y_2 \end{pmatrix} = \begin{pmatrix} y_1 y_2^3 - y_0^4 - x_3 y_0 y_2 z \\ y_2^4 - y_0^2 z \\ y_2^2 z - y_0 w \\ y_2^2 w - y_0 z^2 \\ w^2 - z^3 \end{pmatrix}.
$$

The model B_3 is the symbolic blow-up of X along C.

4.4 Construction by a One-Parameter Deformation

Let $\pi \colon Y \to X$ be a threefold divisorial contraction which contracts the divisor to a curve. Work on the germ $x \in X$ and take the general hyperplane section $x \in H$ of $x \in X$. The function $t \in \mathscr{O}_X$ defining H gives a morphism $t \colon X \to \mathbf{A}^1$. We regard π as a one-parameter family $Y \to X \to \mathbf{A}^1$. The central fibre is the morphism $H_Y \to H$ from the strict transform H_Y in Y of H, and Y is the total space of a deformation of the germ of H_Y along $\pi^{-1}(x)$. This perspective is also applicable to a threefold flipping contraction.

The purpose of this section is to construct examples of the analytic germ of π from this point of view. The strategy was formulated in [276, section 11] and [278, section 3]. We work in the analytic category which fits well into the deformation theory. We use the notation $C \subset X$ for $\pi^{-1}(x) \subset Y$ which coordinates with [276].

We review the definition of deformation. The reader may consult [153, chapter II] for example.

Definition 4.4.1 Let $Z \subset V$ be the germ of an analytic space V along a compact analytic subspace Z. A *deformation* of V over the germ $s \in S$ of an analytic space consists of a flat morphism $v \colon \mathfrak{B} \to S$ of analytic spaces and an isomorphism $i \colon V \to \mathfrak{B}_s$ to the central fibre $\mathfrak{B}_s = \mathfrak{B} \times_S s$ of v. We write $(i, v) \colon V \hookrightarrow \mathfrak{B} \to s \in S$ for it. The space $\mathfrak{B}$ is the germ along $i(Z)$. For two deformations $(i, v) \colon V \hookrightarrow \mathfrak{B} \to s \in S$ and $(i', v') \colon V \hookrightarrow \mathfrak{B}' \to s' \in S'$ of V, a *morphism* from (i, v) to (i', v') consists of morphisms $f \colon s \in S \to s' \in S'$ and $F \colon \mathfrak{B} \to \mathfrak{B}'$ such that $v' \circ F = f \circ v$ and $i' = F \circ i$. If f is an isomorphism, then so is F [153, I lemma 1.86].

Definition 4.4.2 A deformation $(i, v) \colon V \hookrightarrow \mathfrak{B} \to s \in S$ of an analytic space V is said to be *versal* if it satisfies the following lifting property: for every deformation $(j, w) \colon V \hookrightarrow \mathfrak{W} \to t \in T$ and every closed analytic subspace

$t \in T' \subset T$, if the base change $(j', w') \colon V \hookrightarrow \mathfrak{W}' = \mathfrak{W} \times_T T' \to t \in T'$ admits a morphism $\alpha' \colon (j', w') \to (i, v)$ of deformations, then it extends to a morphism $\alpha \colon (j, w) \to (i, v)$ of deformations such that $\alpha' = \alpha \circ \iota$ for the induced embedding $\iota \colon (j', w') \hookrightarrow (j, w)$. If (i, v) is versal, then every (j, w) has a morphism to (i, v) by taking $T' = \{t\}$.

Suppose that $(i, v) \colon V \hookrightarrow \mathfrak{B} \to s \in S$ is versal. It is said to be *miniversal* or *semi-universal* if in the lifting property above, the induced map $T_{T,t} \to T_{S,s}$ of tangent spaces is uniquely determined by α'. By the *inverse function theorem*, the miniversal deformation $\mathfrak{B} \to S$ of V is unique up to non-canonical isomorphism if it exists [153, II lemma 1.12]. In this case, the base space S is called the *miniversal deformation space* of V and denoted by Def V.

We remark that if a versal deformation $\mathfrak{B} \to S$ of V exists, then so does the miniversal deformation of V, and $S \simeq \mathrm{Def}\, V \times \mathfrak{D}^n$ for some n [153, II theorem 1.13, proposition 1.14].

Example 4.4.3 Let $o \in V \subset \mathfrak{D}^n$ with coordinates $x_1, \ldots, x_n$ be an isolated hypersurface singularity defined by $f \in \mathscr{O}_{\mathfrak{D}^n}$. Take functions $g_1, \ldots, g_l$ in $\mathscr{O}_{\mathfrak{D}^n}$ which form a basis of $\mathscr{O}_{\mathfrak{D}^n}/(f, \partial f/\partial x_1, \ldots, \partial f/\partial x_n)\mathscr{O}_{\mathfrak{D}^n}$. Then the family $(f + \sum_{i=1}^{l} t_i g_i = 0) \subset \mathfrak{D}^n \times \mathfrak{D}^l \to \mathfrak{D}^l$ with coordinates $t_1, \ldots, t_l$ is the miniversal deformation of $o \in V$ as elucidated in [277, theorem 4.61].

It is straightforward to extend these notions to deformations of morphisms. We shall use the existence of the miniversal deformation space in the following cases. The second item also holds when W is empty, which means the first item. The second with V being a point means the existence of the miniversal deformation space of a compact analytic space [150].

Theorem 4.4.4 *Let $o \in V$ be the germ of an isolated analytic singularity.*

(i) *(Grauert [149]) The miniversal deformation space Def V of the germ V at o exists.*

(ii) *(Bingener [42, V Satz 4.8]) If $\pi \colon W \to V$ is a proper morphism isomorphic outside o, then the miniversal deformation space Def π of the germ of the morphism π at o exists.*

The next result [276, proposition 11.4] connects several deformation spaces arising from a morphism, in which Def π and Def $(p_i \in W)$ exist by the above theorem.

Theorem 4.4.5 *Let $o \in V$ be the germ of an isolated analytic singularity and let $\pi \colon W \to V$ be a proper morphism with connected fibres and isomorphic outside o. If $R^1\pi_*\mathscr{O}_W = 0$, then the miniversal deformation space Def W of* *the germ W about $\pi^{-1}(o)$ exists and the natural morphism* Def $\pi \to$ Def *W is surjective. If in addition $\pi^{-1}(o)$ is of dimension one and W has only finitely many singular points p_i, then the natural morphism* Def *W* $\to \prod_i$ Def $(p_i \in W)$ *is smooth and in particular surjective.*

Our object is an analytic germ of a threefold along a contractible curve.

Definition 4.4.6 Let $C \subset X$ be the germ of a terminal analytic threefold X along a reduced compact analytic space C of dimension one. The germ $C \subset X$ is called an *extremal neighbourhood* if it admits a projective bimeromorphic morphism $\pi \colon C \subset X \to y \in Y$ to the germ of a normal analytic space such that $C = \pi^{-1}(y)$ and such that $-K_X$ is π-ample. We say that it is *irreducible* if C is irreducible, and that it is *isolated* (resp. *divisorial*) if the exceptional locus of π is of dimension one (resp. two).

The extremal neighbourhood $C \subset X$ satisfies the vanishing $H^1(\mathcal{O}_C) = 0$ by the analytic Kawamata–Viehweg vanishing $R^1\pi_*\mathcal{O}_X = 0$ mentioned in Remark 1.4.14. Hence C is a tree of $\mathbf{P}^1$. By the next lemma, $y \in Y$ is a rational singularity.

Lemma 4.4.7 *Let $\pi \colon X \to Y$ be a contraction such that X has rational singularities. Then Y has rational singularities if and only if $R^i\pi_*\mathcal{O}_X = 0$ for all $i \geq 1$.*

Proof Take a resolution $\mu \colon X' \to X$. Then $R^i\mu_*\mathcal{O}_{X'} = 0$ for all $i \geq 1$. Hence $R^i\pi_*\mathcal{O}_X \simeq R^i(\pi \circ \mu)_*\mathcal{O}_{X'}$ from the Leray spectral sequence, which implies the lemma.   □

Let $C \subset X$ be an extremal neighbourhood equipped with a contraction $\pi \colon C \subset X \to y \in Y$. Let H be a principal prime divisor on X containing C. It is defined by a function in $\mathcal{O}_X$ vanishing along C. We do not assume that H is general. By $\pi_*\mathcal{O}_X = \mathcal{O}_Y$, the strict transform $y \in H'$ in Y of H is a hyperplane section of Y such that $H = \pi^*H'$. As introduced at the beginning of the section, one can regard X as a deformation $H \hookrightarrow X \to \mathfrak{D}^1$ of $C \subset H$ by the function $t \in \mathcal{O}_Y$ defining H', which gives a morphism $Y \to \mathfrak{D}^1$. We seek to recover X from $H \to H'$.

Construction 4.4.8 Consider a rational surface singularity $y \in H'$ in the analytic category and a projective bimeromorphic morphism $\varphi \colon H \to H'$ from a normal surface H such that $-K_H$ is φ-ample. Note that $R^1\varphi_*\mathcal{O}_H = 0$ by $R^1\varphi_*\mathcal{O}_H \subset R^1(\varphi \circ \mu)_*\mathcal{O}_{\bar{H}} = 0$ for a resolution $\mu \colon \bar{H} \to H$ from the Leray spectral sequence. We assume that φ is not an isomorphism and define $C = \varphi^{-1}(y)$ with reduced structure. Suppose that the germ at every singular point p_i of H admits a one-parameter deformation $p_i \in H \hookrightarrow X_i \to \mathfrak{D}^1$ such that $p_i \in X_i$ is terminal. Then by Theorem 4.4.5, one can globalise all $X_i \to \mathfrak{D}^1$ compatibly as a deformation $X \to Y \to \mathfrak{D}^1$ of φ. The central fibre $X_o \to Y_o$ is identified with φ and the germ $p_i \in X$ is isomorphic to $p_i \in X_i$ over $\mathfrak{D}^1$. Since $K_H = K_X|_H$, the germ $C \subset X \to y \in Y$ is an extremal neighbourhood.

Remark 4.4.9 In this construction, if $y \in H'$ is Gorenstein, then so is $y \in Y$ because H' is a hyperplane section of Y. In this case, one can define the relative canonical divisor $K = K_{X/Y}$ which is exceptional and anti-ample over Y. Hence $C \subset X$ is divisorial and K is an effective divisor the support of which coincides with the exceptional locus. It follows that the germ $y \in Y$ is terminal and K is reduced. However in general, it is unclear whether $C \subset X$ is divisorial or isolated. When $C \subset X$ is irreducible, it is divisorial if and only if $y \in Y$ is terminal [342, theorem 1.10].

An important class of singularities on H is the class T.

Definition 4.4.10 (Kollár–Shepherd-Barron [278]) A surface singularity $x \in S$ is said to be of *class T* if it is log terminal and admits a one-parameter deformation $x \in S \hookrightarrow X \to \mathfrak{D}^1$ such that $x \in X$ is terminal.

Theorem 4.4.11 ([278, proposition 3.10]) *A surface singularity $x \in S$ is of class T if and only if it is either a Du Val singularity or a cyclic quotient singularity of type $\frac{1}{r^2 s}(1, brs - 1)$ with positive integers b, r and s such that b and r are coprime.*

The cyclic quotient singularity of type $\frac{1}{r^2 s}(1, brs - 1)$ with orbifold coordinates x_1, x_2 which is of class T is isomorphic to $o \in (y_1 y_2 - y_3^{rs} = 0) \subset \mathfrak{D}^3/\mathbf{Z}_r(1, -1, b)$ by $(y_1, y_2, y_3) = (x_1^{rs}, x_2^{rs}, x_1 x_2)$. It has a deformation $(y_1 y_2 - y_3^{rs} + y_4 = 0) \subset \mathfrak{D}^4/\mathbf{Z}_r(1, -1, b, 0)$ and its total space is terminal with a quotient singularity of type $\frac{1}{r}(1, -1, b)$.

The minimal resolution of a cyclic quotient surface singularity is well known [30, section III.5].

Definition 4.4.12 For a finite sequence $a_1, \ldots, a_n$ of integers at least two, we let $[a_1, \ldots, a_n]$ denote the *continued fraction*

$$a_1 - \cfrac{1}{a_2 - \cfrac{1}{a_3 - \cdots}}$$

and call it a *string*. For instance, $[a_1] = a_1$ and $[a_1, a_2] = a_1 - a_2^{-1}$. For coprime positive integers b and r such that $b < r$, the rational number r/b

is uniquely expressed by a string as $r/b = [a_1, \ldots, a_n]$. We say that two strings $[a_1, \ldots, a_n]$ and $[b_1, \ldots, b_m]$ are *conjugate* if $[a_1, \ldots, a_n] = r/b$ and $[b_1, \ldots, b_m] = r/(r - b)$ for some b and r.

For a quotient singularity of type $\frac{1}{r}(1, b)$ with $r/b = [a_1, \ldots, a_n]$, the exceptional locus of the minimal resolution is the chain

$$\underset{-a_1}{\circ}\!\!\!-\!\!\!- \cdots -\!\!\!-\underset{-a_n}{\circ}$$

of smooth rational curves with self-intersection numbers $-a_1, \ldots, -a_n$. By a *chain* of curves, we mean a finite union $\bigcup_{i=1}^{l} C_i$ of curves $C_1, \ldots, C_l$ such that C_i intersects C_j if and only if $|i - j| = 1$. Conversely, a normal surface singularity is a cyclic quotient singularity if the exceptional locus of the minimal resolution is a chain of smooth rational curves which is a tree. Thus a string $[a_1, \ldots, a_n]$ is identified with the corresponding cyclic quotient singularity.

Theorem 4.4.13 ([278, proposition 3.11])   (i) *The string* [4] *and the strings* $[3, 2, \ldots, 2, 3]$, *including* $[3, 3]$, *are of class T.*

(ii) *If* $[a_1, \ldots, a_n]$ *is of class T, then so are* $[2, a_1, \ldots, a_{n-1}, a_n + 1]$ *and* $[a_1 + 1, a_2, \ldots, a_n, 2]$.

(iii) *Every non-Du Val singularity of class T is obtained from one of the strings in* (i) *by iterating the operation in* (ii).

We shall use a classification of lc surface singularities due to Kawamata. The reader can refer to [264, chapter 3] or [277, section 4.1].

Theorem 4.4.14 (Kawamata [236, theorem 9.6])   *Let* (S, Δ) *be an lc pair on the germ* $o \in S$ *of a normal analytic surface. For the minimal resolution of* $o \in S$, *let* E *denote the exceptional locus and let* Γ *denote the dual graph, in which* $\circ$ *stands for an exceptional curve,* $\bullet$ *for a component of the strict transform of* $\lfloor \Delta \rfloor$ *and the number attached to* $\circ$ *for the self-intersection number. Then one of the following holds. Except the case* (i), E *is an snc divisor consisting of smooth rational curves. The germ* $o \in S$ *is lt exactly in* (v) *to* (viii).

(i) $\Delta = 0$ *and* E *is either a smooth elliptic curve or a nodal cubic. Here a nodal cubic means a complete rational curve with one node, which is unique up to isomorphism and realised as a cubic in* $\mathbf{P}^2$.

(ii) $\Delta = 0$ *and* Γ *is a cycle as follows. The number of* $\circ$ *is at least two.*

(iii) $\Delta = 0$ *and* Γ *is as follows. The part* Γ_i *is contracted to a cyclic quotient singularity* $\mathfrak{D}^2/\mathbf{Z}_{r_i}$ *with* $(r_1, r_2, r_3) = (2, 3, 6)$, $(2, 4, 4)$ *or* $(3, 3, 3)$.

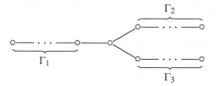

(iv) $\Delta = 0$ *and* Γ *is as follows. The number of* o *is at least five.*

(v) $\lfloor \Delta \rfloor = 0$ *and* $o \in S$ *is a quotient singularity.*

(vi) $(S, \lfloor \Delta \rfloor)$ *is plt and* Γ *is as follows.*

(vii) $\Delta = \lfloor \Delta \rfloor$, (S, Δ) *is not dlt and* Γ *is as follows. The number of* o *is at least three.*

(viii) $\Delta = \lfloor \Delta \rfloor$ *and* Γ *is as follows. The pair* (S, Δ) *is dlt if and only if* S *is smooth and* Δ *is snc.*

Remark 4.4.15 This classification includes the assertion that a normal surface singularity is lt if and only if it is a quotient singularity [232, corollary 1.9]. The dual graph Γ of an lt surface singularity has been classified by Brieskorn [58, Satz 2.10]. It is either a chain or of shape as in Theorem 4.4.14(iii) with $(r_1, r_2, r_3) = (2, 2, r)$, $(2, 3, 3)$, $(2, 3, 4)$ or $(2, 3, 5)$. He also proved that such a singularity is uniquely determined by the dual graph.

In dimension two, pull-backs and intersection numbers are defined for Weil divisors, by which the log discrepancy $a_E(S, \Delta)$ is defined for every couple (S, Δ) of a normal surface S and an effective **R**-divisor Δ. If (S, Δ) is *numerically log canonical* in the sense that $a_E(X, \Delta) \geq 0$ for all divisors E over X, then it is actually log canonical. This is found as [277, proposition 4.11] only in the Japanese edition.

Definition 4.4.16 (Mumford [360, subsection II(b)])   Let D be a Weil divisor on a normal surface S. For every generically finite morphism $\pi \colon T \to S$ from a normal surface, the *pull-back π^*D* by π of D is uniquely defined as a **Q**-divisor on T in such a manner that

- on the smooth locus in S, π^*D coincides with the usual pull-back of D,
- if T is smooth, then $(\pi^*D \cdot E) = 0$ for any curve E contracted by π and
- π^*D is the push-forward of $(\pi \circ \varphi)^*D$ for any proper birational morphism $\varphi \colon T' \to T$.

The *intersection number* $(D \cdot D')$ of Weil divisors D and D' on S is defined as the rational number $(\mu^*D \cdot \mu^*D')_{S'}$ on a resolution $\mu \colon S' \to S$.

Returning to the extremal neighbourhood $C \subset X$ together with $\pi \colon C \subset X \to y \in Y$, we assume that it is irreducible, that is, $C \simeq \mathbf{P}^1$, for simplicity. Then the general elephant S of X is Du Val by Theorem 5.5.1. Let S' be the strict transform in Y of S. The discussion in the previous sections predicts that the structure of π is complicated when the singularity $y \in S'$ is mild. We shall concentrate on the semistable case below and describe the pair $C \subset H$ with a principal divisor H on X containing C. Afterwards we shall provide examples along Construction 4.4.8.

Definition 4.4.17   An irreducible extremal neighbourhood $C \simeq \mathbf{P}^1 \subset X \to y \in Y$ is said to be *semistable* if the strict transform $y \in S'$ in Y of the general elephant of X is Du Val of type A.

The situation is summarised as the following framework. By virtue of Theorem 4.1.9, we assume that X is not Gorenstein.

Framework 4.4.18   Let $C \simeq \mathbf{P}^1 \subset X$ be a semistable extremal neighbourhood such that X is not Gorenstein, together with a contraction $\pi \colon C \subset X \to y \in Y$. Let S be the general elephant of X and let S' be the strict transform in Y. By Theorem 5.5.1, either

(i) $C \not\subset S$ and X has one non-Gorenstein point p or
(ii) $C \subset S$, X has two non-Gorenstein points p_1 and p_2 and X is smooth outside these points.

Let H be a principal prime divisor on X containing C and let H' be the strict transform in Y, for which $H = \pi^*H'$. We assume that H is smooth outside C but do not assume that H is general. We write H^ν for the normalisation of H and write $\bar{H}$ for the minimal resolution of H^ν. We let $\bar{C}$ denote the union of curves in $\bar{H}$ mapped onto C. Note that $\bar{C} \simeq C$ if H is normal.

Lemma 4.4.19 *In Framework 4.4.18 if $(X, S + H)$ is lc, then $y \in H'$ is lt and in particular rational. This is the case if H is taken generally.*

Proof Since π is crepant with respect to $K_X + S + H$, the log canonicity of $(X, S + H)$ is equivalent to that of $(Y, S' + H')$. If $(Y', S' + H')$ is lc, then so is $(H', S'|_{H'})$ by adjunction and H' is lt by Theorem 4.4.14, which is rational by Theorem 1.4.20.

If H is general, then $H'|_{S'}$ is the general hyperplane section of $q \in S'$. It follows from the explicit description $o \in (x_1 x_2 + x_3^{n+1} = 0)$ of $q \in S'$ that $(S', H'|_{S'})$ is an lc pair which belongs to the case (viii) in Theorem 4.4.14. Then $(Y', S' + H')$ is lc by inversion of adjunction in Theorem 1.4.26. ☐

In the rest of the section, the circle ◦ in the dual graph of $\bar{H}$ stands for a curve exceptional over H^ν, the bullet • for a curve in $\bar{C}$ and the number attached to ◦ or • for the self-intersection number.

First we deal with the case when H is normal.

Theorem 4.4.20 (Tziolas [455], Mori–Prokhorov [342]) *In the semistable case in Framework 4.4.18, consider H such that $(X, S + H)$ is lc, which exists by Lemma 4.4.19.*

(i) *Suppose that $C \not\subset S$ and that H is normal. Then H has a singularity of class T at p and possibly has one more singularity z, which is Du Val of type A. The dual graph of $\bar{H}$ is as follows. The parts Γ_p and Γ_z are contracted to p and z respectively, where Γ_z is empty if H is smooth outside p.*

(ii) *Suppose that $C \subset S$. Then H is normal, has singularities of class T at p_1 and p_2 and is smooth outside these points. The dual graph of $\bar{H}$ is as follows. The part Γ_i is contracted to p_i.*

Proof If H is normal, then $\bar{C}$ is irreducible, $(\bar{C}^2) < 0$ and $(K_{\bar{H}} \cdot \bar{C}) \le (K_H \cdot C) = (K_X \cdot C) < 0$. Hence $\bar{C}$ is a (-1)-curve.

We begin with (i). By adjunction, $(H, S|_H)$ is lc. Hence $p \in H$ is lt since S passes through p. The germ $p \in X$ is a one-parameter deformation of $p \in H$ and thus $p \in H$ is of class T by definition.

Take a **Q**-divisor A on H which is the pull-back of an effective **Q**-Cartier **Q**-divisor on H' such that C has coefficient one in A. We consider the pair $(H, S|_H + (1 - \varepsilon)A)$ for small positive ε. Then $K_H + S|_H + (1 - \varepsilon)A \equiv_{H'} 0$ and the non-klt locus N of $(H, S|_H + (1 - \varepsilon)A)$ contains $S|_H$. By the connectedness lemma in Theorem 1.4.23 for $H \to H'$, the locus $N|_C$ must coincide with the point p in $S|_H$ set-theoretically. Hence $(H, S|_H + A)$ is lc along C outside p. In particular by Theorem 4.4.14 and Remark 4.4.15, H has only quotient singularities along C and one can consider the pair (H, C) which is lc outside p. The surface H is Du Val outside p since it is Gorenstein there.

If (H, C) were not plt at a point z other than p, then by Theorem 4.4.14, the dual graph of $\bar{H}$ would contain a part

over z. Then the Du Val singularity $z \in H$ is of type D and $2\bar{C} + 2F_1 + \cdots + 2F_{n-2} + F_{n-1} + F_n$ has self-intersection number zero, which contradicts Grauert's criterion stated as Theorem 4.4.21. Hence (H, C) is plt outside p. Again by Theorem 4.4.14, each singularity $z \in H$ other than p is Du Val of type A and the dual graph of $\bar{H}$ over z is as follows.

$$\bullet\!\!-\!\!-\!\!-\!\!\circ\!\!-\!\!-\!\!\cdots\!-\!\!-\!\!\circ$$

Recall that $y \in H'$ is lt by Lemma 4.4.19. The morphism $\bar{H} \to H'$ factors through the minimal resolution $\bar{H}'$ of H', and $\bar{H}$ is obtained from $\bar{H}'$ by finitely many blow-ups at a point. Hence the (-1)-curve $\bullet$ on $\bar{H}$ intersects at most two curves $\circ$. Thus the number of singular points of H is at most two, which completes (i).

The assertion (ii) can be proved similarly to (i). It follows from the log canonicity of $(X, H + S)$ that H is normal and (H, C) is lc everywhere. By Theorem 4.4.14, the dual graph of $\bar{H}$ over p_1 and p_2 is as in the statement. The section H is smooth outside p_1 and p_2 because the curve $\bullet$ intersects at most two $\circ$. $\qquad\square$

The following theorem does not hold in the algebraic category.

Theorem 4.4.21 (Grauert [148])　*Let $C = \bigcup_i C_i$ be a finite union of compact curves C_i in a smooth analytic surface S. Suppose that C is connected. Then S admits a proper bimeromorphic morphism to a normal analytic surface with* exceptional locus C if and only if the matrix $((C_i \cdot C_j))_{ij}$ of intersection numbers is negative definite.

By Construction 4.4.8, one can obtain an extremal neighbourhood $C \subset X$ from an arbitrary germ $C \subset H \to y \in H'$ in Theorem 4.4.20.

Example 4.4.22 ([455, example 1]) Let $y \in H'$ be an A_2 singularity and take a resolution $\bar{H}$ of it with the following dual graph.

Contract all curves $\circ$ to obtain a surface H. By $[3, 2, 3] = 12/5$, it has a quotient singularity of type $\frac{1}{12}(1, 5)$ which is of class T. The extremal neighbourhood $C \subset X$ obtained from H by Construction 4.4.8 is an example of Theorem 4.4.20(i). It is divisorial by Remark 4.4.9.

Example 4.4.23 ([455, section 6]) Let $y \in H'$ be an A_2 singularity and take a resolution $\bar{H}$ of it with the following dual graph.

$$\underset{-5}{\circ} - \underset{-2}{\circ} - \underset{-1}{\bullet} - \underset{-3}{\circ} - \underset{-2}{\circ} - \underset{-3}{\circ}$$

The extremal neighbourhood $C \subset X$ obtained as in the preceding example is an example of Theorem 4.4.20(ii). It is also divisorial by Remark 4.4.9.

When C is not contained in S, it may occur that every surface H is non-normal. Tziolas, Mori and Prokhorov also obtained a structure theorem in this case, which we quote below without proof. Further by [456] one can construct an extremal neighbourhood $C \subset X$ from the germ $C \subset H$ in this theorem. The relative ampleness of $-K_X$ is derived from the dual graph of $\bar{H}$.

Theorem 4.4.24 ([342], [455]) *In the semistable case in Framework 4.4.18, suppose that every principal divisor containing C is non-normal. Consider H such that $(X, S + H)$ is lc, which exists by Lemma 4.4.19. Then $C \not\subset S$ and the dual graph of $\bar{H}$ is as follows, where $\bar{C} = \bar{C}_1 + \bar{C}_2$. The strings $[a_{11}, \ldots, a_{1n_1}]$ and $[a_{21}, \ldots, a_{2n_2}]$ are conjugate and the corresponding parts Γ_1 and Γ_2 are contracted to the point p in H. It satisfies the inequality $\sum_{i=1}^{m}(b_i - 2) \leq \min\{2, 4 - c\}$.*

$$\overset{\Gamma_1}{\overbrace{\underset{-a_{1n_1}}{\circ} - \cdots - \underset{-a_{11}}{\circ}}} \underset{-1}{\overset{\bar{C}_1}{\bullet}} - \underset{-b_1}{\circ} - \cdots - \underset{-b_m}{\circ} \underset{-c}{\overset{\bar{C}_2}{\bullet}} - \underset{-a_{21}}{\circ} - \overset{\Gamma_2}{\overbrace{\cdots - \underset{-a_{2n_2}}{\circ}}}$$

Remark 4.4.25 Let $\bar{H} \to y \in H'$ be a resolution of a surface singularity with dual graph as in Theorem 4.4.24. Contract all curves $\circ$ in $\bar{H}$ to obtain an lt surface $H^\vee$. One can construct $C \subset H$ as an analytic space [67], [220], [252]. See [123, sections 1.20–21] for a brief exposition. Further following Artin, one can construct H as an algebraic space after algebraisation of $y \in H'$ by Theorem 2.1.8. Here an *algebraic space* consists of a complex scheme U and an equivalence relation R in $U \times U$ such that the two projections $R \to U$ are both étale [20]. It is regarded as the analytic space U_h/R which is the quotient of the analytic space U_h underlying U.

Let $C_i \simeq \mathbf{P}^1 \subset H^\vee$ be the strict transform of $\bar{C}_i$. Let $p_i \in C_i$ be the point to which Γ_i is contracted and let q be the point $C_1 \cap C_2$. Take $C = \mathbf{P}^1$ with two marked points 0 and ∞ and fix a morphism $c \colon C_1 \cup C_2 \to C$ which maps p_1, p_2, q to $\infty, \infty, 0$ respectively. By [21, theorem 6.1], there exists a unique maximal proper morphism $H^\vee \to H$ of algebraic spaces with an embedding $C \subset H$ that induces c on $C_1 \cup C_2$ and an isomorphism elsewhere. The space H is the *fibre coproduct* $H^\vee \amalg_{C_1 \cup C_2} C$ in the category of algebraic spaces.

The assumption in Theorem 4.4.24 is characterised in terms of H.

Lemma 4.4.26 ([342, proposition 6.3])   *In Framework* 4.4.18, *fix H such that $(X, S + H)$ is lc and suppose that H is non-normal. Then every principal divisor on X containing C is non-normal if and only if the fundamental cycle Z on $\bar{H}$ of H' satisfies $2\bar{C} \le Z$.*

Proof Let $\mathfrak{m}$ be the maximal ideal in $\mathcal{O}_Y$ defining y and let $\mathscr{I}$ be the ideal sheaf in $\mathcal{O}_X$ defining C. Since $y \in H'$ is rational by Lemma 4.4.19, it follows from Theorem 2.3.15 that $\mathcal{O}_{\bar{H}}(-Z) = \mathfrak{m}\mathcal{O}_{\bar{H}}$.

If every principal divisor on X containing C is non-normal, then $\mathfrak{m}\mathcal{O}_X \subset \mathscr{I}^{(2)}$ and hence $\mathcal{O}_{\bar{H}}(-Z) \subset \mathscr{I}^{(2)}\mathcal{O}_{\bar{H}} \subset \mathcal{O}_{\bar{H}}(-2\bar{C})$, that is, $2\bar{C} \le Z$. Conversely, suppose the existence of a normal principal divisor G on X containing C. Take a small open analytic subset X° of X meeting C such that X° and $G^\circ = G|_{X^\circ}$ are both smooth. We write $H^\circ = H|_{X^\circ}$ and $\bar{H}^\circ = \bar{H} \times_H H^\circ$. Let B be the blow-up of X° along $C|_{X^\circ}$ and let E be the exceptional divisor. By the log canonicity of (X°, H°), after shrinking X°, one can identify $\bar{H}^\circ$ with the strict transform in B of H° and can assume that $\bar{H}^\circ$ consists of two disjoint components $\bar{H}_1$ and $\bar{H}_2$. Define $\bar{C}_i = \bar{C}|_{\bar{H}_i} = \bar{H}_i \cap E$ by this identification. Since G° is smooth, its strict transform in B does not contain at least one $\bar{C}_i$, say $\bar{C}_1$. Then $\bar{C}_1$ has coefficient one in $\mu_1^* G^\circ|_{H^\circ}$ for $\mu_1 \colon \bar{H}_1 \to H^\circ$. Since $Z|_{\bar{H}_1} \le \mu_1^* G^\circ|_{H^\circ}$, one concludes that $2\bar{C}_1 \not\le Z$ and hence $2\bar{C} \not\le Z$.   □

Example 4.4.27 ([455, example 2]) Let $y \in H'$ be an A_2 singularity and take a resolution $\bar{H}$ of it with the following dual graph. Let $\bar{C} = \bar{C}_1 + \bar{C}_2$.

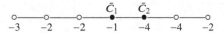

We construct H as well as $H^\vee$ by Remark 4.4.25 and keep the notation $p_i \in H^\vee$ there. The cyclic quotient $p_1 \in H^\vee$ is of type $\frac{1}{7}(1,5)$ and $p_2 \in H^\vee$ is of type $\frac{1}{7}(1,2)$. By a suitable identification of $y \in H'$ with $o \in (x_1x_2 + x_3^3 = 0)$, the divisors $\bar{C}_1$ and $\bar{C}_2$ over H' are obtained by the weighted blow-ups with $\mathrm{wt}(x_1, x_2, x_3) = (7, 26, 11)$ and $(2, 7, 3)$ respectively. By Theorem 2.3.15, the fundamental cycle Z on $\bar{H}$ of H' contains $7\bar{C}_1 + 2\bar{C}_2$. Hence $2\bar{C} \leq Z$. By Lemma 4.4.26, the extremal neighbourhood $C \subset X$ obtained from H is an example of Theorem 4.4.24. It is divisorial by Remark 4.4.9.

Remark 4.4.28 Keep $\pi \colon C \simeq \mathbf{P}^1 \subset X \to y \in Y$ being an irreducible extremal neighbourhood. Let H be a *general* principal prime divisor on X containing C and let $\bar{H}$ be the minimal resolution of the normalisation $H^\vee$ of H. Even if π is not semistable, we have a classification of the dual graph Γ of $\bar{H}$ in several cases. It is obtained in accordance with the classification of π in Theorem 5.5.1 and the isolated case will be discussed in the last section of Chapter 5. Below we provide the list of Γ in the divisorial case for the sake of reference. See also Corollary 5.6.9.

Henceforth we assume that π is *divisorial*. In Γ, $\circ$ stands for a curve exceptional over $H^\vee$ and $\bullet$ for a curve appearing in $H^\vee$. The self-intersection number of $\bullet$ is -1 and that of $\circ$ is -2 unless this number is attached.

(i) Suppose that π belongs to the case (i) in Theorem 5.5.1, in which X has a unique non-Gorenstein point p. If p is $cAx/4$ and H is normal, then Γ is one of the following [344].

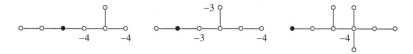

If p is $cAx/4$ and H is not normal, then Γ is as follows [345].

If p is $cD/3$, then H is normal and Γ is one of the following [342].

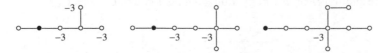

If p is $cAx/2$, $cD/2$ or $cE/2$, then H is normal and Γ is one of the following. The second graph occurs only if p is $cD/2$ or $cE/2$, and the number of $\circ$ in this graph is six when p is $cE/2$. The third graph occurs only if p is $cE/2$ [276, theorem 4.7].

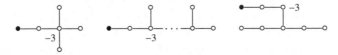

(ii) Suppose that π belongs to the case (iii) in Theorem 5.5.1, in which X has a unique singular point p of type IIB. Then H is normal and Γ is one of the following [343].

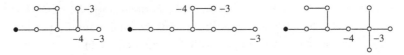

(iii) Suppose that π belongs to the case (iv) in Theorem 5.5.1, in which X has a unique singular point p of type II$^{\vee}$. Then H is normal and Γ is as follows [276, theorem 4.11.2].

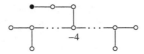

5

Flips

One of the landmarks in birational geometry is the attainment of the existence of threefold flips by Mori. In spite of the conceptual proof by Birkar, Cascini, Hacon and McKernan, the original proof remains important for the reason that it furnishes a geometric classification of flips simultaneously. We elucidate his approach in detail in this chapter.

Passing through the analytic category and constructing the flop of the double cover, we reduce the existence of a threefold flip to the general elephant conjecture on an irreducible extremal neighbourhood. It does not matter in our argument whether it is isolated or divisorial. Hence the classification result is also applicable to threefold divisorial contractions which contract the divisor to a curve.

The study of an extremal neighbourhood is performed with numerical invariants defined in terms of filtrations of the structure sheaf and the dualising sheaf along the exceptional curve. Locally at a point, the inverse image of the curve by the index-one cover turns out to be planar. We divide singular points into types according to this structure. Then we classify the set of singular points in an extremal neighbourhood by deforming the neighbourhood so that every singularity is a quotient singularity or an ordinary double point.

It is easy to prove that the general elephant is Du Val when it does not contain the exceptional curve. In some cases, the anti-bicanonical system has a good member and this is still sufficient to construct the flip. The hardest case is when the general elephant contains the exceptional curve, and this requires a really delicate analysis of how the curve is embedded in the threefold. We need refinements of the filtrations of the structure and dualising sheaves.

As discussed in the preceding chapter, an extremal neighbourhood is considered to be a one-parameter deformation of a principal prime divisor on it. We describe the associated surface morphism and conversely build an example of a threefold flip from the surface morphism.

5.1 Strategy and Flops

We shall make a reduction of the following existence of a threefold flip to the general elephant conjecture in Question 2.4.13.

Theorem 5.1.1 (Mori [335]) *Let* $\pi\colon X \to Y$ *be a threefold flipping contraction in Definition* 1.3.16 *to the algebraic or analytic germ* $y \in Y$. *Then the flip* $\pi^+\colon X^+ \to Y$ *of* π *exists.*

The algebraic statement is equivalent to the analytic one since π is isomorphic outside the isolated singularity $y \in Y$. One can algebraise π by Theorem 2.1.8 as $-K_X$ is π-ample. The *flip* X^+ of X is described as $\mathrm{Proj}_Y \mathscr{R}(Y, K_Y)$ by the graded $\mathscr{O}_Y$-algebra $\mathscr{R}(Y, K_Y) = \bigoplus_{i \in \mathbf{N}} \mathscr{O}_Y(iK_Y)$ in Notation 1.3.17. By the next lemma, the existence of X^+ is equivalent to the finite generation of $\mathscr{R}(Y, K_Y)$. In general by Lemma 1.5.20, $\mathscr{R}(X, D)$ is finitely generated if and only if so is $\mathscr{R}(X, lD)$ for some positive integer l.

Lemma 5.1.2 *Let* D *be a Weil divisor on a normal variety* X. *If the graded* $\mathscr{O}_X$-*algebra* $\mathscr{R}(X, D)$ *is finitely generated, then* $Y = \mathrm{Proj}_X \mathscr{R}(X, D) \to X$ *is a small contraction and the strict transform in* Y *of* D *is a relatively ample* **Q**-*Cartier divisor.*

Proof Working locally and subtracting an effective Cartier divisor from D, we may assume that $N = -D$ is effective. Replacing D by a multiple of it, we may assume that $\mathscr{R}(X, D)$ is generated by the part $\mathscr{O}_X(D)$ of degree one. Then $\pi\colon Y = \mathrm{Proj}_X \mathscr{R}(X, D) \to X$ is the blow-up of X along N.

To see that Y is normal, we shall prove that the $\mathscr{O}_X$-algebra $\mathscr{R}(X, D)_{(f)}$ is normal for non-zero $f \in \mathscr{O}_X(-N)$. Recall that elements in $\mathscr{R}(X, D)_{(f)}$ are of form g/f^i with $g \in \mathscr{O}_X(-iN)$. Consider an element s in the fractional field $\mathscr{K}_X$ of $\mathscr{R}(X, D)_{(f)}$ such that $s^n + a_1 s^{n-1} + \cdots + a_n = 0$ in $\mathscr{K}_X$ with $a_i \in \mathscr{R}(X, D)_{(f)}$. One can write $a_i = b_i/f^l$ with $b_i \in \mathscr{O}_X(-lN)$ for common l, by which $(f^l s)^n + c_1(f^l s)^{n-1} + \cdots + c_n = 0$ with $c_i = b_i f^{(i-1)l} \in \mathscr{O}_X(-ilN)$. This implies that $f^l s$ belongs to $\mathscr{O}_X$ since X is normal, and moreover $f^l s$ belongs to $\mathscr{O}_X(-lN)$ since each prime component of N gives a valuation. Thus $s = f^l s/f^l \in \mathscr{R}(X, D)_{(f)}$, which shows that $\mathscr{R}(X, D)_{(f)}$ is normal.

If π were not small, then Y would have an exceptional prime divisor E. For the π-very ample invertible sheaf $\mathscr{L} = \mathscr{O}_X(-N) \cdot \mathscr{O}_Y$, there exists a positive integer n such that $\mathscr{L}^{\otimes n} \otimes \mathscr{O}_Y(E)$ is relatively generated by global sections. Then $\pi_* \mathscr{L}^{\otimes n}$ is properly contained in $\pi_*(\mathscr{L}^{\otimes n} \otimes \mathscr{O}_Y(E))$. However since N is effective, one has $\pi_* \mathscr{L}^{\otimes n} = \mathscr{O}_X(-nN) = \pi_*(\mathscr{L}^{\otimes n} \otimes \mathscr{O}_Y(E))$ from $\mathscr{L}^{\otimes n} = \mathscr{O}_X(-nN) \cdot \mathscr{O}_Y$, which is absurd. Hence π is small and the strict transform of $D = -N$ is a π-very ample Cartier divisor. $\square$

Lemma 5.1.3 ([236, lemma 3.2])   *Let $\bar{X} \to X$ be a finite surjective morphism of normal varieties. Let D be a Weil divisor on X and let $\bar{D}$ be the unique Weil divisor on $\bar{X}$ that coincides with the pull-back of D on the smooth locus in X. Then $\mathscr{R}(X, D)$ is finitely generated if and only if so is $\mathscr{R}(\bar{X}, \bar{D})$.*

Proof   If $\mathscr{R}(X, D)$ is finitely generated, then we take the small contraction $\pi \colon Y = \mathrm{Proj}_X \mathscr{R}(X, D) \to X$ as in Lemma 5.1.2. The strict transform D_Y in Y of D is π-ample. The induced morphism $\bar{\pi} \colon \bar{Y} \to \bar{X}$ from the normalisation $\bar{Y}$ of $Y \times_X \bar{X}$ is small and the pull-back $\bar{D}_Y$ to $\bar{Y}$ of D_Y is $\bar{\pi}$-ample. Thus $\mathscr{R}(\bar{X}, \bar{D}) = \bigoplus_{i \in \mathbf{N}} \bar{\pi}_* \mathcal{O}_{\bar{Y}}(i\bar{D}_Y)$ and it is finitely generated.

Conversely, suppose that $\mathscr{R}(\bar{X}, \bar{D})$ is finitely generated. Consider the extension $K(X) \subset K(\bar{X})$ of function field and take a finite extension L of the field $K(\bar{X})$ such that $L/K(X)$ is Galois. Let $\mathscr{A}$ be the integral closure in L of $\mathcal{O}_X$, for which $\mathrm{Spec}_X \mathscr{A} \to \bar{X}$ is a finite morphism [308, §33 lemma 1]. Replacing $\bar{X}$ by $\mathrm{Spec}_X \mathscr{A}$ and using what has already been shown, we may assume that $K(\bar{X})/K(X)$ is a Galois extension. Let G be the Galois group.

Take $\bar{Y} = \mathrm{Proj}_{\bar{X}} \mathscr{R}(\bar{X}, \bar{D}) \to \bar{X}$ as in Lemma 5.1.2 and let $\bar{D}_Y$ denote the strict transform in $\bar{Y}$ of $\bar{D}$. The Galois group G acts on $\bar{Y}$. We take the geometric quotient Y of $\bar{Y}$ by G as in Theorem 2.2.3. Then Y is normal as so is $\bar{Y}$. It has the Weil divisor D_Y which is the quotient of $\bar{D}_Y$, and nD_Y is Cartier for the index n of G. The induced morphism $\pi \colon Y \to X$ is small and D_Y is π-ample. Thus $\mathscr{R}(X, D) = \bigoplus_{i \in \mathbf{N}} \pi_* \mathcal{O}_Y(iD_Y)$ and it is finitely generated.   □

In Theorem 5.1.1, the analytic germ $C = \pi^{-1}(y) \subset X$ is an extremal neighbourhood in Definition 4.4.6. It follows from the vanishing $R^1\pi_*\mathcal{O}_X = 0$ that C is a tree of $\mathbf{P}^1$. By Lemma 4.4.7, the germ $y \in Y$ is a rational singularity.

Lemma 5.1.4   *Let $\pi \colon C \subset X \to y \in Y$ be an extremal neighbourhood with irreducible decomposition $C = \bigcup_{i=1}^n C_i$ with $C_i \simeq \mathbf{P}^1$. Then $\mathrm{Pic}\, X \simeq H^2(C, \mathbf{Z}) = \mathbf{Z}^{\oplus n}$. For any subset I of $\{1, \ldots, n\}$, there exists a bimeromorphic morphism $X \to Y'$ to an analytic space projective over Y which contracts exactly the subspace $C' = \bigcup_{i \in I} C_i$ of C, and hence $C' \subset X$ is also an extremal neighbourhood.*

Proof   The argument for the first assertion has appeared in the proof of Theorem 2.3.1. The vanishing $R^1\pi_*\mathcal{O}_X = 0$ yields the isomorphism $R^1\pi_*\mathcal{O}_X^\times \simeq R^2\pi_*\mathbf{Z}$, that is, $\mathrm{Pic}\, X \simeq H^2(X, \mathbf{Z})$. By Theorem 2.3.10, the right-hand side is isomorphic to $H^2(C, \mathbf{Z})$.

For each i, there exists an effective Cartier divisor H_i on the germ X about C such that H_i intersects C_i transversally with $(H_i \cdot C_i) = 1$ and does not intersect any other C_j. These H_i are generators of $\mathrm{Pic}\, X \simeq \mathbf{Z}^{\oplus n}$. Because H_i can be chosen so that $H_i \cap C_i$ is a general point in C_i, the divisor $\sum_{i \notin I} H_i$ is relatively free and defines the morphism $X \to Y'$ over Y which contracts $\bigcup_{i \in I} C_i$.   □

Lemma 5.1.5 *Theorem 5.1.1 follows from the assertion in the case when $\pi^{-1}(y)$ is irreducible.*

Proof We shall verify Theorem 5.1.1 by assuming the statement in the case when $\pi^{-1}(y)$ is irreducible. Work over the analytic germ $y \in Y$ in which $C = \pi^{-1}(y) \subset X$ is an extremal neighbourhood. Take an irreducible component $C_1 \simeq \mathbf{P}^1$ of C. By Lemma 5.1.4, there exists a contraction $X \to Z$ over Y which contracts exactly the curve C_1.

Take the flip $X' \to Z$ of $X \to Z$, which exists by our assumption. Since X' and Y have rational singularities, Lemma 4.4.7 gives the vanishing $R^1 \pi'_* \mathcal{O}_{X'} = 0$ for $\pi' \colon X' \to Y$. From this, one deduces that $C' = (\pi')^{-1}(y)$ is a tree of $\mathbf{P}^1$ and $\operatorname{Pic} X' \simeq H^2(C', \mathbf{Z})$ in the same manner as for $C \subset X$. Hence as far as C' contains an irreducible component C'_1 with $(K_{X'} \cdot C'_1)$ negative, one can construct the contraction $X' \to Z'$ of C'_1 and its flip $X'' \to Z'$. This process generates a sequence $X \dashrightarrow X' \dashrightarrow X'' \dashrightarrow \cdots$ of flips over Y. It terminates by Theorem 1.6.3, the proof of which also works in our analytic setting.

Hence this sequence attains a terminal threefold $\bar{X}$ over Y such that $K_{\bar{X}}$ is relatively nef. The divisor $K_{\bar{X}}$ is further relatively free by the argument for Lemma 5.1.4. In particular, $\mathscr{R}(Y, K_Y) = \bigoplus_{i \in \mathbf{N}} \bar{\pi}_* \mathcal{O}_{\bar{X}}(iK_{\bar{X}})$ for $\bar{\pi} \colon \bar{X} \to Y$ is finitely generated and the flip $X^+ = \operatorname{Proj}_Y \mathscr{R}(Y, K_Y)$ of π exists.   □

We employ the theory of threefold flops for the reduction to the general elephant conjecture. In the book, a flopping contraction is assumed to be extremal.

Definition 5.1.6 A *flopping contraction* $\pi \colon X \to Y$ is a crepant small extremal contraction from a $\mathbf{Q}$-factorial canonical variety. There exists a π-ample $\mathbf{Q}$-divisor $-D$ on X. We call π a *D-flopping contraction*. The *flop*, or *D-flop*, of π is a flopping contraction $\pi^+ \colon X^+ \to Y$ such that the strict transform D^+ in X^+ of D is π^+-ample. In other words, π^+ is a $(-D^+)$-flopping contraction. It does not depend on the choice of D by the observation below. The transformation $X \dashrightarrow X^+$ is also called the *flop* or *D-flop* by abuse of language.

One can make D effective by adding a relatively numerically trivial effective $\mathbf{Q}$-divisor. Then π is associated with the $(K_X + \varepsilon D)$-negative ray $\overline{\operatorname{NE}}(X/Y)$ for the klt pair $(X, \varepsilon D)$ with small positive ε, and π^+ is the log flip of it. Hence $\operatorname{Pic} X / \pi^* \operatorname{Pic} Y \simeq \mathbf{Z}$ by Theorem 1.4.7 and X^+ is $\mathbf{Q}$-factorial by Lemma 1.4.12. Since π is crepant, the canonical divisor K_Y is still $\mathbf{Q}$-Cartier and Y is canonical. This makes it easier to treat flops than flips.

Notation 5.1.7 For a klt pair (X, Δ), we let $e(X, \Delta)$ denote the number of divisors E exceptional over X such that $a_E(X, \Delta) = 1$. It is finite. We write $e(X)$ for $e(X, 0)$ simply.

Theorem 5.1.8 (Kawamata [236]) (i) *For a* **Q**-*factorial strictly canonical threefold X, there exists a crepant extremal contraction $X' \to X$.*

(ii) *For any numbers e and r, there exists a positive integer $b = b(e, r)$ such that if D is a divisor on a* **Q**-*factorial canonical threefold X of index r with $e(X) = e$, then bD is Cartier.*

(iii) *The flop of every threefold flopping contraction $X \to Y$ exists.*

(iv) *Let X be a* **Q**-*factorial canonical threefold and let D be an effective* **Q**-*divisor on X. Then there exists no infinite sequence $X = X_0 \dashrightarrow X_1 \dashrightarrow X_2 \dashrightarrow \cdots$ of D-flops.*

The proof uses a *terminalisation* of a canonical threefold due to Reid.

Theorem 5.1.9 (Reid [399]) *Let X be a canonical threefold. Then there exists a crepant contraction $Y \to X$ from a terminal threefold.*

This relies on a deep analysis of canonical threefold singularities, which pertains to a classification of elliptic singularities by Laufer and Reid in Theorem 2.3.16 and to the existence of the *simultaneous resolution* of a family of Du Val singularities by Brieskorn [57], [59] and Tyurina [452]. The reader can learn the entire proof in [277, section 6.3]. This analysis was applied to the finite generation of the graded algebra on a terminal threefold.

Proposition 5.1.10 ([236, corollary 4.3]) *Let D be a Weil divisor on a terminal threefold X. Then $\mathscr{R}(X, D)$ is finitely generated.*

As a corollary, one can take Y in Theorem 5.1.9 so that it is **Q**-factorial. Such Y is called a **Q**-*factorial terminalisation* of X.

Corollary 5.1.11 *Let X be a terminal threefold. Then there exists a small contraction $Y \to X$ from a* **Q**-*factorial terminal threefold.*

Proof For a terminal threefold X, let $n(X)$ denote the number of divisors E over X with $a_E(X) = 2$ such that $c_X(E)$ is a closed point. This number is finite since X is algebraic and has only isolated singularities. We shall prove the corollary by induction on $n(X)$.

Suppose that a terminal threefold X is not **Q**-factorial and let D be a Weil divisor on X which is not **Q**-Cartier. Take $\pi \colon X' = \mathrm{Proj}_X \mathscr{R}(X, D) \to X$ by Proposition 5.1.10. By Lemma 5.1.2, X' is normal and π is small. In particular, X' is terminal. Note that π is not an isomorphism since the strict transform in X' of D becomes **Q**-Cartier. Consider a curve C in X' contracted by π. The image $\pi(C)$ is a singular point of X. The divisor obtained at the generic point of

C by the blow-up of X' along C is counted in $n(X)$ but not in $n(X')$. Hence the strict inequality $n(X') < n(X)$ holds, from which the procedure of replacement of X by X' must terminate and one eventually attains a **Q**-factorial terminal threefold. □

Theorem 5.1.8 will be proved by induction on $e(X)$. By $(i)_e, \ldots, (iv)_e$, we mean the statements (i), ..., (iv) for X with $e(X) \leq e$. The case $e(X) = 0$ is the terminal case. The statement $(i)_0$ is empty. The statements $(ii)_0$ and $(iv)_0$ are Theorems 2.4.17 and 1.6.3 respectively, and $(iii)_0$ follows from Proposition 5.1.10.

Proof of Theorem 5.1.8 We have observed all the statements when $e(X) = 0$. Let X be a **Q**-factorial strictly canonical threefold with $e(X) = e + 1$. By induction on $e(X)$, it suffices to prove the statements for X by assuming $(i)_e$ to $(iv)_e$. We shall establish them by the four implications in Steps 1 to 4 below, following a simplified proof by Kollár [262].

Step 1 $(iii)_e + (iv)_e \Rightarrow (i)_{e+1}$ Take a **Q**-factorial terminalisation $\bar{X}$ of X by Theorem 5.1.9 and Corollary 5.1.11. Let E be the divisorial part of the exceptional locus in $\bar{X}$. Consider the $(K_{\bar{X}} + \varepsilon E)$-MMP over X for a small positive rational number ε. Since $K_{\bar{X}/X}$ is zero, every morphism in the MMP is crepant. Hence flops in the MMP exist by $(iii)_e$ and sequences of flops terminate by $(iv)_e$. Thus the MMP runs well, and by the negativity lemma in Theorem 1.3.9, it ends with a log divisorial contraction $X' \to X$ to the **Q**-factorial base X, which is a desired crepant contraction.

Step 2 $(i)_{e+1} + (ii)_e \Rightarrow (ii)_{e+1}$ Take a crepant extremal contraction $\varphi \colon X' \to X$ in $(i)_{e+1}$ and let E be the exceptional divisor. Since $K_{X'} = \varphi^* K_X$, it follows from the exact sequence in Theorem 1.4.7 that the index of X' equals that of X. We write r for this common index.

We claim the existence of a curve C in X' contracted by φ such that $(-E \cdot C) \leq 3$. Consider the adjunction $K_{E^\nu} + \Delta^\nu = (K_{X'} + E)|_{E^\nu}$ on the normalisation E^ν of E with the different $\Delta^\nu \geq 0$ in Definition 1.4.24. Let $\mu \colon E' \to E^\nu$ be the minimal resolution of E^ν. We run the $K_{E'}$-MMP over $\varphi(E)$ and produce a surface $F/\varphi(E)$ which is a relative minimal model or admits a structure of a relative Mori fibre space. We write $\alpha \colon E' \to F$. Let C' be a general curve in E' contracted to a point in $\varphi(E)$ and let C be its image in E. From $\alpha^* K_F \leq K_{E'} \leq \mu^*(K_{E^\nu} + \Delta^\nu) = \mu^*(K_{X'} + E)|_{E^\nu}$, one obtains

$$(K_F \cdot \alpha(C')) \leq ((K_{X'} + E)|_{E^\nu} \cdot \mu(C')) = (E \cdot C) < 0.$$

Hence F is a relative Mori fibre space, which is isomorphic to $\mathbf{P}^2$ or a $\mathbf{P}^1$-bundle over a curve. Choose C' so that $\alpha(C')$ is a line in $\mathbf{P}^2$ or a fibre of the $\mathbf{P}^1$-bundle. Then $(-E \cdot C) \leq (-K_F \cdot \alpha(C')) = 3$ or 2, and the claim is completed.

For $b = b(e, r)$ in (ii)$_e$, we shall see that $b(e + 1, r) = (3b)! \cdot b$ satisfies the required property. Let D be an arbitrary divisor on X and let D' be the strict transform in X'. The divisors bD' and bE are Cartier, and $n = (-bE \cdot C)$ is a positive integer at most $3b$. Because the Cartier divisor $nbD' + (bD' \cdot C)bE$ has intersection number zero with C, it is the pull-back of a Cartier divisor on X by Theorem 1.4.7. This Cartier divisor is $\varphi_*(nbD' + (bD' \cdot C)bE) = nbD$. Thus $(3b)! \cdot bD$ is Cartier by $n \le 3b$.

Step 3 (i)$_{e+1}$ + (iii)$_e$ + (iv)$_e$ $\Rightarrow$ (iii)$_{e+1}$ Take a π-ample divisor $-D$ on X for $\pi \colon X \to Y$. We may and shall assume that D is effective. Take $\varphi \colon X' \to X$ in (i)$_{e+1}$ and let E be the exceptional divisor. Let $D' = \varphi^* D$. Run $(K_{X'} + \varepsilon D')/Y$-MMP for small positive ε. This is a two-ray game in Definition 4.2.5, in which every morphism is crepant. By (iii)$_e$ and (iv)$_e$, after a sequence of D'-flops, the MMP produces a $\mathbf{Q}$-factorial canonical threefold X''/Y which satisfies (c) or (n) below, where D'' is the strict transform in X'' of D'.

(c) X'' admits an elementary log divisorial contraction $\varphi^+ \colon X'' \to X^+$ over Y such that $-(K_{X''} + \varepsilon D'')$ is φ^+-ample.

(n) $K_{X''} + \varepsilon D''$ is nef over Y. The base-point free theorem provides a birational contraction $\varphi^+ \colon X'' \to X^+$ over Y such that D'' is the pull-back of a $\mathbf{Q}$-Cartier $\mathbf{Q}$-divisor on X^+ which is ample over Y.

In both cases, we have a diagram $X \leftarrow X' \dashrightarrow X'' \to X^+$ over Y. We write $\pi^+ \colon X^+ \to Y$. Let D^+ be the push-forward to X^+ of D''.

Suppose the case (c). Then π^+ is a flopping contraction and D^+ is the strict transform of D. Since $\pi^+_* D^+ = \pi_* D$ is not $\mathbf{Q}$-Cartier, π^+ is never crepant with respect to D^+. Hence either D^+ or $-D^+$ is π^+-ample. If $-D^+$ were π^+-ample, then $X^+ \simeq \mathrm{Proj}_Y \mathscr{R}(Y, -D)$ and the rational map $X \dashrightarrow X^+$ is an isomorphism. This is a contradiction because $X' \to X$ is crepant with respect to D' but $X' \dashrightarrow X'' \to X^+$ is D'-negative. Thus D^+ is π^+-ample and π^+ is the flop of π.

Suppose the case (n). Then D^+ is π^+-ample and $X^+ \to Y$ is not an isomorphism. Take a general curve C' in E contracted by φ. Let W be a common resolution of X' and X'' with $\mu' \colon W \to X'$ and $\mu'' \colon W \to X''$. By Lemma 1.4.12, $(\mu'')^* D'' \le (\mu')^* D'$. The strict transforms C_W in W and C'' in X'' of C' are well-defined and $(C'' \cdot D'') = (C_W \cdot (\mu'')^* D'') \le (C_W \cdot (\mu')^* D') = (C' \cdot \varphi^* D) = 0$. Hence the strict transform of E must be contracted by φ^+, that is, φ^+ is a log divisorial contraction and π^+ is a flopping contraction. Then D^+ is the strict transform of D and π^+ is the flop of π.

Step 4 (i)$_{e+1}$ + (ii)$_{e+1}$ + (iii)$_e$ + (iv)$_e$ $\Rightarrow$ (iv)$_{e+1}$ Recall that the flop $X \dashrightarrow X^+$ in Step 3 is factorised as $X \leftarrow X' \dashrightarrow X'' \to X^+$. We say that it is of type (c) or (n) in accordance with the case (c) or (n) of X''.

We may assume that D is integral. Let D_i be the strict transform in X_i of D. Following Step 3, we factorise the D_i-flop $X_i \dashrightarrow X_{i+1}$ as $X_i \leftarrow X_i' \dashrightarrow X_i'' \rightarrow X_{i+1}$ inductively in such a way that $X_{i+1}' = X_i''$. Let $D_i' = \varphi_i^* D_i$ with $\varphi_i \colon X_i' \rightarrow X_i$. Let E_i be the exceptional divisor of φ_i and let D_i'' be the strict transform in $X_{i+1}' = X_i''$ of D_i'. Then $D_i'' = D_{i+1}' + a_i E_{i+1}$ where the rational number a_i is positive (resp. zero) if $X_i \dashrightarrow X_{i+1}$ is of type (c) (resp. (n)). Take $b = b(e+1, r)$ in (ii)$_{e+1}$ for the index r of X. One deduces that rK_{X_i} is Cartier inductively, because for the flopping contraction $X_i \rightarrow Y_i$, rK_{X_i} is Cartier if and only if so is rK_{Y_i} by Theorem 1.4.7. It follows that $ba_i E_{i+1} = bD_i'' - bD_{i+1}'$ is integral. Namely, a_i belongs to $b^{-1}\mathbf{N}$.

From $D_i'' = D_{i+1}' + a_i E_{i+1}$, one has $D_0' = D_{n0}' + \sum_{i=0}^{n-1} a_i E_0$ for all n where D_{n0}' is the strict transform in X_0' of D_n'. Hence $\sum_{i=0}^{n-1} a_i E_0 \leq D_0'$ and the sum $\sum_{i=0}^{n-1} a_i$ of $a_i \in b^{-1}\mathbf{N}$ is bounded. Consequently only finitely many a_i are positive. Thus after truncation, one can assume that every flop is of type (n). Then D_i' is the strict transform of D_0' for all i and $X_0' \dashrightarrow X_1' \dashrightarrow X_2' \dashrightarrow \cdots$ is a sequence of D_0'-flops, which terminates by (iv)$_e$. Plainly the original sequence $X_0 \dashrightarrow X_1 \dashrightarrow X_2 \dashrightarrow \cdots$ also terminates.   □

Corollary 5.1.12   *Let X be a canonical threefold. Then there exists a small contraction $X' \rightarrow X$ from a $\mathbf{Q}$-factorial canonical threefold. For every Weil divisor D on X, $\mathscr{R}(X, D)$ is finitely generated.*

Proof   The existence of X' can be proved in the same manner as in Step 1 of the proof of Theorem 5.1.8. Take a $\mathbf{Q}$-factorial terminalisation $\bar{X}$ of X by Theorem 5.1.9 and Corollary 5.1.11 and run the $(K_{\bar{X}} + \varepsilon E)/X$-MMP for small positive ε where E is the divisorial part of the exceptional locus of $\bar{X} \rightarrow X$. Every log flip is an E-flop, and by the negativity lemma, all components of E are contracted. Thus the MMP yields a desired threefold X'.

For the latter statement, we may assume that D is effective. For the $\mathbf{Q}$-factorialisation X' above, let D' be the strict transform in X' of D. Run the $(K_{X'} + \varepsilon D')/X$-MMP for small positive ε to produce a small contraction $\pi^+ \colon X^+ \rightarrow X$ such that the strict transform D^+ of D is π^+-nef and further π^+-semi-ample by the base-point free theorem. Hence $\mathscr{R}(X, D) = \bigoplus_{i \in \mathbf{N}} \pi_*^+ \mathcal{O}_{X^+}(iD^+)$ is finitely generated.   □

Proposition 5.1.13 ([236, theorem 8.7])   *Theorem 5.1.1 holds when the general elephant of X is Du Val.*

Proof   Let S be the general elephant of X, which is Du Val by assumption. By inversion of adjunction in Theorem 1.4.26, the pair (X, S) is plt and further canonical. Let T be the general member of $|-2K_X|$, which is reduced. The pair $(X, T/2)$ is canonical as so is $(X, 2S/2)$ for $2S \in |-2K_X|$. For the push-forward

$T_Y = \pi_* T \in |-2K_Y|$, the pair $(Y, T_Y/2)$ is also canonical by $K_X + T/2 = \pi^*(K_Y + T_Y/2)$.

Using the section in $\Gamma(Y, \mathcal{O}_Y(-2K_Y))$ defining T_Y, one can construct the *double cover* $\mu \colon \bar{Y} = \mathrm{Spec}_Y \mathscr{R} \to Y$ ramified along T_Y as explained before Lemma 2.2.16, where $\mathscr{R} = \mathscr{R}_0 \oplus \mathscr{R}_1$ with $\mathscr{R}_0 = \mathcal{O}_Y$ and $\mathscr{R}_1 = \mathcal{O}_Y(K_Y)$. Then $\bar{Y}$ is normal by Lemma 2.2.16 and $K_{\bar{Y}} = \mu^*(K_Y + T_Y/2)$ by the ramification formula in Theorem 2.2.20. By Corollary 2.2.21, $\bar{Y}$ is canonical.

Let $\bar{D}$ be the divisor on $\bar{Y}$ which coincides with the pull-back of K_Y on the punctured locus $Y \setminus y$. Then $\mathscr{R}(\bar{Y}, \bar{D})$ is finitely generated by Corollary 5.1.12 and so is $\mathscr{R}(Y, K_Y)$ by Lemma 5.1.3. Thus by Lemma 5.1.2, the flip $X^+ = \mathrm{Proj}_Y \mathscr{R}(Y, K_Y)$ exists. □

Remark 5.1.14 By the proof of the proposition, the existence of the flip of π is derived from the existence of $T \in |-2K_X|$ such that the double cover of X ramified along T is canonical.

By virtue of Lemma 5.1.5 and Proposition 5.1.13 with Remark 4.1.6, Theorem 5.1.1 is reduced to the general elephant conjecture on an irreducible extremal neighbourhood. We shall prove it as Theorem 5.5.1 in this chapter.

5.2 Numerical Invariants

In this section, we shall prepare several numerical invariants of an extremal neighbourhood for proving the general elephant conjecture. Some of them have already been introduced in Chapter 3.

Let X be a terminal analytic threefold and let C be a reduced closed analytic subspace of pure dimension one. Let $\mathscr{I}$ denote the ideal in $\mathcal{O}_X$ defining C.

Definition 5.2.1 Let $\mathscr{S}$ be a coherent $\mathcal{O}_X$-module and let $\mathscr{M}$ be a submodule of $\mathscr{S}$. The *saturation* of $\mathscr{M}$ in $\mathscr{S}$ is the maximal submodule $\tilde{\mathscr{M}}$ of $\mathscr{S}$ containing $\mathscr{M}$ such that $\tilde{\mathscr{M}}/\mathscr{M}$ is locally of finite length. For example, the symbolic power $\mathscr{I}^{(n)}$ is the saturation of $\mathscr{I}^n$ in $\mathcal{O}_X$.

For a coherent $\mathcal{O}_X$-module $\mathscr{M}$, we let $F^n \mathscr{M}$ denote the saturation of $\mathscr{I}^n \mathscr{M}$ in $\mathscr{M}$ and define $\mathrm{gr}^n_C \mathscr{M} = F^n \mathscr{M}/F^{n+1} \mathscr{M}$. We write $\mathrm{gr}^n_C \mathcal{O}$ for $\mathrm{gr}^n_C \mathcal{O}_X$ and $\mathrm{gr}^n_C \omega$ for $\mathrm{gr}^n_C \omega_X$ for simplicity. Remark that $\mathrm{gr}^n_C \mathcal{O} = \mathscr{I}^{(n)}/\mathscr{I}^{(n+1)}$. When the pair $C \subset X$ is clear from the context, we just write $\mathrm{gr}^n \mathscr{M}$, $\mathrm{gr}^n \mathcal{O}$ and $\mathrm{gr}^n \omega$ by omitting the subscript C.

We introduce two homomorphisms α_1 and $\beta_0(r)$, which will play fundamental roles in the analysis of an extremal neighbourhood. The maps α_1 and $\beta_0(r)$ pertain to the $\mathcal{O}_C$-modules $\mathrm{gr}^1 \mathcal{O}$ of rank two and $\mathrm{gr}^0 \omega$ of rank one.

Definition 5.2.2 Consider the map $\mathscr{I}/\mathscr{I}^2 \times \mathscr{I}/\mathscr{I}^2 \times \Omega_C \to \Omega_X^3 \otimes \mathscr{O}_C$ defined by $(x, y, zdu) \mapsto zdx \wedge dy \wedge du$. It induces a natural map

$$\alpha_1 \colon \bigwedge^2 \operatorname{gr}^1 \mathscr{O} \to \mathscr{H}om_C(\Omega_C, \operatorname{gr}^0 \omega).$$

For a positive integer r such that rK_X is Cartier, there exists a natural map

$$\beta_0(r) \colon (\operatorname{gr}^0 \omega)^{\otimes r} \to \mathscr{O}_X(rK_X) \otimes \mathscr{O}_C.$$

Let p be a smooth point of C. We define $i_p(1) \in \mathbf{N}$ as the length at p of the cokernel of α_1, and define $w_p(0) \in \mathbf{Q}_{\geq 0}$ so that $r \cdot w_p(0)$ is the length at p of the cokernel of $\beta_0(r)$ for the index r of the germ $p \in X$.

Note that $i_p(1)$ is zero if X is smooth at p, and that $w_p(0)$ is zero if and only if X is Gorenstein at p. We shall also see the converse for $i_p(1)$ in Proposition 5.2.10 after preparation of necessary notation.

Let $p \in X$ be a terminal threefold analytic singularity of index r and let $\mu \colon \bar{p} \in \bar{X} \to p \in X$ be the index-one cover. By Theorem 2.4.8, there exists an identification $\bar{p} \in \bar{X} = (\xi = 0) \subset \mathfrak{D}^4$ with coordinates x_1, x_2, x_3, x_4 by which

$$X = (\xi = 0) \subset \mathfrak{D}^4/\mathbf{Z}_r(w_1, w_2, w_3, w_4)$$

with respect to a fixed primitive character $\chi \colon \mathbf{Z}_r \to \mathbf{C}^\times$. Here ξ is a semi-invariant function in $\mathscr{O}_{\mathfrak{D}^4, \bar{p}}$. We also apply this expression when $p \in X$ is a quotient singularity by choosing as ξ an invariant regular function, which is a special case of cA/r singularity.

Remark 5.2.3 By Theorem 2.4.8, up to permutation of x_i one can assume that $w_4 = \operatorname{wt} \xi$ and that w_1, w_2, w_3 are coprime to r with $w_1 + w_2 \equiv 0 \mod r$. Moreover, either $\operatorname{wt} \xi = 0$ or $(r, \operatorname{wt} \xi) = (4, 2)$.

Let C be a smooth curve in the analytic germ $p \in X$, for which $\operatorname{gr}^1 \mathscr{O}$ and $\operatorname{gr}^0 \omega$ are locally free $\mathscr{O}_C$-modules. Let $\bar{C} = \mu^{-1}(C)$ and let s be the number of irreducible analytic components of $\bar{C}$, which is a divisor of r. It equals the order of the cokernel $\mathbf{Z}_s$ of the natural map $\pi_1(C \setminus p) \to \pi_1(X \setminus p)$ of fundamental groups from $\pi_1(C \setminus p) \simeq \mathbf{Z}$ to $\pi_1(X \setminus p) \simeq \mathbf{Z}_r$, which is in Theorem 2.4.17.

Definition 5.2.4 The number s is called the *splitting degree* of $p \in C \subset X$. The point p in C is said to be *primitive* if $s = 1$ and *imprimitive* if $s \geq 2$. We say that $C \subset X$ is *locally primitive* if all points in C are primitive.

We write $r = r's$ by an integer r'. Elements in $\mathbf{Z}_r$ fixing each irreducible component of $\bar{C}$ form a subgroup $\mathbf{Z}_{r'}$ of order r', and the quotient $p' = \bar{p}/\mathbf{Z}_{r'} \in X' = \bar{X}/\mathbf{Z}_{r'}$ is the covering of $p \in X$ associated with $r'K_X$. The space $C' = \bar{C}/\mathbf{Z}_{r'}$ consists of s irreducible components isomorphic to C.

Take the normalisation $\bar{p}^\nu \in \bar{C}^\nu$ of one of the irreducible components of $\bar{p} \in \bar{C}$ and fix a *uniformising parameter* $t' = t^{1/r'}$ in $\mathscr{O}_{\bar{C}^\nu, \bar{p}^\nu}$, that is, a generator of the maximal ideal in $\mathscr{O}_{\bar{C}^\nu}$ defining $\bar{p}^\nu$. Then $t = (t')^{r'}$ is a uniformising parameter in $\mathscr{O}_{C,p}$. For a semi-invariant function f in $\mathscr{O}_{\bar{X}}$, let $\operatorname{ord}_{\bar{C}} f$ denote the order of the restriction $f|_{\bar{C}^\nu}$ with respect to $(t')^{1/s}$. It equals the local intersection number $((f)_{\bar{X}} \cdot \bar{C})_{\bar{p}}$ with the divisor $(f)_{\bar{X}}$ unless f belongs to the ideal $\bar{\mathscr{I}}$ in $\mathscr{O}_{\bar{X}}$ defining $\bar{C}$. We write $\operatorname{ord} f$ for $\operatorname{ord}_{\bar{C}} f$ simply when there is no confusion. Let

$$\operatorname{ow} f = (\operatorname{ord} f, \operatorname{wt} f) \in s\mathbf{Z} \times \mathbf{Z}_r.$$

Definition 5.2.5 We write $\operatorname{ow} C$ for the submonoid of $\mathbf{Z} \times \mathbf{Z}_r$ which consists of all $\operatorname{ow} f$ for semi-invariant $f \in \mathscr{O}_{\bar{X}} \setminus \bar{\mathscr{I}}$. In like manner, $\operatorname{ord} C$ is the submonoid of $\mathbf{Z}$ which consists of all $\operatorname{ord} f$ for semi-invariant $f \in \mathscr{O}_{\bar{X}} \setminus \bar{\mathscr{I}}$.

Definition 5.2.6 For $w \in \mathbf{Z}_r$, we define $o_p(w)$ as the least integer such that $(o_p(w), w) \in \operatorname{ow} C$. We also define $o'_p(w)$ as the least integer such that $(o'_p(w), w) \in \operatorname{ow} C \setminus \{(0,0)\}$. Hence $o'_p(w) = o_p(w)$ unless $w \equiv 0 \bmod r$. Notice that $o'_p(0) = r$, that is, $(r, 0) \in \operatorname{ow} C$, by the existence of an invariant function u with $u|_C = t$. We write $o(w)$ for $o_p(w)$ and $o'(w)$ for $o'_p(w)$ when the point p is clear from the context.

We remark that the paper [335] defines the order of f as $s^{-1} \operatorname{ord} f$ and uses the notation $R(w) = o(-w)$.

Remark 5.2.7 The complement $s\mathbf{N} \setminus \operatorname{ord} C$ for the inclusion $\operatorname{ord} C \subset s\mathbf{N}$ is a finite set. This follows from the existence of semi-invariant functions f and g in $\mathscr{O}_{\bar{X}} \setminus \bar{\mathscr{I}}$ such that $t' = (f|_{\bar{C}^\nu})(g|_{\bar{C}^\nu})^{-1}$ on the normalisation $\bar{C}^\nu$.

Defining $a_i = o'(w_i)$, one can take a normal form below.

Definition 5.2.8 The expression $p \in X \simeq (\xi = 0) \subset \mathfrak{D}^4 / \mathbf{Z}_r(w_1, w_2, w_3, w_4)$ with orbifold coordinates x_1, x_2, x_3, x_4 is called a *normal form* of $p \in C \subset X$ if $x_i|_{\bar{C}^\nu} = t^{a_i/r}$ with $a_i = o'(w_i)$ for all i, where t is a uniformising parameter in $\mathscr{O}_{C,p}$.

Remark 5.2.9 Via the adjunction $\omega_{\bar{X}} \simeq \omega_{\mathfrak{D}^4} \otimes \mathscr{O}_{\mathfrak{D}^4}(\bar{X})|_{\bar{X}}$, the dualising sheaf $\omega_{\bar{X}}$ is generated by $\theta = (dx_1 \wedge dx_2 \wedge dx_3 \wedge dx_4)/\xi$, which is semi-invariant of weight $\operatorname{wt} x_1 x_2 x_3 x_4 - \operatorname{wt} \xi$. Thus the invariant part ω_X of $\mu_* \omega_{\bar{X}}$ is generated by $f\theta$ for all semi-invariant functions f with $\operatorname{wt} f = w$ where $w = \operatorname{wt} \xi - \sum_{i=1}^4 w_i$. This description shows that $w_p(0) = r^{-1} o(w)$.

Proposition 5.2.10 *The number $i_p(1)$ is zero if and only if X is smooth at p.*

Proof We need to prove the positivity of $i_p(1)$ when X is singular at p. The case $r = 1$ will be settled by Lemma 5.2.11. We shall assume that $r \geq 2$. Take a normal form of $p \in C \subset X$ as in Definition 5.2.8 and fix an invariant monomial u in x_1, x_2, x_3, x_4 such that $u|_C = t$. By Remark 5.2.9, the map $\alpha_1 \otimes \omega_C : \bigwedge^2 \operatorname{gr}^1 \mathscr{O} \otimes \omega_C \to \operatorname{gr}^0 \omega$ is locally given by

$$(g \wedge h) \otimes du \mapsto \frac{dg \wedge dh \wedge du \wedge d\xi}{\xi},$$

and $\operatorname{gr}^0 \omega$ is generated by $v(dx_1 \wedge dx_2 \wedge dx_3 \wedge dx_4)/\xi$ for a semi-invariant monomial v such that $\operatorname{ow} v = (o(w), w)$ where $w = \operatorname{wt} \xi - \sum_{i=1}^4 w_i$. Hence if g and h form a basis of $\operatorname{gr}^1 \mathscr{O}$ at p and $dg \wedge dh \wedge du \wedge d\xi = f \, dx_1 \wedge dx_2 \wedge dx_3 \wedge dx_4$, then the length $i_p(1)$ of the cokernel of α_1 is $r^{-1}(\operatorname{ord} f - \operatorname{ord} v)$. It suffices to verify that $o(w) < \operatorname{ord} f$.

We shall prove this in the case $\operatorname{wt} \xi = 0$ here. The other case $(r, \operatorname{wt} \xi) = (4, 2)$ is similar and easier. Suppose that $\operatorname{wt} \xi = 0$. Then without loss of generality, one can assume that $(w_1, w_2, w_3, w_4) = (1, -1, -w, 0)$ and $u = x_4$. Take g and h as above. Since $dg \wedge dh \wedge du \wedge d\xi = -dg \wedge dh \wedge d\xi \wedge dx_4$, the invariant functions g, h, ξ must contain monomials $u_1 x_1, u_2 x_2, u_3 x_3$ up to permutation for some monomials u_1, u_2, u_3 such that $\operatorname{ord} u_1 u_2 u_3 \leq \operatorname{ord} f$. Note that $\operatorname{wt} u_i = -w_i$, from which $o(w) < o(-1) + o(1) + o(w) \leq \operatorname{ord} u_1 u_2 u_3$. Thus $o(w) < \operatorname{ord} f$. $\qquad \square$

When $r = 1$, the germ $p \in C \subset X$ is always expressed as in the lemma below.

Lemma 5.2.11 *Let $p \in C \subset X$ be a terminal threefold analytic singularity of index r with a smooth curve defined by*

$$p \in C = (x_1\text{-axis}) \subset X = (x_1^a x_2 + f = 0) \subset \mathfrak{D}^4 / \mathbf{Z}_r(w_1, w_2, w_3, w_4)$$

with orbifold coordinates x_1, x_2, x_3, x_4 and $f \in (x_2, x_3, x_4)^2 \mathscr{O}_{\mathfrak{D}^4}$. Then $i_p(1) = a$ when $r = 1$ and $i_p(1) = \lfloor a/r \rfloor + 1$ when $r \geq 2$.

Proof We shall deal with the case of index one for simplicity. The strategy is the same in the case of higher index. Suppose that $r = 1$. We put $\xi = x_1^a x_2 + f$. The image at p of the map $\alpha_1 \otimes \omega_C$ is generated by all $(dx_i \wedge dx_j \wedge dx_1 \wedge d\xi)/\xi$ for $i, j = 2, 3, 4$. Since $d\xi = x_1^a dx_2$ in $\Omega_{\mathfrak{D}^4} \otimes \mathscr{O}_C$, it is eventually generated by the single element $(dx_3 \wedge dx_4 \wedge dx_1 \wedge x_1^a dx_2)/\xi$. This shows that the length $i_p(1)$ of the cokernel is the order a of x_1^a. $\qquad \square$

Definition 5.2.12 We define $u_p = r^{-1}(o(w) + o(-w)) \in \mathbf{N}$ for $w = \operatorname{wt} \xi - \sum_{i=1}^4 w_i$ in the expression $p \in X \simeq (\xi = 0) \subset \mathfrak{D}^4 / \mathbf{Z}_r(w_1, w_2, w_3, w_4)$. This number is positive if and only if $r \geq 2$.

We are interested in the numbers $i_p(1)$ and $w_p(0)$ on an extremal neighbourhood $C \subset X$ equipped with a contraction $\pi \colon C \subset X \to y \in Y$. We shall assume the irreducibility of C in view of Lemma 5.1.5.

Terminology 5.2.13 Henceforth in this chapter, an extremal neighbourhood is assumed to be *irreducible* unless otherwise mentioned. That is, an extremal neighbourhood $C \subset X$ implies that $C \simeq \mathbf{P}^1$.

Proposition 5.2.14 *Let $C \subset X$ be an extremal neighbourhood and hence $C \simeq \mathbf{P}^1$ by Terminology 5.2.13. Then $H^1(\mathrm{gr}^1 \mathscr{O}) = 0$, $\mathrm{gr}^0 \omega \simeq \mathscr{O}_{\mathbf{P}^1}(-1)$, $H^1(\mathrm{gr}^1 \omega) = 0$ and*

$$\deg \mathrm{gr}^1 \mathscr{O} = 1 - \sum_{p \in C} i_p(1), \qquad (K_X \cdot C) = -1 + \sum_{p \in C} w_p(0).$$

In particular $\sum_p i_p(1) \leq 3$ and $\sum_p w_p(0) < 1$. If $\sum_p i_p(1) = 3$, then $\mathrm{gr}^1 \mathscr{O} \simeq \mathscr{O}_{\mathbf{P}^1}(-1)^{\oplus 2}$.

Proof Since $R^1\pi_*\mathscr{O}_X$ is zero, the ideal $\mathscr{I}$ in $\mathscr{O}_X$ defining C has $R^1\pi_*\mathscr{I} = 0$ from the exact sequence $0 \to \mathscr{I} \to \mathscr{O}_X \to \mathscr{O}_C \to 0$. Hence $H^1(\mathrm{gr}^1 \mathscr{O}) = 0$ by the surjection $\mathscr{I} \twoheadrightarrow \mathrm{gr}^1 \mathscr{O} = \mathscr{I}/\mathscr{I}^{(2)}$.

By the Kawamata–Viehweg vanishing $R^1\pi_*\omega_X = 0$, one has $H^1(\mathrm{gr}^0 \omega) = 0$ from the surjection $\omega_X \twoheadrightarrow \mathrm{gr}^0 \omega$. Further the degree of $\mathrm{gr}^0 \omega$ is negative by the existence of the map $\beta_0(r)$ with $(rK_X \cdot C) < 0$. Hence the invertible sheaf $\mathrm{gr}^0 \omega$ is isomorphic to $\mathscr{O}_{\mathbf{P}^1}(-1)$. Then by the exact sequence $0 \to \mathrm{gr}^1 \omega \to \omega_X/F^2\omega_X \to \mathrm{gr}^0 \omega \to 0$, $H^1(\mathrm{gr}^1 \omega)$ is isomorphic to $H^1(\omega_X/F^2\omega_X)$, which is zero by $R^1\pi_*\omega_X = 0$.

The two equations in the statement are evident by the definitions of $i_p(1)$ and $w_p(0)$. Note that $\Omega_C \simeq \mathscr{O}_{\mathbf{P}^1}(-2)$. The last assertion on $\mathrm{gr}^1 \mathscr{O}$ is from Theorem 1.3.5. $\qquad\square$

Proposition 5.2.15 *Let $C \subset X$ be an extremal neighbourhood. Then the inequality $\sum_{p \in C} u_p \leq 2$ holds.*

Proof The natural map $\mathrm{gr}^0 \omega \otimes \mathrm{gr}^0 \omega_X^\vee \to \mathscr{O}_C$ has image $\mathscr{O}_C(-u_p p)$ at p, where $\omega_X^\vee = \mathscr{O}_X(-K_X)$ is the dual of ω_X. Thus

$$\sum_{p \in C} u_p = -(\deg \mathrm{gr}^0 \omega + \deg \mathrm{gr}^0 \omega_X^\vee).$$

We have $\mathrm{gr}^0 \omega \simeq \mathscr{O}_{\mathbf{P}^1}(-1)$ in Proposition 5.2.14. Similarly $H^1(\mathrm{gr}^0 \omega_X^\vee) = 0$ by the Kawamata–Viehweg vanishing $R^1\pi_*\omega_X^\vee = 0$. Hence $\deg \mathrm{gr}^0 \omega_X^\vee \geq -1$ and the proposition follows. $\qquad\square$

Let $p \in C \subset X$ be the germ of an extremal neighbourhood equipped with $\pi: C \subset X \to y \in Y$. In the paragraph following Definition 5.2.4, we constructed the covering $\psi: p' \in C' \subset X' \to p \in C \subset X$ associated with $r'K_X$ where $r = r's$ with index r and splitting degree s. We shall globalise it.

Fix an effective divisor D on the germ $p \in X$ not containing C such that $D \sim r'K_X$ at p. The local intersection number $(D \cdot C)_p$ is an integer because it can be computed on an irreducible component of C' in X'. We regard D as an effective divisor about C such that $(D \cdot C) = (D \cdot C)_p$. Fix a Cartier divisor H on the germ $C \subset X$ along C such that $(H \cdot C) = (D \cdot C)$. Take $E = D - H$ and the strict transform E_Y in Y. By Lemma 5.1.4, the Cartier divisor sE is principal and so is sE_Y. We take the cyclic cover $y' \in Y'$ of $y \in Y$ associated with E_Y as in Definition 2.2.17. It is lifted to the global covering $C' \subset X'$ of $C \subset X$ associated with E. Since $E \sim r'K_Y$ at p, this covering is an extension of ψ. We have the diagram

$$
\begin{array}{ccc}
C' \subset X' & \xrightarrow{\psi} & C \subset X \\
\pi' \downarrow & & \downarrow \pi \\
y' \in Y' & \longrightarrow & y \in Y
\end{array}
$$

and ψ is étale outside p. When p is imprimitive, $C' \subset X'$ is a reducible extremal neighbourhood. One can write $C' = \bigcup_{i=1}^{s} C_i'$ with $C_i' \simeq C \simeq \mathbf{P}^1$.

Definition 5.2.16   The covering $C' \subset X'$ is called the *splitting cover* of $C \subset X$ associated with p.

Lemma 5.2.17   *The natural map* $\mathrm{gr}_{C'}^0 \, \omega_{X'} \to \bigoplus_{i=1}^{s} \mathrm{gr}_{C_i'}^0 \, \omega_{X'}$ *is an isomorphism.*

Proof   The cokernel $\mathscr{Q}$ of the map in the statement fits into the exact sequence $0 \to \mathrm{gr}_{C'}^0 \, \omega_{X'} \to \bigoplus_i \mathrm{gr}_{C_i'}^0 \, \omega_{X'} \to \mathscr{Q} \to 0$. One has $H^1(\mathrm{gr}_{C'}^0 \, \omega_{X'}) = 0$ from the vanishing $R^1\pi_*'\omega_{X'} = 0$ with the surjection $\omega_{X'} \twoheadrightarrow \mathrm{gr}_{C'}^0 \, \omega_{X'}$. One also has the isomorphism $\mathrm{gr}_{C_i'}^0 \, \omega_{X'} \simeq \mathscr{O}_{\mathbf{P}^1}(-1)$ by Proposition 5.2.14 for the extremal neighbourhood $C_i' \subset X'$. Hence the skyscraper sheaf $\mathscr{Q}$ must be zero. $\square$

Corollary 5.2.18   *Let $C \subset X$ be an extremal neighbourhood and let $p \in C$ be an imprimitive point. Then p is the only imprimitive point and the splitting degree s of $p \in C \subset X$ is less than the index r.*

Proof   Keep the notation above in which $C' \subset X'$ is the splitting cover associated with p. By construction, $r' = r/s$ is the index of the germ $p' \in X'$. It follows from Lemma 5.2.17 that $\omega_{X'}$ is never invertible whenever $s \geq 2$. This means that the index r' is greater than one, that is, $s < r$.

Suppose that X had another imprimitive point q. Let t be the splitting degree at q and let $C'' \subset X''$ be the splitting cover associated with q. Then $\tilde{C} = C' \times_C C'' \subset \tilde{X} = X' \times_X X''$ is again a reducible extremal neighbourhood. In particular, $H^1(\mathscr{O}_{\tilde{C}}) = 0$ and $\tilde{C}$ is a tree of $\mathbf{P}^1$. However, $\tilde{C}$ consists of st irreducible components $\tilde{C}_{ij}$ passing through p_i and q_j for $1 \le i \le t$ and $1 \le j \le s$, where $p_1, \ldots, p_t$ and $q_1, \ldots, q_s$ are the points in $\tilde{X}$ above p and q respectively. It cannot be a tree, which is absurd. □

Remark 5.2.19 By the proofs of Lemma 5.2.17 and its corollary, one deduces the general statement that a reducible extremal neighbourhood $C \subset X$ is non-Gorenstein at every singular point of C.

In the delicate analysis of an extremal neighbourhood, we need higher analogues of α_1 and $\beta_0(r)$.

Definition 5.2.20 Consider the natural maps

$$\alpha_n : S^n(\mathrm{gr}^1 \mathscr{O}) \to \mathrm{gr}^n \mathscr{O} \qquad \text{for } n \ge 2,$$
$$\beta_n : \mathrm{gr}^0 \omega \otimes S^n(\mathrm{gr}^1 \mathscr{O}) \to \mathrm{gr}^n \omega \quad \text{for } n \ge 1,$$

where S^n stands for the n-th symmetric power. For $p \in C$, we define $i_p(n)$ (resp. $w_p(n)$) as the length at p of the cokernel of α_n (resp. β_n). We also define

$$w_p^*(n) = \frac{1}{2}n(n+1)i_p(1) - w_p(n) \in \mathbf{Z} \quad \text{for } n \ge 1.$$

Remark 5.2.21 One can make formulae for $i_p(n)$ and $w_p^*(n)$ on an extremal neighbourhood $C \subset X$ analogously to Proposition 5.2.14. For example, the following hold for $w_p^*(n)$. See [335, 2.3] for details.

(i) $\sum_p w_p^*(1) \le 1$.
(ii) Fix $n \ge 1$. If $\sum_p w_p^*(i) = i(i+1)/2$ for all $1 \le i \le n$, then $\mathrm{gr}^n \omega \simeq \mathscr{O}_{\mathbf{P}^1}(-1)^{\oplus(n+1)}$, $H^1(\mathrm{gr}^{n+1} \omega) = 0$ and $\sum_p w_p^*(n+1) \le (n+1)(n+2)/2$.

5.3 Planarity of Covering Curves

Let $p \in C \subset X$ be the germ of an extremal neighbourhood. The purpose of this section is to provide a classification of it with respect to the normal form in Definition 5.2.8. The main theorem is the following. Later the cases IB and IC$^\vee$ will be excluded by Corollary 5.4.15 and Lemma 5.4.9 respectively.

Theorem 5.3.1 ([335, proposition-definition 4.2]) *Let $p \in C \subset X$ be the germ of an extremal neighbourhood and let r be the index of $p \in X$. Then there exists a normal form of $p \in C \subset X$ in Definition 5.2.8 as $p \in X \simeq (\xi = 0) \subset$*

$\mathfrak{D}^4/\mathbf{Z}_r\,(w_1, w_2, w_3, w_4)$ *with orbifold coordinates* x_1, x_2, x_3, x_4 *and* $a_i = o'(w_i)$ *which satisfies* $\operatorname{ow} C = \mathbf{N}\operatorname{ow} x_1 + \mathbf{N}\operatorname{ow} x_2$ *and belongs to the list below.*

If p is a primitive point, then exactly one of the following holds.

(I) $r \geq 2$, $\operatorname{wt}\xi = 0$ *and*

(IA) $a_1 + a_3 \equiv 0 \bmod r$ *and* $a_4 = r \in \mathbf{N}_{>0}a_1 + \mathbf{N}_{>0}a_2$,

(IB) $a_1 + a_3 \equiv 0 \bmod r$, $a_2 = r$ *and* $2 \leq a_1$ *or*

(IC) $a_1 + a_2 = a_3 = r$, $a_4 \neq a_1, a_2$ *and* $2 \leq a_1 < a_2$.

(II) $(r, \operatorname{wt}\xi) = (4, 2)$ *with ξ non-regular and*

(IIA) $(a_1, a_2, a_3, a_4) = (1, 1, 3, 2)$ *or*

(IIB) $(a_1, a_2, a_3, a_4) = (3, 2, 5, 5)$.

(III) $r = 1$.

If p is an imprimitive point, then the splitting degree s of $p \in C \subset X$ satisfies $1 < s < r$ and exactly one of the following holds.

(I$^\vee$) $\operatorname{wt}\xi = 0$ *and*

(IA$^\vee$) $w_1 + w_3 \equiv 0 \bmod r$ *and* $\operatorname{ow} x_4 = (r, 0) \in \mathbf{N}_{>0}\operatorname{ow} x_1 + \mathbf{N}_{>0}\operatorname{ow} x_2$ *or*

(IC$^\vee$) $s = 2$, $r/2$ *is an even integer at least four,* $(a_1, a_2, a_3, a_4) = (2, r-2, r, r+2)$ *and* $(w_1, w_2, w_3, w_4) = (1, -1, 0, r/2 + 1)$.

(II$^\vee$) $(r, \operatorname{wt}\xi) = (4, 2)$ *with ξ non-regular,* $s = 2$, $(a_1, a_2, a_3, a_4) = (2, 2, 2, 4)$ *and* $(w_1, w_2, w_3, w_4) = (1, 3, 3, 2)$.

We write $\bar{p} \in \bar{C} \subset \bar{X}$ for the index-one cover of $p \in C \subset X$. The property $\operatorname{ow} C = \mathbf{N}\operatorname{ow} x_1 + \mathbf{N}\operatorname{ow} x_2$ in the theorem means that the maximal ideal $(x_1, x_2, x_3, x_4)\mathcal{O}_{\bar{C}}$ in $\mathcal{O}_{\bar{C}}$ equals $(x_1, x_2)\mathcal{O}_{\bar{C}}$. Hence $\bar{p} \in \bar{C}$ is of embedding dimension at most two. In other words, it is a planar singularity.

We treat an imprimitive point first.

Proof of Theorem 5.3.1 for imprimitive p　　We write $r = r's$ with the splitting degree s where $1 < s < r$ by Corollary 5.2.18. Take a normal form of $p \in C \subset X$ in Definition 5.2.8 as $p \in X \simeq (\xi = 0) \subset \mathfrak{D}^4/\mathbf{Z}_r\,(w_1, w_2, w_3, w_4)$ with x_1, x_2, x_3, x_4 and $a_i = o'(w_i)$. By Remark 5.2.3, we may and shall assume that $(w_1, w_2, w_3, w_4) = (1, -w, -1, \operatorname{wt}\xi)$. Note that w is coprime to r and all a_i belong to $s\mathbf{Z}$. Let $\bar{\omega}$ be a semi-invariant generator of $\omega_{\bar{X}}$ at $\bar{p}$ and take a monomial u in x_1, x_2, x_3, x_4 such that $u\bar{\omega}$ is a generator of the invertible $\mathcal{O}_C$-module $\operatorname{gr}^0 \omega$. Then $\operatorname{ow} u = (o(w), w)$ and $w_p(0) = r^{-1}o(w)$ as in Remark 5.2.9. In particular, $\operatorname{ord} u = o(w) < r$ by $w_p(0) < 1$ in Proposition 5.2.14.

We divide the proof into three steps, which correspond to IA$^\vee$, IC$^\vee$ and II$^\vee$ eventually.

Step 1 Suppose that $\operatorname{wt}\xi = 0$ and $a_2 \leq r$. In this case, $\operatorname{ow} x_4 = (r, 0)$ since $(r, 0) \in \operatorname{ow} C$. If u were a power of x_2, then by $\operatorname{wt} u = w = -w_2$, one would have $u = x_2^{r-1}$ and hence $r \leq (r-1)s \leq \operatorname{ord} u$, which contradicts the estimate $\operatorname{ord} u < r$. Thus u is divisible by x_1 or x_3. Possibly after permuting x_1 and x_3, we may assume that $u = vx_1$ with some monomial v of weight $w - 1$.

The monomial $ux_2 = vx_1x_2$ of weight zero satisfies $\operatorname{ord} ux_2 < r + a_2 \leq 2r$. Hence $\operatorname{ord} ux_2 = r$ and $\operatorname{ow} vx_1x_2 = (r, 0) = \operatorname{ow} x_4$. In particular, $\operatorname{ord} vx_2 < r$. Then $\operatorname{ow} x_3 = \operatorname{ow} vx_2$ as $w_3 = \operatorname{wt} vx_2$, and v must be a monomial in x_1, x_2. From these observations one concludes that $\operatorname{ow} C = \mathbf{N} \operatorname{ow} x_1 + \mathbf{N} \operatorname{ow} x_2$ and $\operatorname{ow} x_4 \in \mathbf{N}_{>0} \operatorname{ow} x_1 + \mathbf{N}_{>0} \operatorname{ow} x_2$. This is the case IA$^\vee$.

Step 2 Suppose that $\operatorname{wt}\xi = 0$ and $a_2 > r$. Again $\operatorname{ow} x_4 = (r, 0)$. We may and shall assume that $a_1 \leq a_3$. Then $r \leq a_1 + a_3 \leq 2a_3$, where the former inequality comes from the relation $a_1 + a_3 \equiv 0 \bmod r$. Thus the monomials $x_1^{r'}$, x_2, x_1x_3, x_3^2, x_4 have order at least r.

Let $p' \in C' = \bigcup_{i=1}^s C_i' \subset X'$ be the splitting cover of $C \subset X$ associated with p, where p' is the point above p. Take a monomial u_i in x_1, x_2, x_3, x_4 such that $u_i\bar\omega$ is a generator of $\operatorname{gr}^0_{C_i'} \omega_{X'}$ at p'. Then u_i has order less than r for the same reason as does u, and further $\operatorname{ord} u_i = \operatorname{ord} u$ since they are congruent modulo r. The candidates for u_i are thus x_1^λ and x_3, where $u_i = x_1^\lambda$ only if $\lambda a_1 < r$ with $w \equiv \lambda \bmod r'$, and $u_i = x_3$ only if $a_3 < r$ with $w \equiv -1 \bmod r'$. Hence by Lemma 5.2.17, the number s of u_i must be two, and u_1, u_2 are x_1^λ, x_3 up to permutation. It follows that $w \equiv \lambda \equiv -1 \bmod r/2$, $\lambda a_1 < r$ and $a_3 < r$.

From $\lambda \equiv -1 \bmod r/2$, $\lambda a_1 < r$ and $a_1 \in 2\mathbf{N}$, one obtains $\lambda = r/2 - 1$ and $a_1 = 2$. Further $a_3 = r - 2$ from $a_3 < r$ with $a_1 + a_3 \equiv 0 \bmod r$, and $w_2 = 1$ or $r/2 + 1$ from $w_2 = -w \equiv 1 \bmod r/2$. If w_2 were one, then $a_2 = a_1 = 2$ and this contradicts the assumption $a_2 > r$. Hence $w_2 = r/2 + 1$. Then $r/2$ is even by the coprimeness of w_2 and $r = 2r'$. From $a_2 > r$ and $\operatorname{ow} x_1^{r/2+1} = (r + 2, w_2)$, one has $a_2 = r + 2$.

To sum up, $(a_1, a_2, a_3, a_4) = (2, r+2, r-2, r)$ and $(w_1, w_2, w_3, w_4) = (1, r/2 + 1, -1, 0)$. In particular, $\operatorname{ow} C = \mathbf{N} \operatorname{ow} x_1 + \mathbf{N} \operatorname{ow} x_3$. Replacing (x_2, x_3, x_4) by (x_3, x_4, x_2), one attains the case IC$^\vee$ unless $r = 4$. In the case $r = 4$, the coincidence $w_2 = w_3$ contradicts $a_2 \neq a_3$.

Step 3 Suppose that $(r, \operatorname{wt}\xi) = (4, 2)$ with ξ non-regular, in which $s = 2$. We may set $(w_1, w_2, w_3, w_4) = (1, 3, 3, 2)$. Then $\operatorname{wt} u = 1$ and one has $\operatorname{ord} u = 2$ from $\operatorname{ord} u < r = 4$ with $\operatorname{ord} u \in 2\mathbf{N}$. Hence $u = x_1$ and $\operatorname{ow} x_1 = (2, 1)$, in which $\operatorname{ow} x_4 = 2 \operatorname{ow} x_1 = (4, 2)$. Since $(4, 0) \in \operatorname{ow} C$, the element $(2, 3)$ must belong to $\operatorname{ow} C$. Thus $\operatorname{ow} x_2 = \operatorname{ow} x_3 = (2, 3)$ and the case II$^\vee$ holds. □

Remark 5.3.2 When p is of type IA$^\vee$, one can show that $w_p(0) \geq 1/2$ by the argument in Step 2 as follows. Take a monomial u_i such that $u_i\bar\omega$ generates

$\mathrm{gr}^0_{C'_i}\, \omega_{X'}$. Then $\mathrm{ord}\, u_i = \mathrm{ord}\, u < r$. If $w_p(0)$ were less than one-half, then $\mathrm{ord}\, u_i = rw_p(0) < r(1 - w_p(0)) = \mathrm{ord}\, x_2 \le \mathrm{ord}\, x_3$, where the latter inequality follows from ow $x_3 \in \mathbf{N}$ ow $x_1 + \mathbf{N}$ ow x_2 with $w_1 + w_3 \equiv 0$ mod r. Hence x_1^{λ} with the residue λ of w modulo r' is the only candidate for u_i, which contradicts Lemma 5.2.17.

In the primitive case, it follows from Remark 5.2.7 that the structure of the monoid ow C is determined by that of ord C up to the character $\chi \colon \mathbf{Z}_r \to \mathbf{C}^\times$ defining weights. The planarity is thus reduced to the assertion on ord C.

Theorem 5.3.3 *Let $p \in C \subset X$ be the germ of an extremal neighbourhood at a primitive point. Then ord C is generated by at most two elements.*

We take a normal form of $p \in C \subset X$ in Definition 5.2.8 as $p \in X \simeq (\xi = 0) \subset \mathfrak{D}^4/\mathbf{Z}_r(w_1, w_2, w_3, w_4)$ with x_1, x_2, x_3, x_4 and $a_i = o'(w_i)$. Note that Theorem 5.3.3 holds when $2 \in \mathrm{ord}\, C$. If $(r, \mathrm{wt}\, \xi) = (4, 2)$ with ξ non-regular, then the property $4 \in \mathrm{ord}\, C$ implies that some a_i equals one or two and thus $2 \in \mathrm{ord}\, C$. Hence we shall discuss the case $\mathrm{wt}\, \xi = 0$, in which we may assume that $(w_1, w_2, w_3, w_4) = (-w, 1, -1, 0)$ with $0 < w < r$. Note that $a_4 = r$ and $o(w) = rw_p(0) < r$ by Remark 5.2.9 and Proposition 5.2.14.

Lemma 5.3.4 *Suppose that $\mathrm{wt}\, \xi = 0$ and $(w_1, w_2, w_3, w_4) = (-w, 1, -1, 0)$ as above. Then either*

(i) *ord C is generated by at most two of a_1, a_2, a_3, a_4 or*
(ii) *$i_p(1) = 3$, $o(w) = 3w < r$ and $(a_1, a_2, a_3, a_4) = (2r - 3w, 3, kr - 3, r)$ with $a_3 \ge 3$ and $k = 1, 2$ or 3 up to permutation of x_2 and x_3.*

Proof **Step 1** By definition, $ru_p = a_1 + o(w)$. Hence $a_1 = o(-w) < 2r$ by $u_p \le 2$ in Proposition 5.2.15.

If $a_1 < r$, then with $o(w) < r$, one has $u_p = 1$, that is, $a_1 + o(w) = r$. We claim the case (i). Since $o(w) < a_4 = r$ and $a_2 + a_3 \equiv 0$ mod r, the element $o(w) \in \mathrm{ord}\, C$ belongs to $\mathbf{N}a_1 + \mathbf{N}a_2$ possibly after permutation of x_2 and x_3. If $o(w) \in \mathbf{N}a_1$, then $o(w) = (r-1)a_1$ by $w \equiv -w_1$ mod r. Thus $a_1 = 1$ and it generates ord C. If $o(w) \in \mathbf{N}a_1 + \mathbf{N}_{>0}a_2$, then $r - a_2 = a_1 + o(w) - a_2 \in \mathbf{N}a_1 + \mathbf{N}a_2$. Thus $a_3 = r - a_2$ and ord C is generated by a_1 and a_2.

Step 2 Henceforth we assume that $r < a_1 < 2r$, in which $a_1 + o(w) = 2r$. We also assume that $a_i \ge 3$ for all i since the property $2 \in \mathrm{ord}\, C$ implies the case (i). Our aim is to derive the case (ii) from these assumptions.

By Proposition 5.2.14, $i_p(1) \le 3$. As in the proof of Proposition 5.2.10, this number is computed as $i_p(1) = r^{-1}(\mathrm{ord}\, f - o(w))$ in terms of some invariant functions g and h together with ξ with $dg \wedge dh \wedge d\xi \wedge dx_4 = f\, dx_1 \wedge dx_2 \wedge dx_3 \wedge dx_4$. Here g, h, ξ belong to the ideal $\mathcal{I}^A$ in $\mathcal{O}_A$ defining C for $A = \mathfrak{D}^4/\mathbf{Z}_r(w_1, w_2, w_3, w_4)$.

The ideal $\mathscr{I}^A$ is generated by all $x_1^{s_1} x_2^{s_2} x_3^{s_3} - x_4^{s_4}$ with $\sum_{i=1}^{3} a_i s_i = r s_4$. Let

$$S = \{ x_1^{s_1} x_2^{s_2} x_3^{s_3} \mid \textstyle\sum_{i=1}^{3} a_i s_i \equiv 0 \bmod r \} \setminus \{1\}.$$

Let S' be the subset of S which consists of elements not factorised in S, that is, $S' = S \setminus (S \cdot S)$. Then $\mathscr{I}^A$ is generated by all $u - x_4^{r^{-1} \operatorname{ord} u}$ with $u \in S'$. Indeed, $u u' - x_4^{r^{-1} \operatorname{ord} u u'} = u'(u - x_4^{r^{-1} \operatorname{ord} u}) + x_4^{r^{-1} \operatorname{ord} u}(u' - x_4^{r^{-1} \operatorname{ord} u'})$ for $u, u' \in S$. The existence of the triple g, h, ξ with $\operatorname{ord} f = i_p(1) r + o(w)$ guarantees that of the triple u_1, u_2, u_3 in S' such that $d u_1 \wedge d u_2 \wedge d u_3 = \lambda d x_1 \wedge d x_2 \wedge d x_3$ with $\operatorname{ord} \lambda \leq i_p(1) r + o(w)$. Then

$$\operatorname{ord} u_1 u_2 u_3 \leq i_p(1) r + o(w) + a_1 + a_2 + a_3 = (i_p(1) + 2) r + a_2 + a_3.$$

Writing $u_{\sigma(i)} = x_i v_i$ with a monomial v_i of weight $-w_i$ for some permutation σ of $\{1, 2, 3\}$, one has $a_i + o(-w_i) \leq \operatorname{ord} u_{\sigma(i)}$ for $i = 1, 2, 3$. The sum of them becomes $2(r + a_2 + a_3) \leq \operatorname{ord} u_1 u_2 u_3$. Combining it with the inequality $\operatorname{ord} u_1 u_2 u_3 \leq (i_p(1) + 1) r + (r + a_2 + a_3)$ above, one obtains

$$\operatorname{ord} u_1 u_2 u_3 \leq 2(i_p(1) + 1) r.$$

One also has $a_2 + a_3 \leq i_p(1) r \leq 3r$, from which $a_2, a_3 < 3r$.

Step 3 The set S' contains the monomial $x_2 x_3$. Let $S'' = S' \setminus \{x_2 x_3\}$. After permutation of u_1, u_2, u_3, either

(a) $u_1, u_2 \in S''$, $u_3 = x_2 x_3$ and $\operatorname{ord} u_1 u_2 \leq (i_p(1) + 2) r \leq 5r$ by the first estimate of $\operatorname{ord} u_1 u_2 u_3$ in Step 2 or

(b) $u_1, u_2, u_3 \in S''$ and $\operatorname{ord} u_1 u_2 u_3 \leq 2(i_p(1) + 1) r \leq 8r$ by the second one.

Every element in S'' is of form $x_1^{s_1} x_2^{s_2}$ or $x_1^{s_1} x_3^{s_3}$ since it is not factorised in S. Since $a_1 > r$ and $a_2, a_3 \geq 3$, it has order at least $2r$. Further the candidates for $u \in S''$ with $\operatorname{ord} u = 2r$ are $x_1 x_2^w$ and $x_1 x_3^{r-w}$. Since $\operatorname{ord} x_1 x_2^w + \operatorname{ord} x_1 x_3^{r-w} \geq 2a_1 + 3w + 3(r - w) > 5r$, at most one element u in S'' satisfies $\operatorname{ord} u = 2r$.

We have seen that $\operatorname{ord} u_i \geq 2r$ for all $u_i \in S''$, and that $u_i \in S''$ with $\operatorname{ord} u_i = 2r$ is unique if it exists. Hence corresponding to the case (a) or (b) above, up to permutation of u_i we have either

(a) $i_p(1) = 3$ and $\operatorname{ord}(u_1, u_2) = (2, 3)$ or
(b) $i_p(1) = 3$ and $\operatorname{ord}(u_1, u_2, u_3) = (2, 3, 3)$.

Step 4 We shall see that a_2 or a_3 is three. Suppose that $a_2, a_3 \geq 4$ contrarily. Possibly after permuting x_2 and x_3, we may assume that $u_1 = x_1 x_2^w$. Then $2r = \operatorname{ord} u_1 > r + 4w$, that is, $w < r/4$. By $a_1 > r$ and $a_2, a_3 \geq 4$ with $w < r/4$, the candidates for $u_2 \in S''$ with $\operatorname{ord} u_2 = 3r$ are $x_1 x_3^{r-w}$ and $x_1^2 x_3^{r-2w}$. However, the estimate $w < r/4$ yields $\operatorname{ord} x_1 x_3^{r-w} > r + 4(r - w) > 4r$ and $\operatorname{ord} x_1^2 x_3^{r-2w} > 2r + 4(r - 2w) > 4r$, contradicting $\operatorname{ord} u_2 = 3r$.

We may thus assume that $a_2 = 3$ after permutation. Then $a_3 = kr - 3$ with $k \leq 3$ by $a_3 < 3r$. It remains to show that $o(w) = 3w$. We have seen that $u_1 = x_1 x_2^w$ or $x_1 x_3^{r-w}$ in the middle paragraph of Step 3. If $u_1 = x_1 x_2^w$, then $2r = a_1 + 3w$ and hence $o(w) = 2r - a_1 = 3w$. If $u_1 = x_1 x_3^{r-w}$, then $(r - w)a_3 < r$, from which $a_3 = r - 3 \geq 3$ and $w = r - 1$. This yields $w_1 = w_2$ but $a_1 \neq a_2$ by $a_1 > r > 3 = a_2$, which is a contradiction. $\qquad\square$

Proof of Theorem 5.3.3   One has only to disprove the case (ii) in Lemma 5.3.4, where $i_p(1) = 3$, $o(w) = 3w < r$ and $(a_1, a_2, a_3, a_4) = (2r - 3w, 3, kr - 3, r)$ with $a_3 \geq 3$ and $k \leq 3$ for $p \in C \subset X = (\xi = 0) \subset A = \mathfrak{D}^4 / \mathbf{Z}_r(-w, 1, -1, 0)$. Suppose this case. We shall compute the length $w_p(1)$ of the cokernel of the map $\beta_1 \colon \operatorname{gr}^0 \omega \otimes \operatorname{gr}^1 \mathcal{O} \to \operatorname{gr}^1 \omega$ explicitly.

Step 1   By Propositions 5.2.10 and 5.2.14, the equality $i_p(1) = 3$ implies that p is a unique singular point of X along C and $\operatorname{gr}^1 \mathcal{O} \simeq \mathcal{O}_{\mathbf{P}^1}(-1)^{\oplus 2}$. Together with $\operatorname{gr}^0 \omega \simeq \mathcal{O}_{\mathbf{P}^1}(-1)$ in Proposition 5.2.14, one has $\deg(\operatorname{gr}^0 \omega \otimes \operatorname{gr}^1 \mathcal{O}) = -4$. On the other hand, $\deg \operatorname{gr}^1 \omega \geq -2$ by $H^1(\operatorname{gr}^1 \omega) = 0$ in Proposition 5.2.14. Thus β_1 has cokernel of length at least $(-2) - (-4) = 2$.

Let $\bar{\omega}$ be a semi-invariant generator of $\omega_{\bar{X}}$ on the index-one cover $\bar{p} \in \bar{C} \subset \bar{X}$. Locally on the germ at p, the map β_1 fits into the commutative diagram

$$
\begin{array}{ccc}
\operatorname{gr}^0 \omega \otimes \operatorname{gr}^1 \mathcal{O} & \xrightarrow{\beta_1} & \operatorname{gr}^1 \omega \\
{\scriptstyle \delta_1} \uparrow & & \downarrow {\scriptstyle \gamma} \\
\operatorname{gr}^1 \mathcal{O} & \xrightarrow{\delta_2} & \operatorname{gr}^1 \mathcal{O}
\end{array}
$$

with $\delta_1 \colon f \mapsto x_2^w \bar{\omega} \otimes f$ for $f \in \operatorname{gr}^1 \mathcal{O}$ and $\gamma \colon g\bar{\omega} \mapsto x_1 g$ for $g\bar{\omega} \in \operatorname{gr}^1 \omega$. Whilst δ_1 is a local isomorphism, δ_2 is the multiplication by $x_1 x_2^w |_C = t^2$ and thus has cokernel of length $2 \cdot 2 = 4$. Hence γ has cokernel of length at most $4 - 2 = 2$.

Step 2   We shall consider the composite

$$
\bigwedge^2 \operatorname{gr}^1 \omega \xrightarrow{\wedge^2 \gamma} \bigwedge^2 \operatorname{gr}^1 \mathcal{O} \xrightarrow{\alpha_1} \omega_C^{-1} \otimes \operatorname{gr}^0 \omega \simeq \operatorname{gr}^0 \omega,
$$

$$
(g\bar{\omega}) \wedge (h\bar{\omega}) \mapsto \frac{d(x_1 g) \wedge d(x_1 h) \wedge dx_4 \wedge d\xi}{\xi}
$$

of local homomorphisms at p, in which g and h are semi-invariant functions of weight w vanishing along $\bar{C}$. Its cokernel is of length at most $2 + i_p(1) = 5$. The parts $d(x_1 g)$ and $d(x_1 h)$ can be replaced by $x_1 dg$ and $x_1 dh$ respectively in the $\mathcal{O}_C$-module $\operatorname{gr}^0 \omega$ because g and h vanish on C. Recall that $\operatorname{gr}^0 \omega$ is generated by $x_2^w (dx_1 \wedge dx_2 \wedge dx_3 \wedge dx_4)/\xi$ as $\operatorname{ord} x_2^w = o(w)$. Thus the above estimate on the length shows the existence of g and h such that the semi-invariant function f_0 given by $x_1^2 dg \wedge dh \wedge dx_4 \wedge d\xi = f_0 dx_1 \wedge dx_2 \wedge dx_3 \wedge dx_4$

has order at most $5r + o(w)$. Equivalently, the semi-invariant function f with $dg \wedge dh \wedge d\xi \wedge dx_4 = f dx_1 \wedge dx_2 \wedge dx_3 \wedge dx_4$ has

$$\operatorname{ord} f \leq 5r + o(w) - \operatorname{ord} x_1^2 = r + 9w.$$

Step 3 In Step 2 of the proof of Lemma 5.3.4, we saw that ξ belongs to the ideal in $\mathscr{O}_A$ generated by all $u - x_4^{r^{-1} \operatorname{ord} u}$ with $u \in S' = S \setminus (S \cdot S)$ for $S = \{x_1^{s_1} x_2^{s_2} x_3^{s_3} \mid \sum_{i=1}^3 a_i s_i \equiv 0 \bmod r\} \setminus \{1\}$. Let

$$T = \{x_1^{s_1} x_2^{s_2} x_3^{s_3} \mid \sum_{i=1}^3 a_i s_i \equiv w \bmod r\} \setminus \{x_2^w\}, \qquad T' = T \setminus (S \cdot T)$$

and let $p(u) = x_2^w x_4^{r^{-1}(\operatorname{ord} u - 3w)}$ for $u \in T'$. Then the above g and h belong to the $\mathscr{O}_A$-module generated by all $u - p(u)$ with $u \in T'$. Hence there exist $u_1, u_2 \in T'$ and $u_3 \in S'$ such that

$$d(u_1 - p(u_1)) \wedge d(u_2 - p(u_2)) \wedge du_3 \wedge dx_4 = \lambda dx_1 \wedge dx_2 \wedge dx_3 \wedge dx_4$$

with $\operatorname{ord} \lambda \leq r + 9w$. They satisfy

$$\operatorname{ord} u_1 u_2 u_3 \leq (r + 9w) + a_1 + a_2 + a_3 = (k + 3)r + 6w.$$

It follows from $(a_1, a_2, a_3, a_4) = (2r - 3w, 3, kr - 3, r)$ with $r \not\equiv 0 \bmod 3$ that

- for $i = 1, 2$, if $k = 1$, then the candidate for u_i with $\operatorname{ord} u_i \leq r + 3w$ is $x_2^{w+1} x_3$ with order $r + 3w$,
- for $i = 1, 2$, if $k = 2$ or 3, then the candidates for u_i with $\operatorname{ord} u_i \leq 2r + 3w$ are $x_1 x_2^{2w}$ with order $2r + 3w$ and $x_2^{w+1} x_3$ with order $kr + 3w$ and
- the candidates for u_3 with $\operatorname{ord} u_3 \leq 2r$ are $x_1 x_2^w$ with order $2r$ and $x_2 x_3$ with order kr.

By these observations, the triple (u_1, u_2, u_3) has $\operatorname{ord} u_1 u_2 u_3 \leq (k + 3)r + 6w$ only if it is $(u_1, x_2^{w+1} x_3, x_2 x_3)$ with $k = 1$ up to permutation of u_1, u_2. However for this triple, the part $d(u_2 - p(u_2)) \wedge du_3 \wedge dx_4$ in the $\mathscr{O}_C$-module $\operatorname{gr}^0 \omega$ is $x_2^w d(x_2 x_3 - x_4) \wedge d(x_2 x_3) \wedge dx_4 = 0$ because $x_2 x_3 - x_4$ vanishes on C. This implies that $\lambda|_C = 0$, which is absurd. The theorem is completed. $\square$

Proof of Theorem 5.3.1 *for primitive p* This is almost straightforward from Theorem 5.3.3. By this theorem, we may set ow $C = \mathbf{N}$ ow $x_1 + \mathbf{N}$ ow x_2 with respect to a normal form in Definition 5.2.8. For example we consider the case when $r \geq 2$ and $\operatorname{wt} \xi = 0$. If a_1 or a_2 is one, then the case IA occurs after permutation. Suppose that a_1 and a_2 are at least two. Up to permutations of x_1, x_2 and of x_3, x_4, either $a_1 + a_2, a_1 + a_3$ or $a_3 + a_4$ is divisible by r. If $a_1 + a_2 \equiv 0$ $\bmod r$, then $a_1 \neq a_2$ and one can assume that $a_3 = r$. If in addition $a_4 \neq a_1, a_2$, then the case IC occurs by the arrangement of x_1 and x_2 such that $a_1 < a_2$. The case $a_4 = a_1$ or a_2 is reduced to the case $a_1 + a_3 \equiv 0 \mod r$ after permutation. One can handle other cases similarly. Refer to [335, proposition-definition 4.2] for the complete argument. □

We shall close the section by providing computations of $w_p(0)$, $i_p(n)$ and $w_p^*(n)$ according to the type of p in Theorem 5.3.1. The method for computation is essentially found in the proof of Theorem 5.3.1. We omit the proof.

Definition 5.3.5 Let $p \in C \subset X$ be a terminal threefold analytic singularity with a smooth curve. For an element $z = (z_o, z_w)$ in ow C, we define $U(z)$ as the least integer such that $(rU(z), 0) \in z + \text{ow } C$. In other words, $U(z) = r^{-1}(z_o + o(-z_w))$. The definition is extended to the subset ow $C + r\mathbf{Z} \times \{0\}$ of $\mathbf{Z} \times \mathbf{Z}_r$ by $U(z + a(r, 0)) = U(z) + a$ for $z \in \text{ow } C$ and $a \in \mathbf{Z}$. If p is a primitive point, then $U(z)$ is determined by z_o for $z = (z_o, z_w) \in \text{ow } C + r\mathbf{Z} \times \{0\}$ and hence we write $U(z_o)$ for $U(z)$.

Lemma 5.3.6 ([335, theorem 4.9]) *Let $p \in C \subset X$ be as in Theorem 5.3.1.*

(i) *The invariant $w_p(0)$ is as follows.*

type	IA, IA$^\vee$	IB, IC, IC$^\vee$	IIA, IIB	II$^\vee$	III
$w_p(0)$	$(r - a_2)/r$	$o(-w_4)/r$	$3/4$	$1/2$	0

(ii) *If p is a non-smooth quotient singularity, then p is of type IA, IA$^\vee$, IC or IC$^\vee$. In this case, $i_p(1) = U(a_2 \text{ ow } x_1)$, and $i_p(n)$ for $n \geq 2$ and $w_p^*(n)$ for $n \geq 1$ are as follows, where $N_n = \{(i, j) \in \mathbf{N}^2 \mid i + j = n\}$. Note that p is always a quotient singularity if it is of type IC or IC$^\vee$ as in Remark 5.4.1.*

For type IA or IA$^\vee$,

$$i_p(n) = \sum_{(i,j) \in N_n} (iU(a_2 \text{ ow } x_1) - U((ia_2 - j) \text{ ow } x_1)),$$

$$w_p^*(n) = \sum_{(i,j) \in N_n} U((ia_2 - j) \text{ ow } x_1 + \text{ow } x_2) - n - 1.$$

For type IC or IC$^\vee$,

$$i_p(n) = \sum_{(i,j) \in N_n} (iU(a_2 \text{ ow } x_1) + jU(\text{ow } x_4) - U(ia_2 \text{ ow } x_1 + j \text{ ow } x_4)),$$

$$w_p^*(n) = \sum_{(i,j) \in N_n} (U(ia_2 \text{ ow } x_1 + (j + 1) \text{ ow } x_4) - (j + 1)U(\text{ow } x_4)).$$

(iii) *If p is an ordinary double point, then $i_p(1) = 1$, $i_p(n) = \lfloor n^2/4 \rfloor$ for $n \geq 2$ and $w_p^*(n) = \lfloor (n + 1)^2/4 \rfloor$ for $n \geq 1$.*

5.4 Deformations of an Extremal Neighbourhood

Following the local classification obtained in the preceding section, we shall classify the set of singular points in an extremal neighbourhood. The basic tool is deformation of the neighbourhood.

Let $C \subset X$ be an extremal neighbourhood. We have a classification of a singular point $p \in C \subset X$ in Theorem 5.3.1. We begin with constructing a deformation of the germ $p \in C \subset X$ locally. Keep the notation in Theorem 5.3.1 in which $p \in X \simeq (\xi = 0) \subset A = \mathfrak{D}^4/\mathbf{Z}_r(w_1, w_2, w_3, w_4)$ with orbifold coordinates x_1, x_2, x_3, x_4 and $a_i = o'(w_i)$. Since $\mathrm{ow}\, C$ is generated by $\mathrm{ow}\, x_1$ and $\mathrm{ow}\, x_2$, there exists a monomial g_i in x_1, x_2 for $i = 3, 4$ such that $\mathrm{ow}\, x_i = \mathrm{ow}\, g_i$. Put $f_2 = x_1^{a_2} - x_2^{a_1}$, $f_3 = x_3 - g_3$ and $f_4 = x_4 - g_4$. Then

$$p \in C = (f_2 = f_3 = f_4 = 0) \subset X = (\xi = 0) \subset A = \mathfrak{D}^4/\mathbf{Z}_r(w_1, w_2, w_3, w_4).$$

Remark 5.4.1 Every point p of type IC or IC$^\vee$ is a quotient singularity. Indeed in this case, g_3 must be $x_1 x_2$ and one can write $\xi = h_2(x_1^{a_2} - x_2^{a_1}) + h_3(x_3 - x_1 x_2) + h_4(x_4 - g_4)$ with a semi-invariant function h_2 and invariant functions h_3 and h_4. By $r \geq 5$, $p \in X$ is of type cA/r. Since $a_4 \neq a_1, a_2$, either $x_1 x_2$ or x_3 must appear in ξ. Thus h_3 is a unit and $X \simeq \mathfrak{D}^3/\mathbf{Z}_r$.

Fix a small neighbourhood U in A at p. We construct a one-parameter deformation $\mathfrak{C} \subset \mathfrak{X} \subset A \times \mathfrak{D}^1$ over $\mathfrak{D}^1$ of $C \subset X \subset A$ and take the fibre $C_z \subset X_z \subset A$ at general $z \in \mathfrak{D}^1$ close to the origin, for which $C_z|_U$ and $X_z|_U$ are closed subspaces of U.

Construction 5.4.2   (i) Suppose that p is of type IA, IIA, IA$^\vee$ or II$^\vee$. Let $C_z = C \subset X_z = (\xi + z f_4 = 0) \subset A$. Note that X_z is terminal by Theorem 2.4.16. Then p is the only singularity of $X_z|_U$ in $C_z|_U$, which is a quotient singularity. The germ $p \in C_z \subset X_z$ is of type IA (resp. IA$^\vee$) if $p \in C \subset X$ is primitive (resp. imprimitive).

(ii) Suppose that p is of type IB. Let $C_z = C \subset X_z = (\xi + z f_2 = 0) \subset A$. Then p is the only singularity of $X_z|_U$ in $C_z|_U$. The germ $p \in C_z \subset X_z$ is of type IB and for its index-one cover $\bar{p} \in \bar{C}_z \subset \bar{X}_z$, the curve $\bar{C}_z$ is lci in $\bar{X}_z$ defined by f_3 and f_4.

(iii) Suppose that p is of type IB with index-one cover $\bar{p} \in \bar{C} \subset \bar{X}$ such that $\bar{C}$ is lci in $\bar{X}$. Up to unit, one can write $\xi = f_2 + h_3 f_3 + h_4 f_4$ with semi-invariant functions h_3 and h_4. Let $C_z = (f_2 + z x_2 = f_3 = f_4 = 0) \subset X_z = (\xi + z x_2 = 0) \subset A$. Then X_z has a_1 singular points in $C_z|_U$, all of which are quotient singularities of type IA with respect to $C_z \subset X_z$.

(iv) Suppose that p is of type IIB. One can write $\xi = f_2 + h_3 f_3 + h_4 f_4$ up to unit with semi-invariant functions h_3 and h_4 because the monomial x_1^2 appears in ξ by the description in Theorem 2.4.8. Let $C_z = (f_2 + zx_2 = f_3 = f_4 = 0) \subset X_z = (\xi + zx_2 = 0) \subset A$. Then X_z has two singular points in $C_z|_U$, which are quotient singularities of type IA with respect to $C_z \subset X_z$ and of indices four and two.

(v) When p is of type III, we take another expression

$$p \in C = (y_1\text{-axis}) \subset X = (f = 0) \subset A = \mathfrak{D}^4$$

with coordinates y_1, y_2, y_3, y_4 for $f = y_1^a y_2 + g$ with $g \in (y_2, y_3, y_4)^2 \mathcal{O}_{\mathfrak{D}^4}$ as in Lemma 5.2.11. The lemma shows that $a = i_p(1)$. Let $C_z = C \subset X_z = (f + z(y_1 y_2 + y_3 y_4) = 0) \subset A$. Then X_z has $i_p(1)$ singular points $p_{z,i}$ in $C_z|_U$ for $1 \le i \le i_p(1)$, all of which are ordinary double points with $i_{p_{z,i}}(1) = 1$ with respect to $C_z \subset X_z$.

Remark 5.4.3   In the construction above, the deformation in (i) and (ii) preserves the structure ow C of the normal form. In (iii), the monoid ord C_z equals $\mathbf{N}$ at every singular point $p_{z,i}$ of $C_z \subset X_z$ and the invariant $w_{p_{z,i}}(0)$ takes the same value for all i.

Every deformation of $p \in C \subset X$ in Construction 5.4.2 is globalised explicitly as in [335, definition 1b.8] to a deformation $(\mathfrak{C} \subset \mathfrak{X}) \to \mathfrak{D}^1$ of the germ $C \subset X$ which is locally trivial outside U. Further it always extends to a deformation $(\mathfrak{C} \subset \mathfrak{X} \to \mathfrak{Y}) \to \mathfrak{D}^1$ of the associated contraction $C \subset X \to y \in Y$ such that the fibre $C_z \subset X_z \to Y_z$ at general $z \in \mathfrak{D}^1$ close to the origin is an extremal neighbourhood. This follows from Theorem 4.4.5 when $C \subset X$ is isolated, whilst it is proved in [342, theorem 3.2] when $C \subset X$ is divisorial. Notice that in the proof of [342, theorem 3.2], the existence of flips is used only for a reducible extremal neighbourhood.

Definition 5.4.4   The procedure for making an extremal neighbourhood $C_z \subset X_z$ by Construction 5.4.2 followed by its globalisation is referred to as an *L-deformation* of $C \subset X$ at p.

By serial *L*-deformations, one attains an extremal neighbourhood every singular point of which is either a quotient singularity of type IA, IC, IA$^\vee$ or IC$^\vee$ or an ordinary double point of type III.

We shall state the main theorem in this section with the following notion.

Definition 5.4.5   In Theorem 5.3.1, we say that p is *minimal* if $U(a_1 \text{ ow } x_2) = 1$.

Note that $U(a_1 \text{ ow } x_2) = U(a_2 \text{ ow } x_1)$. By Remark 5.4.3, the minimality is preserved by the *L*-deformations in Construction 5.4.2(i) and (ii).

Theorem 5.4.6 ([335, theorem 6.7]) *The set Σ of singular points of an extremal neighbourhood $C \subset X$ is exactly one of the following.*

(i) $\Sigma = \{p\}$ *or* $\{p, z\}$ *where p is of type* IA, IIA *or* IA$^\vee$ *and z is of type* III.

(ii) $\Sigma = \{p\}$ *where p is of type* IC, IIB *or* II$^\vee$.

(iii) $\Sigma = \{p, q\}$ *or* $\{p, q, z\}$ *where p and q are minimal of type* IA, *q is of index two and z is of type* III.

(iv) $\Sigma = \{p, q\}$ *where p and q are minimal of type* IA, IIA *or* IA$^\vee$ *and of index at least three.*

(v) $\Sigma = \emptyset$ *or* $\{z\}$ *where z is of type* III *and $C \subset X$ is divisorial.*

Later in Theorem 5.5.1, we shall prove that p and q in the case (iv) are of type IA.

It is easy to settle (v) in the Gorenstein case by applying the deformation theory of curves. Recall that the *Hilbert scheme* $\mathrm{Hilb}(X/S)$ of a quasi-projective scheme X over S represents the functor from the category of schemes over S to the category of sets given by

$$(T \to S) \mapsto \{Z \subset X \times_S T \mid Z \text{ flat and projective over } T\}.$$

We usually write $\mathrm{Hilb}\, X$ by omitting S when S is the spectrum of the ground field. The *Hom scheme* $\mathrm{Hom}(X, Y)$, which parametrises morphisms from X to Y, is the subscheme of $\mathrm{Hilb}(X \times Y)$ that parametrises graphs of morphisms $X \to Y$, where X is assumed to be projective. The following theorem is found in [266, II theorem 1.3].

Theorem 5.4.7 *Let $f: C \to X$ be a morphism from a projective curve to an lci quasi-projective variety of dimension n such that $f(C)$ meets the smooth locus in X. Then the local dimension of the Hom scheme $\mathrm{Hom}(C, X)$ at $[f]$ is at least $(-K_X \cdot f_*C) + n\chi(\mathscr{O}_C)$.*

Proposition 5.4.8 *If an extremal neighbourhood is Gorenstein, then it is divisorial and has at most one singular point.*

Proof Let $\pi: C \simeq \mathbf{P}^1 \subset X \to y \in Y$ be an extremal neighbourhood such that X is Gorenstein. By Theorem 2.3.4, X has only hypersurface singularities. Suppose that $C \subset X$ were isolated. We consider the Hom scheme $\mathrm{Hom}(C, X)$ after algebraisation of π by Theorem 2.1.8. By Theorem 5.4.7, the local dimension of $\mathrm{Hom}(C, X)$ at the embedding of C is at least $(-K_X \cdot C) + 3 > 3$. Hence C could move in X, but this is impossible since C is the exceptional locus of π. Thus $C \subset X$ is divisorial.

After L-deformations, every singular point p is an ordinary double point and thus has the invariant $w_p^*(1) = 1$ by Lemma 5.3.6(iii). Hence p is unique by Remark 5.2.21(i). □

Next we discuss the case when X has an imprimitive point.

Lemma 5.4.9 *There exists no point of type* $\mathrm{IC}^\vee$ *in an extremal neighbourhood.*

Proof Suppose that an extremal neighbourhood had a point of type $\mathrm{IC}^\vee$. For each singular point of type III, instead of an L-deformation, we apply the deformation in Construction 5.4.2(v) with X_z replaced by the locus $(f+zy_3 = 0)$ and globalise it. Then we obtain an extremal neighbourhood $C \subset X$ which has a point p of type $\mathrm{IC}^\vee$ but has no singular points of type III.

The point p has the numerical data $(a_1, a_2, a_3, a_4) = (2, r - 2, r, r + 2)$ and $(w_1, w_2, w_3, w_4) = (1, -1, 0, r/2 + 1)$. Since $u_p = r^{-1}(o(w_4) + o(-w_4)) = r^{-1}((r+2) + (r-2)) = 2$, Proposition 5.2.15 shows that p is the only singular point. By Lemma 5.3.6(ii),

$$w_p^*(1) = U(3r - 2, r/2 - 1) + U(2r + 4, 2) - 3U(r + 2, r/2 + 1)$$

as a_2 ow $x_1 +$ ow $x_4 = (3r-2, r/2-1)$. It is computed as $w_p^*(1) = 4+4-3\cdot 2 = 2$. This contradicts Remark 5.2.21(i). □

Lemma 5.4.10 *Suppose that $p \in C \subset X$ is a quotient singularity of type* IA *or* $\mathrm{IA}^\vee$ *in Theorem 5.3.1.*

(i) *If $i_p(1) = 1$ and $w_p(0) < 1/2$, then $a_1 = 1$, $w_p^*(1) = 1$ and hence p is primitive.*

(ii) *One has $w_p^*(1) \geq -1$. The equality holds if and only if $a_2 = 1$ and hence p is primitive.*

(iii) *If $a_2 \geq 2$, then $w_p^*(1) \geq i_p(1) - 1$.*

Proof By Lemma 5.3.6(ii), the condition $i_p(1) = 1$ is equivalent to the minimality $U(a_2 \text{ ow } x_1) = 1$, which implies that $a_1 a_2 \leq r$ for the index r. By $w_p(0) = (r - a_2)/r$ in Lemma 5.3.6(i), the condition $w_p(0) < 1/2$ is equivalent to $a_2 > r/2$. From these two one deduces that $a_1 = 1$, in which p is primitive because the splitting degree is a divisor of a_1. When $a_1 = 1$, Lemma 5.3.6(ii) gives $w_p^*(1) = U(2a_2) + U(a_2 - 1) - 2$, which equals one by $a_2 > r/2$. Remark that $U(n) = \lceil n/r \rceil$ when ord $C = \mathbf{N}$. Thus (i) is proved.

For (ii), Lemma 5.3.6(ii) gives

$$w_p^*(1) = U(a_2 \text{ ow } x_1 + \text{ow } x_2) + U(\text{ow } x_2 - \text{ow } x_1) - 2 \geq 1 + 0 - 2 = -1.$$

If the equality holds, then $U(\text{ow } x_2 - \text{ow } x_1) = 0$, which means that ow $x_1 -$ ow $x_2 \in$ ow $C = \mathbf{N}$ ow $x_1 + \mathbf{N}$ ow x_2 and hence ow $x_1 -$ ow $x_2 \in \mathbf{N}$ ow x_2. In this case, ow C is generated by ow x_2 and thus $a_2 = 1$ by $(r, 0) \in$ ow C. It is obvious to derive $w_p^*(1) = -1$ from $a_2 = 1$. Thus (ii) is proved.

If $a_2 \geq 2$, then $U(\text{ow } x_2 - \text{ow } x_1) \geq 1$ by the above argument. Hence $w_p^*(1) \geq U(a_2 \text{ ow } x_1 + \text{ow } x_2) - 1 \geq i_p(1) - 1$, where the latter inequality is from $i_p(1) = U(a_2 \text{ ow } x_1)$ in Lemma 5.3.6(ii). This is (iii). □

Lemma 5.4.11 *If $p \in C \subset X$ is of type IC in Theorem 5.3.1, then $i_p(1) \geq 2$.*

Proof One has $a_2 > r/2$ from the conditions $a_1 + a_2 = r$ and $a_1 < a_2$. Then with Lemma 5.3.6(ii), $i_p(1) = U(a_1 a_2) \geq a_1 a_2/r > a_1/2 \geq 1$. □

Lemma 5.4.12 *If an extremal neighbourhood has an imprimitive point p, then it has at most one singular point other than p. Such a singular point exists only if p is minimal of type $\mathrm{IA}^{\vee}$.*

Proof Let $C \subset X$ denote the extremal neighbourhood. By Corollary 5.2.18, an imprimitive point is unique if it exists. By Lemma 5.4.9, p is of type $\mathrm{IA}^{\vee}$ or $\mathrm{II}^{\vee}$. Remark that a point of type $\mathrm{II}^{\vee}$ is never minimal since it has $U(a_2 \text{ ow } x_1) = U(4, 2) = 2$. Hence after L-deformations with Remark 5.4.3, we may and shall assume that p is a quotient singularity of type $\mathrm{IA}^{\vee}$ and that any other singular point of X is either a quotient singularity of type IA or IC or an ordinary double point of type III. Note that $w_p(0) \geq 1/2$ in Remark 5.3.2.

We shall prove that $C \subset X$ has at most one singular point other than p. If it had three singular points p, q, z, then by Propositions 5.2.10 and 5.2.14, they would exhaust singularities of X with $i_p(1) = i_q(1) = i_z(1) = 1$. By Lemma 5.4.11, q and z are of type IA or III. Further since $w_p(0) \geq 1/2$, it follows from Proposition 5.2.14 that $w_q(0)$, $w_z(0) < 1/2$. One has $w_p^*(1) \geq 0$ from Lemma 5.4.10(ii), and $w_q^*(1) = w_z^*(1) = 1$ from Lemma 5.4.10(i) for type IA and Lemma 5.3.6(iii) for type III. These contradict Remark 5.2.21(i).

It remains to prove that if X has two singular points p and q, then p is minimal or equivalently $i_p(1) = 1$ from Lemma 5.3.6(ii). By Lemma 5.4.10(iii), this is the case if $w_p^*(1) \leq 0$. By virtue of the inequalities $i_p(1) + i_q(1) \leq 3$ in Proposition 5.2.14 and $w_p^*(1) + w_q^*(1) \leq 1$ in Remark 5.2.21(i), it suffices to show that either $i_q(1) \geq 2$ or $w_q^*(1) \geq 1$. Note that $w_q(0) < 1/2$ by Proposition 5.2.14. If q is of type IA, then $i_q(1) \geq 2$ or $w_q^*(1) = 1$ by Lemma 5.4.10(i). If q is of type IC, then $i_q(1) \geq 2$ by Lemma 5.4.11. Finally if q is of type III, then $w_q^*(1) = 1$ by Lemma 5.3.6(iii). □

We proceed to the case when $C \subset X$ is locally primitive. We need a computation of $w_p^*(n)$.

Lemma 5.4.13 *Suppose that $p \in C \subset X$ is a quotient singularity of type* IA *in Theorem* 5.3.1. *If* $a_2 = 1$, *then* $w_p^*(n) = -\lceil n/2 \rceil$ *for* $n < r$ *and* $w_p^*(r) = -\lceil r/2 \rceil + 1$.

Proof Lemma 5.3.6(ii) with $a_2 = 1$ means that

$$w_p^*(n) = \sum_{(i,j) \in N_n} U((i-j)a_1 + 1) - n - 1 = \sum_{0 \le i < n/2} v(i)$$

for $v(i) = U((n-2i)a_1 + 1) + U(-(n-2i)a_1 + 1) - 2$. One computes $v(i)$ as

$$
\begin{aligned}
v(i) &= \left\lceil \frac{(n-2i)a_1 + 1}{r} \right\rceil + \left\lceil \frac{-(n-2i)a_1 + 1}{r} \right\rceil - 2 \\
&= \left\lceil \frac{\overline{(n-2i)a_1} + 1}{r} \right\rceil + \left\lceil \frac{\overline{-(n-2i)a_1} + 1}{r} \right\rceil - 2 =
\begin{cases}
0 & \text{if } n \equiv 2i \bmod r, \\
-1 & \text{if } n \not\equiv 2i \bmod r,
\end{cases}
\end{aligned}
$$

where $\bar{l}$ stands for the residue of l modulo r. This provides the lemma.   □

The next lemma characterises the case when X has three singular points.

Lemma 5.4.14 *If an extremal neighbourhood has three singular points, then they are a point p of type* IA *of odd index, a point q of type* IA *of index two and a point z of type* III.

Proof *Step* 1 Let $C \subset X$ be an extremal neighbourhood which has three singular points p, q and z. By Propositions 5.2.10 and 5.2.14, $i_p(1) = i_q(1) = i_z(1) = 1$. They are all primitive by Lemma 5.4.12, not of type IC by Lemma 5.4.11. If one of p, q, z were of type IB or IIB, then L-deformations at this point would produce an extremal neighbourhood with at least four singular points, which contradicts Propositions 5.2.10 and 5.2.14. Hence p, q, z are of type IA, IIA or III. By serial L-deformations, one attains an extremal neighbourhood $C' \subset X'$ with three singular points, all of which are either a quotient singularity of type IA or an ordinary double point of type III. Note that if one of p, q, z is of type IIA, then $C' \subset X'$ has a singular point of index four. Therefore with Remark 5.4.3, the assertion is reduced to that for $C' \subset X'$.

Thus we may and shall assume that each of p, q, z is a quotient singularity of type IA or an ordinary double point of type III. Up to permutation, either

  (i) p, q, z are of type IA,
 (ii) p, q are of type IA and z is of type III,
(iii) p is of type IA and q, z are of type III or
 (iv) p, q, z are of type III.

Step 2 By Propositions 5.2.15 and 5.4.8, neither the case (i) nor (iv) occurs. We shall also see that the case (iii) does not occur. In the case (iii), $w_q^*(1) =$

$w_z^*(1) = 1$ by Lemma 5.3.6(iii) and hence $w_p^*(1) \le -1$ by Remark 5.2.21(i). Then by Lemma 5.4.10(ii), $a_2 = 1$ and $w_p^*(1) = -1$. Let r be the index of the germ $p \in X$. Using Lemma 5.4.13 for $w_p^*(n)$ and Lemma 5.3.6(iii) for $w_q^*(n)$ and $w_z^*(n)$, one has $\sum_{\theta=p,q,z} w_\theta^*(n) = -\lceil n/2 \rceil + 2\lfloor (n+1)^2/4 \rfloor = n(n+1)/2$ for all $n < r$ and $\sum_\theta w_\theta^*(r) = r(r+1)/2 + 1$. This contradicts Remark 5.2.21(ii).

Step 3 In the remaining case (ii), we shall prove that one of p and q is of odd index and the other is of index two. We write a_{ip}, r_p and a_{iq}, r_q for the notation a_i, r in Theorem 5.3.1 for p and q respectively to distinguish them.

Since $w_p(0) + w_q(0) < 1$ from Proposition 5.2.14, we may assume without loss of generality that $w_p(0) < 1/2$. Then by Lemma 5.4.10(i), $a_{1p} = 1$ and $w_p^*(1) = 1$. Recall the estimate $w_p^*(1) + w_q^*(1) + w_z^*(1) \le 1$ in Remark 5.2.21(i). Together with $w_z^*(1) = 1$ in Lemma 5.3.6(iii), one obtains $w_q^*(1) \le -1$ and hence $a_{2q} = 1$ and $w_q^*(1) = -1$ by Lemma 5.4.10(ii). Then $w_p^*(1) + w_q^*(1) + w_z^*(1) = 1$. By Remark 5.2.21(ii), it induces the estimate $w_p^*(2) + w_q^*(2) + w_z^*(2) \le 3$.

One has $w_z^*(2) = 2$ from Lemma 5.3.6(iii), and

$$w_p^*(2) = \begin{cases} 3 & \text{if } 2r_p/3 < a_{2p}, \\ 2 & \text{if } (r_p + 1)/2 < a_{2p} \le 2r_p/3, \\ 1 \text{ or } 0 & \text{if } a_{2p} = (r_p + 1)/2, \end{cases} \qquad w_q^*(2) = \begin{cases} 0 & \text{if } r_q = 2, \\ -1 & \text{if } r_q \ge 3. \end{cases}$$

The value of $w_q^*(2)$ comes from Lemma 5.4.13 and that of $w_p^*(2)$ from $w_p^*(2) = \lceil 3a_{2p}/r_p \rceil + \lceil (2a_{2p} - 1)/r_p \rceil + \lceil (a_{2p} - 2)/r_p \rceil - 3$ in Lemma 5.3.6(ii). Notice that $a_{2p} > r_p/2$ by $w_p(0) = (r_p - a_{2p})/r_p < 1/2$ in Lemma 5.3.6(i). Since $1 > w_p(0) + w_q(0) = (r_p - a_{2p})/r_p + (r_q - 1)/r_q$ from Lemma 5.3.6(i), one also has

$$\frac{a_{2p}}{r_p} + \frac{1}{r_q} > 1.$$

Thus one concludes that $a_{2p} = (r_p + 1)/2$ and $r_q = 2$ from $w_p^*(2) + w_q^*(2) \le 3 - w_z^*(2) = 1$. The first equality implies that r_p is odd. The lemma is therefore completed. $\square$

Corollary 5.4.15 *There exists no point of type* IB *in an extremal neighbourhood.*

Proof If an extremal neighbourhood had a singular point of type IB, then L-deformations at this point would produce an extremal neighbourhood $C \subset X$ with at least two quotient singularities p and q of type IA, which are of the same index, and by Remark 5.4.3, have the same $a_1 = 1$ for the notation in Theorem 5.3.1 and the same $w_p(0) = w_q(0)$. In particular by Lemma 5.4.14, X has no other singular points. By Proposition 5.2.14, $w_p(0) = w_q(0) < 1/2$.

Further $i_p(1) = i_q(1) = 1$ by Lemma 5.3.6(ii). Hence $w_p^*(1) = w_q^*(1) = 1$ by Lemma 5.4.10(i), which contradicts Remark 5.2.21(i). □

We provide a sufficient condition for a primitive point to be a unique singularity.

Lemma 5.4.16 *If an extremal neighbourhood has a primitive point p of type other than* III *and another singular point, then p is minimal of type* IA *or* IIA.

Proof *Step* 1 Let $C \subset X$ denote the extremal neighbourhood, which has a singular point q besides p. By Corollary 5.4.15, p is not of type IB. If p were of type IIB, then an L-deformation at p would produce an extremal neighbourhood which has at least three singular points two of which are of indices four and two. This contradicts Lemma 5.4.14. Hence after an L-deformation at p with Remark 5.4.3, it suffices to exclude the case when p is a quotient singularity either of type IC or non-minimal of type IA. The proof requires a multitude of delicate computations on numerical invariants, whereas the methods for these computations have already appeared in the preceding arguments in this section. Hence we shall only demonstrate the exclusion of the case of non-minimal type IA here. This case is more interesting for the reason that it involves a global analysis of the $\mathscr{O}_C$-module $\mathrm{gr}^1 \mathscr{O}$ unlike the case of type IC.

Henceforth we shall assume that p is a non-minimal quotient singularity of type IA and derive a contradiction. With Lemma 5.4.9, after L-deformations at q, we may assume that q is a quotient singularity of type IA, IC or IA$^\vee$ or an ordinary double point of type III. By Lemma 5.3.6(ii), the non-minimality of p means that $i_p(1) \geq 2$. Hence by Propositions 5.2.10 and 5.2.14, $i_p(1) = 2$, $i_q(1) = 1$ and X has no other singular points. By Lemma 5.4.11, q is of type IA, IA$^\vee$ or III. We write a_{ip}, r_p and a_{iq}, r_q for the notation a_i, r in Theorem 5.3.1 for p and q respectively.

Step 2 By Lemma 5.3.6(ii), $U(a_{1p}a_{2p}) = i_p(1) = 2$ and in particular a_{1p} and a_{2p} are at least two. We claim that $w_p(0) < 1/2$, $a_{1p} = 2$ and a_{2p}, r_p are odd.

If $w_p(0) \geq 1/2$, then $w_q(0) < 1/2$ by Proposition 5.2.14, and $w_q^*(1) = 1$ by Lemma 5.4.10(i) for type IA or IA$^\vee$ and Lemma 5.3.6(iii) for type III. On the other hand since $i_p(1) = 2$ and $a_{2p} \geq 2$, one has $w_p^*(1) \geq 1$ from Lemma 5.4.10(iii). This contradicts Remark 5.2.21(i). Hence $w_p(0) < 1/2$. By Lemma 5.3.6(i), this is equivalent to $a_{2p} > r_p/2$.

Since $a_{1p}a_{2p} < 2r_p$ from $U(a_{1p}a_{2p}) = 2$, the invariant a_{1p} is two or three. If $a_{1p} = 3$, then $U(3a_{2p}) = 2$ which means that $2r_p - 3a_{2p} \in 3\mathbf{N} + a_{2p}\mathbf{N}$ and hence $2r_p - 3a_{2p} \in 3\mathbf{N}$ as $2r_p - 3a_{2p} < a_{2p}$. Then r_p is a multiple of three, which contradicts the coprimeness of a_{1p} and r_p. Thus $a_{1p} = 2$. Note that r_p

is odd again by the coprimeness to a_{1p}. Since $a_{2p} > r_p/2$, it follows from $r_p \in 2\mathbf{N} + a_{2p}\mathbf{N}$ that $r_p - a_{2p} \in 2\mathbf{N}$, by which a_{2p} is also odd. Then $a_{2p} \geq 3$.

Next we claim that q is of type IA with $a_{2q} = 1$ and $w_p^*(1) + w_q^*(1) = 1$. By $a_{1p} = 2$, Lemma 5.3.6(ii) gives $w_p^*(1) = U(3a_{2p}) + U(a_{2p} - 2) - 2 = U(3a_{2p}) - 1$, where $U(a_{2p} - 2) = 1$ follows from $r_p - a_{2p} \in 2\mathbf{N}$ with $a_{2p} \geq 3$. If $U(3a_{2p}) = 2$, then $2r_p - 3a_{2p} \in 2\mathbf{N} + a_{2p}\mathbf{N}$ and hence $2r_p - 3a_{2p} \in 2\mathbf{N}$, which is absurd since a_{2p} is odd. Hence $U(3a_{2p}) = 3$ and $w_p^*(1) = 2$. Then $w_q^*(1) \leq -1$ by Remark 5.2.21(i). Thus q is not of type III by Lemma 5.3.6(iii) and the claim follows from Lemma 5.4.10(ii).

Step 3 Similarly to Step 3 in the proof of Lemma 5.4.14, one can see that $a_{2p} = (r_p + 1)/2$, $r_q = 2$ and $w_p^*(2) + w_q^*(2) = 3$. Indeed,

$$w_p^*(2) + w_q^*(2) \leq 3, \qquad a_{2p}/r_p + 1/r_q > 1$$

where the former is from Remark 5.2.21(ii) and the latter from $w_p(0) + w_q(0) = (r_p - a_{2p})/r_p + (r_q - 1)/r_q < 1$ in Proposition 5.2.14 and Lemma 5.3.6(i). Lemmata 5.3.6(ii) and 5.4.13 compute $w_p^*(2)$ and $w_q^*(2)$ respectively as

$$w_p^*(2) \begin{cases} \geq 5 & \text{if } a_{2p} > 3r_p/5, \\ \geq 4 & \text{if } a_{2p} > (r_p + 1)/2, \\ = 3 & \text{if } a_{2p} = (r_p + 1)/2, \end{cases} \qquad w_q^*(2) = \begin{cases} 0 & \text{if } r_q = 2, \\ -1 & \text{if } r_q \geq 3. \end{cases}$$

From these the required properties are derived.

Step 4 By Proposition 5.2.14 and Remark 5.2.21(ii), one concludes that $\mathrm{gr}^1 \mathscr{O} \simeq \mathscr{O}_{\mathbf{P}^1}(-1)^{\oplus 2}$ and $\mathrm{gr}^i \omega \simeq \mathscr{O}_{\mathbf{P}^1}(-1)^{\oplus(i+1)}$ for $i = 0, 1, 2$. We shall investigate the global map $\beta_2 \colon \mathrm{gr}^0 \omega \otimes S^2(\mathrm{gr}^1 \mathscr{O}) \to \mathrm{gr}^2 \omega$.

Since $a_{2q} = 1$ and $r_q = 2$, the germ $q \in C \subset X$ is identified with $o \in (x_1\text{-axis}) \subset \mathfrak{D}^3/\mathbf{Z}_2$ with orbifold coordinates x_1, x_2, x_3. Then for $\bar{\omega} = dx_1 \wedge dx_2 \wedge dx_3$, locally at q one has the descriptions

$$\mathrm{gr}^0 \omega = \mathscr{O}_C x_1 \bar{\omega}, \qquad \mathrm{gr}^1 \mathscr{O}_C = \mathscr{O}_C x_1 x_2 \oplus \mathscr{O}_C x_1 x_3,$$

by which $\mathrm{gr}^2 \omega = \mathscr{O}_C x_1 x_2^2 \bar{\omega} \oplus \mathscr{O}_C x_1 x_2 x_3 \bar{\omega} \oplus \mathscr{O}_C x_1 x_3^2 \bar{\omega}$. The image of β_2 at q is $x_1^2 \mathrm{gr}^2 \omega = \mathrm{gr}^2 \omega \otimes \mathscr{O}_C(-q)$. Hence β_2 factors through the map

$$\gamma \colon \mathrm{gr}^0 \omega \otimes S^2(\mathrm{gr}^1 \mathscr{O}) \to \mathrm{gr}^2 \omega \otimes \mathscr{O}_C(-q)$$

which is isomorphic outside p.

From now on, we write $r = r_p$ and $a = a_{2p} = (r + 1)/2 \geq 3$ for simplicity. Recall that a is odd. In like manner at p, we identify the germ $p \in X$ with $o \in \mathfrak{D}^3/\mathbf{Z}_r(2, a, -2)$ with orbifold coordinates y_1, y_2, y_3 such that $(y_1, y_2, y_3)|_C =$

$(t^{2/r}, t^{a/r}, 0)$ with respect to a uniformising parameter t in $\mathcal{O}_{C,p}$. Then for $\bar{\omega} = dy_1 \wedge dy_2 \wedge dy_3$ and $f_2 = y_1^a - y_2^2$, locally at p one has the descriptions

$$\mathrm{gr}^0 \, \omega = \mathcal{O}_C t^{(a-1)/r} \bar{\omega}, \qquad \mathrm{gr}^1 \, \mathcal{O} = \mathcal{O}_C t^{(r-1)/r} f_2 \oplus \mathcal{O}_C t^{2/r} y_3,$$

by which $\mathrm{gr}^2 \, \omega = \mathcal{O}_C t^{(a-3)/r} f_2^2 \bar{\omega} \oplus \mathcal{O}_C t^{a/r} f_2 y_3 \bar{\omega} \oplus \mathcal{O}_C t^{(a+3)/r} y_3^2 \bar{\omega}$. Even when $(r, a) = (5, 3)$, one cannot replace $t^{(a+3)/r} = t^{6/5}$ by $t^{1/5}$ as $1 \notin 2\mathbf{N} + a\mathbf{N}$.

Because $\mathrm{gr}^1 \, \mathcal{O} \simeq \mathcal{O}_{\mathbf{P}^1}(-1)^{\oplus 2}$, the splitting $\mathrm{gr}^1 \, \mathcal{O} \otimes k(p) = \mathbf{C} t^{(r-1)/r} f_2 \oplus \mathbf{C} t^{2/r} y_3$ for the residue field $k(p)$ at p is globalised to a splitting $\mathrm{gr}^1 \, \mathcal{O} = \mathcal{L} \oplus \mathcal{M}$ with $\mathcal{L} \simeq \mathcal{M} \simeq \mathcal{O}_{\mathbf{P}^1}(-1)$ such that $\mathcal{L}$ and $\mathcal{M}$ have at p local generators

$$t^{(r-1)/r} f_2 + ut^{(r+2)/r} y_3, \qquad vt^{(2r-1)/r} f_2 + t^{2/r} y_3$$

respectively with some local functions $u, v \in \mathcal{O}_{C,p}$. Since $t^{(a-1)/r}(t^{(r-1)/r} f_2 + ut^{(r+2)/r} y_3)^2 = t^2 \cdot t^{(a-3)/r}(f_2 + ut^{3/r} y_3)^2$ in $\mathrm{gr}^2 \, \omega$, the map $\mathrm{gr}^0 \, \omega \otimes \mathcal{L}^{\otimes 2} \to \mathrm{gr}^2 \, \omega \otimes \mathcal{O}_C(-q)$ induced by γ factors through the map

$$\mathrm{gr}^0 \, \omega \otimes \mathcal{L}^{\otimes 2} \to \mathrm{gr}^2 \, \omega \otimes \mathcal{O}_C(-2p - q),$$

which is non-zero. This contradicts the isomorphisms $\mathrm{gr}^0 \, \omega \otimes \mathcal{L}^{\otimes 2} \simeq \mathcal{O}_{\mathbf{P}^1}(-3)$ and $\mathrm{gr}^2 \, \omega \otimes \mathcal{O}_C(-2p - q) \simeq \mathcal{O}_{\mathbf{P}^1}(-4)^{\oplus 3}$. $\square$

We arrive at the establishment of Theorem 5.4.6.

Proof of Theorem 5.4.6   By Propositions 5.2.10 and 5.2.14, $C \subset X$ has at most three singular points. They are not of type IB or IC$^\vee$ by Lemma 5.4.9 and Corollary 5.4.15. If X is Gorenstein, then by Proposition 5.4.8, the set Σ belongs to the case (v) in the theorem. We shall assume that X has a singular point of type other than III.

If X has three singular points, then as in Lemma 5.4.14, Σ consists of two points of type IA, which are minimal by Lemma 5.4.16, and a point of type III. Thus Σ belongs to the case (iii). The case when X has a unique singular point is exhausted in the cases (i) and (ii). The remaining case is when X has exactly two singular points p and q, each of which is either minimal of type IA, IIA or IA$^\vee$ or of type III by Lemmata 5.4.12 and 5.4.16.

If p or q is of type III, then Σ belongs to the case (i). Suppose that both are minimal of type IA, IIA or IA$^\vee$. By Lemma 5.3.6(i) and Remark 5.3.2, the invariant $w_\theta(0)$ is at least $1/r$ if θ is of type IA and of index r, and is at least one-half if θ is of type IIA or IA$^\vee$. With $w_p(0) + w_q(0) < 1$ in Proposition 5.2.14, if p or q is of index two and thus of type IA, then the other is also of type IA and of index at least three, in which Σ belongs to the case (iii). If both are of index at least three, then Σ belongs to the case (iv). $\square$

5.5 General Elephants

We shall establish the general elephant conjecture for an extremal neighbourhood $C \subset X$, to which we have reduced Theorem 5.1.1. The approach by Mori with Kollár furnishes a classification of $C \subset X$ simultaneously.

Theorem 5.5.1 (Kollár–Mori [276, theorem 2.2]) *Let $\pi \colon C \simeq \mathbf{P}^1 \subset X \to y \in Y$ be an extremal neighbourhood. Then the general elephant S of X is Du Val. Further, one of the following holds for the set Σ of singular points of X and the dual graph Γ of the minimal resolution of S, in which $\circ$ stands for a curve exceptional over S and $\bullet$ for a curve not exceptional over S but exceptional over $y \in S' = \pi_* S$.*

(i) $\Sigma = \{p\}$ *or* $\{p, z\}$ *where p is of type IA, IIA or IA$^\vee$ and z is of type III. $\Sigma = \{p\}$ and $C \subset X$ is divisorial if $p \in X$ is cAx/2, cD/2 or cE/2. $C \not\subset S$ and $p \in S \simeq S'$ is as follows, where n is the axial multiplicity of $p \in X$ in Definition 2.5.20.*

$p \in X$	cA/r	cAx/4	cAx/2	cD/3	cD/2	cE/2
$p \in S$	A_{rn-1}	D_{n+2}	D_4	E_6	D_{2n}	E_7

(ii) $\Sigma = \{p\}$ *where p is of type IC. $C \subset X$ is isolated. $C \subset S$ and Γ is as below. $p \in X$ is a quotient singularity of odd index r at least five. $p \in S$ is of type A_{r-1}.*

(iii) $\Sigma = \{p\}$ *where p is of type IIB. $C \subset X$ is divisorial. $C \subset S$ and Γ is as below. $p \in S$ is of type D_5.*

(iv) $\Sigma = \{p\}$ *where p is of type II$^\vee$. $C \subset X$ is divisorial. $C \not\subset S$ and $p \in S \simeq S'$ is of type D_{n+2} for the axial multiplicity n of $p \in X$.*

(v) $\Sigma = \{p, q, z\}$ *where p and q are of type IA and z is of type III. $C \subset X$ is divisorial. $C \subset S$ and Γ is as below. $p \in X$ and $q \in X$ are quotient singularities of odd index r and of index two respectively. $p \in S$ is of type A_{r-1} and $q, z \in S$ are of type A_1.*

(vi) $\Sigma = \{p, q\}$ *where p and q are of type* IA. $C \subset S$ *and* Γ *is as below.*
$p \in X$ *is a quotient singularity of odd index r and* $q \in X$ *is* $cA/2$, $cAx/2$
or $cD/2$. $C \subset X$ *is divisorial if either* $r = 3$ *or* $q \in X$ *is* $cAx/2$ *or* $cD/2$.
$p \in S$ *is of type* A_{r-1} *and* $q \in S$ *is of type* D_{2n} *if* $n \geq 2$ *and of type* A_1
if $n = 1$ *for the axial multiplicity n of* $q \in X$. *When* $n = 1$, *S has three*
singular points p, q, z with z of type A_1.

(vii) $\Sigma = \{p_1, p_2\}$ *where* p_1 *and* p_2 *are of type* IA. $C \subset S$ *and* Γ *is as below.*
$p_i \in X$ *is* cA/r_i *with axial multiplicity* n_i. $p_i \in S$ *is of type* $A_{r_i n_i - 1}$.

(viii) $\Sigma = \emptyset$ *or* $\{z\}$ *where z is of type* III. $C \subset X$ *is divisorial.* $C \not\subset S$ *and*
$p \in S \simeq S'$ *is smooth.*

The theorem is proved in accordance with the classification of the set Σ
of singular points in Theorem 5.4.6. The five cases (i) to (v) below mean
those in Theorem 5.4.6. In the case (v), the statement is the analytic version
of Lemma 4.1.11 combined with the subsequent proof of Theorem 4.1.9. The
general elephant conjecture is easy in the case (i) and in the case (ii) with p
imprimitive. In the cases (ii) with p primitive and (iii), one can similarly show
that the double cover of X ramified along the general member of $|-2K_X|$ is
canonical, and as in Remark 5.1.14, this is sufficient for the existence of the
flip. On the other hand, the general elephant conjecture in these cases and (iv)
requires very delicate analysis of how C is embedded in X. We shall pick the
case $\Sigma = \{p\}$ with p of type $\text{II}^\vee$ as a representative in the easy case and then
demonstrate the general elephant conjecture in the hard case (iv).

Proof of Theorem 5.5.1 for $\Sigma = \{p\}$ *with p of type* $\text{II}^\vee$ First of all, we recall
the data $(a_1, a_2, a_3, a_4) = (2, 2, 2, 4)$ and $(w_1, w_2, w_3, w_4) = (1, 3, 3, 2)$ in Theorem 5.3.1. Let S be the effective divisor on the germ $p \in X$ defined by a
general semi-invariant function f in $\mathcal{O}_{\bar{X}}$ of weight three on the index-one cover
$\bar{X}$ of $p \in X$. Then $S \sim -K_X$ at p by Remark 5.2.9 and it is Du Val of type D_{n+2}
by Theorem 2.4.15 for the axial multiplicity n of $p \in X$. The local intersection
number $(S \cdot C)_p$ equals ord $f = o(3)/4 = 2/4$. We regard S as an effective
divisor about C with $(S \cdot C) = (S \cdot C)_p = 1/2$.

The divisor $K_X + S$ is Cartier about C. The formula in Proposition 5.2.14
provides $(K_X \cdot C) = -1 + w_p(0) = -1/2$ with $w_p(0) = 1/2$ in Lemma 5.3.6(i).
Hence $K_X + S$ has intersection number zero with C and $K_X + S \sim 0$ from

Lemma 5.1.4. In this construction, S is the general elephant of X and possesses the properties in (iv) in Theorem 5.5.1. It remains to prove that $C \subset X$ is divisorial.

Let $\psi: C' \subset X' \to C \subset X$ be the splitting cover of $C \subset X$ associated with p. It admits the structure $\pi': C' \subset X' \to y' \in Y'$ of a reducible extremal neighbourhood. The locus C' consists of two curves C_1' and C_2' isomorphic to $C \simeq \mathbf{P}^1$. Note that $(K_{X'} \cdot C_i') = (K_X \cdot C) = -1/2$. The covering X' has a unique singular point p' above p, which is of index two. Let H be an effective Cartier divisor on the germ X along C which intersects C transversally outside p with $(H \cdot C) = 1$. Let H' be the strict transform in X' of H. The Cartier divisor $2K_{X'} + H'$ is principal again from Lemma 5.1.4.

Using the section in $\Gamma(X', \mathscr{O}_{X'}(-2K_{X'}))$ defining H', we construct the double cover W of X' ramified along H' as in the proof of Proposition 5.1.13. This induces the double cover $\pi_W: D \subset W \to z \in Z$ of π' where D is the inverse image of C'. These fit into the diagram

$$
\begin{array}{ccccc}
D \subset W & \xrightarrow{\varphi} & C' \subset X' & \xrightarrow{\psi} & C \subset X \\
\pi_W \downarrow & & \pi' \downarrow & & \downarrow \pi \\
z \in Z & \xrightarrow{\varphi_Z} & y' \in Y' & \longrightarrow & y \in Y \ .
\end{array}
$$

It follows from the ramification formula that $K_W = \varphi^*(K_{X'} + H'/2)$. Further, W is the index-one cover on the germ $p' \in X'$ and it is smooth outside $\varphi^{-1}(p')$. Hence W is terminal of index one. One also has $K_Z = \varphi_Z^*(K_{Y'} + H_Y'/2)$ with the strict transform H_Y' of H', which satisfies $K_W = \pi_W^* K_Z$. By the construction of the double cover, the pull-back K_W of $K_{X'} + H'/2$ becomes principal and so is $K_Z = \pi_{W*}K_W$. Thus $z \in Z$ is canonical of index one. Since $\mathrm{mld}_z Z = \mathrm{mld}_D W = 2$, it is further cDV by Proposition 2.3.17.

The divisor H' consist of two prime divisors, say H_1' and H_2'. Let G_i be the inverse image in W of H_i', which is isomorphic to H_i', and let G_{iZ} be its strict transform in Z. The germ $y' \in Y'$ is the quotient by $\mathbf{Z}_2$ of the hypersurface singularity $z \in Z$. One can linearise this action as in Remark 2.2.5 so that

$$
y' \in Y' = Z/\mathbf{Z}_2 \subset \mathfrak{D}^4/\mathbf{Z}_2(v_1, v_2, v_3, v_4).
$$

In this description, the locus L of fixed points in $\mathfrak{D}^4$ is a linear subspace $\mathfrak{D}^l$ with some l. Plainly $L \cap Z$ has support $G_{1Z} + G_{2Z}$. Hence $l = 3$ and $L|_Z$ is a Cartier divisor of form $a_1 G_{1Z} + a_2 G_{2Z}$ with positive integers a_1 and a_2. One can consider the intersection number $(\pi_W^*(a_1 G_{1Z} + a_2 G_{2Z}) \cdot D) = 0$. Since $(G_i \cdot D)$ is positive, the divisor $\pi_W^*(a_1 G_{1Z} + a_2 G_{2Z})$ contains some extra divisor besides $a_1 G_1 + a_2 G_2$. Thus $C \subset X$ must be divisorial. □

In the rest of the section, we shall discuss the case (iv) in Theorem 5.4.6 in the following framework.

Framework 5.5.2   Let $\pi\colon C \subset X \to y \in Y$ be an extremal neighbourhood with exactly two singular points p_1 and p_2 such that each p_i is minimal of type IA, IIA or IA$^\vee$ and of index r_i at least three. The germ $p_i \in C \subset X$ possesses the data $(a_{i1}, a_{i2}, a_{i3}, a_{i4})$ from its normal form as in Theorem 5.3.1 which satisfies ord $C = \mathbf{N}a_{i1} + \mathbf{N}a_{i2}$ and $a_{i1} + a_{i3} \equiv 0 \bmod r_i$. We let $\mu\colon \bar{p}_i \in \bar{C} \subset \bar{X} \to p_i \in C \subset X$ denote the index-one cover, where we omit the index i for μ, $\bar{C}$ and $\bar{X}$ to avoid intricacy. We write $\mathscr{I}$ for the ideal in $\mathscr{O}_X$ defining C and write $\bar{\mathscr{I}}$ for the ideal in $\mathscr{O}_{\bar{X}}$ defining $\bar{C}$.

For a while we treat the germ at a quotient singularity achieved by an L-deformation at p_i.

Framework 5.5.3   Let $p \in C \subset X$ be the germ of an extremal neighbourhood at a quotient singularity of type IA or IA$^\vee$. Suppose that the index r of $p \in X$ is at least three and that $p \in C \subset X$ is minimal or equivalently $i_p(1) = 1$ from Lemma 5.3.6(ii). Let $\mu\colon \bar{p} \in \bar{C} \subset \bar{X} \to p \in C \subset X$ be the index-one cover. By Theorem 5.3.1, there exist orbifold coordinates x_1, x_2, x_3 with wt $x_1 x_3 = 0$ such that $(x_1, x_2, x_3)|_{\bar{C}} = (t^{a_1/r}, t^{a_2/r}, 0)$ with $a_i = o(\text{wt}\, x_i)$ for a uniformising parameter t in $\mathscr{O}_{C,p}$. We write $\mathscr{I}$ and $\bar{\mathscr{I}}$ for the ideals in $\mathscr{O}_X$ and $\mathscr{O}_{\bar{X}}$ defining C and $\bar{C}$ respectively.

In Framework 5.5.3, one has the description $\bar{p} \in \bar{C} = (f_1 = f_2 = 0) \subset \bar{X} = \mathfrak{D}^3$ with $f_1 = x_3$ and $f_2 = x_1^{a_2} - x_2^{a_1}$. By the minimality $U(a_2\, \text{ow}\, x_1) = 1$, there exists a monomial u in x_1, x_2 such that ow $u x_2^{a_2} = $ ow $u x_2^{a_1} = (r, 0)$. Writing $\text{gr}^1\, \bar{\mathscr{O}}$ for $\text{gr}^1_{\bar{C}}\, \mathscr{O}_{\bar{X}}$, one has the local expressions

$$\text{gr}^1\, \bar{\mathscr{O}} = \mathscr{O}_{\bar{C}} f_1 \oplus \mathscr{O}_{\bar{C}} f_2, \qquad \text{gr}^1\, \mathscr{O} = \mathscr{O}_C x_1 f_1 \oplus \mathscr{O}_C u f_2.$$

The $\mathscr{O}_C$-module $\text{gr}^1\, \mathscr{O}$ is the invariant part of $\mu_* \text{gr}^1\, \bar{\mathscr{O}}$ by the action of $\mathbf{Z}_r$.

Lemma 5.5.4   *In the above setting, either $a_2 = 1$ or x_1 is a factor of u. In the latter case, the inequality $a_1 \le r - a_1 a_2$ holds.*

Proof   If u is a power of x_2, then $a_2 = \text{ord}\, x_2 = 1$ by ow $u x_2^{a_1} = (r, 0)$. If x_1 is a factor of u, then $a_1 = \text{ord}\, x_1 \le \text{ord}\, u = r - a_1 a_2$. $\square$

Lemma 5.5.5   *Working locally in Framework 5.5.3, let $\mathscr{L}$ be an invertible direct summand of the $\mathscr{O}_C$-module $\text{gr}^1\, \mathscr{O}$ on the germ $p \in C \subset X$. Let $\bar{\mathscr{L}}$ denote the saturation in $\text{gr}^1\, \bar{\mathscr{O}}$ of the submodule $\mathscr{O}_{\bar{C}} \cdot \mathscr{L}$. Then $\bar{\mathscr{L}}$ is an invertible direct summand of $\text{gr}^1\, \bar{\mathscr{O}}$ generated by a semi-invariant element and $\mathscr{L}$ is the invariant part of $\mu_* \bar{\mathscr{L}}$. Consider the vector space $\text{gr}^1\, \mathscr{O} \otimes k(p) \simeq \mathbf{C}^2$ for the* *residue field $k(p)$ at p. Then it has a vector subspace $W \simeq \mathbf{C}$ giving a splitting* $\mathrm{gr}^1\, \mathcal{O} \otimes k(p) = \mathcal{L} \otimes k(p) \oplus W$ *such that if an invertible direct summand* $\mathcal{M}$ *of* $\mathrm{gr}^1\, \mathcal{O}$ *satisfies* $\mathcal{M} \otimes k(p) = W$, *then the saturation* $\bar{\mathcal{M}}$ *in* $\mathrm{gr}^1\, \bar{\mathcal{O}}$ *of the submodule* $\mathcal{O}_{\bar{C}} \cdot \mathcal{M}$ *gives a splitting* $\mathrm{gr}^1\, \bar{\mathcal{O}} = \bar{\mathcal{L}} \oplus \bar{\mathcal{M}}$.

Proof We divide the proof in accordance with Lemma 5.5.4. First suppose that $a_2 = 1$. Then $\mathrm{gr}^1\, \mathcal{O} = \mathcal{O}_C x_2^{a_1} f_1 \oplus \mathcal{O}_C x_2^{r-a_1} f_2$ and hence $\mathcal{L}$ has a generator of form either $x_2^{a_1}(f_1 + g_2 f_2)$ or $x_2^{r-a_1}(g_1 f_1 + f_2)$ with a semi-invariant function g_i in $\mathbf{C}\{x_2\}$. If $\mathcal{L} = \mathcal{O}_C x_2^{a_1}(f_1 + g_2 f_2)$, then $\bar{\mathcal{L}} = \mathcal{O}_{\bar{C}} \cdot (f_1 + g_2 f_2)$ in $\mathrm{gr}^1\, \bar{\mathcal{O}}$ and it has the required properties. Take $W = \mathbf{C} x_2^{r-a_1} f_2$. If $\mathcal{M}$ satisfies $\mathcal{M} \otimes k(p) = W$, then $\mathcal{M}$ has a generator of form $h x_2^r \cdot x_2^{a_1} f_1 + x_2^{r-a_1} f_2$ with invariant h. By this description, $\bar{\mathcal{M}} = \mathcal{O}_{\bar{C}} \cdot (h x_2^{2a_1} f_1 + f_2)$ and it gives a splitting $\mathrm{gr}^1\, \bar{\mathcal{O}} = \bar{\mathcal{L}} \oplus \bar{\mathcal{M}}$. Similarly if $\mathcal{L} = \mathcal{O}_C x_2^{r-a_1}(g_1 f_1 + f_2)$, then $\bar{\mathcal{L}} = \mathcal{O}_{\bar{C}} \cdot (g_1 f_1 + f_2)$ and we take $W = \mathbf{C} x_2^{a_1} f_1$.

Next suppose that x_1 is a factor of u. Write $u = x_1 v$ with a monomial v in x_1, x_2. Then $\mathrm{gr}^1\, \mathcal{O} = x_1(\mathcal{O}_C f_1 \oplus \mathcal{O}_C v f_2)$. A generator of $\mathcal{L}$ is of form either $x_1(f_1 + g v f_2)$ with an invariant function g or $x_1(h f_1 + v f_2)$ with an invariant function h which is not a unit. In the former case, $\bar{\mathcal{L}} = \mathcal{O}_{\bar{C}} \cdot (f_1 + g v f_2)$ and we take $W = \mathbf{C} x_1 v f_2$. If $\mathcal{M}$ satisfies $\mathcal{M} \otimes k(p) = W$, then $\mathcal{M}$ has a generator of form $q x_1 f_1 + x_1 v f_2$ with invariant q which is not a unit. Since $t = u x_1^{a_2}|_C$ divides $q|_C$, one can replace q by $u x_1^{a_2} q'$ for some invariant q'. Then $\mathcal{M} = \mathcal{O}_C u(x_1^{a_2+1} q' f_1 + f_2)$. By this description, $\bar{\mathcal{M}} = \mathcal{O}_{\bar{C}} \cdot (x_1^{a_2+1} q' f_1 + f_2)$ and $\mathrm{gr}^1\, \bar{\mathcal{O}} = \bar{\mathcal{L}} \oplus \bar{\mathcal{M}}$. Similarly in the latter case, we replace h by $u x_1^{a_2} h'$ to obtain $\bar{\mathcal{L}} = \mathcal{O}_{\bar{C}} \cdot (x_1^{a_2+1} h' f_1 + f_2)$ and take $W = \mathbf{C} x_1 f_1$. $\qquad\square$

Let $\mathcal{L}$ be an invertible direct summand of $\mathrm{gr}^1\, \mathcal{O}$ at p. Then Lemma 5.5.5 constructs a splitting

$$\mathrm{gr}^1\, \bar{\mathcal{O}} = \bar{\mathcal{L}} \oplus \bar{\mathcal{M}}$$

on the index-one cover such that $\mathcal{L} = (\mu_* \bar{\mathcal{L}})^{\mathbf{Z}_r}$ and such that $\bar{\mathcal{L}} = \mathcal{O}_{\bar{C}} u$ and $\bar{\mathcal{M}} = \mathcal{O}_{\bar{C}} v$ with semi-invariant generators u, v of the ideal $\bar{\mathcal{I}}$. The invariant part $\mathcal{M} = (\mu_* \bar{\mathcal{M}})^{\mathbf{Z}_r}$ provides a local splitting $\mathrm{gr}^1\, \mathcal{O} = \mathcal{L} \oplus \mathcal{M}$ at p. Let

$$b_1 = o(-\mathrm{wt}\, u), \qquad b_2 = o(-\mathrm{wt}\, v).$$

Lemma 5.5.6 *The orders b_1 and b_2 are a_1 and $r - a_1 a_2$ up to permutation.*

Proof This follows from the property that u and v have the same weights up to permutation as $f_1 = x_3$ and $f_2 = x_1^{a_2} - x_2^{a_1}$ have. $\qquad\square$

We take the ideal $\mathcal{J}$ in $\mathcal{O}_X$ on the germ $p \in C \subset X$ such that $\mathcal{I}^{(2)} \subset \mathcal{J} \subset \mathcal{I}$ and such that $\mathcal{L} = \mathcal{J}/\mathcal{I}^{(2)}$ in $\mathrm{gr}^1\, \mathcal{O} = \mathcal{I}/\mathcal{I}^{(2)}$. Consider the decreasing sequence of ideals $\mathcal{I}^{\bar{n}} \mathcal{J}^{\lfloor n/2 \rfloor}$ where $\bar{n} = n - 2\lfloor n/2 \rfloor$, that is,

$$\mathcal{O}_X \supset \mathcal{I} \supset \mathcal{J} \supset \mathcal{I}\mathcal{J} \supset \mathcal{J}^2 \supset \mathcal{I}\mathcal{J}^2 \supset \cdots.$$

Take the saturation $\mathscr{F}_n$ of $\mathscr{I}^{\tilde{n}}\,\mathscr{J}^{\lfloor n/2 \rfloor}$ in $\mathcal{O}_X$. These $\mathscr{F}_n$ also form a decreasing sequence.

The saturation $\bar{\mathscr{J}}$ in $\mathcal{O}_{\bar{X}}$ of the ideal $\mathscr{J} \cdot \mathcal{O}_{\bar{X}}$ is expressed as $\bar{\mathscr{J}} = (u, v^2)\mathcal{O}_{\bar{X}}$ where $\bar{\mathscr{L}} = \mathcal{O}_{\bar{C}}u$ and $\bar{\mathscr{M}} = \mathcal{O}_{\bar{C}}v$. Since u and v form a regular sequence in $\mathcal{O}_{\bar{X},\bar{p}}$, the ideal $\bar{\mathscr{F}}_n = \bar{\mathscr{J}}^{\tilde{n}}\,\bar{\mathscr{J}}^{\lfloor n/2 \rfloor}$ is saturated in $\mathcal{O}_{\bar{X}}$ and $\bar{\mathscr{F}}_n/\bar{\mathscr{F}}_{n+1} = \bigoplus_{i=0}^{\lfloor n/2 \rfloor} \mathcal{O}_{\bar{C}}u^i v^{n-2i}$ as derived from [61, theorem 1.1.8]. This yields a natural isomorphism $\bar{\mathscr{F}}_n/\bar{\mathscr{F}}_{n+1} \simeq \bigoplus_{i=0}^{\lfloor n/2 \rfloor} \bar{\mathscr{L}}^{\otimes i} \otimes \bar{\mathscr{M}}^{\otimes(n-2i)}$. The saturation $\mathscr{F}_n$ equals the invariant part $(\mu_* \bar{\mathscr{F}}_n)^{\mathbf{Z}_r}$ and $\mathscr{F}_n/\mathscr{F}_{n+1} = (\mu_*(\bar{\mathscr{F}}_n/\bar{\mathscr{F}}_{n+1}))^{\mathbf{Z}_r}$. The above isomorphism induces a natural injection

$$i_n : \bigoplus_{i=0}^{\lfloor n/2 \rfloor} \mathscr{L}^{\otimes i} \otimes \mathscr{M}^{\otimes(n-2i)} \hookrightarrow \mathscr{F}_n/\mathscr{F}_{n+1}.$$

Lemma 5.5.7   *The injection i_n has cokernel of length at most $(8r)^{-1}(b_1 + 2b_2)n^2 + O(n)$, where O is Landau's symbol.*

Proof   The component $\mathscr{L}^{\otimes i} \otimes \mathscr{M}^{\otimes(n-2i)} \hookrightarrow (\mu_*(\bar{\mathscr{L}}^{\otimes i} \otimes \bar{\mathscr{M}}^{\otimes(n-2i)}))^{\mathbf{Z}_r}$ of i_n has cokernel of length $r^{-1}(ib_1 + (n-2i)b_2 - o(-\operatorname{wt} u^i v^{n-2i}))$. Hence the length of the cokernel of i_n is at most

$$\sum_{i=0}^{\lfloor n/2 \rfloor} \frac{1}{r}(ib_1 + (n-2i)b_2) = \frac{b_1 + 2b_2}{8r}n^2 + O(n). \qquad \square$$

We apply the argument above to Framework 5.5.2. By L-deformations, one can produce a new extremal neighbourhood $C \subset X$ on which p_1 and p_2 are quotient singularities. By Remark 5.4.3, this process retains the structure of ow C and hence r_i and a_{ij}. Each quotient singularity p_i is minimal of type IA or IA$^{\vee}$. Thus the germ $p_i \in C \subset X$ is as in Framework 5.5.3 and $i_{p_i}(1) = 1$. There exist orbifold coordinates x_{i1}, x_{i2}, x_{i3} at p_i with $\operatorname{wt} x_{i1}x_{i3} = 0$ such that $(x_{i1}, x_{i2}, x_{i3})|_{\bar{C}} = (t^{a_{i1}/r_i}, t^{a_{i2}/r_i}, 0)$ for $a_{ij} = o_{p_i}(\operatorname{wt} x_{ij})$.

From Proposition 5.2.14 with $i_{p_i}(1) = 1$, one has $\deg \operatorname{gr}^1 \mathcal{O} = -1$ and $H^1(\operatorname{gr}^1 \mathcal{O}) = 0$. Hence $\operatorname{gr}^1 \mathcal{O} \simeq \mathcal{O}_{\mathbf{P}^1} \oplus \mathcal{O}_{\mathbf{P}^1}(-1)$ by Theorem 1.3.5. Then a non-zero section in $H^0(\operatorname{gr}^1 \mathcal{O}) \simeq \mathbf{C}$ defines a unique submodule $\mathscr{L} \subset \operatorname{gr}^1 \mathcal{O}$ such that $\mathscr{L} \simeq \mathcal{O}_{\mathbf{P}^1}$. Consider a splitting $\operatorname{gr}^1 \mathcal{O} = \mathscr{L} \oplus \mathscr{M}$ in which $\mathscr{M} \simeq \mathcal{O}_{\mathbf{P}^1}(-1)$. The choice of $\mathscr{M}$ is determined by $\mathcal{H}om(\mathscr{M}, \mathscr{L}) \simeq \mathcal{O}_{\mathbf{P}^1}(1)$, and the evaluation map $H^0(\mathcal{H}om(\mathscr{M}, \mathscr{L})) \to \mathcal{H}om(\mathscr{M}, \mathscr{L}) \otimes (k(p_1) \oplus k(p_2))$ is surjective for the residue field $k(p_i)$ at p_i. Hence by Lemma 5.5.5, one can take $\mathscr{M}$ so that for each i, the splitting is locally at p_i induced by some splitting $\operatorname{gr}^1 \bar{\mathcal{O}} = \bar{\mathscr{L}} \oplus \bar{\mathscr{M}}$ on the index-one cover $\bar{p}_i \in \bar{C} \subset \bar{X}$.

Definition 5.5.8 The splitting $\mathrm{gr}^1 \mathcal{O} = \mathcal{L} \oplus \mathcal{M}$ by invertible submodules $\mathcal{L}$ and $\mathcal{M}$ is called an *l-splitting* if locally at each point p, it is induced by a splitting of $\mathrm{gr}^1 \bar{\mathcal{O}}$ on the index-one cover $\bar{p} \in \bar{C} \subset \bar{X}$ of $p \in C \subset X$.

For an *l*-splitting $\mathrm{gr}^1 \mathcal{O} = \mathcal{L} \oplus \mathcal{M}$, we write $\mathcal{L} = \mathcal{O}_{\bar{C}} u_i$ and $\mathcal{M} = \mathcal{O}_{\bar{C}} v_i$ in $\mathrm{gr}^1 \bar{\mathcal{O}}$ at $\bar{p}_i$ by semi-invariant generators u_i, v_i of $\bar{\mathscr{I}}$. Let $b_{i1} = o_{p_i}(-\mathrm{wt}\, u_i)$ and $b_{i2} = o_{p_i}(-\mathrm{wt}\, v_i)$. By Lemma 5.5.4, b_{i1} and b_{i2} are a_{i1} and $r_i - a_{i1} a_{i2}$ up to permutation.

Proposition 5.5.9 *In Framework 5.5.2 if p_1 and p_2 are quotient singularities, then for each i, $a_{i1} = b_{i1} = 1$ after necessary permutation. In particular, p_i is of type IA.*

The proof consists of two parts. First we prove that p_i is primitive and next deduce that $a_{i1} = b_{i1} = 1$. We shall only demonstrate the former. The latter requires a lengthy discussion on the contrary assumption which will turn out to be false and the techniques there can be learnt in other places in this section.

Proof of primitivity in Proposition 5.5.9 We shall prove the inequality

$$2 \le \frac{b_{11} + 2b_{12}}{r_1} + \frac{b_{21} + 2b_{22}}{r_2}$$

by globalising the result in Lemma 5.5.7. The ideal $\mathscr{J}$ and the saturation $\mathscr{F}_n$ of $\mathscr{I}^{\bar{n}} \mathscr{J}^{\lfloor n/2 \rfloor}$ are defined about C by the equality $\mathcal{L} = \mathscr{J}/\mathscr{J}^{(2)}$. Hence the injection $i_n \colon \bigoplus_{i=0}^{\lfloor n/2 \rfloor} \mathcal{L}^{\otimes i} \otimes \mathcal{M}^{\otimes(n-2i)} \hookrightarrow \mathscr{F}_n/\mathscr{F}_{n+1}$ is defined on the whole C. Note that $\mathcal{L}^{\otimes i} \otimes \mathcal{M}^{\otimes(n-2i)} \simeq \mathcal{O}_{\mathbf{P}^1}(2i - n)$. Thus Lemma 5.5.7 yields the evaluation

$$\chi(\mathscr{F}_n/\mathscr{F}_{n+1}) \le \sum_{i=0}^{\lfloor n/2 \rfloor} \chi(\mathcal{O}_{\mathbf{P}^1}(2i - n)) + \sum_{i=1,2} \frac{b_{i1} + 2b_{i2}}{8r_i} n^2 + O(n)$$

$$= \frac{1}{8}\left(\sum_{i=1,2} \frac{b_{i1} + 2b_{i2}}{r_i} - 2\right) n^2 + O(n)$$

and hence

$$\chi(\mathcal{O}_X/\mathscr{F}_n) = \sum_{l=0}^{n-1} \chi(\mathscr{F}_l/\mathscr{F}_{l+1}) \le \frac{1}{24}\left(\sum_{i=1,2} \frac{b_{i1} + 2b_{i2}}{r_i} - 2\right) n^3 + O(n^2).$$

The left-hand side is positive by the vanishing of $H^1(\mathcal{O}_X/\mathscr{F}_n)$ from $R^1\pi_*\mathcal{O}_X = 0$. This shows the desired inequality.

If one of p_i, say p_2, were imprimitive, then $w_{p_2}(0) \ge 1/2$ by Remark 5.3.2 and hence the other p_1 has $w_{p_1}(0) < 1 - w_{p_2}(0) \le 1/2$ by Proposition 5.2.14.

Hence one obtains $a_{11} = 1$ from Lemma 5.4.10(i) with $i_{p_1}(1) = 1$. Thus it suffices to prove that a_{21} or a_{22} equals one on the assumption $a_{11} = 1$.

From now on, we assume that $a_{11} = 1$. By Lemma 5.5.6 for the point p_1, the orders b_{11} and b_{12} are 1 and $r_1 - a_{12}$ up to permutation. Hence

$$\frac{b_{11} + 2b_{12}}{r_1} \le \frac{3(r_1 - a_{12})}{r_1} < \frac{3a_{22}}{r_2},$$

where the latter inequality follows from $w_{p_1}(0) + w_{p_2}(0) < 1$ in Proposition 5.2.14 together with $w_{p_i}(0) = (r_i - a_{i2})/r_i$ in Lemma 5.3.6(i). By Lemma 5.5.6 for p_2, the orders b_{21} and b_{22} are a_{21} and $r_2 - a_{21}a_{22}$ up to permutation. We are done if $a_{22} = 1$. Hence by Lemma 5.5.4, we may assume that $a_{21} \le r_2 - a_{21}a_{22}$. Then

$$\frac{b_{21} + 2b_{22}}{r_2} \le \frac{a_{21} + 2(r_2 - a_{21}a_{22})}{r_2} = 2 + \frac{a_{21} - 2a_{21}a_{22}}{r_2}.$$

Applying these two estimates to the inequality at the beginning, one obtains $3a_{22} + (a_{21} - 2a_{21}a_{22}) > 0$, that is, $(2a_{21} - 3)(2a_{22} - 1) < 3$. This holds only if a_{21} or a_{22} equals one. $\qquad\square$

We return to the general situation in Framework 5.5.2 where p_i are not necessarily quotient singularities. Whereas p_i is a priori of type IA, IIA or IA$^\vee$, we shall prove that it is actually of type IA.

Proposition 5.5.10 *In Framework 5.5.2 for each i, p_i is of type IA, $a_{i1} = 1$ after necessary permutation and $i_{p_i}(1) = 1$.*

Proof By L-deformations, one obtains a new extremal neighbourhood on which each p_i is a quotient singularity. By Proposition 5.5.9, the point p_i on the new neighbourhood has the invariant $a_{i1} = 1$ after necessary permutation. Because our L-deformations preserve the structure of ow C by Remark 5.4.3, the original germ $p_i \in C \subset X$ also has the invariant $a_{i1} = 1$. In particular, it is of type IA or IIA.

Since $a_{i1} = 1$, there exists an expression $p_i \in C = (x_1\text{-axis}) \subset X = (x_1^a x_2 + f = 0) \subset A = \mathfrak{D}^4/\mathbf{Z}_{r_i}(w_1, w_2, w_3, w_4)$ with $f \in (x_2, x_3, x_4)^2 \mathcal{O}_{\mathfrak{D}^4}$ as in Lemma 5.2.11 in which $i_{p_i}(1) = \lfloor a/r_i \rfloor + 1$. If a were at least r_i, then the deformation $C_z = C \subset X_z = (x_1^{a-r_i}(x_1^{r_i} + z)x_2 + f = 0) \subset A$ followed by its globalisation would produce an extremal neighbourhood which has three singular points of indices r_1, r_2 and 1. This contradicts Lemma 5.4.14 since r_1 and r_2 are at least three. Hence $a < r_i$, that is, $i_{p_i}(1) = 1$.

It remains to exclude the case of type IIA. We shall derive a contradiction assuming that p_1 is of type IIA.

Step 1 By an *L*-deformation at the other point p_2, we can assume that p_2 is a quotient singularity of type IA with $a_{21} = 1$. Since $a_{11} = 1$, the germ at p_1 admits orbifold coordinates x_1, x_2, x_3, x_4 such that

$$p_1 \in C = (x_1\text{-axis}) \subset X = (\xi = 0) \subset A = \mathfrak{D}^4/\mathbf{Z}_4(1,1,3,2)$$

with a semi-invariant function ξ of weight two as in Theorem 2.4.8. Apply the deformation $C_z = C \subset X_z = (\xi + z(x_1x_2 + x_3^2 + x_4^3) = 0) \subset A$ and globalise it like an *L*-deformation. After replacing $C \subset X$ by general $C_z \subset X_z$, the quadratic part ξ_2 of ξ becomes of form $(ax_1 + bx_2)x_2 + cx_3^2$ with $a, c \in \mathbf{C}^\times$ and $b \in \mathbf{C}$. Hence one can change orbifold coordinates so that $\xi_2 = x_1x_2 + x_3^2$. Applying the Weierstrass preparation theorem in Theorem 2.1.10 together with Remark 2.1.11 to ξ with respect to $x_1 \pm \sqrt{-1}x_2$ and then to x_3, one obtains an expression $X = (x_1x_2 + x_3^2 + q(x_4) = 0)$ in which C is contained in the locus $X \cap (x_3 = x_4 = 0)$. Thus one can take $\xi = x_1x_2 + x_3^2 + q(x_4)$ in the above expression and can further take $q = x_4^3$ since the new ξ contains the monomial x_4^3. One eventually attains the expression with $\xi = x_1x_2 + x_3^2 + x_4^3$.

Step 2 Recall that $i_{p_1}(1) = i_{p_2}(1) = 1$. As in the discussion about Definition 5.5.8, one has $\mathrm{gr}^1\,\mathscr{O} \simeq \mathscr{O}_{\mathbf{P}^1} \oplus \mathscr{O}_{\mathbf{P}^1}(-1)$ and there exists a unique submodule $\mathscr{L} \subset \mathrm{gr}^1\,\mathscr{O}$ isomorphic to $\mathscr{O}_{\mathbf{P}^1}$. Moreover, there exists an *l*-splitting $\mathrm{gr}^1\,\mathscr{O} = \mathscr{L} \oplus \mathscr{M}$ by a submodule $\mathscr{M} \simeq \mathscr{O}_{\mathbf{P}^1}(-1)$. We write $\mathrm{gr}^1\,\bar{\mathscr{O}} = \bar{\mathscr{L}} \oplus \bar{\mathscr{M}}$ for the corresponding splitting on the index-one cover $\bar{p}_i \in \bar{C} \subset \bar{X}$.

Recall the description

$$p_1 \in C = (x_1\text{-axis}) \subset X = (x_1x_2 + x_3^2 + x_4^3 = 0) \subset \mathfrak{D}^4/\mathbf{Z}_4(1,1,3,2).$$

At p_1, $\mathrm{gr}^1\,\bar{\mathscr{O}} = \bar{\mathscr{I}}/\bar{\mathscr{I}}^{(2)} = \mathscr{O}_{\bar{C}}x_3 \oplus \mathscr{O}_{\bar{C}}x_4$ and $\mathrm{gr}^1\,\mathscr{O} = \mathscr{O}_C x_1 x_3 \oplus \mathscr{O}_C x_1^2 x_4$. If $\mathscr{L}$ is locally generated by $gx_1^5 x_3 + x_1^2 x_4$ with invariant g, then there exists a deformation of $\mathscr{L}$ on the trivial deformation of the germ $p_1 \in C \subset X$ such that the fibre at $z \in \mathfrak{D}^1$ is $\mathscr{L}_z = \mathscr{O}_C \cdot x_1(gx_1^4 x_3 + x_1 x_4 + zx_3)$, and one can globalise this twisting of $\mathscr{L}$ like an *L*-deformation. Refer to [335, proposition 1b.8.3] for details. Replacing $\mathscr{L}$ by $\mathscr{L}_z$, we may set $\bar{\mathscr{L}} = \mathscr{O}_{\bar{C}}u$ and $\bar{\mathscr{M}} = \mathscr{O}_{\bar{C}}v$ in $\mathrm{gr}^1\,\bar{\mathscr{O}}$ with semi-invariant u and v with $\mathrm{wt}\,u = \mathrm{wt}\,x_3 = 3$ and $\mathrm{wt}\,v = \mathrm{wt}\,x_4 = 2$ in which u is expressed as $x_3 + hx_1 x_4$ up to unit. Replacing x_2 by $x_2 - 2hx_3 x_4 - h^2 x_1 x_4^2$ and x_3 by $x_3 + hx_1 x_4$, we may and shall assume that $u = x_3$, that is, $\bar{\mathscr{L}} = \mathscr{O}_{\bar{C}}x_3$.

On the other hand, the germ at p_2 admits orbifold coordinates y_1, y_2, y_3 with $\mathrm{wt}(y_1, y_2, y_3) = (1, a_{22}, -1)$ such that C is the y_1-axis. Then $\mathrm{gr}^1\,\bar{\mathscr{O}} = \mathscr{O}_{\bar{C}}y_2 \oplus \mathscr{O}_{\bar{C}}y_3$ at $\bar{p}_2$. For the same reason as above, we may and shall assume that $\bar{\mathscr{L}} = \mathscr{O}_{\bar{C}}y_3$ and $\bar{\mathscr{M}} = \mathscr{O}_{\bar{C}}y_2$.

Step 3 We define the ideal $\mathscr{J}$ in $\mathscr{O}_X$ containing $\mathscr{I}^{(2)}$ by $\mathscr{L} = \mathscr{J}/\mathscr{I}^{(2)}$ in $\mathrm{gr}^1\,\mathscr{O} = \mathscr{I}/\mathscr{I}^{(2)}$. Take the saturation $\omega_X = \mathscr{G}_0 \supset \cdots \supset \mathscr{G}_4$ in ω_X of the filtration $\omega_X \supset \mathscr{I}\omega_X \supset \mathscr{J}\omega_X \supset \mathscr{I}\mathscr{J}\omega_X \supset \mathscr{J}^2\omega_X$ in the same manner as for $\mathscr{F}_n$ before Lemma 5.5.7. Then there exist natural injections

$$j_1: \mathscr{M} \otimes \mathrm{gr}^0 \, \omega \hookrightarrow \mathscr{G}_1/\mathscr{G}_2,$$
$$j_2: (\mathscr{L} \oplus \mathscr{M}^{\otimes 2}) \otimes \mathrm{gr}^0 \, \omega \hookrightarrow \mathscr{G}_2/\mathscr{G}_3,$$
$$j_3: (\mathscr{L} \otimes \mathscr{M} \oplus \mathscr{M}^{\otimes 3}) \otimes \mathrm{gr}^0 \, \omega \hookrightarrow \mathscr{G}_3/\mathscr{G}_4,$$

which are isomorphic outside p_1 and p_2. We shall compute the length l_{ij} at p_i of the cokernel of j_j.

We write $\mathrm{gr}^0 \, \bar{\omega}$ for $\mathrm{gr}^0_{\bar{C}} \, \omega_{\bar{X}}$ for simplicity. At p_2, which is a quotient singularity, $\mathrm{gr}^0 \, \bar{\omega}$ is equivariantly isomorphic to $\mathscr{M} = \mathcal{O}_{\bar{C}} y_2$ and we compute l_{2j} as in the proof of Lemma 5.5.7 as

$$l_{21} = \frac{2o_{p_2}(-\mathrm{wt}\, y_2) - o_{p_2}(-2\,\mathrm{wt}\, y_2)}{r_2} = \left\lfloor \frac{2(r_2 - a_{22})}{r_2} \right\rfloor = 0,$$

$$l_{22} = \left\lfloor \frac{1 + (r_2 - a_{22})}{r_2} \right\rfloor + \left\lfloor \frac{3(r_2 - a_{22})}{r_2} \right\rfloor = 0,$$

$$l_{23} = \left\lfloor \frac{1 + 2(r_2 - a_{22})}{r_2} \right\rfloor + \left\lfloor \frac{4(r_2 - a_{22})}{r_2} \right\rfloor = 0.$$

Here we used the estimate $w_{p_2}(0) = (r_2 - a_{22})/r_2 < 1 - w_{p_1}(0) = 1/4$ from Proposition 5.2.14 with Lemma 5.3.6(i).

At p_1, let $\bar{\mathscr{J}}$ be the saturation in $\mathcal{O}_{\bar{X}}$ of the ideal $\mathscr{J} \cdot \mathcal{O}_{\bar{X}}$ on $\bar{p}_1 \in \bar{C} \subset \bar{X}$ and hence $\bar{\mathscr{L}} = \bar{\mathscr{J}}/\bar{\mathscr{J}}^{(2)}$. Recall that $\bar{X} = (x_1 x_2 + x_3^2 + x_4^3 = 0)$, $\bar{\mathscr{L}} = \mathcal{O}_{\bar{C}} x_3$ and $\bar{\mathscr{M}} = \mathcal{O}_{\bar{C}} v$. Hence the saturation $\omega_{\bar{X}} = \bar{\mathscr{G}}_0 \supset \cdots \supset \bar{\mathscr{G}}_4$ of the filtration $\omega_{\bar{X}} \supset \bar{\mathscr{I}}\omega_{\bar{X}} \supset \bar{\mathscr{J}}\omega_{\bar{X}} \supset \bar{\mathscr{I}}\bar{\mathscr{J}}\omega_{\bar{X}} \supset \bar{\mathscr{J}}^2\omega_{\bar{X}}$ is expressed as

$$\bar{\mathscr{G}}_1 = (x_2, x_3, v)\omega_{\bar{X}}, \qquad \bar{\mathscr{G}}_2 = (x_2, x_3, v^2)\omega_{\bar{X}},$$
$$\bar{\mathscr{G}}_3 = (x_2, x_3^2, x_3 v, v^3)\omega_{\bar{X}}, \qquad \bar{\mathscr{G}}_4 = (x_2^2, x_2 x_3, x_2 v, x_3^2, x_3 v^2, v^4)\omega_{\bar{X}}.$$

It follows that $\bar{\mathscr{G}}_1/\bar{\mathscr{G}}_2 = \mathcal{O}_{\bar{C}} v\bar{\omega}$, $\bar{\mathscr{G}}_2/\bar{\mathscr{G}}_3 = \mathcal{O}_{\bar{C}} x_3\bar{\omega} \oplus \mathcal{O}_{\bar{C}} v^2\bar{\omega}$ and $\bar{\mathscr{G}}_3/\bar{\mathscr{G}}_4 = \mathcal{O}_{\bar{C}} x_3 v\bar{\omega} \oplus \mathcal{O}_{\bar{C}} x_2\bar{\omega} = \mathcal{O}_{\bar{C}} x_3 v\bar{\omega} \oplus \mathcal{O}_{\bar{C}} x_1^{-1} v^3\bar{\omega}$ with a semi-invariant generator $\bar{\omega}$ of $\mathrm{gr}^0 \, \bar{\omega}$. That is, $\bar{\mathscr{G}}_1/\bar{\mathscr{G}}_2 \simeq \mathscr{M} \otimes \mathrm{gr}^0 \, \bar{\omega}$, $\bar{\mathscr{G}}_2/\bar{\mathscr{G}}_3 \simeq (\mathscr{L} \oplus \mathscr{M}^{\otimes 2}) \otimes \mathrm{gr}^0 \, \bar{\omega}$ but $\bar{\mathscr{G}}_3/\bar{\mathscr{G}}_4 \simeq (\bar{\mathscr{L}} \otimes \mathscr{M} \oplus \mathscr{M}^{\otimes 3} \otimes \mathcal{O}_{\bar{C}}(\bar{p}_1)) \otimes \mathrm{gr}^0 \, \bar{\omega}$. Hence we compute l_{1j} as

$$l_{11} = \frac{\bar{o}_{p_1}(-\mathrm{wt}\, v) + o_{p_1}(-\mathrm{wt}\, \bar{\omega}) - o_{p_1}(-\mathrm{wt}\, v\bar{\omega})}{4} = \left\lfloor \frac{2+3}{4} \right\rfloor = 1,$$

$$l_{12} = \left\lfloor \frac{1+3}{4} \right\rfloor + \left\lfloor \frac{2 \cdot 2 + 3}{4} \right\rfloor = 2,$$

$$l_{13} = \left\lfloor \frac{1 + 2 + 3}{4} \right\rfloor + \left\lfloor \frac{3 \cdot 2 + o_{p_1}(\mathrm{wt}\, x_1) + 3}{4} \right\rfloor = 3.$$

Step 4 By $\mathscr{L} \simeq \mathcal{O}_{\mathbf{P}^1}$, $\mathscr{M} \simeq \mathcal{O}_{\mathbf{P}^1}(-1)$ and $\mathrm{gr}^0 \, \omega \simeq \mathcal{O}_{\mathbf{P}^1}(-1)$ in Proposition 5.2.14, the source $\mathscr{S}$ of the map $j_1 \oplus j_2 \oplus j_3$ is isomorphic to $\mathcal{O}_{\mathbf{P}^1}(-2) \oplus (\mathcal{O}_{\mathbf{P}^1}(-1) \oplus \mathcal{O}_{\mathbf{P}^1}(-3)) \oplus (\mathcal{O}_{\mathbf{P}^1}(-2) \oplus \mathcal{O}_{\mathbf{P}^1}(-4))$. Hence

$$\chi(\mathcal{G}_1/\mathcal{G}_4) = \chi(\mathcal{S}) + \sum_{i,j} l_{ij} = (5-12) + 6 = -1$$

and $\chi(\omega_X/\mathcal{G}_4) = \chi(\mathcal{G}_1/\mathcal{G}_4) + \chi(\mathrm{gr}^0 \omega) = -1$ by $\omega_X/\mathcal{G}_1 = \mathrm{gr}^0 \omega$. This contradicts the vanishing of $H^1(\omega_X/\mathcal{G}_4)$ from $R^1\pi_*\omega_X = 0$. □

By Proposition 5.5.10, we are now in Framework 5.5.2 with p_i of type IA, $a_{i1} = 1$ and $i_{p_i}(1) = 1$. We shall complete the general elephant conjecture in the restrictive case when p_i are quotient singularities. This restriction makes the stream of the proof transparent and still covers all the arguments for the general case after the description $p_i \in C = (x_1\text{-axis}) \subset X = (x_1 x_3 + f = 0) \subset A = \mathfrak{D}^4/\mathbf{Z}_{r_i}(1, a_{i2}, -1, 0)$ for $f \in (x_2, x_3, x_4)^2 \mathcal{O}_{\mathfrak{D}^4}$ with $\mathcal{L} = \mathcal{O}_{\bar{C}} x_4$.

Henceforth we shall treat an extremal neighbourhood $C \subset X$ in Framework 5.5.2 such that each p_i is a quotient singularity of type IA with $a_{i1} = 1$ and $i_{p_i}(1) = 1$. The germ $p_i \in C \subset X$ admits orbifold coordinates x_{i1}, x_{i2}, x_{i3} such that $p_i \in C = (x_{i1}\text{-axis}) \subset X = \mathfrak{D}^3/\mathbf{Z}_{r_i}(1, a_{i2}, -1)$.

We have $\mathrm{gr}^1 \mathcal{O} \simeq \mathcal{O}_{\mathbf{P}^1} \oplus \mathcal{O}_{\mathbf{P}^1}(-1)$ by $\deg \mathrm{gr}^1 \mathcal{O} = -1$ with $H^1(\mathrm{gr}^1 \mathcal{O}) = 0$ in Proposition 5.2.14. Let $\mathcal{L}$ be the unique submodule of $\mathrm{gr}^1 \mathcal{O}$ isomorphic to $\mathcal{O}_{\mathbf{P}^1}$. As in the discussion about Definition 5.5.8, there exists an l-splitting $\mathrm{gr}^1 \mathcal{O} = \mathcal{L} \oplus \mathcal{M}$ by a submodule $\mathcal{M} \simeq \mathcal{O}_{\mathbf{P}^1}(-1)$. We write $\mathrm{gr}^1 \bar{\mathcal{O}} = \bar{\mathcal{L}} \oplus \bar{\mathcal{M}}$ for the corresponding splitting on $\bar{p}_i \in \bar{C} \subset \bar{X}$. Since $b_{i1} = 1$ in Proposition 5.5.9, the orbifold coordinates can be taken so that

$$\bar{\mathcal{L}} = \mathcal{O}_{\bar{C}} x_{i3}, \qquad \bar{\mathcal{M}} = \mathcal{O}_{\bar{C}} x_{i2}.$$

Then $\mathcal{L} = \mathcal{O}_C x_{i1} x_{i3}$ and $\mathcal{M} = \mathcal{O}_C x_{i1}^{r_i - a_{i2}} x_{i2}$ at p_i.

Lemma 5.5.11 $\mathcal{M} \simeq \mathrm{gr}^0 \omega$ *globally on* C, *and* $\bar{\mathcal{M}} \simeq \mathrm{gr}^0 \bar{\omega}$ *equivariantly on the germ* $\bar{p}_i \in \bar{C} \subset \bar{X}$.

Proof The first isomorphism holds since both are isomorphic to $\mathcal{O}_{\mathbf{P}^1}(-1)$. The second follows from the coincidence $\mathrm{wt}\, x_{i2} = \mathrm{wt}(dx_{i1} \wedge dx_{i2} \wedge dx_{i3})$. □

By this lemma, locally on $\bar{p}_i \in \bar{C} \subset \bar{X}$, one has an equivariant isomorphism $\mathrm{gr}^1_{\bar{C}}(\omega_{\bar{X}}^\vee) \simeq \mathrm{gr}^1 \bar{\mathcal{O}} \otimes (\mathrm{gr}^0 \bar{\omega})^\vee \simeq \bar{\mathcal{L}} \otimes \bar{\mathcal{M}}^{-1} \oplus \mathcal{O}_{\bar{C}}$, from which

$$\mathrm{gr}^1(\omega_X^\vee) = (\mu_* \mathrm{gr}^1_{\bar{C}}(\omega_{\bar{X}}^\vee))^{\mathbf{Z}_{r_i}} \simeq (\mu_*(\bar{\mathcal{L}} \otimes \bar{\mathcal{M}}^{-1}))^{\mathbf{Z}_{r_i}} \oplus \mathcal{O}_C.$$

It follows that the direct summand $\mathcal{M} \otimes \mathcal{M}^{-1} = \mathcal{O}_C$ in $\mathrm{gr}^1 \mathcal{O} \otimes (\mathrm{gr}^0 \omega)^\vee \simeq \mathcal{L} \otimes \mathcal{M}^{-1} \oplus \mathcal{O}_C$ is still a direct summand in $\mathrm{gr}^1(\omega_X^\vee)$ after the natural injection $\mathrm{gr}^1 \mathcal{O} \otimes (\mathrm{gr}^0 \omega)^\vee \hookrightarrow \mathrm{gr}^1(\omega_X^\vee)$. Writing $\mathcal{L} \tilde{\otimes} \mathcal{M}^{-1}$ for the kernel of the projection $\mathrm{gr}^1(\omega_X^\vee) \to \mathcal{O}_C$, one obtains the splitting

$$F^1 \omega_X^\vee / F^2 \omega_X^\vee = \mathrm{gr}^1(\omega_X^\vee) = (\mathcal{L} \tilde{\otimes} \mathcal{M}^{-1}) \oplus \mathcal{O}_C.$$

The space $H^0(F^1\omega_X^\vee/F^2\omega_X^\vee)$ of global sections equals $H^0(\omega_X^\vee/F^2\omega_X^\vee)$ since $\omega_X^\vee/F^1\omega_X^\vee = \mathrm{gr}^0(\omega_X^\vee) \simeq \mathscr{O}_{\mathbb{P}^1}(-1)$, where the last isomorphism follows from $\mathrm{gr}^0\,\omega \otimes \mathrm{gr}^0(\omega_X^\vee) \simeq \mathscr{O}_C(-p_1 - p_2)$. Thus $H^0(\omega_X^\vee/F^2\omega_X^\vee) = H^0(\mathscr{L} \tilde{\otimes} \mathscr{M}^{-1}) \oplus H^0(\mathscr{O}_C)$, and the projection to $H^0(\mathscr{O}_C)$ yields the natural map

$$\theta \colon H^0(\omega_X^\vee) \to H^0(\omega_X^\vee/F^2\omega_X^\vee) = H^0(F^1\omega_X^\vee/F^2\omega_X^\vee) \to H^0(\mathscr{O}_C) \simeq \mathbf{C}.$$

Lemma 5.5.12 *The map θ is surjective.*

Proof *Step* 1 Starting with $(\mathscr{I}_1, \mathscr{L}_1, \mathscr{M}_1) = (\mathscr{I}, \mathscr{L}, \mathscr{M})$, we shall construct a decreasing sequence of ideals $\mathscr{I}_n$ in $\mathscr{O}_X$ for $n \geq 1$ together with l-splittings $\mathrm{gr}^0\,\mathscr{I}_n = \mathscr{L}_n \oplus \mathscr{M}_n$ by invertible submodules $\mathscr{L}_n$ and $\mathscr{M}_n$ of $\mathrm{gr}^0\,\mathscr{I}_n$. The notion of l-splitting means as in Definition 5.5.8 the compatibility with a splitting of $\mathrm{gr}^0\,\bar{\mathscr{I}}_n$ on the index-one cover $\bar{X}$ at each p_i, where $\bar{\mathscr{I}}_n$ is the saturation of $\mathscr{I}_n \cdot \mathscr{O}_{\bar{X}}$ in $\mathscr{O}_{\bar{X}}$. The construction will satisfy the properties for all $l \geq 1$ that

(i) $\mathscr{I}^l \subset \mathscr{I}_l$ and this induces an injection $\mathscr{L}^{\otimes l} \hookrightarrow \mathscr{L}_l$,

(ii) $F^1\mathscr{I}_l \subset \mathscr{I}_{l+1} \subset \mathscr{I}_l$ and $\mathscr{M}_l = \mathscr{I}_{l+1}/F^1\mathscr{I}_l$ in $\mathrm{gr}^0\,\mathscr{I}_l$ and

(iii) $\mathrm{gr}^0\,\mathscr{I}_{l+1} \to \mathrm{gr}^0\,\mathscr{I}_l$ induces an isomorphism $\mathscr{M}_{l+1} \simeq \mathscr{M}_l$ of submodules.

Note that the inclusion $\mathscr{I}^l \subset \mathscr{I}_l$ in (i) follows from $\mathscr{I}\mathscr{I}_{l-1} \subset F^1\mathscr{I}_{l-1} \subset \mathscr{I}_l$ in (ii) inductively. We take $\mathrm{gr}^1\,\mathscr{O}_X = \mathscr{L} \oplus \mathscr{M}$ as the l-splitting $\mathrm{gr}^0\,\mathscr{I}_1 = \mathscr{L}_1 \oplus \mathscr{M}_1$.

Suppose that we have constructed $\mathscr{I}_l$, $\mathscr{L}_l$ and $\mathscr{M}_l$ for all $l \leq n$ which satisfy (i) for $l \leq n$ and (ii), (iii) for $l < n$. We define the ideal $\mathscr{I}_{n+1}$ containing $F^1\mathscr{I}_n$ by the equality $\mathscr{M}_n = \mathscr{I}_{n+1}/F^1\mathscr{I}_n$. The inclusion $\mathscr{M}_n \subset \mathrm{gr}^0\,\mathscr{I}_n$ induces $\mathscr{I}_{n+1} \subset \mathscr{I}_n$ as claimed in (ii). Let $\mathscr{L}_{n+1}$ be the kernel of the composite $\mathrm{gr}^0\,\mathscr{I}_{n+1} \to \mathrm{gr}^0\,\mathscr{I}_n \to \mathscr{M}_n$ of the natural map and the projection, which fits into the exact sequence

$$0 \to \mathscr{L}_{n+1} \to \mathrm{gr}^0\,\mathscr{I}_{n+1} \to \mathscr{M}_n \to 0.$$

By $\mathrm{gr}^0\,\mathscr{I}_{n+1} = \mathscr{I}_{n+1}/F^1\mathscr{I}_{n+1}$ and $\mathscr{M}_n = \mathscr{I}_{n+1}/F^1\mathscr{I}_n$, one can write $\mathscr{L}_{n+1} = F^1\mathscr{I}_n/F^1\mathscr{I}_{n+1}$. Together with $\mathscr{L}_n \simeq \mathrm{gr}^0\,\mathscr{I}_n/\mathscr{M}_n \simeq \mathscr{I}_n/\mathscr{I}_{n+1}$, one has the natural map

$$\mathscr{L}^{\otimes n} \otimes (\mathscr{L} \oplus \mathscr{M}) \to (\mathscr{I}_n/\mathscr{I}_{n+1}) \otimes (\mathscr{I}/\mathscr{I}^{(2)}) \to F^1\mathscr{I}_n/F^1\mathscr{I}_{n+1} = \mathscr{L}_{n+1}$$

which is generically surjective on C. The image of the direct summand $\mathscr{L}^{\otimes n} \otimes \mathscr{M}$ is contained in $(\mathscr{I}^n\mathscr{I}_2 + F^1\mathscr{I}_{n+1})/F^1\mathscr{I}_{n+1}$, which is zero by $\mathscr{I}^n\mathscr{I}_2 \subset \mathscr{I}^{n-1}\mathscr{I}_3 \subset \cdots \subset \mathscr{I}\mathscr{I}_{n+1}$ from $\mathscr{I}\mathscr{I}_l \subset F^1\mathscr{I}_l \subset \mathscr{I}_{l+1}$ for $l \leq n$ in (ii). Hence the injection $\mathscr{L}^{\otimes(n+1)} \hookrightarrow \mathscr{L}_{n+1}$ in (i) holds.

From the injection $\mathscr{O}_{\mathbb{P}^1} \simeq \mathscr{L}^{\otimes(n+1)} \hookrightarrow \mathscr{L}_{n+1}$ and the isomorphism $\mathscr{M}_n \simeq \mathscr{M} \simeq \mathscr{O}_{\mathbb{P}^1}(-1)$, one has $\mathscr{H}om(\mathscr{M}_n, \mathscr{L}_{n+1}) \simeq \mathscr{O}_{\mathbb{P}^1}(a)$ with positive a. This implies the splitting of $\mathscr{L}_{n+1} \subset \mathrm{gr}^0\,\mathscr{I}_{n+1}$ and further the existence of a submodule

$\mathcal{M}_{n+1}$ of $\mathrm{gr}^0\,\mathcal{I}_{n+1}$ which provides an l-splitting $\mathrm{gr}^0\,\mathcal{I}_{n+1} = \mathcal{L}_{n+1} \oplus \mathcal{M}_{n+1}$ as in the argument about Definition 5.5.8. Then $\mathcal{M}_{n+1} \simeq \mathrm{gr}^0\,\mathcal{I}_{n+1}/\mathcal{L}_{n+1} \simeq \mathcal{M}_n$, which is (iii) for $l = n$. We thus obtain a sequence of triples $(\mathcal{I}_n, \mathcal{L}_n, \mathcal{M}_n)$ by induction.

Step 2 Let C_n denote the analytic subspace of X defined by the ideal $\mathcal{I}^{(n)}$. Locally at p_i where $\mathcal{L} = \mathcal{O}_{\bar{C}}x_{i3}$ and $\mathcal{M} = \mathcal{O}_{\bar{C}}x_{i2}$, we take the generator $\lambda = (dx_{i1} \wedge dx_{i2} \wedge dx_{i3})^{-1}$ of $\omega_{\bar{X}}^{\vee}$ on $\bar{p}_i \in \bar{C} \subset \bar{X}$. Then $\bar{\mathcal{I}}_n = (x_{i2}, x_{i3}^n)\mathcal{O}_{\bar{X}}$ after replacement of x_{i2}, and

$$\bar{\mathcal{I}}_n\omega_{\bar{X}}^{\vee}/\bar{\mathcal{I}}^n\omega_{\bar{X}}^{\vee} = (x_{i2}, x_{i3}^n)\lambda/(x_{i2}, x_{i3})^n\lambda = (\mathcal{O}_{\bar{X}}/(x_{i2}, x_{i3})^{n-1}\mathcal{O}_{\bar{X}}) \cdot x_{i2}\lambda.$$

Recall that $F^n\omega_X^{\vee}$ is the saturation of $\mathcal{I}^n\omega_X^{\vee}$ in $\omega_X^{\vee}$. We let $G^n\omega_X^{\vee}$ denote the saturation of $\mathcal{I}_n\omega_X^{\vee}$ in $\omega_X^{\vee}$. Because $x_{i2}\lambda$ is invariant, $G^n\omega_X^{\vee}/F^n\omega_X^{\vee} = (\mu_*(\bar{\mathcal{I}}_n\omega_{\bar{X}}^{\vee}/\bar{\mathcal{I}}^n\omega_{\bar{X}}^{\vee}))^{Z_{r_i}}$ is locally isomorphic to $\mathcal{O}_{C_{n-1}}$. Hence $G^n\omega_X^{\vee}/F^n\omega_X^{\vee}$ is an invertible sheaf on C_{n-1}.

Suppose that $n \geq 2$. On $\bar{p}_i \in \bar{C} \subset \bar{X}$, one has the equivariant map

$$\bar{\mathcal{I}}_n\omega_{\bar{X}}^{\vee}/\bar{\mathcal{I}}^n\omega_{\bar{X}}^{\vee} \twoheadrightarrow \bar{\mathcal{I}}_n\omega_{\bar{X}}^{\vee}/(F^1\bar{\mathcal{I}}_{n-1})\omega_{\bar{X}}^{\vee} = \bar{\mathcal{M}}_{n-1} \otimes \mathrm{gr}_{\bar{C}}^0(\omega_{\bar{X}}^{\vee})$$

$$\simeq \bar{\mathcal{M}} \otimes \bar{\mathcal{M}}^{-1} \simeq \mathcal{O}_{\bar{C}},$$

where $\bar{\mathcal{M}}_{n-1}$ is the direct summand of $\mathrm{gr}^0\,\bar{\mathcal{I}}_{n-1}$ corresponding to $\mathcal{M}_{n-1} \subset \mathrm{gr}^0\,\mathcal{I}_{n-1}$. This yields a global surjective map $k_n\colon G^n\omega_X^{\vee}/F^n\omega_X^{\vee} \twoheadrightarrow \mathcal{O}_C$ compatible with θ. Namely, the map κ_n of global sections induced by k_n fits into the commutative diagram

$$
\begin{array}{ccc}
H^0(G^n\omega_X^{\vee}/F^n\omega_X^{\vee}) & \xrightarrow{\kappa_n} & H^0(\mathcal{O}_C) \\
\big\downarrow & & \big\| \\
\theta\colon\ H^0(\omega_X^{\vee}) \longrightarrow H^0(\omega_X^{\vee}/F^n\omega_X^{\vee}) & \xrightarrow{\theta_n} & H^0(\mathcal{O}_C)\ .
\end{array}
$$

The surjection k_n tensored with $\mathcal{O}_C$ is an isomorphism $(G^n\omega_X^{\vee}/F^n\omega_X^{\vee}) \otimes \mathcal{O}_C \simeq \mathcal{O}_C$. Since $H^1(\mathcal{O}_{C_{n-1}}) = 0$ from $R^1\pi_*\mathcal{O}_X = 0$, one has $\mathrm{Pic}\,C_{n-1} \simeq \mathrm{Pic}\,C$ from Lemma 4.1.12. Thus the invertible $\mathcal{O}_{C_{n-1}}$-module $G^n\omega_X^{\vee}/F^n\omega_X^{\vee}$ is globally isomorphic to $\mathcal{O}_{C_{n-1}}$ on C_{n-1}.

Step 3 By virtue of the isomorphism $G^n\omega_X^{\vee}/F^n\omega_X^{\vee} \simeq \mathcal{O}_{C_{n-1}}$, the map κ_n is surjective and so is $\theta_n\colon H^0(\omega_X^{\vee}/F^n\omega_X^{\vee}) \to H^0(\mathcal{O}_C)$. The inverse system of the kernels of θ_n, which are finite dimensional vector spaces, satisfies the Mittag-Leffler condition. Recall that an inverse system $(A_i, \pi_{ij}\colon A_j \to A_i)$ indexed by a directed set I satisfies the *Mittag-Leffler condition* if for any $i \in I$, there exists $j \geq i$ such that $\pi_{ik}(A_k) = \pi_{ij}(A_j)$ for all $k \geq j$. This condition implies the surjectivity of the induced map $\varprojlim H^0(\omega_X^{\vee}/F^n\omega_X^{\vee}) \to H^0(\mathcal{O}_C)$ by [178, II proposition 9.1], through which θ factors.

The space $\varprojlim H^0(\omega_X^\vee/F^n\omega_X^\vee)$ coincides with $\varprojlim H^0(\omega_X^\vee/\mathscr{I}^n\omega_X^\vee)$ because at p_i,

$$
\begin{aligned}
F^{(2n+1)r_1r_2}\omega_X^\vee &\subset F^{(2n+1)r_i}\omega_X^\vee = (\mu_*\bar{\mathscr{I}}^{(2n+1)r_i}\omega_{\bar{X}}^\vee)^{\mathbf{Z}_{r_i}}\\
&\subset (\mu_*(\bar{\mathscr{I}}^{2nr_i}x_{i2}\lambda + \bar{\mathscr{I}}^{2nr_i}x_{i3}^{r_i-a_{i2}}\lambda))^{\mathbf{Z}_{r_i}}\\
&\subset (\mu_*\bar{\mathscr{I}}^{2nr_i})^{\mathbf{Z}_{r_i}}\omega_X^\vee \subset (x_{i2}^{r_i})^n\omega_X^\vee + (x_{i3}^{r_i})^n\omega_X^\vee \subset \mathscr{I}^n\omega_X^\vee.
\end{aligned}
$$

Hence applying [178, III theorem 11.1], which also holds in the analytic category by its proof, one obtains the surjection

$$
\pi_*\omega_X^\vee \otimes \hat{\mathscr{O}}_{Y,y} \simeq \varprojlim H^0(\omega_X^\vee/\mathscr{I}^n\omega_X^\vee) \twoheadrightarrow H^0(\mathscr{O}_C)
$$

for the completion $\hat{\mathscr{O}}_{Y,y}$ of $\mathscr{O}_{Y,y}$. This shows the surjectivity of θ. □

Proof of Theorem 5.5.1 in Framework 5.5.2   We keep the assumption that p_i are quotient singularities with $a_{i1} = 1$. Let S be the general elephant of the extremal neighbourhood $C \subset X$. It is defined by the general section s in $H^0(\omega_X^\vee)$. Notice that $C \subset S$ since $H^0(\omega_X^\vee/F^2\omega_X^\vee) = H^0(F^1\omega_X^\vee/F^2\omega_X^\vee)$.

By Lemma 5.5.12, the image $\theta(s)$ is non-zero. In particular for an arbitrary closed point z in C other than p_1 or p_2, the section s is non-zero in the germ $\mathrm{gr}^1(\omega_X^\vee)_z \simeq (\mathscr{I}/\mathscr{I}^2)_z$. Hence S is locally at z defined by a function in $\mathscr{I} \setminus \mathscr{I}^2$, which shows that S is smooth at z. On the other hand at the quotient singularity p_i where $\bar{\mathscr{L}} = \mathscr{O}_{\bar{C}}x_{i3}$ and $\bar{\mathscr{M}} = \mathscr{O}_{\bar{C}}x_{i2}$, the condition $\theta(s) \neq 0$ means that S is defined at p_i by a semi-invariant function of form $x_{i2} + g(x_{i1})x_{i3} + h$ with $h \in (x_{i2},x_{i3})^2\mathscr{O}_{\bar{X}}$. It follows that S is Du Val of type A_{r_i-1} at p_i and $p_i \in S \simeq \mathbb{D}^2/\mathbf{Z}_{r_i}(1,-1)$ has orbifold coordinates x_{i1},x_{i3} for which C is defined by x_{i3}. Thus the case (vii) in Theorem 5.5.1 holds. □

5.6 Examples

We shall make examples of threefold flips by means of Construction 4.4.8. The simplest example has already been explained.

Example 5.6.1   The extremal neighbourhood $C' \subset X'$ in Example 1.3.19 belongs to the case (i) in Theorem 5.5.1. It has a unique singular point of type IA, which is a quotient singularity of type $\frac{1}{2}(1,1,1)$.

We are interested in an isolated extremal neighbourhood $C \subset X$. See Section 4.4 for the divisorial case. We keep the assumption of the irreducibility $C \simeq \mathbf{P}^1$ in Terminology 5.2.13. The neighbourhood $C \subset X$ belongs to one of the cases (i), (ii), (vi) and (vii) in Theorem 5.5.1. Note that $p \in X$ is cA/r, $cAx/4$ or $cD/3$ in (i). Our strategy is to build $C \subset X$ from a principal prime divisor H containing C. We describe $C \subset H$ and the flip $C^+ \subset X^+$ for this purpose. By the following result, the flipped locus C^+ is again irreducible.

Theorem 5.6.2 ([276, theorem 13.5]) *Let $C \subset X \rightarrow y \in Y$ be an isolated extremal neighbourhood which is not assumed to be irreducible. Let $C^+ \subset X^+ \rightarrow y \in Y$ be its flip. Then the number of irreducible components of C^+ is at most that of irreducible components of C.*

In this section, we let $\pi \colon C \simeq \mathbf{P}^1 \subset X \rightarrow y \in Y$ be an isolated extremal neighbourhood. Let H be a principal prime divisor on X containing C and let H' be the strict transform in Y. We assume that H is smooth outside C.

In the cases (i) with a cA/r point p and (vii) in Theorem 5.5.1, the neighbourhood $C \subset X$ is semistable in Definition 4.4.17. Hence Theorems 4.4.20 and 4.4.24 are applicable. From the surface morphism $C \subset H \rightarrow y \in H'$, one can construct an extremal neighbourhood $C \subset X$ in the corresponding case by Construction 4.4.8. As will be explained in Example 5.6.3, one may use Lemma 5.6.4 for finding an isolated neighbourhood. Further analyses are in [168], [336]. For example, the paper [168] constructs the universal family of semistable extremal neighbourhoods which have a normal principal divisor containing the exceptional curve.

We shall treat the remaining cases (i) with a $cAx/4$ or $cD/3$ point p, (ii) and (vi) in Theorem 5.5.1, in which H will turn out to be normal if it is general. From now on, we assume the principal divisor H containing C to be *general*, in which H' is the general hyperplane section of $y \in Y$. Let Γ and Γ' denote the dual graphs of the minimal resolutions of H and H' respectively. In Γ and Γ', the circle ○ stands for a curve exceptional over H, the bullet • stands for a curve not exceptional over H but exceptional over H' and the number attached to ○ or • is the self-intersection number. By Theorem 5.6.2, the exceptional locus C^+ in the flip $C^+ \subset X^+$ is also irreducible. The vertex ○ marked with + in Γ' means that C^+ is this curve. The edge marked with + in Γ' means that C^+ is obtained by the blow-up of the minimal resolution of H' at the intersection of the two curves which are vertices on this edge.

Example 5.6.3 ([276, theorems 13.12, 13.17]) Consider an isolated neighbourhood $C \subset X$ with a $cAx/4$ point p in the case (i) in Theorem 5.5.1. Then H is normal and $\Gamma \rightarrow \Gamma'$ is one of the following. If it is the first or third graph, then p is the only singular point of X.

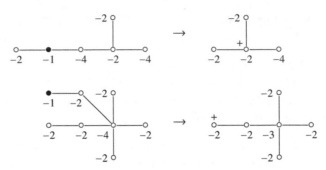

Conversely, start with $C \subset H \to y \in H'$ equipped with one of these dual graphs. Every singularity on H is recovered from the graph Γ by a result of Laufer [286] as in [276, theorem 13.8.2], and it is realised as a hyperplane section of a terminal threefold singularity. Thus one obtains an extremal neighbourhood $C \subset X \to y \in Y$ by Construction 4.4.8. It belongs to (i) in Theorem 5.5.1 with a $cAx/4$ point p. Further $C \subset X$ is isolated. If it were divisorial, then by Lemma 5.6.4, $y \in Y$ would be terminal of index one. But then $y \in H'$ would be Du Val by Theorem 2.3.4, which contradicts the description of Γ'.

Lemma 5.6.4 *Let $C \simeq \mathbf{P}^1 \subset X \to y \in Y$ be a divisorial extremal neighbourhood, in which Y is terminal by Remark 4.4.9. Suppose that $C \subset X$ is locally primitive. If the index of Y has a prime factor l, then there exist two points p_1 and p_2 in X such that l divides the index of each germ $p_i \in X$.*

Proof The pull-back $\pi^* K_Y$ by $\pi \colon X \to Y$ is an integral divisor. Let r_Y denote the index of Y. The divisor $(r_Y/l)K_Y$ produces the cyclic cover $y' \in Y'$ of $y \in Y$ lifted to the cyclic cover $C' \subset X'$ of $C \subset X$ associated with $(r_Y/l)\pi^* K_Y$. Because $C \subset X$ is locally primitive, C' is irreducible and $C' \subset X' \to y' \in Y'$ is an extremal neighbourhood with $C' \simeq \mathbf{P}^1$. It follows from Hurwitz's theorem below that the induced morphism $C' \to C$ is ramified over two points in C. By Theorem 2.4.17, this occurs only if l divides the local index of X at each of the two points. $\qquad \square$

Theorem 5.6.5 (Hurwitz's theorem [178, IV corollary 2.4]) *Let $C \to D$ be a finite separable morphism of degree d of smooth projective curves which is only tamely ramified. Then $2g_C - 2 = d(2g_D - 2) + \sum_{x \in C}(e_x - 1)$ for the genera g_C and g_D of C and D and the ramification index e_x at $x \in C$.*

Example 5.6.6 ([276, theorems 13.11, 13.17]) Consider an isolated neighbourhood $C \subset X$ with a $cD/3$ point p in the case (i) in Theorem 5.5.1. Then H is normal and $\Gamma \to \Gamma'$ is one of the following. The point p is the only singularity of X.

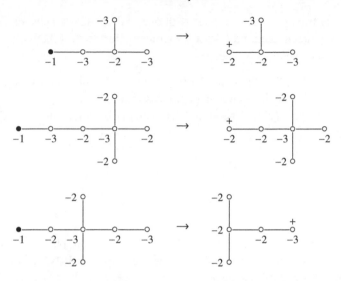

In the same manner as in Example 5.6.3, conversely from $C \subset H \to y \in H'$ equipped with one of these dual graphs, one obtains an extremal neighbourhood $C \subset X$ by Construction 4.4.8. It belongs to (i) in Theorem 5.5.1 with a $cA/3$ or $cD/3$ point p and it is isolated.

For the last two cases, we use the notation Δ_n which stands for the graph

$$\underset{-4}{\circ} \quad \text{or} \quad \underset{-3}{\circ}\text{---}\underset{-2}{\circ}\text{---} \cdots \text{---}\underset{-2}{\circ}\text{---}\underset{-3}{\circ}$$

such that the number of $\circ$ equals n. This is the dual graph for a surface singularity [4] or $[3, 2, \dots, 2, 3]$ of class T in Theorem 4.4.13(i).

Example 5.6.7 ([276, theorems 13.9, 13.18])   Consider an isolated neighbourhood $C \subset X$ with a point p of type IC in the case (ii) in Theorem 5.5.1. Then H is normal and $\Gamma \to \Gamma'$ is one of the following where r is the index of $p \in X$.

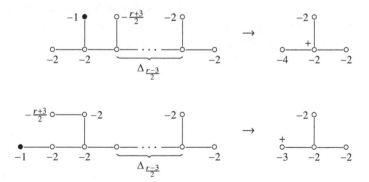

Conversely from $C \subset H \to y \in H'$ with one of these dual graphs, one obtains an extremal neighbourhood $C \subset X$ by Construction 4.4.8. It belongs to (ii) in Theorem 5.5.1 and it is isolated.

Example 5.6.8 ([276, theorems 13.10, 13.18])   Consider an isolated neighbourhood $C \subset X$ with a quotient singularity p of odd index $r \geq 5$ and a $cA/2$ point q in the case (vi) in Theorem 5.5.1. Then H is normal and $\Gamma \to \Gamma'$ is the following.

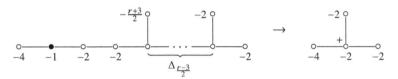

Conversely from $C \subset H \to y \in H'$ with this dual graph, one obtains an extremal neighbourhood $C \subset X$ by Construction 4.4.8. It belongs to (vi) in Theorem 5.5.1 and it is isolated.

Corollary 5.6.9   *Let $C \simeq \mathbf{P}^1 \subset X \to y \in Y$ be an extremal neighbourhood. Then the general hyperplane section $y \in H'$ of $y \in Y$ is a rational singularity. Either it is log terminal or the dual graph of the minimal resolution is one of the following. The first graph occurs in the divisorial case in Theorem 5.5.1(iv) and the second occurs only in the isolated case with a $cAx/4$ or $cD/3$ point in Theorem 5.5.1(i).*

Proof   Almost all cases are treated in Lemma 4.4.19, Remark 4.4.28 and the above four examples. The remaining case is when $C \subset X$ is divisorial and belongs to (v) or (vi) in Theorem 5.5.1. In this case, X has two non-Gorenstein points of odd index and of index two. Then $y \in Y$ is terminal of index one by Lemma 5.6.4 and $y \in H'$ is Du Val by Theorem 2.3.4.   □

6

The Sarkisov Category

The output of the minimal model program is either a minimal model or a Mori fibre space depending on whether the canonical divisor is pseudo-effective or not. We shall proceed to the explicit study of threefold Mori fibre spaces. Unlike a minimal model, a Mori fibre space can be birational to another space with a rather different structure. This chapter is an introduction to the Sarkisov program as a tool for understanding birational maps of Mori fibre spaces.

The Sarkisov program factorises an arbitrary birational map of Mori fibre spaces as a composite of elementary maps called Sarkisov links. Corti established the program in dimension three after the proposal of Sarkisov. By means of the Noether–Fano inequality, the program twists the birational map into an isomorphism by a series of Sarkisov links which decrease the Sarkisov degree. Whereas Hacon and McKernan obtained a new program in an arbitrary dimension using the finiteness of ample models, we stick to the traditional approach which fits better into the explicit study of Sarkisov links.

The Sarkisov program stems from the pioneer work of Iskovskikh and Manin on the irrationality of a smooth quartic threefold. It is fundamental to ask which varieties are rational. We review this problem and discuss the notions of rationality, stable rationality and unirationality. We exhibit examples of threefolds revealing that these are all different notions, whilst they coincide in dimension two.

A rational variety has a large number of Mori fibre spaces in its birational class. Oppositely, a variety birational to essentially only one Mori fibre space is said to be birationally rigid. Iskovskikh and Manin actually obtained the irrationality by proving the birational rigidity. We shall explain a general strategy for applying the Sarkisov program to the rationality and birational rigidity problem. Then we demonstrate this by an example of a terminal quartic threefold due to Corti and Mella which has exactly two birational structures of a Mori fibre space.

6.1 Links of Mori Fibre Spaces

Recall that a *Mori fibre space* $\varphi\colon X \to S$ is by Definition 1.3.21 an extremal contraction of fibre type from a $\mathbf{Q}$-factorial terminal projective variety to a normal projective variety such that $-K_X$ is relatively ample. In particular, φ is projective with connected fibres, $\rho(X/S) = 1$ and $\dim S < \dim X$. By Lemma 1.4.13, the base S is also $\mathbf{Q}$-factorial. By Theorem 1.4.7, $\operatorname{Pic} X/\varphi^* \operatorname{Pic} S \simeq \mathbf{Z}$.

Definition 6.1.1 Let $\varphi\colon X \to S$ and $\psi\colon Y \to T$ be Mori fibre spaces. A birational map $f\colon X \dashrightarrow Y$ is called a *link* from φ to ψ, in which no condition is imposed on the relation between S and T. We often write $f\colon X/S \dashrightarrow Y/T$. When there exists a rational map $g\colon S \dashrightarrow T$ such that $\psi \circ f = g \circ \varphi$, which is unique if it exists, we call g a *compatible* base map of f. We say that f is *fibrewise* if it admits a birational compatible base map $g\colon S \dashrightarrow T$. A fibrewise link f is called a *Sarkisov isomorphism* if both f and g are isomorphisms.

The *Sarkisov category* is the category the objects of which are Mori fibre spaces and the morphisms of which are links. The Sarkisov program, which will be established in the next section, factorises an arbitrary link as a composite of Sarkisov links defined below.

Definition 6.1.2 A link $X/S \dashrightarrow Y/T$ of Mori fibre spaces is called a *Sarkisov link* if it is of one of the following four types.

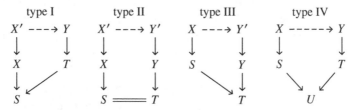

Each diagram is the result by the two-ray game in Definition 4.2.5 from the top-left variety $L = X$ or X' over the bottom variety $B = S$, $S(= T)$, T or U. The morphisms $X' \to X$ and $Y' \to Y$ are elementary divisorial contractions, in which X' and Y' are $\mathbf{Q}$-factorial and terminal, and $S \to U$ and $T \to U$ are extremal contractions to a projective variety. The rational map from L to the top-right variety Y or Y' is a sequence of the $(K_L + \Delta)/B$-flips for some canonical pair (L, Δ). See also Terminology 6.2.11.

Sarkisov links appear in a classical example of rational surfaces.

Example 6.1.3 The *Hirzebruch surface* $\Sigma_n = \mathbf{P}(\mathscr{O}_{\mathbf{P}^1} \oplus \mathscr{O}_{\mathbf{P}^1}(-n)) \to \mathbf{P}^1$ in Definition 1.3.6 has a section $C_n = \mathbf{P}(\mathscr{O}_{\mathbf{P}^1}(-n))$ with self-intersection number $-n$ defined by the surjection $\mathscr{O}_{\mathbf{P}^1} \oplus \mathscr{O}_{\mathbf{P}^1}(-n) \twoheadrightarrow \mathscr{O}_{\mathbf{P}^1}(-n)$, which is unique unless n is zero. Let Σ'_n be the blow-up of Σ_n at a point x in C_n. Then the contraction of the (-1)-curve in Σ'_n which is the strict transform of the fibre in Σ_n through x produces a $\mathbf{P}^1$-bundle isomorphic to Σ_{n+1}. The map $\Sigma_n \dashrightarrow \Sigma_{n+1}$ and its inverse are Sarkisov links of type II.

The blow-up of the projective plane $\mathbf{P}^2$ at a point admits a structure of the Hirzebruch surface Σ_1. The induced map $\mathbf{P}^2 \dashrightarrow \Sigma_1$ is a Sarkisov link of type I and the inverse is a Sarkisov link of type III. The surface $\Sigma_0 \simeq \mathbf{P}^1 \times \mathbf{P}^1$ has two structures of a $\mathbf{P}^1$-bundle, which is an example of a Sarkisov link of type IV. These are illustrated in the diagrams below.

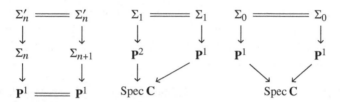

By the next lemma, every threefold Sarkisov link of type IV endowed with a compatible base map is a Sarkisov isomorphism.

Lemma 6.1.4 *Let $f \colon X/S \dashrightarrow Y/T$ be a link of Mori fibre spaces such that $X \dashrightarrow Y$ is small. Suppose that f has a compatible base map $g \colon S \dashrightarrow T$ which is a morphism outside finitely many points in S. Then f is a Sarkisov isomorphism.*

Proof We set $\varphi \colon X \to S$ and $\psi \colon Y \to T$. Let A be the general hyperplane section of T. The pull-back A_S of A by g is well-defined since g is defined in codimension one in S. The divisor A_S is mobile and further nef by the assumption on g. Recall that S is $\mathbf{Q}$-factorial. Let G be the general hyperplane section of X. The strict transform G_Y in Y of G is ψ-ample. By Kleiman's ampleness criterion in Theorem 1.2.8, $\psi^*A + \varepsilon G_Y$ is ample for a small positive rational number ε. By the same criterion, $\varphi^*A_S + \varepsilon G$ is also ample. These ample $\mathbf{Q}$-divisors are transformed to each other by the small map $f \colon X \dashrightarrow Y$. Hence f is an isomorphism by Lemma 3.1.2.

Then $G = f^*G_Y$ and $\varphi^*A_S = (\psi \circ f)^*A$. It follows that every curve in X contracted by φ is also contracted by $\psi \circ f$. Since both φ and $\psi \circ f$ are extremal contractions, they must be associated with the same extremal ray. Thus g is also an isomorphism. □

We shall prove the *Noether–Fano inequality* on certain effective thresholds. It detects the locus where the link fails to be an isomorphism. In preparation for this, we generalise the notion of the strict transform of a divisor to a mobile linear system.

Definition 6.1.5　Let $X \dashrightarrow Y$ be a birational map of normal varieties. Let $\mathscr{H}_Y$ be a mobile linear system on Y. It defines a rational map $Y \dashrightarrow \mathbf{P}^d$ where d is the dimension of $\mathscr{H}_Y$. Let U be the locus in X where the composite $X \dashrightarrow Y \dashrightarrow \mathbf{P}^d$ is defined as a morphism $\varphi_U \colon U \to \mathbf{P}^d$. Note that $X \setminus U$ is of codimension at least two. One can write $\varphi_U^* \mathscr{O}_{\mathbf{P}^d}(1) \simeq \mathscr{O}_U(H|_U)$ by the strict transform H in X of the general member of $\mathscr{H}_Y$, and the image of $H^0(\mathscr{O}_{\mathbf{P}^d}(1)) \to H^0(U, \mathscr{O}_U(H|_U)) = H^0(X, \mathscr{O}_X(H))$ defines a mobile linear subsystem $\mathscr{H}$ of $|H|$. The system $\mathscr{H}$ is called the *homaloidal transform* in X of $\mathscr{H}_Y$. The divisor H is nothing but the general member of $\mathscr{H}$.

Homaloidal transforms are transitive in the following sense. For a mobile linear system $\mathscr{H}_Y$ on Y and a composite $X' \dashrightarrow X \dashrightarrow Y$ of birational maps, the homaloidal transform in X' of the homaloidal transform in X of $\mathscr{H}_Y$ is the homaloidal transform in X' of $\mathscr{H}_Y$.

Definition 6.1.6　Let $\mathscr{H}$ be a mobile linear system on X for a Mori fibre space $\varphi \colon X \to S$. Recall that $\operatorname{Pic} X / \varphi^* \operatorname{Pic} S \simeq \mathbf{Z}$. Define the non-negative rational number μ by $-\mu K_X \sim_{\mathbf{Q},S} H$ for the general member H of $\mathscr{H}$. It is the greatest number such that $\mu K_X + H$ is relatively pseudo-effective. We call μ the *Sarkisov threshold* of $\mathscr{H}$ with respect to φ. We just say that μ is the Sarkisov threshold of $\mathscr{H}$ when the structure φ of the Mori fibre space is clear in the context.

The following is a consequence of the boundedness of $\mathbf{Q}$-Fano varieties due to Birkar.

Theorem 6.1.7　*The Sarkisov threshold μ in Definition* 6.1.6 *has bounded denominator when the dimension of X is fixed.*

Proof　The general fibre X_s of φ is a terminal projective variety with $-K_{X_s}$ ample. Express μ as $(H|_{X_s} \cdot C)/(-K_{X_s} \cdot C)$ by a complete curve C contained in the smooth locus in X_s. The boundedness of the denominator $(-K_{X_s} \cdot C)$ follows from that of X_s in Theorem 9.1.17.　□

Definition 6.1.8　Let $\psi \colon Y \to T$ be a Mori fibre space. A *standard* linear system for ψ is a complete linear system $\mathscr{H}_Y = |-\mu_Y K_Y + \psi^* A|$ on Y with a positive integer μ_Y and a very ample divisor A on T such that $-\mu_Y K_Y + \psi^* A$ is very ample. In this situation, μ_Y is the Sarkisov threshold of $\mathscr{H}_Y$.

Let $f \colon X/S \dashrightarrow Y/T$ be a link of Mori fibre spaces. The Noether–Fano inequality compares a standard linear system $\mathscr{H}_Y$ for Y/T and its homaloidal transform $\mathscr{H}$ in X in terms of Sarkisov threshold. The Sarkisov thresholds of $\mathscr{H}_Y$ and $\mathscr{H}$ coincide if f is square as defined below. We shall see the converse in Theorem 6.1.12(i).

Definition 6.1.9 Let $f: X/S \dashrightarrow Y/T$ be a fibrewise link with a compatible base map $g: S \dashrightarrow T$. We say that f is *square* if f and g induce an isomorphism between the generic fibres of $X \to S$ and $Y \to T$. This is equivalent to the existence of dense open subsets S° and T° of S and T such that $S^\circ \simeq T^\circ$ via g and such that $X \times_S S^\circ \simeq Y \times_T T^\circ$ via f.

Example 6.1.10 Consider the Sarkisov link $\mathbf{P}^2/\operatorname{Spec} \mathbf{C} \dashrightarrow \Sigma_1/\mathbf{P}^1$ of type I in Example 6.1.3. Then $K_{\Sigma_1} \sim -2C - 3f$ with the negative section C and a fibre f of $\Sigma_1 \to \mathbf{P}^1$, and $\mathscr{H}_{\Sigma_1} = |-K_{\Sigma_1} + f|$ is a standard linear system for $\Sigma_1/\mathbf{P}^1$. See [178, section V.2], which also explains Theorem 6.1.11. The homaloidal transform $\mathscr{H}$ in $\mathbf{P}^2$ of $\mathscr{H}_{\Sigma_1}$ is a subsystem of $|\mathcal{O}_{\mathbf{P}^2}(4)|$ and hence $\mathscr{H}$ has Sarkisov threshold $\mu = 4/3$.

Theorem 6.1.11 *Let $\pi: S \to C$ be a $\mathbf{P}^1$-bundle over a smooth projective curve. It is normalised as $S = \mathbf{P}(\mathscr{E})$ by a locally free sheaf $\mathscr{E}$ on C of rank two such that $H^0(\mathscr{E}) \neq 0$ but $H^0(\mathscr{E} \otimes \mathscr{L}) = 0$ for any invertible sheaf $\mathscr{L}$ on C of negative degree. Then π has a section C_0 with $\mathcal{O}_S(C_0) \simeq \mathcal{O}_S(1)$, and $\omega_S \simeq \mathcal{O}_S(-2) \otimes \pi^*(\omega_C \otimes \det \mathscr{E})$. In particular, $K_S \equiv -2C_0 + (2g - 2 - e)f$ with the genus g of C, the integer $e = -(C_0^2) = -\deg \mathscr{E}$ and a fibre f of π.*

We note that the expression of ω_S follows from the relation $\omega_S = \pi^*\omega_C \otimes \Omega_{S/C}$ with the *relative Euler sequence*

$$0 \to \Omega_{S/C} \to \pi^*\mathscr{E} \otimes \mathcal{O}_S(-1) \to \mathcal{O}_S \to 0.$$

Theorem 6.1.12 (Noether–Fano inequality) *Let $f: X/S \dashrightarrow Y/T$ be a link of Mori fibre spaces $\varphi: X \to S$ and $\psi: Y \to T$. Fix a standard linear system $\mathscr{H}_Y = |-\mu_Y K_Y + \psi^* A|$ with A very ample. Let $\mathscr{H}$ be the homaloidal transform in X of $\mathscr{H}_Y$ and let μ be the Sarkisov threshold of $\mathscr{H}$ with respect to φ.*

(i) *One has $\mu_Y \leq \mu$. The equality holds only if and only if f is square.*
(ii) *If $(X, \mu^{-1}H)$ is canonical and $K_X + \mu^{-1}H$ is nef for the general member H of $\mathscr{H}$, then f is a Sarkisov isomorphism.*

Proof Take a common resolution W of X and Y with $p: W \to X$ and $q: W \to Y$ where $q = f \circ p$. Let E_p (resp. E_q) denote the sum of p-exceptional (resp. q-exceptional) prime divisors and let E_0 denote the common part of E_p and E_q. Write H_Y for the general member of $\mathscr{H}_Y$ which is the strict transform of H. Since H_Y is general in $\mathscr{H}_Y$, the strict transform H_W in W is the pull-back $q^* H_Y$. We compare the pairs $(X, \mu_Y^{-1}H)$ and $(Y, \mu_Y^{-1}H_Y)$ by writing

$$K_W + \mu_Y^{-1}H_W = p^*(K_X + \mu_Y^{-1}H) + P = q^*(K_Y + \mu_Y^{-1}H_Y) + Q$$

with $\mathbf{Q}$-divisors P and Q supported in E_p and E_q respectively. Because Y is terminal, $Q = K_{W/Y}$ is effective and has support E_q. Let C be a general relative curve in X/S and let C_W be the strict transform in W. In like manner, let D be a general relative curve in Y/T and let D_W be the strict transform in W. Then C_W avoids E_p and D_W avoids E_q.

Step 1 From $K_Y + \mu_Y^{-1} H_Y \sim_{\mathbf{Q}} \mu_Y^{-1} \psi^* A$, one has $((K_W + \mu_Y^{-1} H_W) \cdot C_W) = \mu_Y^{-1} (\psi^* A \cdot q(C_W)) + (Q \cdot C_W) \ge 0$. Since C_W avoids E_p, it equals $((K_X + \mu_Y^{-1} H) \cdot C) = \mu_Y^{-1}(\mu - \mu_Y)(-K_X \cdot C)$. Hence $\mu_Y \le \mu$ as stated in (i).

Suppose that $\mu_Y = \mu$. Then $(\psi^* A \cdot q(C_W)) = (Q \cdot C_W) = 0$ by the above observation. By $(\psi^* A \cdot q(C_W)) = 0$, $\psi \circ q$ contracts the curve C_W. Then the image of $(\varphi \circ p, \psi \circ q) \colon W \to S \times T$ is birational to S and hence the projection $W \to T$ is factorised as $W \to S \dashrightarrow T$ with a rational map $g \colon S \dashrightarrow T$. Namely, f admits a compatible base map g. By $(Q \cdot C_W) = 0$, the curve C_W avoids not only E_p but also the support E_q of Q.

Consider the pull-back $V = \varphi^* V_S$ of the general hyperplane section V_S of S and its strict transform $V_Y = q_* p^* V$ in Y. Then $(q^* V_Y \cdot C_W) = (p^* V \cdot C_W)$ as C_W avoids E_q. That is, $(V_Y \cdot q(C_W)) = (V \cdot C) = 0$. Since $\rho(Y/T) = 1$, D also has $(V_Y \cdot D) = 0$. Hence $(V \cdot p(D_W)) = (p^* V \cdot D_W) = (V_Y \cdot D) = 0$ as D_W avoids E_q. Consequently the inverse link f^{-1} also admits a compatible base map. In other words, f is fibrewise and g is birational.

Take the generic fibres $X_K = X \times_S \operatorname{Spec} K$ and $Y_K = Y \times_T \operatorname{Spec} K$ for the common function field K of S and T. The induced map $f_K \colon X_K \dashrightarrow Y_K$ is isomorphic in codimension one because C avoids the locus $p(E_p \cup E_q)$. Lemma 6.1.4 also holds for f_K by its proof, showing that f_K is an isomorphism, that is, f is square. This completes (i).

Step 2 To prove (ii), henceforth we assume that $(X, \mu^{-1} H)$ is canonical and $K_X + \mu^{-1} H$ is nef. Then $K_X + \mu^{-1} H \sim_{\mathbf{Q}} \varphi^* B$ with a nef $\mathbf{Q}$-divisor B on S and

$$K_W + \mu^{-1} H_W = p^*(K_X + \mu^{-1} H) + R = q^*(K_Y + \mu^{-1} H_Y) + Q$$

with an effective $\mathbf{Q}$-divisor R supported in E_p. Apply to this the argument at the beginning of Step 1 and obtain $\mu^{-1}(\mu_Y - \mu)(-K_Y \cdot D) \ge 0$ by which $\mu \le \mu_Y$. Thus with (i), $\mu = \mu_Y$ and f is square. Let $g \colon S \dashrightarrow T$ be the compatible base map. Let $c = \mu^{-1} = \mu_Y^{-1}$ so that $K_Y + cH_Y \sim_{\mathbf{Q}} c\psi^* A$ and $K_X + cH \sim_{\mathbf{Q}} \varphi^* B$.

We write $P = P_0 + P_p$ with $\mathbf{Q}$-divisors P_0 and P_p supported in E_0 and $E_p - E_0$ respectively. Likewise $Q = Q_0 + Q_q$ with Q_0 and Q_q supported on E_0 and $E_q - E_0$. Notice that $P = R$ is effective. By the negativity lemma for p, the p-exceptional $\mathbf{Q}$-divisor

$$P_0 - Q_0 + P_p = -p^*(K_X + cH) + q^*(K_Y + cH_Y) + Q_q$$

is negative. Hence $P_0 \le Q_0$ and $P_p \le 0$. In like manner, the negativity lemma for q applied to $Q_0 - P_0 + Q_q$ provides $Q_0 \le P_0$ and $Q_q \le 0$. Together with the effectivity of P and Q, one concludes that $P_0 = Q_0$ and $P_p = Q_q = 0$. Therefore $p^*(K_X + cH) = q^*(K_Y + cH_Y)$ and further $E_q = E_0 \leq E_p$ since $E_q - E_0$ supports Q_q. Thus $f^{-1} : Y \dashrightarrow X$ is a birational contraction map.

Step 3 Let G be the general hyperplane section of X and let $G_Y = q_* p^* G$ be the strict transform in Y. We write $q^* G_Y = p^* G + F$ with an effective q-exceptional $\mathbf{Q}$-divisor F. Since G_Y is ψ-ample, Kleiman's criterion shows that $K_Y + cH_Y + \varepsilon G_Y \sim_{\mathbf{Q}} c\psi^* A + \varepsilon G_Y$ is ample for small positive ε. On the other hand, $K_X + cH + \varepsilon G \sim_{\mathbf{Q}} \varphi^* B + \varepsilon G$ is also ample. Since $q^*(K_Y + cH_Y + \varepsilon G_Y) = p^*(K_X + cH + \varepsilon G) + \varepsilon F$, f^{-1} is the ample model of $K_Y + cH_Y + \varepsilon G_Y$ by Proposition 1.5.4. Thus f^{-1} is an isomorphism by the uniqueness of the ample model in Lemma 1.5.3.

The isomorphism f sends $K_X + cH \sim_{\mathbf{Q}} \varphi^* B$ to $K_Y + cH_Y \sim_{\mathbf{Q}} c\psi^* A$. One sees that g is also an isomorphism by the same argument as in the last paragraph of the proof of Lemma 6.1.4. This completes (ii). □

Remark 6.1.13 In the theorem if $(X, \mu^{-1} H)$ is canonical and $K_X + \mu^{-1} H$ is pseudo-effective, then one still obtains the equality $\mu = \mu_Y$ by the argument at the beginning of Step 2 with the aid of the inequality $((K_X + \mu^{-1} H) \cdot p(D_W)) \geq 0$. In particular, f is square.

6.2 The Sarkisov Program

The *Sarkisov program* is the machinery of the following theorem.

Theorem 6.2.1 (Corti [91], Hacon–Mckernan [173]) *Every link of Mori fibre spaces is a composite of finitely many Sarkisov links.*

Corti formulated the program by following the proposal of Sarkisov [411] and established it in dimension three. After the attainment of the existence of flips, Hacon and McKernan founded a new program in an arbitrary dimension. We adopt the original approach by Corti because the way of building Sarkisov links is explicit in his framework, in which our study of threefold Mori fibre spaces will be performed. The approach by Hacon and McKernan will be explained at the end of the section.

The program of Corti uses the Noether–Fano inequality for the construction of Sarkisov links. For the termination of the program, we track the triple of invariants called the Sarkisov degree.

Definition 6.2.2 Let $\mathcal{H}$ be a mobile linear system on X for a Mori fibre space $\varphi : X \to S$. Assume that $\mathcal{H}$ has Sarkisov threshold μ at least one, and write H

for the general member of $\mathcal{H}$. The *Sarkisov degree* of $\mathcal{H}$ with respect to φ is the triple (μ, c, e) consisting of

- the Sarkisov threshold μ of $\mathcal{H}$ with respect to φ,
- the *canonical threshold* c of H on X in Definition 6.2.4 and
- the number $e = e(X, cH)$ of divisors E exceptional over X such that $a_E(X, cH) = 1$ as in Notation 5.1.7, where we define $e = \infty$ if $\lceil cH \rceil \neq 0$.

We mention the Sarkisov degree of $\mathcal{H}$ without reference to φ when it is clear. In the set of Sarkisov degrees, we introduce a total order in such a manner that $(\mu, c, e) \leq (\mu', c', e')$ if and only if either

- $\mu < \mu'$,
- $\mu = \mu'$ and $c > c'$ or
- $\mu = \mu'$, $c = c'$ and $e \leq e'$.

Remark 6.2.3   We added the natural assumption $\mu \geq 1$ in view of Theorem 6.1.12(i) in order to make arguments lucid. In general unless μ is zero, we define c and e by resetting $H = n^{-1}(H_1 + \cdots + H_n)$ for the general members $H_1, \ldots, H_n$ of $\mathcal{H}$ with $n = \lceil \mu^{-1} \rceil$. By Theorem 6.1.7, the number n is bounded as far as the dimension of X is fixed.

The canonical threshold is defined in parallel to Definition 1.6.10.

Definition 6.2.4   Let X be a canonical variety and let H be a non-zero effective $\mathbf{Q}$-Cartier $\mathbf{Q}$-divisor on X. The *canonical threshold* of H on X is the greatest rational number c such that (X, cH) is canonical.

Thanks to Theorem 6.1.7, the set of Sarkisov degrees satisfies the DCC provided that the ACC for canonical thresholds holds. This is settled in dimension three by Stepanov on smooth threefolds and by J.-J. Chen on terminal threefolds. They applied the classification of threefold divisorial contractions in Chapter 3.

Theorem 6.2.5 (Stepanov [432], J.-J. Chen [77])   *The set of canonical thresholds of integral divisors on terminal threefolds satisfies the ACC.*

Proof   For simplicity, we shall only prove the ACC on smooth threefolds. Let c be the canonical threshold of an effective divisor H on a smooth threefold X. By Lemma 6.2.6 below, there exists a divisorial contraction $\pi \colon Y \to X$ such that the exceptional divisor E has log discrepancy $a_E(X, cH) = 1$. If $\pi(E)$ is a curve C, then by Proposition 4.1.3, E is obtained at the generic point of C by the blow-up of X along C, in which $c = (\mathrm{ord}_E H)^{-1} \in \{1/n \mid n \in \mathbf{N}_{>0}\}$. Thus for the ACC for canonical thresholds c, one has only to take account of c for which E is contracted to a closed point x.

In this case by Theorem 3.5.1, π is a weighted blow-up with respect to a regular system x_1, x_2, x_3 of parameters in $\mathcal{O}_{X,x}$. The ideal $\mathscr{I} = \pi_* \mathcal{O}_Y(-(\mathrm{ord}_E H)E)$ in $\mathcal{O}_X$ contains $\mathcal{O}_X(-H)$ and $\mathrm{ord}_E \mathscr{I} = \mathrm{ord}_E H$. Thus c is still the canonical threshold *of the ideal $\mathscr{I}$* on X. This means the canonical threshold of the divisor $(f)_X$ defined by a general function f in $\mathscr{I}$. Take the étale morphism $x \in X \to o \in \mathbf{A}^3 = \mathrm{Spec}\, \mathbf{C}[x_1, x_2, x_3]$ defined by the system x_1, x_2, x_3. By the definition of the weighted blow-up, the ideal $\mathscr{I}$ is the pull-back of a monomial ideal I in $\mathbf{C}[x_1, x_2, x_3]$. Then c is also the canonical threshold of I on $\mathbf{A}^3$.

Thus it suffices to consider a non-decreasing sequence $\{c_i\}_{i \in \mathbf{N}}$ of canonical thresholds c_i of monomial ideals I_i on $\mathbf{A}^3$. We want to show that c_i is constant passing to a subsequence. Following Mustaţă and Nakamura's argument [365, p.300], we shall prove that $I_{i+1} \subset I_i$ for all i after passing to a subsequence.

We may assume that all I_i are distinct. Then we apply a result of Maclagan [298] that every infinite set of monomial ideals contains two ideals I and J such that $I \subset J$. Her result implies the finiteness of the number of I_i which are maximal with respect to inclusion. Let S be the finite set of indices i of such I_i. It follows from the noetherian property that for any j, there exists $i \in S$ such that $I_j \subset I_i$. Hence one can find an index i_1 in S such that $I_j \subset I_{i_1}$ for infinitely many j. Only take account of these I_j with $i_1 < j$ and repeat the same procedure to find i_2. Inductively one obtains an infinite sequence $i_1 < i_2 < \cdots$ such that $I_{i_{j+1}} \subset I_{i_j}$ for all j, which is a desired subsequence.

The inclusion $I_{i+1} \subset I_i$ implies the inequality $c_{i+1} \leq c_i$. Now the sequence $\{c_i\}_i$ is not only non-decreasing but also non-increasing. It is thus constant. $\square$

Lemma 6.2.6 *Let (X, Δ) be a strictly canonical pair on a $\mathbf{Q}$-factorial terminal variety X. Then there exists an elementary divisorial contraction $Y \to X$ which is crepant with respect to $K_Y + \Delta_Y$ for the strict transform Δ_Y of Δ.*

Proof Fix a log resolution $\mu \colon X' \to X$ of (X, Δ) equipped with an effective exceptional divisor F such that $-F$ is μ-very ample. Let $\{E_i\}_{i \in I}$ be the set of exceptional prime divisors on X' such that $a_{E_i}(X, \Delta) = 1$. We may assume that $I \neq \emptyset$. After slight perturbation of a multiple of F, only one index in I attains the minimum of $\{\mathrm{ord}_{E_i} \Delta / \mathrm{ord}_{E_i} F\}_{i \in I}$. Let $0 \in I$ denote this index. The minimum $m = \mathrm{ord}_{E_0} \Delta / \mathrm{ord}_{E_0} F$ will be assumed to be less than one by multiplication of F.

We shall find a divisorial contraction $\pi \colon Y \to X$ such that E_0 appears as the unique exceptional divisor on Y. Since X is $\mathbf{Q}$-factorial, π is extremal and it is a desired contraction. The existence of π is a local problem by the description $Y = \mathrm{Proj}_X \bigoplus_{l \in \mathbf{N}} \mu_* \mathcal{O}_{X'}(-lE_0)$. Thus we may assume that X is affine.

Since $-F$ is μ-very ample, there exists a reduced divisor H on the affine variety X such that $\mu^*H = H' + F$ for the strict transform H' of H and such that μ is a log resolution of $(X, \Delta + H)$. It satisfies $a_{E_i}(X, mH) = a_{E_i}(X, \Delta) + (\mathrm{ord}_{E_i} \Delta - m \, \mathrm{ord}_{E_i} F) > 1$ for all $i \in I \setminus \{0\}$, whilst $a_{E_0}(X, mH)$ remains equal to one. Take the boundary $\Delta' = (1 - \varepsilon)\Delta + \varepsilon mH$ for small positive ε. By $a_{E_i}(X, \Delta') = (1 - \varepsilon)a_{E_i}(X, \Delta) + \varepsilon a_{E_i}(X, mH)$, the pair (X, Δ') is canonical and E_0 is the only divisor exceptional over X such that $a_{E_0}(X, \Delta') = 1$.

By Proposition 1.5.17, there exists a birational contraction $\pi \colon Y \to X$ from a **Q**-factorial normal variety such that E_0 appears on Y as the unique exceptional divisor E_Y. Then $K_Y + \Delta_Y' = \pi^*(K_X + \Delta')$ for the strict transform Δ_Y' of Δ' and the pair (Y, Δ_Y') becomes terminal. In particular, Y itself is also terminal. Since X is **Q**-factorial, $-E_Y$ of π-ample. Since X is terminal, one can write $K_Y = \pi^*K_X + dE_Y$ with positive discrepancy d. Thus $-K_Y$ is π-ample and π is a divisorial contraction. □

In Theorem 6.1.12 unless $f \colon X/S \dashrightarrow Y/T$ is a Sarkisov isomorphism, either $(X, \mu^{-1}H)$ fails to be canonical or $K_X + \mu^{-1}H$ fails to be nef. The Sarkisov program of Corti builds a Sarkisov link by a two-ray game as in Proposition 6.2.8 or 6.2.9. We take the framework of Definition 6.2.2 and assume the mobile linear system $\mathscr{H}$ to be big, by which the Sarkisov threshold of $\mathscr{H}$ is positive. Here a linear system $\Lambda = \mathbf{P}V^\vee$ on a normal variety X is said to be *big* if the rational map $X \dashrightarrow \mathbf{P}V$ defined by Λ is birational to the image.

Framework 6.2.7 Let $\varphi \colon X \to S$ be a Mori fibre space. Let $\mathscr{H}$ be a mobile and big linear system on X with Sarkisov degree (μ, c, e). We write H for the general member of $\mathscr{H}$ if $\mu \geq 1$ but reset it as in Remark 6.2.3 if $\mu < 1$.

Proposition 6.2.8 *In Framework 6.2.7 if $(X, \mu^{-1}H)$ is not canonical, then there exists a Sarkisov link $g \colon X/S \dashrightarrow X'/S'$ of type* I *or* II *which decreases the Sarkisov degree strictly. That is, the Sarkisov degree of the homaloidal transform in X' of $\mathscr{H}$ is less than (μ, c, e). The divisorial contraction $Z \to X$ which initiates the Sarkisov link g is crepant with respect to $K_Z + cH_Z$ for the strict transform H_Z of H.*

Proof The assumption is nothing but $c < \mu^{-1}$, in which $\lfloor cH \rfloor = 0$. For the pair (X, cH), we take a divisorial contraction $p \colon Z \to X$ as in Lemma 6.2.6. Then $K_Z + cH_Z = p^*(K_X + cH)$ for the strict transform H_Z of H.

The **Q**-divisor $K_Z + cH_Z = p^*(K_X + cH) \sim_{\mathbf{Q},S} -(\mu^{-1} - c)p^*H$ is not pseudo-effective over S since a general relative curve in X/S has negative intersection number with $-H$. We run the $(K_Z + cH_Z)/S$-MMP, which is a two-ray game. It works by Corollary 1.5.13 because the direction of the MMP is unique in the two-ray game. By Proposition 1.4.16, the MMP never ends with a log minimal model. Thus after a finite sequence $Z \dashrightarrow Z'$ of log flips, it produces either a log divisorial contraction $Z' \to X'$ or a log Mori fibre space $X' = Z' \to S'$. Since this MMP is also the $(K_Z + (c - \varepsilon)H_Z)$-MMP for small positive ε in which $(Z, (c - \varepsilon)H_Z)$ is terminal, the pair $(X', (c - \varepsilon)H')$ is terminal for the strict transform H' of H. In particular, X' is terminal.

In the case of a log divisorial contraction $Z' \to X'$, it is actually a divisorial contraction and the MMP ends with the contraction $\varphi' \colon X' \to S$ of fibre type. Since $-K_{X'} = -(K_{X'} + cH') + cH'$ is φ'-ample, φ' is a Mori fibre space and $g \colon X/S \dashrightarrow X'/S$ is a Sarkisov link of type II. In the case of a log Mori fibre space $X' = Z' \to S'$, the contraction $\varphi' \colon X' \to S'$ is also a Mori fibre space for the same reason and $g \colon X/S \dashrightarrow X'/S'$ is a Sarkisov link of type I.

It remains to confirm that the Sarkisov degree (μ', c', e') of the homaloidal transform in X' of $\mathscr{H}$ is less than (μ, c, e). Let E denote the exceptional divisor of p and write

$$K_Z + \mu^{-1}H_Z + aE = p^*(K_X + \mu^{-1}H) \sim_{\mathbf{Q},S} 0$$

with the positive rational number $a = (\mu^{-1} - c) \operatorname{ord}_E H$. Then $K_{X'} + \mu^{-1}H' + aE' \sim_{\mathbf{Q},S} 0$ for the push-forward E' of E and hence a general relative curve C' in X'/S has intersection number $((K_{X'} + \mu^{-1}H') \cdot C') = -(aE' \cdot C') \le 0$. Thus $\mu' \le \mu$. Since (X', cH') is the result by the $(K_Z + cH_Z)$-MMP, it is canonical and hence $c' \ge c$. Finally if $c' = c \, (< \mu^{-1})$, then by Lemma 1.4.12, $e' = e(X', cH') \le e(Z, cH_Z) = e(X, cH) - 1 = e - 1$. From these one concludes that $(\mu', c', e') < (\mu, c, e)$. □

Proposition 6.2.9 *In Framework* 6.2.7 *if* $(X, \mu^{-1}H)$ *is canonical but* $K_X + \mu^{-1}H$ *is not nef, then there exists a Sarkisov link* $g \colon X/S \dashrightarrow X'/S'$ *of type* III *or* IV *which does not increase the Sarkisov threshold. That is, the Sarkisov threshold* μ' *of the homaloidal transform in* X' *of* $\mathscr{H}$ *satisfies* $\mu' \le \mu$. *One can write* $K_X + \mu^{-1}H \sim_{\mathbf{Q}} \varphi^*B$ *with a* $\mathbf{Q}$-*divisor* B *on* S *such that* $-B$ *is ample over the base* U *of the two-ray game* g. *The pair* $(X', \mu^{-1}H')$ *is canonical for the strict transform* H' *of* H. *If* X *is of dimension at most three, then either* g *is of type* III *or* μ' *is less than* μ.

Proof First we shall find $\pi \colon S \to U$ to a projective base U. We write $K_X + \mu^{-1}H \sim_{\mathbf{Q}} \varphi^*B$ with a $\mathbf{Q}$-Cartier $\mathbf{Q}$-divisor B on S. By assumption, B is not nef. Take an ample $\mathbf{Q}$-divisor L on S generally so that $B + L$ is nef but not ample. Then the locus $\{z \in \overline{\mathrm{NE}}(S) \mid ((B + L) \cdot z) = 0\}$ is an extremal ray of the closed cone $\overline{\mathrm{NE}}(S)$ of curves in S, and the locus F in $\overline{\mathrm{NE}}(X)$ perpendicular to $K_X + \mu^{-1}H + \varphi^*L = \varphi^*(B + L)$ is a two-dimensional extremal face. The face F is negative with respect to $K_X + (\mu^{-1} - \varepsilon)H$ for small positive ε since $K_X + (\mu^{-1} - \varepsilon)H$ equals $-(\varepsilon H + \varphi^*L)$ as a function on F. By the contraction theorem for the terminal pair $(X, (\mu^{-1} - \varepsilon)H)$, there exists a contraction $X \to U$ associated with F, which factors through φ, and the induced contraction $\pi: S \to U$ is extremal. Alternatively one may apply Theorem 1.4.28.

We run the $(K_X + \mu^{-1}H)/U$-MMP as a two-ray game by Corollary 1.5.13. Even if μ^{-1} is an integer in which $(X, \mu^{-1}H)$ is not klt, the MMP works by replacing H by $2^{-1}(H + H_1)$ for general $H_1 \sim_{\mathbf{Q}} H$. Remark that $K_X + \mu^{-1}H \equiv_U -\varphi^*L$ is pseudo-effective over U if and only if π is birational.

Suppose that π is of fibre type. In this case by Proposition 1.4.16, the MMP never ends with a log minimal model. After a finite sequence $X \dashrightarrow Z'$ of log flips, it produces either a log divisorial contraction $Z' \to X'$ followed by $\varphi': X' \to S' = U$ or a log Mori fibre space $\varphi': X' = Z' \to S'$. As in the proof of Proposition 6.2.8, φ' is a Mori fibre space and $g: X/S \dashrightarrow X'/S'$ is a Sarkisov link of type III in the former case and of type IV in the latter case. Since φ' is the output of the $(K_X + \mu^{-1}H)$-MMP, the pair $(X', \mu^{-1}H')$ is canonical and $-(K_{X'} + \mu^{-1}H')$ is φ'-ample. In particular, $\mu' < \mu$.

Suppose that π is birational. Then the MMP ends with a log minimal model X'/U, that is, $K_{X'} + \mu^{-1}H'$ is nef over U. The model X' is attained by either

(i) a finite sequence $X \dashrightarrow X'$ of log flips or
(ii) a log divisorial contraction $Z' \to X'$ following a finite sequence $X \dashrightarrow Z'$ of log flips.

In the case (i), $X \dashrightarrow X'$ is not an isomorphism since $K_{X'} + \mu^{-1}H'$ is nef over U. Considering the two-ray game to be the $(K_X + (\mu^{-1} - \varepsilon)H)/U$-MMP for small positive ε, one can continue this game with an extremal contraction $\varphi': X' \to S'$ over U and φ' is the ample model of $K_{X'} + \mu^{-1}H'$ over U in Definition 1.5.2. Indeed, $K_{X'} + \mu^{-1}H'$ is nef but not ample over U since $K_X + \mu^{-1}H \equiv_U -\varphi^*L$ is not big over U. It must be zero on the extremal ray of $\overline{NE}(X'/U)$ corresponding to φ' since it is positive on the other ray. Then $K_{X'} + \mu^{-1}H' \not\equiv_U 0$ is the pull-back of a $\mathbf{Q}$-divisor on S' which is ample over U, which means that φ' is its ample model. Note that φ' is of fibre type since $K_{X'} + \mu^{-1}H'$ is not big over U. In the case (ii), we set $\varphi': X' \to S' = U$. Again $K_{X'} + \mu^{-1}H' \sim_{\mathbf{Q},S'} 0$.

Similarly to the case when π is of fibre type, φ' is a Mori fibre space and $g: X/S \dashrightarrow X'/S'$ is a Sarkisov link of type IV in (i) and of type III in (ii). The pair $(X', \mu^{-1}H')$ is canonical. Further $\mu' = \mu$ by $K_{X'} + \mu^{-1}H' \sim_{\mathbf{Q},S'} 0$.

To complete the proposition, we shall disprove the case (i) in dimension at most three. In this case, $X \dashrightarrow X'$ is small but not an isomorphism, and $g: X/S \dashrightarrow X'/S'$ is square by Theorem 6.1.12(i). If X were of dimension at most three, then by Lemma 6.1.4, g would be a Sarkisov isomorphism, which is absurd. $\qquad\square$

Note that unlike Proposition 6.2.8, the decrease in Sarkisov degree is not claimed in Proposition 6.2.9.

Definition 6.2.10 Let $X \to S$ be a Mori fibre space and let $\mathcal{H}$ be a mobile and big linear system on X. The *Sarkisov program of Corti* is the program of producing a sequence $X/S = X_0/S_0 \dashrightarrow X_1/S_1 \dashrightarrow \cdots$ of Sarkisov links $g_i \colon X_i/S_i \dashrightarrow X_{i+1}/S_{i+1}$ constructed from the homaloidal transform $\mathcal{H}_i$ in X_i of $\mathcal{H}$ by Propositions 6.2.8 and 6.2.9, until $(X_n, \mu_n^{-1} H_n)$ is canonical and $K_{X_n} + \mu_n^{-1} H_n$ is nef for the Sarkisov threshold μ_n of $\mathcal{H}_n$ and the general member H_n of $\mathcal{H}_n$ if $\mu_n \geq 1$, which should be reset as in Remark 6.2.3 if $\mu_n < 1$.

Let $f \colon X/S \dashrightarrow Y/T$ be a link of Mori fibre spaces and let $\mathcal{H}$ be the homaloidal transform in X of a standard linear system $\mathcal{H}_Y$ for Y/T. In this case, the induced link $f \circ g_0^{-1} \circ \cdots \circ g_{n-1}^{-1} \colon X_n/S_n \dashrightarrow Y/T$ is a Sarkisov isomorphism by the Noether–Fano inequality, Theorem 6.1.12(ii).

Terminology 6.2.11 In the book, except in the explanation of the program of Hacon and McKernan, the *Sarkisov program* means the Sarkisov program of Corti, and accordingly a *Sarkisov link* is a link constructed by Proposition 6.2.8 or 6.2.9.

Theorem 6.2.1 follows from the termination of this program. One has to deal with the sequences of Sarkisov links in Lemmata 6.2.12 and 6.2.13.

Lemma 6.2.12 *Keep in Framework* 6.2.7. *There exists no infinite sequence* $X/S = X_0/S_0 \dashrightarrow X_1/S_1 \dashrightarrow \cdots$ *of Sarkisov links* $g_i \colon X_i/S_i \dashrightarrow X_{i+1}/S_{i+1}$ *of type* I *or* II *constructed from* $\mathcal{H}$ *by Proposition* 6.2.8 *such that the homaloidal transform* $\mathcal{H}_i$ *in* X_i *of* $\mathcal{H}$ *has constant Sarkisov threshold* $\mu_i = \mu$ *for all* i.

Proof *Step* 1 The lemma follows from the ACC for canonical thresholds which is only known up to dimension three as in Theorem 6.2.5. We shall take a diversion using the ACC for lc thresholds by Hacon, McKernan and Xu.

Let H_i be the strict transform in X_i of H and let c_i be the canonical threshold of H_i on X_i. Then $c_i < \mu_i^{-1} = \mu^{-1}$. By Proposition 6.2.8, c_i form a non-decreasing sequence and the strict increase $c_i < c_{i+1}$ occurs for infinitely many i. Take the limit $\bar{c} \in \mathbf{R}$ of c_i, which exists by $c_i < \mu^{-1}$ and satisfies $\bar{c} \leq \mu^{-1}$.

Let a_i be the lc threshold of H_i on X_i in Definition 1.6.10. Note that $c_i < a_i$ since the canonical pair $(X_i, c_i H_i)$ is klt by $c_i < \mu^{-1}$. If $c_i < a_i < \bar{c}$ for infinitely many i, then there would exist a strictly increasing subsequence of $\{a_i\}_{i \in \mathbf{N}}$, which contradicts the ACC for lc thresholds in Theorem 1.6.11. Thus after truncation, we may and shall assume that $\bar{c} \leq a_i$ for all i. In other words, $(X_i, \bar{c} H_i)$ is lc for all i.

Step 2 Recall that g_i is the result by the $(K_{Z_i} + c_i H_{Z_i})/S_i$-MMP from a divisorial contraction $p_i \colon Z_i \to X_i$ crepant with respect to $K_{Z_i} + c_i H_{Z_i}$ where H_{Z_i} is the strict transform of H_i. Let E_i be the exceptional divisor of p_i and write $K_{Z_i} + \mu^{-1} H_{Z_i} + a_i E_i = p_i^*(K_{X_i} + \mu^{-1} H_i) \sim_{\mathbf{Q}, S_i} 0$ with $a_i > 0$. Since $K_{Z_i} + \bar{c} H_{Z_i} \sim_{\mathbf{R}, S_i} -((\mu^{-1} - \bar{c}) H_{Z_i} + a_i E_i)$ is not nef over S_i but ample over X_i, it must be relatively anti-ample with respect to the first contraction in the $(K_{Z_i} + c_i H_{Z_i})/S_i$-MMP. Inductively, the strict transform of $K_{Z_i} + \bar{c} H_{Z_i}$ is relatively anti-ample with respect to every new contraction. Hence our MMP is also a part of the $(K_{Z_i} + \bar{c} H_{Z_i})$-MMP. Further p_i is positive with respect to $K_{Z_i} + \bar{c} H_{Z_i}$. Therefore it follows from Lemma 1.4.12 that for the sequence $X/S = X_0/S_0 \dashrightarrow X_1/S_1 \dashrightarrow \cdots$, every divisor E over X satisfies

$$a_E(X, \bar{c} H) \le a_E(X_i, \bar{c} H_i)$$

and the equality holds if and only if $X \dashrightarrow X_i$ is isomorphic at the generic point of the centre $c_X(E)$.

Step 3 For a closed subvariety Γ of H, we consider the canonical threshold c_Γ of H at the generic point η_Γ of X. By definition, c_Γ is the greatest rational number such that $(X, c_\Gamma H)$ is canonical at η_Γ. Observe the finiteness of the set of all c_Γ for $\Gamma \subset H$ by taking a log resolution of (X, H). Fix a rational number c' less than $\bar{c}$ such that there exists no c_Γ with $c' < c_\Gamma < \bar{c}$.

We claim that the lc pair $(X_i, \bar{c} H_i)$ is plt whenever $c' < c_i$. Indeed if a divisor E exceptional over X had $a_E(X_i, \bar{c} H_i) = 0$, then $a_E(X, \bar{c} H) = 0$ by the estimate $a_E(X, \bar{c} H) \le a_E(X_i, \bar{c} H_i)$. Moreover, $X \dashrightarrow X_i$ is isomorphic at the generic point of the centre $\Gamma = c_X(E)$. Hence c_Γ can be computed on X_i and it should be at least the global canonical threshold c_i. Further $c_\Gamma < \bar{c}$ from $a_E(X, \bar{c} H) = 0$. Thus $c' < c_i \le c_\Gamma < \bar{c}$, which contradicts the choice of c'.

Step 4 Thus after truncation, we may assume that $(X, \bar{c} H)$ is plt. Then the set A of divisors E over X with $a_E(X, \bar{c} H) < 1$ is finite. The divisor E_i belongs to A for all i since

$$a_{E_i}(X, \bar{c} H) \le a_{E_i}(X_i, \bar{c} H_i) < a_{E_i}(X_i, c_i H_i) = 1.$$

Because every prime divisor on X_i that is exceptional over X is realised as some E_j with $j \le i$, the number of g_i of type I is bounded by the cardinality of A. Thus by truncation, we may assume that all g_i are of type II, in which $S_i = S$ for all i.

Let A_i be the subset of A which consists of all $E \in A$ appearing as a divisor on X_i. Take two indices i and j with $i < j$ such that $A_i = A_j$. If there existed a prime divisor on X_j which does not appear on X_i, then it would be realised as E_k for some $i \le k < j$. Then $E_k \in A_j \setminus A_i$, contradicting $A_i = A_j$. Hence $X_i \dashrightarrow X_j$ is small by $\rho(X_i) = \rho(X_j)$. Then by Lemma 6.1.4, the link $X_i/S \dashrightarrow X_j/S$ is a

Sarkisov isomorphism. This is impossible since the Sarkisov degree of $\mathcal{H}_j$ is less than that of $\mathcal{H}_i$ by Proposition 6.2.8.   □

Lemma 6.2.13   *Keep in Framework 6.2.7. Let n be the dimension of X. Assume either $n \le 3$ or the termination of canonical n-fold log flips. Then there exists no infinite sequence $X/S = X_0/S_0 \dashrightarrow X_1/S_1 \dashrightarrow \cdots$ of Sarkisov links $g_i \colon X_i/S_i \dashrightarrow X_{i+1}/S_{i+1}$ of type III or IV constructed from $\mathcal{H}$ by Proposition 6.2.9 such that the homaloidal transform $\mathcal{H}_i$ in X_i of $\mathcal{H}$ has constant Sarkisov threshold $\mu_i = \mu$ for all i.*

Proof   If $n \le 3$, then by Proposition 6.2.9, all g_i are of type III and the assertion follows from the decrease $\rho(X_{i+1}) = \rho(X_i) - 1$ in Picard number. For an arbitrary n, just notice that the sequence $X = X_0 \dashrightarrow X_1 \dashrightarrow \cdots$ is a $(K_X + \mu^{-1}H)$-MMP for the constant $\mu = \mu_i$.   □

Corollary 6.2.14   *The Sarkisov program of Corti terminates in dimension up to four. It terminates in an arbitrary dimension on the assumption of the termination of canonical log flips.*

Proof   By Propositions 6.2.8 and 6.2.9, the Sarkisov threshold does not increase in the course of the Sarkisov program of Corti. If the program generates a Sarkisov link of type III or IV which does not decrease the Sarkisov threshold strictly, then by Proposition 6.2.9, the subsequent Sarkisov link is again of type III or IV. By Lemma 6.2.13 and [127], the Sarkisov threshold decreases strictly after finitely many Sarkisov links of type III or IV. From this and Lemma 6.2.12, one deduces that the Sarkisov degree decreases strictly after finitely many steps during the program. Thus the program must terminate by Theorem 6.1.7.   □

We move to the approach by Hacon and McKernan to Theorem 6.2.1 in an arbitrary dimension. Let $f \colon X/S \dashrightarrow Y/T$ be a link of Mori fibre spaces $\varphi \colon X \to S$ and $\psi \colon Y \to T$. Take a common resolution W of X and Y with $p \colon W \to X$ and $q \colon W \to Y$. One can write $K_X + G \sim_{\mathbf{Q}} \varphi^*C$ and $K_Y + H \sim_{\mathbf{Q}} \psi^*D$ with general ample effective $\mathbf{Q}$-divisors G, H, C, D on X, Y, S, T respectively. Then X is a log minimal model of (W, p^*G) and S is the ample model of $K_W + p^*G$. Likewise Y is a log minimal model of (W, q^*H) and T is the ample model of $K_W + q^*H$. The Sarkisov program of Hacon and McKernan makes use of the following variant of the finiteness of ample models in Theorem 1.5.8(i).

Theorem 6.2.15   *Let X be a normal projective variety and let A be an ample effective $\mathbf{Q}$-divisor on X. Let $V = V_{\mathbf{Q}} \otimes_{\mathbf{Q}} \mathbf{R}$ be the extension of a finite dimensional vector subspace $V_{\mathbf{Q}}$ of the rational vector space $Z^1(X) \otimes \mathbf{Q}$ of $\mathbf{Q}$-divisors on X. Suppose that (X, B_0) is klt for some $B_0 \in V$. Define the closed convex*

$$\mathscr{E}_A = \{\Delta = A + B \mid B \in V, (X, B) \text{ lc}, K_X + \Delta \text{ pseudo-effective}\}$$

in the real affine space $V_A = A + V$. For a rational map $\pi\colon X \dashrightarrow Y$, define

$$\mathscr{A}_{A,\pi} = \{\Delta \in \mathscr{E}_A \mid \pi \text{ the ample model of } K_X + \Delta\}.$$

Then there exist finitely many rational maps $\pi_i\colon X \dashrightarrow Y_i$ such that $\mathscr{E}_A = \bigsqcup_i \mathscr{A}_{A,\pi_i}$. These π_i and $\mathscr{A}_{A,\pi_i}$ have the same properties as g_i and $\mathscr{A}_i$ in Theorem 1.5.8(i) have.

The statement is local on the compact set $\mathscr{E}_A$. It is reduced to Theorem 1.5.8(i) as follows. For a fixed boundary $\Delta = A + B \in \mathscr{E}_A$, we take a small positive real number ε such that $A' = (1 - \varepsilon)A + \varepsilon(B - B_0)$ is ample. Let $B_1 = (1 - \varepsilon)B + \varepsilon(A_0 + B_0)$ with a general effective $\mathbf{Q}$-divisor $A_0 \sim_{\mathbf{Q}} A$. Then (X, B_1) is klt and $K_X + A' + B_1 \sim_{\mathbf{R}} K_X + A + B$. The pair $(X, \Delta_1 = A_1 + B_1)$ is klt for a general effective $\mathbf{R}$-divisor $A_1 \sim_{\mathbf{R}} A'$. The affine map λ which sends $\Delta' \in \mathscr{E}_A$ to $\Delta_1 + (\Delta' - \Delta)$ satisfies $\Delta' \sim_{\mathbf{R}} \lambda(\Delta')$ and $(X, \lambda(\Delta'))$ is klt if Δ' is sufficiently close to Δ, for which one can use Theorem 1.5.8(i).

We apply Theorem 6.2.15 to some two-dimensional affine subspace $V_A = A + V$ of $Z^1(W) \otimes \mathbf{R}$. We write $\mathscr{A}_\pi$ for $\mathscr{A}_{A,\pi}$ simply. By choosing V_A suitably, one can show that $\mathscr{A}_p$ and $\mathscr{A}_q$ are of dimension two and that $\mathscr{A}_{\varphi \circ p}$ and $\mathscr{A}_{\psi \circ q}$ are of dimension one. The loci $\mathscr{A}_{\varphi \circ p}$ and $\mathscr{A}_{\psi \circ q}$ lie on the boundary of the polytope $\mathscr{E}_A$ as in Figure 6.1 since neither $\varphi \circ p$ nor $\psi \circ q$ is birational. Take general points $\Delta_P \in \mathscr{A}_{\varphi \circ p}$ and $\Delta_Q \in \mathscr{A}_{\varphi \circ q}$.

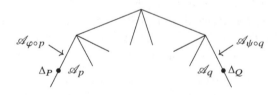

Figure 6.1 The polytope $\mathscr{E}_A$

There exist two paths from Δ_P to Δ_Q along the boundary of $\mathscr{E}_A$. We choose the path γ such that $K_W + \Delta$ is not big for all $\Delta \in \gamma$. Only finitely many points $\Delta_1, \ldots, \Delta_n$ on γ are contained in more than two closures of form $\overline{\mathscr{A}_\pi}$. Look at a point $\Delta_i \in \gamma$ as in Figure 6.2.

The two edges $\mathscr{A}_{\varphi_i \circ p_i}$ and $\mathscr{A}_{\varphi_{i+1} \circ p_{i+1}}$ correspond to the ample models $W \dashrightarrow S_i$ and $W \dashrightarrow S_{i+1}$. The two-dimensional chambers $\mathscr{A}_{p_i}$ and $\mathscr{A}_{p_{i+1}}$ adjacent to them correspond to the ample models $p_i\colon W \dashrightarrow X_i$ and $p_{i+1}\colon W \dashrightarrow X_{i+1}$. By the general choice of A_V, the associated contractions $\varphi_i\colon X_i \to S_i$ and $\varphi_{i+1}\colon X_{i+1} \to S_{i+1}$ are Mori fibre spaces. Let $W \dashrightarrow X_{ij}$ be the ample model associated with $\mathscr{A}_i(j)$ for the two-dimensional chambers $\mathscr{A}_i(1), \ldots, \mathscr{A}_i(n_i)$ between $\mathscr{A}_{p_i}$ and $\mathscr{A}_{p_{i+1}}$ locally about Δ_i. These admit natural contractions to

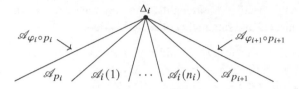

Figure 6.2 The neighbourhood at a point Δ_i

the ample model $W \dashrightarrow U_i$ of $K_W + \Delta_i$ as in the diagram below. Note that n_i may be zero and even $\mathscr{A}_{p_i}$ may equal $\mathscr{A}_{p_{i+1}}$.

$$
\begin{array}{ccccccc}
X_i & \xleftarrow{\alpha_i} & X_{i1} & \dashrightarrow & X_{in_i} & \xdashrightarrow{\beta_i} & X_{i+1} \\
{\scriptstyle\varphi_i}\big\downarrow & & \searrow & & \swarrow & & \big\downarrow{\scriptstyle\varphi_{i+1}} \\
S_i & \longrightarrow & & U_i & \longleftarrow & & S_{i+1}
\end{array}
$$

Hacon and McKernan showed that X_{ij} are **Q**-factorial and terminal with $\rho(X_{ij}/U_i) = 2$ and that $\alpha_i \colon X_{i1} \dashrightarrow X_i$ and $\beta_i \colon X_{in_i} \dashrightarrow X_{i+1}$ are a divisorial contraction or a (converse of) log flip, crepant with respect to the push-forward of $K_W + \Delta_i$. Further every $X_{ij} \dashrightarrow X_{i,j+1}$ is a log flip of the same kind. The pair $(\rho(S_i/U_i), \rho(S_{i+1}/U_i))$ of relative Picard numbers is $(0,1)$, $(0,0)$, $(1,0)$ or $(1,1)$. Accordingly, the link from φ_i to φ_{i+1} is a Sarkisov link of type I, II, III or IV. The sequence of log flips is in fact that of suitable canonical log flips.

In this manner, the link $f \colon X/S \dashrightarrow Y/T$ from $\varphi = \varphi_1$ to $\psi = \varphi_{n+1}$ is the composite of the n Sarkisov links.

6.3 Rationality and Birational Rigidity

We shall provide a framework of how to apply the Sarkisov program to the rationality problem. Prior to this, we briefly review the rationality problem of complex varieties. The reader may refer to [36] and [266, chapter IV].

A *rational* variety means a variety birational to the projective space $\mathbf{P}^n$. It is a fundamental problem to determine which varieties are rational. The origin of the Sarkisov program is a refinement of the work of Iskovskikh and Manin on the irrationality of a smooth quartic threefold, which is now an immediate consequence of Theorem 6.4.3 in the next section.

Corollary 6.3.1 (Iskovskikh–Manin [211]) *Every smooth quartic threefold in* $\mathbf{P}^4$ *is irrational.*

From the perspective of rationality, one may regard a Mori fibre space as a family of uniruled and rationally connected varieties.

Definition 6.3.2   A variety X is said to be *uniruled* if there exists a generically finite dominant rational map $\mathbf{P}^1 \times Z \dashrightarrow X$ with some variety Z. The variety X is said to be *ruled* if X is birational to $\mathbf{P}^1 \times Z$.

We have a numerical criterion for uniruledness.

Theorem 6.3.3 (Miyaoka–Mori [328])   *A smooth projective complex variety X is uniruled if and only if there exists a dense open subset U of X such that for any $x \in U$, there exists a curve C in X through x with $(K_X \cdot C) < 0$.*

The proof uses uncountability of the ground field $\mathbf{C}$. The if part is a consequence of the effective bend and break based on Theorem 6.3.4. The original assumption of the irrationality of $f(C)$ is removable as in [266, II corollary 5.6]. The *bend and break* technique produces a rational curve in a given variety from a curve which possesses enough deformations inside the variety. Whereas the bend and break still holds in positive characteristic, even a variety of general type can be uniruled in positive characteristic. It is thus natural to consider a *separably* uniruled variety which satisfies the additional condition that the map $\mathbf{P}^1 \times Z \dashrightarrow X$ is separable.

Theorem 6.3.4 ([328, theorem 4])   *Let X be a projective variety of dimension n in any characteristic with an ample divisor H. Let $f \colon C \to X$ be a nonconstant morphism from a smooth projective curve such that the image $f(C)$ is contained in the smooth locus in X. Suppose that the local dimension of the Hom scheme $\mathrm{Hom}(C, X)$ at $[f]$ is at least $ln + 1$ where l is a positive integer. Then for any point x in $f(C)$, there exists a rational curve D through x such that $(H \cdot D) \leq 2(H \cdot f_* C)/l$.*

Proposition 6.3.5   *Let $(X/S, \Delta)$ be a pair with a contraction $X \to S$ of fibre type such that $-(K_X + \Delta)$ is relatively ample. Then X is uniruled.*

Proof   Let U be the smooth locus in X. Shrinking S, we may assume that every fibre $F = X \times_S s$ at a closed point is a variety of the same dimension $n \geq 1$ and that the complement in F of $U|_F$ is of codimension at least two. Take any point $x \in U$ and the fibre F containing x. The intersection of $n-1$ general hyperplane sections of F through x contains a complete curve C in $U|_F$ through x. As far as x is away from the support of Δ, it has $(K_X \cdot C) \leq ((K_X + \Delta) \cdot C) < 0$ by the ampleness of $-(K_X + \Delta)|_F$. Thus by Theorem 6.3.3, X is uniruled.   □

Theorem 6.3.6 (Boucksom–Demailly–Păun–Peternell [55])   *A smooth projective variety X is uniruled if and only if K_X is not pseudo-effective.*

Proof   The only-if part follows from Theorem 6.3.3. Conversely if K_X is not pseudo-effective, then by Corollary 1.5.13, the MMP with scaling produces a Mori fibre space $X' \to S$ after a birational map $X \dashrightarrow X'$. By Proposition 6.3.5, the total space X' is uniruled and so is X. Note that the original proof is independent of the work of Birkar, Cascini, Hacon and McKernan.   □

Roughly speaking, the uniruledness means the existence of a rational curve through a general point. We make the notion of rational connectedness which means the existence of a rational curve through two general points.

Definition 6.3.7 (Campana [64], Kollár–Miyaoka–Mori [273])   A complete complex variety X is said to be *rationally connected* if there exists a dense open subset U of $X \times X$ such that for any $(x, y) \in U$, there exists a rational curve in X through x and y. This is actually equivalent to the stronger condition that for any finitely many points in X, there exists a rational curve in X through all these points.

The rational curve in the definition is irreducible in our terminology. When X is smooth, it is rationally connected if and only if it is *rationally chain connected*, meaning that for any $(x, y) \in U$, x and y are connected by a chain of rational curves [273, theorem 2.1]. This equivalence will be proved effectively in Theorem 9.1.14.

Example 6.3.8   Let X be the surface in $\mathbf{P}^3$ defined by $x_0^3 + x_1^3 + x_2^3$ for the homogeneous coordinates x_0, x_1, x_2, x_3 of $\mathbf{P}^3$, which is the *classical projective cone* over the elliptic curve E in $\mathbf{P}^2$ defined by $x_0^3 + x_1^3 + x_2^3$. The surface X is rationally chain connected but not rationally connected. Indeed, the blow-up B of X at the vertex $[0, 0, 0, 1]$ admits the structure of a $\mathbf{P}^1$-bundle $B \to E$. Fibres of $B \to E$ exhaust all rational curves in B since E is never the image of a rational curve by Hurwitz's theorem in Theorem 5.6.5.

The following statement for Fano varieties will be proved in Theorems 9.1.8 and 9.1.14.

Theorem 6.3.9 ([64], [274], [477])   *Let (X, Δ) be a klt projective pair such that $-(K_X + \Delta)$ is nef and big. Then X is rationally connected.*

For a contraction $X \to S$ from a klt pair (X, Δ), the next result shows that the total space X is rationally connected if and only if so is the base S.

Theorem 6.3.10 (Graber–Harris–Starr [147])   *Let $X \to Y$ be a surjective morphism of complete varieties such that the general fibre is rationally connected. Then X is rationally connected if and only if so is Y.*

We introduce two intermediate notions between rational connectedness and rationality.

Definition 6.3.11   (i) We say that a variety X is *unirational* if there exists a generically finite dominant rational map $\mathbf{P}^n \dashrightarrow X$.

(ii) We say that a variety X is *stably rational* if $X \times \mathbf{P}^m$ is rational for some m.

A stably rational variety is unirational since for a birational map $\mathbf{P}^{n+m} \dashrightarrow X \times \mathbf{P}^m$, a general n-dimensional linear subspace of $\mathbf{P}^{n+m}$ is projected to X dominantly. By definition, a unirational complete variety is rationally connected. The converse is unknown in dimension at least three.

Conjecture 6.3.12   *There exist complex varieties which are rationally connected but not unirational.*

A unirational (or even rationally connected) surface is rational by *Castelnuovo's criterion*, namely that a smooth complete surface S is rational if and only if the *irregularity* $q = h^1(\mathscr{O}_S)$ and the second *plurigenus* $P_2 = h^0(\mathscr{O}_S(2K_S))$ are both zero [35, chapter V]. By the minimal model program, every rationally connected projective surface admits a contraction to a Hirzebruch surface or the projective plane. The question of whether a unirational variety is rational is known as *Lüroth's problem*. As will be seen below, the notions of rationality, stable rationality and unirationality are all different in dimension at least three.

Example 6.3.13   Let $X = X_3 \subset \mathbf{P}^4$ be a general cubic threefold. Clemens and Griffiths [88] showed that X is irrational by means of the *intermediate Jacobian*. They proved Theorem 6.3.15 and confirmed that X does not satisfy the property in the theorem. Here we shall demonstrate the unirationality of X.

The *Grassmannian* $G = \mathrm{Gr}(2, 5)$ parametrises lines in the ambient space $\mathbf{P}^4$. By the following standard argument, one sees that the family of lines in X is of dimension two in G. Let H be the Hilbert scheme which parametrises cubic hypersurfaces in $\mathbf{P}^4$. The *incidence* subscheme I of $G \times H$ consists of pairs of a line l and a cubic X such that $l \subset X$. The projection $I \to G$ is a $\mathbf{P}^{30}$-bundle and hence I is a variety of dimension $30 + \dim G = 36$. Since H is of dimension 34, it suffices to check the surjectivity of the projection $I \to H$. This is equivalent to the existence of a two-dimensional fibre of $I \to H$, that is, the existence of a cubic X with exactly two-dimensional family of lines in X. The cone over the cubic surface defined by $x_0^3 + x_1^3 + x_2^3 + x_3^3$ is such an example of X.

Fix a general line l in X. Let B be the blow-up of X along l. The exceptional divisor E is a Hirzebruch surface by the structure of the $\mathbf{P}^1$-bundle $E \to l$. Planes in $\mathbf{P}^4$ containing l form a family $\bar{B} \to \mathbf{P}^2$ for the blow-up $\bar{B}$ of $\mathbf{P}^4$ along l, and for a general plane Π amongst them, the intersection $X \cap \Pi$ consists of l and a conic C in Π. These C form a conic bundle $B \to \mathbf{P}^2$. The induced morphism $E \to \mathbf{P}^2$ is generically two-to-one, where the fibre at $[\Pi]$ consists of the two points at which the strict transform of C intersects E. The normalisation Y of the dominant component of the base change $B \times_{\mathbf{P}^2} E$ has the structure $Y \to E$ of a conical fibration in Definition 7.1.1 with a section $E \to Y$. The fibre at the generic point is a projective line $\mathbf{P}_K^1$ over the function field K of E. Thus Y is birational to $\mathbf{P}^1 \times E$, which is rational as so is E, and X is unirational by the existence of the generically two-to-one morphism $Y \to B \to X$.

Recall the *Hodge decomposition* for compact Kähler manifolds. A good textbook is [466] by Voisin.

Theorem 6.3.14 (Hodge decomposition)   *Let V be a compact Kähler manifold. Define the Hodge cohomology group $H^{p,q}(V) = H^q(V, \Omega_V^p)$. Then there exists a natural isomorphism*

$$H^i(V, \mathbf{C}) \simeq \bigoplus_{p+q=i} H^{p,q}(V)$$

independent of the choice of Kähler metric. Via this isomorphism, $H^{p,q}(V)$ is complex conjugate to $H^{q,p}(V)$.

The *Hodge number* $h^{p,q}(X)$ is defined as the dimension of the Hodge cohomology group $H^{p,q}(X)$. The Hodge decomposition provides the equalities $b_i(X) = \sum_{p+q=i} h^{p,q}(X)$ and $h^{p,q}(X) = h^{q,p}(X)$, where $b_i(X)$ is the Betti number.

For a smooth projective threefold X, we let $W = H^{1,2}(X) \oplus H^{0,3}(X)$ and let Λ be the image of $H^3(X, \mathbf{Z})$ by the projection $H^3(X, \mathbf{C}) \to W$. The *intermediate Jacobian* of X is the complex torus $J_2(X) = W/\Lambda$. If $H^{0,1}(X) = H^{0,3}(X) = 0$, which holds for a rational threefold, then $J_2(X) = H^{1,2}(X)/\Lambda$ admits a positive definite hermitian form $(\alpha, \beta) = 2\sqrt{-1} \int_X \alpha \wedge \bar{\beta}$, by which $J_2(X)$ is a *principally polarised abelian variety*. This notion is an extension of that of the *Jacobian* $J(C) = H^{0,1}(C)/\Lambda' \simeq \mathrm{Pic}^0 C$ for a smooth projective curve C where Λ' is the image of $H^1(C, \mathbf{Z}) \to H^{0,1}(C)$.

Theorem 6.3.15 (Clemens–Griffiths)   *If X is a smooth rational projective threefold, then $J_2(X)$ is isomorphic as a principally polarised abelian variety to a product of Jacobians of smooth curves.*

Proof   By Hironaka's resolution, a birational map $\mathbf{P}^3 \dashrightarrow X$ is resolved by successive blow-ups along smooth centres. To be precise, starting from $Y_0 = \mathbf{P}^3$, there exists a finite sequence of blow-ups $Y_{i+1} \to Y_i$ along a point or a smooth curve Z_i for $0 \leq i < l$ such that the induced map from Y_l to X is a morphism. Then there exists a natural isomorphism

$$H^3(Y_{i+1}, \mathbf{Z}) \simeq H^3(Y_i, \mathbf{Z}) \oplus H^1(Z_i, \mathbf{Z})$$

of Hodge structures as stated in [466, theorem 7.31]. Hence $H^3(Y_l, \mathbf{Z}) \simeq \bigoplus_{i \in I} H^1(Z_i, \mathbf{Z})$ for the set I of indices i such that Z_i is a curve. Notice that $H^3(\mathbf{P}^3, \mathbf{Z}) = 0$ as mentioned in Example 2.5.3. This provides an isomorphism $J_2(Y_l) \simeq \prod_{i \in I} J(Z_i)$ of principally polarised abelian varieties.

By Lemma 2.5.4, the morphism $Y_l \to X$ induces an injection $H^3(X, \mathbf{Z}) \hookrightarrow H^3(Y_l, \mathbf{Z})$ in such a way that $H^3(X, \mathbf{Z})$ is a direct summand of $H^3(Y_l, \mathbf{Z})$. It is also compatible with the Hodge structures. Thus one can write $J_2(Y_l) \simeq J_2(X) \times A$ with some principally polarised abelian variety A.

We have obtained $J_2(X) \times A \simeq \prod_{i \in I} J(Z_i)$. This implies that $J_2(X)$ is isomorphic to the product of some of $J(Z_i)$ because a principally polarised abelian variety has a unique decomposition into the product of irreducible ones [88, corollary 3.23]. Note that the Jacobian of a smooth curve is irreducible since the theta divisor is the image of the Abel–Jacobi map.   □

Example 6.3.16 Artin and Mumford [22] constructed a unirational but not stably rational threefold. Let A be the conic in $\mathbf{P}^2$ defined by the quadric $\alpha = x_0 x_1 - x_2^2$. Take smooth cubics D_1 and D_2 in $\mathbf{P}^2$ meeting at nine points such that $D_1 + D_2$ meets A tangentially at six points outside $D_1 \cap D_2$. There exists a cubic B in $\mathbf{P}^2$ such that $(D_1 + D_2)|_A = 2B|_A$. We choose cubic forms δ_1, δ_2, β in x_0, x_1, x_2 defining D_1, D_2, B respectively so that α divides $\delta_1 \delta_2 - \beta^2$, and write $\delta_1 \delta_2 = \beta^2 - 4\alpha\gamma$ with a quartic form γ.

Let K be the quartic surface in $\mathbf{P}^3$ defined by $\alpha x_3^2 + \beta x_3 + \gamma$. It has 10 ordinary double points including $p = [0, 0, 0, 1]$. We construct the *double solid* $X' = \mathrm{Proj}_{\mathbf{P}^3}(\mathcal{O}_{\mathbf{P}^3} \oplus \mathcal{O}_{\mathbf{P}^3}(-2))$, namely the double cover of $\mathbf{P}^3$, ramified along K by the relation $\mathcal{O}_{\mathbf{P}^3}(2)^{\otimes 2} = \mathcal{O}_{\mathbf{P}^3}(K)$ as explained before Lemma 2.2.16. Explicitly, X' is given in the weighted projective space $\mathbf{P}(1, 1, 1, 1, 2)$ by $y^2 - (\alpha x_3^2 + \beta x_3 + \gamma)$ with a new coordinate y of degree two. Let X be the blow-up of X' at the 10 ordinary double points.

The smooth threefold X is unirational. Indeed, the restriction U of X to the affine locus $(x_0 \neq 0)$ is given in $\mathbf{A}^4$ by $y^2 - (x_1 - x_2^2)x_3^2 - \beta(1, x_1, x_2)x_3 - \gamma(1, x_1, x_2)$. Setting $z^2 = x_1 - x_2^2$ with a new coordinate z, there exists a finite morphism to U from the affine hypersurface given by

$$(y + zx_3)(y - zx_3) - \beta(1, x_2^2 + z^2, x_2)x_3 - \gamma(1, x_2^2 + z^2, x_2)$$

in $\mathbf{A}^4$, which is birational to $\mathbf{A}^3$ with coordinates $x_2, z, y - zx_3$.

Let B the blow-up of $\mathbf{P}^3$ at the 10 ordinary double points of K. The threefold X is alternatively realised as the double cover of B ramified along the strict transform of K. Thus the projection $\mathbf{P}^3 \dashrightarrow \mathbf{P}^2$ from p induces a morphism $X \to B \to \mathbf{P}^2$. The general fibre of $X \to \mathbf{P}^2$ is a conic given by $y^2 - (\alpha' x_3^2 + \beta' x_0 x_3 + \gamma' x_0^2)$ with homogeneous coordinates x_0, x_3, y, where $(\alpha', \beta', \gamma') = (\alpha x_0^{-2}, \beta x_0^{-3}, \gamma x_0^{-4})$. It degenerates over the locus where the discriminant $\beta^2 - 4\alpha\gamma = \delta_1 \delta_2$ vanishes, and the fibre at a general point in D_i consists of two lines l_i and l_i'.

Computing a certain linking number [413, section 77], Artin and Mumford checked that the cycle $[l_1] - [l_2]$ defines a two-torsion in $H_2(X, \mathbf{Z})$. Thus $H^3(X, \mathbf{Z})$ has a torsion since it contains $\mathrm{Ext}(H_2(X, \mathbf{Z}), \mathbf{Z})$ by the universal coefficient theorem. Consequently X is not stably rational by Theorem 6.3.17. See Remark 7.4.3 for an alternative perspective.

Theorem 6.3.17 *Every smooth stably rational projective variety X has no torsions in $H^3(X, \mathbf{Z})$.*

Proof Recall the description $H^\bullet(\mathbf{P}^m, \mathbf{Z}) = \mathbf{Z}[\xi]/(\xi^{m+1})$ in Example 2.5.3. Hence $H^3(X \times \mathbf{P}^m, \mathbf{Z}) \simeq H^3(X, \mathbf{Z}) \oplus H^1(X, \mathbf{Z})$ by the Künneth formula, which reduces the theorem to the case when X is rational. As in the proof of Theorem 6.3.15, we resolve a birational map $\mathbf{P}^n \dashrightarrow X$ by a finite sequence of blow-ups $Y_{i+1} \to Y_i$ along a smooth centre Z_i for $0 \le i < l$ where $Y_0 = \mathbf{P}^n$. Then $H^3(Y_l, \mathbf{Z}) \simeq \bigoplus_i H^1(Z_i, \mathbf{Z})$. This group is torsion-free since any variety V has no torsions in $H^1(V, \mathbf{Z}) \simeq \mathrm{Hom}(H_1(V, \mathbf{Z}), \mathbf{Z})$ from the universal coefficient theorem. By Lemma 2.5.4, $Y_l \to X$ induces an injection $H^3(X, \mathbf{Z}) \hookrightarrow H^3(Y_l, \mathbf{Z})$. Thus $H^3(X, \mathbf{Z})$ is also torsion-free. $\square$

Voisin [468] proved that the double solid ramified along a very general quartic surface with at most seven ordinary double points is not stably rational. She used the criterion that a smooth stably rational projective variety has a *Chow-theoretic decomposition of the diagonal*. It is remarkable that its desingularisation by the usual blow-up has no two-torsions in the integral cohomology groups as proved by Endrass [122].

Example 6.3.18 Beauville, Colliot-Thélène, Sansuc and Swinnerton-Dyer [37] constructed a stably rational but irrational threefold. Let C be a smooth trigonal curve, where we say that C is *trigonal* if it admits a finite morphism $C \to \mathbf{P}^1$ of degree three. Suppose that C has genus at least three and has no points of ramification index three. There exists an affine chart of C in $\mathbf{A}^2 = \mathrm{Spec}\,\mathbf{C}[t, x]$ given by a polynomial $P(t, x)$ of degree three with respect to x on which $C \to \mathbf{P}^1$ is defined by the inclusion $\mathbf{C}[t] \subset \mathbf{C}[t, x]$. Let $\delta(t) \in \mathbf{C}[t]$ be the *discriminant* of the polynomial $P(t, x)$ in x over $\mathbf{C}[t]$. For discriminant, see [283, section IV.6] for example. Let X be the hypersurface in $\mathbf{A}^4$ defined by $y_1^2 - \delta(t) y_2^2 - P(t, x)$. They proved that $X \times \mathbf{A}^3$ is rational and hence X

is stably rational, and also checked that X is irrational using Theorem 6.3.15. Later, Shepherd-Barron [418] showed that $X \times \mathbf{A}^2$ is already rational.

As to the rationality problem, consider a very general hypersurface X in $\mathbf{P}^{n+1}$ of degree d. Kollár [265] proved that X is not ruled if $d \geq 2\lceil (n+3)/3 \rceil$. After the work of Voisin [468] and a generalisation [90] of it, Totaro [448] proved that X is not stably rational if $d \geq 2\lceil (n+2)/3 \rceil$ and $n \geq 2$. Schreieder [412] improved this by showing the stable irrationality for $d \geq \log_2 n + 2$ and $n \geq 3$.

The properties discussed above behave well under deformation.

Theorem 6.3.19 *Let* $X \to B$ *be a smooth proper morphism of complex varieties with connected fibres. In the following cases, the set of points* $b \in B$ *such that the fibre* $X_b = X \times_B b$ *has property* $\mathscr{P}$ *is either B or empty.*

(i) (Fujiki [126], Levine [290]) $\mathscr{P}$ *is being uniruled.*
(ii) (Kollár–Miyaoka–Mori [273]) $\mathscr{P}$ *is being rationally connected.*

In the following cases, the set of $b \in B$ *such that* X_b *has property* $\mathscr{Q}$ *is a countable union of closed subsets of B.*

(iii) (Matsusaka [309]) $\mathscr{Q}$ *is being ruled.*
(iv) (Nicaise–Shinder [372]) $\mathscr{Q}$ *is being stably rational.*
(v) (Kontsevich–Tschinkel [280]) $\mathscr{Q}$ *is being rational.*

Levine [291] constructed an example of a family X/B which needs infinitely many closed subsets of B to parametrise ruled fibres. Hassett, Pirutka and Tschinkel [180] proved that for the family X/B of hypersurfaces in $\mathbf{P}^2 \times \mathbf{P}^3$ of bidegree $(2,2)$, the set of $b \in B$ at which the fibre $X \times_B b$ is rational is Euclidean dense in B but the fibre at the very general point in B is not stably rational.

We return to the geometry of Mori fibre spaces. We characterise the rationality of a variety by the property of having a large number of Mori fibre spaces birational to it. For example, a standard conic bundle over $\mathbf{P}^2$ with suitable quintic discriminant divisor has rational total space as will be seen in Theorem 7.4.18. The *Cremona group*, namely the group of birational self-maps of $\mathbf{P}^n$, needs uncountably many generators besides automorphisms when n is at least three [192], [378]. From this standpoint, we define the following notion opposite to rationality for Mori fibre spaces.

Definition 6.3.20 A Mori fibre space $\varphi \colon X \to S$ is said to be *birationally rigid* (resp. *birationally rigid over the base*) if for any link $f \colon X/S \dashrightarrow Y/T$ from φ to a Mori fibre space $Y \to T$, there exists a self-link (resp. fibrewise self-link) $\sigma \colon X/S \dashrightarrow X/S$ such that $f \circ \sigma \colon X/S \dashrightarrow Y/T$ is square. We say that

φ is *birationally superrigid* if every link $X/S \dashrightarrow Y/T$ from φ to a Mori fibre space is square.

The birational rigidity over the base is trivially equivalent to the birational rigidity if the base of a Mori fibre space is a point. For a Mori fibre space of relative dimension one, the birational rigidity over the base is equivalent to the birational superrigidity.

Example 6.3.21 We have seen the irrationality of a general cubic threefold X in Example 6.3.13. It has Pic $X \simeq H^2(X, \mathbf{Z}) \simeq \mathbf{Z}$, where the former isomorphism follows from the Kodaira vanishing $H^i(\mathcal{O}_X) = 0$ for $i \geq 1$ and the latter from Theorem 6.3.22. Hence $X \to \operatorname{Spec} \mathbf{C}$ is a Mori fibre space. It is not birationally rigid since it is linked to the conic bundle $B \to \mathbf{P}^2$.

The reader can learn the following famous theorem in [467, chapter 1] and [319, corollary 7.3].

Theorem 6.3.22 (Lefschetz hyperplane theorem) *Let X be a projective complex variety of dimension n. Let Y be a hyperplane section of X such that $X \setminus Y$ is smooth. Then the restriction map $H^i(X, \mathbf{Z}) \to H^i(Y, \mathbf{Z})$ is an isomorphism for $i < n - 1$ and is injective for $i = n - 1$. The map $H_i(Y, \mathbf{Z}) \to H_i(X, \mathbf{Z})$ is an isomorphism for $i < n - 1$ and is surjective for $i = n - 1$.*

Remark 6.3.23 The Lefschetz hyperplane theorem is reduced to the vanishing of homology groups of an affine variety. More generally, Goresky and MacPherson established the following vanishing. Let $\pi \colon V \to A$ be a proper morphism from an equi-dimensional complex scheme to an affine complex scheme. Stratify A into the disjoint union of constructible subsets A_i in such a manner that every $a \in A_i$ has fibre $\pi^{-1}(a)$ of dimension i. Let I be the maximum of $2i + \dim A_i$. Then $H_n(V, \mathbf{Z}) = 0$ for all $n > I$. See [145, introduction 2.3].

Obviously the birational rigidity of $\varphi \colon X \to S$ implies the irrationality of X. We shall explain a general method for proving birational rigidity. The Sarkisov program constructs a Sarkisov link $f \colon X/S \dashrightarrow X'/S'$ when φ admits a link to another Mori fibre space. We seek to show that $f \circ \sigma$ becomes square by an appropriate self-link σ of X/S.

If f is a Sarkisov link of type III or IV, then it is uniquely determined as a two-ray game. Our task is to track the game explicitly. This case never occurs when X is a **Q**-Fano variety with Picard number one. If f is of type I or II, then it starts with a divisorial contraction $\pi \colon Y \to X$. One can apply the structural classification of π in Chapters 3 and 4. Once π is fixed, the link is determined again by a two-ray game. The contraction π is crepant with respect to $K_Y + cH_Y$ for a strictly canonical pair (X, cH) and the strict transform H_Y of H, where H

is the general member of the homaloidal transform $\mathscr{H}$ of some standard linear system. The centre Γ on X of the exceptional divisor of π is thus a maximal centre in the following sense.

Definition 6.3.24 Let $X \to S$ be a Mori fibre space. Let $\mathscr{H}$ be a mobile linear system on X with positive Sarkisov threshold μ. Let $H \sim_{\mathbb{Q},S} -\mu K_X$ be the general member of $\mathscr{H}$ if $\mu \geq 1$, which should be reset as in Remark 6.2.3 if $\mu < 1$. Suppose that $(X, \mu^{-1}H)$ is not canonical. Let c be the canonical threshold of H on X, which satisfies $c < \mu^{-1}$ and $\lfloor cH \rfloor = 0$. A closed subset Γ of X is called a *maximal centre* in X with respect to $\mathscr{H}$ if there exists a divisorial contraction $E \subset Y \to \Gamma \subset X$ which contracts the divisor E to Γ and which is crepant with respect to $K_Y + cH_Y$ for the strict transform H_Y of H.

By $\lfloor cH \rfloor = 0$, the number of maximal centres is finite. The existence of a maximal centre follows from Lemma 6.2.6. We seek to exclude the possibility for a locus Γ in X to be a maximal centre except a few special loci. In dimension three, the next application is useful for excluding smooth points. It was proved in [93, corollary 3.4] and generalised to higher dimensions in [103].

Proposition 6.3.25 *Let $x \in X$ be the germ of a smooth threefold. Let $\mathscr{H}$ be a mobile linear system on X and let H and H' be the general members of $\mathscr{H}$. If $\mathrm{mld}_x(X, cH) = 1$ with a positive rational number c, then $\mathrm{mult}_x(H \cap H') \geq 4c^{-2}$.*

Proof We may assume that $c < 1$. We shall provide a proof which uses the classification in Chapter 3. Fix a divisor F over X which computes $\mathrm{mld}_x(X, cH)$ in the sense that $a_F(X, cH) = 1$ and $c_X(F) = x$. Since H is general, the ideal $\{f \in \mathscr{O}_X \mid \mathrm{ord}_F H \leq \mathrm{ord}_F f\}$ in $\mathscr{O}_X$ contains all h such that $(h)_X \in \mathscr{H}$. Take the divisor $D = (f)_X$ defined by a general function f in this ideal. Then (X, cD) is terminal outside x but has $\mathrm{mld}_x(X, cD) = 1$. Further if a divisor E over X has $a_E(X, cD) = 1$, then $a_E(X, cH) = 1$. Applying Lemma 6.2.6 to the pair (X, cD), one finds a divisorial contraction $\pi \colon E \subset Y \to x \in X$ which contracts the divisor E to x and which has $a_E(X, cH) = 1$.

By Theorem 3.5.1, π is the weighted blow-up with $\mathrm{wt}(x_1, x_2, x_3) = (1, r_1, r_2)$ for a suitable regular system x_1, x_2, x_3 of parameters in $\mathscr{O}_{X,x}$. Note that $E^3 = 1/r_1 r_2$. It follows from $a_E(X, cH) = 1$ that $c \, \mathrm{ord}_E H = r_1 + r_2$ and hence $\pi^*H = H_Y + c^{-1}(r_1 + r_2)E$ with the strict transform H_Y of H. Likewise $\pi^*H' = H'_Y + c^{-1}(r_1 + r_2)E$. On the other hand, the general hyperplane section $x \in A$ of $x \in X$ has $\pi^*A = A_Y + E$ with the strict transform A_Y. The multiplicity $\mathrm{mult}_x(H \cap H') = (H \cdot H' \cdot A)_x$ is expressed as $(\pi^*H \cdot \pi^*H' \cdot \pi^*A)_U = (H_Y \cdot$

$H'_Y \cdot A_Y)_U + c^{-2}(r_1 + r_2)^2 E^3$ on a small neighbourhood U in Y about E. It is estimated as

$$c^2 \operatorname{mult}_x(H \cap H') \geq (r_1 + r_2)^2 E^3 = \frac{(r_1 + r_2)^2}{r_1 r_2} \geq 4. \qquad \square$$

Corollary 6.3.26 *In Definition 6.3.24 if X is a threefold and the maximal centre Γ is a smooth point x of X, then $\operatorname{mult}_x(H \cap H') > 4\mu^2$ for the general members H and H' of $\mathscr{H}$.*

If a curve C in a threefold X is a maximal centre, then by Proposition 4.1.3, the divisorial contraction $E \subset Y \to C \subset X$ is the symbolic blow-up of X along C. Set $m = \operatorname{mult}_C H$ for the general member H of $\mathscr{H}$ with the notation in Definition 6.3.24. Then $m = \operatorname{ord}_E H = c^{-1}$. In particular, the following holds.

Lemma 6.3.27 *In Definition 6.3.24 if X is a threefold and the maximal centre Γ is a curve C, then $\operatorname{mult}_C H > \mu$ for the general member H of $\mathscr{H}$.*

In order to exclude the possibility for a curve C to be a maximal centre, we shall study the restriction $H|_S$ to a suitable normal surface S containing C, called a *test surface*. We choose S so that $\operatorname{ord}_C H|_S = \operatorname{mult}_C H = m$, but the restriction $\mathscr{H}|_S$ may have fixed components other than C. Here for the system $\mathscr{H} = \mathbf{P}V^\vee$ with a subspace V of $H^0(\mathscr{O}_X(H))$, the *restriction* $\mathscr{H}|_S$ to S of $\mathscr{H}$ means the linear subsystem $\mathbf{P}V_S^\vee$ of $|H|_S|$ defined by the image V_S of V by the restriction map $H^0(\mathscr{O}_X(H)) \to H^0(\mathscr{O}_S(H|_S))$. We have an expression $H|_S = mC + F + L$ with a mobile Weil divisor L and the fix part $mC + F$ of $\mathscr{H}|_S$.

6.4 Pliability

Corti introduced the set of substantially distinct Mori fibre spaces birational to a given variety.

Definition 6.4.1 (Corti)   Given a uniruled variety Z, we consider the set $\mathscr{S}$ of Mori fibre spaces $X \to S$ such that X is birational to Z. We introduce an equivalence relation in $\mathscr{S}$ in such a way that $X \to S$ is equivalent to $Y \to T$ if and only if there exists a square link $X/S \dashrightarrow Y/T$. The quotient of $\mathscr{S}$ by this relation is called the *pliability* of Z and denoted by $\mathscr{P}(Z)$.

By definition, a Mori fibre space $X \to S$ is birationally rigid if and only if the pliability $\mathscr{P}(X)$ consists of just one element. One motivation for introducing pliability is to grasp the behaviour of birational rigidity under deformation by detecting an algebraic structure in the pliability. Contrary to expectations,

Cheltsov and Grinenko [75] pointed out that birational rigidity is not an open condition. Neither is birational superrigidity [76, example 6.3], [478].

This section is devoted to the example by Corti and Mella of a threefold with pliability of cardinality two. This will reveal how the Sarkisov program works in practice.

Theorem 6.4.2 (Corti–Mella [95])   *Let $X = X_4 \subset \mathbf{P}^4$ be a general quartic threefold having one singular point analytically isomorphic to $o \in (x_1 x_2 + x_3^3 + x_4^3 = 0) \subset \mathfrak{D}^4$. Then $X \to \operatorname{Spec} \mathbf{C}$ is a Mori fibre space with X factorial and the pliability $\mathscr{P}(X)$ consists of exactly two elements.*

The element in $\mathscr{P}(X)$ other than X itself is provided in Construction 6.4.7. The assumption on the singularity of X is crucial. This contrasts with the original result of Iskovskikh and Manin.

Theorem 6.4.3 (Iskovskikh–Manin [211])   *Every smooth quartic threefold in $\mathbf{P}^4$ is birationally superrigid.*

It has been generalised to higher dimensions. For $n \geq 4$, every smooth hypersurface in $\mathbf{P}^n$ of degree n is birationally superrigid [102], [393].

We shall clarify the meaning of 'general' in Theorem 6.4.2. Let p denote the unique singular point of X. We set the homogeneous coordinates $x_0, \dots, x_4$ of $\mathbf{P}^4$ so that $p = [1, 0, 0, 0, 0]$ and

$$X = (x_0^2 x_1 x_2 + x_0 a_3(x_1, \dots, x_4) + b_4(x_1, \dots, x_4) = 0) \subset \mathbf{P}^4$$

with some cubic and quartic forms a_3 and b_4. The generality condition means that the choices of a_3 and b_4 are general in the Zariski topology.

Lemma 6.4.4   *The quartic X is a factorial terminal threefold with $\operatorname{Pic} X = \mathbf{Z}[\mathscr{O}_X(1)]$ and $\mathscr{O}_X(1) = \mathscr{O}_X(-K_X)$. In particular, $X \to \operatorname{Spec} \mathbf{C}$ is a Mori fibre space.*

Proof   The threefold X is terminal with one cA_2 point p and $K_X = (K_{\mathbf{P}^4} + X)|_X \sim -H$ for the hyperplane section H of X inside $\mathbf{P}^4$. We shall prove that the divisor class group $\operatorname{Cl} X$ of X equals $\mathbf{Z}[H]$ by means of the *Grothendieck–Lefschetz theory*. Then X is factorial with $\operatorname{Pic} X = \mathbf{Z}[H]$ and this is sufficient for the lemma.

Remark that $\operatorname{Cl} X \simeq \operatorname{Pic}(X \setminus p)$. Let S be the surface in $X = (x_0^2 x_1 x_2 + x_0 a_3 + b_4 = 0)$ defined by x_0. It is the general quartic in $\mathbf{P}^3$ defined by b_4 and hence $\operatorname{Pic} S = \mathbf{Z}[H|_S]$ by the *Noether–Lefschetz theorem* [467, theorem 3.5]. Let $\hat{X}$ be the *formal completion* of X along S [178, section II.9]. It is the locally ringed space on the topological space $\hat{X} = S$ such that $\mathcal{O}_{\hat{X}} = \varprojlim \mathcal{O}_X/\mathcal{O}_X(-nS)$. Consider the composite

$$\mathrm{Pic}(X \setminus p) \to \mathrm{Pic}\,\hat{X} \to \mathrm{Pic}\,S$$

of restriction maps. The pair $S \subset X$ satisfies the *Lefschetz condition* $\mathrm{Lef}(X, S)$ below as in [161, X exemple 2.2] and this implies the injectivity of the first map $\mathrm{Pic}(X \setminus p) \to \mathrm{Pic}\,\hat{X}$ by [161, XI proposition 2.1]. The second map $\mathrm{Pic}\,\hat{X} \to \mathrm{Pic}\,S$ is also injective by [161, XI proposition 1.1] since $H^1(\mathcal{O}_S(-n))$, dual to $H^1(\omega_S \otimes \mathcal{O}_S(n))$, is zero for $n \geq 1$ by Kodaira vanishing. These induce the injection $\mathrm{Cl}\,X \hookrightarrow \mathrm{Pic}\,S = \mathbf{Z}[H|_S]$ and hence $\mathrm{Cl}\,X = \mathbf{Z}[H]$. □

Definition 6.4.5 ([161, exposé X]) Let X be an algebraic scheme and let Y be a closed subset of X. Let $\hat{X}$ denote the formal completion of X along Y. We say that $Y \subset X$ satisfies the *Lefschetz condition* $\mathrm{Lef}(X, Y)$ if for every locally free coherent sheaf $\mathcal{E}$ on an open neighbourhood U in X about Y, the restriction map $\Gamma(U, \mathcal{E}) \to \Gamma(\hat{X}, \mathcal{E}|_{\hat{X}})$ is an isomorphism. We say that $Y \subset X$ satisfies the *effective Lefschetz condition* $\mathrm{Leff}(X, Y)$ if it satisfies $\mathrm{Lef}(X, Y)$ and in addition every locally free coherent sheaf on $\hat{X}$ is the restriction of some locally free sheaf on an open neighbourhood about Y.

The factoriality of X is a subtle condition. The analytic singularity given by $x_1x_2+x_3^3+x_4^3$ is not $\mathbf{Q}$-factorial. For instance, the divisor $(x_1 = x_3+x_4 = 0)$ is not $\mathbf{Q}$-Cartier. A quartic threefold with only ordinary double points is factorial if and only if the singular points impose linearly independent conditions on cubics in $\mathbf{P}^4$ [87], [105], [471]. In this case, Mella [315] proved that it is birationally rigid. By the work of Cheltsov [71], [74], a quartic threefold with at most eight ordinary double points is factorial.

We construct Sarkisov links from X with the following notation.

Notation 6.4.6 The notation $\mathbf{P}(a_1^{m_1}, \ldots, a_n^{m_n})$ stands for the weighted projective space $\mathbf{P}(a_1, \ldots, a_1, \ldots, a_n, \ldots, a_n)$ of dimension $\sum_{i=1}^{n} m_i - 1$ in which a_i appears m_i times.

Construction 6.4.7 By a new coordinate $y_1 = x_0x_1$ of degree two, we let $\mathbf{P}(1^4, 2) = \mathbf{P}(1, 1, 1, 1, 2)$ be the weighted projective space with coordinates x_1, x_2, x_3, x_4, y_1. The rational map $\mathbf{P}^4 \dashrightarrow \mathbf{P}(1^4, 2)$ defines a birational map from X to the threefold

$$Z = Z_5 = (x_2y_1^2 + a_3y_1 + b_4x_1 = 0) \subset \mathbf{P}(1^4, 2).$$

The germ $p \in X \subset \mathbf{P}^4$ has local coordinates z_1, z_2, z_3, z_4 with $z_i = x_ix_0^{-1}$. The weighted blow-up $\pi\colon W \to X$ with $\mathrm{wt}(z_1, z_2, z_3, z_4) = (2, 1, 1, 1)$ is a divisorial contraction and it resolves the indeterminacy of $X \dashrightarrow Z$ with a crepant small contraction $\alpha : W \to Z$ which contracts the strict transforms of the 12 lines $(x_1 = a_3 = b_4 = 0)$ in X. The generality condition is used implicitly here to guarantee that the locus $(x_1 = a_3 = b_4 = 0)$ consists of 12 lines.

Set $\mathbf{P}(1^4, 2^2)$ with coordinates $x_1, x_2, x_3, x_4, y_1, y_2$ by adding a new coordinate y_2 of degree two. The projection $\mathbf{P}(1^4, 2^2) \dashrightarrow \mathbf{P}(1^4, 2)$ induces a birational map from the threefold

$$Y = Y_{3,4} = (x_2 y_1 + x_1 y_2 + a_3 = y_1 y_2 - b_4 = 0) \subset \mathbf{P}(1^4, 2^2)$$

to Z. The model Y has two singular points $q_1 = [0, 0, 0, 0, 1, 0]$ and $q_2 = [0, 0, 0, 0, 0, 1]$, which are quotient singularities of type $\frac{1}{2}(1, 1, 1)$. The blow-up $\pi^+ : W^+ \to Y$ at q_2 is a divisorial contraction and it resolves the indeterminacy of $Y \dashrightarrow Z$ with a crepant small contraction $\alpha^+ : W^+ \to Z$ which contracts the strict transforms of the 12 curves $(y_1 = x_1 = a_3 = b_4 = 0)$ in Y.

Observe that the transformation

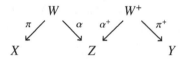

is a two-ray game which is a Sarkisov link of type II. Let D be the strict transform in W of the hyperplane section in X defined by x_1. Then $W \dashrightarrow W^+$ is the D-flop and D is transformed to the exceptional divisor of π^+. The end product $Y \to \operatorname{Spec} \mathbf{C}$ is a Mori fibre space. The inverse $Y \dashrightarrow X$ is a Sarkisov link of type II from Y. One can make another Sarkisov link by interchanging the roles of x_1 and x_2, but it ends with the same threefold Y as an abstract variety.

Construction 6.4.8 Choose a line l in X through p. It is in the locus $(x_1 x_2 = a_3 = b_4 = 0)$. By coordinate change, we attain the expression $l = (x_2 = x_3 = x_4 = 0)$ in $X = (x_0^2 x_1 x_2 + x_0 a_3 + b_4 = 0)$. The general hypersurface in X containing l is Du Val of type A_2 at p. Hence by Theorem 4.2.10, there exists a divisorial contraction $\pi : W \to X$ which contracts the divisor to l and it is the blow-up of X along the third symbolic power $\mathcal{I}^{(3)}$ of the ideal $\mathcal{I}$ in $\mathcal{O}_X$ defining l.

We write $a_3 = x_1^2 c_1 + x_1 c_2 + c_3$ and $b_4 = x_1^3 d_1 + x_1^2 d_2 + x_1 d_3 + d_4$ with homogeneous polynomials c_i and d_i in x_2, x_3, x_4 of degree i. Then $\mathcal{I}^{(3)}$ is generated by y and cubics in x_2, x_3, x_4 where

$$y = x_0^2 x_2 + x_0 (x_1 c_1 + c_2) + (x_1^2 d_1 + x_1 d_2 + d_3),$$

which satisfies the relation $x_1 y + x_0 c_3 + d_4 = 0$ on X. Hence considering y to be a coordinate of degree three, we embed W into $X \times \mathbf{P}(1^3, 3)$ with the coordinates x_2, x_3, x_4, y of $\mathbf{P}(1^3, 3)$.

It follows from the relation $x_0x_2(x_1y + x_0c_3 + d_4) = 0$ on X that

$$x_0x_1x_2y + (y - x_0(x_1c_1 + c_2) - (x_1^2d_1 + x_1d_2 + d_3))c_3 + x_0x_2d_4 = 0,$$

which is rewritten as $x_1z + x_0(x_2d_4 - c_2c_3) + c_3(y - d_3) = 0$ with

$$z = x_0x_2y - x_0c_1c_3 - x_1c_3d_1 - c_3d_2 \in \mathscr{I}^{(5)}.$$

These can be arranged neatly as $A\,{}^t(x_1, x_0, 1) = 0$ by

$$A = \begin{pmatrix} y & c_3 & d_4 \\ z & x_2d_4 - c_2c_3 & c_3(y - d_3) \\ c_3d_1 & c_1c_3 - x_2y & z + c_3d_2 \end{pmatrix}.$$

The determinant $\det A$ is divided by c_3, and the polynomial $-c_3^{-1}\det A$ is of form $z^2 + (c_2y + e_5)z - (x_2y^3 + e_4y^2 + e_7y + e_{10})$ with homogeneous polynomials e_i in x_2, x_3, x_4 of degree i. It defines a threefold Z in $\mathbf{P}(1^3, 3, 5)$ with coordinates x_2, x_3, x_4, y, z which admits a double cover $\beta: Z \to \mathbf{P}(1^3, 3)$. There exists a contraction $\alpha: W \to Z$ fitting into the commutative diagram

$$
\begin{array}{ccc}
W & \subset & X \times \mathbf{P}(1^3, 3) \\
{\scriptstyle \alpha}\downarrow & & \downarrow \\
Z & \xrightarrow{\ \beta\ } & \mathbf{P}(1^3, 3)
\end{array}
\quad .
$$

From the generality condition, α is a crepant small contraction.

Since Z is defined by a monic quadratic in z, there exists an involution ι on Z compatible with β, that is, $\beta = \beta \circ \iota$. Explicitly it multiplies $z + (c_2y + e_5)/2$ by -1. The transformation

$$
\begin{array}{ccccc}
 & W & & W & \\
{\scriptstyle \pi}\swarrow & & {\scriptstyle \alpha}\searrow \quad {\scriptstyle \alpha}\swarrow & & \searrow{\scriptstyle \pi} \\
X & & Z \xrightarrow{\ \iota\ } Z & & X
\end{array}
$$

is a Sarkisov link of type II.

Actually, all Sarkisov links from X are obtained as above.

Theorem 6.4.9 *Let $X = X_4 \subset \mathbf{P}^4$ be the general quartic threefold in Theorem 6.4.2 and let $Y = Y_{3,4} \subset \mathbf{P}(1^4, 2^2)$ be the threefold defined from X in Construction 6.4.7.*

 (i) *Every Sarkisov link from X is obtained as in Construction 6.4.7 or 6.4.8.*
 (ii) *Every Sarkisov link from Y is the inverse of the link $X \dashrightarrow Y$ obtained as in Construction 6.4.7.*

The Sarkisov program, Corollary 6.2.14, reduces Theorem 6.4.2 to this theorem. We shall demonstrate the first statement. The idea for the second is the same.

Since X has Picard number one, every Sarkisov link from X is of type I or II and it starts with a divisorial contraction $\pi\colon W \to X$. As discussed prior to Definition 6.3.24, the centre Γ on X of the exceptional divisor of π is a maximal centre with respect to some mobile and big linear system $\mathscr{H}$. Since $\operatorname{Pic} X = \mathbf{Z}[\mathscr{O}_X(1)]$ by Lemma 6.4.4, one can write

$$\mathscr{H} \subset |-\mu K_X| = |\mathscr{O}_X(\mu)|$$

with a positive integer μ.

The case when Γ is the singular point p or a line through p is settled by the classification of π.

Lemma 6.4.10 *Let Γ be either the point p or a line in X through p. Then every Sarkisov link from X starting with a divisorial contraction which contracts the divisor to Γ is obtained as in Construction 6.4.7 or 6.4.8.*

Proof Recall that a Sarkisov link of type I or II is uniquely determined by the initial divisorial contraction as far as it exists. Hence one has only to verify that $\pi\colon W \to X$ in Constructions 6.4.7 and 6.4.8 exhaust all the divisorial contractions that contract the divisor to Γ. This follows from Theorem 6.4.11 when $\Gamma = p$ and from Proposition 4.1.3 when Γ is a line through p. □

Theorem 6.4.11 *Let $\pi\colon E \subset Y \to x \in X$ be a divisorial contraction which contracts the divisor E to the cA_2 point x analytically given by $x_1 x_2 + x_3^3 + x_4^3$. Then π is the weighted blow-up with $\operatorname{wt}(x_1, x_2, x_3, x_4) = (2, 1, 1, 1)$ or $(1, 2, 1, 1)$.*

Proof This was first obtained by Corti and Mella in the preprint version of [95] based on the approach to Theorem 3.1.10 by the connectedness lemma. We derive it from the classification result of divisorial contractions in Chapter 3. By Theorems 3.5.5 and 3.5.7, one has only to exclude the case of type e3 in Table 3.4. In general if there exists a divisorial contraction of type e3 to a cA_2 point $o \in (x_1 x_2 + g(x_3, x_4) = 0)$, then the cubic part of g becomes x_3^3 after coordinate change [223, theorem 1.13]. This is impossible when $g = x_3^3 + x_4^3$. □

For Theorem 6.4.9(i), we need to exclude the possibility for any other locus to be a maximal centre with respect to $\mathscr{H} \subset |\mathscr{O}_X(\mu)|$.

Lemma 6.4.12 *A smooth point of X is not a maximal centre.*

Proof Let x be a smooth point of X. For the general members H and H' of $\mathscr{H}$, one has $\text{mult}_x(H \cap H') \le \deg(H \cap H') = 4\mu^2$. By Corollary 6.3.26, x is not a maximal centre. $\qquad\square$

From now on, we let H denote the general member of $\mathscr{H}$. Let C be a curve in X and let $m = \text{mult}_C H$. By Lemma 6.3.27, C is a maximal centre with respect to $\mathscr{H}$ only if $m > \mu$. In order to exclude C as a maximal centre, one has only to derive the inequality $m \le \mu$.

Lemma 6.4.13 *A curve C in X of degree at least four is not a maximal centre.*

Proof Take another general member H' of $\mathscr{H}$ and observe the estimate $4m^2 \le (\text{mult}_C H)^2 \deg C \le \deg H \cap H' = 4\mu^2$, that is, $m \le \mu$. $\qquad\square$

Hence any curve C that is a maximal centre is of degree at most three and consequently it is a *space curve*, namely a curve contained in a three-dimensional linear subspace $\mathbf{P}^3$.

Lemma 6.4.14 *Let C be a curve in $\mathbf{P}^n$ of degree d which is not contained in any hyperplane $\mathbf{P}^{n-1}$. Then $d \ge n$. The equality $d = n$ holds if and only if C is the n-tuple embedding of $\mathbf{P}^1$ up to automorphism of $\mathbf{P}^n$, which is called the rational twisted curve.*

Proof Take a hyperplane Π in $\mathbf{P}^n$ through n general points in C. Since C is not in Π entirely, one has $d = (\Pi \cdot C) \ge n$.

Suppose that $d = n$. If C were singular, then the hyperplane through a singular point and $n - 1$ general points in C would have intersection number greater than n with C. Thus C is smooth. Fix a linear subspace $\Sigma = \mathbf{P}^{n-2}$ through $n - 1$ general points in C. For the same reason, $C \cap \Sigma$ consists of exactly these $n - 1$ points, and the pencil of hyperplanes containing Σ gives a birational morphism $C \setminus \Sigma \to \mathbf{P}^1$ which associates a point $x \in C \setminus \Sigma$ with the hyperplane spanned by Σ and x. Thus the smooth curve C is rational. The restriction map

$$H^0(\mathbf{P}^n, \mathscr{O}_{\mathbf{P}^n}(1)) \to H^0(C, \mathscr{O}_C(1)) \simeq H^0(\mathbf{P}^1, \mathscr{O}_{\mathbf{P}^1}(n))$$

is injective since no hyperplane contains C. Hence it is an isomorphism of vector spaces of dimension $n + 1$ and $C \subset \mathbf{P}^n$ is the rational twisted curve. $\qquad\square$

Remark 6.4.15 As a higher dimensional analogue, let X be a closed subvariety of $\mathbf{P}^n$ of dimension m and degree d not contained in any hyperplane $\mathbf{P}^{n-1}$. Then one has $d \ge n - m + 1$ by applying Lemma 6.4.14 to the curve $X \cap \Sigma$ cut out by a general linear subspace $\Sigma = \mathbf{P}^{n-m+1}$ of $\mathbf{P}^n$. The subvariety X with $d = n - m + 1$ is either a quadric hypersurface, a rational scroll, the Veronese surface, the rational twisted curve or the cone over one of them [407, section 1].

Let C be a space curve of degree at most three. It may be a *plane curve*, namely a curve contained in a plane $\mathbf{P}^2$. Postponing the treatment of this case, we consider a non-planar space curve C of degree three, which is the triple embedding of $\mathbf{P}^1$ by Lemma 6.4.14.

Lemma 6.4.16 *A non-planar space curve C in X is not a maximal centre.*

Proof By what we have seen, we have only to consider a smooth cubic C which is a rational twisted curve. Recall the notation of the general member H of $\mathscr{H}$. We want to show that $m = \mathrm{mult}_C H \le \mu$.

By the definition of a rational twisted curve, the ideal sheaf $\mathscr{I}$ in $\mathscr{O}_{\mathbf{P}^4}$ defining C is generated by smooth quadrics. Hence by Bertini's theorem, the intersection $S = S_{2,4} = Q \cap X$ with a general quadric Q in $\mathbf{P}^4$ containing C is a normal surface of degree eight and it is smooth outside p. Indeed outside the inverse image of p on the blow-up of X along C, the strict transform of S is free and isomorphic to $S \setminus p$. The restriction $\mathscr{H}|_S$ of the linear system $\mathscr{H}$ is mobile outside C and one can write $H|_S = mC + L$ with a mobile Weil divisor L on S. We take S as a *test surface* explained at the end of the preceding section. Note that $K_S = (K_{\mathbf{P}^4} + X + Q)|_S \sim A$ for the general hyperplane section A on S with $\mathscr{O}_S(A) \simeq \mathscr{O}_S(1)$.

If $p \notin C$, then $p \notin S$ and S is smooth. One has $-2 = \deg_C K_C = ((K_S + C) \cdot C) = 3 + (C^2)$ on S and hence $(C^2) = -5$. Since $L \sim \mu A - mC$,

$$0 \le (L^2) = (\mu A - mC)^2 = 8\mu^2 - 6\mu m - 5m^2 = (2\mu + m)(4\mu - 5m).$$

Thus $m \le 4\mu/5 < \mu$.

From now on, we assume that $p \in C$. Let $\bar{B} \to \mathbf{P}^4$ be the blow-up at p and let $\bar{E} \simeq \mathbf{P}^3$ be the exceptional divisor. The strict transform B in $\bar{B}$ of X is the blow-up of X at p. Its exceptional locus $\bar{E}|_B$ consists of two planes $E_1 = \mathbf{P}^2$ and $E_2 = \mathbf{P}^2$ in $\bar{E}$, and B has three ordinary double points lying on the line $E_1 \cap E_2$. Take the blow-up Y of B at these three singular points and let F_1, F_2, F_3 be the exceptional divisors. Note that Y is smooth. Since C is smooth at p, its strict transform C_Y in Y intersects exactly one of $E_{1Y}, E_{2Y}, F_1, F_2, F_3$ transversally, where E_{iY} is the strict transform of E_i. Without loss of generality, we assume that C_Y intersects E_{1Y} or F_1. Since $\mathscr{I}$ is generated by smooth quadrics, the strict transform T in Y of $S = Q \cap X$ is smooth again by Bertini's theorem. One has $(C_Y^2)_T = -5$ as in the case when $p \notin C$. We write $\alpha \colon T \to S$ and let $e_i = E_{iY}|_T$ and $f_i = F_i|_T$.

In the case when C_Y intersects e_1, the exceptional locus of α is $e_1 + e_2$ and thus $p \in S$ is Du Val of type A_2. The surface S is $\mathbf{Q}$-factorial and $\alpha^* C = C_Y + (2/3)e_1 + (1/3)e_2$, from which $(C^2)_S = (C_Y \cdot \alpha^* C)_T = -5 + 2/3 < -4$. In the case when C_Y intersects f_1, the exceptional locus of α is $e_1 + f_1 + e_2$ and

$p \in S$ is Du Val of type A_3. In like manner, $\alpha^*C = C_Y + (1/2)e_1 + f_1 + (1/2)e_2$ and $(C^2)_S = -4$. In both cases, $(C^2) \le -4$ and

$$0 \le (L^2) = (\mu A - mC)^2 \le 8\mu^2 - 6\mu m - 4m^2 \le 4(2\mu + m)(\mu - m).$$

Thus $m \le \mu$. □

Finally we treat a plane curve C.

Lemma 6.4.17 *A plane curve C in X is not a maximal centre unless C is a line through p.*

Proof The outline is similar to that of the proof of Lemma 6.4.16. We shall work on a certain test surface S containing C. Again we write A for the general hyperplane section on S with $\mathcal{O}_S(A) \simeq \mathcal{O}_S(1)$. By Lemma 6.4.13, we may assume that the degree d of C is at most three. Note that C may be a nodal cubic.

If $p \notin C$, then we take a general hypersurface K in $\mathbf{P}^4$ of degree d which contains C. By Proposition 4.1.14, the blow-up of X along the plane curve C has only isolated singularities. Hence by Bertini's theorem, $S = S_{d,4} = K \cap X$ is a smooth surface of degree $4d$ with $K_S \sim (d - 1)A$. Then

$$(d - 3)d = \deg_C \omega_C = ((K_S + C) \cdot C) = (d - 1)d + (C^2)$$

and hence $(C^2) = -2d$. Writing $H|_S = mC + L$ with a mobile divisor L on S, one obtains

$$0 \le (L^2) = (\mu A - mC)^2 = 2d(2\mu^2 - \mu m - m^2) = 2d(2\mu + m)(\mu - m).$$

Thus $m \le \mu$.

Now suppose that $p \in C$ with $d = 2$ or 3. Let $\Pi = \mathbf{P}^2$ be the plane in $\mathbf{P}^4$ which contains C. Because X is factorial with $\operatorname{Pic} X = \mathbf{Z}[\mathcal{O}_X(1)]$ by Lemma 6.4.4, the plane Π is never contained in X. Hence for general hyperplanes G and G' in $\mathbf{P}^4$ containing C, the scheme $D = G \cap G' \cap X = \Pi \cap X$ is of dimension one. Outside p, it is the pull-back of a line by the projection $X \dashrightarrow \mathbf{P}^3$ from p. Hence D is reduced from the generality condition on a_3 and b_4. Then D consists either of a cubic and a line, of a conic and two lines or of two conics. Since the argument is similar in every case, we shall only discuss the first case.

We take the normal surface $S = S_{1,4} = G \cap X$. It is smooth outside p because the blow-up of X along $D = \Pi \cap X$ has only isolated singularities outside the inverse image of p. We are assuming that the Cartier divisor D on S is the sum $D = C + l$ of the nodal cubic C and a line l.

Consider the intersection number $(C \cdot l)_S \in \mathbf{Q}$ in Definition 4.4.16. We shall prove that it is at least one. Since C and l intersect inside $\Pi = \mathbf{P}^2$, the estimate is obvious unless the support of $C \cap l$ equals the unique singular point p of S. If $C \cap l$ is supported on p, then the local intersection number $(C \cdot l)_{p \in \Pi}$ on the plane Π is three. It follows that the line l is tangent at p to a component of the analytic germ $p \in C$. Note that $p \in C$ has two analytic components when it is nodal. Take the blow-up B of X at p and write $E_1 + E_2$ for the exceptional locus as in the proof of Lemma 6.4.16. By the above observation on tangency, the strict transforms C_B and l_B of C and l intersect at one point, say b, in B. The point b is away from $E_1 \cap E_2$ since the generality condition again guarantees that no line in X has strict transform intersecting $E_1 \cap E_2$. Thus the strict transform S_B of S is smooth at b as so is the section $b \in S_B \cap (E_1 + E_2)$, by which $(C \cdot l)_S \geq (C_B \cdot l_B)_{b \in S_B} \geq 1$.

Recall that $D = C + l \sim A$. One can write $H|_S = mC + nl + L \sim \mu A$ with a mobile Weil divisor L on S and an integer n. Because $D = S \cap G'$ is reduced, the coincidence $n = \text{mult}_l H$ follows from the general choice of G. If C is a maximal centre, then $(X, m^{-1}H)$ is canonical and in particular $n \leq m$. Thus one can assume that $n \leq m$. Then from $(m - n)C + L \sim (\mu - n)A$ and $(C \cdot l)_S \geq 1$, one obtains

$$0 \leq (L \cdot l)_S = (\mu - n)(A \cdot l)_S - (m - n)(C \cdot l)_S \leq (\mu - n) - (m - n).$$

Thus $m \leq \mu$. $\square$

Proof of Theorem 6.4.9(i)   Combine Lemmata 6.4.10, 6.4.12, 6.4.13, 6.4.16 and 6.4.17. $\square$

The arguments above are sufficient for Theorem 6.4.3.

Proof of Theorem 6.4.3   Remark that by Lemma 9.5.2, every smooth quartic threefold X has Picard number one and thus $X \to \text{Spec } \mathbf{C}$ is a Mori fibre space. No point in X is a maximal centre as in Lemma 6.4.12. No curve in X is a maximal centre by the proofs of Lemmata 6.4.13, 6.4.16 and 6.4.17 in the case when $p \notin C$. Thus there exists no Sarkisov link from X, which means the birational superrigidity of X. $\square$

7

Conical Fibrations

A Mori fibre space of relative dimension one is a conical fibration. It possesses about the generic fibre the structure of a family of conics in a projective plane bundle. Every conical fibration has a good representative in the equivalence class, called a standard conic bundle. The principal object in this chapter is a Mori fibre space from a threefold to a surface. It is termed a $\mathbf{Q}$-conic bundle since all fibres are of dimension one.

The analytic germ along a fibre of a $\mathbf{Q}$-conic bundle is analogous to an extremal neighbourhood except for the crucial point that the higher direct image of the canonical sheaf may not vanish. Paying careful attention to the non-birationality, one can develop arguments similar to those for the study of an extremal neighbourhood. In fact, Mori and Prokhorov achieved a structural classification of $\mathbf{Q}$-conic bundle germs and the general elephant conjecture in the case when the central fibre is irreducible. This implies Iskovskikh's conjecture that the base surface has only Du Val singularities of type A.

We shall also discuss the rationality problem of threefold standard conic bundles. We assume the rationality of the base surface since it is derived from that of the total threefold. Then the conic bundle is essentially recovered from the associated double cover of the discriminant divisor along which conics are degenerate. Beauville, after Mumford, proved that the intermediate Jacobian is isomorphic to the Prym variety of this cover. By virtue of Shokurov's analysis of the Prym variety, we have a complete criterion for rationality when the base surface is relatively minimal.

The linear systems of suitable combinations of the canonical divisor and the discriminant divisor are invariant under birational transformation. The pseudo-effective threshold is defined in terms of them. Sarkisov proved that a standard conic bundle is birationally superrigid if the threshold is at least four. In contrast, the rationality in dimension three is nearly paraphrased as having the threshold less than two.

7.1 Standard Conic Bundles

We shall furnish the general theory of conic bundles and construct a standard model of a Mori fibre space of relative dimension one.

Definition 7.1.1   A *conic bundle* $X \to S$ is a contraction between smooth varieties with one-dimensional fibres such that $-K_X$ is relatively ample. The conic bundle $X \to S$ is said to be *standard* if it has relative Picard number $\rho(X/S) = 1$. A contraction $X \to S$ is called a *conical fibration* if the restriction to some dense open subset of S is a conic bundle. Two conical fibrations $\pi \colon X \to S$ and $\pi' \colon X' \to S'$ are said to be *equivalent* if there exist birational maps $f \colon X \dashrightarrow X'$ and $g \colon S \dashrightarrow S'$ such that $\pi' \circ f = g \circ \pi$. In this case, f and g induce an isomorphism between the generic fibres of π and π'.

Notice that a conic bundle is flat by [308, theorem 23.1]. Every Mori fibre space $X \to S$ with $\dim X = \dim S + 1$ is a conical fibration. It is a standard conic bundle outside a locus in S of codimension least two. Since X is **Q**-factorial and $\rho(X/S) = 1$, fibres are of dimension one outside a locus in S of codimension at least three. The equivalence of Mori fibre spaces as conical fibrations means nothing but the existence of a square link between them.

Definition 7.1.2   Let S be a smooth variety. A conical fibration $X \to S$ is called an *embedded conical fibration* if X is embedded into a projective space bundle $\mathbf{P}(\mathscr{E})$ over S with a locally free sheaf $\mathscr{E}$ on S of rank three as $X \subset \mathbf{P}(\mathscr{E}) \to S$ in such a way that the general fibre $X_s \subset \mathbf{P}^2$ is a conic.

Let $\pi \colon X \to S$ be a conic bundle. It satisfies the Kawamata–Viehweg vanishing $R^1\pi_*\mathscr{O}_X(-K_X) = 0$ and hence $H^1(\mathscr{O}_X(-K_X) \otimes \mathscr{O}_{X_s}) = 0$ for the fibre $X_s = X \times_S s$ at any closed point s. If s is general, then X_s is a smooth rational curve with $(-K_X \cdot X_s) = 2$. By cohomology and base change in Theorem 1.1.5, the direct image $\mathscr{E} = \pi_*\mathscr{O}_X(-K_X)$ is locally free of rank three. Further by Remark 4.1.13, $-K_X$ is relatively very ample. Thus it yields an embedding $X \subset \mathbf{P}(\mathscr{E})$ over S with $\mathscr{O}_X(1) = \mathscr{O}_X(-K_X)$. By this embedding, each fibre X_s is a (not necessarily smooth) conic in $\mathbf{P}^2$. In particular, π is an embedded conical fibration. This is the origin of the terminology of conic bundle.

Proposition 7.1.3   *Let $\pi \colon X \to S$ be a conic bundle, which is embedded in $\bar{\pi} \colon P = \mathbf{P}(\mathscr{E}) \to S$ with $\mathscr{E} = \pi_*\mathscr{O}_X(-K_X)$. Then X as a divisor on P satisfies $\mathscr{O}_P(X) \simeq \mathscr{O}_P(2) \otimes \bar{\pi}^*((\det \mathscr{E}^\vee) \otimes \omega_S^{-1})$.*

Proof   Since $X \subset P$ is a family of conics, one can write $\mathscr{O}_P(X) \simeq \mathscr{O}_P(2) \otimes \bar{\pi}^*\mathscr{L}$ with some invertible sheaf $\mathscr{L}$ on S, from which $\pi^*\mathscr{L} \simeq \mathscr{O}_X(-2) \otimes \mathscr{O}_P(X)|_X$. Using the adjunction $\mathscr{O}_X(-1) = \omega_X = \omega_P \otimes \mathscr{O}_P(X)|_X$ and the relation $\omega_P = \bar{\pi}^* \omega_S \otimes \det \Omega_{P/S}$, one obtains $\pi^* \mathscr{L} \simeq \mathscr{O}_X(-3) \otimes \det \Omega_{P/S}^{\vee}|_X \otimes \pi^* \omega_S^{-1}$. The *relative Euler sequence*

$$0 \to \Omega_{P/S} \to \bar{\pi}^* \mathscr{E} \otimes \mathscr{O}_P(-1) \to \mathscr{O}_P \to 0$$

yields $\det \Omega_{P/S}^{\vee} \simeq \bar{\pi}^* \det \mathscr{E}^{\vee} \otimes \mathscr{O}_P(3)$. Hence $\pi^* \mathscr{L} \simeq \pi^* (\det \mathscr{E}^{\vee} \otimes \omega_S^{-1})$ and thus $\mathscr{L} \simeq \det \mathscr{E}^{\vee} \otimes \omega_S^{-1}$ by the projection formula. $\qquad \square$

More generally, let $\pi \colon X \to S$ be an embedded conical fibration in the $\mathbf{P}^2$-bundle $\bar{\pi} \colon P = \mathbf{P}(\mathscr{E}) \to S$ with a locally free sheaf $\mathscr{E}$ of rank three, by which S is assumed to be smooth. Then X is defined by a section σ of $H^0(\mathscr{O}_P(2) \otimes \bar{\pi}^* \mathscr{L})$ for some invertible sheaf $\mathscr{L}$ on S. By the inclusion

$$\bar{\pi}_*(\mathscr{O}_P(2) \otimes \bar{\pi}^* \mathscr{L}) = S^2 \mathscr{E} \otimes \mathscr{L} \subset \mathscr{H}om(\mathscr{E}^{\vee}, \mathscr{E} \otimes \mathscr{L})$$

which sends $ab \otimes l$ to $(f \mapsto 2^{-1}(f(a)b + f(b)a) \otimes l)$ for $a, b \in \mathscr{E}, l \in \mathscr{L}$ and $f \in \mathscr{E}^{\vee}$, where $S^2 \mathscr{E}$ is the second symmetric power of $\mathscr{E}$, the section σ defines a homomorphism

$$q(\sigma) \colon \mathscr{E}^{\vee} \to \mathscr{E} \otimes \mathscr{L}$$

such that $q(\sigma)^{\vee} \otimes \mathscr{L} = q(\sigma)$. The induced map $q_0(\sigma) = \bigwedge^3 q(\sigma) \colon \det \mathscr{E}^{\vee} \to \det \mathscr{E} \otimes \mathscr{L}^{\otimes 3}$ is considered to be a global section of $\mathscr{H}om(\det \mathscr{E}^{\vee}, \det \mathscr{E} \otimes \mathscr{L}^{\otimes 3}) \simeq (\det \mathscr{E})^{\otimes 2} \otimes \mathscr{L}^{\otimes 3}$.

Definition 7.1.4 Let $\pi \colon X \to S$ be an embedded conical fibration. With notation as above, the effective divisor Δ on S defined by the section $q_0(\sigma) \in H^0(S, (\det \mathscr{E})^{\otimes 2} \otimes \mathscr{L}^{\otimes 3})$ is called the *discriminant divisor* of π.

Locally on a germ of S, the $\mathbf{P}^2$-bundle $\bar{\pi}$ is trivial as $\mathbf{P}(\mathscr{E}) \simeq \mathbf{P}^2 \times S$ by $\mathscr{E} \simeq \mathscr{O}_S^{\oplus 3}$ with basis x_0, x_1, x_2 of $\mathscr{E}$. Then σ is represented by a quadratic form $Q = \sum_{i,j=0,1,2} a_{ij} x_i x_j$ with homogeneous coordinates x_0, x_1, x_2 of $\mathbf{P}^2$ and a symmetric matrix $A = (a_{ij})$ with coefficients in $\mathscr{O}_S$. In this description, X is defined by Q and the discriminant divisor Δ is defined by $\det A$.

Lemma 7.1.5 *Let $\pi \colon X \to S$ be an embedded conical fibration endowed with $q(\sigma) \colon \mathscr{E}^{\vee} \to \mathscr{E} \otimes \mathscr{L}$ as above. Let Δ be the discriminant divisor. For $s \in S$, let r_s denote the rank of $q(\sigma) \otimes k(s) \colon \mathbf{C}^3 \to \mathbf{C}^3$ for the residue field $k(s)$ at s. Then Δ has multiplicity at least $3 - r_s$ at s. Further, $r_s = 3$ if and only if $s \notin \Delta$.*

Proof Locally X and Δ are defined by $Q = \sum_{i,j} a_{ij} x_i x_j$ and the determinant of $A = (a_{ij})$ respectively as above. Let $\mathfrak{m}$ be the maximal ideal in $\mathscr{O}_S$ defining s. The assertion is clear if $r_s = 3$ or 2. If $r_s = 1$, then Q is expressed as $x_0^2 + ax_1^2 + 2bx_1x_2 + cx_2^2$ with $a, b, c \in \mathfrak{m}$ up to unit after change of coordinates x_0, x_1, x_2, for which $\det A = ac - b^2 \in \mathfrak{m}^2$. If $r_s = 0$, then $\det A \in \mathfrak{m}^3$. $\qquad \square$

Lemma 7.1.6   *The embedded conical fibration π in Lemma 7.1.5 is a conic bundle if and only if for all $s \in S$,*

(i) $r_s = 3$ *if and only if $s \notin \Delta$,*
(ii) $r_s = 2$ *if and only if s is a smooth point of Δ and*
(iii) $r_s = 1$ *if and only if s is a singular point of Δ, and in this case $s \in \Delta$ is given by a function of form $z_1 z_2 + h(z_3, \dots, z_n)$ for suitable coordinates $z_1, \dots, z_n$ of the analytic germ $s \in S \simeq \mathfrak{D}^n$.*

In particular, the discriminant divisor of a conic bundle is reduced and normal crossing in codimension two.

Proof   The fibration π is a conic bundle if and only if it is a flat morphism from a smooth variety. We shall work on the germ $s \in S$ at a point and use the local expressions $Q = \sum_{i,j} a_{ij} x_i x_j$ and $\det A$ for X and Δ where $A = (a_{ij})$. Let $\mathfrak{m}$ denote the maximal ideal in $\mathscr{O}_S$ defining s.

By the characterisation of flatness in terms of the Hilbert polynomial [178, III theorem 9.9], π is flat at s if and only if r_s is positive. Hence we may assume that $r_s \geq 1$. If $r_s = 3$, then $Q = x_0^2 + u x_1^2 + v x_2^2$ with $u, v \in \mathscr{O}_S^\times$ up to unit after change of coordinates, in which X is smooth about the fibre X_s at s. If $r_s = 2$, then $Q = x_0^2 + u x_1^2 + a x_2^2$ with $u \in \mathscr{O}_S^\times$ and $a \in \mathfrak{m}$ up to unit after coordinate change. In this case, X is smooth about X_s if and only if $a \in \mathfrak{m} \setminus \mathfrak{m}^2$, which means that the divisor Δ defined by a is smooth at s.

Suppose that $r_s = 1$. We write $Q = x_0^2 + a x_1^2 + 2b x_1 x_2 + c x_2^2$ with $a, b, c \in \mathfrak{m}$ up to unit suitably, for which $\det A = ac - b^2$. Take analytic coordinates $z_1, \dots, z_n$ of S at s. We claim that X is smooth about X_s if and only if the rank q of the quadratic part of $\det A$ in $z_1, \dots, z_n$ is at least two. By the Jacobian criterion, the former condition is equivalent to the non-existence of non-trivial solutions to the equation $M\, {}^t(x_1^2, x_1 x_2, x_2^2) = 0$ for the $n \times 3$ matrix M the i-th row of which is $((\partial a / \partial z_i)_s, 2(\partial b / \partial z_i)_s, (\partial c / \partial z_i)_s)$, where $(\partial f / \partial z_i)_s$ denotes the value of $\partial f / \partial z_i$ at s. We write r_M for the rank of M.

Let v be the dimension of the vector subspace of $\mathfrak{m}/\mathfrak{m}^2$ spanned by a, b, c. If $v \leq 1$, then $q \leq 1$, $r_M \leq 1$ and the equation $M\, {}^t(x_1^2, x_1 x_2, x_2^2) = 0$ has a non-trivial solution. If $v = 3$, then one can choose $(z_1, z_2, z_3) = (a, b, c)$ by which $q = r_M = 3$ and $M\, {}^t(x_1^2, x_1 x_2, x_2^2) = 0$ has no non-trivial solutions. Finally if $v = 2$, then we may assume that $(z_1, z_2) = (a, c)$ after coordinate change. Write $b = b_1 z_1 + b_2 z_2 + d$ with $b_1, b_2 \in \mathbf{C}$ and $d \in \mathfrak{m}^2$. Taking the first two rows of M, we can write our equation as

$$
\begin{pmatrix} 1 & 2b_1 & 0 \\ 0 & 2b_2 & 1 \end{pmatrix} \begin{pmatrix} x_1^2 \\ x_1 x_2 \\ x_2^2 \end{pmatrix} = 0.
$$

It has no non-trivial solutions if and only if $4b_1b_2 \neq 1$. This is equivalent to $q = 2$ for the rank q of the quadric $z_1z_2 - (b_1z_1 + b_2z_2)^2$. The claim is now completed.

The lemma follows from the above properties and Lemma 7.1.5. $\quad\square$

Remark 7.1.7 A conic bundle on a germ of a smooth surface is analytically given by a quadratic form $Q = x_0^2 + x_1^2 + x_2^2$, $x_0^2 + x_1^2 + z_1x_2^2$ or $x_0^2 + z_1x_1^2 + z_2x_2^2$ with local coordinates z_1, z_2. This is obvious if the discriminant divisor Δ is smooth. When Δ is singular, the above proof gives the expression $Q = x_0^2 + z_1x_1^2 + 2(b_1z_1 + b_2z_2)x_1x_2 + z_2x_2^2$ modulo $\mathfrak{m}^2$ with $4b_1b_2 \neq 1$. Taking a solution $\lambda \in \mathbf{C}$ to $b_1b_2\lambda^2 - \lambda + 1 = 0$, one computes $(1 - \lambda^4 b_1^2 b_2^2)(Q - x_0^2)$ to be $(z_1 - \lambda^2 b_2^2 z_2)(x_1 + \lambda b_1 x_2)^2 + (z_2 - \lambda^2 b_1^2 z_1)(x_2 + \lambda b_2 x_1)^2$ modulo $\mathfrak{m}^2$, from which one derives a desired form of Q using $4b_1b_2 \neq 1$.

Sarkisov constructed a good representative of each equivalence class of conical fibrations as follows. The theorem in dimension three was obtained by Miyanishi [321] after Zagorskii [476].

Theorem 7.1.8 (Sarkisov [410, theorem 1.13]) *For any conical fibration $\pi\colon X \to S$, there exists a standard conic bundle $\pi'\colon X' \to S'$ equipped with a birational contraction $g\colon S' \to S$ such that π and π' are equivalent via g, that is, there exists a birational map $f\colon X' \dashrightarrow X$ with $\pi \circ f = g \circ \pi'$.*

Proof Step 1 To begin with, taking a resolution S_1 of S and replacing π by the contraction $X_1 \to S_1$ from the normalisation X_1 of the dominant component of $X \times_S S_1$, we may assume that S is smooth. Take a dense open subset U of S such that the restriction $X \times_S U \to U$ is a conic bundle. Shrinking U, we may assume that $X \times_S U$ is embedded into the trivial bundle $\mathbf{P}^2 \times U \to U$ and that X is defined by a diagonal form $Q_U = a_0x_0^2 + a_1x_1^2 + a_2x_2^2$ with $a_i \in \mathcal{O}_U^\times$ with homogeneous coordinates x_0, x_1, x_2 of $\mathbf{P}^2$. As rational functions on S, these a_i define principal divisors $(a_i)_S$ on S. Let B be the least divisor on S such that $(a_i)_S + B \geq 0$ for all i. It is supported in $S \setminus U$.

The endomorphism of $\mathcal{O}_U^{\oplus 3}$ which defines Q_U is extended to the whole S as a symmetric form $q\colon \mathcal{E}^\vee \to \mathcal{E} \otimes \mathcal{L}$ for $\mathcal{E} = \mathcal{O}_S^{\oplus 3}$ and $\mathcal{L} = \mathcal{O}_S(B)$. It provides a compactification $X' \to S$ in $\mathbf{P}(\mathcal{E})$ of $X \times_S U \to U$. Replacing X by X', we may and shall assume that π is an embedded conical fibration in $\mathbf{P}(\mathcal{E}) \to S$ defined by a section σ of $H^0(S^2\mathcal{E} \otimes \mathcal{L})$ such that $\mathcal{E}$ is the direct sum of invertible sheaves $\mathcal{E}_0, \mathcal{E}_1, \mathcal{E}_2$ and such that the associated map $q(\sigma)\colon \mathcal{E}^\vee \to \mathcal{E} \otimes \mathcal{L}$ is the direct sum of $\mathcal{E}_i^\vee \to \mathcal{E}_i \otimes \mathcal{L}$ for $i = 0, 1, 2$.

Let Δ be the discriminant divisor on S. Replacing S by a log resolution of (S, Δ), we may and shall assume that Δ has snc support. One can describe

X locally on S by a diagonal form $a_0 x_0^2 + a_1 x_1^2 + a_2 x_2^2$ compatible with the decomposition $\mathscr{E} = \bigoplus_i \mathscr{E}_i$, in which Δ is given by $a_0 a_1 a_2$.

Step 2 We shall make replacement of X over the same base S after which Δ is reduced and $q(\sigma)$ is of rank two at the generic point of any prime component of Δ. Let D be a prime divisor in Δ and let r_D be the rank of the map $q(\sigma)|_D \colon \mathscr{E}^\vee \otimes \mathscr{O}_D \to \mathscr{E} \otimes \mathscr{L} \otimes \mathscr{O}_D$ on D. Note that r_D is positive since X' is irreducible. If either $r_D = 1$ or $2D \le \Delta$, then locally on the germ V of S at a general point in D, the restriction $X \times_S V$ is given up to unit and up to permutation of $\mathscr{E}_i$ by a diagonal form

$$x_0^2 + a x_1^2 + b x_2^2 \quad \text{if } r_D = 1, \qquad x_0^2 + u x_1^2 + c x_2^2 \quad \text{if } r_D = 2$$

with $a, b \in \mathscr{O}_V(-D|_V)$, $u \in \mathscr{O}_V^\times$ and $c \in \mathscr{O}_V(-2D|_V)$.

Let

$$\mathscr{E}' = \begin{cases} (\mathscr{E}_0 \otimes \mathscr{O}_S(mD)) \oplus \mathscr{E}_1 \oplus \mathscr{E}_2 & \text{if } r_D = 1, \\ (\mathscr{E}_0 \otimes \mathscr{O}_S(D)) \oplus (\mathscr{E}_1 \otimes \mathscr{O}_S(D)) \oplus \mathscr{E}_2 & \text{if } r_D = 2 \end{cases}$$

with the greatest integer m such that $a, b \in \mathscr{O}_V(-mD|_V)$ and let

$$\mathscr{L}' = \begin{cases} \mathscr{L} \otimes \mathscr{O}_S(-mD) & \text{if } r_D = 1, \\ \mathscr{L} \otimes \mathscr{O}_S(-2D) & \text{if } r_D = 2. \end{cases}$$

The natural composite $(\mathscr{E}')^\vee \hookrightarrow \mathscr{E}^\vee \xrightarrow{q(\sigma)} \mathscr{E} \otimes \mathscr{L} \hookrightarrow \mathscr{E}' \otimes \mathscr{L}$ factors through $(\mathscr{E}')^\vee \to \mathscr{E}' \otimes \mathscr{L}'$ and defines a section σ' of $H^0(S^2\mathscr{E}' \otimes \mathscr{L}')$. It defines a new conical fibration $X' \subset \mathbf{P}(\mathscr{E}')$, given on V by

$$d^m x_0^2 + a d^{-m} x_1^2 + b d^{-m} x_2^2 \quad \text{if } r_D = 1, \qquad x_0^2 + u x_1^2 + c d^{-2} x_2^2 \quad \text{if } r_D = 2$$

with the function $d \in \mathscr{O}_V$ defining D. Thus $X' \to S$ has discriminant divisor $\Delta - mD$ if $r_D = 1$ and $\Delta - 2D$ if $r_D = 2$. In particular, the replacement of X by X' decreases discriminant divisor strictly. Repeating this procedure, one attains $\pi \colon X \to S$ with Δ reduced and with $r_D = 2$ for any D.

Step 3 We write the snc divisor Δ as a sum $\Delta = \sum_i D_i$ of prime divisors D_i. Let $D_{ij} = D_i \cap D_j$ and $D_{ijk} = D_i \cap D_j \cap D_k$ for distinct i, j (and k).

Suppose that $D_{ij} \ne \emptyset$. If the restriction $q(\sigma)|_{D_{ij}}$ to D_{ij} is of rank two, then locally at a general point in D_{ij}, X is given up to unit and up to permutation by $x_0^2 + u x_1^2 + d_i d_j x_2^2$ with a unit u and the functions d_i, d_j defining D_i, D_j. Let $g \colon S' \to S$ be the blow-up along D_{ij} and let E be the exceptional divisor. Let

$$\mathscr{E}' = g^* \mathscr{E}_0 \oplus g^* \mathscr{E}_1 \oplus (g^* \mathscr{E}_2 \otimes \mathscr{O}_{S'}(-E)).$$

The pull-back $(g^*\mathscr{E})^\vee \to g^*\mathscr{E} \otimes g^*\mathscr{L}$ of $q(\sigma)$ factors through $q(\sigma') \colon (\mathscr{E}')^\vee \to \mathscr{E}' \otimes g^*\mathscr{L}$. The map $q(\sigma')$ defines an embedded conical fibration $X' \subset \mathbf{P}(\mathscr{E}')$

over S' the discriminant divisor of which is the strict transform of Δ. Hence repeating the procedure of replacement of X/S by X'/S', one attains a model $\pi\colon X \to S$ such that $q(\sigma)|_{D_{ij}}$ is of rank one as far as $D_{ij} \neq \emptyset$.

After this reduction, if $D_{ijk} \neq \emptyset$, then X is given at a general point in D_{ijk} by $d_i x_0^2 + d_j x_1^2 + d_k x_2^2$ with the functions d_i, d_j, d_k defining D_i, D_j, D_k up to permutation. Let $g\colon S' \to S$ be the blow-up along D_{ijk} and let E be the exceptional divisor. The pull-back of $q(\sigma)$ factors through $(g^*\mathcal{E})^\vee \to g^*\mathcal{E} \otimes (g^*\mathcal{L} \otimes \mathcal{O}_{S'}(-E))$, and the latter map defines an embedded conical fibration $X' \subset \mathbf{P}(g^*\mathcal{E})$ over S' the discriminant divisor of which is the strict transform of Δ. Repeating the procedure of replacement of X/S by X'/S', one attains a model such that all D_{ijk} are empty.

Step 4 Now $\pi\colon X \to S$ is an embedded conical fibration to a smooth projective variety with snc discriminant divisor $\Delta = \sum_i D_i$ such that for any i, j, k, $q(\sigma)|_{D_i}$ is of rank two, $q(\sigma)|_{D_{ij}}$ is of rank one as far as $D_{ij} \neq \emptyset$, and $D_{ijk} = \emptyset$. Thus by Lemma 7.1.6, π is a conic bundle. Contracting superfluous divisors on X by Lemma 7.1.9, one finally obtains a standard conic bundle. $\square$

The standard conic bundle $\pi'\colon X' \to S'$ constructed in the above proof is described Zariski locally at each point $s' \in S'$ by a diagonal form $x_0^2 + a_1 x_1^2 + a_2 x_2^2$ such that non-units out of a_1, a_2 form a part of a regular system of parameters in $\mathcal{O}_{S',s'}$. Compare this with Remark 7.1.7. Further $\pi'_*\mathcal{O}_{X'}(-K_{X'})$ splits into a direct sum of invertible sheaves compatible with this diagonalisation.

Lemma 7.1.9 *A conic bundle $\pi\colon X \to S$ is standard if and only if the pull-back π^*D of any prime divisor D on S is again a prime divisor on X. If π^*D is not prime, then it is the sum $E_1 + E_2$ of two prime divisors. For each $i = 1, 2$, π factors through a divisorial contraction $X \to Y_i$ which contracts E_i and the induced morphism $Y_i \to S$ is a conic bundle.*

Proof The pull-back π^*D may fail to be prime only if D is a component of the discriminant divisor Δ of π. If $E = \pi^*D$ is prime for $D \subset \Delta$, then the normalisation $E^\vee$ admits the Stein factorisation $E^\vee \to \bar{D} \to D$ by a finite morphism $\bar{D} \to D$ of degree two. Every line l in a fibre of $E \to D$ is numerically equivalent since it is the image of a fibre of $E^\vee \to \bar{D}$. Note that $2[l] \equiv [f]$ for the general fibre f of π. Thus if π^*D is prime for all D, then $\rho(X/S) = 1$ and hence π is standard.

Conversely suppose that π^*D is not prime for some $D \subset \Delta$. Then $\pi^*D = E_1 + E_2$ with prime divisors E_i and the general fibre of $E_1 + E_2 \to D$ consists of two lines l_1 and l_2 with $l_i \subset E_i$, for which $(E_2 \cdot l_1)$ is positive but $(E_2 \cdot l_2) = -(E_1 \cdot l_2)$ is negative. Thus l_1 and l_2 are not numerically proportional. In particular $\rho(X/S) \neq 1$ and hence π is not standard.

Since the line l_i is numerically equivalent to any fibre of $E_i \to D$ with $(E_i \cdot l_i) < 0$, it spans an extremal ray $\mathbf{R}_{\geq 0}[l_i]$ of $\overline{\mathrm{NE}}(X/S)$ and the associated contraction $\varphi_i \colon X \to Y_i$ is a divisorial contraction which contracts E_i. It follows from Theorem 7.1.10(i) that Y_i is smooth. The induced morphism $\pi_i' \colon Y_i \to S$ is flat by [308, theorem 23.1] and $-K_{Y_i}$ is π_i'-ample by the relation $-\varphi_i^* K_{Y_i} = -K_X + E_i$. Thus π_i' is a conic bundle. □

Thanks to Ando, we have a structure theorem for non-small extremal contractions from a smooth variety with fibres of dimension at most one. His proof supplies the following local statement. See also [15].

Theorem 7.1.10 (Ando [14]) *Let $\pi \colon X \to Y$ be an extremal contraction from a smooth variety X to the germ $y \in Y$ of a normal variety such that $-K_X$ is relatively ample. Suppose that the central fibre $\pi^{-1}(y)$ is of dimension one.*

(i) *If π is a divisorial contraction, then it is the blow-up of a smooth variety along a smooth subvariety of codimension two.*

(ii) *If π is of fibre type, then it is a standard conic bundle.*

The theorem in dimension three was obtained by Mori and extended to the case when X is Gorenstein by Cutkosky. Whilst the result in the first case is Theorem 4.1.9, that in the second is the following.

Theorem 7.1.11 (Mori [332], Cutkosky [98]) *Let $\pi \colon X \to S$ be a relative Mori fibre space from a Gorenstein terminal threefold to a surface. Then S is smooth and π is a flat embedded conical fibration.*

Recall the terminology of the relative Mori fibre space $\pi \colon X \to S$ by which π is assumed to be extremal and X is to be $\mathbf{Q}$-factorial. Once S turns out to be smooth, the theorem follows along the argument in the paragraph after Definition 7.1.2. Below we provide the original algebraic proof. The assertion actually holds without π being extremal as will be seen in Proposition 7.2.5.

Algebraic proof of Theorem 7.1.11 The total space X is factorial by Theorem 2.3.18 and then S is also factorial by Lemma 1.4.13. We shall work on the germ $s \in S$ at a point. Similarly to Lemma 4.1.11, one can show the relative freedom of $-K_X$ by taking the intersection $F = \pi^* H \cap \pi^* H'$ of the pull-backs of the general hyperplane sections H and H' of $s \in S$. It follows that the general elephant T of X is a smooth surface. We analyse the finite morphism $\pi_T \colon T \to S$ of degree two.

Take an element f in $\pi_{T*} \mathcal{O}_T \setminus \mathcal{O}_S$. It satisfies a relation $f^2 + af + b = 0$ with some a, b in the function field of S. We may assume that $a, b \in \mathcal{O}_S$ by multiplying f by a function in $\mathcal{O}_S$ if necessary. We may also assume that

$a = 0$ replacing f by $f + a/2$. Then $f^2 + b = 0$. Since S is factorial, one can write $b = c^2 b'$ with $b', c \in \mathcal{O}_S$ so that b' defines a reduced divisor on S. The quotient f/c belongs to the integral closure $\pi_{T*} \mathcal{O}_T$ of $\mathcal{O}_S$ in the function field of T. Hence replacing f by f/c and b by b', we may and shall assume that b defines a reduced divisor. Then by Serre's criterion, the algebra $\mathcal{O}_S[f] \simeq \mathcal{O}_S[F]/(F^2 + b)$ with indeterminate F is normal. By Zariski's main theorem, one has $T = \operatorname{Spec}_S \mathcal{O}_S[f]$. Hence π_T is flat by [308, theorem 22.6] and S is smooth by [308, theorem 23.7]. Remark that the factoriality of S is crucial for the flatness of π_T. For instance, $\mathbf{A}^2 \to \mathbf{A}^2/\mathbf{Z}_2(1, 1)$ is not flat. □

From Lemma 7.1.9, one can see that the discriminant divisor of a standard conic bundle is determined by the generic structure.

Proposition 7.1.12 *Let $\pi \colon X \to S$ and $\pi' \colon X' \to S$ be standard conic bundles which are equivalent via a birational map $f \colon X \dashrightarrow X'$ such that $\pi = \pi' \circ f$. Then π and π' have the same discriminant divisor.*

Proof Let D be a prime divisor on S. We shall derive a contradiction by assuming that D is contained in the discriminant divisor of π but not in that of π'. Shrinking S, we may assume that D is smooth. Let $E = \pi^* D$ and $E' = (\pi')^* D$. By Lemma 7.1.9, E is a non-normal prime divisor. By Stein factorisation, the morphism $E^\nu \to D$ from the normalisation E^ν of E factors through a finite morphism $\bar{D} \to D$ of degree two.

Take a common log resolution W of (X, E) and (X', E') with $\mu \colon W \to X$ and $\mu' \colon W \to X'$ where $\pi \circ \mu = \pi' \circ \mu'$. Let F and F' denote the strict transforms in W of E and E' respectively. Whilst $F \to D$ factors through $\bar{D}$, the fibres of $F' \to E' \to D$ are connected. In particular, $F \ne F'$.

Let $Z = \mu(F')$ be the centre on X of E', which lies on E. The morphism $Z \to D$ is surjective with connected fibres as so is $F' \to D$. Because $E^\nu \to D$ factors through $\bar{D}$, the centre Z must coincide with the non-normal locus in E. Then F' would have coefficient at least two in $\mu^* E$, which contradicts the equality $\mu^* E = (\pi \circ \mu)^* D = (\pi' \circ \mu')^* D = (\mu')^* E'$. □

We relate the discriminant divisor to the canonical divisors. See also Remark 7.5.5.

Proposition 7.1.13 *Let $\pi \colon X \to S$ be a conic bundle over a projective base S. Then $\pi_* K_X^2 \equiv -(4K_S + \Delta)$ for the discriminant divisor Δ. Namely, $(\pi_* K_X^2 \cdot C) = (-(4K_S + \Delta) \cdot C)$ for any curve C in S.*

Proof We refer the reader to [141] or [178, appendix A] for the intersection theory. Let n be the dimension of S. If $n = 1$, then contracting one of the two (-1)-curves of X at each point in Δ, one obtains a $\mathbf{P}^1$-bundle $X' \to S$. Then

$K_X^2 = K_{X'}^2 - \deg \Delta$ whilst $K_{X'}^2 = -4 \deg K_S$ by Theorem 6.1.11. These give the assertion.

Suppose that $n \geq 2$. We use a result of Kleiman [41, XIII théorème 4.6] that a Cartier divisor D on a projective variety of dimension n is numerically equivalent to zero if and only if $D^2 H^{n-2} = DH^{n-1} = 0$ for an ample divisor H. Hence it suffices to prove that $\pi_* K_X^2|_T \equiv -(4K_S + \Delta)|_T$ for the intersection $T = H_1 \cap \cdots \cap H_{n-2}$ of $n-2$ general hyperplane sections of S. This means the equality $(\pi_* K_X^2 \cdot A)_S = -((4K_S + \Delta) \cdot A)_S$ for any very ample curve A in T because every curve in T is linearly equivalent to the difference between two very ample curves.

We take $\varphi \colon Y = \pi^* H_1 \cap \cdots \cap \pi^* H_{n-2} \to T$ and $B = \varphi^* A$. Then $B \to A$ is a conic bundle with discriminant divisor $\Delta|_A$ after replacement of A by the general member of $|A|$. We write $H_T = (\sum_{i=1}^{n-2} H_i)|_T + A$ formally. By the adjunction formula, $K_A = (K_S|_T + H_T)|_A$ and $K_B = (K_X|_Y + \varphi^* H_T)|_B$. By the projection formula, $(\pi_* K_X^2 \cdot A)_S$ equals

$$(K_X^2 \cdot B)_X = K_B^2 - ((2K_X|_Y + \varphi^* H_T) \cdot \varphi^* H_T \cdot \varphi^* A)_Y = K_B^2 + 4(H_T \cdot A)_T.$$

On the other hand, $-((4K_S + \Delta) \cdot A)_S$ equals $-\deg(4K_A + \Delta|_A) + 4(H_T \cdot A)_T$. They are the same number from the result in $n = 1$.   □

7.2 Q-Conic Bundles

In this and the next section, we shall study the analytically local structure of a Mori fibre space from a threefold to a surface, following a series of works of Mori and Prokhorov [338], [339], [340]. The object is a **Q**-conic bundle germ.

Definition 7.2.1 ([338])   A **Q**-*conic bundle* $X \to S$ in the algebraic or analytic category is a contraction from a terminal threefold to a normal surface such that $-K_X$ is relatively ample and such that all fibres are of dimension one. It is called a *Gorenstein conic bundle* if X is Gorenstein. A **Q**-*conic bundle germ* $\pi \colon C \subset X \to o \in S$ means a **Q**-conic bundle over the analytic germ $o \in S$, in which the notation $C = \pi^{-1}(o)$ stands for the support of the central fibre. We say that π is *irreducible* if C is irreducible.

A **Q**-conic bundle germ $\pi \colon C \subset X \to o \in S$ has many features in common with an extremal neighbourhood $C \subset X \to y \in Y$ in Definition 4.4.6. Most of the study by Mori and Prokhorov follows the approach to an extremal neighbourhood discussed in Chapters 4 and 5. For example, π satisfies the vanishing $R^1 \pi_* \mathcal{O}_X = 0$ and this shows that C is a tree of $\mathbf{P}^1$. However, the vanishing of $R^1 \pi_* \omega_X$ may not hold for the **Q**-conic bundle germ. We shall concentrate on the properties which stem from the non-birationality of π, without repeating the arguments in the previous chapters.

Remark 7.2.2 For a **Q**-conic bundle germ $\pi\colon C \subset X \to o \in S$, every one-dimensional connected subspace D of C is contractible for the same reason as for Lemma 5.1.4. If $D \neq C$, then the contraction is birational and hence $D \subset X$ is an extremal neighbourhood.

Lemma 7.2.3 *The base S of a **Q**-conic bundle $X \to S$ is lt and thus has only quotient singularities.*

Proof Locally on S, we take the general hyperplane section H of X, which is smooth, and consider the finite morphism $\pi_H\colon H \to S$. By Theorem 2.2.20, one can write $K_H = \pi_H^*(K_S + \Delta)$ in the sense of Definition 4.4.16 with some effective **Q**-divisor Δ on S. A priori, $K_S + \Delta$ may not be **Q**-Cartier. In spite of this, one can apply Corollary 2.2.21 to the above relation by its proof and it shows that (S, Δ) is *numerically klt*, that is, $a_E(S, \Delta) > 0$ for all divisors E over S. Then (S, Δ) is actually klt as mentioned before Definition 4.4.16. Further by Theorem 4.4.14, S itself is lt and has quotient singularities. □

By virtue of the lemma, we make the following framework.

Framework 7.2.4 Let $\pi\colon C \subset X \to o \in S$ be a **Q**-conic bundle germ. Write $o \in S = \mathfrak{D}^2/G$ with a finite subgroup G of $\mathrm{GL}(2, \mathbf{C})$ by Theorem 2.2.4, which still holds in the analytic category. By Remark 2.2.7, we assume that G is small. Let $\bar{X}$ be the normalisation of the base change $X \times_S \mathfrak{D}^2$ of X by $\mu\colon \mathfrak{D}^2 \to S = \mathfrak{D}^2/G$. We set $\bar{\pi}\colon \bar{X} \to \bar{o} \in \mathfrak{D}^2$ with origin $\bar{o}$, $\bar{C} = \bar{\pi}^{-1}(\bar{o})$ and $\mu_X\colon \bar{X} \to X$ as in the diagram

$$
\begin{array}{ccc}
\bar{C} \subset \bar{X} & \xrightarrow{\ \mu_X\ } & C \subset X \\
\bar{\pi} \downarrow & & \downarrow \pi \\
\bar{o} \in \mathfrak{D}^2 & \xrightarrow{\ \mu\ } & o \in S \ .
\end{array}
$$

The group G acts on $\bar{X}$ by which $X = \bar{X}/G$. From $K_{\bar{X}} = \mu_X^* K_X$, the space $\bar{X}$ is terminal and $\bar{\pi}$ is again a **Q**-conic bundle germ.

The structure of a Gorenstein conic bundle is simple.

Proposition 7.2.5 *If $\pi\colon X \to S$ is a Gorenstein conic bundle, then S is smooth and π is an embedded conical fibration.*

Proof As discussed prior to the proof of Theorem 7.1.11, it suffices to show that S is smooth. We consider a Gorenstein conic bundle germ $\pi\colon C \subset X \to o \in S$ and take $\bar{\pi}\colon \bar{C} \subset \bar{X} \to \bar{o} \in \mathfrak{D}^2$ as in Framework 7.2.4. We want $G = 1$.

The covering $\bar{X}$ is also Gorenstein and $\bar{\pi}$ is an embedded conical fibration in $\mathbf{P}(\bar{\pi}_* \mathscr{O}_{\bar{X}}(-K_{\bar{X}})) \rightarrow \mathfrak{D}^2$. In particular, $\bar{C}$ consists of one or two smooth rational curves. By Theorems 2.3.4 and 2.3.19, every closed point in X has simply connected punctured neighbourhood. It follows that G acts on $\bar{C}$ freely. If $\bar{C} \simeq \mathbf{P}^1$, then by Hurwitz's theorem in Theorem 5.6.5 for $\bar{C} \rightarrow \bar{C}/G$, the group G acting on $\mathbf{P}^1$ freely must be trivial. If $\bar{C}$ is a union of two lines, then G fixes the unique singular point of $\bar{C}$ and hence it is trivial. □

Remark 7.2.6 Using this proposition, Prokhorov [385] earlier classified a $\mathbf{Q}$-conic bundle germ which is a non-trivial cyclic quotient of a Gorenstein conic bundle. It belongs to (i), (ii) or (iv) in Theorem 7.3.1.

Henceforth we shall study an *irreducible* $\mathbf{Q}$-conic bundle germ unless otherwise mentioned.

Lemma 7.2.7 *In Framework 7.2.4 if π is irreducible, then $\bar{C}$ has at most one singular point.*

Proof Since $\bar{C}$ is a tree of $\mathbf{P}^1$, there exists an irreducible component $\bar{C}_1$ of $\bar{C}$ which intersects the closure of $\bar{C} \setminus \bar{C}_1$ at only one point. This property is shared with all the components $\bar{C}_i$ of $\bar{C}$ because G acts on the set of $\bar{C}_i$ transitively. Consequently all $\bar{C}_i$ meet at only one point, which is the assertion. □

Lemma 7.2.8 *The base $o \in S$ of an irreducible $\mathbf{Q}$-conic bundle germ $C \simeq \mathbf{P}^1 \subset X \rightarrow o \in S$ is a cyclic quotient singularity.*

Proof We write $\pi \colon C \subset X \rightarrow o \in S = \mathfrak{D}^2/G$ and $\bar{\pi} \colon \bar{C} \subset \bar{X} \rightarrow \bar{o} \in \mathfrak{D}^2$ as in Framework 7.2.4. First we consider the case $\bar{C} \simeq \mathbf{P}^1$. By the action of G on $\bar{C} \simeq \mathbf{P}^1$, one can regard G as a subgroup of the projective general linear group $\mathrm{PGL}(2, \mathbf{C})$. Then G is conjugate to a finite subgroup of the projective special unitary group $\mathrm{PSU}(2, \mathbf{C})$, which is either a cyclic group C_n, a dihedral group D_n, the tetrahedral group T, the octahedral group O or the icosahedral group I. See [431, section 4.4]. Hence if G were not cyclic, then it would have at least two elements of order two. One of them must be a reflection as an element in $G \subset \mathrm{GL}(2, \mathbf{C})$, contradicting the assumption that G is small. Thus G is cyclic.

If $\bar{C} \neq \mathbf{P}^1$, then $\bar{C}$ has a unique singular point $\bar{p}$ by Lemma 7.2.7, which is fixed by G. By Theorem 2.4.17, there exists a neighbourhood U in X at $p = \mu_X(\bar{p})$ such that $\pi_1(U \setminus p) \simeq \mathbf{Z}_r$ for the index r of $p \in X$. Since μ_X induces the surjection $\pi_1(U \setminus p) \twoheadrightarrow G$, the image G is also cyclic. □

Definition 7.2.9 The *topological index* of an irreducible $\mathbf{Q}$-conic bundle germ $C \subset X \rightarrow o \in S$ is the order d of the fundamental group $\pi_1(S \setminus o)$. It is characterised by the isomorphism $S \simeq \mathfrak{D}^2/\mathbf{Z}_d$.

Let $\pi\colon C \simeq \mathbf{P}^1 \subset X \to o \in S$ be an irreducible **Q**-conic bundle germ. We apply Definition 5.2.4 to the germ $p \in C \subset X$ at each point p in C. The *splitting degree* of $p \in C \subset X$ is the order s of the cokernel of the natural map $\pi_1(C \setminus p) \to \pi_1(X \setminus p)$, and we say that p is *primitive* (resp. *imprimitive*) if $s = 1$ (resp. $s \geq 2$). We say that $C \subset X$ is *locally primitive* if all points in C are primitive.

If $C \subset X$ has an imprimitive point p, then one can construct the *splitting cover* $\pi'\colon C' \subset X' \to o' \in S'$ as in Definition 5.2.16. An imprimitive point is unique if it exists by the proof of the first assertion of Corollary 5.2.18. The cover $\mu'_X\colon X' \to X$ is locally at p the covering of $p \in X$ associated with $r'K_X$, where $r = r's$ with the index r and the splitting degree s of the germ $p \in C \subset X$. It is étale outside p. In Framework 7.2.4, $\mu\colon \bar{o} \in \mathfrak{D}^2 \to o \in S = \mathfrak{D}^2/G$ factors through $o' \in S' \to o \in S$ because $\mathfrak{D}^2 \setminus \bar{o}$ is the universal cover of $S \setminus o$. By Lemma 7.2.7, $\bar{C}$ has a unique singular point, which is above p.

The following summarises what we have seen.

Framework 7.2.10 Let $\pi\colon C \simeq \mathbf{P}^1 \subset X \to o \in S$ be an irreducible **Q**-conic bundle germ of topological index d. It fits into the diagram

$$
\begin{array}{ccccc}
\bar{C} \subset \bar{X} & \xrightarrow{\ \mu''_X\ } & C' \subset X' & \xrightarrow{\ \mu'_X\ } & C \subset X \\
{\scriptstyle\bar{\pi}}\downarrow & & {\scriptstyle\pi'}\downarrow & & \downarrow{\scriptstyle\pi} \\
\bar{o} \in \mathfrak{D}^2 & \xrightarrow{\ \mu''\ } & o' \in S' = \mathfrak{D}^2/\mathbf{Z}_{d'} & \xrightarrow{\ \mu'\ } & o \in S = \mathfrak{D}^2/\mathbf{Z}_d
\end{array}
$$

with the normalisation $\bar{X}$ of $X \times_S \mathfrak{D}^2$ and $d = d's$. If $C \subset X$ is locally primitive, then $s = 1$ and $\pi' = \pi$. If $C \subset X$ has a unique imprimitive point, then s is its splitting degree and π' is the associated splitting cover. The spaces $\bar{C}$ and C' have s irreducible components. The germ $C'_i \subset X'$ along each component C'_i of C' is locally primitive.

Lemma 7.2.11 *Let $\pi\colon C \subset X \to o \in S$ be an irreducible* **Q**-*conic bundle germ of topological index d. Let $p_1, \ldots, p_n$ be all the non-Gorenstein points of X and let r_i be the index of $p_i \in X$.*

(i) *There exists a* **Q**-*Cartier integral divisor D on X such that $(D \cdot C) = d/\prod_{i=1}^n r_i$. One can take $D = -K_X$ unless the morphism $\bar{\pi}\colon \bar{X} \to \mathfrak{D}^2$ in Framework 7.2.10 is smooth.*

(ii) *Assume that $p_2, \ldots, p_n$ are primitive and set $r_1 = sr'_1$ by the splitting degree s of $p_1 \in C \subset X$. Then $(\prod_{i=1}^n r_i)/d$ equals the lcm of $r'_1, r_2, \ldots, r_n$.*

Proof Work in Framework 7.2.10. Since $R^1\pi_*\mathcal{O}_X = 0$, the exponential sequence with Theorem 2.3.10 yields Pic $X \simeq H^2(X, \mathbf{Z}) \simeq H^2(C, \mathbf{Z}) = \mathbf{Z}$.

Step 1 Let G be the subgroup of the divisor class group $\mathrm{Cl}\,X$ generated by $\mathbf{Q}$-Cartier integral divisors on X. Every numerically trivial $\mathbf{Q}$-Cartier integral divisor on X is a torsion in G and thus defines a finite cover of π globally by Definition 2.2.17, through which the cover $\bar{\pi}$ factors. Hence the kernel K of the degree map $\delta\colon G \to \mathbf{Q}$ defined by $\delta(D) = (D \cdot C)$ equals $\pi_1(S \setminus o) \simeq \mathbf{Z}_d$. The image of δ is of form $u^{-1}\mathbf{Z}$ in $\mathbf{Q}$ with some positive integer u, fitting into the exact sequence

$$0 \to \mathbf{Z}_d \to G \xrightarrow{\delta} u^{-1}\mathbf{Z} \to 0.$$

Since $\mathrm{Pic}\,X \cap K = 0$ by $\mathrm{Pic}\,X \simeq \mathbf{Z}$, one obtains the induced sequence

$$0 \to K \to G/\mathrm{Pic}\,X \to u^{-1}\mathbf{Z}/\mathbf{Z} \to 0.$$

In particular, the group $G/\mathrm{Pic}\,X$ is of order ud.

Let G_i be the subgroup of the divisor class group of the germ $p_i \in X$ generated by $\mathbf{Q}$-Cartier integral divisors on $p_i \in X$, which is isomorphic to $\mathbf{Z}_{r_i}$ by Theorem 2.4.17. From the exact sequence

$$0 \to \mathrm{Pic}\,X \to G \to \prod_{i=1}^{n} G_i \to 0,$$

one has $G/\mathrm{Pic}\,X \simeq \prod_{i=1}^{n} \mathbf{Z}_{r_i}$. Thus the order ud of $G/\mathrm{Pic}\,X$ equals $\prod_{i=1}^{n} r_i$. Hence $u = (\prod_{i=1}^{n} r_i)/d$, which is the first part of (i).

Step 2 Set $r_i' = r_i$ for $i \geq 2$ for convenience and write L for the lcm of $r_1', \ldots, r_n'$. The statement (ii) means that $u = L$. For each prime number l, let $a(l)$ be the greatest integer such that $r_{i(l)}'$ is divisible by $l^{a(l)}$ for some $1 \leq i(l) \leq n$. Similarly let $b(l)$ be the greatest integer such that u is divisible by $l^{b(l)}$.

The group G_i is generated by a divisor D_i with $(D_i \cdot C) = 1/r_i'$ such that the support of D_i intersects C exactly at p_i. Since $G/\mathrm{Pic}\,X \simeq \prod_{i=1}^{n} G_i \simeq \prod_{i=1}^{n} \mathbf{Z}_{r_i}$, the group $(G/(\mathrm{Pic}\,X + K)) \otimes \mathbf{Z}_{l^N}$ for $N \geq a(l)$ is the cyclic group $\mathbf{Z}_{l^{a(l)}}$ generated by the class of $D_{i(l)}$. On the other hand since $G/(\mathrm{Pic}\,X + K) \simeq \mathbf{Z}_u$, it is also isomorphic to $\mathbf{Z}_u \otimes \mathbf{Z}_{l^N} \simeq \mathbf{Z}_{l^{b(l)}}$ as far as $N \geq b(l)$. Hence $a(l) = b(l)$ and

$$L = \prod_{l} l^{a(l)} = \prod_{l} l^{b(l)} = u,$$

where the product is taken over all prime numbers l. This is (ii).

Step 3 Write $-K_X \equiv kD$ with a positive integer k. Then $2 = (-K_X \cdot f) = k(D \cdot f)$ for the general fibre f of π, from which $k = 1$ or 2. For the second part of (i), we shall assume that $(D \cdot f) = 1$ and derive the smoothness of $\bar{\pi}$ from this assumption.

Since the base of $\bar{\pi}$ is smooth, the pull-back $\bar{D} = (\mu'_X \circ \mu''_X)^*D$ satisfies $(\bar{D} \cdot \bar{f}) = (D \cdot f) = 1$ for every scheme-theoretic fibre $\bar{f}$ of $\bar{\pi}$. Locally at each point $\bar{p} \in \bar{C} \subset \bar{X}$, the sheaf $\mathscr{O}_{\bar{X}}(\bar{D})$ is a direct summand of the direct image of the invertible sheaf $\mathscr{O}_{\tilde{X}}(\gamma^*\bar{D})$ for the index-one cover $\gamma \colon \bar{p} \in \tilde{X} \to \bar{p} \in \bar{X}$. Hence $\mathscr{O}_{\bar{X}}(\bar{D})$ is flat over $\mathfrak{D}^2$ by [308, theorem 23.1]. Since $R^1\bar{\pi}_*\mathscr{O}_{\bar{X}}(\bar{D}) = 0$ by Kawamata–Viehweg vanishing, it follows from cohomology and base change that the direct image $\bar{\pi}_*\mathscr{O}_{\bar{X}}(\bar{D})$ is locally free of rank two.

The $\mathbf{P}^1$-bundle $\bar{\pi}_Y \colon \bar{Y} = \mathbf{P}(\pi_*\mathscr{O}_{\bar{X}}(\bar{D})) \to \mathfrak{D}^2$ is isomorphic to $\bar{\pi}$ outside $\bar{o}$. The induced bimeromorphic map $\bar{X} \dashrightarrow \bar{Y}$ over $\mathfrak{D}^2$ transforms the $\bar{\pi}$-ample divisor $\bar{D}$ to a $\bar{\pi}_Y$-ample divisor. It is actually an isomorphism for the same reason as for Lemma 3.1.2. Thus $\bar{\pi} \simeq \bar{\pi}_Y$ is smooth. $\qquad\square$

Recall the notation $\mathrm{gr}^0_C \omega$ in Definition 5.2.1, which stands for the quotient of $\omega_X \otimes \mathscr{O}_C$ by torsion. Without the vanishing of $R^1\pi_*\omega_X$, the first cohomology $H^1(\mathrm{gr}^0_C \omega)$ may not be zero. We shall investigate this case. The next theorem is in fact a corollary to a stronger result [338, theorem 4.4].

Theorem 7.2.12 *Let $C \subset X \to o \in S$ be a not necessarily irreducible* **Q***-conic bundle germ over a smooth base S. If $H^1(\mathrm{gr}^0_C \omega) \neq 0$, then the reduced space C coincides with the scheme-theoretic fibre $X \times_S o$.*

Proof We write $\pi \colon X \to S$ and $D = X \times_S o$. Since S is smooth, D is the intersection $\pi^*H \cap \pi^*H'$ of the pull-backs of the general hyperplane sections H and H' of the germ $o \in S$. In particular, D is Cohen–Macaulay and $\omega_D \simeq \omega_X \otimes \mathscr{O}_D$ by adjunction. Since π is flat and $R^1\pi_*\mathscr{O}_X = 0$, the cohomology and base change theorem provides the isomorphism $H^0(\mathscr{O}_D) \simeq \pi_*\mathscr{O}_X \otimes k(o) \simeq \mathbf{C}$ for the residue field $k(o)$ at o.

Let $\mathscr{I}$ denote the ideal sheaf in $\mathscr{O}_D$ defining C. If the Cohen–Macaulay scheme D were not reduced, then there would exist an irreducible component $C_1 \simeq \mathbf{P}^1$ of C along which D is not reduced. It corresponds to an associated prime of the $\mathscr{O}_D$-module $\mathscr{I}$. In other words, $\mathscr{H}om_D(\mathscr{O}_{C_1}, \mathscr{I}) \neq 0$. The functor $\mathscr{H}om_D(\mathscr{O}_{C_1}, \cdot)$ applied to the exact sequence $0 \to \mathscr{I} \to \mathscr{O}_D \to \mathscr{O}_C \to 0$ yields the exact sequence

$$0 \to \mathscr{H}om_D(\mathscr{O}_{C_1}, \mathscr{I}) \to \mathscr{H}om_D(\mathscr{O}_{C_1}, \mathscr{O}_D) \to \mathscr{H}om_C(\mathscr{O}_{C_1}, \mathscr{O}_C).$$

Since C is reduced, $\mathscr{H}om_C(\mathscr{O}_{C_1}, \mathscr{O}_C)$ is an invertible $\mathscr{O}_{C_1}$-module. Since D is Cohen–Macaulay, $\mathscr{H}om_D(\mathscr{O}_{C_1}, \mathscr{O}_D)$ is a torsion-free $\mathscr{O}_{C_1}$-module. One has $\omega_{C_1} = \mathscr{H}om_D(\mathscr{O}_{C_1}, \omega_D)$ from Lemma 1.1.11. Hence $\mathscr{H}om_D(\mathscr{O}_{C_1}, \mathscr{O}_D)$ is invertible. It follows that the injection $\mathscr{H}om_D(\mathscr{O}_{C_1}, \mathscr{I}) \hookrightarrow \mathscr{H}om_D(\mathscr{O}_{C_1}, \mathscr{O}_D)$ is an isomorphism. The surjection $\mathscr{O}_D \twoheadrightarrow \mathscr{O}_{C_1}$ induces an inclusion

$$\mathscr{H}om_D(\mathscr{O}_{C_1}, \mathscr{O}_D) \simeq \mathscr{H}om_D(\mathscr{O}_{C_1}, \mathscr{I}) \hookrightarrow \mathscr{H}om_D(\mathscr{O}_D, \mathscr{I}) \simeq \mathscr{I}.$$

The surjection $\mathscr{O}_D \twoheadrightarrow \mathscr{O}_{C_1}$ also induces the *trace map*

$$\iota_* \omega_{C_1} = \iota_* \mathscr{H}om_D(\mathscr{O}_{C_1}, \omega_D) \hookrightarrow \mathscr{H}om_D(\mathscr{O}_D, \omega_D) \simeq \omega_D$$

for $\iota \colon C_1 \hookrightarrow D$. Let $\mathscr{S}$ be the saturation of $\iota_* \omega_{C_1}$ in ω_D in Definition 5.2.1. Since $H^1(\omega_{C_1}) \to H^1(\omega_D)$, dual to $H^0(\mathscr{O}_D) \simeq \mathbf{C} \to H^0(\mathscr{O}_{C_1})$, is an isomorphism, one has $H^1(\omega_D / \mathscr{S}) = H^1(\omega_D / \iota_* \omega_{C_1}) = 0$. On the other hand, $\mathscr{S}$ is also the saturation of $\mathscr{H}om_D(\mathscr{O}_{C_1}, \mathscr{O}_D) \omega_D$ in ω_D. It is contained in the saturation of $\mathscr{I} \omega_D$ in ω_D because of the inclusion $\mathscr{H}om_D(\mathscr{O}_{C_1}, \mathscr{O}_D) \subset \mathscr{I}$. Thus there exists a surjection $\omega_D / \mathscr{S} \twoheadrightarrow \mathrm{gr}^0_C \, \omega_D \simeq \mathrm{gr}^0_C \, \omega$. This contradicts the assumption $H^1(\mathrm{gr}^0_C \, \omega) \neq 0$. The fibre D is therefore reduced and $C = D$. $\square$

Now we provide a classification of the germ $p \in C \subset X$ at a point for an irreducible $\mathbf{Q}$-conic bundle germ $C \subset X \to o \in S$. The case when $H^1(\mathrm{gr}^0_C \, \omega) \neq 0$ is added in comparison with Theorem 5.3.1 for an extremal neighbourhood. As is the case with an extremal neighbourhood, the cases IB and $\mathrm{IC}^\vee$ in the following theorem will be excluded by Theorems 7.3.1 and 7.3.3.

Theorem 7.2.13 ([338, section 5]) *Let $C \subset X \to o \in S$ be an irreducible $\mathbf{Q}$-conic bundle germ and let $p \in C \subset X$ be the germ at a point of index r. Then there exists a normal form of $p \in C \subset X$ in Definition 5.2.8 as $p \in X \subset \mathfrak{D}^4 / \mathbf{Z}_r(w_1, w_2, w_3, w_4)$ with $a_i = o'(w_i)$ which belongs to exactly one of* (IA), (IB), (IC), (IIA), (IIB), (III), $(\mathrm{IA}^\vee)$, $(\mathrm{IC}^\vee)$, $(\mathrm{II}^\vee)$ *in Theorem 5.3.1 and* $(\mathrm{ID}^\vee)$, $(\mathrm{IE}^\vee)$ *below, where s is the splitting degree.*

$(\mathrm{ID}^\vee)$ $r = 2$, $s = 2$, $(a_1, a_2, a_3, a_4) = (2, 2, 2, 2)$, $(w_1, w_2, w_3, w_4) = (1, 1, 1, 0)$.
$(\mathrm{IE}^\vee)$ $r = 8$, $s = 4$, $(a_1, a_2, a_3, a_4) = (4, 4, 4, 8)$, $(w_1, w_2, w_3, w_4) = (5, 1, 3, 0)$.

Proof The proof in the case when $H^1(\mathrm{gr}^0_C \, \omega) = 0$ is substantially the same as that of Theorem 5.3.1. Here we concentrate on the case when $H^1(\mathrm{gr}^0_C \, \omega) \neq 0$. We shall work in Framework 7.2.10 with the irreducible decomposition $C' = \bigcup_{i=1}^s C'_i$. We remark that Steps 2 and 4 below modifies the arguments (5.3.4) and (5.3.8) in [338] respectively.

Step 1 Henceforth we assume the non-vanishing $H^1(\mathrm{gr}^0_C \, \omega) \neq 0$. This implies that $H^1(\mathrm{gr}^0_{\bar{C}} \, \omega_{\bar{X}}) \neq 0$ since $\mathrm{gr}^0_C \, \omega$ is the invariant part of the direct image of $\mathrm{gr}^0_{\bar{C}} \, \omega_{\bar{X}}$. Then by Theorem 7.2.12, $\bar{C} = \bar{X} \times_{\mathfrak{D}^2} \bar{o}$ and $(-K_{\bar{X}} \cdot \bar{C}) = 2$.

If $C \subset X$ is locally primitive, then $\bar{X} \times_{\mathfrak{D}^2} \bar{o} = \bar{C} \simeq \mathbf{P}^1$ and hence $\bar{X} \to \mathfrak{D}^2$ is a trivial $\mathbf{P}^1$-bundle. This case was studied by Prokhorov as mentioned in Remark 7.2.6. We thus assume that $C \subset X$ has an imprimitive point q. Then by Remark 7.2.2, every $C'_i \subset X'$ is an extremal neighbourhood. If $p \neq q$, then for the point $p'_i \in C'_i$ above p, one can apply Theorem 5.3.1 to the germ $p'_i \in C'_i \subset X'$, which is analytically isomorphic to $p \in C \subset X$. Thus we shall assume that $p = q$.

Now p is a unique imprimitive point of $C \subset X$. Let p' and $\bar{p}$ denote the points above p in X' and $\bar{X}$ respectively. Write $r = r's$ and $d = d's$. If $r' = 1$, then by Lemma 7.2.14 below, X' is Gorenstein everywhere as so is it at p'. In this case, $d = s = r$ by Lemma 7.2.11(ii). Thanks to Proposition 7.2.5, the number s of irreducible components of C' has to be two and $(-K_{X'} \cdot C_i') = 1$. Since C' is a tree, $a_i = 2$ for $i \neq 4$ and the case $(\mathrm{ID}^\vee)$ holds. From now on, we assume that $r' \geq 2$.

Step 2 We claim that p is the only non-Gorenstein point of X. By $r' \geq 2$, X' is not Gorenstein at p'. By Theorem 5.4.6, the extremal neighbourhood $C_i' \subset X'$ has at most two non-Gorenstein points. Because $X' \to X$ is étale outside p, $C \subset X$ has at most one non-Gorenstein point besides p.

Suppose that X had two non-Gorenstein points p and p_2. Let r_2 be the index of $p_2 \in X$. By Lemma 7.2.11(ii), $d' = \gcd(r', r_2)$. By Lemma 7.2.7, $\bar{p}$ is the only singular point of $\bar{C}$. Hence the scheme-theoretic expression $\bar{C} = \bar{X} \times_{\mathfrak{D}^2} \bar{o}$ implies that $\bar{X}$ is smooth outside $\bar{p}$. In particular, r_2 divides d' and thus $r_2 = d'$. Write $r' = d'r''$ with an integer r''. By Lemma 7.2.11(i), $(-K_X \cdot C) = 1/r'$. Hence $2 = (-K_{\bar{X}} \cdot \bar{C}) = d(-K_X \cdot C) = d/r' = s/r''$, that is, $s = 2r''$.

If $r'' \geq 2$, then by Remark 7.2.2, the quotient $C'' \subset X''$ of $C' \subset X'$ by $\mathbf{Z}_{r''}$ has an extremal neighbourhood $C_i'' \subset X''$ for an irreducible component C_i'' of C''. It would have an imprimitive point of index $r/2 = r'r''$ and a primitive point of index $r_2 = d'$, which contradicts the classification in Theorem 5.5.1. If $r'' = 1$, then $d = r$ and $\bar{X}$ is Gorenstein at $\bar{p}$. Then the whole $\bar{X}$ is Gorenstein by Lemma 7.2.14 and the existence of p_2 contradicts a result of Prokhorov in Remark 7.2.6. The claim is now completed.

Step 3 Since p is the only non-Gorenstein point of X, one has $d' = 1$ from Lemma 7.2.11(ii) and thus $(-K_X \cdot C) = 1/r'$ from Lemma 7.2.11(i). Thus $2 = (-K_{\bar{X}} \cdot \bar{C}) = d(-K_X \cdot C) = s/r'$. Hence $s = 2r' \geq 4$ and $r = r' \cdot 2r' \geq 8$. In particular, $p \in X$ is of type cA/r from the classification in Theorem 2.4.8.

Take a normal form of $p \in C \subset X$ in Definition 5.2.8 as $p \in X \subset \mathfrak{D}^4 / \mathbf{Z}_r(w_1, w_2, w_3, w_4)$ with orbifold coordinates x_1, x_2, x_3, x_4 and $a_i = o'(w_i)$. Note that $a_i \in s\mathbf{Z}$. Since $p \in X$ is cA/r, we may and shall assume that $(w_1, w_2, w_3, w_4) = (1, -w, -1, 0)$ with w coprime to r. Then $a_4 = r$ and $a_1 + a_3 \equiv 0 \bmod r$. It follows from the intersection number $(-K_X \cdot C) = s/r$ that $a_2 \equiv s \bmod r$. We may assume that $a_1 \leq a_3$, in which $r/2 \leq a_3$.

On $X' = \bar{X}$, the cokernel $\mathscr{Q}$ of the map $\mathrm{gr}_{C'}^0 \, \omega \to \bigoplus_{i=1}^s \mathrm{gr}_{C_i'}^0 \, \omega$ fits into the exact sequence

$$0 \to \mathrm{gr}_{C'}^0 \, \omega \to \bigoplus_{i=1}^s \mathrm{gr}_{C_i'}^0 \, \omega \to \mathscr{Q} \to 0.$$

From $C' = X' \times_{S'} o'$ with S' smooth, one has $\omega_{X'} \otimes \mathcal{O}_{C'} \simeq \omega_{C'}$ and hence $\mathrm{gr}^0_{C'} \omega \simeq \omega_{C'}$. Then $H^1(\mathrm{gr}^0_{C'} \omega) \simeq H^1(\omega_{C'}) = \mathbf{C}$ dually to $H^0(\mathcal{O}_{C'})$. On the other hand, $\mathrm{gr}^0_{C'_i} \omega \simeq \mathcal{O}_{\mathbf{P}^1}(-1)$ by Proposition 5.2.14 for the extremal neighbourhood $C'_i \subset X'$. Thus $H^0(\mathscr{Q}) \simeq H^1(\mathrm{gr}^0_{C'} \omega) \simeq \mathbf{C}$.

We tensor the above sequence with the residue field $k(p')$ at p' and derive the exact sequence

$$\mathrm{gr}^0_{C'} \omega \otimes k(p') \xrightarrow{\alpha} \mathbf{C}^{\oplus s} \to \mathbf{C} \to 0.$$

Let $\tilde{\omega}$ be a semi-invariant generator of $\omega_{\tilde{X}}$ on the index-one cover $\tilde{p} \in \tilde{C} \subset \tilde{X}$ of $p \in C \subset X$. Then $\mathrm{gr}^0_{C'} \omega$ is generated by all $u\tilde{\omega}$ for monomials u in x_1, x_2, x_3, x_4 of weight w modulo r'. Since $C'_i \subset X'$ is an extremal neighbourhood, the invariant $w_{p'}(0)$ defined from $p' \in C'_i \subset X'$ as in Definition 5.2.2 satisfies $w_{p'}(0) < 1$ by Proposition 5.2.14. Thus if the monomial u has $o_p(u) \geq r$, then $u\tilde{\omega}$ vanishes in $\mathrm{gr}^0_{C'_i} \omega \otimes k(p')$.

Step 4 Take the residue $w' = w - \lfloor w/r' \rfloor r'$ of w modulo r'. If $a_2 \geq r$, then the monomials $x_1^{r'}$, x_2, $x_1 x_3$, x_3^2, x_4 have order at least r with respect to o_p. Hence the map $\alpha \colon \mathrm{gr}^0_{C'} \omega \otimes k(p') \to \mathbf{C}^{\oplus s}$ factors through the space $V = (\mathcal{O}_{\tilde{C}} \tilde{\omega}/(x_1^{r'}, x_2, x_1 x_3, x_3^2, x_4) \mathcal{O}_{\tilde{C}} \tilde{\omega})^{\mathbf{Z}_{r'}}$, which is spanned by $x_1^{w'} \tilde{\omega}$ and possibly $x_3 \tilde{\omega}$. Then $s \leq \dim V + 1 \leq 3$, which contradicts $s \geq 4$.

Thus $a_2 < r$ and then $a_2 = s$ by $(-K_X \cdot C) = s/r$. Since monomials $x_1^{r'}$, $x_2^{r'}$, $x_1 x_3$, x_3^2, x_4 have order at least r, the map α factors through the space $W = (\mathcal{O}_{\tilde{C}} \tilde{\omega}/(x_1^{r'}, x_2^{r'}, x_1 x_3, x_3^2, x_4) \mathcal{O}_{\tilde{C}} \tilde{\omega})^{\mathbf{Z}_{r'}}$. The space W is spanned by $x_1^{c_i} x_2^i \tilde{\omega}$ with $0 \leq i < r'$ and $x_3^c x_3 \tilde{\omega}$, where c_i is the residue of $(i+1)w'$ modulo r' and c is the integer with $0 \leq c < r'$ such that $-cw' - 1 \equiv w' \mod r'$. It follows that $2r' = s \leq \dim W + 1 \leq r' + 2$. This occurs only if $r' = 2$, $w' = 1$ and W has basis $x_1 \tilde{\omega}, x_2 \tilde{\omega}, x_3 \tilde{\omega}$. Therefore $r = 8$, $(a_1, a_2, a_3, a_4) = (4, 4, 4, 8)$ and $(w_1, w_2, w_3, w_4) = (1, 1, 7, 0)$ or $(1, 5, 7, 0)$ up to permutation of x_1 and x_3.

If $w_1 = w_2$, then $x_1|_{\tilde{C}} = x_2|_{\tilde{C}}$ and so $x_1 \tilde{\omega}$ and $x_2 \tilde{\omega}$ coincide in W. This contradicts $\dim W = 3$. Thus $(w_1, w_2, w_3, w_4) = (1, 5, 7, 0)$ and the case $(\mathrm{IE}^\vee)$ holds. $\qquad\square$

Lemma 7.2.14 (Kollár [268, proposition 4.2]) *Let* $C \subset X \to o \in S$ *be a reducible* $\mathbf{Q}$-*conic bundle germ. If* C *has a singular point at which* X *is Gorenstein, then* X *is Gorenstein everywhere.*

Proof Let p be a singular point of C at which X is Gorenstein. Since C is a tree of $\mathbf{P}^1$, there exist irreducible components C_1 and C_2 of C which intersect at p. If C had more than two components, then $C_1 \cup C_2 \subset X$ would be an extremal neighbourhood by Remark 7.2.2 but this contradicts Remark 5.2.19. Thus $C = C_1 \cup C_2$ and p is the unique singular point of C.

First we prove the assertion by assuming that S is smooth. Because the $\mathscr{O}_C$-module $\mathrm{gr}^0_C \, \omega$ is invertible at the Gorenstein point p of X, the cokernel $\mathscr{Q}$ of the injection $\mathrm{gr}^0_C \, \omega \hookrightarrow \mathrm{gr}^0_{C_1} \, \omega \oplus \mathrm{gr}^0_{C_2} \, \omega$ is non-trivial at p, fitting into

$$0 \to \mathrm{gr}^0_C \, \omega \to \mathrm{gr}^0_{C_1} \, \omega \oplus \mathrm{gr}^0_{C_2} \, \omega \to \mathscr{Q} \to 0.$$

Since $\mathrm{gr}_{C_i} \, \omega \simeq \mathscr{O}_{\mathbf{P}^1}(-1)$ by Proposition 5.2.14, one obtains $H^1(\mathrm{gr}^0_C \, \omega) \simeq H^0(\mathscr{Q}) \neq 0$. Hence $C = X \times_S o$ by Theorem 7.2.12. When S is smooth, this shows that C is the scheme-theoretic intersection of two Cartier divisors on X. Consequently X is smooth outside p and the assertion holds.

To complete the lemma, we prove that S is actually smooth. In Framework 7.2.10, the morphism $\mu \colon \bar{X} \to X$ is étale at p because the punctured neighbourhood in X at p is simply connected by Theorems 2.3.4 and 2.3.19. Applying the above result to $\bar{C} \subset \bar{X} \to \bar{o} \in \mathfrak{D}^2$, we see that $\bar{X}$ is Gorenstein everywhere. Then by Proposition 7.2.5, $\bar{C}$ has at most one singular point. Thus μ, which is étale at p, must be an isomorphism. $\qquad\square$

7.3 Classification

Based on the local description of singularities in Theorem 7.2.13, Mori and Prokhorov established a structural classification of $\mathbf{Q}$-conic bundle germs simultaneously with the general elephant conjecture. The method is principally the same as that for extremal neighbourhoods fully studied in Chapter 5. It is not reasonable to reproduce the entire discussion here. Instead, we state the results and provide examples along the classification.

We have a satisfactory classification when the base is singular.

Theorem 7.3.1 ([338, theorem 1.2]) *Let $\pi \colon C \simeq \mathbf{P}^1 \subset X \to o \in S$ be an irreducible $\mathbf{Q}$-conic bundle germ of topological index d greater than one. Then one of the following holds, where x_0, x_1, x_2, y are the homogeneous coordinates of $\mathbf{P}^1$, $\mathbf{P}^2$ or $\mathbf{P}(1, 1, 1, 2)$, t_1, t_2 are the coordinates of $\mathfrak{D}^2$, g is a generator of $\mathbf{Z}_d$ and $\zeta = \chi(g)$ for a primitive character $\chi \colon \mathbf{Z}_d \to \mathbf{C}^\times$. The germ $C \subset X$ is locally primitive in (i) and (ii), whilst it has a unique imprimitive point p in (iii) to (vi). Every case has an example.*

(i) *π is the quotient of the $\mathbf{P}^1$-bundle $\bar{X} = \mathbf{P}^1 \times \mathfrak{D}^2 \to \mathfrak{D}^2$ by $\mathbf{Z}_d$ such that $g(x_0, x_1, t_1, t_2) = (\zeta^a x_0, x_1, \zeta t_1, \zeta^{-1} t_2)$ where a is coprime to d.*

(ii) *π is the quotient of the conic bundle*

$$\bar{X} = (x_0^2 + t_1 x_1^2 + t_2 x_2^2 = 0) \subset \mathbf{P}^2 \times \mathfrak{D}^2 \to \mathfrak{D}^2$$

by $\mathbf{Z}_d$ such that $g(x_0, x_1, x_2, t_1, t_2) = (\zeta^a x_0, \zeta^{-1} x_1, x_2, \zeta t_1, \zeta^{-1} t_2)$ with $d = 2a + 1$.

(iii) *p is of type* IA$^\vee$, *d = 2 and π is the quotient of the* **Q**-*conic bundle*

$$\bar{X} = \left(\begin{array}{c} x_0^2 - x_1^2 = t_1 f_1 + t_2 f_2 \\ x_2^2 = t_1 g_1 + t_2 g_2 \end{array} \right) \subset \mathbf{P}(1,1,1,2) \times \mathfrak{D}^2 \to \mathfrak{D}^2$$

by $\mathbf{Z}_2$ *such that* $g(x_0, x_1, x_2, y, t_1, t_2) = (-x_0, x_1, x_2, -y, -t_1, -t_2)$ *with* $f_i, g_i \in \mathbf{C}\{t_1, t_2\}[x_0, x_1, x_2, y]$ *where* $t_i f_i$ *and* $t_i g_i$ *are invariant. p is a quotient singularity of type* $\frac{1}{4}(1, 3, 3)$.

(iv) *p is of type* ID$^\vee$, *d = 2 and π is the quotient of the Gorenstein conic bundle*

$$\bar{X} = (x_0^2 + x_1^2 + f(t_1, t_2)x_2^2 = 0) \subset \mathbf{P}^2 \times \mathfrak{D}^2 \to \mathfrak{D}^2$$

by $\mathbf{Z}_2$ *such that* $g(x_0, x_1, x_2, t_1, t_2) = (-x_0, x_1, x_2, -t_1, -t_2)$ *where f is an invariant non-unit without multiple factors. p is* cA/2 *or* cAx/2 *and it is a unique singular point of X.*

(v) *p is of type* IE$^\vee$, *d = 4 and π is the quotient of the* **Q**-*conic bundle*

$$\bar{X} = \left(\begin{array}{c} x_0^2 - x_1^2 = t_1 f_1 + t_2 f_2 \\ x_0 x_1 - x_2^2 = t_1 g_1 + t_2 g_2 \end{array} \right) \subset \mathbf{P}(1,1,1,2) \times \mathfrak{D}^2 \to \mathfrak{D}^2$$

by $\mathbf{Z}_4$ *such that* $g(x_0, x_1, x_2, y, t_1, t_2) = (-\zeta x_0, \zeta x_1, -x_2, \zeta y, \zeta t_1, -\zeta t_2)$ *with* $f_i, g_i \in \mathbf{C}\{t_1, t_2\}[x_0, x_1, x_2, y]$ *where* $t_i^3 f_i$ *and* $t_i g_i$ *are invariant. p is a quotient singularity of type* $\frac{1}{8}(5, 1, 3)$ *and it is a unique singular point of X.*

(vi) *p is of type* II$^\vee$, *d = 2 and π is the quotient of the* **Q**-*conic bundle by* $\mathbf{Z}_2$ *which is of the same form as in* (iii).

Observe that the base $o \in S$ is Du Val of type A_{d-1} in every case in the theorem. Running the MMP, one obtains the following deep result conjectured by Iskovskikh [209]. See the remark on Conjecture 7.4.19.

Corollary 7.3.2   *The base of a* **Q**-*conic bundle has only Du Val singularities of type A.*

Proof   Let $X \to S$ be a **Q**-conic bundle and take an arbitrary point o in S. The K_X-MMP over the *analytic* germ $o \in S$ by [368] outputs an irreducible **Q**-conic bundle germ on $o \in S$. Then the assertion follows from Theorem 7.3.1.   □

When the base is smooth, we have a list of possible singularities like that in Theorem 5.4.6.

Theorem 7.3.3 ([338, theorem 1.3])   *Let* $\pi\colon C \subset X \to o \in S$ *be an irreducible* **Q**-*conic bundle germ such that S is smooth. Then* $C \subset X$ *is locally primitive and the set* Σ *of singular points of X in terms of type is one of the following.*

(i) $\Sigma = \emptyset$, {III}, {III, III}, {IA}, {IA, III}, {IA, III, III}, {IIA} *or* {IIA, III}.
(ii) $\Sigma = $ {IC}, {IIB}, {IA, IA} *or* {IA, IA, III}.

We have an explicit description of the general elephant comparable to Theorem 5.5.1.

Theorem 7.3.4 ([338], [340]) *Let $\pi : C \simeq \mathbf{P}^1 \subset X \to o \in S$ be an irreducible* **Q***-conic bundle germ. Then the general elephant T of X is Du Val. Further if π belongs to one of* (i), (iii) *to* (vi) *in Theorem 7.3.1 or* (i) *in Theorem 7.3.3, then $C \not\subset T$ and $x \in T$ is the general elephant of the germ $x \in X$ at each point x in $T \cap C$. If π belongs to* (ii) *in Theorem 7.3.1 or* (ii) *in Theorem 7.3.3, then $C \subset T$ and one of the following holds for the set Σ of singular points of X and the dual graph Γ of the minimal resolution of the Du Val surface $\operatorname{Spec}_S \pi_* \mathscr{O}_T$, where $\circ$ stands for a curve exceptional over T and $\bullet$ for the curve C.*

(i) $\Sigma = \{p_1, p_2\}$ *in Theorem 7.3.1(ii). Each p_i is a quotient singularity of type $\frac{1}{d}(1, -1, a)$. $p_i \in T$ is of type A_{d-1} and Γ is as follows.*

(ii) $\Sigma = \{IC\}$ *in Theorem 7.3.3. Σ consists of a quotient singularity p of type $\frac{1}{5}(2, 3, 1)$ by [343, theorem 1.1]. $p \in T$ is of type A_4 and Γ is as follows.*

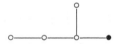

(iii) $\Sigma = \{IIB\}$ *in Theorem 7.3.3. Σ consists of a $cAx/4$ point p. $p \in T$ is of type D_5 and Γ is as follows.*

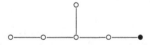

(iv) $\Sigma = \{IA, IA\}$ *in Theorem 7.3.3. Σ consists of a quotient singularity p of odd index r and a point q of type $cA/2$, $cAx/2$ or $cD/2$. $p \in T$ is of type A_{r-1} and $q \in T$ is of type D_{2n} if $n \geq 2$ and of type A_1 if $n = 1$ for the axial multiplicity n of $q \in X$. When $n = 1$, T has three singular points p, q, z with z of type A_1. Γ is as follows.*

(v) $\Sigma = \{IA, IA, III\}$ *in Theorem 7.3.3. Σ consists of a quotient singularity p of odd index r, a quotient singularity q of index two and a Gorenstein* *point z. $p \in T$ is of type A_{r-1}, $q \in T$ is of type A_1 and $z \in T$ is of type A_1 or A_2 where the type A_2 occurs only if $r = 3$. Γ is one of the following.*

Mori and Prokhorov [339] also classified reducible **Q**-conic bundle germs over singular surfaces and verified the general elephant conjecture for them.

We proceed to examples. Examples 7.3.5 to 7.3.8 are in the imprimitive cases in Theorem 7.3.1. We use the notation in Theorem 7.3.1.

Example 7.3.5   The quotient of

$$\bar{X} = \begin{pmatrix} x_0^2 - x_1^2 = t_1 y \\ x_2^2 = t_1^2 x_0^2 + \lambda t_1 x_0 x_1 + t_2 y \end{pmatrix} \subset \mathbf{P}(1,1,1,2) \times \mathfrak{D}^2 \to \mathfrak{D}^2$$

by $\mathbf{Z}_2$ such that $g(x_0, x_1, x_2, y, t_1, t_2) = (-x_0, x_1, x_2, -y, -t_1, -t_2)$ with $\lambda \in \mathbf{C}$ is a **Q**-conic bundle with p of type IA$^\vee$ in the case (iii) in Theorem 7.3.1. If $\lambda = 0$, then it also has one ordinary double point.

Example 7.3.6   The quotient of

$$\bar{X} = (x_0^2 + x_1^2 + (t_1^{2e} + t_2^{2e}) x_2^2 = 0) \subset \mathbf{P}^2 \times \mathfrak{D}^2 \to \mathfrak{D}^2$$

by $\mathbf{Z}_2$ such that $g(x_0, x_1, x_2, t_1, t_2) = (-x_0, x_1, x_2, -t_1, -t_2)$ with $e \geq 1$ is a **Q**-conic bundle with p of type ID$^\vee$ in the case (iv) in Theorem 7.3.1. The point p is $cA/2$ if $e = 1$ and $cAx/2$ if $e \geq 2$.

Example 7.3.7   The quotient of

$$\bar{X} = \begin{pmatrix} x_0^2 - x_1^2 = t_1 y \\ x_0 x_1 - x_2^2 = t_2 y \end{pmatrix} \subset \mathbf{P}(1,1,1,2) \times \mathfrak{D}^2 \to \mathfrak{D}^2$$

by $\mathbf{Z}_4$ such that $g(x_0, x_1, x_2, y, t_1, t_2) = (-\zeta x_0, \zeta x_1, -x_2, \zeta y, \zeta t_1, -\zeta t_2)$ is a **Q**-conic bundle with p of type IE$^\vee$ in the case (v) in Theorem 7.3.1.

Example 7.3.8   The quotient of

$$\bar{X} = \begin{pmatrix} x_0^2 - x_1^2 = t_1^3 y + t_2 y \\ x_2^2 = t_1^2 x_0^2 + \lambda t_1 x_0 x_1 + t_2 y \end{pmatrix} \subset \mathbf{P}(1,1,1,2) \times \mathfrak{D}^2 \to \mathfrak{D}^2$$

by $\mathbf{Z}_2$ such that $g(x_0, x_1, x_2, y, t_1, t_2) = (-x_0, x_1, x_2, -y, -t_1, -t_2)$ with $\lambda \in \mathbf{C}$ is a **Q**-conic bundle with p of type II$^\vee$ in the case (vi) in Theorem 7.3.1. If $\lambda = 0$, then it also has one ordinary double point.

Next we treat the case when the base is smooth in accordance with the classification of the set Σ of singular points in Theorem 7.3.3.

Example 7.3.9 The fibrations

$$X_1 = (x_0^2 + (t_1 + t_2)x_1^2 + t_1 t_2 x_2^2 = 0) \subset \mathbf{P}^2 \times \mathfrak{D}^2 \to \mathfrak{D}^2,$$
$$X_2 = (x_0^2 + t_1 x_1 x_2 + t_2^2(x_1^2 + x_2^2) = 0) \subset \mathbf{P}^2 \times \mathfrak{D}^2 \to \mathfrak{D}^2$$

are Gorenstein conic bundles. The space X_1 has one singular point of type cA_1 and X_2 has two ordinary double points.

Example 7.3.10 ([342, section 7]) The fibration

$$X = \begin{pmatrix} x_0^2 = (\lambda t_1 + t_2)x_2^2 + (t_1^e - t_2^3)y \\ x_1^2 = t_1 x_2^2 + t_2 y \end{pmatrix} \subset \mathbf{P}(1,1,1,2) \times \mathfrak{D}^2 \to \mathfrak{D}^2$$

with $\lambda = 1, 0$ and $e = 1, 2, 3$ is a **Q**-conic bundle with the following Σ. Every singular point of type III is an ordinary double point.

(λ, e)	$(1,1), (0,1)$	$(1,2)$	$(0,2)$	$(1,3)$	$(0,3)$
Σ	IA, III, III	IA, III	IA	IA, III	IA
IA point	$\frac{1}{2}(1,1,1)$	$cAx/2$	$cAx/2$	$cD/2$	$cE/2$

Let $\pi\colon C \subset X \to o \in S$ be an irreducible **Q**-conic bundle germ such that S is smooth and let Σ be the set of singular points of X. The general principal prime divisor H on X containing C is defined by an element in $\pi_* \mathcal{O}_X = \mathcal{O}_S$ and hence the image $\pi(H)$ is a smooth curve in S. The dual graph Γ of the minimal resolution of the normalisation H^ν of H is classified in several cases. In Γ, we let $\circ$ stand for a curve exceptional over H^ν, $\bullet$ for a curve appearing in H^ν and the number attached to $\circ$ or $\bullet$ for the self-intersection number.

Theorem 7.3.11 *Let* $\pi\colon C \subset X \to o \in S$ *be as in Theorem 7.3.3.*

(i) *([342]) Suppose that* $\Sigma = \{$IA$\}$, $\{$IA, III$\}$ *or* $\{$IA, III, III$\}$ *and p is the point of type* IA *and of index r. If $r = 2$, then X is embedded in* $\mathbf{P}(1,1,1,2) \times S$ *over S. If $r \geq 3$, then p is of type cA/r, H is not normal and Γ is as follows, where the strings $[a_{11}, \dots, a_{1n_1}]$ and $[a_{21}, \dots, a_{2n_2}]$ are conjugate.*

$$\underset{-a_{1n_1}}{\circ} \text{---} \cdots \text{---} \underset{-a_{11}}{\circ} \text{---} \underset{-1}{\bullet} \text{---} \underset{-a_{21}}{\circ} \text{---} \cdots \text{---} \underset{-a_{2n_2}}{\circ}$$

(ii) *([344], [345]) Suppose that* $\Sigma = \{$IIA$\}$ *or* $\{$IIA, III$\}$*. If H is normal, then* $\Sigma = \{$IIA$\}$ *and Γ is as follows.*

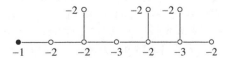

If H is not normal, then Γ is as follows.

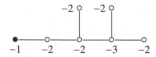

(iii) ([343]) *If $\Sigma = \{IC\}$ or $\{IIB\}$, then H is normal and Γ is as follows.*

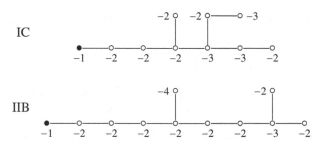

Example 7.3.12 ([345, example 4.8])   Let

$$C = (x_1 = y = z = 0) \subset X = (wf = x_0^6 g) \subset \mathbf{P}(1, 1, 2, 3, 4)$$

with homogeneous coordinates x_0, x_1, y, z, w, where

$$f = x_0^4 y + x_1^2 w + y^3 + z^2, \qquad g = \lambda x_0^3 x_1 + \mu x_0^2 y + y^2 + \xi x_1 z$$

with $\lambda \in \mathbf{C}$ and general $\mu, \xi \in \mathbf{C}$ with respect to λ. The germ $C \subset X$ has a point $[0, 0, 0, 0, 1]$ of type IIA. If $\lambda = 0$, then it also has one ordinary double point.

The rational function $f/x_0^6 = g/w$ on X defines a rational map $\alpha \colon X \dashrightarrow A = \mathbf{P}^1$ which is regular on a neighbourhood U in X about C. It contracts $H = (f = g = 0)$ to a point a in A with $U \times_A a = H|_U$. The surface H is rational since it has global coordinates $x_0 x_1^{-1}$ and $y x_1^{-2}$ on the locus $(x_1 \neq 0)$. It is normal if $\lambda \neq 0$ and non-normal if $\lambda = 0$. The normalisation H^ν of H has a unique singular point and only one curve C^ν in H^ν is mapped to C in H. The explicit computations in [344], [345] verify that the minimal resolution $\bar{H}$ of H^ν has dual graph Γ as in Theorem 7.3.11(ii).

By contracting all curves $\circ$ and $\bullet$ in $\bar{H}$ but the rightmost (-2)-curve $\circ$ in Γ, one obtains a smooth rational surface G. The strict transform l in G of the surviving $\circ$ has $(l^2) = 0$. Since G is rational, the K_G-MMP contracting only curves perpendicular to the nef divisor l produces a fibration $G \to B \simeq \mathbf{P}^1$. It follows from $(l^2) = 0$ that l is contracted in the last step of the MMP. The composite $\bar{H} \to G \to B$ factors through H^ν. The induced morphism $\beta^\nu \colon C^\nu \subset H^\nu \to b \in B$ for $C^\nu = (\beta^\nu)^{-1}(b)$ factors through H and descends to $\beta \colon C \subset H \to b \in B$. Indeed when H is not normal, β is defined outside the ramification locus of $C^v \to C$ and it is extended to the whole H because H is Cohen–Macaulay.

Let $\hat{X}$ and $\hat{H}$ be the formal completions of X and H along C. Mori and Prokhorov proved the vanishing $H^1(\mathscr{O}_{\hat{X}}) = 0$. Then the exact sequence $0 \to \mathscr{O}_X \to \mathscr{O}_X(H) \to \mathscr{O}_H(H|_H) \to 0$ yields the surjection $H^0(\mathscr{O}_{\hat{X}}) \twoheadrightarrow H^0(\mathscr{O}_{\hat{H}})$ since the divisor H on X is principal about C. Hence the regular function on a neighbourhood in H about C defining β extends to a function $\hat{v}$ in $H^0(\mathscr{O}_{\hat{X}})$. Together with the function $\hat{u}$ in $H^0(\mathscr{O}_{\hat{X}})$ defining $\alpha|_U$, the pair $(\hat{u}, \hat{v})$ produces a fibration $\hat{\pi} \colon \hat{X} \to \hat{\mathbf{A}}^2$ to the formal completion of $\mathbf{A}^2$ at origin. The central fibre $\hat{X} \times_{\hat{\mathbf{A}}^2} \operatorname{Spec} \mathbf{C}$ is isomorphic to $F = H \times_B b \subset H|_U$.

Analogously to the Hilbert scheme, the *Douady space* is the analytic space which parametrises compact analytic subspaces of a given analytic space [109], [383]. Let $o \in D$ be the germ of the relative Douady space of U/A at the point $o = [F]$. The fibre $o \in D \times_A a$ is isomorphic to $b \in B$, and the existence of $\hat{\pi}$ implies that D dominates the analytic germ $a \in A$. Hence $D \simeq \mathfrak{D}^2$ and the universal family $\mathfrak{F} \subset U \times_A \mathfrak{D}^2 \to \mathfrak{D}^2$ is an extension of $\hat{\pi}$. The projection $\mathfrak{F} \to U$ is isomorphic about $F = \mathfrak{F} \times_{\mathfrak{D}^2} o$ and thus one can consider $\mathfrak{F} \to \mathfrak{D}^2$ to be a $\mathbf{Q}$-conic bundle germ $C \subset X \to o \in \mathfrak{D}^2$. It has $\Sigma = \{\text{IIA}\}$ if $\lambda \neq 0$ and $\Sigma = \{\text{IIA}, \text{III}\}$ if $\lambda = 0$.

7.4 Rationality

We begin with a basic lemma on transformations of threefold standard conic bundles. Recall that by Lemma 7.1.6, the discriminant divisor of a threefold conic bundle is reduced and normal crossing.

Lemma 7.4.1 *Let $s \in S$ be the germ of a smooth surface. Let $g \colon S' \to S$ be the blow-up at s and let E be the exceptional curve.*

(i) *([409, proposition 2.4]) For any standard conic bundle $\pi \colon X \to S$, there exists a standard conic bundle $\pi' \colon X' \to S'$ which is isomorphic to π outside s.*

(ii) *([208, lemma 4]) For any standard conic bundle $\pi' \colon X' \to S'$ such that $(\Delta' \cdot E) \leq 1$ for the discriminant divisor Δ', there exists a standard conic bundle $\pi \colon X \to S$ to which π' is isomorphic outside s.*

In both statements, E is contained in Δ' if and only if s is a singular point of Δ, where Δ and Δ' are the discriminant divisors of π and π' respectively.

Proof We shall reduce (i) to (ii). We may work analytically on S since a conic bundle has relatively ample anti-canonical divisor. Then $\pi \colon X \to S$ is described by a diagonal quadratic form as in Remark 7.1.7. The assertion (i) is trivial if $s \notin \Delta$. If s is a singular point of Δ, then one can replace the base change $X \times_S S' \to S'$ as in Step 2 of the proof of Theorem 7.1.8 to obtain a required bundle $\pi' : X' \to S'$. To be precise, when the map $q(\sigma) : \mathscr{O}_S^{\oplus 3} \to \mathscr{O}_S^{\oplus 3}$ gives the diagonal form $x_0^2 + ax_1^2 + bx_2^2$ for π, we take $\mathscr{E}' = \mathscr{O}_{S'}(E) \oplus \mathscr{O}_{S'}^{\oplus 2}$ and $\mathscr{L} = \mathscr{O}_{S'}(-E)$. Then $q(\sigma)$ induces a map $(\mathscr{E}')^{\vee} \to \mathscr{E}' \otimes \mathscr{L}$ which defines π'. The remaining case is when s is a smooth point of Δ.

In this case, the base change $X \times_S S' \to S'$ is an embedded conical fibration with discriminant divisor $g^*\Delta$. The total space $X \times_S S'$ has a singular point over the point s' in S' at which E intersects the strict transform of Δ. Take the blow-up $g' : S'' \to S'$ at s'. Replacing $X' \times_{S'} S'' \to S''$ as in Step 3 of the proof of Theorem 7.1.8, one obtains a conic bundle $Y \to S''$ the discriminant divisor of which is the disjoint union of the strict transform E' of E and the strict transform Δ'' of Δ. Contracting one of the two irreducible components of $Y \times_{S''} E'$ by Lemma 7.1.9, one obtains a standard conic bundle $\pi'' : X'' \to S''$ with discriminant divisor Δ''. Then one can make a standard conic bundle π' in (i) by applying (ii) to g' and π''.

Thus it suffices to prove (ii). Given $\pi' : X' \to S'$, we let $\mathscr{E}' = \pi'_* \mathscr{O}_{X'}(-K_{X'})$ by which $X' \subset \mathbf{P}(\mathscr{E}')$. The double dual $\mathscr{E} = (g_*\mathscr{E}')^{\vee\vee}$ of its direct image is reflexive and hence free at s [179, section 1]. We take the closure X in $\mathbf{P}(\mathscr{E})$ of $X' \setminus (\pi')^{-1}(E) \subset \mathbf{P}(\mathscr{E}'|_{S'\setminus E}) \simeq \mathbf{P}(\mathscr{E}|_{S\setminus s})$, which defines an embedded conical fibration $\pi : X \to S$. Using Lemma 7.1.6, we shall verify that π is a conic bundle.

The discriminant divisor Δ of π is $g_*\Delta'$ by construction and it is normal crossing from the assumption on Δ'. The condition in Lemma 7.1.6 at $s \in S$ follows from Lemma 7.1.5 immediately once one excludes the case when Δ is singular at s with $r_s = 2$. Note that r_s is positive by Lemma 7.1.5 since Δ has only nodal singularities. If Δ were singular at s with $r_s = 2$, then Step 3 of the proof of Theorem 7.1.8 would construct a standard conic bundle over S' the discriminant divisor of which is the strict transform of Δ, not Δ'. This contradicts Proposition 7.1.12. Thus π is a conic bundle which is standard. $\quad\square$

In this section, we shall discuss the rationality problem of threefold conical fibrations. By virtue of Theorem 7.1.8, we study a standard conic bundle $\pi : X \to S$ over a smooth projective surface with discriminant divisor Δ.

Artin and Mumford observed that Δ must be connected in order for X to be stably rational. The argument works even if Δ is singular.

Theorem 7.4.2 ([22, §4 proposition 3]) *Let $X \to S$ be a standard conic bundle over a smooth projective surface. If the discriminant divisor is disconnected,* then X has a two-torsion in $H^3(X, \mathbf{Z})$ and in particular X is not stably rational by Theorem 6.3.17.

Remark 7.4.3 The conical fibration $X \to \mathbf{P}^2$ in Example 6.3.16 is equivalent to a standard conic bundle over the blow-up of $\mathbf{P}^2$ at the nine points $D_1 \cap D_2$ such that the discriminant divisor is the disjoint union of the strict transforms of D_1 and D_2. Hence Theorem 7.4.2 shows that X is not stably rational.

Once X is rational, S is unirational and further rational as explained posterior to Conjecture 6.3.12. It is thus harmless to assume the rationality of S. On this assumption, either S admits a conical fibration to $\mathbf{P}^1$ or it is isomorphic to $\mathbf{P}^2$. The next result is a consequence of the classification of standard conic bundles over the projective line defined over a perfect field, due to Iskovskikh. In general for a field k, we add the prefix 'k-' to make the meaning relative to Spec k, such as a k-point, a k-isomorphism or k-rationality, and we write $\mathbf{P}^n_k = \operatorname{Proj} k[x_0, \ldots, x_n]$.

Proposition 7.4.4 *Let $X \to S$ be a standard conic bundle over a smooth projective surface and let Δ be the discriminant divisor. If there exists a conical fibration $S \to \mathbf{P}^1$ such that $(\Delta \cdot f) \leq 3$ for its fibre f, then X is rational.*

Proof We shall use *Tsen's theorem* [266, IV theorem 6.5]. It asserts that for the function field F of a curve defined over an algebraically closed field, every homogeneous polynomial of positive degree d in more than d variables has a non-trivial zero in F.

The base change to the generic point η of $\mathbf{P}^1$ produces a standard conic bundle $X_K = X \times_{\mathbf{P}^1} \eta \to S_K = S \times_{\mathbf{P}^1} \eta$ over the function field K of $\mathbf{P}^1$. It has discriminant divisor $\Delta \times_{\mathbf{P}^1} \eta$ of degree $(\Delta \cdot f)$. Thanks to Tsen's theorem, the conic S_K in $\mathbf{P}^2_K$ has a K-point and hence it is K-isomorphic to $\mathbf{P}^1_K$. Then the general fibre of $X_K \to S_K \simeq \mathbf{P}^1_K$ is defined over K and it has a K-point again by Tsen's theorem. Thus X_K has a K-point.

One obtains $K^2_{X_K} = 8 - (\Delta \cdot f) \geq 5$ from Theorem 6.1.11 applied to the base change of $X_K \to S_K$ to the algebraic closure of K. Then by Theorem 7.4.5, X_K is K-rational. This means that the fibration $X \to S \to \mathbf{P}^1$ is birational over $\mathbf{P}^1$ to the trivial bundle $\mathbf{P}^2 \times \mathbf{P}^1 \to \mathbf{P}^1$. In particular, X is rational. $\square$

Theorem 7.4.5 (Iskovskikh [199], [200], [201], [210]) *Let $S \to \mathbf{P}^1_k$ be a standard conic bundle defined over a perfect field k. Let $d = K^2_S \leq 8$. Then the following hold. In particular, S is k-rational if and only if S has a k-point with $d \geq 5$.*

(i) *If $d \leq 0$, then S is k-birationally superrigid.*
(ii) *If $d = 1$ or 2, then S is k-birationally rigid.*

(iii) *If $d = 3$ or 4, then S is not k-rational.*

(iv) *If $d = 5$, then S admits a k-birational morphism to $\mathbf{P}_k^2$.*

(v) *If $d = 6$, then S admits a k-birational morphism to a non-singular quadric in $\mathbf{P}_k^3$.*

(vi) *The case $d = 7$ never occurs.*

(vii) *If $d = 8$, then S is k-birational to the product of two non-singular geometrically irreducible curves of genus zero.*

Unsurprisingly, this theorem is founded on results in the geometric case when k is algebraically closed. For instance, (vi) is proved as follows.

Proof of Theorem 7.4.5(vi)   If $\pi\colon S \to \mathbf{P}_k^1$ had $K_S^2 = 7$, then the base change $\bar{\pi}\colon \bar{S} \to \mathbf{P}_{\bar{k}}^1$ to the algebraic closure $\bar{k}$ of k would be the blow-up of a Hirzebruch surface at a point. Hence $\bar{\pi}$ has a unique section $\bar{C}$ with negative self-intersection number. The unique degenerate fibre of $\bar{\pi}$ consists of two smooth rational curves $\bar{f}_1$ and $\bar{f}_2$, exactly one of which intersects $\bar{C}$. Thus $\bar{C}$, $\bar{f}_1$ and $\bar{f}_2$ are all invariant by the action of the Galois group of $\bar{k}/k$. Then π also has a fibre which is reducible over k. This contradicts the assumption that it is standard.   □

The work of Clemens and Griffiths in Theorem 6.3.15 addresses the rationality problem of a smooth projective threefold X by means of the intermediate Jacobian $J_2(X)$. When X has a structure of a standard conic bundle, $J_2(X)$ turns out to be isomorphic to the *Prym variety*. We shall briefly review the theory of Prym varieties.

Definition 7.4.6 (Beauville [34, définition 0.3.1])   Let C and $\tilde{C}$ be reduced schemes of pure dimension one with only nodal singularities. A finite morphism $\delta\colon \tilde{C} \to C$ is called a *pseudo-cover* if

- it is étale of degree two at every smooth point of C and
- it is analytically isomorphic to the morphism $(\tilde{x}_1\tilde{x}_2 = 0) \to (x_1x_2 = 0)$ given by $(x_1, x_2) = (\tilde{x}_1^2, \tilde{x}_2^2)$ at every singular point of C.

Let $\delta\colon \tilde{C} \to C$ be a pseudo-cover. It satisfies the relation $p_a(\tilde{C}) = 2p_a(C) - 1$ of the arithmetic genera. The *arithmetic genus* $p_a(X)$ of a projective scheme X of dimension n is defined as $(-1)^n(\chi(\mathcal{O}_X) - 1)$. There exists an involution $\iota\colon \tilde{C} \to \tilde{C}$ compatible with δ, that is, $\delta = \delta \circ \iota$. By this, C is the quotient of $\tilde{C}$ by $\mathbf{Z}_2$. Let $\mathscr{L}$ be the subsheaf of $\delta_*\mathcal{O}_{\tilde{C}}$ which consists of semi-invariant functions of weight one modulo two, that is, functions $f \in \delta_*\mathcal{O}_{\tilde{C}}$ such that $\iota^*f = -f$. Then $\mathscr{L}$ is a torsion-free sheaf of rank one such that $\mathscr{L} \simeq \mathscr{L}^\vee$. One can write $\tilde{C} = \operatorname{Spec}_C(\mathcal{O}_C \oplus \mathscr{L})$ by $\delta_*\mathcal{O}_{\tilde{C}} = \mathcal{O}_C \oplus \mathscr{L}$. Abstractly, $\mathscr{L}$ is isomorphic to the cokernel of the natural inclusion $\mathcal{O}_C \hookrightarrow \delta_*\mathcal{O}_{\tilde{C}}$.

An algebraic definition of the Prym variety was given by Mumford for a double cover of a smooth curve and generalised to a pseudo-cover $\delta\colon \tilde{C} \to C$ by Beauville. Note that the involution ι on $\tilde{C}$ induces $\iota^*\colon J(\tilde{C}) \to J(\tilde{C})$ on the Jacobian $J(\tilde{C})$ of $\tilde{C}$.

Definition 7.4.7 ([33], [363]) Let $\delta\colon \tilde{C} \to C$ be a pseudo-cover with involution $\iota\colon \tilde{C} \to \tilde{C}$. The *Prym variety* of δ is $\mathrm{Pr}(\tilde{C}/C) = (\mathrm{id} - \iota^*)(J(\tilde{C})) \subset J(\tilde{C})$. It is a principally polarised abelian variety in such a manner that the principal polarisation of $J(\tilde{C})$ induces twice that of $\mathrm{Pr}(\tilde{C}/C)$.

Let $\pi\colon X \to S$ be a conic bundle over a smooth surface, which has an embedding $X \subset \mathbf{P}(\mathscr{E})$ by $\mathscr{E} = \pi_* \mathscr{O}_X(-K_X)$. As in [164, I §9.7], the *relative Grassmannian* $\mathrm{Gr}(2, \mathscr{E}) \to S$ represents the functor from the category of schemes over S to the category of sets given by

$$(f\colon T \to S) \mapsto \{f^*\mathscr{E} \twoheadrightarrow \mathscr{Q} \mid \mathscr{Q} \text{ locally free of rank two}\}.$$

Let Δ be the discriminant divisor of π. The family $X \times_S \Delta \to \Delta$ of two lines defines a subscheme $\tilde{\Delta}$ of $\mathrm{Gr}(2, \mathscr{E}) \times_S \Delta$. From the local description in Remark 7.1.7, one sees that the projection $\delta\colon \tilde{\Delta} \to \Delta$ is a pseudo-cover. By Lemma 7.1.9, the conic bundle π is standard if and only if no irreducible component of $\tilde{\Delta}$ is isomorphic to its image by δ.

Definition 7.4.8 The morphism $\delta\colon \tilde{\Delta} \to \Delta$ constructed as above is called the *associated pseudo-cover* of the conic bundle $\pi\colon X \to S$.

Beauville obtained an isomorphism between $J_2(X)$ and $\mathrm{Pr}(\tilde{\Delta}/\Delta)$ when $S = \mathbf{P}^2$ after Mumford. This was extended to surfaces with irregularity zero by Beltrametti.

Theorem 7.4.9 ([33], [38], [88, appendix C]) *Let S be a smooth projective surface with irregularity zero, that is, $h^1(\mathscr{O}_S) = 0$. Let $X \to S$ be a standard conic bundle with associated pseudo-cover $\tilde{\Delta} \to \Delta$. Then the intermediate Jacobian $J_2(X)$ is isomorphic to the Prym variety $\mathrm{Pr}(\tilde{\Delta}/\Delta)$ as a principally polarised abelian variety.*

If S is rational, then the conic bundle π is recovered from δ up to equivalence. This is essentially due to Artin and Mumford [22], and the case when Δ is singular is treated in [410, section 5]. The statement is found in [206, lemma 1(iv)].

Proposition 7.4.10 *Let S be a smooth rational projective surface and let Δ be a reduced normal crossing divisor on S. Then for any pseudo-cover $\delta\colon \tilde{\Delta} \to \Delta$ without components of $\tilde{\Delta}$ isomorphic to the image, there exists a standard conic bundle $X \to S$ with associated pseudo-cover δ. All such conic bundles are equivalent.*

The proof uses the following exact sequence with respect to the *Brauer group* [162] which consists of equivalence classes of *Azumaya algebras*.

Theorem 7.4.11   *Let S be a smooth complete surface with the first étale cohomology $H^1_{\text{ét}}(S, \mathbf{Q}/\mathbf{Z}) = 0$. Then there exists a natural exact sequence*

$$0 \to \operatorname{Br} S \to \operatorname{Br} K(S) \xrightarrow{\alpha} \bigoplus_{C \subset S} H^1_{\text{ét}}(K(C), \mathbf{Q}/\mathbf{Z}) \xrightarrow{\beta} \bigoplus_{s \in S} \mu^{-1} \to \mu^{-1} \to 0,$$

in which the direct sums are taken over all curves C and points s respectively. The symbols $K(S)$ and $K(C)$ stand for the function fields of S and C, Br stands for the Brauer group and $\mu^{-1} = \bigcup_{n \in \mathbf{N}_{>0}} \operatorname{Hom}(\mathbf{Z}_n, \mathbf{Q}/\mathbf{Z})$.

In Proposition 7.4.10, the pseudo-cover δ represents an element $[\delta]$ of order two in $\bigoplus_C H^1_{\text{ét}}(K(C), \mathbf{Q}/\mathbf{Z})$ such that $\beta([\delta]) = 0$. With the property $\operatorname{Br} S = 0$ of the rational surface S, there exists a unique class of an Azumaya algebra A_η of rank four over $K(S)$ such that $\alpha([A_\eta]) = [\delta]$. Then A_η extends to an Azumaya algebra $\mathscr{A}_U$ on the complement $U = S \setminus \Delta$. The algebra $\mathscr{A}_U$ corresponds to a standard conic bundle over S with associated pseudo-cover δ.

Henceforth we focus on the rationality problem in the case when $S \simeq \mathbf{P}^2$. Let $\pi \colon X \to \mathbf{P}^2$ be a standard conic bundle with discriminant divisor Δ. Let d denote the degree of Δ.

Lemma 7.4.12   *If $d \leq 4$, then X is rational.*

Proof   Take the blow-up B of $\mathbf{P}^2$ at a point, which is assumed to lie on Δ when $d = 4$, and construct a standard conic bundle $\pi' \colon X' \to B$ from π by Lemma 7.4.1(i). By the structure $B \simeq \Sigma_1 \to \mathbf{P}^1$ of a Hirzebruch surface, π' satisfies the assumption of Proposition 7.4.4. Hence X as well as X' is rational.   □

On the other hand, X is irrational whenever $d \geq 6$.

Theorem 7.4.13 (Beauville [34, proposition 4.7])   *Let $X \to \mathbf{P}^2$ be a standard conic bundle with associated pseudo-cover $\tilde{\Delta} \to \Delta$. If Δ is of degree at least six, then $\operatorname{Pr}(\tilde{\Delta}/\Delta)$ is never isomorphic as a principally polarised abelian variety to a product of Jacobians of smooth curves. In particular, X is irrational by Theorems 6.3.15 and 7.4.9.*

In the remaining case $d = 5$, we apply Shokurov's detailed analysis of a Prym variety isomorphic to a product of Jacobians of smooth curves. For simplicity, we state his result only for plane quintics. Recall that a pseudo-cover $\delta \colon \tilde{C} \to C$ is expressed as $\tilde{C} = \operatorname{Spec}_C(\mathscr{O}_C \oplus \mathscr{L})$ by the torsion-free sheaf $\mathscr{L}$ of semi-invariant functions of weight one. When C is a plane quintic, namely a reduced divisor on $\mathbf{P}^2$ of degree five, the sheaf $\mathscr{O}_C(1) \otimes \mathscr{L}$ is a *theta characteristic* in the following sense.

Definition 7.4.14 ([33, section 1], [362]) Let C be a reduced scheme of pure dimension one. A torsion-free $\mathscr{O}_C$-module $\mathscr{M}$ of rank one is called a *theta characteristic* on C if $\mathscr{M} \simeq \mathscr{H}om_C(\mathscr{M}, \omega_C)$. A theta characteristic $\mathscr{M}$ is said to be *even* (resp. *odd*) if the dimension $h^0(\mathscr{M})$ is even (resp. odd).

Definition 7.4.15 We say that a pseudo-cover $\tilde{C} \to C$ of a plane quintic C is *defined by* even (resp. odd) theta characteristic if the theta characteristic $\mathscr{O}_C(1) \otimes \mathscr{L}$ above is even (resp. odd). We also say that a standard conic bundle $X \to \mathbf{P}^2$ with quintic discriminant divisor is *defined by* even (or odd) theta characteristic if so is the associated pseudo-cover.

Theorem 7.4.16 (Shokurov [422]) *Let $\delta \colon \tilde{C} \to C$ be a pseudo-cover of a plane quintic C. Then the Prym variety $\mathrm{Pr}(\tilde{C}/C)$ is isomorphic as a principally polarised abelian variety to a product of Jacobians of smooth curves if and only if δ is defined by even theta characteristic.*

We remark that hyperelliptic, trigonal and quasi-trigonal curves in other cases in the main theorem in [422] are excluded because the *canonical map* $C \dashrightarrow PH^0(\omega_C) = \mathbf{P}^5$ of a plane quintic C is an embedding which factors through the Veronese embedding of $\mathbf{P}^2$. The canonical map of a hyperelliptic curve is not an embedding, and the canonical embedding of a trigonal or quasi-trigonal curve is contained in a surface of minimal degree which is a rational scroll or the cone over a smooth rational curve. Refer to [403, 2.10].

By virtue of Theorem 7.4.16, we assume that the standard conic bundle $\pi \colon X \to \mathbf{P}^2$ with quintic discriminant divisor Δ is defined by even theta characteristic. In this case, according to Masiewicki [305] and Tyurin [451], the Prym variety $\mathrm{Pr}(\tilde{\Delta}/\Delta)$ is isomorphic to the Jacobian $J(D)$ of a smooth curve D which is a complete intersection $Q_1 \cap Q_2 \cap Q_3$ of three quadrics in $\mathbf{P}^4$, and further the pseudo-cover $\delta \colon \tilde{\Delta} \to \Delta$ is described explicitly as in the next example. It follows that X is rational [206], [379].

Example 7.4.17 In the above setting where $D = Q_1 \cap Q_2 \cap Q_3 \subset \mathbf{P}^4$, we fix a point o in D. Let $\Pi \simeq \mathbf{P}^2$ be the linear subsystem of $|\mathscr{O}_{\mathbf{P}^4}(2)|$ spanned by the three quadrics Q_1, Q_2, Q_3. The incidence subvariety I of $\Pi \times \mathbf{P}^4$ consists of $(Q, x) \in \Pi \times \mathbf{P}^4$ such that $x \in Q$. Then $\Pi \times o \subset I$ is a section of the projection $I \to \Pi$. One can define the subvariety Y of I which parametrises $(Q, x) \in \Pi \times \mathbf{P}^4$ such that $x \in Q \cap T_{Q,o}$ for the tangent space $T_{Q,o}$ identified with the hyperplane in $\mathbf{P}^4$ tangent to Q at o.

For any point x in $Q \cap T_{Q,o} \setminus o$, the line l joining o and x is contained in Q. Otherwise l would intersect Q at x and tangentially at o, contradicting $(Q \cdot l) = 2$. Hence each fibre $Q \cap T_{Q,o}$ of $Y \to \Pi$ is a cone with vertex o, and $Y \to \Pi$ factors through the image Y' in $\Pi \times \mathbf{P}^3$ of $Y \subset \Pi \times \mathbf{P}^4$ by the base change of the projection $p \colon \mathbf{P}^4 \dashrightarrow \mathbf{P}^3$ from o. By the works of Masiewicki and Tyurin, the induced morphism $\varphi \colon Y' \to \Pi$ is a conic bundle with associated pseudo-cover isomorphic to $\delta \colon \tilde{\Delta} \to \Delta$.

For the rationality of Y', we shall see that the second projection $q \colon Y' \to \mathbf{P}^3$ is birational. For any line l in $\mathbf{P}^4$ through o and for any two points in l other than o, the linear system $\Pi \simeq \mathbf{P}^2$ possesses a quadric Q through these two points and it must contain the whole l. Then $(Q, [l]) \in Y'$. It shows that q is surjective. If $q^{-1}([l])$ contains two quadrics Q'_1 and Q'_2 in Π, then $D = Q'_1 \cap Q'_2 \cap Q'_3$ with a third quadric Q'_3 in Π and $l \cap Q'_3 \subset D$. Hence l is a line either joining o and another point in D or tangent at o to D. It follows that $[l] \in p(D)$. In particular, q is bijective outside $p(D)$ and thus it is birational.

We have constructed the conic bundle $\varphi \colon Y' \to \Pi$ with associated pseudo-cover δ. By Proposition 7.4.10, φ is equivalent to $\pi \colon X \to \mathbf{P}^2$. Therefore X is rational as so is Y'.

We complete the rationality problem of conic bundles over $\mathbf{P}^2$ by summing up Lemma 7.4.12, Theorems 7.4.13 and 7.4.16 and Example 7.4.17.

Theorem 7.4.18 *Let $\pi \colon X \to \mathbf{P}^2$ be a standard conic bundle and let Δ be the discriminant divisor. Let d denote the degree of Δ. Then X is rational if and only if either $d \le 4$ or $d = 5$ with π defined by even theta characteristic.*

The results hitherto obtained lead us to the following conjecture.

Conjecture 7.4.19 (Iskovskikh [206]) *Let $\pi \colon X \to S$ be a conical fibration from a smooth threefold. Then X is rational if and only if π is equivalent to a standard conic bundle $\pi' \colon X' \to S'$ over a smooth projective surface with discriminant divisor Δ' such that either*

(i) *there exists a conical fibration $S' \to \mathbf{P}^1$ with $(\Delta' \cdot f) \le 3$ for its fibre f or*
(ii) *$S' = \mathbf{P}^2$, Δ' is a quintic and π' is defined by even theta characteristic.*

The if part of the conjecture holds by Proposition 7.4.4 and Example 7.4.17. Corollary 7.3.2 settles the only-if part for standard conic bundles with discriminant divisor of arithmetic genus at least 19 [209]. By Theorem 7.4.18, the conjecture holds when $S \simeq \mathbf{P}^2$. Moreover, Shokurov proved it when S is a Hirzebruch surface [422, section 10].

Theorem 7.4.20 *Conjecture 7.4.19 holds when S is the projective plane $\mathbf{P}^2$ or a Hirzebruch surface Σ_n.*

The conic bundle constructed from a cubic threefold is defined by odd theta characteristic.

Example 7.4.21 Let $X = X_3 \subset \mathbf{P}^4$ be a general cubic threefold discussed in Example 6.3.13. We have seen that X is irrational and contains a two-dimensional family of lines. The blow-up B of X along a general line l in X admits a conic bundle structure $\pi \colon B \to \mathbf{P}^2$. It is a standard conic bundle since $\rho(B) = \rho(X) + 1 = 2$ as seen in Example 6.3.21.

Consider the restriction $\pi_L \colon B_L = B \times_{\mathbf{P}^2} L \to L$ to a general line L in $\mathbf{P}^2$. Planes $\Pi = \mathbf{P}^2$ in $\mathbf{P}^4$ parametrised by L span a hyperplane H_L containing l, and B_L is isomorphic to the smooth cubic surface $X_L = X \cap H_L$. Hence π_L is the family of conics C realised in $X_L \cap \Pi = l \cup C$ for $[\Pi] \in L$ in which Π is a plane in $H_L = \mathbf{P}^3$ containing l. It is a classical result [35, lemma IV.15] that C degenerates for exactly five planes Π, which means that π has discriminant divisor of degree five. By Theorem 7.4.18, π is defined by odd theta characteristic.

Finally we provide examples of standard conic bundles over $\mathbf{P}^2$ with discriminant divisor of degree d for $d = 6, 7$ and 8, according to [33, exemples 1.4].

Example 7.4.22 Let $D = D_4$ be a general quartic in $\mathbf{P}^3$ having one ordinary double point p. In the same way as in Example 6.3.16, we take the double solid X' ramified along D and the blow-up X of X' at the ordinary double point above p. The morphism $X \to \mathbf{P}^3$ factors through the blow-up B of $\mathbf{P}^3$ at p and the projection from p induces the morphism $\pi \colon X \to B \to \mathbf{P}^2 = S$.

Without loss of generality, we set $p = [0, 0, 0, 1]$ with homogeneous coordinates x_0, x_1, x_2, x_3. Then the quartic form defining D is written as $a_2 x_3^2 + a_3 x_3 + a_4$ with forms a_d in x_0, x_1, x_2 of degree d. On the chart $U_i = (x_i \neq 0)$ in S for $i = 0, 1, 2$, π is given by $y^2 - (a'_2 x_3^2 + a'_3 x_i x_3 + a'_4 x_i^2)$ in $\mathbf{P}^2 \times U_i$ for $a'_d = a_d x_i^{-d} \in \mathscr{O}_{U_i}$ with homogeneous coordinates x_i, x_3, y of $\mathbf{P}^2$. Hence $\pi \colon X \to S$ is a standard conic bundle and its discriminant divisor is globally defined by $a_3^2 - 4a_2 a_4$, which is of degree six.

Example 7.4.23 Let $X = X_{2,2,2} \subset \mathbf{P}^6$ be the intersection of three general quadrics Q_1, Q_2, Q_3 in $\mathbf{P}^6$. By Theorem 6.3.22, $\rho(X) = 1$ and $X \to \operatorname{Spec} \mathbf{C}$ is a Mori fibre space. Let Π be the linear subsystem of $|\mathscr{O}_{\mathbf{P}^6}(2)|$ spanned by these three quadrics. A standard argument using Grassmannians, such as in Example 6.3.13, shows that X contains a one-dimensional family of lines [451, lemma 5.2]. Fix a general line l in X and let $\Sigma \simeq \mathbf{P}^4$ parametrise planes P in $\mathbf{P}^6$ containing l.

Take the incidence subvariety I of $\Pi \times \Sigma$ consisting of $(Q, P) \in \Pi \times \Sigma$ such that $P \subset Q$. We shall see that the projection $\pi \colon I \to \Pi \simeq \mathbf{P}^2$ is a conic bundle.

We may set $l = (x_0 = \cdots = x_4 = 0)$ with homogeneous coordinates $x_0, \ldots, x_6$. Then $\Sigma \simeq \mathbf{P}^4$ has homogeneous coordinates $x_0, \ldots, x_4$, by which Σ is the target of the projection $p \colon \mathbf{P}^6 \dashrightarrow \Sigma$ from l. A quadric Q containing l is given by $ax_5 + bx_6 + c$ with linear forms a and b and a quadratic form c in $x_0, \ldots, x_4$. The fibre $\pi^{-1}(Q)$ is the plane conic $(a = b = c = 0)$ in Σ, which makes π a conic bundle. The locus $(a = b = 0)$ is certainly a plane in Σ since Q is smooth about X.

Write $a_i x_5 + b_i x_6 + c_i$ for the quadratic form defining Q_i. Then p maps X birationally to the quartic Y in Σ defined by the determinant of the 3×3 matrix M which consists of the three vectors (a_i, b_i, c_i) for $i = 1, 2, 3$. Let B be the blow-up of X along l. The induced morphism $p_B \colon B \to Y$ is isomorphic on the locus where M is of rank two. On the other hand, the projection $I \to \Sigma$ sends a general point $(Q, P) \in I$ to $p_B(B \cap P_B) \in Y$ for the strict transform P_B of P in the blow-up of $\mathbf{P}^6$ along l, where $B \cap P_B$ in $P_B \simeq P$ is the intersection of the two lines l_2' and l_3' for $Q_i' \cap P = l \cup l_i'$ with $Q_i' \in \Pi$ such that $X = Q \cap Q_2' \cap Q_3'$. In particular, it induces a birational morphism $I \to Y$. One can further verify that $B \to Y \leftarrow I$ is a flop and that the diagram

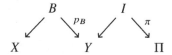

is a Sarkisov link of type I from $X \to \operatorname{Spec} \mathbf{C}$ to $\pi \colon I \to \Pi$ as mentioned in [97, p.17 example 5.5.1].

Without loss of generality, we express two general points in Π as $x_0 x_5 + x_1 x_6 + c$ and $x_2 x_5 + x_3 x_6 + d$ with quadratic forms c and d in $x_0, \ldots, x_4$. Then a point on the line joining them is expressed as $(x_0 + \lambda x_2)x_5 + (x_1 + \lambda x_3)x_6 + (c + \lambda d)$ with $\lambda \in \mathbf{P}^1$, and the corresponding conic given by $(c + \lambda d)(-\lambda x_2, -\lambda x_3, x_2, x_3, x_4)$ degenerates for seven values of λ. Thus π has discriminant divisor of degree seven.

Example 7.4.24   Let $X = X_4 \subset \mathbf{P}^4$ be a general quartic which is singular along a fixed line l. The projection from l induces a standard conic bundle $\pi \colon B \to \mathbf{P}^2 = S$ from the blow-up B of X along l. Setting $l = (x_0 = x_1 = x_2 = 0)$ with homogeneous coordinates $x_0, \ldots, x_4$, one can express the quartic form defining X as $a_{33}x_3^2 + 2a_{34}x_3 x_4 + a_{44}x_4^2 + 2a_{35}x_3 + 2a_{45}x_4 + a_{55}$ with forms a_{mn} in x_0, x_1, x_2. Then on the chart $U_i = (x_i \neq 0)$ in S for $i = 0, 1, 2$, π is given by $\sum_{m,n=3}^{5} a_{mn}' x_m x_n$ in $\mathbf{P}^2 \times U_i$ for $a_{mn}' = a_{mn} x_i^{-\deg a_{mn}}$ and the ad hoc notation $x_5 = x_i$. The discriminant divisor of π is globally defined by $\det(a_{mn})$, which is of degree eight.

7.5 Birational Rigidity

Let $\pi\colon X \to S$ be a threefold standard conic bundle and let Δ be the discriminant divisor. We shall study the linear system of form $|aK_S + b\Delta|$ for the birational rigidity problem. A good survey is [93, chapter 4].

Lemma 7.5.1 *Let $X \to S$ and $X' \to S'$ be equivalent standard conic bundles over smooth projective surfaces and let Δ and Δ' be the discriminant divisors. Then for any positive integers a and b such that $a \geq b$, there exists a natural identification of the linear systems $|aK_S + b\Delta|$ and $|aK_{S'} + b\Delta'|$. For any real number μ at least one, $\mu K_S + \Delta$ is pseudo-effective if and only if so is $\mu K_{S'} + \Delta'$.*

Proof We shall demonstrate the first assertion. The proof of the second is substantially the same. Take a common resolution T of S and S'. By Proposition 7.1.12, equivalent standard conic bundles over T have common discriminant divisor Δ_T and hence have common $|aK_T + b\Delta_T|$. Thus for $g\colon T \to S$, it suffices to construct one standard conic bundle $Y \to T$ with discriminant divisor Δ_T such that $g_* \mathscr{O}_T(aK_T + b\Delta_T) = \mathscr{O}_S(aK_S + b\Delta)$. This is reduced to the case when g is the blow-up at a point since g is a finite sequence of such blow-ups.

Assuming that g is the blow-up at a point, we construct Y/T from X/S by Lemma 7.4.1(i). Then Δ_T is $g^*\Delta$ or $g^*\Delta - E$ for the exceptional curve E, and $aK_T + b\Delta_T = g^*(aK_S + b\Delta) + eE$ with $e = a$ or $a - b \geq 0$. The projection formula yields the desired equality $g_* \mathscr{O}_T(aK_T + b\Delta_T) = \mathscr{O}_S(aK_S + b\Delta)$. □

The main theorem in this section is the following result on birational rigidity in terms of $4K_S + \Delta$.

Theorem 7.5.2 (Sarkisov [409], [410]) *Let $\pi\colon X \to S$ be a standard conic bundle over a smooth projective variety and let Δ be the discriminant divisor. If $4K_S + \Delta$ is pseudo-effective, then π is birationally superrigid.*

We prove it in the framework of the Sarkisov program. We extend the notion of discriminant divisor to Mori fibre spaces of relative dimension one.

Definition 7.5.3 Let $\pi\colon X \to S$ be a relative Mori fibre space, in which X is **Q**-factorial and $\rho(X/S) = 1$, such that $\dim X = \dim S + 1$. Take the maximal open subset S° of S such that the restriction $X^\circ = X \times_S S^\circ \to S^\circ$ is a standard conic bundle. The complement $S \setminus S^\circ$ is of codimension at least two. We call the closure in S of the discriminant divisor of X°/S° the *divisorial part of discriminant locus* of π.

Remark 7.5.4 In the definition, consider a standard conic bundle $\pi'\colon X' \to S'$ equivalent to π via a birational morphism $g\colon S' \to S$, which exists by

Theorem 7.1.8. Then the divisorial part of discriminant locus of π is the push-forward of the discriminant divisor of π' because they coincide on any open subset U of S° with $g^{-1}(U) \simeq U$ by Proposition 7.1.12.

Remark 7.5.5 If $\pi\colon X \to S$ is a Mori fibre space with $\dim X = \dim S + 1$ and with divisorial part Δ of discriminant locus, then the relation $\pi_* K_X^2 \equiv -(4K_S + \Delta)$ holds similarly to Proposition 7.1.13 by its proof, where the projective base S is **Q**-factorial by Lemma 1.4.13. Notice that $\varphi\colon Y \to T$ in the proof becomes a **Q**-conic bundle and hence T is **Q**-factorial by Lemma 7.2.3.

Proof of Theorem 7.5.2   Let $f\colon X/S \dashrightarrow Y/T$ be an arbitrary link from π to a Mori fibre space $Y \to T$. We want to show that f is square. Fixing a standard linear system $\mathscr{H}_Y$ for Y/T in Definition 6.1.8, we run the Sarkisov program as in Definition 6.2.10 to produce a sequence $X/S = X_0/S_0 \dashrightarrow X_1/S_1 \dashrightarrow \cdots$ of Sarkisov links $g_i\colon X_i/S_i \dashrightarrow X_{i+1}/S_{i+1}$. Let $\mathscr{H}_i$ be the homaloidal transform in X_i of $\mathscr{H}_Y$ and let H_i be its general member. Let μ_i be the Sarkisov threshold of $\mathscr{H}_i$ defined by $\mu_i K_{X_i} + H_i \sim_{\mathbf{Q},S_i} 0$.

As far as g_i is a Sarkisov link of type I or II from a conical fibration X_i/S_i, it is necessarily square and $\mu_i = \mu_{i+1}$. By Lemma 6.2.12, the program attains a model X_i/S_i square to X/S such that either $X_i/S_i \simeq Y/T$ or $(X_i, \mu_i^{-1} H_i)$ is canonical. Then Remark 6.1.13 reduces the assertion to the pseudo-effectivity of $K_{X_i} + \mu_i^{-1} H_i$.

We write $\pi_i\colon X_i \to S_i$ with divisorial part Δ_i of discriminant locus and write $K_{X_i} + \mu_i^{-1} H_i \sim_{\mathbf{Q}} \pi_i^* B_i$ with a **Q**-divisor B_i on S_i. Since H_i is mobile,

$$\mu_i^{-2} \pi_{i*} H_i^2 \equiv \pi_{i*} (\pi_i^* B_i - K_{X_i})^2 = 4B_i + \pi_{i*} K_{X_i}^2$$

is pseudo-effective. Hence together with Remark 7.5.5, it suffices to prove that $4K_{S_i} + \Delta_i$ is pseudo-effective. By Theorem 7.1.8, we have a standard conic bundle $\pi'\colon X' \to S'$ on a common resolution of S and S_i via which π' admits square links to π and π_i. Let Δ' be the discriminant divisor of π'. Because the pair $(S, \Delta/4)$ is terminal [229, proposition 6] from the description of Δ in Lemma 7.1.6, $4K_{S'} + \Delta'$ is at least the pull-back of $4K_S + \Delta$. Thus it is pseudo-effective and so is the push-forward $4K_{S_i} + \Delta_i$ as in Remark 7.5.4. $\square$

Whilst Theorem 7.5.2 pertains to $4K_S + \Delta$, Shokurov obtained the vanishing of $|2K_S + \Delta|$ in the study of Theorem 7.4.20.

Theorem 7.5.6 (Shokurov [422, theorem 10.2]) *Let* $X \to S$ *be a standard conic bundle over the projective plane or a Hirzebruch surface. If X is rational, then $|2K_S + \Delta| = \emptyset$ for the discriminant divisor Δ.*

He further conjectured the same vanishing for any standard conic bundle from a rational threefold. We state this in the style of Iskovskikh.

Conjecture 7.5.7 (Shokurov [422, conjecture 10.3]) *Let $\pi\colon X \to S$ be a standard conic bundle over a smooth rational projective surface. Then X is rational if and only if*

(i) *$|2K_S + \Delta| = \emptyset$ for the discriminant divisor Δ and*
(ii) *every standard conic bundle $X' \to \mathbf{P}^2$ with quintic discriminant divisor that is equivalent to π is defined by even theta characteristic.*

The following will be explained at the end of the section.

Theorem 7.5.8 (Iskovskikh [208]) *The if part of Conjecture 7.5.7 holds. Conjectures 7.4.19 and 7.5.7 are equivalent.*

From these observations, we predict that the pseudo-effective threshold μ below measures how close the total space X is to being rational or to being birationally rigid.

Definition 7.5.9 Let $\pi\colon X \to S$ be a standard conic bundle over a smooth rational projective surface and let Δ be the discriminant divisor. The *pseudo-effective threshold* μ of π is the greatest real number such that $\mu K_S + \Delta$ is pseudo-effective. It is actually a rational number as will be seen in Proposition 7.5.13.

Lemma 7.5.10 *Notation and assumptions as in Definition 7.5.9.*

(i) *If $\Delta = 0$, then $\mu = 0$ and X is rational.*
(ii) *If $\Delta \neq 0$, then $|K_S + \Delta| \neq \emptyset$ and in particular $\mu \geq 1$.*
(iii) *If $2K_S + \Delta$ is nef, then $|2K_S + \Delta| \neq \emptyset$.*

Proof Note that K_S is not pseudo-effective. If $\Delta = 0$, then X is rational by Proposition 7.4.4. We shall assume that $\Delta \neq 0$.

By Lemma 7.5.11 below, the normal crossing divisor Δ is never a tree of $\mathbf{P}^1$. In other words, $H^1(\mathcal{O}_\Delta) \neq 0$. One has $H^1(\mathcal{O}_S) = H^2(\mathcal{O}_S) = 0$ for the rational surface S. Hence the exact sequence $0 \to \mathcal{O}_S(-\Delta) \to \mathcal{O}_S \to \mathcal{O}_\Delta \to 0$ yields $H^2(\mathcal{O}_S(-\Delta)) \neq 0$. By Serre duality, $H^0(\mathcal{O}_S(K_S + \Delta)) \neq 0$, which is (ii).

Take an effective divisor D linearly equivalent to $K_S + \Delta$. Suppose that $2K_S + \Delta \sim K_S + D$ is nef. Then $D > 0$. By the Riemann–Roch formula, $\chi(\mathcal{O}_S(K_S + D)) = (K_S + D)D/2 + \chi(\mathcal{O}_S) \geq \chi(\mathcal{O}_S) = 1$. Thus $H^0(\mathcal{O}_S(K_S + D)) \neq 0$ since $H^2(\mathcal{O}_S(K_S + D))$ is dual to $H^0(\mathcal{O}_S(-D)) = 0$. This is (iii). $\square$

Lemma 7.5.11 *Let $X \to S$ be a standard conic bundle over a smooth projective surface and let Δ be the discriminant divisor. If Δ contains a smooth rational curve C, then C intersects $\Delta - C$ at more than one point.*

Proof If $C \simeq \mathbf{P}^1$ were contained in Δ with $((\Delta - C) \cdot C) \leq 1$, then the associated pseudo-cover $\tilde{\Delta} \to \Delta$ of the conic bundle would induce a finite morphism $\tilde{C} \to C$ of degree two from a smooth curve $\tilde{C}$ which ramifies possibly at only one point in C. This contradicts Hurwitz's theorem. □

Unless μ is zero, there exists a representative X/S in the equivalence class such that $K_S + \Delta$ is nef.

Lemma 7.5.12 *Let* $\pi\colon X \to S$ *be a standard conic bundle over a smooth rational projective surface. Suppose that* π *has positive pseudo-effective threshold* μ, *in which* $\mu \geq 1$ *by Lemma 7.5.10. Then there exist a birational contraction* $g\colon S \to S'$ *and a standard conic bundle* $\pi'\colon X' \to S'$ *with discriminant divisor* Δ' *such that* π *and* π' *are equivalent via* g *and such that* $K_{S'} + \Delta'$ *is nef. The bundle* π' *has the same pseudo-effective threshold* μ.

Proof We write Δ for the discriminant divisor of π. By Lemma 7.5.10(ii), there exists an effective divisor D linearly equivalent to $K_S + \Delta$.

Unless $K_S + \Delta$ is nef, there exists an irreducible component C of D such that $((K_S + \Delta) \cdot C) < 0$. Then $c(C^2) = ((K_S + \Delta) \cdot C) - ((D - cC) \cdot C) < 0$ for the coefficient c of C in D. If C were in Δ, then $((K_S + C) \cdot C) = ((K_S + \Delta) \cdot C) - ((\Delta - C) \cdot C) < 0$ and hence C would be a smooth rational curve with $((\Delta - C) \cdot C) \leq 1$, which contradicts Lemma 7.5.11. Thus $C \not\subset \Delta$. Then $(K_S \cdot C) < -(\Delta \cdot C) \leq 0$ and hence C is a (-1)-curve disjoint from Δ.

Take the contraction $S \to S_1$ of the (-1)-curve C. From Lemma 7.4.1(ii), one obtains a standard conic bundle $\pi_1\colon X_1 \to S_1$ equivalent to π. Repeating this procedure, one eventually attains a standard conic bundle $\pi'\colon X' \to S'$ as in the statement. This π' also has pseudo-effective threshold $\mu' \geq 1$ and then the equality $\mu = \mu'$ follows from Lemma 7.5.1. □

It turns out that μ belongs to a discrete subset of the rational numbers.

Proposition 7.5.13 *The pseudo-effective threshold of a standard conic bundle over a smooth rational projective surface belongs to the set* $(2^{-1}\mathbf{N} \cup 3^{-1}\mathbf{N}) \setminus \{1/2, 1/3, 2/3\}$.

Proof Still working in Definition 7.5.9, we treat $\pi\colon X \to S$ with discriminant divisor Δ and pseudo-effective threshold μ. By Lemma 7.5.12, we may assume that $K_S + \Delta$ is nef, in which $\mu \geq 1$. Run the K_S-MMP with scaling of Δ as in Remark 1.5.12. It makes a finite sequence $S = S_0 \to S_1 \to \cdots \to S_n$ of contractions $g_i\colon S_i \to S_{i+1}$ of (-1)-curves which ends with a Mori fibre space $g_n\colon S_n \to U$ with $U \simeq \mathbf{P}^1$ or Spec $\mathbf{C}$. We define the *nef threshold* u_i to be the greatest real number such that $u_i K_{S_i} + \Delta_i$ is nef, where Δ_i is the strict transform of Δ. Note that u_i is the reciprocal of the threshold t_i in Definition 1.5.10. By the construction of the MMP with scaling, g_i is crepant with respect to $u_i K_{S_i} + \Delta_i$ and $u_i \leq u_{i+1}$ for all i.

The push-forward $\mu K_{S_n} + \Delta_n$ of $\mu K_S + \Delta$ is pseudo-effective. One deduces from the estimates $u_i \leq u_{i+1}$ that $u_n K_S + \Delta$ is at least the pull-back of the nef **R**-divisor $u_n K_{S_n} + \Delta_n$. Hence $u_n K_S + \Delta$ is pseudo-effective, that is, $u_n \leq \mu$. Let f be a fibre of g_n if $U \simeq \mathbf{P}^1$ and be a line in $S_n \simeq \mathbf{P}^2$ if $U = \operatorname{Spec} \mathbf{C}$. Then

$$0 \leq ((\mu K_{S_n} + \Delta_n) \cdot f) \leq ((u_n K_{S_n} + \Delta_n) \cdot f) = 0,$$

which shows that $\mu = u_n = (\Delta_n \cdot f)/(-K_{S_n} \cdot f)$. Since $(-K_{S_n} \cdot f) = 2$ or 3, μ belongs to $2^{-1}\mathbf{N}$ or $3^{-1}\mathbf{N}$. □

Proof of Theorem 7.5.8 For a standard conic bundle $\psi \colon Y \to T$ with discriminant divisor Γ, Lemma 7.5.1 shows that the property $|2K_T + \Gamma| = \emptyset$ is invariant in the equivalence class. The bundle $\pi' \colon X' \to S'$ in Conjecture 7.4.19 satisfies $|2K_{S'} + \Delta'| = \emptyset$. When $T = \mathbf{P}^2$, this property means that the degree d of Γ is at most five. If $d \leq 4$, then by the proof of Lemma 7.4.12, ψ is equivalent to π' in Conjecture 7.4.19(i). If $d = 5$, then by Theorem 7.4.18, Y is rational if and only if ψ is defined by even theta characteristic.

Let $\pi \colon X \to S$ be an arbitrary standard conic bundle over a smooth rational projective surface and let Δ be the discriminant divisor. By the above remarks, the theorem is reduced to the following statement: if $|2K_S + \Delta| = \emptyset$, then π is equivalent to a standard conic bundle $X' \to S'$ either as in Conjecture 7.4.19(i) or with $S' \simeq \mathbf{P}^2$. This can be verified by Lemma 7.4.1(ii) if $\Delta = 0$. Thus we may and shall assume that $\Delta \neq 0$. Then by Lemmata 7.5.10(ii) and 7.5.12, we may also assume that $K_S + \Delta$ is nef with $|K_S + \Delta| \neq \emptyset$ but $|2K_S + \Delta| = \emptyset$. Let $\mu \geq 1$ be the pseudo-effective threshold of π.

Run the K_S-MMP with scaling of Δ as in the proof of Proposition 7.5.13. We keep the notation $g_i \colon S_i \to S_{i+1}$, Δ_i, u_i and $g_n \colon S_n \to U$ with $U \simeq \mathbf{P}^1$ or $\operatorname{Spec} \mathbf{C}$ there. We shall inductively prove the existence of a standard conic bundle $\pi_i \colon X_i \to S_i$ with discriminant divisor Δ_i for all $i \leq n$. This is trivial for $i = 0$ by taking $\pi_0 = \pi$. For $i < n$, suppose the existence of π_i. By Lemma 7.5.1, $|K_{S_i} + \Delta_i| \neq \emptyset$ but $|2K_{S_i} + \Delta_i| = \emptyset$. By Lemma 7.5.10(iii), $2K_{S_i} + \Delta_i$ is not nef, that is, $u_i < 2$. Since g_i contracts a (-1)-curve C_i such that $((u_i K_{S_i} + \Delta_i) \cdot C_i) = 0$, one has $u_i = (\Delta_i \cdot C_i) \in \mathbf{Z}$ and further $u_i = 1$ from $1 \leq u_0 \leq u_i < 2$. Then Lemma 7.4.1(ii) constructs a desired bundle π_{i+1}.

Therefore we have a standard conic bundle $\pi_n \colon X_n \to S_n$ equivalent to π. Again by Lemma 7.5.10(iii), $2K_{S_n} + \Delta_n$ is not nef, that is, $u_n < 2$. We are done if $S_n \simeq \mathbf{P}^2$. If g_n is a $\mathbf{P}^1$-bundle, then the fibre f has $(\Delta_n \cdot f) = 2u_n < 4$ since $((u_n K_{S_n} + \Delta_n) \cdot f) = 0$. Thus $(\Delta_n \cdot f) \leq 3$ and π_n satisfies the condition (i) in Conjecture 7.4.19. This completes the theorem. □

8

Del Pezzo Fibrations

A Mori fibre space of relative dimension two is a del Pezzo fibration. It is considered to be a compactification of the del Pezzo surface defined over the function field of the base. The most fundamental invariant is the degree defined as the self-intersection number of the canonical divisor on this surface.

We address the problem of finding a good representative in each square equivalence class of threefold del Pezzo fibrations. In the case of degree at least three, Corti realised a standard model embedded anti-canonically into a projective space bundle. We explain the construction by Kollár for degree three as a semistable model of a cubic hypersurface and the reduction to it. The stability theory completes the existence of standard models for degree two and one. Further we have uniqueness for certain models.

Mori and Prokhorov listed all the possibilities for the singularities along a multiple fibre by means of the singular Riemann–Roch formula. As a result, the multiplicity of a fibre is bounded by six.

We shall also discuss the rationality problem of del Pezzo fibrations to a rational curve. It follows from the classification of del Pezzo surfaces over a perfect field that the total space is always rational when the degree is at least five. We exhibit Alexeev's work for degree four, in which the Sarkisov program produces a Sarkisov link of type I to a standard conic bundle. He established a criterion for rationality in terms of topological Euler characteristic by applying the criterion for that of conic bundles in the preceding chapter.

The lower degree case is treated from the point of view of birational rigidity. Pukhlikov proposed the K^2-condition and proved that it is sufficient for birational rigidity. The K^2-condition means that the square of the canonical divisor lies outside the interior of the cone of curves. We compare it with Grinenko's K-condition that the anti-canonical divisor lies outside the interior of the mobile cone. For a fibration of degree one from a smooth threefold, the K-condition turns out to be equivalent to the birational superrigidity.

8.1 Standard Models

Definition 8.1.1 A *del Pezzo fibration* is a contraction $X \to B$ from a terminal variety such that $\dim X = \dim B + 2$ and such that $-K_X$ is relatively ample. The del Pezzo fibration $X \to B$ is said to be *elementary* if X is **Q**-factorial with $\rho(X/B) = 1$. In other words, an elementary del Pezzo fibration is a relative Mori fibre space of relative dimension two.

Let $X \to B$ be a del Pezzo fibration. For the function field K of B, the generic fibre $X_K = X \times_B \operatorname{Spec} K$ is a del Pezzo surface over K. By definition, a *del Pezzo surface* S over a field is a geometrically irreducible smooth projective scheme of dimension two such that the anti-canonical divisor $-K_S$ is ample. The self-intersection number $d = K_S^2$ is called the *degree* of the del Pezzo surface S. It satisfies $1 \le d \le 9$. The description of S is well-known as below. See [266, section III.3] or [302, section 24] for example. The if part of the last statement in Theorem 8.1.2 is proved in [205, p.31 remark].

Theorem 8.1.2 *A del Pezzo surface of degree d over an algebraically closed field is isomorphic to either a blow-up of $\mathbf{P}^2$ at $9 - d$ general points or $\mathbf{P}^1 \times \mathbf{P}^1$ with $d = 8$. The blow-up of $\mathbf{P}^2$ at $9 - d$ points for $1 \le d \le 9$ is a del Pezzo surface if and only if no three points lie on a line, no six on a conic and no eight on a cubic which is singular at one of them.*

Theorem 8.1.3 *Let S be a del Pezzo surface of degree d over a field k. Set the anti-canonical k-algebra $R = \bigoplus_{i \in \mathbf{N}} R_i$ with $R_i = H^0(\mathscr{O}_S(-iK_S))$.*

(i)   *If $d \ge 3$, then R is generated by R_1.*

(ii) *If $d = 2$, then R is generated by R_1 and R_2 and $S = \operatorname{Proj} R$ is described as a quartic $S_4 \subset \mathbf{P}(1, 1, 1, 2)$. To be precise, $R \simeq k[x_0, x_1, x_2, y]/(f)$ with $x_0, x_1, x_2 \in R_1$, $y \in R_2$ and a weighted homogeneous polynomial $f = f(x_0, x_1, x_2, y)$ of weighted degree four.*

(iii) *If $d = 1$, then R is generated by R_1, R_2 and R_3 and $S = \operatorname{Proj} R$ is described as a sextic $S_6 \subset \mathbf{P}(1, 1, 2, 3)$, that is, $R \simeq k[x_0, x_1, y, z]/(g)$ with $x_0, x_1 \in R_1$, $y \in R_2$, $z \in R_3$ and g of weighted degree six.*

Remark 8.1.4 Theorem 8.1.3 still holds when S is normal and Gorenstein as shown by Hidaka and Watanabe [185]. It holds even for a (not necessarily normal) Gorenstein complex surface S with ample ω_S^{-1} as derived from the works of Reid [402] and Abe and Furushima [2].

Definition 8.1.5 The *degree* of a del Pezzo fibration $X \to B$ is the degree $K_{X_K}^2$ of the generic fibre X_K as a del Pezzo surface.

We are interested in a threefold del Pezzo fibration $X \to B$, in which B is a smooth curve. The K-points in the generic fibre X_K form a dense subset.

Theorem 8.1.6 *Let* $\pi \colon X \to B$ *be a threefold del Pezzo fibration so that B is a smooth curve. Let U be the open locus in B such that $\pi^{-1}(U) \to U$ is smooth. Then for any finitely many points $x_1, \ldots, x_n \in \pi^{-1}(U)$ lying on distinct fibres, there exists a section of π passing through all $x_1, \ldots, x_n$.*

Proof The existence of a section $s \colon B \to X$ of π is due to Colliot-Thélène [89, proposition 2]. The existence of s with $x_1, \ldots, x_n \in s(B)$ is due to Kollár, Miyaoka and Mori [273, theorem 2.13].   □

Remark 8.1.7 For an *elementary* threefold del Pezzo fibration $X \to B$, the fibre $X_b = X \times_B b$ at any closed point $b \in B$ has irreducible support F. Write $X_b = mF$ as a divisor on X. The multiplicity m is the least positive integer such that mF is Cartier because of the intersection number $(mF \cdot C) = 1$ with a section C of $X \to B$ given by Theorem 8.1.6. By Theorem 2.4.17, m divides the index of X about X_b. In particular if X is Gorenstein, then X_b is integral.

Given a del Pezzo surface X_K over the function field K of a smooth curve B, it is a fundamental problem to find a simple del Pezzo fibration $X \to B$ with generic fibre X_K analogous to a standard conic bundle. This is formulated on the germ $b \in B$ at each point b in B.

Notation 8.1.8 In this section for the germ $b \in B$ of a smooth curve, we let K denote the function field of B and let $\mathscr{O}$ denote the local ring $\mathscr{O}_{B,b}$, which is a discrete valuation ring. We let t denote a uniformising parameter in $\mathscr{O}$, namely a generator of the maximal ideal in $\mathscr{O}$.

If the degree d of the del Pezzo surface X_K is at least three, then by Theorem 8.1.3, X_K is embedded by $-K_{X_K}$ into $\mathbf{P}_K^d = \mathbf{P}H^0(\mathscr{O}_{X_K}(-K_{X_K}))$ as a projectively normal variety. It is thus natural to require a compactification of $X_K \subset \mathbf{P}_K^d$.

Definition 8.1.9 Let $b \in B$ be the germ of a smooth curve with function field K. Let X_K be a del Pezzo surface of degree d at least three over K. A del Pezzo fibration $X \to B$ with generic fibre X_K is called a *standard model* of X_K if the central fibre $X \times_B b$ is integral and the total threefold X is Gorenstein. By the lemma below, it is always embedded by $-K_X$ into $\mathbf{P}^d \times B$.

Lemma 8.1.10 *Let* $\pi \colon X \to B$ *be a del Pezzo fibration of degree d form a Gorenstein terminal threefold X to the germ $b \in B$ of a smooth curve. Suppose that the central fibre $X \times_B b$ is integral. Set the anti-canonical $\mathscr{O}_B$-algebra $\mathscr{R} = \bigoplus_{i \in \mathbf{N}} \mathscr{R}_i$ with $\mathscr{R}_i = \pi_* \mathscr{O}_X(-iK_X)$.*

(i) *If $d \geq 3$, then $\mathscr{R}$ is generated by $\mathscr{R}_1$ and $\mathscr{R}_1$ is a free $\mathscr{O}_B$-module.*
(ii) *If $d = 2$, then $\mathscr{R}$ is generated by $\mathscr{R}_1$ and $\mathscr{R}_2$ and $X = \mathrm{Proj}_B \mathscr{R}$ is described as a quartic in $\mathbf{P}(1, 1, 1, 2) \times B$.*
(iii) *If $d = 1$, then $\mathscr{R}$ is generated by $\mathscr{R}_1$, $\mathscr{R}_2$ and $\mathscr{R}_3$ and $X = \mathrm{Proj}_B \mathscr{R}$ is described as a sextic in $\mathbf{P}(1, 1, 2, 3) \times B$.*

Proof Take the integral fibre $F = X \times_B b$ and the **C**-algebra $R = \bigoplus_{i \in \mathbf{N}} R_i$ with $R_i = H^0(\mathscr{O}_X(-iK_X) \otimes \mathscr{O}_F) \simeq H^0((\omega_F^{-1})^{\otimes i})$. The corresponding statement for R holds by Remark 8.1.4. The multiplication map $t \colon \mathscr{R} \to \mathscr{R}$ by the function t in $\mathscr{R}_0 = \mathscr{O}_B$ defining b induces the exact sequence $0 \to \mathscr{R} \to \mathscr{R} \to R \to 0$ by the Kawamata–Viehweg vanishing $R^1\pi_*\mathscr{O}_X(-iK_X) = 0$ for $i \geq 0$. Then the assertion follows from the structure of R with the aid of Nakayama's lemma. □

The same description actually holds without the integral condition on the central fibre provided that the general elephant conjecture holds for π [92, proposition 3.13]. We have the following result, in which the singularities of the fibre are of class T in Definition 4.4.10.

Theorem 8.1.11 (Hacking–Prokhorov [167], [389]) *Let $X \to B$ be a del Pezzo fibration to the germ $b \in B$ of a smooth curve. If the central fibre $X \times_B b$ is log terminal, then the general elephant of X has only Du Val singularities of type A.*

When the degree of X_K is three, the standard model $X \to B$ is a cubic hypersurface in $\mathbf{P}^3 \times B$ and hence the problem amounts to finding a good cubic form $f(x_0, x_1, x_2, x_3)$ over $\mathscr{O}_B$ which defines X_K over K. Corti built an algorithm for constructing it and obtained the existence of standard models by the reduction to the case of degree three.

Theorem 8.1.12 (Corti [92]) *A standard model of a del Pezzo surface X_K of degree at least three exists.*

The purpose of this section is to prove it. For degree three, we have a sufficient condition for a cubic hypersurface $X \subset \mathbf{P}^3 \times B$ to be a standard model.

Proposition 8.1.13 *Let $b \in B$ be the germ of a smooth curve with function field K. Let X be a prime divisor on $\mathbf{P}^3 \times B$ such that the generic fibre $X_K = X \times_B \mathrm{Spec}\, K$ is a del Pezzo surface of degree three. Suppose that*

- *the central fibre $X_0 = X \times_B b$ is a prime divisor on $\mathbf{P}^3$,*
- *X is smooth at the generic point of any line in $X_0 \subset \mathbf{P}^3$ and*
- *X has multiplicity at most two at any point in X_0.*

Then X has only isolated cDV singularities of type cA_n with $n \leq 5$, cD_4, cD_5 or cE_6. In particular, X is a standard model of X_K.

Proof   Once X turns out to be terminal, it is evident from $\mathscr{O}_X(-K_X) \simeq \mathscr{O}_X(1)$ that X is a standard model of X_K.

If X_0 were singular along a curve other than lines, then every line l joining two points in this curve would be contained in X_0 since otherwise the intersection number $(X_0 \cdot l)_{\mathbf{P}^3}$ exceeds the degree three of X_0. This occurs only if X_0 contains a plane, which contradicts the first assumption. From this and the second assumption, X has only isolated singularities.

Let x be a singular point of X, which is isolated. Take the homogeneous coordinates x_0, x_1, x_2, x_3 of $\mathbf{P}^3$ so that $x = ([0,0,0,1], b)$. Then the central fibre $X_0 = X \times_B b$ is given in $\mathbf{P}^3$ by an irreducible cubic form $\bar{f}_2(x_0, x_1, x_2)x_3 + \bar{f}_3(x_0, x_1, x_2)$. If X_0 is normal and $\bar{f}_2$ is non-zero, then it follows from a classification by Bruce and Wall [60] that $x \in X_0$ is Du Val of type A_n with $n \leq 5$, D_4, D_5 or E_6. Thus one has only to deal with the cases when X_0 is non-normal and when $\bar{f}_2$ is zero. In the former case, X_0 is singular along a line l by the above exclusion of a curve as a singular locus and further X_0 is smooth outside l by the same argument. In particular, x lies on l. Thus without loss of generality, we may assume that either X_0 is singular along $l = (x_0 = x_1 = 0)$ or $\bar{f}_2$ is zero.

The total space X is given in $\mathbf{P}^3 \times B$ by a cubic form $t^2 f_0 x_3^3 + t f_1 x_3^2 + f_2 x_3 + f_3$ over $\mathscr{O}_B$, where $t \in \mathscr{O}_B$ defines the point b and $f_i = f_i(x_0, x_1, x_2)$ is homogeneous of degree i with coefficients in $\mathscr{O}_B$. We write $\bar{f}_i = \bar{f}_i(x_0, x_1, x_2)$ for the reduction of f_i modulo t as a form with coefficients in $\mathbf{C}$ via the isomorphism $\mathscr{O}_B/t\mathscr{O}_B \simeq \mathbf{C}$. The germ $x \in X$ is given by the function

$$ F = t^2 f_0 + t f_1 + f_2 + f_3 $$

in x_0, x_1, x_2, t. By the third assumption, it is an isolated singularity of multiplicity two. If $\bar{f}_2 \neq 0$, then X_0 is singular along l whilst X is smooth at the generic point of l. Hence tx_2, tx_2^2 or tx_2^3 appears in F in this case.

We look at the quadratic part $q = t^2 \bar{f}_0 + t\bar{f}_1 + \bar{f}_2$ of F, which is non-zero by $\mathrm{mult}_x X = 2$. If q has rank four or three, then x is cA_1. Suppose that q has rank two. If $\bar{f}_2 \neq 0$, then X_0 is singular along l in which q is a form of rank two in x_0, x_1, t. One can check that x is cA_2 or cA_3 using the irreducibility of $\bar{f}_2 + \bar{f}_3$ and the appearance of tx_2^2 or tx_2^3. If $\bar{f}_2 = 0$, then $\bar{f}_3$ is irreducible and x is cA_2. Suppose that q has rank one. If $\bar{f}_2 \neq 0$, then we may assume that $\bar{f}_2 = x_0^2$, in which x_1^3 or $x_1^2 x_2$ appears in F besides the appearance of tx_2^2 or tx_2^3. In this case, x is cD_4, cD_5 or cE_6. If $\bar{f}_2 = 0$, then $\bar{f}_1 = 0$ and $\bar{f}_0 \neq 0$. In this case, x is cD_4 from the irreducibility of $\bar{f}_3$. $\qquad\qquad\square$

Hence for degree three, it suffices to construct a hypersurface X as in the proposition. Following Kollár [267], we shall obtain it as a semistable model of a hypersurface. We explain this in a general setting by treating hypersurfaces in $\mathbf{P}^n$ of degree d over an algebraically closed field k. They are parametrised by the complete linear system $P = \mathbf{P}H^0(\mathscr{O}_{\mathbf{P}^n}(d))^\vee$ which is a projective space.

Lemma 8.1.14 *The locus in $P = \mathbf{P}H^0(\mathscr{O}_{\mathbf{P}^n}(d))^\vee$ which parametrises singular hypersurfaces is a hypersurface in P.*

Proof We take the incidence subscheme I of $\mathbf{P}^n \times P$ consisting of pairs of a point x and a hypersurface H such that H is singular at x. The hypersurface H defined by $f \in H^0(\mathscr{O}_{\mathbf{P}^n}(d))$ is singular at x if and only if $\partial f/\partial x_0, \ldots, \partial f/\partial x_n$ vanish at x for the homogeneous coordinates $x_0, \ldots, x_n$ of $\mathbf{P}^n$. Notice that the latter condition implies that $x \in H$ since $f = d^{-1} \sum_i x_i \partial f/\partial x_i$. Thus the first projection $I \to \mathbf{P}^n$ is a $\mathbf{P}^{p-n-1}$-bundle for the dimension p of P, and I is a variety of dimension $p - 1$.

The image D of the second projection $I \to P$ is projective and irreducible. As far as a hypersurface H has only isolated singularities, the fibre of $I \to D$ at $[H]$ is finite. Hence $I \to D$ is generically finite and D has the same dimension $p - 1$ as I has. Thus D is a hypersurface in $P \simeq \mathbf{P}^p$. $\qquad\square$

Let $U = U_{n,d}$ denote the open subset of the affine space $H^0(\mathscr{O}_{\mathbf{P}^n}(d)) \simeq \mathbf{A}^h$ with $h = h^0(\mathscr{O}_{\mathbf{P}^n}(d))$ which parametrises homogeneous polynomials of degree d defining smooth hypersurfaces. By the lemma above, U is the complement of a divisor on $H^0(\mathscr{O}_{\mathbf{P}^n}(d))$ and in particular it is an affine variety. In fixing homogeneous coordinates of $\mathbf{P}^n$, there exists a natural right action of $G = \mathrm{GL}(n + 1, k)$ on $H^0(\mathscr{O}_{\mathbf{P}^n}(1))$. This induces actions of G on $H^0(\mathscr{O}_{\mathbf{P}^n}(d))$ and on U. Assuming that d is at least three, we shall explain the stability of every point in U with respect to this action in the geometric invariant theory.

Theorem 8.1.15 *If $d \geq 3$, then every point in $U_{n,d}$ has finite stabiliser with respect to the action of $\mathrm{GL}(n + 1, k)$.*

Proof We follow [357, theorem 5.23]. Keep the notation $U = U_{n,d}$ and $G = \mathrm{GL}(n + 1, k)$. For $f \in U$, the morphism $G \simeq f \times G \to U$ defined by the group action induces a linear map $g : \mathfrak{g} \to T_{U,f}$ from the Lie algebra $\mathfrak{g}$ of G to the tangent space $T_{U,f}$ at f, where $\mathfrak{g}$ is the k-algebra of $(n + 1) \times (n + 1)$ matrices with entries in k. It suffices to show the injectivity of g.

The map g is expressed as

$$A = (a_{ij}) \mapsto \sum_{i,j} a_{ij} x_j \frac{\partial f}{\partial x_i}$$

by the isomorphism $T_{U,f} \simeq H^0(\mathcal{O}_{\mathbf{P}^n}(d))$. Write $f_i = \partial f/\partial x_i$ and $l_i = \sum_j a_{ij}x_j$ for brevity, which are homogeneous of degree $d-1$ and 1 unless zero. Assuming that $\sum_i l_i f_i = 0$ as a polynomial, we want to have $l_i = 0$ for all i.

Consider the ideal $I = (f_0, \ldots, f_n)R$ in $R = k[x_0, \ldots, x_n]$, which contains $f = d^{-1} \sum_i x_i f_i$. Because f defines a smooth hypersurface, I is $(x_0, \ldots, x_n)R$-primary. Hence $f_0, \ldots, f_n$ form a regular sequence in R. If $\sum_i l_i f_i = 0$ and $l_0 \neq 0$, say, then $l_0 \notin I$ by $\deg l_0 = 1 < d - 1 = \deg f_i$, and the expression $l_0 f_0 = -\sum_{i \neq 0} l_i f_i$ would imply that f_0 is a zero-divisor in $R/(f_1, \ldots, f_n)R$, which is absurd. Thus g is injective. $\qquad\square$

Suppose that $d \geq 3$. By the above theorem, the orbit O_f by G of any closed point $f \in U$ is of the same dimension $(n+1)^2 = \dim G$. Then O_f is closed. Otherwise a point g in $\bar{O}_f \setminus O_f$ would have orbit $O_g \subset \bar{O}_f \setminus O_f$, contradicting $\dim O_f = \dim O_g$. Hence by Theorem 2.2.3, the geometric quotient U/G exists as an affine variety. This also holds for the action by $\mathrm{SL}(n+1, k)$. One has

$$U = U_{n,d} \twoheadrightarrow U/\mathrm{SL}(n+1, k) \twoheadrightarrow U/\mathrm{GL}(n+1, k).$$

Fix $f \in U_{n,d}$, where we keep the assumption $d \geq 3$. By the existence of $U/\mathrm{SL}(n+1, k)$ and $U/\mathrm{GL}(n+1, k)$, there exists an $\mathrm{SL}(n+1, k)$-invariant non-constant function ξ in $\mathcal{O}_{U_{n,d}}$ such that $\xi(f) \neq 0$. Indeed, just take an element ξ in $H^0(\mathcal{O}_{U/\mathrm{SL}(n+1,k)}) \setminus H^0(\mathcal{O}_{U/\mathrm{GL}(n+1,k)})$ which does not vanish at the image of f. One can choose ξ so that it is a homogeneous function on the affine space $H^0(\mathcal{O}_{\mathbf{P}^n}(d))$ by multiplying ξ by a regular function and picking a homogeneous part of it.

Lemma 8.1.16 *Suppose that $d \geq 3$ and fix $f \in U_{n,d}$. As above, let ξ be an $\mathrm{SL}(n+1, k)$-invariant homogeneous function on $H^0(\mathcal{O}_{\mathbf{P}^n}(d))$ of positive degree r such that $\xi(f) \neq 0$. Then $n+1$ divides dr and one has*

$$\xi(c \cdot f(Ax)) = c^r (\det A)^{dr/(n+1)} \xi(f(x))$$

for all $A \in \mathrm{GL}(n+1, k)$ and $c \in k$, where $x = {}^t(x_0, \ldots, x_n)$ is the column vector of homogeneous coordinates of $\mathbf{P}^n$.

Proof Write $A = aB$ with $a \in k^\times$ and $B \in \mathrm{SL}(n+1, k)$, where $a^{n+1} = \det A$. Then $\xi(f(Ax)) = \xi(f(aBx)) = \xi(f(ax)) = \xi(a^d f(x)) = a^{dr} \xi(f(x))$. This should be independent of the choice of a solution a to $a^{n+1} = \det A$. Hence the exponent dr is divisible by $n+1$. $\qquad\square$

Consider the situation in Notation 8.1.8 where $\mathcal{O}$ is the local ring at a point of a smooth complex curve and K is its function field. Let X_K be a smooth hypersurface in $\mathbf{P}^n_K$ of degree d at least three. We fix an embedding of $\mathbf{P}^n_K$ into $\mathbf{P}^n_{\mathcal{O}} = \mathbf{P}^n \times_{\mathrm{Spec}\,\mathbf{C}} \mathrm{Spec}\,\mathcal{O}$ by specifying a K-isomorphism $\mathbf{P}^n_K \simeq$

$\mathbf{P}^n \times_{\mathrm{Spec}\,\mathbf{C}} \mathrm{Spec}\,K$. The compactification X in $\mathbf{P}^n_{\mathcal{O}}$ of X_K is flat over $\mathrm{Spec}\,\mathcal{O}$ and it is defined by an element f in $H^0(\mathcal{O}_{\mathbf{P}^n_{\mathcal{O}}}(d))$. As discussed above, there exists an $\mathrm{SL}(n+1, \bar{K})$-invariant homogeneous function ξ on $H^0(\mathcal{O}_{\mathbf{P}^n_{\bar{K}}}(d))$ of positive degree r such that $\xi(f) \neq 0$, where $\bar{K}$ is the algebraic closure of K. One can take as ξ a function defined over K by replacing it by the norm of ξ with respect to a finite Galois extension of K over which ξ is defined. Further, multiplying ξ by a regular function, we may and shall assume that ξ is the extension of a function

$$\xi \colon H^0(\mathcal{O}_{\mathbf{P}^n_{\mathcal{O}}}(d)) \to \mathcal{O}.$$

In the *geometric invariant theory*, a point f in the affine space A endowed with an action of G is said to be *stable* if there exists an invariant function ξ with $\xi(f) \neq 0$ such that all orbits in $A_\xi = (\xi \neq 0)$ are closed in A_ξ [364, definition 1.7]. This is the case with points in $U_{n,d}$. The following definition is in the spirit of the *Hilbert–Mumford criterion* [364, theorem 2.1].

Definition 8.1.17 Keep Notation 8.1.8 for $\mathcal{O}$, K and t and let $x_0, \dots, x_n$ be the homogeneous coordinates of $\mathbf{P}^n$. Let f be a homogeneous polynomial in $x_0, \dots, x_n$ of degree d with coefficients in $\mathcal{O}$, that is, a non-zero element in $A = H^0(\mathcal{O}_{\mathbf{P}^n_{\mathcal{O}}}(d))$. We say that f is *semistable* over $\mathcal{O}$ if

$$\mathrm{ord}_t(\sigma_g f)(t^{w_0}x_0, \dots, t^{w_n}x_n) \leq \frac{d}{n+1}\sum_i w_i$$

for all $g \in \mathrm{SL}(n+1, \mathcal{O})$ and all $w = (w_0, \dots, w_n) \in \mathbf{N}^{n+1}$, where $\mathrm{ord}_t\, h$ stands for the greatest integer m such that $h \in t^m \mathcal{O}[x_0, \dots, x_n]$ and $\sigma_g f$ stands for the image of (f, g) by the action $A \times G \to A$. If further the inequality is strict for all $w \neq 0$, then we say that f is *stable* over $\mathcal{O}$.

Let $X \subset \mathbf{P}^n_{\mathcal{O}}$ be the hypersurface defined by f and let $X_K \subset \mathbf{P}^n_K$ be the generic fibre. We call X a *semistable* (resp. *stable*) *model* of X_K over $\mathcal{O}$ if f is semistable (resp. stable).

Theorem 8.1.18 (Kollár [267]) *Let $b \in B$ be the germ of a smooth curve with function field K. Let X_K be a smooth hypersurface in $\mathbf{P}^n_K$ of degree at least three. Then there exists a semistable model $X \subset \mathbf{P}^n_{\mathcal{O}}$ of X_K over $\mathcal{O} = \mathcal{O}_{B,b}$.*

Proof We write $d \geq 3$ for the degree of X_K. Take an embedding $\mathbf{P}^n_K \subset \mathbf{P}^n_{\mathcal{O}}$ and let $f \in H^0(\mathcal{O}_{\mathbf{P}^n_{\mathcal{O}}}(d))$ define the compactification in $\mathbf{P}^n_{\mathcal{O}}$ of X_K. We have seen the existence of an $\mathrm{SL}(n+1, \mathcal{O})$-invariant homogeneous function $\xi \colon H^0(\mathcal{O}_{\mathbf{P}^n_{\mathcal{O}}}(d)) \to \mathcal{O}$ of positive degree r such that $\xi(f) \neq 0$.

Let t be a uniformising parameter in $\mathcal{O}$ and let $x_0, \dots, x_n$ be the homogeneous coordinates of $\mathbf{P}^n$. Unless f is semistable, after changing coordinates by an element in $\mathrm{SL}(n + 1, \mathcal{O})$, there exists $(w_0, \ldots, w_n) \in \mathbb{N}^{n+1}$ such that $m = \mathrm{ord}_t f(t^{w_0} x_0, \ldots, t^{w_n} x_n)$ is greater than $d \sum_i w_i/(n+1)$. Then by Lemma 8.1.16, the new polynomial $f_1 = t^{-m} f(t^{w_0} x_0, \ldots, t^{w_n} x_n)$ satisfies

$$\xi(f_1) = t^{-mr}(t^{\sum_i w_i})^{dr/(n+1)} \xi(f) = t^{-(m-d \sum_i w_i/(n+1))r} \xi(f).$$

Hence $t\xi(f_1)$ divides $\xi(f) \neq 0$. This induces the strict inclusion $\xi(f)\mathcal{O} \subsetneq \xi(f_1)\mathcal{O}$ of ideals in $\mathcal{O}$. Thus after the replacement of f by f_1 finitely many times, we eventually attain a semistable polynomial f. $\square$

The existence of a standard model of degree three is an easy consequence of this theorem and Proposition 8.1.13.

Proof of Theorem 8.1.12 *for degree three* We keep the notation where the del Pezzo surface X_K of degree three is defined over the function field K of the germ $b \in B$ of a smooth curve. By Theorem 8.1.18, there exists a semistable model $X \subset \mathbf{P}^3_{\mathcal{O}}$ with $\mathcal{O} = \mathcal{O}_{B,b}$, which extends to a cubic hypersurface $X \subset \mathbf{P}^3 \times B$. The semistability of the cubic form f with coefficients in $\mathcal{O}_B$ defining X implies the following for $w = (w_0, w_1, w_2, w_3)$ where $X_0 = X \times_B b$ is the central fibre.

- By $\mathrm{ord}_t f \leq \lfloor 3/4 \rfloor = 0$ for $w = (0, 0, 0, 1)$, X_0 contains no planes and hence it is a prime divisor on $\mathbf{P}^3$.
- By $\mathrm{ord}_t f \leq \lfloor 6/4 \rfloor = 1$ for $w = (0, 0, 1, 1)$, X is smooth at the generic point of any line in $X_0 \subset \mathbf{P}^3$.
- By $\mathrm{ord}_t f \leq \lfloor 9/4 \rfloor = 2$ for $w = (0, 1, 1, 1)$, X has multiplicity at most two at any point in X_0.

Thus X satisfies the assumptions in Proposition 8.1.13, by which $X \to B$ is a standard model of X_K. $\square$

Example 8.1.19 ([267, example 6.4.3]) Let $b \in B$ be the germ at origin of $\mathbf{A}^1 = \mathrm{Spec}\,\mathbf{C}[t]$. The cubic form $f = t^6 x_0^3 + x_1^3 + x_2^3 + x_0 x_3^2$ satisfies the semistability condition in Definition 8.1.17 for all $g \in \mathrm{SL}(4, \mathcal{O}_{B,b})$ and $w = (w_0, w_1, w_2, w_3) = (0, 0, 0, 1)$, $(0, 0, 1, 1)$ and $(0, 1, 1, 1)$. In particular by the proof above, it defines a standard model of the generic fibre. However, it is not semistable with respect to $w = (0, 2, 2, 3)$. The replacement of f by f_1 as in the proof of Theorem 8.1.18 yields a semistable form $x_0^3 + x_1^3 + x_2^3 + x_0 x_3^2$.

In the general case, we construct a standard model as the ample model of the anti-canonical divisor on a suitable fibration.

Proof of Theorem 8.1.12 We shall prove the theorem by induction on the degree d of X_K. It has already been proved when $d = 3$. Suppose that $d \geq 4$.

By Theorem 8.1.6, K-points in X_K form a dense subset of X_K. The blow-up Y_K of X_K at a general K-point is again a del Pezzo surface. This follows from Theorem 8.1.2 after the base change to the algebraic closure of K. By inductive hypothesis, there exists a standard model $Y \to B$ of Y_K on the germ $b \in B$ of a smooth curve with function field K. On the punctured neighbourhood $B^\circ = B \setminus b$, X_K is the generic fibre of a smooth projective morphism $X^\circ \to B^\circ$ and $Y^\circ = Y \times_B B^\circ$ is the blow-up of X° along a section of $X^\circ \to B^\circ$. Let E be the closure in Y of the exceptional divisor of $Y^\circ \to X^\circ$.

Take a $\mathbf{Q}$-factorialisation Y' of Y and the strict transform E' in Y' of E. Running the $(K_{Y'} + \varepsilon E')/Y$-MMP over Y for small positive ε, we may assume that $E' \equiv_Y \varepsilon^{-1}(K_{Y'} + \varepsilon E')$ is nef over Y. Then for any $z \in \overline{\mathrm{NE}}(Y'/B)$, the equality $(K_{Y'} \cdot z) = 0$ implies that $(E' \cdot z) \geq 0$ since $-K_{Y'}$ is the pull-back of the ample divisor $-K_Y$. Further the general fibre f of $E' \to B$ has $(K_{Y'} \cdot f) = (E' \cdot f) = -1$. In this situation, the cone theorem supplies a $K_{Y'}$-negative extremal ray R of $\overline{\mathrm{NE}}(Y'/B)$ which is at the same time E'-negative.

Let $\pi \colon Y' \to X'$ be the extremal contraction associated with R. Since Y' is Gorenstein, Theorem 5.5.1 shows that it is a divisorial contraction which contracts the divisor E' to a curve. Then by Theorem 4.1.9, X' is smooth and $K_{Y'} = \pi^* K_{X'} + E'$. By construction, X' is a compactification of X°, and the closed fibre $X' \times_B b$ is integral as so is $Y \times_B b$. Further $-K_{X'}$ is nef over B. Indeed for any relative curve C in X'/B, the projection formula gives $(-K_{X'} \cdot C) = ((-K_{Y'} + E') \cdot C') \geq 0$ for the strict transform C' in Y' of C. Note that $-K_{Y'}$ is nef over B and C' does not lie on E'.

Recall that $-K_{X^\circ}$ is ample over B°. By the base-point free theorem, $-K_{X'}$ is semi-ample over B and a multiple of it defines the ample model $\varphi \colon X' \to X$ of $-K_{X'}$ over B. It is isomorphic over B° and hence small, and X is terminal. One can apply the exact sequence in Theorem 1.4.7 to φ by finding a small boundary Δ' on X' such that $-\Delta'$ is φ-ample. It follows that X is Gorenstein as so is X'. Thus X is a standard model of X_K and the theorem is completed by induction. □

8.2 Simple Models and Multiple Fibres

Let X_K be a del Pezzo surface over the function field of a smooth curve. If the degree d is at least three, then the standard model $X \to B$ of X_K in Definition 8.1.9 is embedded by $-K_X$ into $\mathbf{P}^d \times B$. In view of Lemma 8.1.10, we adopt the following definition of a standard model of degree at most two.

Definition 8.2.1 Let $b \in B$ be the germ of a smooth curve with function field K. Let X_K be a del Pezzo surface of degree two (resp. one) over K. A del Pezzo fibration $\pi \colon X \to B$ with generic fibre X_K is called a *standard model* of X_K

if the central fibre $X \times_B b$ is integral and $-2K_X$ (resp. $-6K_X$) is Cartier and relatively very ample.

Corti [92] first constructed a standard model of degree two by the same idea as in his original proof for degree three. The existence of a standard model was completed by refining the notion of Kollár's stability.

Theorem 8.2.2 (Abban–Fedorchuk–Krylov [1]) *Let $b \in B$ be the germ of a smooth curve with function field K. Let X_K be a del Pezzo surface of degree d at most two over K. Then there exists a standard model $X \to B$ of X_K such that*

(i) *if $d = 2$, then X is described as an intersection $X_{4,2}$ of a quartic and a quadric in $\mathbf{P}(1, 1, 1, 2, 2) \times B$ and*

(ii) *if $d = 1$, then X is described as an intersection $X_{6,3}$ of a sextic and a cubic in $\mathbf{P}(1, 1, 2, 3, 3) \times B$.*

By Lemma 8.1.10, a Gorenstein del Pezzo fibration $X \to B$ with integral central fibre is a standard model of the generic fibre. It is an ideal model because of the realisation in $P \times B$ for $P = \mathbf{P}^d$, $\mathbf{P}(1, 1, 1, 2)$ or $\mathbf{P}(1, 1, 2, 3)$ according to the degree. However, there may exist no Gorenstein standard model when the degree is at most two.

Example 8.2.3 ([92, section 5]) Consider $\mathbf{P}(1, 1, 1, 2) \times \mathbf{A}^1$ with homogeneous coordinates x_0, x_1, x_2, y of $\mathbf{P}(1, 1, 1, 2)$ and coordinate t of $\mathbf{A}^1$, and work on the germ $o \in \mathbf{A}^1$ at origin. Let X be the divisor on $\mathbf{P}(1, 1, 1, 2) \times \mathbf{A}^1$ defined by

$$ty^2 + a(x_0, x_1, x_2, t)y + b(x_0, x_1, x_2, t)$$

with general quadratic and quartic forms a and b in x_0, x_1, x_2 with coefficients in $\mathcal{O}_{\mathbf{A}^1}$. The projection $\pi \colon X \to \mathbf{A}^1$ is a standard model of the generic fibre X_K which is a del Pezzo surface of degree two. The threefold X has a quotient singularity of type $\frac{1}{2}(1, 1, 1)$. We shall see that π is a unique standard model of X_K up to isomorphism over $\mathbf{A}^1$.

Let $\pi' \colon X' \to \mathbf{A}^1$ be any standard model of X_K. There exists a birational map $f \colon X \dashrightarrow X'$ such that $\pi = \pi' \circ f$. The general elephant $S' \in |-K_{X'}|$ is transformed by f^{-1} into a member $S \in |-K_X|$ smooth outside $X \times_{\mathbf{A}^1} o$. After coordinate change over the germ $o \in \mathbf{A}^1$, S is defined in X by x_2. The surface S has only Du Val singularities of type A. Indeed, it has a quotient singularity of type $\frac{1}{2}(1, 1)$ at $([0, 0, 0, 1], o)$ and the fibre $S \times_{\mathbf{A}^1} o$ has only nodes. The latter is deduced from the generality condition that $a(x_0, x_1, x_2, 0)$ and $b(x_0, x_1, x_2, 0)$ define smooth curves in $\mathbf{P}^2$ which intersect transversally, as it implies the non-existence of a multiple factor of the polynomial $a(x_0, x_1, 0, 0)y + b(x_0, x_1, 0, 0)$ defining the fibre. Hence the pair (X, S) is plt and canonical by inversion of adjunction in Theorem 1.4.26. The pair (X, S) with $K_X + S \sim 0$ satisfies the Noether–Fano inequality in Theorem 6.1.12(ii) with respect to the map f over the germ $o \in \mathbf{A}^1$ by its proof. It asserts that f is an isomorphism.

Even a standard model $X \to B$ of degree two is not necessarily realised as a quartic in $\mathbf{P}(1, 1, 1, 2) \times B$.

Example 8.2.4 ([92, example 4.11]) Consider $P \times \mathbf{A}^1$ for $P = \mathbf{P}(1, 1, 1, 2, 2)$ with homogeneous coordinates x_0, x_1, x_2, y_0, y_1 of P and coordinate t of $\mathbf{A}^1$. Let A be the hypersurface in $P \times \mathbf{A}^1$ defined by $ty_1 - x_0^2 + x_1 x_2$. Set $(y_2, \ldots, y_6) = (x_0 x_1, x_0 x_2, x_1^2, x_1 x_2, x_2^2)$. Then A is isomorphic to the subvariety of $\mathbf{P}^6 \times \mathbf{A}^1$ defined by the 2×2 minors of the matrix

$$\begin{pmatrix} ty_1 + y_5 & y_2 & y_3 \\ y_2 & y_4 & y_5 \\ y_3 & y_5 & y_6 \end{pmatrix}.$$

The general fibre of $A \to \mathbf{A}^1$ is isomorphic to $\mathbf{P}(1, 1, 1, 2)$, whilst the fibre A_o at origin o of $\mathbf{A}^1$ is isomorphic to $\mathbf{P}(1, 1, 4, 4)$ with homogeneous coordinates z_0, z_1, y_0, y_1 by $(x_0, x_1, x_2) = (z_0 z_1, z_0^2, z_1^2)$.

Let X be the hypersurface in A defined by

$$y_0^2 + y_1 a(y_2, \ldots, y_6) + b(y_2, \ldots, y_6)$$

with general linear and quadratic forms a and b. Write $X_o \subset A_o \subset P \times o$ for the fibre at o of $X \subset A \subset P \times \mathbf{A}^1 \to \mathbf{A}^1$. From the generality condition, X_o is smooth outside the point $x = ([0, 0, 0, 0, 1], o)$ and then X is smooth outside x. A direct computation shows that X is terminal of type $cA/2$ at x. Thus the projection $X \subset P \times \mathbf{A}^1 \to \mathbf{A}^1$ is a del Pezzo fibration of degree two which is a standard model as in Theorem 8.2.2. The embedding $X_o \subset A_o \simeq \mathbf{P}(1, 1, 4, 4)$ is not compatible with any embedding $X_o \subset \mathbf{P}(1, 1, 1, 2)$ as a quartic.

As to the uniqueness of a standard model, Park obtained the following result in the case of degree at most four.

Theorem 8.2.5 (Park [381]) *Let $b \in B$ be the germ of a smooth curve with function field K. Let X_K be a del Pezzo surface of degree at most four over K. Then a standard model $X \to B$ of X_K which is a smooth morphism is unique up to isomorphism over B as far as it exists.*

If the degree is at most three, then the uniqueness still holds for a standard model from a smooth threefold. It is due to Pukhlikov for degree three and to Grinenko for degree two and one. They directly studied the polynomial which defines the embedding into a weighted projective space bundle.

Theorem 8.2.6 (Pukhlikov [394], Grinenko [156], [157]) *Let $b \in B$ be the germ of a smooth curve with function field K. Let X_K be a del Pezzo surface of degree at most three over K. Then a standard model $X \to B$ of X_K from a smooth threefold is unique up to isomorphism over B as far as it exists.*

We shall explain the conceptual proof of Theorem 8.2.5 and then the direct approach to Theorem 8.2.6. The assumption of degree $d \le 4$ comes from the estimate $c_d > 1/2$ in the next lemma. One can check it using Theorems 8.1.2 and 8.1.3. See also Example 9.5.17. We give a proof for $d = 3$ as an example.

Lemma 8.2.7 *Let S be a del Pezzo surface of degree d. Then $(S, c_d C)$ is lc for all $C \in |-K_S|$ where $c_1 = 5/6$, $c_2 = 3/4$, $c_3 = c_4 = 2/3$, $c_5 = c_6 = 1/2$ and $c_7 = c_8 = c_9 = 1/3$.*

Proof for $d = 3$ By Theorem 8.1.3, a del Pezzo surface S of degree three is a cubic surface in $\mathbf{P}^3$ which is projectively normal. Hence every member C of $|-K_S|$ is a hyperplane section $S \cap \Pi$ with $\Pi = \mathbf{P}^2 \subset \mathbf{P}^3$. If C contains a line, then it consists either of a line and a conic or of three lines [35, lemma IV.15]. Thus C is always reduced and any singularity of C is analytically given by $x_1 x_2$, $x_1 x_2 (x_1 + x_2)$, $x_1 (x_1 + x_2^2)$ or $x_1^2 + x_2^3$. Thus $(S, (2/3)C)$ is lc. $\square$

Proof of Theorem 8.2.5 We write $d \le 4$ for the degree of the del Pezzo surface X_K. Let $\pi \colon X \to B$ and $\pi' \colon X' \to B$ be standard models of X_K as in the statement. Then there exists a birational map $f \colon X \dashrightarrow X'$ such that $\pi = \pi' \circ f$ and such that f is isomorphic over $B \setminus b$. The theorem means that f is an isomorphism. By the smoothness of π and π', the central fibres $F = X \times_B b$ and $F' = X' \times_B b$ are del Pezzo surfaces of degree d. We shall prove that f is small. In other words, the divisors F on X and F' on X' are transformed to each other. This implies that f is an isomorphism similarly to Lemma 6.1.4.

Step 1 Take a common resolution W of X and X' with $\mu \colon W \to X$ and $\mu' \colon W \to X'$ where $\mu' = f \circ \mu$. Let $S \in |-K_X|$ be the general elephant of X. Since $\pi_* \mathcal{O}_X(-K_X) \to H^0(\mathcal{O}_F(-K_X|_F))$ is surjective by the vanishing $R^1 \pi_* \mathcal{O}_X(-K_X) = 0$, the restriction $S|_F \in |-K_F|$ for $K_F \sim K_X|_F$ is the general elephant of F.

We shall see that the general elephant $S|_F$ of F is smooth, which implies the log canonicity of the pair $(X, F + S)$. This is obvious if $d \ge 2$ because $-K_F$ is free in this case by Theorem 8.1.3. If $d = 1$, then by the same theorem, the smooth surface F is defined in $\mathbf{P}(1, 1, 2, 3)$ by $z^2 + y^3 + g(x_0, x_1)y + h(x_0, x_1)$ with suitable homogeneous coordinates x_0, x_1, y, z. The system $|-K_F|$ has one base point $p = [0, 0, -1, 1]$ but every member of $|-K_F|$ is smooth at p.

Step 2 Let $T' \in |-K_{X'}|$ be the general elephant of X', for which $(X', F' + T')$ is also lc by symmetry. Then $\mu_* (\mu')^* (K_{X'} + F' + T') = K_X + (1 - a')F + T$ for the log discrepancy $a' = a_F(X', F' + T') \geq 0$ and the strict transform T of T'. One obtains

$$\mu^*(K_X + (1 - a')F + T) = (\mu')^*(K_{X'} + F' + T')$$

from $K_{X'} + F' + T' \sim 0$ using the negativity lemma for μ.

We claim that S does not contain the centre $Z = c_X(F')$ of F'. By the observation in Step 1, it suffices to exclude the case when Z is the base point p of $|-K_F|$ for $d = 1$. The above equality gives $a_{F'}(X, (1 - a')F + T) = a_{F'}(X', F'+T') = 0$. Hence Z is a non-klt centre of $(X, F+T)$ by $a' \geq 0$. It is also a non-klt centre of $(F, T|_F)$ by precise inversion of adjunction in Theorem 1.6.9. Thus if $Z = p$, then $T|_F$ would be singular at p, which contradicts the property that every member of $|-K_F|$ is smooth at p.

Step 3 By symmetry, one has

$$\mu^*(K_X + F + S) = (\mu')^*(K_{X'} + (1 - a)F' + S')$$

for $a = a_{F'}(X, F+S) \geq 0$ and the strict transform S' of S. Take the discrepancies $e = a_{F'}(X) - 1 \geq 0$ and $e' = a_F(X') - 1 \geq 0$. Note that $\mathrm{ord}_F F' = \mathrm{ord}_{F'} F = 1$ by $\mu^* F = (\pi \circ \mu)^* b = (\pi' \circ \mu')^* b = (\mu')^* F'$. Then

$$0 = a_F(X, F + S) = a_F(X', (1 - a)F' + S') = (1 + e') - (1 - a) - \mathrm{ord}_F S',$$

that is, $\mathrm{ord}_F S' = a + e'$, and

$$e = a_{F'}(X, F) = a_{F'}(X, F + S) = a_{F'}(X', (1 - a)F' + S') = a,$$

where the second equality comes from $Z \not\subset S$. Therefore $\mathrm{ord}_F S' = e + e'$.

By Lemma 8.2.7 and inversion of adjunction, the pair $(X', F' + c_d S')$ is lc. In particular $0 \leq a_F(X', F' + c_d S') = e' - c_d(e + e')$, that is, $c_d(e + e') \leq e'$. By symmetry, $c_d(e + e') \leq e$. Summing up these two inequalities, one obtains $2c_d(e + e') \leq e + e'$. Hence $e = e' = 0$ from $c_d > 1/2$. This occurs only if $F = F'$ as a divisor over X, which is what we want to prove. $\square$

We demonstrate Theorem 8.2.6 in the case of degree one as an instance.

Proof of Theorem 8.2.6 for degree one We assume that X_K is of degree one. Let $\pi \colon X \to B$ be a standard model of X_K from a smooth threefold. By Lemma 8.1.10, it is realised as a sextic in $\mathbf{P}(1, 1, 2, 3) \times B$ given by

$$z^2 + y^3 + g(x_0, x_1)y + h(x_0, x_1)$$

with homogeneous coordinates x_0, x_1, y, z where $g, h \in \mathcal{O}_B[x_0, x_1]$.

Let $\pi' \colon X' \to B$ be another standard model of X_K from a smooth threefold. The corresponding birational map $f \colon X \dashrightarrow X'$ over B transforms the coordinate system x_0, x_1, y, z into that of the weighted projective space $\mathbf{P}(1, 1, 2, 3) \times$

Spec K over K in which the generic fibre of π' is embedded. Take a uniformising parameter t in $\mathscr{O}_{B,b}$. One can take a coordinate system x'_0, x'_1, y', z' of $\mathbf{P}(1, 1, 2, 3) \times B$ over the germ $b \in B$ (not a system of $\mathbf{P}(1, 1, 2, 3)$) such that

$$(f^{-1})^*(x_0, x_1, y, z) = (t^{a_0}x'_0, t^{a_1}x'_1, t^b y', t^c z')$$

for some $(a_0, a_1, b, c) \in \mathbf{N}^4 \setminus (v + \mathbf{N}^4)$ where $v = (1, 1, 2, 3)$. Then X' is given by

$$(z')^2 + (y')^3 + t^{-2b}g(t^{a_0}x'_0, t^{a_1}x'_1)y' + t^{-3b}h(t^{a_0}x'_0, t^{a_1}x'_1)$$

for which $2c = 3b$, t^{2b} divides $g(t^{a_0}x'_0, t^{a_1}x'_1)$ and t^{3b} divides $h(t^{a_0}x'_0, t^{a_1}x'_1)$.

If we start from π' instead of π and obtain a quartuple (a'_0, a'_1, b', c'), then the sum $(a_0+a'_0, a_1+a'_1, b+b', c+c')$ is a multiple of v. Hence interchanging π and π' if necessary, and x_0 and x_1 likewise, we may assume that $(a_0, a_1, b, c) = (a_0, a_1, 2e, 3e)$ with $a_0 \leq a_1$ and $a_0 + a_1 \leq 2e$.

Unless $(a_0, a_1) = (e, e)$, the number a_0 would be less than e. In this case, $g \in (t^4x_0^4, t^2x_0^3x_1, x_1^2)\mathscr{O}_B[x_0, x_1] \subset (t^2, x_1^2)\mathscr{O}_B[x_0, x_1]$ since t^{4e} divides $g(t^{a_0}x'_0, t^{a_1}x'_1)$. In like manner, $h \in (t^2, x_1^3)\mathscr{O}_B[x_0, x_1]$. But then X would be singular at $([1, 0, 0, 0], b)$, which is absurd. Thus $(a_0, a_1, b, c) = ev$ and it must be zero. This shows that f is an isomorphism over the germ $b \in B$. □

Grinenko provided interesting examples of pairs of del Pezzo fibrations $\pi \colon X \to B$ and $\pi \colon X' \to B$ of degree one which are linked by a birational map $f \colon X \dashrightarrow X'$ over B isomorphic outside b.

Example 8.2.8 Let π and π' be the fibrations defined by

$$X \colon z^2 + y^3 + x_0^5 x_1 + x_0 x_1^5, \qquad X' \colon z^2 + y^3 + x_0^5 x_1 + t^{24}x_0 x_1^5,$$

for which f is given by the data $(a_0, a_1, b, c) = (0, 6, 2, 3)$ defined in the proof of Theorem 8.2.6. Then X is smooth and X' is terminal with a cE_8 point.

Example 8.2.9 Let π and π' be the fibrations defined by

$$X \colon z^2 + y^3 + x_0 x_1^5 + t^4 x_0^5 x_1, \qquad X' \colon z^2 + y^3 + t^4 x_0 x_1^5 + x_0^5 x_1,$$

for which f is given by $(a_0, a_1, b, c) = (0, 2, 2, 3)$. Then X and X' are abstractly isomorphic, which are terminal with a cE_8 point.

He also provided examples of degree two in which X and X' are quartics in $\mathbf{P}(1, 1, 1, 2) \times B$. Take the coordinate system x_0, x_1, x_2, y of $\mathbf{P}(1, 1, 1, 2)$ and consider its transformation into $t^{a_0}x_0, t^{a_1}x_1, t^{a_2}x_2, t^b y$.

Example 8.2.10 Let π and π' be the fibrations defined by

$$X \colon y^2 + x_0^3 x_2 + x_1^4 + x_2^4, \qquad X' \colon y^2 + x_0^3 x_2 + x_1^4 + t^{12}x_2^4,$$

for which f is given by $(a_0, a_1, a_2, b) = (0, 1, 4, 2)$. Then X is smooth and X' is terminal with a cE_6 point.

Example 8.2.11 Let π and π' be the fibrations defined by

$$X\colon y^2 + x_1^4 + x_0 x_2^3 + t^2 x_0^3 x_2, \qquad X'\colon y^2 + x_1^4 + t^2 x_0 x_2^3 + x_0^3 x_2,$$

for which f is given by $(a_0, a_1, a_2, b) = (0, 1, 2, 2)$. Then X and X' are abstractly isomorphic, which are terminal with a cD_4 point.

In general, a fibre of a del Pezzo fibration is not always reduced. We shall introduce the work of Mori and Prokhorov on multiple fibres. The result holds for a *weak del Pezzo fibration* $X \to B$ which satisfies the conditions for a del Pezzo fibration except that the relative ampleness of $-K_X$ is relaxed to the condition that $-K_X$ is relatively nef and big. Its *degree* is still defined as the self-intersection number $K_{X_K}^2$ on the generic fibre X_K.

For a prime divisor F on X which supports a fibre of $X \to B$, we shall use the same notation as in the proof of Theorem 2.6.6. Notice that F is **Q**-Cartier. Let $\{q_\iota$ of type $\frac{1}{r_\iota}(1, -1, b_\iota)\}_{\iota \in I_0}$ be the basket of fictitious singularities in Definition 2.5.22 from the singularities of X about F. Take an integer e_ι such that $F \sim e_\iota K_X$ at q_ι by Theorem 2.4.17. We may assume that $v_\iota = \overline{e_\iota b_\iota} \le r_\iota/2$, where $\bar{l}$ is the residue of l modulo r_ι as in Notation 2.2.14. Let $I = \{\iota \in I_0 \mid \bar{e}_\iota \ne 0\}$ and $J = \{(r_\iota, v_\iota)\}_{\iota \in I}$.

Theorem 8.2.12 (Mori–Prokhorov [341]) *Let $\pi\colon X \to B$ be a weak del Pezzo fibration of degree d from a terminal threefold X to the germ $b \in B$ of a smooth curve. Suppose that the central fibre $X \times_B b$ is a multiple mF of a prime divisor F on X. Construct $J = \{(r_\iota, v_\iota)\}_{\iota \in I}$ as above from the basket of fictitious singularities at which F is not Cartier. Then π belongs to one of the cases in Table 8.1. Every case in Table 8.1 has an example.*

Table 8.1 *The multiple fibre*

m	J	d	m	J	d
1	$\emptyset$		3	$(3, 1), (3, 1), (3, 1)$	$\equiv 0\ (3)$
2	$(2, 1), (2, 1), (2, 1), (2, 1)$	even	3	$(3, 1), (6, 2)$	$\equiv 0\ (3)$
2	$(2, 1), (2, 1), (4, 2)$	odd	3	$(9, 3)$	$\not\equiv 0\ (3)$
2	$(4, 2), (4, 2)$	even	4	$(2, 1), (4, 1), (4, 1)$	$4, 8$
2	$(2, 1), (6, 3)$	even	5	$(5, 1), (5, 2)$	5
2	$(8, 4)$	odd	6	$(2, 1), (3, 1), (6, 1)$	6

Table 8.1 is almost the same as Table 2.1, and as one can predict, the strategy is similar to that for Theorem 2.6.6, which uses the singular Riemann–Roch

formula. There may exist a non-Gorenstein point at which F is Cartier. It makes no contribution to our basket J like a hidden non-Gorenstein point in Definition 3.2.4.

Proof of Theorem 8.2.12 It follows from $\pi_*\mathcal{O}_X = \mathcal{O}_B$ and $\pi^*b = mF$ that $\pi_*\mathcal{O}_X(iF) = \mathcal{O}_B(\lfloor i/m \rfloor b)$ in the constant sheaf $\mathcal{K}_B$ of the function field of B for all integers i. Thus the periodic function $\delta(i) = \lfloor i/m \rfloor - \lfloor (i-1)/m \rfloor$ is realised as $\delta(i) = \dim \pi_*\mathcal{O}_X(iF)/\pi_*\mathcal{O}_X((i-1)F)$. Since $R^p\pi_*\mathcal{O}_X(iF) = 0$ for $p \geq 1$ by Kawamata–Viehweg vanishing, it is expressed as $\delta(i) = h^0(\mathcal{O}_F(iF|_F)) = \chi(\mathcal{O}_F(iF|_F))$ by the $\mathcal{O}_F$-module $\mathcal{O}_F(iF|_F)$ given by Proposition 2.2.23. Compactifying X to a terminal projective threefold, one obtains

$$\delta(i) = \chi(\mathcal{O}_X(iF)) - \chi(\mathcal{O}_X((i-1)F)).$$

Let $G \sim mF$ denote the general fibre of π. Then $FK_X^2 = GK_X^2/m = d/m$. The number $Fc_2(X)$ defined in Theorem 2.5.23 equals $c_2(G)/m = (12-d)/m$. The last equality follows from *Noether's formula*

$$\chi(\mathcal{O}_G) = \frac{K_G^2 + c_2(G)}{12},$$

which is the special case of the Hirzebruch–Riemann–Roch theorem in Theorem 2.5.6 with $n = 2$ and $\mathscr{E} = \mathcal{O}_X$. Hence Theorem 2.5.23 provides

$$\delta(i) = \frac{1}{m} + \sum_{\iota \in I}(A_\iota(ie_\iota) - A_\iota((i-1)e_\iota))$$

with the contribution $A_\iota(l)$ in Definition 2.6.8. Using Lemma 2.6.9(ii), one has

$$\delta(i) - \delta(1-i) = \sum_{\iota \in I}(B_\iota((i-1)v_\iota) - B_\iota(iv_\iota)).$$

Assume that $m \geq 2$. Then $\delta(0) = 1$ and $\delta(1) = 0$. The above formula for $i = 0$ is $\sum_{\iota \in I} B_\iota(v_\iota) = 1$. The same equation has appeared in the proof of Theorem 2.6.6 and the solutions $J = \{(r_\iota, v_\iota)\}_{\iota \in I}$ are listed in Table 2.1, with the column for r ignored.

For each candidate for the set $\tilde{J} = \{(r_\iota, v_\iota, b_\iota)\}_{\iota \in I}$, the explicit computation of the right-hand side of $\delta(i) = 1/m + \sum_{\iota \in I}(A_\iota(ie_\iota) - A_\iota((i-1)e_\iota))$ determines the multiplicity m as in Table 8.1 and excludes the case $J = \{(2,1),(8,2)\}$ in Table 2.1. For example in the case when $\tilde{J} = \{(9,3,b)\}$, one computes $A_\iota(e_\iota)$ to be $-1, -2/3, -2/3, -1/3, -1/3, 0$ if $b = 1, 2, 4, 5, 7, 8$ respectively. From the equality $0 = \delta(1) = 1/m + A_\iota(e_\iota)$, one concludes that $m = 3$ and $b = 5$ or 7.

In order to determine the degree d, we shall use an expression of $\delta_1 = \dim \pi_*\mathcal{O}_X(-K_X)/\pi_*\mathcal{O}_X(-K_X - F)$. Similarly to $\delta(i)$, Theorem 2.5.23 with

$FK_X^2 = d/m$ and $Fc_2(X) = (12-d)/m$ yields $\delta_1 = (d+1)/m + \sum_{\iota \in I}(A_\iota(-1) - A_\iota(-1 - e_\iota))$. Using Lemma 2.6.9(i), one has

$$\delta_1 - \delta(0) = \frac{d}{m} + \sum_{\iota \in I}(B_\iota(b_\iota + v_\iota) - B_\iota(b_\iota)).$$

One can derive the condition on d stated in Table 8.1 from the integral property $\delta_1 - \delta(0) \in \mathbf{Z}$. For example in the above case $\tilde{J} = \{(9, 3, b)\}$ with $b = 5$ or 7, $\delta_1 - \delta(0) = d/3 + B_\iota(b + 3) - B_\iota(b)$. One computes $B_\iota(b + 3) - B_\iota(b)$ to be $-2/3, -1/3$ if $b = 5, 7$ respectively. Hence $d \equiv 2, 1 \bmod 3$ respectively. $\qquad \square$

Let $\pi \colon X \to B$ be a del Pezzo fibration as in Theorem 8.2.12. Take the covering $\varphi \colon b' \in B' = \mathrm{Spec}_B \, \mathcal{O}_B[U]/(U^m - t) \to b \in B$ of degree m ramified at b by a uniformising parameter t in $\mathcal{O}_{B,b}$. One has the diagram

$$
\begin{array}{ccc}
F' \subset X' & \longrightarrow & mF \subset X \\
\pi' \downarrow & & \downarrow \pi \\
b' \in B' & \xrightarrow{\ \varphi\ } & b \in B
\end{array}
$$

with the normalisation X' of $X \times_B B'$, in which $F' = (\pi')^* b'$ is reduced. The induced morphism $X' \to X$ is étale outside the points in X at which F is not Cartier. In particular, $\pi' \colon X' \to B'$ is again a del Pezzo fibration.

We shall construct several examples of π from π' assuming that X' is Gorenstein. Let $\pi' \colon F' \subset X' \to b' \in B'$ be a del Pezzo fibration from a Gorenstein terminal threefold such that the central fibre $F' = X' \times_{B'} b'$ is integral and Du Val. Suppose that the cyclic group $\mathbf{Z}_m$ acts on X' and B' equivariantly in such a manner that

- the action on B' is free outside b',
- the action on F' is free outside finitely many points and
- the quotient $F = F'/\mathbf{Z}_m$ is Du Val, which exists by the ampleness of $-K_{F'}$ as explained in the paragraph after the proof of Lemma 2.6.12.

Then the quotient $\pi \colon X = X'/\mathbf{Z}_m \to B = B'/\mathbf{Z}_m$ is again a del Pezzo fibration. For this, we need to verify that X is terminal. The cyclic quotient $X' \to X$ is étale outside finitely many points. Hence K_X is $\mathbf{Q}$-Cartier and X has only isolated lt singularities by Corollary 2.2.21. The surjections $\mathcal{O}_X(\pm(K_X + F)) \twoheadrightarrow \mathcal{O}_F(\pm K_F)$ in Proposition 2.2.23 induce the surjection $\mathcal{O}_X(K_X + F) \otimes \mathcal{O}_X(-(K_X + F)) \twoheadrightarrow \mathcal{O}_F(K_F) \otimes \mathcal{O}_F(-K_F) = \mathcal{O}_F$, from which $K_X + F$ is Cartier. Then the pair (X, F) is canonical by inversion of adjunction, and in particular X is terminal.

Example 8.2.13 ([341, example 5.2]) Let F' be a projective surface with only Du Val singularities such that $-K_{F'}$ is ample. Suppose that the cyclic group $\mathbf{Z}_m$ acts on F' freely outside finitely many points in such a way that $F = F'/\mathbf{Z}_m$ is still Du Val. By Theorem 8.1.3 and Remark 8.1.4, F' is embedded in the (weighted) projective space $P = \mathbf{P}^d$, $\mathbf{P}(1, 1, 1, 2)$ or $\mathbf{P}(1, 1, 2, 3)$. The action of $\mathbf{Z}_m$ extends to that on P with appropriate semi-invariant coordinates. Let $\mathbf{Z}_m$ act on the germ $o \in \mathbf{A}^1$ by multiplication by a primitive m-th root of unity. Then one can make an invariant Gorenstein terminal threefold X' in $P \times \mathbf{A}^1$ by extending $F' \subset P \simeq P \times o$. The quotient $X = X'/\mathbf{Z}_m \to \mathbf{A}^1/\mathbf{Z}_m$ is an example of a del Pezzo fibration explained above. In Table 8.2, we pick some cases from the list of the fundamental group of the smooth locus in F in [322].

Table 8.2 *Some examples*

m	K_F^2	sing. of F	$K_{F'}^2$	F'	J
2	4	A_1, A_1, A_3	8	$\mathbf{P}(1, 1, 2)$	$(2, 1), (2, 1), (2, 1), (2, 1)$
3	3	A_2, A_2, A_2	9	$\mathbf{P}^2$	$(3, 1), (3, 1), (3, 1)$
4	2	A_1, A_3, A_3	8	$\mathbf{P}^1 \times \mathbf{P}^1$	$(2, 1), (4, 1), (4, 1)$
5	1	A_4, A_4	5	smooth	$(5, 1), (5, 2)$
6	1	A_1, A_2, A_5	6	smooth	$(2, 1), (3, 1), (6, 1)$

Example 8.2.14 Mori and Prokhorov gave simple explicit examples of del Pezzo fibrations for $m = 2, 3$ and 4. Let $\mathbf{Z}_m$ act on $\mathbf{A}^1$ by multiplication by a primitive m-th root of unity.

(i) Let $\mathbf{Z}_2$ act on two copies of $\mathbf{P}^1$ by multiplication by -1. The projection $(\mathbf{P}^1 \times \mathbf{P}^1 \times \mathbf{A}^1)/\mathbf{Z}_2 \to \mathbf{A}^1/\mathbf{Z}_2$ is an example for $m = 2$ with $J = \{(2, 1), (2, 1), (2, 1), (2, 1)\}$ in Table 8.1.

(ii) Let $\mathbf{Z}_3$ act on $\mathbf{P}^2$ so that a generator sends $[x_0, x_1, x_2]$ to $[x_0, \omega x_1, \omega^2 x_2]$ where $\omega = (-1 + \sqrt{-3})/2$. The projection $(\mathbf{P}^2 \times \mathbf{A}^1)/\mathbf{Z}_3 \to \mathbf{A}^1/\mathbf{Z}_3$ is an example for $m = 3$ in Table 8.2.

(iii) Let $\mathbf{Z}_4$ act on $\mathbf{P}^1 \times \mathbf{P}^1$ so that a generator sends $([x_0, x_1], [y_0, y_1])$ to $([y_0, y_1], [x_0, -x_1])$. The projection $(\mathbf{P}^1 \times \mathbf{P}^1 \times \mathbf{A}^1)/\mathbf{Z}_4 \to \mathbf{A}^1/\mathbf{Z}_4$ is an example for $m = 4$ in Table 8.2.

Finally we provide an example with a quotient singularity of index nine.

Example 8.2.15 ([341, example 5.5]) Let $\mathbf{Z}_3$ act on

$$X' = (tz^2 + y^3 + x_0^6 + x_1^6 = 0) \subset \mathbf{P}(1, 1, 2, 3) \times \mathbf{A}^1$$

so that a generator sends $([x_0, x_1, y, z], t)$ to $([x_0, \omega x_1, \omega y, \omega z], \omega t)$ where $\omega = (-1 + \sqrt{-3})/2$. The quotient $X'/\mathbf{Z}_3 \to \mathbf{A}^1/\mathbf{Z}_3$ is a del Pezzo fibration for $m = 3$ with $J = \{(9, 3)\}$ in Table 8.1. It has a quotient singularity of type $\frac{1}{9}(1, -1, 7)$.

8.3 Rationality

Iskovskikh [210] listed all the possible Sarkisov links over a perfect field from a del Pezzo surface with Picard number one and from a standard conic bundle over a smooth rational curve. We extract its consequence on a del Pezzo surface. Compare it with Theorem 7.4.5 for a conic bundle. It is based on the works of Manin [300], [301], [302] and Iskovskikh [205]. The existence of a k-point for degree five discussed by Enriques was verified by Swinnerton-Dyer [436].

Theorem 8.3.1 *Let S be a del Pezzo surface of degree d with Picard number one defined over a perfect field k. Then the following hold. In particular, S is k-rational if and only if S has a k-point with $d \geq 5$.*

 (i) *If $d = 1$, then S is k-birationally superrigid.*
 (ii) *If $d = 2$ or 3, then S is k-birationally rigid.*
 (iii) *If $d = 4$, then S is not k-rational.*
 (iv) *If $d = 5$, then S is k-rational.*
 (v) *If $d = 6, 8$ or 9, then S is k-rational if and only if S has a k-point.*
 (vi) *The case $d = 7$ never occurs.*

We shall address the rationality problem of the total threefold X of a del Pezzo fibration $X \rightarrow \mathbf{P}^1$. The following is an immediate consequence of the above theorem with the existence of a section.

Corollary 8.3.2 *Let $X \rightarrow \mathbf{P}^1$ be an elementary del Pezzo fibration of degree at least five. Then X is rational.*

Proof By Theorem 8.1.6, the generic fibre X_K over the function field K of $\mathbf{P}^1$ has a K-point. Then by Theorem 8.3.1, X_K is K-rational. This means that X is birational to $\mathbf{P}^2 \times \mathbf{P}^1$ over $\mathbf{P}^1$. In particular, X is rational. □

In this section, we consider an elementary threefold del Pezzo fibration $X \rightarrow B$ of degree four. The lower degree case will be discussed in the next section from the point of view of birational rigidity. Alexeev constructed a link from X/B to a standard conic bundle over a surface with a structure of a $\mathbf{P}^1$-bundle. We explain this by the Sarkisov program. The original statement assumes that X/B has normal fibres.

Theorem 8.3.3 (Alexeev [6]) *Let $X \to B$ be an elementary del Pezzo fibration of degree four from a smooth projective threefold. Then there exists a Sarkisov link $X/B \dashrightarrow Y/S$ of type I to a standard conic bundle $Y \to S$ for which $S \to B$ is a $\mathbf{P}^1$-bundle.*

Proof We write $\pi\colon X \to B$ and let X_K be the generic fibre of π defined over the function field K of B. By Theorem 8.1.6, K-points in X_K form a dense subset. Hence by Theorem 8.1.2 over the algebraic closure of K, the blow-up W_K of X_K at a general K-point C_K is again a del Pezzo surface. This is the generic fibre of the blow-up $p\colon W \to X$ along the general section C of π which is the closure of C_K. Let E denote the exceptional divisor of p.

By Remark 8.1.7, the fibre $F = X \times_B b$ at any $b \in B$ is integral. It follows from Lemma 8.1.10 that $-K_X$ is relatively very ample. The restriction map $\pi_* \mathscr{O}_X(-K_X) \to H^0(\mathscr{O}_X(-K_X) \otimes \mathscr{O}_F)$ is surjective since $R^1 \pi_* \mathscr{O}_X(-K_X) = 0$. Hence $-K_W = -p^* K_X - E$ is free over B.

Further $-K_W$ is relatively ample over a dense open subset B° of B. Indeed if $(-K_W \cdot l_W) = 0$ for a curve l_W in the fibre $W \times_B b$ at general $b \in B$, then the multiplicity of the image l in $F = X \times_B b$ at the point $C|_F$ would equal the degree $(-K_X \cdot l)$ of l with respect to the embedding $F \subset \mathbf{P}^4$ by $-K_F$. Hence l must be a line by a standard argument as in the proof of Lemma 6.4.14. This contradicts the general choice of C_K because the del Pezzo surface F of degree four contains only 16 lines [35, proposition IV.12].

We shall perform the two-ray game in Definition 4.2.5 over B starting from the smooth threefold W. If $-K_W$ is ample over B, then we set $Y = W$ and take the extremal contraction $\psi\colon Y \to S$ over B other than p. If $-K_W$ is nef but not ample over B, then the two-ray game commences with a crepant contraction $W \to Z$, which is isomorphic over B°. This is a flopping contraction since $W \to B$ has irreducible fibres. After the flop $Y \to Z$ of it, Y is again smooth by Theorem 10.3.4. The anti-canonical divisor $-K_Y$ is still nef and big over B, and the two-ray game generates an extremal contraction $\psi\colon Y \to S$ over B.

In both cases, $-K_Y$ is ψ-ample, and by Theorem 5.5.1, $\psi\colon Y \to S$ is not small. Thus ψ is non-trivial over the generic point $\operatorname{Spec} K$ of B. The base change $Y_K \simeq W_K \to S_K$ to $\operatorname{Spec} K$ is nothing but the standard conic bundle which is the projection of the cubic surface Y_K in $\mathbf{P}^3_K$ from the line $E \times_B \operatorname{Spec} K$. In particular, ψ is a Mori fibre space of relative dimension one and it is defined by $-K_Y - E_Y$ where E_Y is the strict transform of E. For $\psi'\colon Y \to S \to B$, one has the Kawamata–Viehweg vanishing $R^i \psi'_* \mathscr{O}_Y(-K_Y - E_Y) = 0$ for $i \geq 1$. Hence by cohomology and base change in Theorem 1.1.5, the direct image $\mathscr{E} = \psi'_* \mathscr{O}_Y(-K_Y - E_Y)$ is locally free of rank two and it makes a $\mathbf{P}^1$-bundle $\mathbf{P}(\mathscr{E}) \to B$. The induced morphism $S \to \mathbf{P}(\mathscr{E})$ is finite and birational. Thus

$S = \mathbf{P}(\mathscr{E})$ and ψ is a standard conic bundle, to which a Sarkisov link from π has been constructed. □

Suppose that $B = \mathbf{P}^1$ in the theorem. Then Alexeev almost furnished the following rationality criterion in terms of the topological Euler characteristic $\chi(X) = \sum_i (-1)^i b_i(X)$ in Remark 2.5.5, which was completed by Shramov.

Theorem 8.3.4 (Alexeev, Shramov) *Let $X \to \mathbf{P}^1$ be an elementary del Pezzo fibration of degree four from a smooth projective threefold. Then X is rational if and only if the topological Euler characteristic $\chi(X)$ equals 0, −4 or −8.*

Proof We construct a standard conic bundle $\psi : Y \to S$ from $X \to B = \mathbf{P}^1$ as in Theorem 8.3.3. It is obtained from (the flop of) the blow-up W of X along a section $C \simeq B$, and the base S/B is a Hirzebruch surface $\Sigma_n / \mathbf{P}^1$. Let Δ be the discriminant divisor of ψ, which is reduced and normal crossing by Lemma 7.1.6. Let Z denote the union of singular points of Δ.

The conic bundle $Y \times_B b \to S \times_B b$ for general $b \in B$ is ramified at five points [35, lemma IV.15]. Hence one can write $\Delta \sim 5e + mf$ with some $m \in \mathbf{Z}$ for the section e of $S \to B$ with self-intersection number $e^2 = -n$ and a fibre f of $S \to B$. Since $K_S \sim -2e - (n+2)f$ by Theorem 6.1.11, one has $K_S + \Delta \sim 3e + (m - n - 2)f$ and hence

$$\deg \omega_\Delta = (3e + (m - n - 2)f) \cdot (5e + mf) = 8m - 20n - 10.$$

The definition of topological Euler characteristic is extended to an arbitrary reduced complex scheme X by adopting the alternating sum of the ranks of cohomology groups $H_c^i(X, \mathbf{Z})$ *with compact support*. It satisfies the excision property $\chi(X) = \chi(X \setminus Z) + \chi(Z)$ for a closed subscheme Z of X and the relation $\chi(X) = \chi(F)\chi(B)$ for a proper fibre bundle $X \to B$ with fibre F. See [140, note 13 to chapter 4] and [430, 9.3 theorem 1]. Thus one has

$$\chi(W) = \chi(X) - \chi(C) + \chi(\mathbf{P}^1)\chi(C) = \chi(X) + 2,$$
$$\chi(Y) = \chi(\mathbf{P}^1)(\chi(S) - \chi(\Delta)) + \chi(\mathbf{P}^1 \vee \mathbf{P}^1)(\chi(\Delta) - \chi(Z)) + \chi(\mathbf{P}^1)\chi(Z)$$
$$= 2(4 - \chi(\Delta)) + 3(\chi(\Delta) - \chi(Z)) + 2\chi(Z) = 8 - \deg \omega_\Delta,$$

using $\chi(\mathbf{P}^1) = 2$, $\chi(\Sigma_n) = 4$, $\chi(\mathbf{P}^1 \vee \mathbf{P}^1) = 3$ for the bouquet $\mathbf{P}^1 \vee \mathbf{P}^1$ and $\deg \omega_\Delta = \chi(Z) - \chi(\Delta)$. These are the same number thanks to Theorem 10.3.4. Hence

$$\chi(X) = 6 - \deg \omega_\Delta = 6 - (8m - 20n - 10) = -8m + 20n + 16.$$

By Lemma 7.5.11, the section e is never a connected component of the reduced divisor $\Delta \sim 5e + mf$. This implies that $4n + 1 \le m$. Suppose that X and thus W are rational. Then by Theorem 7.5.6, $2K_S + \Delta \sim e + (m - 2n - 4)f$

defines an empty linear system and hence $m - 2n - 4 \leq -1$, that is, $m \leq 2n + 3$. The solutions (n, m) to the system $4n + 1 \leq m \leq 2n + 3$ of these two inequalities are $(0, 1)$, $(0, 2)$, $(0, 3)$ and $(1, 5)$, in which $\chi(X) = 8, 0, -8, -4$ respectively. Lemma 7.5.11 excludes the case $(n, m) = (0, 1)$.

Conversely if $\chi(X) = 0, -8$ or -4, then (n, m) is uniquely determined as above by the expression $\chi(X) = -8m + 20n + 16$ with $4n + 1 \leq m$. By Proposition 7.4.4, W is rational if $(n, m) = (0, 2)$ or $(0, 3)$. The rationality of W in the last case $(n, m) = (1, 5)$ is a result of Shramov [428].   □

Ahmadinezhad, renamed Abban, constructed a del Pezzo fibration $X \to B = \mathbf{P}^1$ of degree four from an irrational threefold which is linked to a del Pezzo fibration of degree two besides the conic bundle in Theorem 8.3.3. In particular, the pliability $\mathscr{P}(X)$ in Definition 6.4.1 contains at least three elements.

Example 8.3.5 (Ahmadinezhad [4])   Define the $\mathbf{P}^4$-bundle $\pi_P \colon P = \mathbf{P}(\mathscr{E}) \to \mathbf{P}^1$ by $\mathscr{E} = \mathscr{O}_{\mathbf{P}^1} \oplus \mathscr{O}_{\mathbf{P}^1}(1) \oplus \mathscr{O}_{\mathbf{P}^1}(2) \oplus \mathscr{O}_{\mathbf{P}^1}(3)^{\oplus 2}$. Let h be a divisor on P with $\mathscr{O}_P(h) = \mathscr{O}_P(1)$ and let f be the general fibre of π_P. We take $X = Q \cap Q'$ for the general members Q and Q' of $|2h - 3f|$ and $|2h - 2f|$ respectively.

The ambient space P is the toric variety defined from the homogeneous coordinate ring $\mathbf{C}[x_0, x_1, y_0, \ldots, y_4]$ and the *irrelevant ideal* $(x_0, x_1) \cap (y_0, \ldots, y_4)$, endowed with the bigrading with respect to weights

$$w_h = (0, 0, 1, 1, 1, 1, 1), \qquad w_f = (1, 1, 0, -1, -2, -3, -3).$$

This is an example of a *Cox ring*, which the reader can learn from [23]. The space P is the quotient of $(\mathbf{A}^2 \setminus o) \times (\mathbf{A}^5 \setminus o)$ by the action of $(\mathbf{G}_m)^2$ with respect to weights w_h and w_f. The linear system $|ah + bf|$ is generated by monomials of bidegree (a, b). For example, Q is given by a general linear combination of the monomials listed below. It is straightforward to see that X is smooth.

degree in x_0, x_1	0	1	2	3
part in $y_0, \ldots, y_4$	$y_0 y_3, y_0 y_4, y_1 y_2$	$y_1 y_3, y_1 y_4, y_2^2$	$y_2 y_3, y_2 y_4$	$y_3^2, y_3 y_4, y_4^2$

Let $r_{-1} = f$ and $r_i = h - if$ for $0 \leq i \leq 3$, which correspond to the coordinates $x_0, x_1, y_0, \ldots, y_4$. Take a divisor D_i on P which belongs to the interior of the chamber in $N^1(P) \simeq \mathbf{R}^2$ spanned by $\mathbf{R}_{\geq 0}[r_{i-1}]$ and $\mathbf{R}_{\geq 0}[r_i]$. Its ample model $P_i = \operatorname{Proj} \bigoplus_{l \in \mathbf{N}} H^0(\mathscr{O}_P(lD_i))$ is the toric variety defined with the irrelevant ideal $(x_0, \ldots, y_{i-1}) \cap (y_i, \ldots, y_4)$, which is the quotient of $(\mathbf{A}^{2+i} \setminus o) \times (\mathbf{A}^{5-i} \setminus o)$ by $(\mathbf{G}_m)^2$. It admits contractions $P_i \to Z_{i-1}$ and $P_i \to Z_i$ where $Z_i = \operatorname{Proj} \bigoplus_{l \in \mathbf{N}} H^0(\mathscr{O}_P(lr_i))$ denotes the ample model of r_i. We have the diagram

$$\mathbf{P}^1 \leftarrow P = P_0 \dashrightarrow P_1 \dashrightarrow P_2 \dashrightarrow P_3 \to \mathbf{P}^1,$$

where the last $\mathbf{P}^1 = Z_3$ has homogeneous coordinates y_3, y_4 and $P_3 \to \mathbf{P}^1$ is a $\mathbf{P}(1, 1, 3, 2, 1)$-bundle by $3w_h + w_f = (1, 1, 3, 2, 1, 0, 0)$. The birational map $P_i \dashrightarrow P_{i+1}$ for $i = 0, 1, 2$ is factorised as $P_i \to Z_i \leftarrow P_{i+1}$, in which $P_i \to Z_i$ and $P_{i+1} \to Z_i$ are small.

The restrictions to Q and to $X = Q'|_Q$ yield diagrams $Q_0 \dashrightarrow Q_1 \dashrightarrow Q_2 \dashrightarrow Q_3$ and $X_0 \dashrightarrow X_1 \dashrightarrow X_2 \dashrightarrow X_3$ respectively. From the explicit expressions of the polynomials defining Q and Q', one deduces that X_3 has one quotient singularity of type $\frac{1}{3}(1, 1, 2)$. Notice that the strict transform in P_3 of Q' contains the locus $(x_0 = x_1 = y_1 = y_2 = 0)$. Each $\alpha_i \colon X_i \dashrightarrow X_{i+1}$ is investigated in [4]. The map α_0 is the inverse of a flip from a terminal threefold X_1, α_1 is an isomorphism and α_2 is the Atiyah flop in Example 1.3.18.

The morphisms $X \to \mathbf{P}^1$ and $X_3 \to \mathbf{P}^1$ are del Pezzo fibrations of degree four and two respectively. In fact they are elementary by Lemma 8.3.6. Thus the pliability $\mathscr{P}(X)$ contains $X/\mathbf{P}^1$, $X_3/\mathbf{P}^1$ and the standard conic bundle Y/S in Theorem 8.3.3. As will be seen in Lemma 8.3.7, X is irrational.

Lemma 8.3.6 Pic $X \simeq \mathbf{Z}^{\oplus 2}$.

Proof If the Picard number of X is two, then the contraction $\pi \colon X \to \mathbf{P}^1$ is extremal and the lemma follows from the exact sequence $0 \to \pi^* \mathrm{Pic}\, \mathbf{P}^1 \to \mathrm{Pic}\, X \to \mathbf{Z}$ given by the contraction theorem. Thus it suffices to prove that $\rho(X) = 2$. By virtue of $\rho(P_2) = \rho(P) = 2$ and the description of $X \dashrightarrow X_2$, we shall prove the equivalent assertion $\rho(P_2) = \rho(X_2)$.

By the definition of rational singularity and the Kawamata–Viehweg vanishing $R^i \pi_{P*} \mathscr{O}_P = 0$ for $i \geq 1$, the Leray spectral sequence via a common resolution W of P and P_2 provides $H^i(\mathscr{O}_{P_2}) \simeq H^i(\mathscr{O}_W) \simeq H^i(\mathscr{O}_P) \simeq H^i(\mathscr{O}_{\mathbf{P}^1}) = 0$ for $i \geq 1$. Hence Pic $P_2 = H^1(\mathscr{O}_{P_2}^\times) \simeq H^2(P_2, \mathbf{Z})$. In like manner, Pic $Q_2 \simeq H^2(Q_2, \mathbf{Z})$ and Pic $X_2 \simeq H^2(X_2, \mathbf{Z})$. The assertion is thus reduced to the isomorphism of $H^2(P_2, \mathbf{Z}) \to H^2(Q_2, \mathbf{Z}) \to H^2(X_2, \mathbf{Z})$.

First we consider the restriction of P_2 to Q_2. We use the ampleness of the divisor Q_2 on P_2, from which the complement $P_2 \setminus Q_2$ is affine. Applying the variant of the Lefschetz hyperplane theorem in Remark 6.3.23 to the identity of $P_2 \setminus Q_2$, one obtains $H_i(P_2 \setminus Q_2, \mathbf{Z}) = 0$ for $i \geq 6$. Poincaré duality holds on the homology manifold $P_2 \setminus Q_2$. A *homology manifold* M of dimension n is characterised by the isomorphism $H_i(M, M \setminus x, \mathbf{Z}) \simeq H_i(\mathbf{R}^n, \mathbf{R}^n \setminus o, \mathbf{Z})$ for all i at any $x \in M$, for which the reader can refer to [284]. It follows that $H_c^{10-i}(P_2 \setminus Q_2, \mathbf{Z}) = 0$ for $i \geq 6$ for cohomology groups with compact support. Thus the long exact sequence yields the isomorphism $H^2(P_2, \mathbf{Z}) \simeq H^2(Q_2, \mathbf{Z})$.

The argument from Q_2 to X_2 is similar but not the same. We explain how to change the above one. The divisor X on Q is not transformed to an ample divisor on any Q_i. Instead we take the inverse $Q_1 \to R_1 \leftarrow Q_2$ of a flip, where R_1 is the image in Z_1 of Q_1. This is isomorphic about X_1. Hence the image Y_1 in R_1 of X_1 is a **Q**-Cartier divisor, which is ample by $\rho(R_1) = 1$. Thus the complement $R_1 \setminus Y_1$ turns out to be affine. Applying Remark 6.3.23 to $Q_2 \setminus X_2 \to R_1 \setminus Y_1$, one obtains $H_i(Q_2 \setminus X_2, \mathbf{Z}) = 0$ for $i \geq 5$. The rest is the same, and the isomorphism $H^2(Q_2, \mathbf{Z}) \simeq H^2(X_2, \mathbf{Z})$ follows. We remark that Poincaré duality fails on $R_1 \setminus Y_1$. □

Lemma 8.3.7 *The threefold X is irrational.*

Proof The topological Euler characteristic $\chi(X)$ is regarded as the top Chern class $c_3(X)$ as in Remark 2.5.5. We shall compute it by the axioms in Theorem 2.5.1. We use the same notation h and f for $c_1(\mathscr{O}_P(h))$ and $c_1(\mathscr{O}_P(f))$ respectively. Note that $h^5 = 9$, $fh^4 = 1$ and $f^2 = 0 \in H^4(X, \mathbf{Z})$.

The total Chern class $c(P)$ of P equals the product of $c(\pi_P^* \mathscr{E}^\vee \otimes \mathscr{O}_P(1))$ and $c(\pi_P^* \mathscr{T}_{\mathbf{P}^1})$ from the exact sequence $0 \to \mathscr{T}_{P/\mathbf{P}^1} \to \mathscr{T}_P \to \pi_P^* \mathscr{T}_{\mathbf{P}^1} \to 0$ and the dual $0 \to \mathscr{O}_P \to \pi_P^* \mathscr{E}^\vee \otimes \mathscr{O}_P(1) \to \mathscr{T}_{P/\mathbf{P}^1} \to 0$ of the relative Euler sequence, where $\mathscr{T}_{P/\mathbf{P}^1} = \Omega^\vee_{P/\mathbf{P}^1}$. Hence

$$c(P) = (1+h)(1+h-f)(1+h-2f)(1+h-3f)^2 \cdot (1+2f)$$
$$= (1+2f)(1+h)^5 - 9f(1+h)^4.$$

On the other hand from the exact sequences $0 \to \mathscr{T}_Q \to \mathscr{T}_P|_Q \to \mathscr{O}_P(Q)|_Q \to 0$ and $0 \to \mathscr{T}_X \to \mathscr{T}_Q|_X \to \mathscr{O}_Q(X)|_X \to 0$, one has

$$c(P)|_X = c(X)(1 + 2\bar{h} - 3\bar{f})(1 + 2\bar{h} - 2\bar{f})$$

in the cohomology ring $H^\bullet(X, \mathbf{Z})$, where $\bar{h} = h|_X$ and $\bar{f} = f|_X$. The class $c(X)$ is uniquely determined by these two equalities as

$$c(X) = 1 + (\bar{h} - 2\bar{f}) + (2\bar{h}^2 - 3\bar{f}\bar{h}) + (-2\bar{h}^3 + 6\bar{f}\bar{h}^2).$$

Using $\bar{h}^3 = h^3(2h-2f)(2h-3f) = 26$ and $\bar{f}\bar{h}^2 = fh^2(2h-2f)(2h-3f) = 4$, one obtains $\chi(X) = -2\bar{h}^3 + 6\bar{f}\bar{h}^2 = -2 \cdot 26 + 6 \cdot 4 = -28$. Thus X is irrational by Theorem 8.3.4. □

8.4 Birational Rigidity

For a del Pezzo fibration of degree at most three, Pukhlikov proposed the notion of K^2-condition as a sufficient condition for birational rigidity.

Definition 8.4.1 (Pukhlikov) We say that an elementary del Pezzo fibration $X \to B$ from a projective threefold satisfies the K^2-*condition* if the class of the one-cycle K_X^2 lies outside the interior of the closed cone $\overline{\mathrm{NE}}(X)$ of curves in X.

Theorem 8.4.2 (Pukhlikov [392]) *If an elementary del Pezzo fibration of degree two (resp. one) from a smooth projective threefold satisfies the K^2-condition, then it is birationally rigid over the base (resp. birationally super-rigid).*

In fact Pukhlikov also proved the birational rigidity over the base for a general del Pezzo fibration of degree three which satisfies the K^2-condition. We remark that for an elementary del Pezzo fibration $X \to B$ of degree one, every fibrewise self-link $X/B \dashrightarrow X/B$ is square because the generic fibre is birationally superrigid over the function field of B by Theorem 8.3.1. In particular, the birational rigidity over the base for X/B is equivalent to the birational superrigidity.

We shall prove the theorem by the Sarkisov program. In a general setting, let $\pi\colon X \to B$ be an elementary del Pezzo fibration from a smooth projective threefold. Recall that every fibre $F = X \times_B b$ is integral by Remark 8.1.7. The Sarkisov program in Definition 6.2.10 runs in fixing a mobile and big linear system $\mathscr{H}$ on X. Let μ denote the Sarkisov threshold of $\mathscr{H}$ in Definition 6.1.6. Let H be the general member of $\mathscr{H}$ if $\mu \geq 1$, which should be reset as in Remark 6.2.3 if $\mu < 1$. The Sarkisov threshold μ is defined by the relation $K_X + \mu^{-1} H \sim_{\mathbf{Q}} \pi^* L$ with some **Q**-divisor L on B.

Let d denote the degree of π as a del Pezzo fibration. When d is at most three, the pair $(X, \mu^{-1} H)$ becomes horizontally canonical after twisting of π by a fibrewise self-link.

Lemma 8.4.3 *If $d = 1$, then $(X, \mu^{-1} H)$ is canonical outside finitely many fibres of π. If $d = 2$ or 3, then the same assertion holds for the homaloidal transform of $\mathscr{H}$ in the target of some fibrewise self-link $X/B \dashrightarrow X/B$.*

Proof Let K be the function field of B and take the generic fibre $X_K = X \times_B \operatorname{Spec} K$. The lemma will follow from the birational (super)rigidity of X_K in Theorem 8.3.1 by a slight modification of Proposition 6.2.8.

Let c_K be the canonical threshold of the (**Q**-)divisor $H \times_B \operatorname{Spec} K$ on X_K. It is the greatest rational number such that $(X, c_K H)$ is canonical about X_K. As far as $c_K < \mu^{-1}$, there exists a Sarkisov link g_K of type I or II over K starting from $X_K \to \operatorname{Spec} K$ by the corresponding statement over K of Proposition 6.2.8. By Theorem 8.3.1, such a link does not exist if $d = 1$, and ends with the same space $X_K \to \operatorname{Spec} K$ if $d = 2$ or 3.

Thus the canonicity of $(X, \mu^{-1} H)$ about X_K holds for $d = 1$. When $d = 2$ or 3, the link g_K extends to a fibrewise self-link $g\colon X/B \dashrightarrow X/B$. Defining the Sarkisov degree over $\operatorname{Spec} K$ by taking only divisors over X dominating B into account, one deduces that it decreases strictly by the replacement of $\mathscr{H}$ by the homaloidal transform in the target of g, similarly to Proposition 6.2.8. Hence by Theorems 6.1.7 and 6.2.5, the canonical threshold c_K becomes at least μ^{-1} after finitely many fibrewise self-links $X/B \dashrightarrow X/B$, in which $(X, \mu^{-1}H)$ is canonical about X_K.   □

When d is at most two, the pair $(X, \mu^{-1}H)$ is canonical at the generic point of a relative curve.

Lemma 8.4.4   *If $d = 1$ or 2, then $(X, \mu^{-1}H)$ is canonical at the generic point of an arbitrary curve in X contracted by π.*

Proof   Fix a curve C in the fibre $F = X \times_B b$ at some $b \in B$. Recall that F is integral. For the same reason as for Lemma 6.3.27, it suffices to prove the inequality $\mathrm{mult}_C H \le \mu$. From the estimate

$$(\mathrm{mult}_C H)(-K_X \cdot C) \le (\mathrm{mult}_C H|_F)(-K_X \cdot C) \le (-K_X \cdot H \cdot F) = d\mu,$$

one has only to consider the case when $d = 2$ and $(-K_X \cdot C) = 1$. In this case by Remark 8.1.4, the Gorenstein surface F is embedded as a quartic in $\mathbf{P}(1, 1, 1, 2)$, which does not contain the point $[0, 0, 0, 1]$. By Lemma 8.1.10, $-K_X$ is a relatively free Cartier divisor.

Take any point x in C. Since $-K_X$ is relatively free, there exists a member $G_U \in |-K_U|$ on a neighbourhood U in X about F such that $x \in G_U$ and such that F, G_U and the support of H intersect properly. Then $(\mathrm{mult}_x H)(\mathrm{mult}_x F) \le (G_U \cdot H \cdot F)_U = 2\mu$. It follows that $\mathrm{mult}_C H \le \mathrm{mult}_x H \le \mu$ unless $\mathrm{mult}_x F = 1$. Thus we may and shall assume that F is smooth about C.

Consider the projection $\varphi \colon F \to \mathbf{P}(1, 1, 1) \simeq \mathbf{P}^2$, by which $\mathcal{O}_F(-K_X|_F) = \varphi^* \mathcal{O}_{\mathbf{P}^2}(1)$. The double cover φ is ramified along the quartic Δ in $\mathbf{P}^2$ defined by f for the expression $y^2 + f(x_0, x_1, x_2)$ of F with a suitable weighted coordinate system. This induces an involution $\iota \colon F \to F$ over $\mathbf{P}^2$. It follows from $(-K_X \cdot C) = 1$ that C is mapped to a line in $\mathbf{P}^2$ isomorphically. Let $C' = \iota(C)$, for which $C + C' \sim -K_X|_F$. Then F is smooth about C' as well as about C, from which $\varphi(C) \not\subset \Delta$ and hence $C \ne C'$. Otherwise F would be singular above the intersection of $\varphi(C)$ and $\Delta - \varphi(C)$.

Intersection numbers are defined on smooth germs of F along C and C'. Write $m = \mathrm{mult}_C H|_F$ and $m' = \mathrm{mult}_{C'} H|_F$ for brevity. Using $((C + C') \cdot C)_F = ((C + C') \cdot C')_F = 1$, one has

$$\mu = (H|_F \cdot C)_F \ge ((mC + m'C') \cdot C)_F = m + (m' - m)(C' \cdot C)_F,$$

$$\mu = (H|_F \cdot C')_F \ge ((mC + m'C') \cdot C')_F = m + (m' - m)(C')_F^2.$$

Because C and C' intersect at $C \cap \varphi^{-1}(\Delta)$, one has $(C \cdot C')_F \ge 1$ and $(C')_F^2 = 1 - (C \cdot C')_F \le 0$. Consequently either $(m' - m)(C' \cdot C)_F$ or $(m' - m)(C')_F^2$

is non-negative, and the inequality $\mu \geq m$ follows from the corresponding one of the above two inequalities. Then $\mathrm{mult}_C H \leq m \leq \mu$. □

Now we follow the simplified argument by Corti in [93, theorem 5.1]. The most technical part is the formulation of Lemma 8.4.6. In the next two lemmata, we extend the notions of log discrepancy and canonical singularity to pairs (X, Δ) without assuming that Δ is effective. We say that (X, Δ) is *canonical* if $a_E(X, \Delta) \geq 1$ for all divisors E exceptional over X.

Lemma 8.4.5 *Let $x \in S$ be the germ of a smooth surface. Let $\mathscr{H}$ be a mobile linear system on S and let H and G be the general members of $\mathscr{H}$. Let μ be a rational number at least one and let A be an effective $\mathbf{Q}$-divisor on S. If $(S, \mu^{-1}H - A)$ is not canonical, then the local intersection number $(H \cdot G)_x$ at x satisfies $(H \cdot G)_x > 4 \mathrm{mult}_x A \cdot \mu^2$.*

Proof We write $h = \mathrm{mult}_x H = \mathrm{mult}_x G$ and $a = \mathrm{mult}_x A$. The pair $(S, \mu^{-1}H)$ is canonical in codimension one as $\mu \geq 1$. Hence the assumption means the existence of a divisor E over S with centre $c_S(E) = x$ and with $a_E(S, \mu^{-1}H - A) < 1$. We shall prove the inequality on $(H \cdot G)_x$ by induction on the log discrepancy $a_E(S) \in \mathbf{Z}_{>0}$.

Take the blow-up $\pi \colon S' \to S$ at x and let F be the exceptional curve. Then $K_{S'} + \mu^{-1}H' - A' - fF = \pi^*(K_S + \mu^{-1}H - A)$ for $f = 1 - \mu^{-1}h + a$ and the strict transforms H', G', A' of H, G, A. If $f < 0$, that is, $(1 + a)\mu < h$, then

$$(H \cdot G)_x = ((H' + hF) \cdot (G' + hF))_{S'} \geq h^2 > (1 + a)^2 \mu^2 \geq 4a\mu^2.$$

If $f \geq 0$, then the centre $x' = c_{S'}(E)$ is a point in F. Apply the inductive hypothesis to the pair $(S', \mu^{-1}H' - A' - fF)$ at x', where $a_E(S') < a_E(S)$. This provides $(H' \cdot G')_{x'} > 4(\mathrm{mult}_{x'} A' + f)\mu^2 \geq 4f\mu^2$. Combining it with the equality $(H \cdot G)_x = (H' \cdot G')_{S'} + h^2$, one has $(H \cdot G)_x \geq (H' \cdot G')_{x'} + h^2 > 4f\mu^2 + h^2$. Thus

$$(H \cdot G)_x > 4(1 - \mu^{-1}h + a)\mu^2 + h^2 = 4a\mu^2 + (2\mu - h)^2 \geq 4a\mu^2.$$

These estimates are what we want. □

Lemma 8.4.6 *Let $x \in X$ be the germ of a smooth threefold and let $\{S_i\}_{i \in I}$ be a finite set of prime divisors on X. Let $\mathscr{H}$ be a mobile linear system on X and let H and G be the general members of $\mathscr{H}$. Choose a cycle-theoretic expression $H \cdot G = Z_h + \sum_{i \in I} Z_i$ with effective one-cycles Z_h and Z_i such that Z_h has no components in S_i and Z_i is supported in S_i for all i. Let $\mu \geq 1$ and $\lambda_i > 0$ be rational numbers. If $(X, \mu^{-1}H)$ is canonical outside $\sum_{i \in I} S_i$ and $(X, \mu^{-1}H - \sum_{i \in I} \lambda_i S_i)$ is not canonical, then there exist rational numbers*

$t_i > 0$ *such that*

$$\text{mult}_x \, Z_h + \sum_{i \in I} t_i \, \text{mult}_x \, Z_i > 4\Big(1 + \sum_{i \in I} t_i \lambda_i\Big)\mu^2.$$

Proof The idea is essentially the same as that for the preceding lemma. The *multiplicity* of a one-cycle on a germ equals the local intersection number with the general hyperplane section because a curve is always Cohen–Macaulay. By assumption, $I \neq \emptyset$ and there exists a divisor E over X with centre $c_X(E)$ in $\sum_i S_i$ such that $a_E(X, \mu^{-1}H - \sum_i \lambda_i S_i) < 1$. The centre $c_X(E)$ is a curve or the point x.

Step 1 Suppose that $c_X(E)$ is a curve. Take a general closed point y in $c_X(E)$ and the general hyperplane section T of $y \in X$. The restriction $(T, \mu^{-1}H|_T - \sum_i \lambda_i S_i|_T)$ is not canonical. Applying Lemma 8.4.5 to this pair, one obtains

$$\sum_{j \in J} \text{mult}_x \, Z_j \geq \sum_{j \in J} \text{mult}_y \, Z_j > 4 \sum_{j \in J} \lambda_j \, \text{mult}_y \, S_j \cdot \mu^2 \geq 4 \sum_{j \in J} \lambda_j \cdot \mu^2$$

for the subset J of I consisting of indices j such that $y \in S_j$. Note that $J \neq \emptyset$. Set $t_i = 1$ for $i \in I \setminus J$ and $t_j = t$ for $j \in J$ with a rational number t. If t is chosen sufficiently large, then the inequality in the statement holds.

Step 2 Thus we shall assume that $c_X(E) = x$. We shall prove the inequality by induction on $a_E(X)$. Again we write $h = \text{mult}_x \, H = \text{mult}_x \, G$ and $s_i = \text{mult}_x \, S_i$ for brevity. We also write $z_h = \text{mult}_x \, Z_h$ and $z_i = \text{mult}_x \, Z_i$.

Take the blow-up $\pi: X' \to X$ at x and let F be the exceptional divisor. Write H', G', S_i' for the strict transforms in X' of H, G, S_i respectively and take the expression $H' \cdot G' = Z_h' + \sum_i Z_i' + Z_f$ with the strict transforms Z_h', Z_i' of Z_h, Z_i and an effective one-cycle Z_f supported in F. For a point x' in X', the multiplicities $z_h' = \text{mult}_{x'} \, Z_h'$, $z_i' = \text{mult}_{x'} \, Z_i'$ and $z_f = \text{mult}_{x'} \, Z_f$ satisfy

$$z_h \geq z_h', \qquad z_i \geq z_i', \qquad z_h + \sum_i z_i \geq z_f + h^2$$

as in the surface case. For example, the last one follows from the equality $(H \cdot G \cdot A)_x = (H' \cdot G' \cdot A')_{x'} + h^2$ for the general hyperplane section A of $x \in X$ and its strict transform A'.

One has $K_{X'} + \mu^{-1}H' - \sum_i \lambda_i S_i' - fF = \pi^*(K_X + \mu^{-1}H - \sum_i \lambda_i S_i)$ for $f = 2 - \mu^{-1}h + \sum_i \lambda_i s_i$. If $f \leq 0$, then

$$z_h + \sum_i z_i \geq h^2 \geq \Big(2 + \sum_i \lambda_i s_i\Big)^2 \mu^2 \geq \Big(2 + \sum_i \lambda_i\Big)^2 \mu^2 > 4\Big(1 + \sum_i \lambda_i\Big)\mu^2$$

and hence we may just take $t_i = 1$. We need to discuss the case $f > 0$.

Step 3 Suppose that $f > 0$. By this assumption, the centre $c_{X'}(E)$ is a curve or a point. Take a point x' in $c_{X'}(E)$ and set the multiplicities z_h', z_i' and z_f at

x' as above. To the pair $(X', \mu^{-1}H' - \sum_i \lambda_i S_i' - fF)$ at x', we apply Step 1 if $c_{X'}(E)$ is a curve, and apply the inductive hypothesis if it is a point. It follows that $z_h' + \sum_i t_i' z_i' + t_f z_f > 4(1 + \sum_i t_i' \lambda_i + t_f f)\mu^2$ for some positive rational numbers t_i' and t_f. A priori, the summation in this inequality is over i with $x' \in S_i'$. However, one can make the inequality with summation over all $i \in I$ by choosing t_i' sufficiently small whenever $x' \notin S_i'$.

We shall verify that $t_i = (t_i' + t_f)/(1 + t_f)$ satisfy the inequality. We have

$$z_h + \sum_i t_i z_i = \frac{t_f(z_h + \sum_i z_i)}{1 + t_f} + \frac{z_h + \sum_i t_i' z_i}{1 + t_f}$$

$$\geq \frac{t_f(z_f + h^2)}{1 + t_f} + \frac{z_h' + \sum_i t_i' z_i'}{1 + t_f} \geq 4 \cdot \frac{1 + \sum_i t_i' \lambda_i + t_f f}{1 + t_f}\mu^2 + \frac{t_f h^2}{1 + t_f}.$$

By $f = 2 - \mu^{-1}h + \sum_i \lambda_i s_i$, the right-hand side equals

$$4\left(1 + \sum_i \frac{t_i' + t_f s_i}{1 + t_f}\lambda_i\right)\mu^2 + \frac{t_f(2\mu - h)^2}{1 + t_f},$$

which is at least $4(1 + \sum_i t_i \lambda_i)\mu^2$. Thus $z_h + \sum_i t_i z_i \geq 4(1 + \sum_i t_i \lambda_i)\mu^2$ and this completes the lemma. □

Proof of Theorem 8.4.2 Let $f: X/B \dashrightarrow Y/T$ be an arbitrary link from π to a Mori fibre space $\psi: Y \to T$. We fix a standard linear system $\mathcal{H}_Y = |-\mu_Y K_Y + \psi^* A|$ as in Definition 6.1.8. Let $\mathcal{H}$ be the homaloidal transform in X of $\mathcal{H}_Y$ and let H be the general member of $\mathcal{H}$, which is the strict transform of the general member H_Y of $\mathcal{H}_Y$. We have a pair $(X, \mu^{-1}H)$ for the Sarkisov threshold μ of $\mathcal{H}$, which satisfies $1 \leq \mu_Y \leq \mu$ by Theorem 6.1.12(i). Then $K_X + \mu^{-1}H \sim_{\mathbb{Q}} \pi^* L$ with some $\mathbb{Q}$-divisor L on B. By Lemmata 8.4.3 and 8.4.4, we may assume that $(X, \mu^{-1}H)$ is canonical outside finitely many points.

Let $\sum_{i \in I} F_i$ be the sum of all fibres F_i of π about which $(X, \mu^{-1}H)$ is not canonical. For each $i \in I$, we take the least rational number λ_i such that $(X, \mu^{-1}H - \lambda_i F_i)$ is canonical about F_i. Then $(X, \mu^{-1}H - \lambda_i F_i)$ is strictly canonical at some point x_i in F_i.

Step 1 We write $K_X + \mu^{-1}H - \sum_i \lambda_i F_i \equiv cF$ with $c = \deg L - \sum_i \lambda_i$ and the general fibre F of π. We shall prove that f is square if $c \geq 0$.

Take a common resolution W of X and Y with $p: W \to X$ and $q: W \to Y$ where $q = f \circ p$. Since $(X, \mu^{-1}H - \sum_i \lambda_i F_i)$ is canonical, one can write

$$K_W + \mu^{-1}H_W - \sum_i \lambda_i F_{Wi} = p^*\left(K_X + \mu^{-1}H - \sum_i \lambda_i F_i\right) + P \equiv cp^*F + P$$

with some effective p-exceptional $\mathbf{Q}$-divisor P and the strict transforms H_W and F_{Wi} of H and F_i. Then

$$K_Y + \mu^{-1} H_Y = q_*(K_W + \mu^{-1} H_W) \equiv cF_Y + q_*\left(P + \sum_i \lambda_i F_{Wi}\right) \geq cF_Y$$

for $F_Y = q_* p^* F$, which is the strict transform of the general fibre F.

Since $K_Y + \mu_Y^{-1} H_Y \sim_{\mathbf{Q},T} 0$, a general relative curve D in Y/T satisfies

$$(\mu^{-1} - \mu_Y^{-1})(H_Y \cdot D) = ((K_Y + \mu^{-1} H_Y) \cdot D) \geq c(F_Y \cdot D).$$

Note that $\mu \geq \mu_Y$, $(H_Y \cdot D) > 0$ and $(F_Y \cdot D) \geq 0$. If $c \geq 0$, then the above inequality implies that $\mu = \mu_Y$. By Theorem 6.1.12(i), f is square.

Step 2 By Step 1, it suffices to prove that $\sum_i \lambda_i \leq \deg L$. For the general members H and G of $\mathcal{H}$, we express $H \cdot G = Z + \sum_i Z_i$ with effective one-cycles Z and Z_i so that Z has no components in F_i and Z_i is supported in F_i for all i.

We claim the estimate $\text{mult}_{x_i} Z_i \geq 4\lambda_i \mu^2$. Applying Lemma 8.4.6 to the pair $(X, \mu^{-1} H - (\lambda_i - \varepsilon) F_i)$ at x_i for small positive ε, one finds a positive rational number t_i such that $\text{mult}_{x_i} Z + t_i \, \text{mult}_{x_i} Z_i > 4(1 + t_i(\lambda_i - \varepsilon))\mu^2$. Together with the bound $\text{mult}_{x_i} Z \leq (H \cdot G \cdot F_i) = d\mu^2$, one has

$$t_i \, \text{mult}_{x_i} Z_i > 4(1 + t_i(\lambda_i - \varepsilon))\mu^2 - d\mu^2 > 4t_i(\lambda_i - \varepsilon)\mu^2$$

and hence $\text{mult}_{x_i} Z_i > 4(\lambda_i - \varepsilon)\mu^2$. Since ε can be taken arbitrarily small, the claim $\text{mult}_{x_i} Z_i \geq 4\lambda_i \mu^2$ follows.

The numerical class l of the rational one-cycle $(-K_X \cdot F)/d$ is characterised by $(-K_X \cdot l) = 1$ and $(F \cdot l) = 0$. Write $\sum_i Z_i \equiv nl$ with $n \in \mathbf{Q}_{\geq 0}$. Since $-K_X|_{F_i}$ is mobile, one obtains the inequality

$$4\mu^2 \sum_i \lambda_i \leq \sum_i \text{mult}_{x_i} Z_i \leq \sum_i (-K_X \cdot Z_i) = n.$$

Step 3 It follows from the expression $K_X + \mu^{-1} H \sim_{\mathbf{Q}} \pi^* L$ that

$$\mu^2 K_X^2 \equiv (\mu\pi^* L - H)^2 \equiv -2\mu \deg L(H \cdot F) + Z + \sum_i Z_i$$

$$\equiv (n - 2d\mu^2 \deg L)l + Z.$$

Note that $[Z] \notin \mathbf{R}_{\geq 0}[l]$ in $N_1(X)$ by $(F \cdot Z) = (F \cdot H \cdot G) = d\mu^2 \neq 0$. Because of the K^2-condition that K_X^2 is outside the interior of $\overline{\text{NE}}(X)$, the coefficient $n - 2d\mu^2 \deg L$ of l is not positive. This and the inequality on n in Step 2 are combined as

$$4\mu^2 \sum_i \lambda_i \leq n \leq 2d\mu^2 \deg L \leq 4\mu^2 \deg L$$

and therefore $\sum_i \lambda_i \leq \deg L$. The theorem follows from Step 1. $\qquad \square$

Inspired by Pukhlikov's work, Grinenko introduced a condition weaker than the K^2-condition and proved that it is equivalent to the birational superrigidity in the case of degree one.

Definition 8.4.7 Let $X \to S$ be a proper morphism from a normal variety to a variety. The *mobile cone* $\mathrm{Mb}(X/S)$ is the convex cone in $N^1(X/S)$ spanned by the classes of relatively mobile Cartier divisors. By convention as in Section 1.2, we write $\mathrm{Mb}(X)$ for $\mathrm{Mb}(X/S)$ when S is the spectrum of the ground field.

Definition 8.4.8 (Grinenko) We say that an elementary del Pezzo fibration $X \to B$ from a projective threefold satisfies the *K-condition* if the class of $-K_X$ lies outside the interior of the mobile cone $\mathrm{Mb}(X)$.

Lemma 8.4.9 *K^2-condition implies K-condition.*

Proof We shall prove the contrapositive. Let $X \to B$ be an elementary del Pezzo fibration of degree d from a projective threefold. Suppose that $-K_X$ belongs to the interior of $\mathrm{Mb}(X)$. Then $-nK_X - F$ still belongs to $\mathrm{Mb}(X)$ for a sufficiently large integer n and the general fibre F, which means that $-mnK_X - mF$ is mobile for some positive integer m. We write $\mu = mn$ for brevity. Then for the general members H and H' of $|-\mu K_X - mF|$, the one-cycle

$$H \cdot H' \equiv (-\mu K_X - mF)^2 = \mu^2 K_X^2 - 2\mu m(-K_X \cdot F)$$

belongs to the closed cone $\overline{\mathrm{NE}}(X)$ of curves in X. The ray $\mathbf{R}_{\geq 0}[-K_X \cdot F]$ belongs to $\overline{\mathrm{NE}}(X)$ and does not contain $[H \cdot H']$ by $(H \cdot H' \cdot F) = d\mu^2 \neq 0$. Thus $\mu^2 K_X^2 \equiv (H \cdot H') + 2\mu m(-K_X \cdot F)$ lies in the interior of $\overline{\mathrm{NE}}(X)$. $\square$

Theorem 8.4.10 (Grinenko [155], [156]) *An elementary del Pezzo fibration $X \to \mathbf{P}^1$ of degree one from a smooth projective threefold is birationally superrigid if and only if it satisfies the K-condition.*

Replacing superrigidity by rigidity over the base, Grinenko [155], [158] also proved the correspondent equivalence for a general del Pezzo fibration of degree two. In the singular case, Ahmadinezhad [3] constructed a counter-example to the if part for a Gorenstein del Pezzo fibration of degree two, using a Cox ring in the same manner as in Example 8.3.5.

Let $X \to \mathbf{P}^1$ be a fibration as in Theorem 8.4.10. It has integral fibres by Remark 8.1.7 and is embedded into a $\mathbf{P}(1,1,2,3)$-bundle $P \to \mathbf{P}^1$ by Lemma 8.1.10. By the construction as in Example 8.3.5, P is the toric variety defined by the homogeneous coordinate ring $\mathbf{C}[x_0, x_1, y_0, \ldots, y_3]$ and the irrelevant ideal $(x_0, x_1) \cap (y_0, \ldots, y_3)$. One can normalise its bigrading so that it has weights $(0, 0, 1, 1, 2, 3)$ and $(1, 1, 0, a, 2b, 3b)$ with some $a \in \mathbf{Z}_{\geq 0}$ and $b \in \mathbf{Z}$, where the relation between the last two entries $2b$ and $3b$ comes from the assumption that X is smooth. The divisor X is of bidegree $(6, 6b)$ with respect to these weights.

Lemma 8.4.11　*The integer b is positive. Either $a \leq b$ or $5a = 6b$.*

Proof　The polynomial f in $x_0, x_1, y_0, \ldots, y_3$ defining X is of bidegree $(6, 6b)$. Hence if $b < a$, then X has the section $D = (y_0 = y_2 = y_3 = 0)$. In this case, X is smooth about D only if f contains the monomial $y_0 y_1^5$. Then $0 < 5a = 6b$. It remains to exclude the case $a = b = 0$. In this case, X would be the product of $\mathbf{P}^1$ and a del Pezzo surface of degree one, which contradicts $\rho(X) = 2$.　□

The if part of the theorem is proved along Pukhlikov's approach to Theorem 8.4.2 according to the numerical data (a, b). In fact, one can characterise the K^2 and K-conditions in terms of (a, b).

Lemma 8.4.12　*Let $X \to \mathbf{P}^1$ be as in Theorem 8.4.10.*

(i) *It satisfies the K^2-condition if and only if $(a, b) \neq (0, 1)$, $(1, 1)$, $(1, 2)$, $(2, 2)$, $(3, 3)$ or $(6, 5)$.*

(ii) *It satisfies the K-condition if and only if $(a, b) \neq (0, 1)$ or $(1, 1)$.*

Proof　Recall that $a \geq 0$ and $b \geq 1$. We write h and f for the restrictions to X of the divisors on P of bidegree $(1, 0)$ and $(0, 1)$ respectively, in which f is linearly equivalent to a fibre of $X/\mathbf{P}^1$. Then $-K_P|_X = 7h + (2 + a + 5b)f$ and $-K_X = -(K_P + X)|_X = h + (2 + a - b)f$. The fibration $X/\mathbf{P}^1$ has the section $C = (y_0 = y_1 = \lambda y_2^3 + y_3^2 = 0)$ with some $\lambda \neq 0$.

In the case $a \leq b$ in Lemma 8.4.11, the closed cone $\overline{\mathrm{NE}}(X)$ of curves is generated by $C \equiv h(h + af)$ and hf. Notice that C has intersection number zero with the nef divisor $h + bf$. Hence $K_X^2 \equiv C + (4 + a - 2b)hf$ belongs to its interior if and only if $2a \leq 2b \leq 3 + a$, which corresponds to (i). On the other hand, the closure of the mobile cone $\mathrm{Mb}(X)$ is generated by $h + af$ and f. Hence $-K_X = (h + af) + (2 - b)f$ belongs to its interior if and only if $b = 1$, which corresponds to (ii).

The other case $5a = 6b$ is similar. In this case, $\overline{\mathrm{NE}}(X)$ is generated by the section $D = (y_0 = y_2 = y_3 = 0) \equiv h(h + bf)$ and hf, and $\mathrm{Mb}(X)$ is generated by $h + bf$ and f.　□

The only-if part of the theorem holds even if the total space is singular.

Theorem 8.4.13　*Let $\pi\colon X \to B$ be an elementary del Pezzo fibration of degree one from a terminal projective threefold. If π is birationally superrigid, then it satisfies the K-condition.*

Proof Let F be the general fibre of π. If π does not satisfy the K-condition, then the linear system $\mathscr{H} = |-\mu K_X - mF|$ is mobile and big for some positive integers μ and m. On this assumption, we shall construct a link from π which is not square.

Let H be the general member of $\mathscr{H}$. If the pair $(X, \mu^{-1}H)$ is not canonical, then by Proposition 6.2.8, one can construct a Sarkisov link $g\colon X/B \dashrightarrow X_1/B_1$ of type I or II. This is the two-ray game over B starting with a divisorial contraction $Y \to X$ over a maximal centre in X with respect to $\mathscr{H}$. By the birational superrigidity of the generic fibre $X \times_B \operatorname{Spec} K$ over the function field K of B in Theorem 8.3.1, the link g is isomorphic over $\operatorname{Spec} K$, which means that it is a square Sarkisov link of type II over $B = B_1$. The strict transform H_1 in X_1 of H satisfies $\mu K_{X_1} + H_1 \sim_{\mathbf{Q},B} 0$ for the same integer μ.

In general for a square link $X/B \dashrightarrow X_i/B$, let H_i be the general member of the homaloidal transform in X_i of $\mathscr{H}$. As far as $(X_i, \mu^{-1}H_i)$ is not canonical, one obtains a square Sarkisov link $X_i/B \dashrightarrow X_{i+1}/B$ of type II in the same manner as above. Hence by Lemma 6.2.12, after finitely many Sarkisov links, we attain a del Pezzo fibration $\pi_n\colon X_n \to B$ square to π such that $(X_n, \mu^{-1}H_n)$ is canonical.

We have $H_n \sim_{\mathbf{Q}} -\mu K_{X_n} - m_n F_n$ for some $m_n \in \mathbf{Q}$ and the general fibre F_n of π_n. We shall prove that m_n is positive. Take a common resolution W of X and X_n with $p\colon W \to X$ and $q\colon W \to X_n$. Let C_W be the strict transform in W of a general curve C in X, which avoids the indeterminacy of p^{-1}. Then $((\mu K_W + H_W) \cdot C_W) = ((\mu K_X + H) \cdot C) = -m(F \cdot C) < 0$ for the strict transform H_W of H. Since $(X_n, \mu^{-1}H_n)$ is canonical, $\mu K_W + H_W$ is at least the pull-back $q^*(\mu K_{X_n} + H_n)$ and hence

$$-m_n(F_n \cdot C_n) = ((\mu K_{X_n} + H_n) \cdot C_n) \le ((\mu K_W + H_W) \cdot C_W) < 0$$

for $C_n = q(C_W)$, which shows that $m_n > 0$.

Thus by replacing π by π_n and $\mathscr{H}$ by $|-l(\mu K_{X_n} + m_n F_n)|$ for some positive integer l, we may assume that $(X, \mu^{-1}H)$ is canonical. Since $K_X + \mu^{-1}H \sim_{\mathbf{Q}} -\mu^{-1}mF$ is not pseudo-effective, one can run the $(K_X + \mu^{-1}H)$-MMP as the two ray game over $\operatorname{Spec}\mathbf{C}$. It is also the $(K_X + (\mu^{-1} - \varepsilon)H)$-MMP for small positive ε. Thus it ends with a Mori fibre space $Y \to T$ by a Sarkisov link $h\colon X/B \dashrightarrow Y/T$ of type III or IV.

We shall verify that h is not square. If h is of type III, then T is a point and hence h is never square. If it is of type IV, then the map $X \dashrightarrow Y$ is small. If h were square, then by Lemma 6.1.4, it would be a Sarkisov isomorphism and would transform $K_X + \mu^{-1}H \equiv_B 0$ to $K_Y + \mu^{-1}h_*H$ anti-ample over T, which is absurd. $\qquad\square$

Finally we give an example of a del Pezzo fibration of degree one which is birationally rigid but not birationally superrigid.

Example 8.4.14 Consider an elementary del Pezzo fibration $\pi\colon X \to \mathbf{P}^1$ of degree one from a smooth projective threefold with the data $(a, b) = (0, 1)$ defined before Lemma 8.4.11. The divisor X is of bidegree $(6, 6)$ with respect to weights $(0, 0, 1, 1, 2, 3)$ and $(1, 1, 0, 0, 2, 3)$. Grinenko constructed a flop $X \dashrightarrow X^+$ to another elementary del Pezzo fibration $\pi^+\colon X^+ \to \mathbf{P}^1$ of degree one with the same data $(a, b) = (0, 1)$ and proved that the pliability $\mathscr{P}(X)$ consists of $X/\mathbf{P}^1$ and $X^+/\mathbf{P}^1$ [155, proposition 2.12]. The flop $X \dashrightarrow X^+$ is never fibrewise since $-K_X$ is π-ample. In particular, π is not birationally superrigid. This can also be derived from Theorem 8.4.13 and Lemma 8.4.12(ii).

The ambient $\mathbf{P}(1, 1, 2, 3)$-bundle $P \to \mathbf{P}^1$ has homogeneous coordinates $x_0, x_1, y_0, \ldots, y_3$, and X is defined by a polynomial f of bidegree $(6, 6)$. Okada exhibited the flop $X \dashrightarrow X^+$ as the interchange of the two pairs (x_0, x_1) and (y_0, y_1) [376, section 5]. In terms of Cox ring, X^+ is defined with the irrelevant ideal $(y_0, y_1) \cap (x_0, x_1, y_2, y_3)$. He verified that π is certainly an elementary del Pezzo fibration when f is general amongst those which are invariant by this interchange. In this case, X and X^+ are abstractly isomorphic and π and π^+ are joined by a Sarkisov isomorphism. Thus π is birationally rigid.

9

Fano Threefolds

A Fano variety is defined by the ampleness of the anti-canonical divisor. The projective line is the only Fano curve, and a Fano surface is nothing but a del Pezzo surface. As an extension of these low dimensional varieties, Fano threefolds form an extremely fruitful class for classification.

Fano varieties are rationally connected. Estimating the degree of a rational curve joining two general points, Kollár, Miyaoka and Mori proved that Fano varieties of fixed dimension form a bounded family. In the singular case, Birkar recently settled the boundedness of ε-lc $\mathbf{Q}$-Fano varieties known as the Borisov–Alexeev–Borisov conjecture. Once the restriction by ε is relaxed, the boundedness no longer holds even birationally in dimension three.

As is the case with threefold contractions of any other type, the general elephant conjecture is an important step towards a classification of Fano threefolds. This holds in the Gorenstein case thanks to Shokurov and Reid. We provide a unified proof focusing on the non-vanishing. Without the Gorenstein condition, there exist counter-examples.

Iskovskikh established a systematic classification of Fano threefolds with Picard number one. His approach is founded upon the work of Fano, who studied an anti-canonically embedded Fano threefold by projecting it doubly from a line. We perform the classification using the MMP. Then we explain Mukai's biregular description by means of vector bundles. He realised a Fano threefold as a linear section of a subvariety of some Grassmannian.

There exist 95 families of terminal $\mathbf{Q}$-Fano weighted hypersurfaces in dimension three. By the Sarkisov program, Corti, Pukhlikov and Reid concluded that a general $\mathbf{Q}$-Fano threefold in each of these families is birationally rigid. This is considered to be a generalisation of the irrationality of a quartic threefold due to Iskovskikh and Manin. The Sarkisov program also constructs several interesting examples of $\mathbf{Q}$-Fano threefolds.

Finally we describe the relation between birational rigidity and K-stability. The K-stability was introduced by Tian in the context of differential geometry for the problem of the existence of a Kähler–Einstein metric on a Fano variety. It can be formulated in terms of birational geometry. If a **Q**-Fano threefold in one of the 95 families is birationally superrigid, then it is K-stable.

9.1 Boundedness

A Mori fibre space over a point is a **Q**-Fano variety. It is characterised by the ampleness of the anti-canonical divisor. This section is devoted to the boundedness of **Q**-Fano varieties.

Definition 9.1.1 Let X be a smooth projective variety. We call X a *Fano* variety if the anti-canonical divisor $-K_X$ is ample. We call X a *weak Fano* variety if $-K_X$ is nef and big. Allowing singularities, let X be a log terminal projective variety. We call X a **Q**-*Fano* (resp. *weak* **Q**-*Fano*) variety if $-K_X$ is ample (resp. nef and big). We say that X is *Gorenstein Fano* (resp. *Gorenstein weak Fano*) if it is Gorenstein and **Q**-Fano (resp. weak **Q**-Fano).

We have observed in Theorem 6.3.9 that a **Q**-Fano variety is rationally connected. For Fano varieties, the boundedness is proved simultaneously with the rational connectedness.

Theorem 9.1.2 (Kollár–Miyaoka–Mori [272], [274]) *Fano varieties of fixed dimension form a bounded family. More precisely, given a positive integer n, there exists a morphism $V \to S$ of algebraic schemes such that every Fano variety of dimension n is isomorphic to the fibre $V_s = V \times_S s$ at some closed point s in S.*

We shall work over an algebraically closed field k in general. The characteristic of k is assumed to be zero unless otherwise mentioned from the perspective of deformation theory.

Closed subschemes Z of the projective space $\mathbf{P}^N$ with fixed *Hilbert polynomial* $\chi(\mathscr{O}_Z(l)) \in \mathbf{Q}[l]$ are parametrised by the *Hilbert scheme*. Let X be a Fano variety such that $-mK_X$ is very ample for a positive integer m. By Kodaira vanishing, the dimension $h = h^0(\mathscr{O}_X(-mK_X))$ equals $\chi(\mathscr{O}_X(-mK_X))$. Hence the embedding $X \subset \mathbf{P}^{h-1}$ belongs to a bounded family once the multiplier m and the Hilbert polynomial $\chi(\mathscr{O}_X(-lK_X)) \in \mathbf{Q}[l]$ are fixed. Thus thanks to *Matsusaka's big theorem* below, Theorem 9.1.2 is reduced to the boundedness of $\chi(\mathscr{O}_X(-lK_X))$.

Theorem 9.1.3 (Matsusaka's big theorem [310]) *Fix a polynomial $P = P(l)$ in $\mathbf{Q}[l]$. Then there exists an integer m depending only on P such that if H is an ample divisor on a smooth projective variety X with Hilbert polynomial $\chi(\mathscr{O}_X(lH)) = P(l)$, then lH is very ample for all $l \geq m$.*

For a Fano variety X of dimension n, the boundedness of the polynomial $\chi(\mathscr{O}_X(-lK_X))$ is in fact derived from that of the *volume* $\mathrm{vol}(-K_X) = (-K_X)^n$ in Definition 1.7.7.

Proposition 9.1.4 (Kollár–Matsusaka [271]) *Fix integers n, d and e. Then there exist only finitely many possibilities for the polynomial $\chi(\mathscr{O}_X(lH)) \in \mathbf{Q}[l]$ defined by a nef and big divisor H on a smooth projective variety X of dimension n with $(H^n) = d$ and $(K_X \cdot H^{n-1}) = e$.*

Proof We shall explain the proof along [266, VI exercise 2.15.8]. The polynomial $\chi(\mathscr{O}_X(lH))$ is determined by the $n + 1$ values at $l = -1, \ldots, -n - 1$. By Serre duality and Kawamata–Viehweg vanishing, $(-1)^n \chi(\mathscr{O}_X(lH)) = \chi(\mathscr{O}_X(K_X - lH)) = h^0(\mathscr{O}_X(K_X - lH))$ for $l < 0$. Thus it suffices to bound $h^0(\mathscr{O}_X(K_X - lH))$ for $l = -1, \ldots, -n - 1$ in terms of n, (H^n) and $(K_X \cdot H^{n-1})$. More generally for a divisor D and a nef and big divisor H on X, we shall bound $h = h^0(\mathscr{O}_X(D))$ in terms of n, (H^n) and $(D \cdot H^{n-1})$.

We may assume that D is mobile, in which the linear system $|D|$ defines a rational map $\varphi \colon X \dashrightarrow \mathbf{P}^{h-1}$. We shall mean by D the general member of the system $|D|$. Take a resolution $\mu \colon X' \to X$ which resolves the indeterminacy of φ. Namely, the linear system $|D'|$ of the strict transform D' of D defines a morphism $\varphi' \colon X' \to \mathbf{P}^{h-1}$ which factors through φ as $\varphi' = \varphi \circ \mu$. Then D' is free and $h = h^0(\mathscr{O}_{X'}(D'))$. Replacing (X, H, D) by (X', μ^*H, D'), we may assume that D is free and $\varphi \colon X \to \mathbf{P}^{h-1}$ is a morphism.

Let m be the dimension of the image $\varphi(X)$ and let d be the degree of $\varphi(X)$ in $\mathbf{P}^{h-1}$. Since $\varphi(X)$ is not contained in any hyperplane in $\mathbf{P}^{h-1}$, one has $h - m \leq d$ from Remark 6.4.15. The degree d appears in the relation $D^m \equiv dF$ of $(n-m)$-cycles on X where F is the general fibre of φ. Since H is big, Kodaira's lemma finds an ample $\mathbf{Q}$-divisor A on X such that $A \leq H$. Then the number $(H^{n-m} \cdot F)$ satisfies $0 < (A^{n-m} \cdot F) \leq (A^{n-m-1} \cdot H \cdot F) \leq \cdots \leq (H^{n-m} \cdot F)$ since H is nef. Hence

$$h - m \leq d \leq d(H^{n-m} \cdot F) = (D^m \cdot H^{n-m}).$$

In general for ample divisors A and B on a smooth projective variety of dimension n, the *Hodge index theorem* stated below yields the inequality

$$\frac{(A^i \cdot B^{n-i})}{(A^{i-1} \cdot B^{n-i+1})} \leq \frac{(A^{i-1} \cdot B^{n-i+1})}{(A^{i-2} \cdot B^{n-i+2})}$$

for $2 \leq i \leq n$. Hence

$$\frac{(A^i \cdot B^{n-i})}{(B^n)} = \prod_{j=1}^{i} \frac{(A^j \cdot B^{n-j})}{(A^{j-1} \cdot B^{n-j+1})} \leq \left(\frac{(A \cdot B^{n-1})}{(B^n)} \right)^i$$

for $0 \leq i \leq n$. It follows that $(A^i \cdot B^{n-i}) \cdot (B^n)^{i-1} \leq (A \cdot B^{n-1})^i$ for any nef divisors A and B. In particular, the right-hand side above is bounded as

$$(D^m \cdot H^{n-m}) \leq \frac{(D \cdot H^{n-1})^m}{(H^n)^{m-1}}.$$

Thus h is bounded in terms of n, (H^n) and $(D \cdot H^{n-1})$. □

Theorem 9.1.5 (Hodge index theorem)   *The self-intersection number (D^2) of a divisor D on a smooth projective surface S has signature $(1, \rho(S) - 1)$ as a quadratic form on $N^1(S)$. In other words, for any ample divisor H on S, the self-intersection number is negative-definite on the subspace $H^{\perp}$ of $N^1(S)$ perpendicular to H.*

   The rational connectedness of a Fano variety X means that two general points in X are joined by a rational curve C. The volume $(-K_X)^n$ where $n = \dim X$ turns out to be bounded in terms of $(-K_X \cdot C)$.

Lemma 9.1.6   *Let X be a Fano variety of dimension n and let c be a positive number. If two general points in X are joined by a curve C with $(-K_X \cdot C) \leq c$, then $(-K_X)^n \leq c^n$. Notice that a curve is irreducible in our terminology.*

Proof   We write $v = ((-K_X)^n)^{1/n}$ for brevity. We want the estimate $v \leq c$. Let $\mathfrak{m}$ denote the maximal ideal in $\mathcal{O}_X$ defining a general point x in X. Consider the exact sequence

$$0 \to \mathcal{O}_X(-lK_X) \otimes \mathfrak{m}^e \to \mathcal{O}_X(-lK_X) \to \mathcal{O}_X(-lK_X) \otimes \mathcal{O}_X/\mathfrak{m}^e \to 0$$

for positive integers l and e. The dimension of $H^0(\mathcal{O}_X(-lK_X))$, which equals $\chi(\mathcal{O}_X(-lK_X))$ by Kodaira vanishing, is expressed as $(v^n/n!)l^n + O(l^{n-1})$ with Landau's symbol O. On the other hand, the dimension of $\mathcal{O}_X(-lK_X) \otimes \mathcal{O}_X/\mathfrak{m}^e \simeq \mathcal{O}_X/\mathfrak{m}^e$ equals $\binom{n+e-1}{n} = (1/n!)e^n + O(e^{n-1})$. Thus if l and e are sufficiently large with $e < vl$, then $h^0(\mathcal{O}_X(-lK_X) \otimes \mathcal{O}_X/\mathfrak{m}^e)$ is less than $h^0(\mathcal{O}_X(-lK_X))$ and hence $H^0(\mathcal{O}_X(-lK_X) \otimes \mathfrak{m}^e) \neq 0$.

   Fixing a positive real number ε, we choose l and e so that $(v - \varepsilon)l < e < vl$. A non-zero section of $H^0(\mathcal{O}_X(-lK_X) \otimes \mathfrak{m}^e)$ defines an effective divisor $D \sim -lK_X$ with $\mathrm{mult}_x D \geq e > (v-\varepsilon)l$. Take a general point y in X, which is outside the support of D. Then x and y are joined by a curve C with $(-K_X \cdot C) \leq c$. By $x \in C \not\subset D$, one has

$$(v - \varepsilon)l < \text{mult}_x D \leq (D \cdot C) = (-lK_X \cdot C) \leq cl.$$

Hence $v - \varepsilon < c$ for any $\varepsilon > 0$, that is, $v \leq c$. $\qquad\qquad\square$

Theorem 9.1.2 is now reduced to the existence of a rational curve C with bounded intersection number $(-K_X \cdot C)$ which joins two general points in a Fano variety X. We shall construct C by two steps. Firstly we join two points in X by a chain Γ of rational curves with bounded $(-K_X \cdot \Gamma)$. Secondly we attach extra rational curves D_i to Γ and deform the union $\Gamma \cup \bigcup_i D_i$ to a single curve C with bounded $(-K_X \cdot C)$.

Let X be a smooth projective variety. A *covering family* $\mathfrak{C}$ of rational curves in X means a flat projective morphism $\mathfrak{C} \subset X \times S \to S$ to a variety S such that the fibre $\mathfrak{C}_s \subset X$ at every $s \in S$ is a rational curve and such that the projection $\mathfrak{C} \to X$ has dense image. Such a family exists if and only if X is uniruled. By the theory of Hilbert schemes, there exist countably many closed proper subsets $Z_i \subsetneq X$ such that every rational curve C meeting the complement $X \setminus \Sigma$ of the union $\Sigma = \bigcup_i Z_i$ is a member of a covering family of rational curves in X. In other words, a rational curve through a *very* general point belongs to a covering family.

On a Fano variety, the *bend and break* produces a family of rational curves of bounded degree. The following theorem in any characteristic is found in [274, lemma 3.1].

Theorem 9.1.7 *Let $X \dashrightarrow Y$ be a dominant rational map from a Fano variety of dimension n to a smooth projective variety in any characteristic. Let $\mu \colon X' \to X$ be a resolution which resolves the indeterminacy as a morphism $\pi \colon X' \to Y$. Then there exists a covering family $\mathfrak{C} \subset X' \times S$ of rational curves in X' such that $(-\mu^* K_X \cdot \mathfrak{C}_s) \leq n + 1$ for any curve $\mathfrak{C}_s$ belonging to $\mathfrak{C}$ and such that $\pi_* \mathfrak{C} = (\pi, \text{id}_S)(\mathfrak{C}) \subset Y \times S$ is a covering family of rational curves in Y.*

Let $\mathfrak{C} \subset X \times S$ be a covering family of rational curves in a smooth projective variety X with projections $p \colon \mathfrak{C} \to X$ and $q \colon \mathfrak{C} \to S$. For a point x in X, the locus $L_1(\mathfrak{C}, x) = pq^{-1}qp^{-1}(x)$ in X consists of points joined to x by a single curve $\mathfrak{C}_s$ belonging to $\mathfrak{C}$. In like manner, one can inductively define the locus $L_i(\mathfrak{C}, x) = pq^{-1}qp^{-1}(L_{i-1}(\mathfrak{C}, x))$ which consists of points joined to x by a chain of at most i curves in $\mathfrak{C}$. When x is a general point in X, $L_{i-1}(\mathfrak{C}, x) \neq L_i(\mathfrak{C}, x)$ if and only if $\dim L_{i-1}(\mathfrak{C}, x) < \dim L_i(\mathfrak{C}, x)$. In particular, the locus

$$L(\mathfrak{C}, x) = \bigcup_i L_i(\mathfrak{C}, x)$$

coincides with $L_i(\mathfrak{C}, x)$ for some i at most the dimension of $L(\mathfrak{C}, x)$. Thus if a point in X is connected to the general point x by a chain of rational curves $C_1, \ldots, C_i$ in $\mathfrak{C}$, then one can choose it so that $i \leq \dim L(\mathfrak{C}, x)$.

Theorem 9.1.8 ([274, theorem 3.3]) *Let be X a Fano variety of dimension n over an algebraically closed field in any characteristic. Then any two points in X are joined by a chain Γ of rational curves such that $(-K_X \cdot \Gamma) \leq (2^n - 1)(n+1)$. In particular, X is rationally chain connected.*

Proof Note that Campana [64] also proved the rational chain connectedness independently. We give a proof in characteristic zero. Setting $X_0 = Y_0 = X$, we construct a finite sequence $Y_0 \dashrightarrow Y_1 \dashrightarrow \cdots \dashrightarrow Y_r$ of dominant rational maps of smooth projective varieties inductively in the following manner.

Suppose that we have a dominant map from X to a smooth projective variety Y_i and a resolution $\mu_i \colon X_i \to X$ which resolves its indeterminacy as $\pi_i \colon X_i \to Y_i$. For $X \dashrightarrow Y_i$, we take a covering family $\mathfrak{C}_i$ of rational curves in X_i as in Theorem 9.1.7. Then $(-K_X \cdot \mu_i(C)) \leq n+1$ for any curve C belonging to $\mathfrak{C}_i$, and $\mathfrak{D}_i = \pi_{i*}\mathfrak{C}_i$ is a covering family in Y_i. As above, for a general point $y \in Y_i$, let $L(\mathfrak{D}_i, y) = \bigcup_j L_j(\mathfrak{D}_i, y)$ denote the locus of points in Y_i joined to y by a chain of rational curves in the family $\mathfrak{D}_i$. Consider the rational map $Y_i \dashrightarrow \operatorname{Hilb} Y_i$ to the Hilbert scheme of Y_i which sends y to the closure of $L(\mathfrak{D}_i, y)$, and take a resolution Y_{i+1} of the image of Y_i in $\operatorname{Hilb} Y_i$. We obtain a dominant rational map $Y_i \dashrightarrow Y_{i+1}$. Resolve $X \dashrightarrow Y_{i+1}$ as $X \leftarrow X_{i+1} \to Y_{i+1}$ so that it induces a morphism $X_{i+1} \to X_i$.

Since the dimension n_i of Y_i decreases strictly as far as $n_i > 0$, we attain $n_r = 0$ for some $r \leq n$. For two points x and y in the same fibre of $\pi_i \colon X_i \to Y_i$, we claim that $\mu_i(x)$ and $\mu_i(y)$ are joined by a chain Γ of rational curves in X with $(-K_X \cdot \Gamma) \leq (2^{n-n_i} - 1)(n+1)$. The claim for $i = r$ is what we want.

This is trivial for $i = 0$. We shall prove the claim for $i + 1$ assuming that for i. Because the property of $(x, y) \in X_{i+1} \times_{Y_{i+1}} X_{i+1}$ that $\mu_{i+1}(x)$ and $\mu_{i+1}(y)$ are joined by Γ as in the claim is a closed condition, we may assume that x is a general point in X_{i+1}. Write $x', y' \in X_i$ for the images of x, y by $X_{i+1} \to X_i$.

The condition $\pi_{i+1}(x) = \pi_{i+1}(y)$ means that $L(\mathfrak{D}_i, \pi_i(x')) = L(\mathfrak{D}_i, \pi_i(y'))$, which is of dimension $n_i - n_{i+1}$. Prior to the theorem, we have seen that $L(\mathfrak{D}_i, \pi_i(x')) = L_{n_i - n_{i+1}}(\mathfrak{D}_i, \pi_i(x'))$. Hence $\pi_i(x')$ and $\pi_i(y')$ are joined by a chain of images $\pi_i(C_1), \ldots, \pi_i(C_l)$ of rational curves $C_1, \ldots, C_l$ in $\mathfrak{C}_i$ with $l \leq n_i - n_{i+1}$. Precisely with $s_0 = x'$ and $t_l = y'$, there exist points t_{j-1} and s_j in C_j such that $\pi_i(s_j) = \pi_i(t_j)$ for all j. Remark that $(-K_X \cdot \mu_i(C_j)) \leq n+1$.

By inductive hypothesis, the two points $\mu_i(s_j)$ and $\mu_i(t_j)$ are joined by a chain Γ_j of rational curves with $(-K_X \cdot \Gamma_j) \leq (2^{n-n_i} - 1)(n+1)$. Then the union $\Gamma = \Gamma_0 \cup \mu_i(C_1) \cup \Gamma_1 \cup \cdots \cup \mu_i(C_l) \cup \Gamma_l$ connects $\mu_{i+1}(x) = \mu_i(s_0)$ and $\mu_{i+1}(y) = \mu_i(t_l)$ with

$$\frac{1}{n+1}(-K_X \cdot \Gamma) = \frac{1}{n+1}\sum_{j=1}^{l}(-K_X \cdot \mu_i(C_j)) + \frac{1}{n+1}\sum_{j=0}^{l}(-K_X \cdot \Gamma_j)$$
$$\leq l + (l+1)(2^{n-n_i} - 1) \leq 2^l \cdot 2^{n-n_i} - 1 \leq 2^{n-n_{i+1}} - 1.$$

This is the claim for $i + 1$, and the theorem is completed. $\qquad\qquad\square$

The last step will need the notion of freedom of a rational curve.

Definition 9.1.9 Let C be a rational projective curve in a smooth variety X with normalisation $\nu \colon \mathbf{P}^1 \to C$. Decompose the pull-back of the tangent sheaf $\mathscr{T}_X$ as $\nu^*\mathscr{T}_X|_C \simeq \bigoplus_i \mathscr{O}_{\mathbf{P}^1}(a_i)$ with $a_i \in \mathbf{Z}$ by Theorem 1.3.5. We say that C is *free* in X if $a_i \geq 0$ for all i, and say that C is *very free* in X if $a_i \geq 1$ for all i.

We refer the reader to [266, section II.3] for the basic properties of free curves. The freedom of C is an open condition since it is equivalent to the vanishing $H^1(\nu^*\mathscr{T}_X|_C \otimes \mathscr{O}_{\mathbf{P}^1}(-1)) = 0$. This notion behaves well in the deformation theory. In particular, a free rational curve belongs to a covering family.

Proposition 9.1.10 *Let C be a rational projective curve in a smooth variety X in any characteristic. Let $\nu \colon \mathbf{P}^1 \to C$ be the normalisation. Then C is free if and only if the evaluation morphism $e \colon \mathbf{P}^1 \times \mathrm{Hom}(\mathbf{P}^1, X) \to X$ is smooth about $\mathbf{P}^1 \times [\nu]$. In fact C is free if the induced map $T_{\mathbf{P}^1,p} \oplus T_{\mathrm{Hom}(\mathbf{P}^1,X),[\nu]} \to T_{X,\nu(p)}$ of tangent spaces is surjective for some $p \in \mathbf{P}^1$.*

In characteristic zero, the proposition implies that the general member of a covering family of rational curves is free. It follows that a rational curve through a very general point is free. More precisely, there exists a countable union Σ of closed proper subsets of X such that an arbitrary rational curve meeting $X \setminus \Sigma$ is free. Similar results hold for very free curves. For example for a fixed point x in X, the very general member of a covering family of rational curves through x is very free. See [266, II theorem 3.11].

Let C be a projective subscheme of a smooth variety X. Let $\mathfrak{C} \subset X \times T \to T$ be a one-parameter deformation in X of C over the algebraic germ $o \in T$ of a smooth curve. For the fibre $\mathfrak{C}_t$ at $t \in T$, we say that $C = \mathfrak{C}_o$ is *deformed* to $\mathfrak{C}_t$. If all $\mathfrak{C}_t$ pass through a prescribed point c in C, then we say that C is *deformed* to $\mathfrak{C}_t$ with c *fixed*. We introduce some results from deformation theory in any characteristic.

Proposition 9.1.11 ([266, II theorem 7.6]) *Let Γ be a tree of rational curves $C_1, \ldots, C_r$ in a smooth variety X in any characteristic such that Γ has only nodal singularities. For each i, let x_i be a smooth point of Γ lying on C_i. If all*

C_i are free, then Γ is deformed to a free rational curve with x_1 fixed. If all C_i are very free, then Γ is deformed to a very free rational curve with all x_i fixed.

Consider the union $E = C \cup \bigcup_{i=1}^{r} D_i$ of smooth rational projective curves $C, D_1, \ldots, D_r$ such that C intersects D_i transversally at one point x_i and such that E is smooth outside $x_1, \ldots, x_r$, where $x_i \neq x_j$ if $i \neq j$.

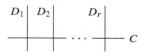

The scheme E is called a *comb* with r teeth. By Theorem 4.4.5, it has the miniversal deformation space $\mathrm{Def}\, E$ smooth over $\prod_i \mathrm{Def}(x_i \in E) \simeq \mathfrak{D}^r$, where $\mathrm{Def}(x_i \in E) \simeq \mathfrak{D}^1$ is given by Example 4.4.3. The relative dimension of $\mathrm{Def}\, E \to \mathfrak{D}^r$ is the maximum of $r - 3$ and 0 since the central fibre comes from deformations of $\mathbf{P}^1$ with r marked points.

The *gluing lemma* [266, II theorem 7.9], [274, theorem 4.2] on a birational image of a comb will play a central role in our last step.

Theorem 9.1.12 (Gluing lemma) *Let X be a smooth projective variety of dimension n in any characteristic. Let C be a rational curve in X and let $x_1, \ldots, x_r$ be distinct points in C. Let $D_1, \ldots, D_r$ be free rational curves in X such that D_i intersects C at x_i for all i. Take the normalisation $\nu \colon \mathbf{P}^1 \to C$ and fix two points p and q in $\mathbf{P}^1$ outside $\bigcup_{i=1}^{r} \nu^{-1}(x_i)$. Suppose that $r \geq n + 2 + (K_X \cdot C)$ and that the scheme $\mathrm{Hom}(\mathbf{P}^1, X, \nu|_{\{p,q\}})$ parametrising morphisms $f \colon \mathbf{P}^1 \to X$ with $f|_{\{p,q\}} = \nu|_{\{p,q\}}$ has local dimension one at ν. Then there exists a non-empty subset J of $\{1, \ldots, r\}$ such that $C \cup \bigcup_{i \in J} D_i$ is deformed to a single rational curve with $\nu(p)$ and $\nu(q)$ fixed.*

Remark 9.1.13 In the theorem if $\mathrm{Hom}(\mathbf{P}^1, X, \nu|_{\{p,q\}})$ has local dimension at least two at ν, then the bend and break asserts that the one-cycle C is algebraically equivalent to a sum $C' = \sum_i C_i'$ of plural rational curves C_i' such that the support of C' contains $\nu(p)$ and $\nu(q)$. See [101, proposition 3.2].

Theorem 9.1.14 ([274, theorem 4.3]) *Let X be a smooth projective variety of dimension n in characteristic zero. Let H be an ample divisor on X such that $H - K_X$ is nef. Suppose that any two points x and y in X are joined by a chain Γ of rational curves with $(H \cdot \Gamma) \leq d$. Then there exists a constant $c = c(n, d)$ depending only n and d such that any x and y are joined by a single very free rational curve C with $(H \cdot C) \leq c$. In particular, X is rationally connected.*

Proof We follow [266, IV complement 3.10.1]. The ground field may be assumed to be uncountable, which guarantees the existence of a very general point.

Let y be an arbitrary point in X. We shall prove that y is joined to a very general point z in X by a rational curve C with bounded $(H \cdot C)$. This C is very free as explained posterior to Proposition 9.1.10. For any point y' in X, y' and z are likewise joined by a very free rational curve C' with bounded $(H \cdot C')$. By Proposition 9.1.11, the union $C \cup C'$ is deformed to a single very free curve with y and y' fixed, which will establish the theorem.

Let x be a very general point in X. We join x and y by a chain of rational curves $C_1, \ldots, C_l$ with $\sum_{i=1}^{l}(H \cdot C_i) \le d$ endowed with a sequence of distinct points $x = x_0, x_1, \ldots, x_{l-1}, x_l = y$ such that $x_{i-1}, x_i \in C_i$. We may choose this chain maximally in the sense that no C_i is algebraically equivalent to a sum Σ of plural rational curves with $x_i, x_{i+1} \in \Sigma$. Then by Remark 9.1.13, the local dimension of $\mathrm{Hom}(\mathbf{P}^1, C_i, v_i|_{\{0,\infty\}})$ at the normalisation $v_i \colon \mathbf{P}^1 \to C_i$ with $v_i(0) = x_{i-1}$ and $v_i(\infty) = x_i$ is one.

Set $c_1 = d$ and define $c_1, \ldots, c_d$ inductively by the recurrence relation

$$c_{i+1} = d + (n + 2 + d)c_i.$$

We shall inductively prove the existence of a rational curve C_i' through x_i, x_{i+1} and a very general point in X such that $(H \cdot C_i') \le c_i$. Note that such C_i' is free. We can take $C_1' = C_1$ since x is a very general point.

Suppose that we have obtained C_i'. It follows from Proposition 9.1.10 that C_i' is deformed to a free rational curve D through a very general point in C_{i+1} and a very general point in X. Attach to C_{i+1} such curves $D_1, \ldots, D_{n+2+d}$ and obtain the union $C_{i+1} \cup \bigcup_{j=1}^{n+2+d} D_j$, where $d \ge (H \cdot C_{i+1}) \ge (K_X \cdot C_{i+1})$ since $H - K_X$ is nef. By Theorem 9.1.12, there exists a subscheme $C_{i+1} \cup \bigcup_{j \in J} D_j$ with non-empty J deformed to a single rational curve C_{i+1}' with x_i, x_{i+1} fixed. The deformation C_{i+1}' passes through a very general point if it is chosen very generally, and it is a desired curve. Indeed, the intersection number

$$(H \cdot C_{i+1}') = (H \cdot C_{i+1}) + \sum_{j \in J}(H \cdot D_j) = (H \cdot C_{i+1}) + N(H \cdot C_i'),$$

where N is the cardinality of J, is at most $d + (n + 2 + d)c_i = c_{i+1}$.

Let $c = \sum_{i=1}^{d} c_i$. Then x and y are joined by the chain Γ of free rational curves $C_1', \ldots, C_l'$ with $(H \cdot \Gamma) \le c$. By Proposition 9.1.11, Γ is deformed to a single free rational curve C with y fixed. It passes through a very general point and satisfies $(H \cdot C) \le c$. This completes the theorem. $\square$

Proof of Theorem 9.1.2 By Theorem 9.1.3 and Proposition 9.1.4, it suffices to bound the volume $(-K_X)^n$ for a Fano variety X of dimension n. This follows from Theorems 9.1.8 and 9.1.14 with Lemma 9.1.6. $\square$

The boundedness of $\mathbf{Q}$-Fano varieties stated as Theorem 9.1.17 had been known as the *BAB conjecture* until it was proved by Birkar. Henceforth we work over the field $\mathbf{C}$ of complex numbers.

Definition 9.1.15 Let ε be a positive real number. A klt pair (X, Δ) is said to be *ε-lc* if $a_E(X, \Delta) \geq \varepsilon$ for all divisors E over X.

Definition 9.1.16 A normal projective variety X is said to be *of Fano type* if it admits a klt pair (X, Δ) such that $-(K_X + \Delta)$ is nef and big. If one can choose this Δ so that (X, Δ) is ε-lc, then X is said to be *of ε-lc Fano type*.

Theorem 9.1.17 (Birkar [46], [47]) *Fix a positive integer n and a positive real number ε. Then projective varieties of dimension n and of ε-lc Fano type form a bounded family.*

This was conjectured by Alexeev and the Borisov brothers, which is the origin of the name of the BAB conjecture. It was proved for toric $\mathbf{Q}$-Fano varieties by the Borisov brothers [54] and in dimension two by Alexeev [8]. The boundedness of canonical $\mathbf{Q}$-Fano threefolds was achieved in [239], [275].

We shall describe the outline of the proof of Theorem 9.1.17. By a result of Hacon and Xu [176], the theorem is reduced to finding a positive integer r such that every projective variety X of dimension n and of ε-lc Fano type has a boundary B such that (X, B) is klt with $r(K_X + B) \sim 0$. Such B is called a klt *r-complement* of X.

By the MMP, we may assume that $-K_X$ is nef and big. Weakening the klt condition, we prove the existence of an lc *r-complement* B which forms an lc pair (X, B) with $r(K_X + B) \sim 0$. This is shown simultaneously with the effective birationality stated as Theorem 9.1.18. By the ACC for lc thresholds in Theorem 1.6.11 and the induction on dimension, it is enough to deal with an *exceptional* pair (X, Δ) in the sense that $(X, \Delta + B)$ is klt for any effective $\mathbf{R}$-divisor $B \sim_{\mathbf{R}} -(K_X + \Delta)$. In this situation, the existence of an lc complement is derived from Theorem 9.1.18.

Theorem 9.1.18 ([46, theorem 1.2]) *Fix a positive integer n and a positive real number ε. Then there exists an integer m depending only n and ε such that if X is an ε-lc weak $\mathbf{Q}$-Fano variety of dimension n, then the linear system $|-mK_X|$ defines a birational map from X to the image.*

Take a positive integer l such that $(2n)^n < (-lK_X)^n$. By an argument as in the proof of Lemma 9.1.6, we cover X by a family of subvarieties G with the following property: for general points x and y in X, there exists an effective $\mathbf{Q}$-divisor $\Delta \sim_{\mathbf{Q}} -(l+1)K_X$ such that (X, Δ) is lc at x with a unique lc centre G and such that (X, Δ) is not klt at y. Taking a family of G of minimal dimension

n_G, one can bound the volume $(-lK_X|_G)^{n_G}$ from above. Actually one also bounds it from below, by which m/l is bounded for the least integer m such that $|-mK_X|$ defines a birational map. We obtain the boundedness of m by bounding X birationally.

Besides Theorem 9.1.18, the volume $(-K_X)^n$ is also bounded from above. From these the log birational boundedness of pairs of X and an lc r-complement B follows. Precisely speaking, there exist a smooth projective morphism $V \to S$ of smooth complex schemes and a divisor W on V relatively snc over S such that for every (X, B), the fibre (V_s, W_s) at some $s \in S$ has a birational map $V_s \dashrightarrow X$ for which W_s contains all exceptional divisors of $V_s \dashrightarrow X$ and the support of the strict transform of B. Take the (not necessarily effective) **Q**-divisor C_s on V_s such that the pull-backs of $K_{V_s} + C_s$ and $K_X + B$ coincide on a common resolution. Using Theorem 9.1.19 below, one can perturb C_s to make a **Q**-divisor the push-forward of which becomes an klt r-complement of X, after replacing r uniformly. This completes Theorem 9.1.17.

Theorem 9.1.19 ([47, theorem 1.6]) *Fix a positive integer n and a positive real number ε. Then there exists a positive real number t depending only n and ε such that if (X, Δ) is an ε-lc projective pair of dimension n with $-(K_X + \Delta)$ nef and big, then $(X, \Delta + tB)$ is lc for any effective **R**-divisor $B \sim_{\mathbf{R}} -(K_X + \Delta)$.*

After a slight modification of the statement, this theorem is reduced to the boundedness of the multiplicity of a divisor E over X which computes the minimal log discrepancy $\varepsilon/2$ of $(X, \Delta + sB)$ with an appropriate real number s. Bringing this to the setting where X is smooth and Δ has snc support, one can assume that E is obtained by a sequence of toroidal blow-ups. Then the assertion is a consequence of the Borisov brothers' BAB in the toric case.

In the BAB theorem, the restriction by ε is indispensable.

Example 9.1.20 Take the Hirzebruch surface $\Sigma_n \to \mathbf{P}^1$ for $n \geq 1$ and contract the negative section e with $e^2 = -n$ to obtain a surface $S_n \simeq \mathbf{P}(1, 1, n)$. Then $K_{\Sigma_n/S_n} = (2/n - 1)e$ and hence S_n is a **Q**-Fano surface which is $(2/n)$-lc if $n \geq 2$. The surfaces S_n for $n \geq 1$ never form a bounded family because S_n has a quotient singularity of type $\frac{1}{n}(1, 1)$.

Actually without the restriction by ε, the statement fails even if one replaces boundedness by birational boundedness. We explain an example of a birationally unbounded family of **Q**-Fano threefolds due to Lin [294]. Okada [375] constructed an example in dimension at least six.

Example 9.1.21 Fix a positive integer n. Following Lin, we shall construct a **Q**-Fano threefold $Y = Y_n$ which is $(2/n)$-lc if $n \geq 2$. It will be obtained by a contraction $\varphi \colon X \to Y$ of the total space of a standard conic bundle $\pi \colon X \to \mathbf{P}^2$ with smooth discriminant divisor Δ of degree $2n + 3$.

Define the $\mathbf{P}^2$-bundle $\pi_P \colon P = \mathbf{P}(\mathscr{E}) \to \mathbf{P}^2$ by $\mathscr{E} = \mathscr{O}_{\mathbf{P}^2}^{\oplus 2} \oplus \mathscr{O}_{\mathbf{P}^2}(n)$. Let h be a divisor on P with $\mathscr{O}_P(h) = \mathscr{O}_P(1)$ and let f be a fibre of π_P. The projection $\mathscr{E} \twoheadrightarrow \mathscr{O}_{\mathbf{P}^2}^{\oplus 2}$ defines a subvariety E_P of P which is a trivial $\mathbf{P}^1$-bundle over $\mathbf{P}^2$. Similarly to Example 8.3.5, P is the toric variety defined from the homogeneous coordinate ring $\mathbf{C}[x_0, x_1, x_2, y_0, y_1, y_2]$ and the irrelevant ideal $(x_0, x_1, x_2) \cap (y_0, y_1, y_2)$ with the bigrading with respect to $w_h = (0, 0, 0, 1, 1, 1)$ and $w_f = (1, 1, 1, 0, 0, -n)$.

Recall that for a smooth subvariety Z of a smooth variety X, the *normal sheaf* $\mathscr{N}_{Z/X}$ of Z in X is defined as $\mathscr{H}om_Z(\mathscr{I}/\mathscr{I}^2, \mathscr{O}_Z)$ with the ideal sheaf $\mathscr{I}$ in $\mathscr{O}_X$ defining Z, which fits into the exact sequence

$$0 \to \mathscr{I}_Z \to \mathscr{I}_X \otimes \mathscr{O}_Z \to \mathscr{N}_{Z/X} \to 0.$$

From the relative Euler sequence, $\mathscr{N}_{E_P/P} \simeq (\mathscr{I}_{P/\mathbf{P}^2}|_{E_P})/\mathscr{I}_{E_P/\mathbf{P}^2}$ is isomorphic to $p^*\mathscr{O}_{\mathbf{P}^2}(-n) \otimes q^*\mathscr{O}_{\mathbf{P}^1}(1)$ for the projections $p = \pi_P|_{E_P}$ and q to $\mathbf{P}^2$ and $\mathbf{P}^1$ from $E_P \simeq \mathbf{P}^1 \times \mathbf{P}^2$. The Picard number of P is two and the closed cone $\overline{\mathrm{NE}}(P)$ of curves is generated by the classes of a line in $f \simeq \mathbf{P}^2$ and a line in a fibre of q. The linear system $|h|$ gives the contraction $\varphi_P \colon P \to Y_P$ associated with the latter ray.

Let X be the general member of $|2h + f|$. The restriction $\pi = \pi_P|_X \colon X \to \mathbf{P}^2$ is a conic bundle given by $\sum_{i,j=0}^2 a_{ij} y_i y_j$ with homogeneous polynomials $a_{ij} = a_{ji}$ in x_0, x_1, x_2 where $a_{00}, a_{01}, a_{11}, a_{02}, a_{12}, a_{22}$ are of degree 1, 1, 1, $n + 1, n + 1, 2n + 1$ respectively. The discriminant divisor Δ in $\mathbf{P}^2$ is defined by the determinant of the matrix (a_{ij}), which is a smooth curve of degree $2n + 3$ from the generality of X. Since X is an ample divisor on P, one can show that $\rho(X) = 2$ in the same manner as in the proof of Lemma 8.3.6. Thus π is a standard conic bundle.

The restriction $\varphi = \varphi_P|_X \colon X \to Y = \varphi_P(X)$ is a birational contraction which contracts $E = E_P|_X$ to $\mathbf{P}^1$. One has $K_X = (K_P + X)|_X = -\bar{h} + (n-2)\bar{f}$ and $E \sim \bar{h} - n\bar{f}$ for $\bar{h} = h|_X$ and $\bar{f} = f|_X$. Hence $\varphi^* K_Y = K_X + (1 - 2/n)E \sim_{\mathbf{Q}} -(2/n)\bar{h}$. It follows that Y is $(2/n)$-lc $\mathbf{Q}$-Fano for $n \geq 2$.

Theorem 9.1.22 (Lin) *The $\mathbf{Q}$-Fano threefolds Y_n for $n \geq 1$ in Example 9.1.21 form a birationally unbounded family. More precisely, there exists no morphism $V \to S$ of complex schemes such that every Y_n is birational to the fibre $V_s = V \times_S s$ at some $s \in S$.*

Proof We write $\pi_n \colon X_n \to \mathbf{P}^2$ for the standard conic bundle in Example 9.1.21 endowed with the log divisorial contraction $X_n \to Y_n$. We shall use the birational superrigidity of π_n for $n \geq 5$ derived from Theorem 7.5.2. Assuming the existence of $v\colon V \to S$ such that Y_n is birational to the fibre $V_n = V \times_S s_n$ at some $s_n \in S$ for infinitely many n, we shall derive a contradiction. We may assume that the points s_n form a dense subset of S. Replacing V and shrinking S, we may assume that v is a smooth projective morphism of smooth quasi-projective varieties.

By Theorem 6.3.9, the fibre V_n birational to the **Q**-Fano threefold Y_n is rationally connected. Then so are all fibres of v by Theorem 6.3.19(ii). We apply the result [276, theorem 12.4.2] on the MMP for a family of threefolds. It asserts that for our family v of uniruled threefolds, after a replacement of S by an étale cover of it, v is birational over S to a flat projective morphism $W \to S$ admitting a contraction $\psi\colon W \to T$ over S such that at any $s \in S$, it induces a birational map $V_s \dashrightarrow W_s$ of fibres and a Mori fibre space $W_s \to T_s$. Consider the fibre $V_n \dashrightarrow W_n \to T_n$ at s_n for $n \geq 5$. Because V_n is birational to the total space X_n of the birationally superrigid Mori fibre space π_n, there exists a square link from $\pi_n\colon X_n \to \mathbf{P}^2$ to the Mori fibre space $\psi_n\colon W_n \to T_n$. In particular, ψ_n is a **Q**-conic bundle and ψ is a conical fibration.

Recall that the discriminant divisor Δ_n of π_n is a smooth curve of degree $2n + 3$. Take a resolution B_n of $\mathbf{P}^2$ which resolves the compatible base map $\mathbf{P}^2 \dashrightarrow T_n$ as a morphism $B_n \to T_n$. By Lemma 7.4.1, there exists a standard conic bundle $X_n' \to B_n$ equivalent to π_n such that the discriminant divisor Δ_n' is isomorphic to the smooth curve Δ_n via $B_n \to \mathbf{P}^2$. Then by Remark 7.5.4, the image Γ_n in T_n of Δ_n' is the divisorial part of discriminant locus of ψ_n. The induced morphism $\Delta_n \simeq \Delta_n' \to \Gamma_n$ is considered to be the normalisation of Γ_n. Hence $h^1(\mathscr{O}_{\Gamma_n}) \geq h^1(\mathscr{O}_{\Delta_n}) = (n + 1)(2n + 1)$.

Let Γ be the divisorial part of the locus in T at which ψ is not smooth. Shrinking S, we may assume that Γ is flat over S and the fibre $\Gamma \times_S s_n$ coincides with Γ_n. Then the arithmetic genus $p_a(\Gamma_n) = h^1(\mathscr{O}_{\Gamma_n})$ must be independent of n, which contradicts the estimate $h^1(\mathscr{O}_{\Gamma_n}) \geq (n + 1)(2n + 1)$. $\qquad\square$

9.2 General Elephants

In this section, we prove the general elephant conjecture in Question 2.4.13 for Gorenstein weak Fano threefolds.

Theorem 9.2.1 (Shokurov [420], Reid [400, theorem 0.5]) *The general elephant of a Gorenstein weak Fano threefold is irreducible and Du Val.*

We prepare two numerical invariants of a Fano threefold, called the Fano index and the genus. They will play fundamental roles in the classification.

Lemma 9.2.2 *Let X be a variety of Fano type. Then $\chi(\mathscr{O}_X) = 1$, $\operatorname{Pic} X \simeq H^2(X, \mathbf{Z})$ and it is torsion-free.*

Proof By Kawamata–Viehweg vanishing, $H^i(\mathscr{O}_X) = 0$ for $i \geq 1$. Hence $\chi(\mathscr{O}_X) = 1$ and $\operatorname{Pic} X = H^1(\mathscr{O}_X^\times) \simeq H^2(X, \mathbf{Z})$. If $\operatorname{Pic} X$ had a torsion of index $r > 1$, then it would define an étale covering $\bar{X} \to X$ of degree r by Definition 2.2.17. By Corollary 2.2.21, $\bar{X}$ is also of Fano type and thus $\chi(\mathscr{O}_{\bar{X}}) = 1$. This contradicts the equality $\chi(\mathscr{O}_{\bar{X}}) = r\chi(\mathscr{O}_X)$ in Theorem 9.2.3.   □

The following is in [141, example 18.3.9]. The reader may refer to [288, example 1.1.30] for the proof in the projective case.

Theorem 9.2.3 *Let $\pi\colon Y \to X$ be an étale morphism of degree d of complete schemes. Then $\chi(\pi^*\mathscr{F}) = d\chi(\mathscr{F})$ for any coherent sheaf $\mathscr{F}$ on X.*

Definition 9.2.4 The *Fano index* of a $\mathbf{Q}$-Fano variety X is the greatest integer m such that $-K_X \sim mH$ by some integral divisor H. When X is factorial, one may replace $-K_X \sim mH$ by $-K_X \equiv mH$ by the above lemma.

Let X be a Fano threefold. The Riemann–Roch formula gives $\chi(\mathscr{O}_X(K_X)) = K_X c_2(X)/12 + \chi(\mathscr{O}_X)$. One has $-K_X c_2(X) = 24$ using the Serre duality $\chi(\mathscr{O}_X(K_X)) = -\chi(\mathscr{O}_X)$ and the equality $\chi(\mathscr{O}_X) = 1$ in Lemma 9.2.2. The dimension $h^0(\mathscr{O}_X(-iK_X))$ for $i \geq 0$, which equals $\chi(\mathscr{O}_X(-iK_X))$ by Kodaira vanishing, is thus computed by the Riemann–Roch formula as

$$h^0(\mathscr{O}_X(-iK_X)) = \frac{1}{12}i(i+1)(2i+1)(-K_X^3) + 2i + 1.$$

In particular, $-K_X^3$ is an even integer.

Definition 9.2.5 The *genus* g of a Fano threefold X is defined by the equality $-K_X^3 = 2g - 2$. It is an integer at least two and satisfies $h^0(-K_X) = g + 2$.

The intersection C of two general elephants S and S' of X has genus g by $\deg K_C = ((K_X + S + S') \cdot S \cdot S') = -K_X^3$, provided that it is a curve. This is the origin of the definition of genus of X.

Shokurov first proved the following assertion in the smooth case and then Reid generalised it to Theorem 9.2.1.

Theorem 9.2.6 (Shokurov [420]) *Let X be a Fano threefold of Fano index m. Take a divisor H on X such that $-K_X \sim mH$. Then the general member of $|H|$ is irreducible and smooth.*

These results are unified in the log Fano case. The final form was obtained by Ambro.

Theorem 9.2.7 (Ambro [11]) *Let (X, Δ) be a klt pair on a projective threefold X. Let H be a nef and big Cartier divisor on X such that $-(K_X + \Delta) \equiv mH$ for a positive real number m. Then H is mobile and $(X, \Delta + S)$ is plt for the general member S of the linear system $|H|$.*

Proof of Theorems 9.2.1 and 9.2.6 from Theorem 9.2.7 We shall derive Theorem 9.2.1 from Theorem 9.2.7. The same argument finds an irreducible Du Val member of $|H|$ in Theorem 9.2.6. Its smoothness will be obtained later in the course of the proof of Proposition 9.2.11.

Let X be a Gorenstein weak Fano threefold. We apply Theorem 9.2.7 to X with $\Delta = 0$ and $H = -K_X$. It asserts that (X, S) is plt for the general elephant S of X. Since $K_X + S$ is Cartier, (X, S) is canonical. In particular, X and S are smooth at the generic point of every curve in S. By Proposition 2.2.23, S is Cohen–Macaulay. Thus S is normal and $K_S = (K_X + S)|_S$. By adjunction, S is canonical and hence Du Val.

The irreducibility of S is equivalent to the connectedness. The cohomology $H^1(\mathscr{O}_X(-S))$, dual to $H^2(\mathscr{O}_X(K_X + S))$, is zero by Kawamata–Viehweg vanishing. Thus $H^0(\mathscr{O}_X) \to H^0(\mathscr{O}_S)$ is surjective and S is connected. □

In Theorem 9.2.7, the non-vanishing of $H^0(\mathscr{O}_X(H))$ is a consequence of the Riemann–Roch theorem.

Lemma 9.2.8 $|H| \neq \emptyset$.

Proof Let $d = H^3$ and $e = -(K_X \cdot H^2) = ((\Delta + mH) \cdot H^2)$, both of which are positive integers. We use the Riemann–Roch formula for the polynomial $p(l) = \chi(\mathscr{O}_X(lH))$. It shows that the top and second-top terms of $p(l)$ are $(d/6)l^3$ and $(e/4)l^2$ respectively.

One has $p(l) = h^0(\mathscr{O}_X(lH))$ for $l \geq 0$ from the Kawamata–Viehweg vanishing for $lH \equiv (K_X + \Delta) + (m + l)H$. Thus $p(0) = 1$ and $p(1) = h^0(\mathscr{O}_X(H))$. On the other hand, $p(-1) = -\chi(\mathscr{O}_X(K_X + H))$ by Serre duality, which equals $-h^0(\mathscr{O}_X(K_X + H))$ by the vanishing on a **Q**-factorialisation of X. In particular, $p(-1) \leq 0$.

Using $p(0) = 1$, one can express the polynomial $p(l) = (d/6)l^3 + \cdots$ as

$$p(l) = \frac{d}{6}l(l + 1)(l + a) + l(1 - p(-1)) + 1$$

with some $a \in \mathbf{Q}$. The coefficient $(1 + a)d/6$ of l^2 equals $e/4$. Hence $1 + a = 3e/2d$ and $h^0(\mathscr{O}_X(H)) = p(1) = e/2 + 2 - p(-1) > 0$. □

By the lemma, there exists a member S of $|H|$. Theorem 9.2.7 will be proved by restricting H to an lc centre Z of $(X, \Delta + tS)$ for the lc threshold t of S. We shall use a result of the non-vanishing on Z.

Theorem 9.2.9 (Kawamata [245, theorem 3.1])   *Let* (X, Δ) *be a klt projective pair of dimension at most two. Let* D *be a nef Cartier divisor on* X *such that* $D - (K_X + \Delta)$ *is nef and big. Then* $H^0(\mathscr{O}_X(D)) \neq 0$.

Proof   By Kodaira's lemma, there exists a sufficiently small effective **R**-divisor B such that $D - (K_X + \Delta + B)$ is ample. Adding B to Δ and perturbing Δ slightly, we may assume that Δ is a **Q**-divisor and that $D - (K_X + \Delta)$ is ample. Take an effective **Q**-divisor $H \sim_{\mathbf{Q}} D - (K_X + \Delta)$ such that $(X, \Delta + H)$ is klt and such that $\varepsilon D \leq H$ for a small positive rational number ε. We may also assume that $(X, \Delta + H - \varepsilon D)$ is klt.

The locus F in $\overline{\mathrm{NE}}(X)$ perpendicular to D is a $(K_X + \Delta)$-negative extremal face. The contraction theorem gives a contraction $\varphi \colon X \to Z$ associated with F, by which D is the pull-back of an ample Cartier divisor D_Z on Z. Theorem 1.4.28 applied to the pair $(X, \Delta + H - \varepsilon D)$ shows that Z admits a klt pair (Z, Δ_Z) with $K_X + \Delta + H \sim_{\mathbf{Q}} \varphi^*(K_Z + \Delta_Z + \varepsilon D_Z) \sim_{\mathbf{Q}} \varphi^* D_Z$. It follows from the exact sequence in Theorem 1.4.7 that $K_Z + \Delta_Z + \varepsilon D_Z \sim_{\mathbf{Q}} D_Z$. Replacing (X, Δ) by (Z, Δ_Z) and D by D_Z, we shall assume that D is ample with $K_X + \Delta + \varepsilon D \sim_{\mathbf{Q}} D$.

There is nothing to prove if X is a point. When X is a curve, D and $D - K_X$ are ample. The assertion is obvious if the genus g of X is zero. If $g \geq 1$, then $h^0(\mathscr{O}_X(D)) \geq \chi(\mathscr{O}_X(D)) > \chi(\mathscr{O}_X(K_X)) = g - 1 \geq 0$, where the strict inequality follows from the ampleness of $D - K_X$.

We shall assume that X is a surface. We replace $K_X + \Delta$ and D by their pull-backs to the minimal resolution of X. This makes X smooth but abandons the ampleness of D. Our D is a semi-ample and big Cartier divisor such that $K_X + \Delta + \varepsilon D \sim_{\mathbf{Q}} D$. By the vanishing and the Riemann–Roch formula,

$$h^0(\mathscr{O}_X(D)) = \chi(\mathscr{O}_X(D)) = \frac{1}{2} D(D - K_X) + \chi(\mathscr{O}_X).$$

Hence $h^0(\mathscr{O}_X(D)) > \chi(\mathscr{O}_X)$ since $D(D - K_X) = D(\Delta + \varepsilon D) \geq \varepsilon D^2 > 0$. We need to discuss the case when $\chi(\mathscr{O}_X)$ is negative.

Assume that $\chi(\mathscr{O}_X) < 0$ henceforth. Then the irregularity $q = h^1(\mathscr{O}_X)$ is at least two. Since the Euler characteristic $\chi(\mathscr{O}_X)$ is a birational invariant of a smooth variety, we deduce from the classification in Theorem 1.3.7 that the MMP from X never ends with a minimal surface. It ends with a Mori fibre space $\pi_Y \colon Y \to C$. In particular, $h^2(\mathscr{O}_X) = h^0(\mathscr{O}_X(K_X)) = 0$ and hence $\chi(\mathscr{O}_X) = 1 - q$. The Leray spectral sequence yields $H^1(\mathscr{O}_C) \simeq H^1(\mathscr{O}_Y) \simeq H^1(\mathscr{O}_X)$. Thus C is a smooth curve of genus $q \geq 2$. We write $\pi \colon X \to Y \to C$.

We have seen that $h^0(\mathscr{O}_X(D)) = D(D - K_X)/2 + \chi(\mathscr{O}_X)$. Take an effective **Q**-divisor $H \sim_{\mathbf{Q}} \varepsilon D$ such that $(X, \Delta + H)$ is klt. Recall that $D \sim_{\mathbf{Q}} K_X + \Delta + H$

is semi-ample. One sees that $D - \pi^* K_C$ is nef by the *semi-positivity theorem* quoted as Theorem 9.2.10. With $D - K_X \sim_{\mathbf{Q}} \Delta + H \geq 0$, one obtains

$$D(D - K_X) \geq (\pi^* K_C)(D - K_X) \geq (\pi^* K_C)(-K_X) = 2 \deg K_C = 4(q - 1)$$

and thus $h^0(\mathcal{O}_X(D)) \geq 2(q - 1) + (1 - q) = q - 1 \geq 1$. □

We used the following logarithmic version of Kawamata's semi-positivity theorem. We say that a locally free coherent sheaf $\mathcal{E}$ on a complete variety X is *nef* if for any morphism $\varphi \colon C \to X$ from a smooth projective curve, every invertible quotient sheaf of the pull-back $\varphi^* \mathcal{E}$ has non-negative degree. We remark that Ambro's original proof of Theorem 9.2.7 is elementary and does not use the semi-positivity theorem.

Theorem 9.2.10 ([132, theorem 1.11]) *Let $\pi \colon X \to C$ be a surjective morphism from a smooth projective variety to a smooth projective curve. Let Δ be an effective $\mathbf{Q}$-divisor on X such that (X, Δ) is lc outside a finite union of closed fibres of π. Suppose that $l(K_X + \Delta)$ is integral and π-free with a positive integer l. Then the locally free sheaf $\pi_* \mathcal{O}_X(l(K_{X/C} + \Delta))$ is nef for $K_{X/C} = K_X - \pi^* K_C$. In particular, $K_{X/C} + \Delta$ is a nef $\mathbf{Q}$-divisor.*

Proof of Theorem 9.2.7 Note that $-(K_X + \Delta) \sim_{\mathbf{R}} mH$ by Lemma 9.2.2. By the base-point free theorem, $H \sim_{\mathbf{R}} -m^{-1}(K_X + \Delta)$ is semi-ample and there exists a birational contraction $\varphi \colon X \to Y$ such that H is the pull-back of an ample Cartier divisor H_Y on Y. Then $-(K_Y + \Delta_Y) \sim_{\mathbf{R}} mH_Y$ for $\Delta_Y = \varphi_* \Delta$. Replacing (X, Δ) by (Y, Δ_Y) and H by H_Y, we may and shall assume that H is ample. Take the general member S of $|H|$ by Lemma 9.2.8.

Let t be the lc threshold of S on (X, Δ) in Definition 1.6.10 and let Z be a *minimal lc centre* of $(X, \Delta + tS)$ introduced before Theorem 1.4.27. Then there exist an effective $\mathbf{Q}$-divisor $A \sim_{\mathbf{Q}} H$ and a positive constant c such that $(X, B = \Delta + \varepsilon A + (t - c\varepsilon)S)$ is lc and has a unique lc centre Z for any sufficiently small positive ε. By the subadjunction formula, Theorem 1.4.27, Z admits a klt pair (Z, B_Z) such that $K_Z + B_Z \sim_{\mathbf{R}} (K_X + B)|_Z$.

One has $H \sim_{\mathbf{R}} K_X + B + \lambda H$ with $\lambda = 1 + m - t + (c - 1)\varepsilon > 0$. The unique lc centre Z is given by the *multiplier ideal* $\mu_* \mathcal{O}_{X'}(\lceil K_{X'} - \mu^*(K_X + B) \rceil)$ in $\mathcal{O}_X$ for a log resolution $\mu \colon X' \to X$ of (X, B). It follows from *Nadel vanishing* [131, theorem 3.4.2] that the restriction map $H^0(\mathcal{O}_X(H)) \to H^0(\mathcal{O}_Z(H|_Z))$ is surjective. Since $H|_Z \sim_{\mathbf{R}} K_Z + B_Z + \lambda H|_Z$, Theorem 9.2.9 provides $H^0(\mathcal{O}_Z(H|_Z)) \neq 0$. Hence from the above surjection, Z is not contained in the fixed locus of $|H|$. This implies that $t = 1$ and Z is an irreducible component of S because $(X, \Delta + S)$ is plt outside the fixed locus of $|H|$ by Bertini's theorem.

Taking account of all minimal lc centres Z of $(X, \Delta + S)$, one concludes that $(X, \Delta + S)$ is plt. The mobility of H follows from the surjection $H^0(\mathscr{O}_X(H)) \twoheadrightarrow H^0(\mathscr{O}_Z(H|_Z)) \neq 0$ for every component Z of S. $\qquad\square$

Whereas Theorem 9.2.6 does not claim the freedom of the linear system $|H|$, it turns out that H is free except a few special cases. A complete classification will be supplied in the next section.

Proposition 9.2.11 *Let X be a Fano threefold of Fano index m. Take a divisor H on X such that $-K_X \sim mH$. Then H is free except the following cases.*

(i) $m = 2$, $H^3 = 1$ and $|H|$ has a unique base point. $-K_X \sim 2H$ is free.
(ii) $m = 1$ and $|H|$ has base locus $C \simeq \mathbf{P}^1$ scheme-theoretically. The restriction to the general member S of $|H|$ satisfies $H|_S \sim C + gE$ and $(C^2)_S = -2$ where g is the genus of X and E is an elliptic curve such that $|E|$ defines a fibration $S \to \mathbf{P}^1$ with a section C.

Proof Let S be the general member of $|H|$. Recall that the smoothness of S in Theorem 9.2.6 has not yet been obtained. We shall prove it simultaneously with the proposition.

We have seen that S is irreducible and Du Val. By Bertini's theorem, S is smooth outside the base locus of $|H|$. The Kodaira vanishing $H^1(\mathscr{O}_X(H - S)) = 0$ yields the surjection $H^0(\mathscr{O}_X(H)) \twoheadrightarrow H^0(\mathscr{O}_S(H_S))$ for $H_S = H|_S$. In particular, the base locus of $|H|$ coincides with that of $|H_S|$ on S.

One has $K_S = (K_X + S)|_S \sim (1 - m)H_S$. If $m \geq 2$, then $-K_S$ is ample. In this case by Remark 8.1.4, if H_S is not free, then $K_S^2 = 1$, $H_S \sim -K_S$, $|H_S|$ has a unique base point, at which S is smooth, and $2H_S$ is free. Then $m = 2$, $H^3 = K_S^2 = 1$ and S is smooth. This is the case (i).

Now suppose that $m = 1$. Then $K_S \sim 0$. By Kodaira vanishing, $H^1(\mathscr{O}_X) = 0$ and $H^2(\mathscr{O}_X(-S)) = H^1(\mathscr{O}_X(K_X + S))^\vee = 0$. Hence $H^1(\mathscr{O}_S) = 0$. We take the minimal resolution $\mu \colon S' \to S$. Then $H^1(\mathscr{O}_{S'}) = H^1(\mathscr{O}_S) = 0$ since S has rational singularities. Thus S' is a *K3 surface* in Theorem 1.3.7(i), characterised by $K_{S'} \sim 0$ and $q = h^1(\mathscr{O}_{S'}) = 0$.

We have the surjection $H^0(\mathscr{O}_X(H)) \twoheadrightarrow H^0(\mathscr{O}_S(H_S)) \simeq H^0(\mathscr{O}_{S'}(\mu^*H_S))$. If μ^*H_S is not free, then the result [419, lemma 2.1] based on Saint-Donat's work [407] gives the expression $\mu^*H_S \sim C' + bE'$, where $C' \simeq \mathbf{P}^1$ is the base locus of $|\mu^*H_S|$, $(C')_{S'}^2 = -2$, $b \geq 2$ and E' is an elliptic curve such that $|E'|$ defines a fibration $S' \to \mathbf{P}^1$ with a section C'. The image $\mu(C')$ supports the base locus of $|H_S|$ and $C' = \mu^{-1}\mu(C')$. Either μ is isomorphic about C' or $\mu(C')$ is a point at which S is of type A_1.

If μ is isomorphic about C', then S is smooth as so is it about the base locus $\mu(C')$ of $|H|$. From $2g - 2 = (H_S^2)_S = 2b - 2$, one obtains $b = g$ as in the case (ii). If C' were μ-exceptional, then $|H_S|$ would have a unique base point $x = \mu(C')$, at which S is singular, and $b = 2$ by $(\mu^* H_S \cdot C') = 0$. For another general elephant T of X and the general hyperplane section $x \in G$ of X through x, the local intersection number $(S \cdot T \cdot G)_x$ is at least $(\mathrm{mult}_x S)(\mathrm{mult}_x T) = 4$. This contradicts the computation

$$(S \cdot T \cdot G)_x = (\mu^* T|_S \cdot \mu^* G|_S)_{U'} = (2E' \cdot \mu^* G|_S)_{U'} = 2$$

on a neighbourhood U' in S' about C'. □

The general elephant conjecture can fail for **Q**-Fano threefolds.

Example 9.2.12 (Iano-Fletcher [195, example 16.1]) Let $X = X_{12,14} \subset P = \mathbf{P}(2, 3, 4, 5, 6, 7)$ be the intersection of two general hypersurfaces of degrees 12 and 14 in P. It has 10 quotient singularities of types $7 \times \frac{1}{2}(1, 1, 1)$, $2 \times \frac{1}{3}(1, 2, 1)$ and $1 \times \frac{1}{5}(3, 1, 2)$. The threefold X is terminal **Q**-Fano with $\mathscr{O}_X(-K_X) = \mathscr{O}_X(1)$ and $-K_X^3 = 12 \cdot 14/(2 \cdot 3 \cdot 4 \cdot 5 \cdot 6 \cdot 7) = 1/30$. From the vanishing in Definition 2.2.13, the restriction map $H^0(\mathscr{O}_P(1)) \to H^0(\mathscr{O}_X(-K_X))$ is an isomorphism. Thus $|-K_X|$ is empty. Alternatively one may compute $h^0(\mathscr{O}_X(-K_X))$ to be zero by the singular Riemann–Roch formula.

Sano [408, section 4] constructed several examples as below for which the anti-canonical system has a unique member.

Example 9.2.13 As observed in [9, 4.8.3], let $X = X_{14}$ be the hypersurface

$$X = (x^{14} + x^2 y_1^6 + y_1^3 y_2^4 + y_2^7 + y_1 z^4 + w^2 = 0) \subset P = \mathbf{P}(1, 2, 2, 3, 7)$$

with homogeneous coordinates x, y_1, y_2, z, w. It is terminal **Q**-Fano with four quotient singularities of types $3 \times \frac{1}{2}(1, 1, 1)$ and $1 \times \frac{1}{3}(1, 2, 1)$ and one $cA/2$ point $(x^2 + y_2^4 + z^4 + w^2 = 0) \subset \mathscr{D}^4/\mathbf{Z}_2(1, 0, 1, 1)$. One has $\mathbf{C} \simeq H^0(\mathscr{O}_P(1)) \simeq H^0(\mathscr{O}_X(-K_X))$ in the same way as in the preceding example. The unique member S of $|-K_X|$ is given by x as $S = (y_1^3 y_2^4 + y_2^7 + y_1 z^4 + w^2 = 0) \subset \mathbf{P}(2, 2, 3, 7)$. It has an elliptic singularity $(y_2^4 + z^4 + w^2 = 0) \subset \mathscr{D}^3/\mathbf{Z}_2(0, 1, 1)$.

Example 9.2.14 Let $X = X_{15}$ be the hypersurface

$$X = (x^{15} + xy^7 + z^5 + w_1^3 + w_2^3 = 0) \subset \mathbf{P}(1, 2, 3, 5, 5)$$

with homogeneous coordinates x, y, z, w_1, w_2. It is terminal **Q**-Fano with four quotient singularities of types $1 \times \frac{1}{2}(1, 1, 1)$ and $3 \times \frac{1}{5}(1, 2, 3)$. It satisfies $H^0(\mathscr{O}_X(-K_X)) \simeq \mathbf{C}$. The unique member $S = (z^5 + w_1^3 + w_2^3 = 0) \subset \mathbf{P}(2, 3, 5, 5)$ of $|-K_X|$ has a non-lc singularity $(z^5 + w_1^3 + w_2^3 = 0) \subset \mathscr{D}^3/\mathbf{Z}_2(1, 1, 1)$.

Example 9.2.15 Let $X = X_{16}$ be the hypersurface

$$X = (x^{16} + xy^6 z + xz^5 + w^4 + yv^2 = 0) \subset \mathbf{P}(1, 2, 3, 4, 7)$$

with homogeneous coordinates x, y, z, w, v. It is terminal $\mathbf{Q}$-Fano with two quotient singularities of types $\frac{1}{3}(1, 2, 1)$ and $\frac{1}{7}(1, 3, 4)$, two cA_1 points $(xz + w^4 + v^2 = 0) \subset \mathfrak{D}^4$ and one $cA/2$ point $(xz + w^4 + v^2 = 0) \subset \mathfrak{D}^4/\mathbf{Z}_2(1, 1, 0, 1)$. It satisfies $H^0(\mathscr{O}_X(-K_X)) \simeq \mathbf{C}$. The unique member $S = (w^4 + yv^2 = 0) \subset \mathbf{P}(2, 3, 4, 7)$ of $|-K_X|$ is not normal.

9.3 Classification in Special Cases

In this and the next section, we shall provide the classification of Fano three-folds by Iskovskikh after Fano and by Mori and Mukai. The book [212] is an encyclopaedia of this classification. Let X be a Fano threefold of Fano index m, in which $-K_X \sim mH$ by some divisor H on X. This section deals with the case when H is special.

Based on Proposition 9.2.11, we proceed with the classification of X with non-free H in [202, theorems 3.3, 4.2(iv)] and [204, I theorem 6.3].

Theorem 9.3.1 (Iskovskikh)   *Let X be a Fano threefold of Fano index m with genus g and Picard number ρ. Take a divisor H on X such that $-K_X \sim mH$. If H is not free, then one of the following holds.*

(i) *In the case* (i) *in Proposition 9.2.11, $m = 2$, $g = 5$ and X is realised as a sextic in $\mathbf{P}(1, 1, 1, 2, 3)$. $\rho = 1$.*

(ii) *In the case* (ii) *in Proposition 9.2.11, $m = 1$ and $g = 3$ or 4.*

If $g = 3$, then X is isomorphic to the blow-up of a Fano threefold Y in (i) *along an elliptic curve $T \cap T'$ for $2T \sim 2T' \sim -K_Y$. $\rho = 2$.*

If $g = 4$, then $X \simeq S \times \mathbf{P}^1$ with a del Pezzo surface S of degree one. $\rho = 10$.

Let S be the general member of $|H|$, which is a smooth surface by Theorem 9.2.6. If X belongs to the case (i) in Proposition 9.2.11, then S is a del Pezzo surface of degree one with $K_S \sim -H|_S$. Using the surjection $H^0(\mathscr{O}_X(iH)) \twoheadrightarrow H^0(\mathscr{O}_S(-iK_S))$ for $i \geq 0$ and the computation of $h^0(\mathscr{O}_X(iH))$ by the Riemann–Roch formula, one can recover the algebra $\bigoplus_{i \in \mathbf{N}} H^0(\mathscr{O}_X(iH))$ from the anti-canonical ring of S in Theorem 8.1.3. The description in Theorem 9.3.1(i) is derived at once from its structure. The assertion $\rho = 1$ follows from Lemma 9.5.2.

Henceforth we shall address the case (ii) in Proposition 9.2.11. Note that $g \geq 3$ by $0 < (-K_X \cdot C) = ((C + gE) \cdot C)_S = g - 2$. Take the exact sequence

$$0 \to \mathscr{N}_{C/S} \to \mathscr{N}_{C/X} \to \mathscr{N}_{S/X}|_C \to 0$$

of normal sheaves. One has $\mathcal{N}_{C/S} \simeq \mathcal{O}_{\mathbf{P}^1}(-2)$ from $(C^2)_S = -2$ and $\mathcal{N}_{S/X}|_C \simeq \mathcal{O}_{\mathbf{P}^1}(g-2)$ from $(S \cdot C) = g - 2$. Hence with Theorem 1.3.5, one can write

$$\mathcal{N}_{C/X} \simeq \mathcal{O}_{\mathbf{P}^1}(a) \oplus \mathcal{O}_{\mathbf{P}^1}(b)$$

with $-2 \le a \le b$ and $a + b = g - 4$, in which $b \ge \lceil (g-4)/2 \rceil \ge 0$.

Let $\pi \colon Y \to X$ be the blow-up along C and let F be the exceptional divisor. The restriction $\pi_F \colon F \simeq \mathbf{P}(\mathcal{N}_{C/X}^{\vee}) \to C$ is the Hirzebruch surface Σ_{b-a}. We write e for the section with $e^2 = -(b-a)$ and write f for its general fibre. Then

$$-K_Y|_F \sim e + (b+2)f, \qquad -F|_F \sim e - af.$$

The divisor $-K_Y = \pi^*(-K_X) - F$ is free because C is the scheme-theoretic base locus of $|-K_X|$. Let $\varphi \colon Y \to Z$ be the contraction defined by a multiple of $-K_Y$. Since $-K_Y^3 = (gE)_S^2 = 0$, it is of fibre type. From $-K_Y|_F \sim e + (b+2)f$, the restriction $\varphi_F \colon F \to Z$ is either an isomorphism or the contraction of e, where e is contracted if and only if $a = -2$.

We write

$$-K_Y = \varphi^*(\varphi_{F*}(e + (b+2)f)) = T + (b+2)U,$$

where U consists of all divisors U_i in $\varphi^*(\varphi(f))$ such that $\varphi(U_i) = \varphi(f)$ and $T = \varphi^*(\varphi_{F*}(e + (b+2)f)) - (b+2)U$ consists of divisors T_i such that $\varphi(T_i) \subset \varphi(e)$. These T and U are effective and integral. The divisor U may differ from $\varphi^*(\varphi(f))$ only if φ contracts e, in which $\Delta = \varphi^*(\varphi(f)) - U$ may be a $\mathbf{Q}$-divisor. They satisfy the scheme-theoretic equalities $T \cap F = e$ and $U \cap F = f$. In particular, T and U are smooth about F.

$$F \subset Y \supset T$$
$$\pi \qquad \qquad \varphi$$
$$C \subset X \qquad \qquad Z \supset \varphi(e)$$

Lemma 9.3.2 (i) *The divisor T is prime. If T is normal, then $b = 0$.*
(ii) *An effective divisor $U_f \sim U$ with $U_f \cap F = f$ is naturally defined for every fibre f of π_F. If $f \neq f'$, then $U_f \cap U_{f'} = \emptyset$. In particular, U is free.*

Proof If T were not prime, then by $T \cap F = e$, it would contain a divisor disjoint from F. Such a divisor contains a curve l contracted by φ. Then $(-K_Y \cdot l) = 0$ and the curve $\pi(l)$ has $(-K_X \cdot \pi(l)) = ((-K_Y + F) \cdot l) = 0$, which contradicts the ampleness of $-K_X$. Thus T is a prime divisor.

The difference $\Delta = \varphi^*(\varphi(f)) - U$ is independent of the choice of the general fibre f. One can define the effective divisor $U_f = \varphi^*(\varphi(f)) - \Delta \sim U$ for every fibre f. It is integral as so is $U_f - U_{f_1} = \varphi^*(\varphi(f) - \varphi(f_1))$ for general f_1. If

U_f intersected $U_{f'}$ for $f \neq f'$, then $U_f \cap U_{f'}$ would contain a curve disjoint from F which is contracted by φ. This is again a contradiction and hence (ii) holds. Moving f, the subsystem of $|U|$ defines a pencil $\psi: Y \to \mathbf{P}^1$.

Assume that T is normal. We want $b = 0$ to complete (i). The restriction $\psi|_T: T \to \mathbf{P}^1$ has connected fibres. Otherwise from $T \cap U \cap F = e \cap f$, one could derive the existence of a curve disjoint from F and contracted by φ, which contradicts the ampleness of $-K_X$ as above. The general fibre of $\psi|_T$ is an elliptic curve since $K_T = (K_Y + T)|_T = -(b+2)U|_T \sim_{\mathbf{P}^1} 0$. For the minimal resolution T' of T, the $K_{T'}$-MMP over $\mathbf{P}^1$ produces a fibration $\bar{T} \to \mathbf{P}^1$ with $K_{\bar{T}} \sim_{\mathbf{Q},\mathbf{P}^1} 0$. It has $K_{\bar{T}}^2 = 0$ and $H^0(\omega_{\bar{T}}) = H^0(\omega_{T'}) \subset H^0(\omega_T) = 0$. Hence $h^1(\mathscr{O}_{T'}) = h^1(\mathscr{O}_{\bar{T}}) \leq 1$ by the classification result [35, lemma VI.1(a)]. Then $h^1(\mathscr{O}_T) \leq 1$ by the inclusion $H^1(\mathscr{O}_T) \subset H^1(\mathscr{O}_{T'})$ from the Leray spectral sequence.

We have $H^i(\mathscr{O}_Y) = H^i(\mathscr{O}_X) = 0$ for $i \geq 1$. By $-K_Y - T = (b+2)U \sim U_1 + \cdots + U_{b+2}$ with $U_i = U_{f_i}$ for distinct general fibres f_i of π_F, there exists an exact sequence

$$0 \to \mathscr{O}_Y(K_Y + T) \to \mathscr{O}_Y \to \sum_{i=1}^{b+2} \mathscr{O}_{U_i} \to 0.$$

Hence $b + 1 = h^1(\mathscr{O}_Y(K_Y + T))$, which equals $h^2(\mathscr{O}_Y(-T))$ by Serre duality. Further $h^2(\mathscr{O}_Y(-T)) = h^1(\mathscr{O}_T)$ from the exact sequence $0 \to \mathscr{O}_Y(-T) \to \mathscr{O}_Y \to \mathscr{O}_T \to 0$. Consequently $b + 1 = h^1(\mathscr{O}_T) \leq 1$ and thus $b = 0$. $\square$

Corollary 9.3.3 *The image $\varphi(e)$ is a curve. In other words, $a \geq -1$.*

Proof If $\varphi(e)$ were a point, then $a = -2$ and $b = g - 2 \geq 1$. Hence by Lemma 9.3.2(i), T must be singular along some curve. This curve is disjoint from F and contracted by φ, which contradicts the ampleness of $-K_X$. $\square$

It seems that Iskovskikh's original proof ignores the possibility for $T \to \varphi(e)$ to be a cuspidal fibration, in which the estimate $h^1(\mathscr{O}_T) \leq 1$ may fail as explained in [400, counterexample 4.11]. We shall adopt the argument by Jahnke and Radloff [217] which uses the theory of extremal contractions. They classified Gorenstein Fano threefolds with non-free anti-canonical divisor.

Proof of Theorem 9.3.1(ii) By virtue of Corollary 9.3.3, one has the isomorphism $\varphi_F: F \simeq Z \simeq \Sigma_{b-a}$ with $a \geq -1$. Recall that $-K_Y|_F = e + (b+2)f$ and $-F|_F = e - af$. We write $T_X = \pi(T)$ and $U_X = \pi(U)$. Then $T \simeq T_X$, $\pi^* T_X = T + F$ and $\pi^* U_X = U$. Lemma 9.3.2(ii) gives a pencil $\psi: Y \to \mathbf{P}^1$.

Step 1 First we treat the case $a < b$. The divisor $T = \varphi^*(\varphi(e))$ is trivial on the locus in $\overline{\mathrm{NE}}(Y)$ perpendicular to the free divisor $-K_Y$. Further $(K_Y \cdot e) =$

$-a - 2 < 0$ and $(T \cdot e) = a - b < 0$. Hence by the cone theorem, there exists an extremal ray of $\overline{\mathrm{NE}}(Y)$ which is negative with respect to both K_Y and T. Consider the associated extremal contraction. It is not small by Theorem 5.5.1. Since $-T$ is relatively ample, it is a divisorial contraction which contracts T. It does not contract a fibre λ of $\psi|_T : T \to \mathbf{P}^1$ since $(T \cdot \lambda) = (-K_Y \cdot \lambda) \geq 0$. Hence T is contracted to a curve. Then T is smooth by Theorem 4.1.9, and $(a, b) = (-1, 0)$ by Lemma 9.3.2(i). In this case, $g = 3$, $F \simeq \Sigma_1$, $-K_Y = T + 2U$ and $-F|_F = e + f$.

The divisor $T + U$ is the pull-back of the free divisor $\varphi(e) + \varphi(f)$ on Z. With $(T + U + F)|_F \sim 0$, one sees that $T + U + F$ is nef. It is big by $(T + U + F)^3 = (T + U)^2 F = 1$. Then $T_X + U_X$ is also nef and big by $\pi^*(T_X + U_X) = T + U + F$. By the base-point free theorem, $T_X + U_X$ is semi-ample and a multiple of it defines a contraction α_0 from X.

Since $(T_X + U_X) T_X U_X = (T + F + U) TU = 1$, α_0 does not contract the curve $D_X = T_X \cap U_X$. Since $T_X - K_X = 2(T_X + U_X)$, every curve l in X contracted by α_0 has $(T_X \cdot l) = (K_X \cdot l) < 0$. Hence α_0 is isomorphic outside T_X and factors through an extremal contraction $\alpha : X \to W$ such that $-K_X$ and $-T_X$ are both α-ample. This α is a divisorial contraction which contracts T_X to the curve $D = \alpha(D_X)$. It follows that $\alpha_0 = \alpha$. Again by Theorem 4.1.9, W and D are smooth and X is the blow-up of W along D.

Let $U_W = \alpha(U_X)$. Then $\alpha^* U_W = T_X + U_X$ and U_W is ample. From $-K_W \sim 2U_W$ and $U_W^3 = (T + U + F)^3 = 1$, W is a Fano threefold which belongs to the case (i) in Proposition 9.2.11. Using $U_X \cap U_X' = \emptyset$ for $U_X' = \pi(U')$ with $U \sim U'$ from Lemma 9.3.2(ii), one obtains $D = U_W \cap U_W'$ for $U_W' = \alpha(U_X')$. Therefore X is as in Theorem 9.3.1(ii) with $g = 3$.

Step 2 Second we treat the case $a = b$. Then $F \simeq \mathbf{P}^1 \times \mathbf{P}^1$. One can choose the section e generally so that $T = \varphi^* \varphi(e)$ is smooth. Thus $(a, b) = (0, 0)$ by Lemma 9.3.2(i). In this case, $g = 4$, $-K_Y = T + 2U$ and $-F|_F = e$.

By Lemma 1.1.1, $\pi_F \circ \varphi_F^{-1} \circ \varphi : Y \to Z \simeq F \to C \simeq \mathbf{P}^1$ factors through X. Let $\beta : X \to \mathbf{P}^1$ be the induced morphism. Its general fibre is U_X. One can write $C = T_X \cap T_X'$ by the image T_X' of T' obtained from another section e' in the same manner as T. The divisor T_X is nef by $(T_X \cdot C) = ((T + F) \cdot e) = 0$, and it is further semi-ample by the base-point free theorem. Let $\gamma : X \to V$ be the contraction defined by a multiple of T_X, which is of fibre type as $T_X^3 = (T_X \cdot C) = 0$.

Because $-K_X \sim T_X + 2U_X$ is ample, no curves are contracted by both β and γ. Hence $(\beta, \gamma) : X \to \mathbf{P}^1 \times V$ is finite. Since $\dim V \leq 2$, it is surjective. The fibres U_X of β and C of γ have intersection number $(U_X \cdot C) = 1$. Thus $X \simeq \mathbf{P}^1 \times V$ and $U_X \simeq V$. Here U_X is a smooth del Pezzo surface of degree $(K_X + U_X)^2 U_X = 1$. Therefore X is as in Theorem 9.3.1(ii) with $g = 4$. $\square$

Next we classify Fano threefolds of higher Fano index.

Theorem 9.3.4 (Iskovskikh [202, theorem 4.2], T. Fujita [137])   *Let X be a Fano threefold of Fano index $m \geq 2$ with Picard number ρ. Take a divisor H with $-K_X \sim mH$ and let $d = H^3$. Then one of the following holds. Except the case* (iii), *H is free and defines a morphism $\varphi \colon X \to \mathbf{P}^{h-1} = \mathbf{P}H^0(\mathcal{O}_X(H))$ with $h = h^0(\mathcal{O}_X(H))$. Except the cases* (iii) *and* (iv), *φ is an embedding and $X \simeq \varphi(X) \subset \mathbf{P}^{h-1}$ is projectively normal as described below. In* (i), (ii), (vii) *to* (x), *$X \subset \mathbf{P}^{h-1}$ is unique up to automorphism of $\mathbf{P}^{h-1}$.*

   (i) $(m, d) = (4, 1)$. $X = \mathbf{P}^3$. $\rho = 1$.
  (ii) $(m, d) = (3, 2)$. $X = Q \subset \mathbf{P}^4$ *is a quadric.* $\rho = 1$.
 (iii) $(m, d) = (2, 1)$. *X is realised as a sextic in* $\mathbf{P}(1, 1, 1, 2, 3)$ *as in Theorem* 9.3.1(i). $\rho = 1$.
  (iv) $(m, d) = (2, 2)$. $\varphi \colon X \to \mathbf{P}^3$ *is a double cover ramified along a smooth quartic, in which X is realised as a quartic in* $\mathbf{P}(1, 1, 1, 1, 2)$. $\rho = 1$.
   (v) $(m, d) = (2, 3)$. $X = X_3 \subset \mathbf{P}^4$ *is a cubic.* $\rho = 1$.
  (vi) $(m, d) = (2, 4)$. $X = X_{2,2} \subset \mathbf{P}^5$ *is a complete intersection of two quadrics.* $\rho = 1$.
 (vii) $(m, d) = (2, 5)$. $X \subset \mathbf{P}^6$ *is a codimension-three linear section of the Plücker embedding* $\mathrm{Gr}(2, 5) \subset \mathbf{P}^9$ *of the Grassmannian.* $\rho = 1$.
(viii) $(m, d) = (2, 6)$. $X \subset \mathbf{P}^7$ *is a hyperplane section of the Segre embedding* $\mathbf{P}^2 \times \mathbf{P}^2 \subset \mathbf{P}^8$, *in which each projection $X \to \mathbf{P}^2$ is isomorphic to the $\mathbf{P}^1$-bundle* $\mathbf{P}(\mathcal{T}_{\mathbf{P}^2})$. $\rho = 2$.
  (ix) $(m, d) = (2, 6)$. $X \subset \mathbf{P}^7$ *is the Segre embedding* $\mathbf{P}^1 \times \mathbf{P}^1 \times \mathbf{P}^1 \subset \mathbf{P}^7$. $\rho = 3$.
   (x) $(m, d) = (2, 7)$. $X \subset \mathbf{P}^8$ *is the projection of the Veronese embedding* $\mathbf{P}^3 \subset \mathbf{P}^9$ *from a point in* $\mathbf{P}^3$, *in which X is isomorphic to the blow-up of* $\mathbf{P}^3$ *at this point.* $\rho = 2$.

We explain the description in the case (vii).

Example 9.3.5   Fix a complex vector space A of dimension n. The *Grassmannian* $G = \mathrm{Gr}(r, n) = \mathrm{Gr}(r, A)$ parametrises vector subspaces $V \subset A$ of dimension r. A good introduction is [120, chapters 3, 4]. It is a homogeneous space for the group $\mathrm{GL}(n, \mathbf{C})$. The *Plücker embedding* $G \subset \mathbf{P}(\bigwedge^r A^\vee) = \mathbf{P}^N$ with $N = \binom{n}{r} - 1$ is given by $\bigwedge^r V \subset \bigwedge^r A$. This is projectively normal since the representation $H^0(\mathcal{O}_G(l))$ of $\mathrm{GL}(n, \mathbf{C})$ is irreducible for $l \geq 0$ by the *Borel–Weil theory*, for which the reader may refer to [68] or [256]. It satisfies $\mathrm{Pic}\, G = \mathbf{Z}[\mathcal{O}_G(1)]$. In fact, a subspace W of A of codimension r defines a hyperplane section H of G where $G \setminus H \simeq \mathbf{A}^{r(n-r)}$ consists of $[V]$ such that

$V \cap W = 0$. *Schubert calculus* [141, example 14.7.11] computes the degree d of $G \subset \mathbf{P}^N$ as

$$d = (r(n-r))! \prod_{i=0}^{r-1} \frac{i!}{(n-r+i)!}.$$

There exists a *universal* exact sequence $0 \to \mathcal{E} \to A \otimes \mathcal{O}_G \to \mathcal{Q} \to 0$ on G with locally free sheaves $\mathcal{E}$ and $\mathcal{Q}$ of ranks r and $n - r$, which satisfies $V = \mathcal{E} \otimes k([V])$ in A for the residue field $k([V])$ at $[V] \in G$. Then $\mathcal{T}_G \simeq \mathcal{H}om(\mathcal{E}, \mathcal{Q}) \simeq \mathcal{E}^{\vee} \otimes \mathcal{Q}$ [120, theorem 3.5]. One has $\bigwedge^r \mathcal{E} = \mathcal{O}_G(-1)$ and hence $\bigwedge^{n-r} \mathcal{Q} = \mathcal{O}_G(1)$. Thus

$$\omega_G \simeq \left(\bigwedge^r \mathcal{E}\right)^{\otimes(n-r)} \otimes \left(\bigwedge^{n-r} \mathcal{Q}^{\vee}\right)^{\otimes r} \simeq \mathcal{O}_G(-n).$$

When $G = \mathrm{Gr}(2, 5) \subset \mathbf{P}^9$, which is of degree five by the above formula for d, we let $X \subset \mathbf{P}^6$ be a transversal linear section of G of codimension three, namely a smooth threefold $G \cap \Sigma$ with a linear subspace $\Sigma = \mathbf{P}^6$ of $\mathbf{P}^9$. Plainly X is of degree five and $\omega_X = \omega_G \otimes \mathcal{O}_X(3) \simeq \mathcal{O}_X(-2)$. By the Lefschetz hyperplane theorem, $\mathrm{Pic}\, X = \mathbf{Z}[\mathcal{O}_X(1)]$. Thus X is a Fano threefold of Fano index two with $-K_X^3 = 2^3 \cdot 5$.

In preparation for Theorem 9.3.4, we restrict possibilities for the numerical invariants.

Lemma 9.3.6 *Let X be a Fano threefold of Fano index m. Take a divisor H with $-K_X \sim mH$ and let $d = H^3$. Suppose that H is free and hence it defines a morphism $\varphi: X \to \mathbf{P}^{h-1}$ with $h = h^0(\mathcal{O}_X(H))$, which is finite since H is ample. Let D be the degree of $\varphi: X \to \varphi(X)$. Then (m, D, d) is $(4, 1, 1)$, $(3, 1, 2)$, $(2, 2, 2)$, $(2, 1, d)$, $(1, 2, d)$ or $(1, 1, d)$.*

Proof We have $\chi(\mathcal{O}_X) = 1$ in Lemma 9.2.2 and $-K_X c_2(X) = 24$ as computed before Definition 9.2.5. Hence

$$h = h^0(\mathcal{O}_X(H)) = \frac{1}{12}(1+m)(2+m)d + \frac{2}{m} + 1$$

by the Riemann–Roch formula with Kodaira vanishing. The degree of $\varphi(X)$ in $\mathbf{P}^{h-1}$ is d/D. The inequality in Remark 6.4.15 for $\varphi(X) \subset \mathbf{P}^{h-1}$ is written as

$$\frac{d}{D} \geq h - 3 = \frac{1}{12}(1+m)(2+m)d + \frac{2}{m} - 2.$$

The lemma is elementarily derived from this inequality with $d/D, h \in \mathbf{N}$. □

The classification in the first three cases in Lemma 9.3.6 is easy. If $(m, D, d) = (4, 1, 1)$, then $h = 4$ and $\varphi: X \to \mathbf{P}^3$ is bijective, which is an isomorphism by Zariski's main theorem. If $(m, D, d) = (3, 1, 2)$, then $\varphi \colon X \to \varphi(X)$ is bijective and $\varphi(X) \subset \mathbf{P}^4$ is a quadric. Hence $\varphi(X)$ is normal and $X \simeq \varphi(X)$. If $(m, D, d) = (2, 2, 2)$, then $\varphi \colon X \to \mathbf{P}^3$ is finite of degree two. Since X is smooth, φ is ramified along a smooth surface T, the degree of which is four by the ramification formula $K_X = \varphi^*(K_{\mathbf{P}^3} + T/2)$ in Theorem 2.2.20.

We need to study the case $(m, D) = (2, 1)$. In this case, $h = d + 2$ and $\varphi \colon X \to \varphi(X) \subset \mathbf{P}^{d+1}$ is bijective to the image. In particular, $d \geq 3$. On the other hand, one has $d \leq 9$ since the general member S of $|H|$ is a del Pezzo surface of degree d with $-K_S = H|_S$. From Kodaira vanishing, one has the surjection $H^0(\mathcal{O}_X(H)) \twoheadrightarrow H^0(\mathcal{O}_S(H|_S))$. By Theorem 8.1.3, $S \simeq \varphi(S)$ and $\varphi(S) \subset \bar{H} = \mathbf{P}^d$ is projectively normal for the hyperplane $\bar{H}$ in $\mathbf{P}^{d+1}$ such that $S = X \cap \varphi^{-1}(\bar{H})$. Using the surjection above, one derives the projective normality of $\varphi(X) \subset \mathbf{P}^{d+1}$ from that of $\varphi(S)$ as in the proof of Lemma 8.1.10. Thus $X \simeq \varphi(X) \subset \mathbf{P}^{d+1}$.

Lemma 9.3.7 *If $(m, D) = (2, 1)$, then $3 \leq d \leq 7$.*

Proof One has only to exclude the case when $d = 8$ or 9, in which $S \simeq \mathbf{P}^1 \times \mathbf{P}^1$, Σ_1 or $\mathbf{P}^2$. Here we shall discuss the simplest case $d = 9$ when $S \simeq \mathbf{P}^2$ with triple embedding into $\mathbf{P}^9$. The Lefschetz hyperplane theorem provides the surjection $H_2(S, \mathbf{Z}) \twoheadrightarrow H_2(X, \mathbf{Z})$, by which $H_2(S, \mathbf{Z}) = \mathbf{Z} \simeq H_2(X, \mathbf{Z})$. Then $\operatorname{Pic} X \simeq H^2(X, \mathbf{Z}) \simeq H^2(S, \mathbf{Z}) = \mathbf{Z}$ by Poincaré duality with Lemma 9.2.2. Hence there would exist a divisor G on X such that $\mathcal{O}_S(G|_S) \simeq \mathcal{O}_{\mathbf{P}^2}(1)$. It would satisfy $H \sim 3G$ by $H|_S \sim 3G|_S$. This contradicts the Fano index two of X. $\qquad\qquad\square$

We are treating a Fano threefold X of Fano index two. The description in Theorem 9.3.4 for $d = 3$ and 4 is derived from that of the corresponding del Pezzo surface $S \subset \mathbf{P}^d$ similarly to Theorem 9.3.1(i). In the remaining cases $d = 5$, 6 and 7, the proof of the theorem uses properties of lines in $X \subset \mathbf{P}^{d+1}$. By a *line* and a *plane* respectively, we mean a curve and a surface of degree one in the ambient projective space.

Proposition 9.3.8 *Let $X = X_d \subset \mathbf{P}^{d+1}$ be a Fano threefold of Fano index two with $\mathcal{O}_X(-K_X) = \mathcal{O}_X(2)$, in which d is the degree of X with $3 \leq d \leq 7$.*

(i) *For any point x in X, there exists a line in X through x.*
(ii) *If $3 \leq d \leq 6$, then there exists no plane in X. If $d = 7$, then X contains a unique plane.*
(iii) *A line l through a general point in X has normal sheaf $\mathcal{N}_{l/X} \simeq \mathcal{O}_{\mathbf{P}^1}^{\oplus 2}$.*

Proof First of all, there exists a line in X since the general hyperplane section of X is a del Pezzo surface of degree d, which contains a line. Let T be the

Hilbert scheme parametrising lines in X and let $U \subset X \times T$ be the universal family. The fibre of $U \to T$ at a point $[l] \in T$ is the corresponding line l.

We write H for the general hyperplane section of X with $-K_X \sim 2H$. Let l be an arbitrary line in X. The linear system $\mathbf{P}H^0(\mathscr{O}_X(1) \otimes \mathscr{I})^\vee \subset |H|$ for the ideal $\mathscr{I}$ in $\mathscr{O}_X$ defining l is transformed into a free linear system on the blow-up of X along l. Hence the general hyperplane section S of X containing l is smooth. By $(l^2)_S = -2 - (K_S \cdot l)_S = -1$ and $(S \cdot l)_X = 1$, the exact sequence $0 \to \mathscr{N}_{l/S} \to \mathscr{N}_{l/X} \to \mathscr{N}_{S/X}|_l \to 0$ is expressed as

$$0 \to \mathscr{O}_{\mathbf{P}^1}(-1) \to \mathscr{N}_{l/X} \to \mathscr{O}_{\mathbf{P}^1}(1) \to 0,$$

by which $H^0(\mathscr{N}_{l/X}) \simeq \mathbf{C}^2$ and $H^1(\mathscr{N}_{l/X}) = 0$. By virtue of Theorem 9.3.9 below, T is smooth of dimension two at $[l]$. In particular, every connected component T_0 of T is a smooth projective surface and $U_0 = U \times_T T_0$ is a smooth projective threefold.

We write $p: U_0 \to X$ and $q: U_0 \to T_0$. Either p is surjective or $\Pi = p(U_0)$ is a plane. Indeed if p is not surjective, then Π is a surface, and for any $x \in \Pi$, the locus $q(p^{-1}(x))$ in T_0 is of positive dimension. It follows that two general points in Π are joined by a line in Π, which occurs only if Π is a plane.

We shall prove the assertions (i) and (ii) by induction on d. These for $d = 7$ are actually obtained after the description in Theorem 9.3.4(x). When $3 \le d \le 6$, if X contained a plane Π, then a line l in Π would have $(-\Pi \cdot l)_X = (-K_\Pi \cdot l)_\Pi + (K_X \cdot l)_X = 3 - 2 = 1$. Thus lines in Π form an extremal ray of $\overline{\mathrm{NE}}(X)$. The associated contraction is a divisorial contraction $\pi: X \to Y$ which contracts Π to a point. By Theorem 3.1.1, Y is smooth and π is the blow-up at $y = \pi(\Pi)$. From $\pi^*(-K_Y) = -K_X + 2\Pi$, one deduces that Y is Fano with $-K_Y \sim 2\pi_*H$ and $(\pi_*H)^3 = d + 1$, which is of Fano index two by Lemma 9.3.6 with Theorem 9.3.1. By the inductive assumption (i) for Y, there exists a line l_Y in Y through y. However, its strict transform l in X has $(-K_X \cdot l) = (-K_Y \cdot l_Y) - 2 = 0$, which contradicts the property that X is Fano. Thus (ii) holds for X. Then $p: U_0 \to X$ is surjective and this implies (i) for X.

The above exact sequence on a line l together with Theorem 1.3.5 shows that $\mathscr{N}_{l/X}$ is isomorphic to $\mathscr{O}_{\mathbf{P}^1}^{\oplus 2}$ or $\mathscr{O}_{\mathbf{P}^1}(-1) \oplus \mathscr{O}_{\mathbf{P}^1}(1)$. On the other hand by (i) and Proposition 9.1.10, every line l through a general point in X is a free curve. In other words, $\mathscr{T}_X|_l$ splits into invertible sheaves of non-negative degrees. So does the image $\mathscr{N}_{l/X}$ of the surjection $\mathscr{T}_X|_l \twoheadrightarrow \mathscr{N}_{l/X}$. Thus $\mathscr{N}_{l/X} \simeq \mathscr{O}_{\mathbf{P}^1}^{\oplus 2}$, which is (iii). $\qquad\square$

We used the following result in the deformation theory. Compare it with Theorem 5.4.7 and see [266, section I.2].

Theorem 9.3.9 *Let X be a projective variety and let Y be a closed subvariety of X. Let $\mathcal{N}_{Y/X} = \mathcal{H}om_Y(\mathcal{I}/\mathcal{I}^2, \mathcal{O}_Y)$ by the ideal sheaf $\mathcal{I}$ in $\mathcal{O}_X$ defining Y. Then the tangent space of the Hilbert scheme Hilb X at $[Y]$ is isomorphic to $H^0(\mathcal{N}_{Y/X})$. If $Y \subset X$ is locally defined by a regular sequence in $\mathcal{O}_X$, then the local dimension of Hilb X at $[Y]$ is at least $h^0(\mathcal{N}_{Y/X}) - h^1(\mathcal{N}_{Y/X})$.*

We demonstrate the description in the most interesting case when $d = 5$.

Proof of Theorem 9.3.4(vii)   *Step* 1   Recall that $X \subset \mathbf{P}^6$ is not contained in any hyperplane and that it is projectively normal. Fix a line l in X through a general point, which exists by Proposition 9.3.8(i). Let $\pi \colon Y \to X$ be the blow-up along l and let F be the exceptional divisor. Then $F \simeq \mathbf{P}(\mathcal{N}_{l/X}^\vee) \simeq \mathbf{P}^1 \times \mathbf{P}^1$ by Proposition 9.3.8(iii). We write e for a section of $F \to l$ with $e^2 = 0$ and write f for a fibre of it. Note that $-F|_F \simeq e$. The divisor $H_Y = \pi^* H - F$ defines a morphism $\varphi \colon Y \to \mathbf{P}^4$ with $\varphi^* \mathcal{O}_{\mathbf{P}^4}(1) = \mathcal{O}_Y(H_Y)$, which resolves the projection $X \dashrightarrow \mathbf{P}^4$ from l. Since $-K_Y = \pi^*(-K_X) - F \sim \pi^* H + H_Y$, Y is Fano.

The image $Z = \varphi(Y) \subset \mathbf{P}^4$ is not contained in any hyperplane. We shall write $\varphi \colon Y \to Z$ using the same symbol φ. Since $H_Y^3 = \pi^* H^3 + 3(\pi^* H \cdot F^2) = 5 - 3 = 2$, φ is birational and Z is a quadric. The restriction $\varphi|_F \colon F \to Q = \varphi(F)$ is the isomorphism defined by $H_Y|_F \sim e + f$ since $H^0(\mathcal{O}_Y(H_Y)) \to H^0(\mathcal{O}_F(H_Y|_F))$ is surjective by the Kawamata–Viehweg vanishing for $H_Y - F \sim K_Y + 3H_Y$.

Step 2   Let E denote the exceptional locus of $\varphi \colon Y \to Z$. We apply the property [202, proposition 4.4] that X is the scheme-theoretic intersection of quadrics in $\mathbf{P}^6$ containing X, which comes from the same property of the elliptic curve $X \cap \Sigma$ in $\Sigma = \mathbf{P}^4$ due to Saint-Donat [405] for a general codimension-two linear subspace Σ of $\mathbf{P}^6$. This shows that for any plane Π in $\mathbf{P}^6$ containing l, the divisorial part in Π of the intersection $X \cap \Pi$ is either l or $l + l'$ with one more line l' (which may equal l). Notice that $\Pi \not\subset X$ by Proposition 9.3.8(ii). Hence every fibre of $E \to C = \varphi(E)$ is the strict transform of a line which meets l. Thus φ is an extremal contraction about each connected component of E. By Theorems 4.1.9 and 5.5.1, Z and C are smooth and φ is the blow-up of Z along the union C of smooth curves.

Write $C \sim a\varphi(e) + b\varphi(f)$ with $a, b \in \mathbf{N}$ as a divisor on $Q \simeq \mathbf{P}^1 \times \mathbf{P}^1$. We claim that $(a, b) = (1, 2)$ or $(2, 1)$. For a general hyperplane $\bar{H}$ in $\mathbf{P}^6$ through l, the surface $S = X \cap \bar{H}$ is a del Pezzo surface of degree five as seen in the proof of Proposition 9.3.8. Hence by Theorem 8.1.2 or [35, proposition IV.12], l intersects exactly three lines in S. These lines are transformed to the union $E|_{S_Y}$ of fibres of φ, where S_Y is the strict transform of S with $S_Y \sim H_Y$. Thus $\deg C = 3$, that is, $a + b = 3$. If a or b were zero, then there would exist a line m in $Q \subset \mathbf{P}^4$ intersecting C transversally at three points. Then the curve $m_Y = (\varphi|_F)^{-1}(m)$ has $(K_Y \cdot m_Y) = (K_Z \cdot m) + (E \cdot m_Y) = -3 + 3 = 0$. This contradicts the property that Y is Fano, and the claim is completed.

Step 3 By what we have seen, the pair $C \subset Z$ in $\mathbf{P}^4$ consists of a rational twisted cubic curve and a smooth quadric threefold such that the hyperplane $\mathbf{P}^3$ containing C cuts out the smooth quadric surface Q as $Z \cap \mathbf{P}^3$. It is uniquely determined up to automorphism of $\mathbf{P}^4$. Indeed, $Q \subset Z$ is expressed as $(x_0 x_1 + x_2 x_3 = x_4 = 0) \subset (x_0 x_1 + x_2 x_3 + x_4^2 = 0)$ by suitable homogeneous coordinates. It thus suffices to verify the uniqueness of $C \subset Q$ in $\mathbf{P}^3$. Up to involution, C is of bidegree $(2, 1)$ in $Q \simeq \mathbf{P}^1 \times \mathbf{P}^1$. The subgroup $G = \operatorname{Aut} \mathbf{P}^1 \times \operatorname{Aut} \mathbf{P}^1$ of automorphisms of $\mathbf{P}^3$ acts on the linear system $|C| \simeq \mathbf{P}^5$ on Q. The stabiliser of C is of dimension one because it fixes the set of two ramification points of the double cover $C \to \mathbf{P}^1$. By dimension count, the orbit of C by G consists of all smooth curves in $|C|$.

Consequently the blow-up Y of Z along C and the target X of the other contraction from Y are also unique. The unique X is isomorphic to the linear section of $\operatorname{Gr}(2, 5)$ in Example 9.3.5. □

Finally we consider a Fano threefold X of Fano index one such that $-K_X$ is free but not very ample. By Lemma 9.3.6, the finite morphism $\varphi \colon X \to \varphi(X) \subset \mathbf{P}^{g+1}$ defined by $|-K_X|$ is of degree one or two, where g is the genus of X. If φ is of degree one, then $-K_X$ is very ample by the following lemma.

Lemma 9.3.10 *Let X be a Fano threefold of Fano index one with genus g. Suppose that $-K_X$ is free and that it defines a bijection $\varphi \colon X \to \varphi(X) \subset \mathbf{P}^{g+1}$ to the image. Then $\varphi(X) \subset \mathbf{P}^{g+1}$ is projectively normal.*

Proof We may assume that $g \geq 3$. Since $-K_X$ defines a bijection, the intersection $C = S \cap S'$ of two general elephants S and S' in $|-K_X|$ is irreducible and smooth. By the surjection $H^0(\mathscr{O}_X(-K_X)) \twoheadrightarrow H^0(\mathscr{O}_S(-K_X|_S)) \twoheadrightarrow H^0(\mathscr{O}_C(-K_X|_C))$, the projective normality of X is reduced to that of C as in the discussion for Theorem 9.3.4 prior to Lemma 9.3.7. By $K_C = -K_X|_C$, the smooth curve C has genus g and $\varphi|_C \colon C \to \mathbf{P}^{g-1}$ is the canonical map. It is never hyperelliptic since $\varphi|_C$ is bijective, and the projective normality of $\varphi(C)$ is a classical result of M. Noether [406]. □

If X is of Fano index one and $\varphi \colon X \to Z = \varphi(X) \subset \mathbf{P}^{g+1}$ is of degree two, then Z is of degree $g - 1$ in $\mathbf{P}^{g+1}$ and $Z \subset \mathbf{P}^{g+1}$ is completely classified as in Remark 6.4.15. Starting with the description of Z, one obtains the following classification of X.

Theorem 9.3.11 (Iskovskikh [202, theorem 7.2]) *Let X be a Fano threefold of Fano index one with genus g and Picard number ρ. If $-K_X$ is not free, then X is as in Theorem 9.3.1(ii). If $-K_X$ is free but not very ample, then $|-K_X|$ defines a double cover $\varphi \colon X \to Z \subset \mathbf{P}^{g+1}$ of a smooth threefold Z ramified along a* smooth divisor D. It is one of the following and X is uniquely recovered from the pair of Z and D.

(i) $g = 2$. $Z = \mathbf{P}^3$ and D is a sextic. $\rho = 1$.
(ii) $g = 3$. $Z \subset \mathbf{P}^4$ is a smooth quadric and $D = \bar{D} \cap Z$ is cut out by a quartic $\bar{D}$ in $\mathbf{P}^4$. $\rho = 1$.
(iii) $g = 4$. $Z = \mathbf{P}^2 \times \mathbf{P}^1 \subset \mathbf{P}^5$ is the Segre embedding and D is of bidegree $(4, 2)$. $\rho = 2$.
(iv) $g = 5$. Z is the $\mathbf{P}^2$-bundle $\mathbf{P}(\mathscr{E})$ over $\mathbf{P}^1$ for $\mathscr{E} = \mathscr{O}_{\mathbf{P}^1}(1)^{\oplus 2} \oplus \mathscr{O}_{\mathbf{P}^1}(2)$ embedded as $Z \subset \mathbf{P}^6 = \mathbf{P}H^0(\mathscr{E})$ and $D = \bar{D} \cap Z$ is cut out by a quartic $\bar{D}$ in $\mathbf{P}^6$. $\rho = 2$.
(v) $g = 7$. $Z = \mathbf{P}^2 \times \mathbf{P}^1 \subset \mathbf{P}^8 = \mathbf{P}H^0(\mathscr{O}_Z(1, 2))$ and D is of bidegree $(4, 0)$. Here $\mathscr{O}_Z(1, 2)$ stands for $p^* \mathscr{O}_{\mathbf{P}^2}(1) \otimes q^* \mathscr{O}_{\mathbf{P}^1}(2)$ with projections $p \colon Z \to \mathbf{P}^2$ and $q \colon Z \to \mathbf{P}^1$. In this case, X is isomorphic to the product $S \times \mathbf{P}^1$ with a del Pezzo surface S of degree two. $\rho = 9$.

We only explain the uniqueness of X for a given pair $D \subset Z$. The double cover $\varphi \colon X \to Z$ is determined by the analytic structure of the restriction to $U = Z \setminus D$. Indeed if another cover $\varphi' \colon X' \to Z$ satisfies $\varphi^{-1}(U) \simeq (\varphi')^{-1}(U)$ over U analytically, then it induces a bimeromorphic map $X \dashrightarrow X'$ over Z, which turns out to be bijective via $\varphi^{-1}(D) \simeq D \simeq (\varphi')^{-1}(D)$. By the GAGA principle, Theorem 1.1.16, it is an algebraic isomorphism $X \simeq X'$.

Thus the uniqueness is reduced to the isomorphism $\operatorname{Hom}(\pi_1(U), \mathbf{Z}_2) \simeq \mathbf{Z}_2$ for the fundamental group $\pi_1(U)$ of U. The existence of φ implies that $\operatorname{Hom}(\pi_1(U), \mathbf{Z}_2) \neq 0$. By the universal coefficient theorem, it is isomorphic to $H^1(U, \mathbf{Z}_2)$. This cohomology fits into the *Kummer sequence*

$$1 \to H^0(\mathscr{O}_U^\times)/H^0(\mathscr{O}_U^\times)^2 \to H^1(U, \mathbf{Z}_2) \to \operatorname{Pic}_2 U \to 0,$$

where $\operatorname{Pic}_2 U$ is the group of two-torsions in $\operatorname{Pic} U$. Because D is irreducible by the description of $D \subset Z$ in Theorem 9.3.11, no principal divisor on X but zero is supported in D. Thus $H^0(\mathscr{O}_U^\times) \simeq \mathbf{C}^\times$ and hence $\operatorname{Hom}(\pi_1(U), \mathbf{Z}_2) \simeq \operatorname{Pic}_2 U$. Since $\operatorname{Pic} Z \simeq \mathbf{Z}$ or $\mathbf{Z}^{\oplus 2}$ by the description of Z, the group $\operatorname{Pic}_2 U$ is zero or $\mathbf{Z}_2$ from the exact sequence $0 \to \mathbf{Z}[D] \to \operatorname{Pic} Z \to \operatorname{Pic} U \to 0$. Thus $\operatorname{Hom}(\pi_1(U), \mathbf{Z}_2) \simeq \mathbf{Z}_2$.

9.4 Classification of Principal Series

We shall complete a classification of Fano threefolds. After Theorems 9.3.4 and 9.3.11, we are interested in an anti-canonically embedded Fano threefold $X = X_{2g-2} \subset \mathbf{P}^{g+1}$ of Fano index one, in which $\mathscr{O}_X(-K_X) = \mathscr{O}_X(1)$ and g is the genus of X. The *anti-canonical embedding* means the embedding defined by the complete linear system $|-K_X|$. By Lemma 9.3.10, $X \subset \mathbf{P}^{g+1}$ is projectively normal. Iskovskikh said that X is of *principal series* and classified it when the Picard number is one. In the case when $g \geq 7$, he investigated the *double projection* from a line, developing the approach by Fano. We shall explain this using the MMP and then introduce a biregular description by means of vector bundles due to Mukai.

Iskovskikh's study relies on the existence of a line due to Shokurov. A *line* in $X \subset \mathbf{P}^{g+1}$ is a curve of degree one as a subvariety of $\mathbf{P}^{g+1}$.

Theorem 9.4.1 (Shokurov [421], Reid [397]) *Let $X \subset \mathbf{P}^{g+1}$ be an anti-canonically embedded Fano threefold of Fano index one. Then there exists a line in X unless $X \simeq \mathbf{P}^1 \times \mathbf{P}^2$.*

The line has several properties similar to those in Proposition 9.3.8. Consult [203, section 3] and [207, section 1] for the proof.

Proposition 9.4.2 *Let $X \subset \mathbf{P}^{g+1}$ be an anti-canonically embedded Fano threefold of Fano index one with genus g and Picard number one.*

(i) *The normal sheaf $\mathcal{N}_{l/X}$ of a line l in X is isomorphic to $\mathcal{O}_{\mathbf{P}^1}(-1) \oplus \mathcal{O}_{\mathbf{P}^1}$ or $\mathcal{O}_{\mathbf{P}^1}(-2) \oplus \mathcal{O}_{\mathbf{P}^1}(1)$.*

(ii) *The Hilbert scheme parametrising lines in X is of pure dimension one. It is smooth at $[l]$ if and only if $\mathcal{N}_{l/X} \simeq \mathcal{O}_{\mathbf{P}^1}(-1) \oplus \mathcal{O}_{\mathbf{P}^1}$.*

(iii) *If $g \geq 4$, then every point in X lies on only finitely many lines. If $g \geq 7$, then every line in X intersects only finitely many lines.*

It can occur that every line has normal sheaf isomorphic to $\mathcal{O}_{\mathbf{P}^1}(-2) \oplus \mathcal{O}_{\mathbf{P}^1}(1)$. This case was omitted in the original work of Iskovskikh, and an example was found by Mukai and Umemura.

Example 9.4.3 (Mukai–Umemura [359]) Let $V \simeq \mathbf{C}^{13}$ be the vector space of homogeneous polynomials in x_1, x_2 of degree 12. The group $\mathrm{SL}(2, \mathbf{C})$ acts on $\mathbf{P}(V^\vee) = \mathbf{P}^{12}$ naturally. Consider the point $h = x_1 x_2 (x_1^{10} + 11 x_1^5 x_2^5 - x_2^{10})$ in $\mathbf{P}^{12}$. Its stabiliser in $\mathrm{SL}(2, \mathbf{C})$ is the binary icosahedral group $2\mathfrak{I}$ [255, I §II.13]. Let X be the closure of the orbit of h by $\mathrm{SL}(2, \mathbf{C})$, which is a compactification of the quotient $\mathrm{SL}(2, \mathbf{C})/2\mathfrak{I}$. Mukai and Umemura proved that X is a Fano threefold with genus 12 and $\mathrm{Pic}\, X = \mathbf{Z}[\mathcal{O}_X(1)]$. The embedding $X \subset \mathbf{P}^{12}$ is the projection of the anti-canonical embedding $X \subset \mathbf{P}^{13}$ from a point. They also proved that all lines l in X are disjoint. This implies that $\mathcal{N}_{l/X} \simeq \mathcal{O}_{\mathbf{P}^1}(-2) \oplus \mathcal{O}_{\mathbf{P}^1}(1)$ by [203, theorem 3.4(iv)].

Let $X = X_{2g-2} \subset \mathbf{P}^{g+1}$ be an anti-canonically embedded Fano threefold of Fano index one with Picard number one. We shall assume that $g \geq 7$. Fix a line l, which exists by Theorem 9.4.1, and write $\mathscr{I}$ for the ideal in $\mathscr{O}_X$ defining l. Our strategy is to study the rational map defined by the linear system $\mathbf{P}H^0(\mathscr{O}_X(1) \otimes \mathscr{I}^2)^{\vee}$ which consists of hyperplane sections singular along l.

Definition 9.4.4 Let $X \subset \mathbf{P}^{g+1}$ be a Fano threefold as above with genus $g \geq 7$. The *double projection* of X from a line l is the dominant rational map $p_{2l} \colon X \dashrightarrow W \subset \mathbf{P}^{g-6} = \mathbf{P}H^0(\mathscr{O}_X(1) \otimes \mathscr{I}^2)$. The equality $h^0(\mathscr{O}_X(1) \otimes \mathscr{I}^2) = g - 5$ will be proved in Theorem 9.4.7.

We shall realise the double projection p_{2l} by a two-ray game. Let $\pi \colon Y \to X$ be the blow-up along l and let F be the exceptional divisor. By Proposition 9.4.2(i), $\pi|_F \colon F \simeq \mathbf{P}(\mathscr{N}_{l/X}^{\vee}) \to l$ is a Hirzebruch surface Σ_n with $n = 1$ or 3. We write e for the negative section of $\pi|_F$ and write f for a fibre of it. Then $-K_Y|_F = e + ((n+3)/2)f$ and $-F|_F = e + ((n+1)/2)f$. The blow-up π resolves the usual projection $p_l \colon X \subset \mathbf{P}^{g+1} \dashrightarrow \mathbf{P}^{g-1}$ from l. In particular, $-K_Y = \pi^*(-K_X) - F$ is free. It is big as $-K_Y^3 = 2g - 6 > 0$.

The double projection p_{2l} makes sense by the next lemma.

Lemma 9.4.5 $h^0(\mathscr{O}_X(1) \otimes \mathscr{I}^2) \geq g - 5 \geq 2$.

Proof By the projection formula $\mathscr{O}_X(1) \otimes \mathscr{I}^2 = \pi_* \mathscr{O}_Y(-K_Y - F)$, it suffices to show that $h^0(\mathscr{O}_Y(-K_Y - F)) \geq g - 5$. Consider the exact sequence

$$0 \to \mathscr{O}_Y(-K_Y - F) \to \mathscr{O}_Y(-K_Y) \to \mathscr{O}_F(-K_Y|_F) \to 0.$$

One has $H^i(\mathscr{O}_Y(-K_Y)) = H^i(\mathscr{O}_F(-K_Y|_F)) = 0$ for $i \geq 1$ by Kawamata–Viehweg vanishing because $-K_Y$ and $-K_Y|_F - (K_F + e/3)$ are nef and big. Hence $H^i(\mathscr{O}_Y(-K_Y - F)) = 0$ for $i \geq 2$, by which $h^0(\mathscr{O}_Y(-K_Y - F)) \geq \chi(\mathscr{O}_Y(-K_Y - F)) = \chi(\mathscr{O}_Y(-K_Y)) - \chi(\mathscr{O}_F(-K_Y|_F))$. One obtains $\chi(\mathscr{O}_Y(-K_Y)) = -K_Y^3/2 + 3 = g$ and $\chi(\mathscr{O}_F(-K_Y|_F)) = 5$ from the Riemann–Roch formula on the weak Fano threefold Y and on the Hirzebruch surface F. $\square$

Construction 9.4.6 We perform the two-ray game starting from the blow-up Y which has Picard number two. Let $\alpha \colon Y \to Z$ be the contraction defined by a multiple of the free divisor $-K_Y$. This is the Stein factorisation of $Y \to p_l(X)$ for the projection $p_l \colon X \dashrightarrow \mathbf{P}^{g-1}$ from l. It is either an isomorphism or the contraction of the ray of $\overline{\mathrm{NE}}(Y)$ other than $\mathbf{R}_{\geq 0}[f]$.

We claim that α is small. By $(-K_Y|_F)^2 = 3$, it does not contract F. It suffices to see that $p_l \colon X \setminus l \to \mathbf{P}^{g-1}$ contracts only finitely many curves. Each fibre of it is expressed as $X \cap \Pi \setminus l$ by a plane Π in $\mathbf{P}^{g+1}$ containing l. Note that $\Pi \not\subset X$ from $\mathrm{Pic}\, X = \mathbf{Z}[-K_X]$. If $X \cap \Pi$ contains a curve m other than l, then

m is also a line because X is an intersection of quadrics [203, proposition 1.7, theorem 2.5]. By Proposition 9.4.2(iii), the number of such m is finite.

Thus α is a flopping contraction or an isomorphism. We take the flop $\alpha^+ \colon Y^+ \to Z$ of α in the former case and take $\alpha^+ = \alpha$ in the latter case. By Theorem 10.3.4, Y^+ is again smooth and it is weak Fano. The two-ray game generates an extremal contraction $\psi \colon Y^+ \to W$ with respect to which $-K_{Y^+}$ and the strict transform F^+ of F are relatively ample.

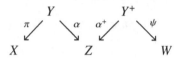

Since $Y \dashrightarrow Y^+$ is crepant, one has

$$K_{Y^+}^3 = K_Y^3 = 6 - 2g, \quad K_{Y^+}^2 F^+ = K_Y^2 F = 3, \quad K_{Y^+} F^{+2} = K_Y F^2 = 2.$$

Theorem 9.4.7 (Iskovskikh [203], [207]) *Let X be a Fano threefold of Fano index one with genus g and Picard number one. Then one of the following holds. Except the cases* (i) *and* (ii), $-K_X$ *is very ample and defines a projectively normal embedding $X \subset \mathbf{P}^{g+1}$. If $g \geq 7$, then the double projection $p_{2l} \colon X \dashrightarrow W \subset \mathbf{P}^{g-6}$ from a line l is obtained by the two-ray game in Construction 9.4.6. We keep the notation $\psi \colon Y^+ \to W$ there.*

   (i) *$g = 2$. X is a double cover of $\mathbf{P}^3$ as in Theorem 9.3.11(i).*
  (ii) *$g = 3$. X is a double cover of a quadric in $\mathbf{P}^4$ as in Theorem 9.3.11(ii).*
 (iii) *$g = 3$. $X = X_4 \subset \mathbf{P}^4$ is a quartic.*
  (iv) *$g = 4$. $X = X_{2,3} \subset \mathbf{P}^5$ is a complete intersection of a quadric and a cubic.*
   (v) *$g = 5$. $X = X_{2,2,2} \subset \mathbf{P}^6$ is a complete intersection of three quadrics.*
  (vi) *$g = 6$. $X \subset \mathbf{P}^7$ is a quadratic section of either a codimension-two linear section $\bar{X} \subset \mathbf{P}^7$ of the Plücker embedding $\mathrm{Gr}(2,5) \subset \mathbf{P}^9$ or the cone over a linear section of $\bar{X} \subset \mathbf{P}^7$.*
 (vii) *$g = 7$. ψ is a del Pezzo fibration of degree five over $W = \mathbf{P}^1$.*
(viii) *$g = 8$. ψ is a conic bundle over $W = \mathbf{P}^2$ with quintic discriminant divisor.*
  (ix) *$g = 9$. $W = \mathbf{P}^3$ and ψ is the blow-up of W along a smooth curve of genus three and degree seven.*
   (x) *$g = 10$. $W \subset \mathbf{P}^4$ is a smooth quadric and ψ is the blow-up of W along a smooth curve of genus two and degree seven.*
  (xi) *$g = 12$. $W \subset \mathbf{P}^6$ is a Fano threefold as in Theorem 9.3.4(vii) and ψ is the blow-up of W along a smooth rational curve of degree five.*

Corollary 9.4.8 *A Fano threefold of Fano index one with Picard number one is rational if it has genus 7, 9, 10 or 12.*

Proof In Theorem 9.4.7, W is rational if $g = 9$, 10 or 12. When $g = 7$, the rationality of the total space Y^+ of the del Pezzo fibration ψ is obtained from Corollary 8.3.2. □

We prove the theorem for $g \geq 7$ as Lemmata 9.4.9 and 9.4.10 according to the type of ψ. We remark that Takeuchi [439] investigated the two-ray game from the blow-up of a Fano threefold along a conic or a point by the method below.

Lemma 9.4.9 *If ψ is a Mori fibre space, then $W \simeq \mathbf{P}^1$ or $\mathbf{P}^2$, and X is as in Theorem 9.4.7(vii) or (viii).*

Proof The base W is not a point by $\rho(W) = 1$, and it is rational since X and hence Y^+ are rationally connected by Theorem 6.3.9 or 9.1.14. If W is a surface, then by Theorem 7.1.11, it is smooth with $\rho(W) = 1$. Hence $W \simeq \mathbf{P}^1$ or $\mathbf{P}^2$.

One has Pic $Y^+ = \mathbf{Z}[K_{Y^+}] \oplus \mathbf{Z}[F^+]$ via Pic $Y^+ \simeq$ Pic Y. Define the divisor $L = -aK_{Y^+} - bF^+$ with $a, b \in \mathbf{Z}$ by the isomorphism $\psi^* \mathscr{O}_W(1) \simeq \mathscr{O}_{Y^+}(L)$. Then $\mathbf{Z}[L]$ is a direct summand of Pic Y^+ from the exact sequence $0 \to$ Pic $W \to$ Pic $Y^+ \to \mathbf{Z}$ in Theorem 1.4.7. We remark that $X \dashrightarrow W$ is the double projection from l if and only if $(a, b) = (1, 1)$. The integer a is positive since a general curve γ in Y^+ is away from F^+ and has $(K_{Y^+} \cdot \gamma) < 0$.

Suppose that $\psi \colon Y^+ \to W = \mathbf{P}^1$ is a del Pezzo fibration of degree $1 \leq d \leq 9$. The general fibre of ψ contains a curve λ with $(-K_{Y^+} \cdot \lambda) = 1, 2$ or 3 depending on the degree d. Since $-(F^+ \cdot \lambda)K_{Y^+} - (-K_{Y^+} \cdot \lambda)F^+$ belongs to $\mathbf{Z}[L]$, the pair (b, d) is $(1, d)$, $(2, 8)$ or $(3, 9)$. The system of Diophantine equations

$$\begin{cases} 0 = -L^2 K_{Y^+} = (2g - 6)a^2 - 6ab - 2b^2, \\ d = L K_{Y^+}^2 \quad = (2g - 6)a - 3b \end{cases}$$

has a unique solution $(a, b, g, d) = (1, 1, 7, 5)$ and ψ is as in Theorem 9.4.7(vii).

Suppose that $\psi \colon Y^+ \to W = \mathbf{P}^2$ is a standard conic bundle with discriminant divisor of degree d. Since $-(F^+ \cdot \lambda)K_{Y^+} - (-K_{Y^+} \cdot \lambda)F^+$ belongs to $\mathbf{Z}[L]$ for a line or a conic λ contracted by ψ, either $b = 1$ or $(b, d) = (2, 0)$. The system

$$\begin{cases} 2 = -L^2 K_{Y^+} = (2g - 6)a^2 - 6ab - 2b^2, \\ 12 - d = L K_{Y^+}^2 \quad = (2g - 6)a - 3b, \end{cases}$$

where the second equation follows from Proposition 7.1.13, has a unique solution $(a, b, g, d) = (1, 1, 8, 5)$ as in Theorem 9.4.7(viii). □

Lemma 9.4.10 *If ψ is birational, then it is a divisorial contraction which contracts the divisor to a curve, and X is as in Theorem 9.4.7(ix), (x) or (xi).*

Proof By Theorem 5.5.1, ψ is divisorial. Let E denote its exceptional divisor and write $E \sim -aK_{Y^+} - bF^+$ with $a, b \in \mathbf{Z}$.

If $\psi(E)$ were a point, then by Theorem 3.1.1, the discrepancy e of ψ given by $K_{Y^+} = \psi^* K_W + eE$ is $1/2$, 1 or 2, and $eE^3 = 2$ as in Table 3.1. One would have the system

$$\begin{cases} -2 = -E^2 K_{Y^+} = (2g-6)a^2 - 6ab - 2b^2, \\ 2e = EK_{Y^+}^2 \quad = (2g-6)a - 3b, \end{cases}$$

but it has no solutions. Thus ψ contracts E to a curve. By Theorem 4.1.9, it is the blow-up of a smooth threefold W along a smooth curve $\Gamma = \psi(E)$ with $K_{Y^+} = \psi^* K_W + E$.

By $\rho(W) = 1$, W is a Fano threefold. Let m be the Fano index of W and write $-K_W = mG$ with a divisor G. Then $\psi^* G \equiv (-(a+1)K_{Y^+} - bF^+)/m$ generates a direct summand $\mathbf{Z}[\psi^* G]$ of $\mathrm{Pic}\, Y^+ = \mathbf{Z}[K_{Y^+}] \oplus \mathbf{Z}[F^+]$. A general curve γ in Y^+ is away from F^+ with $(K_{Y^+} \cdot \gamma) < 0$, by which $a + 1$ is positive. Since $-(F^+ \cdot \lambda)K_{Y^+} - F^+$ belongs to $\mathbf{Z}[\psi^* G]$ for a fibre λ of $E \to \Gamma$, one obtains $(a+1, b) = (cm, m)$ with $c = (F^+ \cdot \lambda) \geq 1$.

Let $d = G^3$ and $d_\Gamma = (G \cdot \Gamma)$. Let g_Γ be the genus of Γ. Consider the system

$$\begin{cases} md = -(K_{Y^+} - E)^2 K_{Y^+}/m^2 = (2g-6)c^2 - 6c - 2, \\ d_\Gamma = (K_{Y^+} - E)EK_{Y^+}/m \quad = (2g-6)ac - 3(2a+1) - 2m, \\ 2g_\Gamma - 2 = -E^2 K_{Y^+} \quad\quad = (2g-6)a^2 - 6am - 2m^2, \end{cases}$$

where the last equation comes from the expression $\deg K_{\Gamma^+} = (K_{Y^+} + \psi^* A)(-E + \psi^* A)E = -E^2 K_{Y^+}$ for a section $\Gamma^+ \sim (-E + \psi^* A)|_E$ of $E \to \Gamma$. If $m \geq 2$, then by Theorem 9.3.4, $(m, d) = (4, 1), (3, 2)$ or $(2, d)$ with $1 \leq d \leq 5$. In this case, the above system has solutions $(m, d, g, c, d_\Gamma, g_\Gamma) = (4, 1, 9, 1, 7, 3)$, $(3, 2, 10, 1, 7, 2)$ and $(2, 5, 12, 1, 5, 0)$ as in Theorem 9.4.7(ix)-(xi). In every case, G is very ample, $\psi^* G = -K_{Y^+} - F^+$ and $h^0(\mathscr{O}_W(G)) = g - 5$. Thus $X \dashrightarrow W \subset \mathbf{P}^{g-6}$ is the double projection from l.

It remains to exclude the case $m = 1$. Notice that $a \geq 1$ from $d_\Gamma \geq 1$. One has $H^0(\mathscr{O}_{Y^+}(E)) \simeq \mathbf{C}$ because E is exceptional. This space is isomorphic to $H^0(\mathscr{O}_Y(-aK_Y - mF))$ via $E \sim -aK_{Y^+} - mF^+$. Since $-K_Y$ is free, one can choose an injection $H^0(\mathscr{O}_Y(-K_Y - mF)) \hookrightarrow H^0(\mathscr{O}_Y(-aK_Y - mF)) \simeq \mathbf{C}$ using $a \geq 1$. This would contradict Lemma 9.4.5 if $m = 1$. □

The description by the double projection of X with $g \geq 7$ has the advantage that it clarifies whether X is rational or not. In contrast to this, Mukai described $X = X_{2g-2} \subset \mathbf{P}^{g+1}$ biregularly. The approach was initiated by Gushel' [165].

Theorem 9.4.11 (Mukai [352]) *Let X be a Fano threefold of Fano index one with Picard number one. If X has genus $g \geq 7$, then the anti-canonical embedding $X = X_{2g-2} \subset \mathbf{P}^{g+1}$ is described as a linear section of $G \subset \mathbf{P}^n$ as in Constructions 9.4.12 to 9.4.16, where $n = 15, 14, 13, 13, 13$ if $g = 7, 8, 9, 10, 12$ respectively.*

The case $g = 8$ is similar to Example 9.3.5.

Construction 9.4.12 ($g = 8$ [166], [354])   By the discussion in Example 9.3.5, the Grassmannian $G = \mathrm{Gr}(2, 6) \subset \mathbf{P}^{14}$ is of dimension 8 and degree 14 with $\omega_G \simeq \mathcal{O}_G(-6)$. Its transversal linear section of codimension five is a Fano threefold with genus eight.

For $g = 7, 9$ and 10, one also finds a Fano threefold as a linear section of a homogeneous space.

Construction 9.4.13 ($g = 7$ [355])   Let A be a vector space of dimension $2n$ equipped with a non-degenerate symmetric bilinear form b. A subspace $V \subset A$ of dimension n is called a *Lagrangian* if $b(V, V) = 0$. We fix Lagrangians V_0 and V_∞ such that $V_0 \cap V_\infty = 0$, which are dual via b.

The subscheme of the Grassmannian $\mathrm{Gr}(n, 2n)$ which parametrises Lagrangians $V \subset A$ consists of two components L^+ and L^-, which are homogeneous spaces of dimension $n(n - 1)/2$ for the *special orthogonal group* $\mathrm{SO}(2n, \mathbf{C})$. We choose them so that $[V_0] \in L^+$. The variety $\mathrm{OGr}(n, 2n) = L^+$ is called the *orthogonal Grassmannian*. The point $[V_0]$ in $\mathrm{OGr}(n, 2n)$ has an open neighbourhood isomorphic to the kernel K of $\mathrm{Hom}(V_0, V_\infty) \simeq V_\infty^{\otimes 2} \twoheadrightarrow S^2 V_\infty$ which associates $f \in K$ with the graph $\Gamma_f \subset V_0 \times V_\infty \simeq A$ as a Lagrangian. The Plücker embedding is expressed as $f \mapsto 1 + f + (f \wedge f) + \cdots \in \bigoplus_i \mathrm{Hom}(\bigwedge^i V_0, \bigwedge^i V_\infty) \simeq \bigwedge^n A$.

Decompose the exterior algebra $S = \bigwedge^\bullet V_\infty$ into $S^+ = \bigoplus_{i \in 2\mathbf{N}} \bigwedge^i V_\infty$ and $S^- = \bigoplus_{i \in 1 + 2\mathbf{N}} \bigwedge^i V_\infty$ according to parity of degree. The *Clifford map* $A \to \mathrm{Hom}(S, S)$, $a \mapsto \varphi_a$, is the linear map such that φ_a for $a \in V_\infty$ is the wedge product by a and such that φ_a for $a \in V_0 = V_\infty^\vee$ is the contraction by a. For a Lagrangian V, there exists a non-zero element $s_V \in S^+ \cup S^-$, unique up to multiplication, such that $\varphi_v(s_V) = 0$ for all $v \in V$. For example, $s_{V_0} = 1$ and s_{V_∞} is the volume element in $\bigwedge^n V_\infty$. The *Spinor embedding* $L^\pm \subset \mathbf{P}(S^{\pm \vee}) = \mathbf{P}^{2^{n-1}-1}$ is defined by sending $[V] \in L^\pm$ to $[s_V] \in \mathbf{P}(S^{\pm \vee})$. Its Veronese embedding corresponds to the Plücker embedding.

Take the universal exact sequence $0 \to \mathcal{E} \to A \otimes \mathcal{O}_{\mathrm{Gr}(n,2n)} \to \mathcal{Q} \to 0$ on $\mathrm{Gr}(n, 2n)$. The section of $S^2 \mathcal{E}^\vee$ given by b defines $L^+ \sqcup L^-$. By Remark 2.5.5, the top Chern class of $S^2 \mathcal{E}^\vee$ is Poincaré dual to the class $[L^+] + [L^-]$. By

Schubert calculus [444] with [141, examples 14.5.1, 14.7.11], the embedding $OGr(n, 2n) \subset \mathbf{P}(S^{+\vee})$ is of degree

$$d_n = \frac{1}{2 \cdot 2^{\frac{n(n-1)}{2}}} \cdot c_{\frac{n(n+1)}{2}} (S^2 \mathscr{E}^{\vee}) \cdot c_1(\mathscr{E}^{\vee})^{\frac{n(n-1)}{2}} = \binom{n}{2}! \prod_{i=1}^{n-1} \frac{(i-1)!}{(2i-1)!}.$$

Since the normal sheaf of $OGr(n, 2n)$ in $Gr(n, 2n)$ is the restriction of $S^2 \mathscr{E}^{\vee}$, $OGr(n, 2n) \subset \mathbf{P}(S^{+\vee})$ has canonical sheaf $\mathscr{O}(-2n + 2)$.

When $n = 5$, $G = OGr(5, 10) \subset \mathbf{P}(S^{+\vee}) = \mathbf{P}^{15}$ is of dimension 10 and degree 12 with $\omega_G \simeq \mathscr{O}_G(-8)$. Its transversal linear section of codimension seven is a Fano threefold with genus seven.

Construction 9.4.14 ($g = 9$ [358])   Let A be a vector space of dimension $2n$ equipped with a non-degenerate skew-symmetric bilinear form b. A subspace $V \subset A$ of dimension n is called a *Lagrangian* if $b(V, V) = 0$. The subvariety $SpGr(n, 2n)$ of $Gr(n, 2n)$ which parametrises Lagrangians $V \subset A$ is called the *symplectic Grassmannian*. It is a homogeneous space of dimension $n(n + 1)/2$ for the *symplectic group* $Sp(2n, \mathbf{C})$. For Lagrangians V_0 and V_∞ with $V_0 \cap V_\infty = 0$, the point $[V_0] \in SpGr(n, 2n)$ has an open neighbourhood isomorphic to the kernel of $\mathrm{Hom}(V_0, V_\infty) \twoheadrightarrow \bigwedge^2 V_\infty$.

Let M be the quotient space of $\bigwedge^n A^{\vee}$ by $b \wedge (\bigwedge^{n-2} A^{\vee})$, which is of dimension $N = \binom{2n}{n} - \binom{2n}{n-2}$. Then $SpGr(n, 2n)$ is embedded in $\mathbf{P}(M) = \mathbf{P}^{N-1}$. By Schubert calculus, the embedding $SpGr(n, 2n) \subset \mathbf{P}(M)$ is of degree $2^{\binom{n}{2}} d_{n+1}$, where d_n is the degree in Construction 9.4.13. It has canonical sheaf $\mathscr{O}(-n - 1)$.

When $n = 3$, $G = SpGr(3, 6) \subset \mathbf{P}(M) = \mathbf{P}^{13}$ is of dimension 6 and degree 16 with $\omega_G \simeq \mathscr{O}_G(-4)$. Its transversal linear section of codimension three is a Fano threefold with genus nine.

Construction 9.4.15 ($g = 10$ [351])   Let A be a vector space of dimension seven equipped with a skew-symmetric quadrilinear form f which belongs to the open orbit in $\bigwedge^4 A^{\vee}$ by $GL(7, \mathbf{C})$. The subvariety G of $Gr(5, 7)$ which parametrises five-dimensional subspaces $V \subset A$ such that $f(V, V, V, V) = 0$ is a G_2-*manifold*. It is a homogeneous space of dimension five for an exceptional group of type G_2. Let M be the quotient space of $\bigwedge^5 A^{\vee}$ by $f \wedge A^{\vee}$, which is of dimension 14. The embedding $G \subset \mathbf{P}(M) = \mathbf{P}^{13}$ is of degree 18 with $\omega_G \simeq \mathscr{O}_G(-3)$. Its transversal linear section of codimension two is a Fano threefold with genus 10.

One can still describe a Fano threefold with $g = 12$ as a subvariety of some Grassmannian.

Construction 9.4.16 ($g = 12$ [353, section 3], [356, section 5]) Let A be a vector space of dimension seven equipped with a three-dimensional vector subspace $N \subset \bigwedge^2 A^\vee$ of skew-symmetric bilinear forms such that $\mathrm{Gr}(3, A^\vee) \cap$ $\mathbf{P}((N \wedge A^\vee)^\vee) = \emptyset$ in $\mathbf{P}(\bigwedge^3 A)$. Let G be the subvariety of $\mathrm{Gr}(3, 7) = \mathrm{Gr}(3, A)$ which parametrises three-dimensional subspaces $V \subset A$ such that $b(V, V) = 0$ for all $b \in N$. It is a smooth threefold. Let M be the quotient space of $\bigwedge^3 A^\vee$ by $N \wedge A^\vee$, which is of dimension 14. The embedding $G \subset \mathbf{P}(M) = \mathbf{P}^{13}$ is of degree 22 with $\omega_G \simeq \mathcal{O}_G(-1)$. It is a Fano threefold with genus 12.

Let $X = X_{2g-2} \subset \mathbf{P}^{g+1}$ be an anti-canonically embedded Fano threefold of Fano index one with genus $g \geq 7$ and Picard number one. Let $S \subset \mathbf{P}^g$ be the general hyperplane section. It satisfies $H^1(\mathcal{O}_S) = 0$ by the Kodaira vanishing $H^1(\mathcal{O}_X) = 0$ and its dual $H^2(\mathcal{O}_X(-S)) = 0$. Hence S is a *polarised* K3 surface with polarisation $\mathcal{O}_S(1)$, where *polarisation* means specification of an ample invertible sheaf. By a result of Moishezon [330], $\mathrm{Pic}\, S = \mathbf{Z}[-K_X|_S]$.

The proof of Theorem 9.4.11 uses the theory of vector bundles introduced below. We refer the reader to [193, chapters 9, 10] or [350]. A locally free coherent sheaf is alternatively called a *vector bundle*.

Definition 9.4.17 Let S be a smooth projective surface. The *Mukai vector* $v(\mathscr{E})$ of a vector bundle $\mathscr{E}$ on S is the topological invariant

$$v(\mathscr{E}) = \mathrm{ch}(\mathscr{E}) \cdot \sqrt{\mathrm{td}(\mathscr{T}_S)} \in \bigoplus_{i=0,2,4} H^i(S, \mathbf{Q}) \simeq \mathbf{Q} \oplus H^2(S, \mathbf{Q}) \oplus \mathbf{Q},$$

where the square root of $\mathrm{td}(\mathscr{T}_S)$ is chosen so that the entry in $H^0(S, \mathbf{Q})$ is one. The *Mukai pairing* $\langle v, w \rangle = v_2 w_2 - v_0 w_4 - v_4 w_0 \in \mathbf{Q}$ is defined for $v = (v_0, v_2, v_4)$ and $w = (w_0, w_2, w_4)$ in $\bigoplus_{i=0,2,4} H^i(S, \mathbf{Q})$.

When S is a K3 surface, the Mukai vector and the Mukai pairing are expressed by the Hirzebruch–Riemann–Roch theorem as

$$v(\mathscr{E}) = (\mathrm{rk}\, \mathscr{E}, \det \mathscr{E}, \chi(\mathscr{E}) - \mathrm{rk}\, \mathscr{E}) \in \mathbf{Z} \oplus \mathrm{Pic}\, S \oplus \mathbf{Z}$$

and $\langle v(\mathscr{E}), v(\mathscr{F}) \rangle = -\chi(\mathscr{E}^\vee \otimes \mathscr{F})$, where $\mathrm{rk}\, \mathscr{E}$ denotes the rank of $\mathscr{E}$.

Theorem 9.4.18 ([352, theorem 3]) *Let S be a K3 surface polarised by an ample divisor H such that $\mathrm{Pic}\, S = \mathbf{Z}[H]$. Let r and s be positive integers such that $v = (r, H, s)$ satisfies $\langle v, v \rangle = -2$. Then there exists a unique Gieseker stable (as in Remark 10.1.5) vector bundle $\mathscr{E}$ with Mukai vector $v(\mathscr{E}) = v$. This $\mathscr{E}$ is generated by global sections, $H^i(\mathscr{E}) = 0$ for $i \geq 1$ and the induced map $\bigwedge^r H^0(\mathscr{E}) \to H^0(\bigwedge^r \mathscr{E})$ is surjective.*

We shall outline the proof of Theorem 9.4.11. Take the general elephant S of X, which is a K3 surface polarised by $-K_X|_S$. For $g = 8, 9, 10, 12$, we apply Theorem 9.4.18 with $(r, s) = (2, 4), (3, 3), (5, 2), (3, 4)$ respectively. It provides a unique stable vector bundle $\mathscr{E}_S$ with Mukai vector $v(\mathscr{E}_S) = (r, -K_X|_S, s)$. This extends to a unique stable vector bundle $\mathscr{E}$ on X with the same properties as $\mathscr{E}_S$ has. That is, $\mathscr{E}$ is generated by global sections, $H^i(\mathscr{E}) = 0$ for $i \geq 1$ and $\bigwedge^r H^0(\mathscr{E}) \to H^0(\bigwedge^r \mathscr{E})$ is surjective.

It follows from $\operatorname{Pic} X \simeq \operatorname{Pic} S$ that $\bigwedge^r \mathscr{E} \simeq \mathscr{O}_X(-K_X)$ and this defines an embedding $X \hookrightarrow \mathbf{P}H^0(\bigwedge^r \mathscr{E})$. The evaluation map $H^0(\mathscr{E}) \twoheadrightarrow \mathscr{E} \otimes k(x)$ at each $x \in X$ with residue field $k(x)$ defines a morphism $X \to \operatorname{Gr}(r, H^0(\mathscr{E})^\vee)$ which sends x to the point $[(\mathscr{E} \otimes k(x))^\vee]$. These fit into the commutative diagram

$$
\begin{array}{ccc}
X & \longrightarrow & \operatorname{Gr}(r, H^0(\mathscr{E})^\vee) \\
\downarrow & & \downarrow \\
\mathbf{P}^{g+1} = \mathbf{P}H^0(\bigwedge^r \mathscr{E}) & \longhookrightarrow & \mathbf{P}(\bigwedge^r H^0(\mathscr{E})) \ ,
\end{array}
$$

where the right vertical morphism is the Plücker embedding and the bottom horizontal morphism is given by the surjection $\bigwedge^r H^0(\mathscr{E}) \twoheadrightarrow H^0(\bigwedge^r \mathscr{E})$.

Suppose that $g = 8$. Then $h^0(\mathscr{E}) = r + s = 6$. The Fano threefold $X \subset \mathbf{P}^9$ is the intersection of quadrics [203, proposition 1.7, theorem 2.5] given by the kernel K of $H^0(\mathscr{O}_{\mathbf{P}^9}(2)) \twoheadrightarrow H^0(\mathscr{O}_X(-2K_X))$. On the other hand, $\operatorname{Gr}(2, 6) = \operatorname{Gr}(2, H^0(\mathscr{E})^\vee)$ is the intersection of all *Plücker quadrics* ($f(2^{-1}x \wedge x) = 0$) for $x \in \bigwedge^2 H^0(\mathscr{E})^\vee$ given by $f \in \operatorname{Gr}(4, H^0(\mathscr{E}))$ [354, proposition 1.3]. This may also be described by using all $f \in I = \bigwedge^4 H^0(\mathscr{E})$. The vector spaces K and I are of the same dimension 15. Mukai proved that $I \to K$ is an isomorphism. This means that the above diagram is cartesian. Thus $X = \operatorname{Gr}(2, 6) \cap \mathbf{P}^9$, which is the theorem for $g = 8$.

The strategy for $g = 9, 10, 12$ is similar. For example when $g = 9$, we observe that the kernel of $\bigwedge^2 H^0(\mathscr{E}) \to H^0(\bigwedge^2 \mathscr{E})$ is generated by a non-degenerate form b, which factorises the above diagram into

$$
\begin{array}{ccccc}
X & \longrightarrow & \operatorname{SpGr}(3, 6) & \longhookrightarrow & \operatorname{Gr}(3, H^0(\mathscr{E})^\vee) \\
\downarrow & & \downarrow & & \downarrow \\
\mathbf{P}^{10} & \longhookrightarrow & \mathbf{P}^{13} = \mathbf{P}(M) & \longhookrightarrow & \mathbf{P}(\bigwedge^3 H^0(\mathscr{E}))
\end{array}
$$

as in Construction 9.4.14. Then we prove that the left square is cartesian. Finally when $g = 7$, we begin with a vector bundle $\mathscr{E}_S$ with $v(\mathscr{E}_S) = (5, -2K_X|_S, 5)$.

We close the section with the classification of Fano threefolds with higher Picard number due to Mori and Mukai. It is based upon the investigation of extremal contractions from a Fano threefold.

Theorem 9.4.19 (Mori–Mukai [337]) *There exist exactly* 88 *types of Fano threefolds X with Picard number ρ at least two up to deformation, as listed in* [337]. *In particular, the following hold.*

(i) $\rho \leq 10$. *If $\rho \geq 6$, then X is the product of $\mathbf{P}^1$ and a del Pezzo surface.*

(ii) *Except 13 types, X is the blow-up of some Fano threefold along a smooth curve. If X is of one of the excluded 13 types, then X admits a structure of a standard conic bundle over $\mathbf{P}^2$ or $\mathbf{P}^1 \times \mathbf{P}^1$ and hence $\rho = 2$ or 3.*

As a result, a Fano threefold X has bounded volume $-K_X^3 \leq 64$, and it attains the maximum 64 if and only if $X \simeq \mathbf{P}^3$. By deformation, this bound remains the case with a terminal Gorenstein Fano threefold [370]. Prokhorov [386] obtained the bound $-K_X^3 \leq 72$ for a Gorenstein Fano threefold X and proved that the equality holds if and only if $X \simeq \mathbf{P}(1, 1, 1, 3)$ or $\mathbf{P}(1, 1, 4, 6)$. He also proved that a $\mathbf{Q}$-factorial terminal non-Gorenstein $\mathbf{Q}$-Fano threefold X with Picard number one has $-K_X^3 \leq 125/2$ and that the equality holds if and only if $X \simeq \mathbf{P}(1, 1, 1, 2)$ [387]. Suzuki [435] showed that a $\mathbf{Q}$-factorial terminal $\mathbf{Q}$-Fano threefold X with Picard number one is of Fano index $1, \ldots, 9, 11, 13, 17$ or 19, and Prokhorov [388] confirmed that it attains the maximum 19 if and only if $X \simeq \mathbf{P}(3, 4, 5, 7)$.

9.5 Birational Rigidity and K-Stability

In Section 6.4, we demonstrated how the Sarkisov program works in the study of birational rigidity. The approach is applicable to $\mathbf{Q}$-Fano threefolds of lower codimension.

Let $P = \mathbf{P}(a_0, \ldots, a_n)$ be a weighted projective space. It is the quotient of $\mathbf{A}^{n+1} \setminus o$ by the action of $\mathbf{G}_m$. Recall that $\mathbf{P}(a_0, \ldots, a_n) \simeq \mathbf{P}(la_0, \ldots, la_n) \simeq \mathbf{P}(la_0, \ldots, a_i, \ldots, la_n)$. We say that $P = \mathbf{P}(a_0, \ldots, a_n)$ is *well-formed* if any n integers out of $a_0, \ldots, a_n$ have no common factors. Let X be a closed subvariety of P. The closure C_X in $\mathbf{A}^{n+1}$ of the inverse image $\pi^{-1}(X)$ by $\pi : \mathbf{A}^{n+1} \setminus o \to P$ is called the *classical affine cone* over X.

Definition 9.5.1 A closed subvariety $X \subset \mathbf{P}(a_0, \ldots, a_n)$ is said to be *quasi-smooth* if the classical affine cone C_X is smooth outside the vertex o.

We are mostly interested in hypersurfaces. Reid produced the list of all families $S \subset \mathbf{P}(a_1, a_2, a_3, a_4)$ of K3 surfaces [396, p.300], which are 95 families in well-formed weighted projective spaces. They correspond bijectively to families of terminal quasi-smooth threefolds $X \subset \mathbf{P}(1, a_1, a_2, a_3, a_4)$ with $\omega_X \simeq \mathcal{O}_X(-1)$. The list of these families is found in [195, 16.6] and [96, chapter 6]. We call them the 95 *families of Reid–Fletcher*. The list begins with the simplest one, a quartic $X_4 \subset \mathbf{P}(1,1,1,1,1)$ of no 1, and ends with $X_{66} \subset \mathbf{P}(1,5,6,22,33)$ of no 95. Such $X \to \operatorname{Spec} \mathbf{C}$ is a Mori fibre space as below.

Lemma 9.5.2 *Let P be a well-formed weighted projective space of dimension at least four. Let X be a quasi-smooth hypersurface in P, which is normal. Then X has divisor class group* $\operatorname{Cl} X = \mathbf{Z}[\mathcal{O}_X(1)]$.

Proof We use a result of Grothendieck [161, XI corollaire 3.14] that a complete intersection singularity is factorial if so is it in codimension three. It shows that the classical affine cone $C_X \subset \mathbf{A}^{n+1}$ of X is factorial, where $n = \dim P \geq 4$. Thus every Weil divisor on X is the quotient of an invariant principal divisor on the affine variety C_X. □

Corti, Pukhlikov and Reid applied the Sarkisov program to the study of **Q**-Fano threefold weighted hypersurfaces and established the following birational rigidity. Relaxing the generality condition, Cheltsov and Park [76] extended it to any terminal quasi-smooth member in the 95 families.

Theorem 9.5.3 (Corti–Pukhlikov–Reid [96]) *Let $X \subset \mathbf{P}(1, a_1, a_2, a_3, a_4)$ be a general member of any of the 95 families of Reid–Fletcher. Then X is birationally rigid.*

The theorem includes the quartic case $X_4 \subset \mathbf{P}^4 = \mathbf{P}(1,1,1,1,1)$ of no 1. The strategy is the same as that explained in Section 6.4 for the **Q**-Fano threefold with pliability of cardinality two. If a **Q**-Fano threefold X with Picard number one admits a Sarkisov link, then it is of type I or II and starts from a divisorial contraction $Y \to X$. The centre Γ in X of the exceptional divisor is a maximal centre in Definition 6.3.24 with respect to some mobile and big linear system $\mathcal{H}$. For each of the 95 families, Corti, Pukhlikov and Reid excluded the possibility for a locus Γ to be a maximal centre unless Γ is a singular point, and verified that every Sarkisov link from a singular point is a self-link which is either a *quadratic involution* or an *elliptic involution* below. The construction shares some features in common with Constructions 6.4.7 and 6.4.8.

Construction 9.5.4 (quadratic involution) Suppose the expression

$$X = (x_1 x_0^2 + a x_0 + b = 0) \subset \mathbf{P}(1, a_1, a_2, a_3, a_4)$$

with permutation $x_0, \ldots, x_4$ of the homogeneous coordinates, where $a, b \in \mathbf{C}[x_1, \ldots, x_4]$. We assume that x_0 has weight greater than one, by which X has a quotient singularity at $p = [1, 0, 0, 0, 0]$.

Taking $y = x_0 x_1$, one defines a birational map from X to the hypersurface

$$Z = (y^2 + ay + bx_1 = 0)$$

in the weighted projective space with homogeneous coordinates $x_1, \ldots, x_4, y$. We assume that it is the composite of the inverse of the Kawamata blow-up $\pi: Y \to X$ at p in Definition 3.1.4 and a flopping contraction $\alpha: Y \to Z$. Then Z is a terminal $\mathbf{Q}$-Fano threefold which is not $\mathbf{Q}$-factorial.

There exists an involution $\iota: Z \to Z$ which interchanges the two roots of the quadratic equation $y^2 + ay + bx_1 = 0$. Explicitly it multiplies $y + a/2$ by -1. The transformation

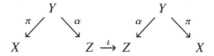

is a Sarkisov link of type II. This is called a *quadratic involution* on X.

Example 9.5.5   Consider a general quintic $X_5 \subset \mathbf{P}(1, 1, 1, 1, 2)$ of no 2 in Reid–Fletcher's list. One can write $X = (x_0 y^2 + a_3 y + b_5 = 0)$ with $a_3, b_5 \in \mathbf{C}[x_0, \ldots, x_3]$ by suitable homogeneous coordinates $x_0, \ldots, x_3, y$, which has a quotient singularity of type $\frac{1}{2}(1, 1, 1)$ at $p = [0, 0, 0, 0, 1]$.

Taking $z = x_0 y$, one defines a birational map from X to the sextic $Z_6 = (z^2 + a_3 z + b_5 x_0 = 0)$ in $\mathbf{P}(1, 1, 1, 1, 3)$ with homogeneous coordinates $x_0, \ldots, x_3, z$. The Kawamata blow-up $Y \to X$ at p resolves the indeterminacy and the morphism $Y \to Z$ is the flopping contraction which contracts the strict transforms of the 15 curves $(x_0 = a_3 = b_5 = 0)$ in X.

Construction 9.5.6 (elliptic involution)   Suppose the expression

$$X = (x_0 x_1^2 + ax_1 + x_2 x_0^3 + bx_0^2 + cx_0 + d = 0) \subset \mathbf{P}(1, a_1, a_2, a_3, a_4)$$

with permutation $x_0, \ldots, x_4$ of the homogeneous coordinates, where $a, b, c, d \in \mathbf{C}[x_2, x_3, x_4]$. We assume that x_0 has weight greater than one, by which X has a quotient singularity at $p = [1, 0, 0, 0, 0]$.

Let

$$y = x_1^2 + x_2 x_0^2 + bx_0 + c, \qquad z = yx_1 - ax_2 x_0 - ab.$$

The equation of X becomes $ax_1 + yx_0 + d = 0$. Cancelling the term x_1^2 in the relation $x_1(ax_1 + yx_0 + d) = 0$, one has $dx_1 + zx_0 + a(y - c) = 0$. Next consider the relation $-(x_2 x_0 + b)(ax_1 + yx_0 + d) = 0$. Cancelling the part $(x_2 x_0 + b)a$ by the definition of z and then the part $x_1^2 + x_2 x_0^2 + bx_0$ by the definition of y, one obtains $zx_1 - y(y - c) - (x_2 x_0 + b)d = 0$. These relations are arranged neatly as

$$\begin{pmatrix} a & y & d \\ d & z & a(y-c) \\ y & -ax_2 & -z - ab \\ z & -dx_2 & -y(y-c) - bd \end{pmatrix} \begin{pmatrix} x_1 \\ x_0 \\ 1 \end{pmatrix} = 0.$$

Multiplying this by the row vector $(dx_2, -ax_2, -z, y)$ from left, one defines a birational map from X to the hypersurface

$$Z = (z^2 + abz - y^3 + cy^2 - (a^2x_2 + bd)y + (a^2c + d^2)x_2 = 0)$$

in the weighted projective space with homogeneous coordinates x_2, x_3, x_4, y, z.

In the same manner as in Construction 9.5.4, we assume that it is the composite of the inverse of the Kawamata blow-up $\pi: Y \to X$ at p and a flopping contraction $\alpha: Y \to Z$. There exists an involution $\iota: Z \to Z$ which interchanges the two roots of the quadratic equation on z. The transformation

$$X \overset{\pi}{\leftarrow} Y \overset{\alpha}{\to} Z \overset{\iota}{\to} Z \overset{\alpha}{\leftarrow} Y \overset{\pi}{\to} X$$

is a Sarkisov link of type II. This is called an *elliptic involution* on X.

Example 9.5.7 Consider a general hypersurface $X_{25} \subset \mathbf{P}(1, 4, 5, 7, 9)$ of degree 25 of no 61 in Reid–Fletcher's list. Arranging the defining polynomial by the coordinate of degree nine and then by that of degree seven, one can write $X = (wv^2 + a_{16}v + yw^3 + b_{11}w^2 + c_{18}w + d_{25} = 0)$ with $a, b, c, d \in \mathbf{C}[x, y, z]$ by suitable homogeneous coordinates x, y, z, w, v, which has a quotient singularity of type $\frac{1}{7}(1, 5, 2)$ at $p = [0, 0, 0, 1, 0]$.

Taking $u = v^2 + yw^2 + bw + c$ and $t = uv - ayw - ab$, one defines a birational map from X to the hypersurface

$$Z_{54} = (t^2 + abt - u^3 + cu^2 - (a^2y + bd)u + (a^2c + d^2)y = 0)$$

of degree 54 in $\mathbf{P}(1, 4, 5, 18, 27)$ with homogeneous coordinates x, y, z, u, t. The Kawamata blow-up $\pi: Y \to X$ at p resolves the indeterminacy and the morphism $\alpha: Y \to Z$ is the flopping contraction which contracts the strict transforms of the 20 curves $(u = a_{16} = d_{25} = 0)$ in X. Indeed, α maps the exceptional divisor $\mathbf{P}(1, 5, 2)$ of π to a surface S by $[x, z, v] \mapsto [x, v^2, z, -av, -dv]$. From the equation in Construction 9.5.6 involving a 4×3 matrix, one derives that α is isomorphic outside S. Further from the definitions of u and t, α is isomorphic outside the locus $(y = b = v^2 + c = av + d = 0)$ besides the images of the 20 curves, which is empty since a, b, c, d are general.

Theorem 9.5.8 *Let $X \subset \mathbf{P}(1, a_1, a_2, a_3, a_4)$ be a general hypersurface in Theorem 9.5.3. Then every Sarkisov link from X is either a quadratic involution in Construction 9.5.4 or an elliptic involution in Construction 9.5.6.*

This was proved by a large number of computations. All the essential ideas have already appeared in the proof of Theorem 6.4.9. We do not reproduce the details.

As another application of the Sarkisov program, one can construct new examples of $\mathbf{Q}$-Fano threefolds.

Example 9.5.9 (Prokhorov–Reid [391])   Consider $X = \mathbf{P}^3$ with a hyperplane H_X. Take a quadric $S_X \sim 2H_X$ in X which is a cone with an ordinary double point x. Then $S_X \simeq \mathbf{P}(1, 1, 2)$ and $\mathrm{Cl}\, S_X = \mathbf{Z}[l]$ for a line l in S_X. Let C be a general non-lci curve in S_X of degree seven which is singular at x with multiplicity three. By Theorem 4.3.13, the symbolic blow-up $\pi \colon Y \to X$ along C is a divisorial contraction. Since C is general, Y has exactly one singular point, which is a quotient singularity of type $\frac{1}{2}(1, 1, 1)$, as seen in the course of the proof of Theorem 4.3.13.

The strict transform S in Y of S_X satisfies $S + E \sim 2H$ with the exceptional divisor E and the pull-back $H = \pi^* H_X$. Via the isomorphism $S \simeq S_X$ obtained from Theorem 4.3.13, $3H + 2S \sim 7H - 2E$ is restricted to S as $7H_X|_{S_X} - 2C \sim 0$. In particular, $3H + 2S$ is nef. By the base-point free theorem, it is semi-ample. Note that $K_Y \sim -2H - S$.

Let $\varphi \colon Y \to Z$ be the contraction defined by a multiple of $3H + 2S$. For a curve γ in Y, the intersection number $((3H + 2S) \cdot \gamma)$ is zero if and only if $\gamma \subset S$. Hence the exceptional locus of φ equals S and it is contracted to a point. Using $K_Y \sim -2H - S \equiv_Z S/3$ and $\rho(Y) = 2$, one sees that φ is a divisorial contraction to a terminal $\mathbf{Q}$-Fano threefold Z. It satisfies $K_Y = \varphi^* K_Z + S/3$, $\mathrm{Cl}\, Z = \mathbf{Z}[H_Z]$ and $-K_Z \sim 2H_Z$ for $H_Z = \varphi_* H$.

Since Y has one quotient singularity of type $\frac{1}{2}(1, 1, 1)$, φ is of type o2 with $r = 2$ in Theorem 3.2.2. By Hayakawa's work [181, section 9], $\varphi(S) \in Z$ is not $cD/3$. Hence by Theorem 3.5.5, φ is the Kawamata blow-up at a quotient singularity of type $\frac{1}{3}(1, 2, 1)$. The linear system $|H_Z|$ is mobile and has the same dimension three as $|H_X|$ has. By computation, $-K_Y^3 = 37/2$ and $H_Z^3 = 7/3$.

Example 9.5.10   Let Z be a terminal non-Gorenstein $\mathbf{Q}$-Fano threefold of Fano index two, in which $-K_Z \sim 2H_Z$ by some divisor H_Z. Assume that every divisor on Z is $\mathbf{Q}$-linearly equivalent to an integral multiple of H_Z. Prokhorov and Reid proved that $|H_Z|$ has dimension at most four and that Z is constructed as follows whenever $\dim |H_Z| = 4$.

Instead of Example 9.5.9, we start with a smooth quadric threefold X in $\mathbf{P}^4$. Take a hyperplane section S_X of X which is a cone with an ordinary double point x. It is the intersection $X \cap T_{X,x}$ with the tangent space $T_{X,x} = \mathbf{P}^3$ as a hyperplane in $\mathbf{P}^4$. Let C be a general non-lci curve in S_X of degree five with multiplicity three at x. Like Example 9.5.9, the symbolic blow-up of X along C is contracted to a terminal $\mathbf{Q}$-Fano threefold Z by a multiple of $3H + 2S$, where H is the pull-back of a hyperplane section and S is the strict transform of S_X. Then $\mathrm{Cl}\, Z = \mathbf{Z}[H_Z]$, $-K_Z \sim 2H_Z$ and $|H_Z|$ is mobile of dimension four for the push-forward H_Z of H.

Example 9.5.11 Ducat [112] found more examples of **Q**-factorial terminal **Q**-Fano threefolds with Picard number one by the method of Prokhorov and Reid. Amongst them is an example of Fano index three, which is constructed from the symbolic blow-up of a quartic X_4 in $\mathbf{P}(1, 1, 2, 2, 3)$ along a curve C of degree nine with a singular point of multiplicity three.

Kaloghiros [219] introduced the notion of a *Sarkisov relation*, which is a loop of Sarkisov links. Ahmadinezhad and Zucconi exhibited an interesting example of a Sarkisov relation.

Example 9.5.12 (Ahmadinezhad–Zucconi [5]) We shall illustrate the example without details. With Notation 6.4.6, let $X \subset \mathbf{P}(1^4, 2^2, 3)$ with homogeneous coordinates $x_0, \ldots, x_3, y_0, y_1, z$ be the **Q**-Fano threefold defined by the maximal Pfaffians of the skew-symmetric 5×5 matrix

$$
\begin{pmatrix}
y_0 & A_3 & y_1 + C_2 & -x_1 \\
 & B_3 & D_2 & x_0 \\
 & & z & -y_1 \\
 & & & x_3
\end{pmatrix},
$$

where A_3, B_3 are general cubic forms and C_2, D_2 are general quadratic forms in $x_0, \ldots, x_3$. It is quasi-smooth with two quotient singularities p_1 of type $\frac{1}{2}(1, 1, 1)$ and p_2 of type $\frac{1}{3}(1, 2, 1)$. It is **Q**-factorial with Picard number one by the Lefschetz hyperplane theorem and Poincaré duality for homology manifolds used in the proof of Lemma 8.3.6. Ahmadinezhad and Zucconi exhibited the following Sarkisov links from the Mori fibre space $X \to \operatorname{Spec} \mathbf{C}$ as summarised in Figure 9.1.

The two-ray game from the Kawamata blow-up $X_1 \to X$ at p_1 produces a Sarkisov link of type II. It ends with a divisorial contraction $Y_1 \to Y$ which is the blow-up at an ordinary double point. The **Q**-Fano threefold Y is defined in $\mathbf{P}(1^5, 2)$ by two cubics and it has a quotient singularity q of type $\frac{1}{2}(1, 1, 1)$. The two-ray game from the Kawamata blow-up $Y_2 \to Y$ at q produces a Sarkisov link of type I which ends with a del Pezzo fibration $Z \to \mathbf{P}^1$ of degree three.

The two-ray game from the Kawamata blow-up $X_2 \to X$ at p_2 produces a Sarkisov link of type I. It ends with a del Pezzo fibration $Z' \to \mathbf{P}^1$ of degree three. The Mori fibre spaces $Z/\mathbf{P}^1$ and $Z'/\mathbf{P}^1$ are joined by a square Sarkisov link of type II. In Figure 9.1, $W \to Z$ is the blow-up at an ordinary double point and $W' \to Z'$ is the Kawamata blow-up at a singularity of type $\frac{1}{2}(1, 1, 1)$.

The birational rigidity of a **Q**-Fano variety is related to K-stability. The notion of K-stability was introduced by Tian [442] in the context of differential geometry to describe when a Fano variety admits a *Kähler–Einstein metric*,

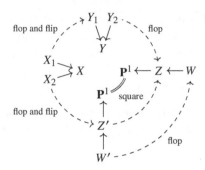

Figure 9.1 The circle

that is, a Kähler form ω satisfying the equation Ric $\omega = \lambda\omega$ with the associated Ricci form Ric ω and a real constant λ. Donaldson [108] reformulated this algebraically in terms of the Donaldson–Futaki invariant.

Definition 9.5.13 Let (X, L) be a *polarised* variety which consists of a normal projective variety X and an ample divisor L on X. A *test configuration* for (X, L) is a projective morphism $(\mathfrak{X}, \mathfrak{L}) \to \mathbf{A}^1$ from a normal variety $\mathfrak{X}$ with a relatively ample invertible sheaf $\mathfrak{L}$ together with a $\mathbf{G}_m$-action on $(\mathfrak{X}, \mathfrak{L})$ such that $(\mathfrak{X}, \mathfrak{L}) \to \mathbf{A}^1$ is $\mathbf{G}_m$-equivariant, where $\mathbf{G}_m$ acts on $\mathbf{A}^1$ multiplicatively, and such that $(\mathfrak{X}, \mathfrak{L})|_{\mathbf{A}^1 \setminus o}$ is $\mathbf{G}_m$-equivariantly isomorphic to $(X, \mathscr{O}_X(rL)) \times (\mathbf{A}^1 \setminus o)$ for some positive integer r, in which rL is Cartier.

The test configuration $(\mathfrak{X}/\mathbf{A}^1, \mathfrak{L})$ is naturally compactified to a contraction $(\bar{\mathfrak{X}}, \bar{\mathfrak{L}}) \to \mathbf{P}^1$. One can write $h^0(\mathscr{O}_X(lL)) = a_0 l^n + a_1 l^{n-1} + O(l^{n-2})$ with $a_0 = (L^n)/n!$ and $a_1 = -2^{-1}(K_X \cdot L^{n-1})/(n-1)!$ for large and divisible l, where $n = \dim X$. Let $\mathfrak{L}_o$ be the restriction of $\mathfrak{L}$ to the central fibre $\mathfrak{X} \times_{\mathbf{A}^1} o$. For a multiple l of r with $H^0(\mathfrak{L}_o^{\otimes l/r}) \neq 0$, the *total weight* $w(l)$ of the action of $\mathbf{G}_m$ on $H^0(\mathfrak{L}_o^{\otimes l/r})$ is the integer such that $t \in \mathbf{G}_m = \mathbf{C}^\times$ defines multiplication by $t^{w(l)}$ on $\det H^0(\mathfrak{L}_o^{\otimes l/r}) \simeq \mathbf{C}$. One can write $w(l) = b_0 l^{n+1} + b_1 l^n + O(l^{n-1})$ with $b_0 = (r^{-1}\bar{\mathfrak{L}})^{n+1}/(n+1)!$ and $b_1 = -2^{-1}(K_{\bar{\mathfrak{X}}/\mathbf{P}^1} \cdot (r^{-1}\bar{\mathfrak{L}})^n)/n!$, where $K_{\bar{\mathfrak{X}}/\mathbf{P}^1}$ is $K_{\bar{\mathfrak{X}}}$ minus the pull-back of $K_{\mathbf{P}^1}$. The *Donaldson–Futaki invariant* DF$(\mathfrak{X}, \mathfrak{L})$ of $(\mathfrak{X}/\mathbf{A}^1, \mathfrak{L})$ is the rational number $(a_1 b_0 - a_0 b_1)/a_0^2$, which appears in the expression

$$\frac{w(l)}{l h^0(\mathscr{O}_X(lL))} = \frac{b_0}{a_0} - \mathrm{DF}(\mathfrak{X}, \mathfrak{L}) l^{-1} + O(l^{-2}).$$

Definition 9.5.14 We say that a polarised variety (X, L) is *K-stable* (resp. *K-semistable*) if the Donaldson–Futaki invariant DF$(\mathfrak{X}, \mathfrak{L})$ is positive (resp. non-negative) for any non-trivial test configuration $(\mathfrak{X}, \mathfrak{L})$ for (X, L). A K-semistable polarised variety (X, L) is said to be *K-polystable* if DF$(\mathfrak{X}, \mathfrak{L})$

becomes zero only when $\mathfrak{X}$ is isomorphic over $\mathbf{A}^1$ to the product $X \times \mathbf{A}^1$. We say that a $\mathbf{Q}$-Fano variety X is *K-stable, K-polystable* or *K-semistable* if so is $(X, -K_X)$.

X. Chen, Donaldson and Sun [84] proved that every K-polystable Fano variety admits a Kähler–Einstein metric. See also [443]. The result has been generalised to $\mathbf{Q}$-Fano varieties [40], [293], [296]. The opposite direction for $\mathbf{Q}$-Fano varieties is due to Berman [39], which has developed from Tian's work [442]. These settle the *Yau–Tian–Donaldson conjecture* for $\mathbf{Q}$-Fano varieties as stated below.

Theorem 9.5.15 *A $\mathbf{Q}$-Fano variety admits a Kähler–Einstein metric if and only if it is K-polystable.*

Tian introduced the alpha invariant in his earlier work [441] on the existence of Kähler–Einstein metric. In the algebraic context, it amounts to the lc threshold of the rational anti-canonical system.

Definition 9.5.16 The *alpha invariant* $\alpha(X)$ of a $\mathbf{Q}$-Fano variety X is the greatest rational number α such that $(X, \alpha D)$ is lc for any effective $\mathbf{Q}$-divisor $D \sim_{\mathbf{Q}} -K_X$.

Example 9.5.17 The projective space $\mathbf{P}^n$ has $\alpha(\mathbf{P}^n) = 1/(n+1)$. Cheltsov [72] computed the alpha invariant of a del Pezzo surface S_d of degree d. The sharp bound $c_d \leq \alpha(S_d)$ holds with the number c_d in Lemma 8.2.7.

Tian's result corresponds to the following.

Theorem 9.5.18 ([374]) *If a $\mathbf{Q}$-Fano variety X of dimension n has $\alpha(X) > n/(n+1)$ (resp. $\geq n/(n+1)$), then X is K-stable (resp. K-semistable).*

One can say that the greater the alpha invariant is, the smaller the rational anti-canonical system is. It is thus tentative to expect K-stability from birational superrigidity as supported by the next observation.

Proposition 9.5.19 *Let X be a $\mathbf{Q}$-factorial terminal $\mathbf{Q}$-Fano variety with Picard number one. Then X is birationally superrigid if and only if $(X, \mu^{-1}H)$ is canonical for the general member H of any mobile linear system $\mathcal{H}$ on X with Sarkisov threshold μ at least one.*

Proof By the definition of the Sarkisov threshold, $K_X + \mu^{-1}H \sim_{\mathbf{Q}} 0$. The if part is implied by the Noether–Fano inequality, Theorem 6.1.12(ii). If $(X, \mu^{-1}H)$ is not canonical, then Proposition 6.2.8, which holds even for mobile $\mathcal{H}$ by its proof, produces a non-trivial link from X. This is the only-if part. □

The K-(semi)stability for **Q**-Fano varieties is characterised in terms of volume.

Definition 9.5.20 (K. Fujita)   Let X be a **Q**-Fano variety of index r. Let E be a divisor over X and take a resolution $\mu\colon X' \to X$ which realises E as a divisor on X'. We say that E is *dreamy* if the bigraded **C**-algebra

$$R = \bigoplus_{(i,j)\in\mathbf{N}^2} H^0(\mathcal{O}_{X'}(-ir\mu^*K_X - jE))$$

is finitely generated. The definition is independent of the choice of μ because the space $H^0(\mathcal{O}_{X'}(-ir\mu^*K_X - jE))$ defines the linear subsystem of $|-irK_X|$ which consists of $D \sim -irK_X$ with $\mathrm{ord}_E D \geq j$.

Theorem 9.5.21 (K. Fujita [136], Li [292])   *Let X be a **Q**-Fano variety. For a divisor E over X, let*

$$\beta(E) = a_E(X)\,\mathrm{vol}(-K_X) - \int_0^\infty \mathrm{vol}(-K_X - tE)\,dt.$$

Here $a_E(X)$ is the log discrepancy of E and $\mathrm{vol}(-K_X - tE)$ is the volume of $-\mu^*K_X - tE$ on a resolution $\mu\colon X' \to X$ realising E as a divisor, which is independent of μ. Then X is K-stable (resp. K-semistable) if and only if $\beta(E) > 0$ (resp. $\beta(E) \geq 0$) for all dreamy divisors E over X. Actually, X is K-semistable if and only if $\beta(E) \geq 0$ for all divisors E over X.

We apply this to birationally superrigid **Q**-Fano varieties.

Theorem 9.5.22 (Stibitz–Zhuang [434])   *Let X be a birationally superrigid **Q**-factorial terminal **Q**-Fano variety with Picard number one. If $\alpha(X) \geq 1/2$ and $X \not\cong \mathbf{P}^1$, then X is K-stable.*

Proof   We may assume that the dimension n of X is at least two. Assuming that X is not K-stable, we shall derive the inequality $\alpha(X) < 1/2$.

Step 1   By Theorem 9.5.21, there exists a dreamy divisor E over X such that

$$b = \frac{1}{\mathrm{vol}(-K_X)} \int_0^\infty \mathrm{vol}(-K_X - tE)\,dt$$

is not less than the log discrepancy $a = a_E(X) \geq 1$. We fix a resolution $\mu\colon X' \to X$ which realises E as a divisor. We write $v(t) = \mathrm{vol}(-K_X - tE)$ for simplicity, which extends to a continuous function on **R** as mentioned in Definition 1.7.7.

Let $|-K_X|_{\mathbf{Q}}$ denote the set of effective **Q**-divisors G on X such that $G \sim_{\mathbf{Q}} -K_X$. Let m be the supremum of $\mathrm{ord}_E D$ for $D \in |-K_X|_{\mathbf{Q}}$. This is attained as $m = \mathrm{ord}_E D$ by some D, because m is the maximum of j/ir for $(i,j) \in \mathbf{N}_{>0}\times\mathbf{N}$

such that $H^0(\mathscr{O}_{X'}(-ir\mu^*K_X - jE)) \neq 0$, which exists by the finite generation of the algebra R in Definition 9.5.20. Then

$$a \leq b = \frac{1}{v(0)} \int_0^m v(t)dt < \frac{1}{v(0)} \int_0^m v(0)dt = m,$$

where the strict inequality comes from the continuity of v with $0 = v(m) < v(0)$.

Since X is $\mathbf{Q}$-factorial with $\rho(X) = 1$, one can write $D = \sum_i t_i D_i$ with $t_i \in \mathbf{Q}_{>0}$ and irreducible $D_i \in |-K_X|_{\mathbf{Q}}$, that is, each D_i is a rational multiple of a prime divisor. Then $\sum_i t_i = 1$, and $\mathrm{ord}_E D_i = m$ for all i by the estimate $\mathrm{ord}_E D_i \leq m$. Thus we may assume that D is an irreducible $\mathbf{Q}$-divisor.

Step 2 We claim that D is a unique irreducible member of $|-K_X|_{\mathbf{Q}}$ with $\mathrm{ord}_E D \geq a$. This is trivial if E appears on X. Suppose that E is exceptional over X. If another irreducible member $\Delta \in |-K_X|_{\mathbf{Q}}$ satisfied $\mathrm{ord}_E \Delta \geq a$, then for a positive integer l with $lD \sim l\Delta$, the linear subsystem $\mathscr{H}$ of $|-lK_X|$ spanned by lD and $l\Delta$ is mobile and would have a general member H with $\mathrm{ord}_E H \geq la = la_E(X)$. The pair $(X, l^{-1}H)$ is not canonical, which contradicts Proposition 9.5.19.

Fix $a \leq t \leq m$. Suppose that $\Gamma \in |-K_X|_{\mathbf{Q}}$ has $\mathrm{ord}_E \Gamma = t$. Write $\Gamma = \Gamma_0 + cD$ with $c \in \mathbf{Q}_{\geq 0}$ so that Γ_0 does not contain the prime support of D. Then $\Gamma_0 \sim_{\mathbf{Q}} (1 - c)(-K_X)$, and the claim above gives $\mathrm{ord}_E \Gamma_0 = t - cm \leq (1 - c)a$, that is, $(t - a)/(m - a) \leq c$. In other words, $(t - a)/(m - a) \cdot D \leq \Gamma$. Thus for the strict transform $D' \sim_{\mathbf{Q}} -\mu^*K_X - mE$ in X' of D, the decomposition

$$-\mu^*K_X - tE \sim_{\mathbf{Q}} \frac{m - t}{m - a}(-\mu^*K_X - aE) + \frac{t - a}{m - a}D'$$

yields the isomorphism

$$H^0(\mathscr{O}_{X'}(l(-\mu^*K_X - tE))) \simeq H^0\left(\mathscr{O}_{X'}\left(l \cdot \frac{m - t}{m - a}(-\mu^*K_X - aE)\right)\right)$$

for any sufficiently large and divisible integer l.

Step 3 We use the expression of $v(t)$ in terms of the restricted volume defined in [115]. For a $\mathbf{Q}$-divisor G on X', let $d_{X'|E}(lG)$ denote the dimension of the image of the restriction map $H^0(\mathscr{O}_{X'}(\lfloor lG \rfloor)) \to H^0(\mathscr{O}_{X'}(\lfloor lG \rfloor) \otimes \mathscr{O}_E)$. The *restricted volume* $\mathrm{vol}_{X'|E}(G)$ of G along E is defined as

$$\mathrm{vol}_{X'|E}(G) = \limsup_{l \to \infty} \frac{d_{X'|E}(lG)}{l^e/e!},$$

where $e = n - 1$ is the dimension of E. We write $r(t) = \mathrm{vol}_{X'|E}(-\mu^*K_X - tE)$. By [115, theorem A] and the explicit description about $t = m$ below, $r(t)$ extends to a continuous function on $\mathbf{R}$. Note that the continuity of restricted volume may fail in general [115, example 5.10].

Using an effective $\mathbf{Q}$-divisor F on X' such that $-\mu^* K_X - F$ is ample and such that the coefficient ε of E in F is at most a, one can write

$$-\mu^* K_X - aE \sim_{\mathbf{Q}} \frac{m - a}{m - \varepsilon}(-\mu^* K_X - F) + \frac{m - a}{m - \varepsilon}(F - \varepsilon E) + \frac{a - \varepsilon}{m - \varepsilon}D'.$$

Hence $r(a)$ is positive. The isomorphism in Step 2 implies the equality

$$r(t) = \left(\frac{m - t}{m - a}\right)^{n-1} r(a)$$

for $a \leq t \leq m$. On the other hand, $r(t)$ satisfies the inequality

$$r(t) \geq \left(\frac{t}{a}\right)^{n-1} r(a)$$

for $0 \leq t \leq a$ since $-\mu^* K_X - tE$ is decomposed into $(t/a)(-\mu^* K_X - aE)$ and semi-ample $(1 - t/a)(-\mu^* K_X)$.

Step 4 One has the expression $v(t) = n \int_t^m r(s)ds$ for $0 \leq t \leq m$ from [289, corollary 4.27(ii)]. The relation $bv(0) = \int_0^m v(t)dt$ is now rewritten as

$$b \int_0^m r(s)ds = \int_0^m \int_t^m r(s)dsdt = \int_0^m sr(s)ds$$

and hence

$$\int_0^a (b - t)r(t)dt = \int_a^m (t - b)r(t)dt.$$

Hence from the estimates in Step 3, one obtains

$$\int_0^a (b - t)\left(\frac{t}{a}\right)^{n-1} dt \leq \int_a^m (t - b)\left(\frac{m - t}{m - a}\right)^{n-1} dt.$$

By calculus, this is computed to be $(n + 1)(b - a) \leq m - 2a$. Therefore $\operatorname{ord}_E D = m \geq 2a = 2a_E(D)$. Thus $(X, D/2)$ is not klt and the weak inequality $\alpha(X) \leq 1/2$ follows.

Step 5 We exclude the possibility of $\alpha(X) = 1/2$. If this were the case, then $m = 2a$ and $b = a$. The arguments in Steps 2 to 4 remain valid even if a is replaced by a rational number a' with $a - 1 < a' < a$, by which one concludes that $(n + 1)(b - a') \leq m - 2a'$. When $m = 2a$ and $b = a$, this means that $(n - 1)(a - a') \leq 0$, which contradicts $n \geq 2$. $\square$

The following K-stability is obtained by verifying the assumption in Theorem 9.5.22.

Theorem 9.5.23 (Cheltsov [73], Kim–Odaka–Won [253])   *Every terminal quasi-smooth $\mathbf{Q}$-Fano threefold in the 95 families of Reid–Fletcher has alpha* *invariant at least one-half. Hence if it is birationally superrigid, then it is K-stable.*

The K-semistability for **Q**-Fano varieties is an open condition.

Theorem 9.5.24 (Blum–Liu–Xu [49], [473])   *Let $X \to B$ be a flat projective morphism from a **Q**-Gorenstein normal variety such that the fibre $X_b = X \times_B b$ is **Q**-Fano for all $b \in B$. Then the set of $b \in B$ such that X_b is K-semistable is open in B.*

10

Minimal Models

The minimal model program outputs a minimal model when the canonical divisor is pseudo-effective. The final chapter treats minimal threefolds.

The abundance conjecture predicts that the nef canonical divisor of a minimal model is semi-ample. We explain the abundance for threefolds due to Miyaoka and Kawamata according to the numerical Kodaira dimension v. The abundance is known in an arbitrary dimension when v equals the dimension or zero. We deal with the intermediate case when v is one or two.

The initial step is to prove the non-vanishing which means the existence of a global section of some pluricanonical divisor. If the threefold has non-zero irregularity, then the Albanese map provides enough geometric information. In the case of irregularity zero, Miyaoka applied the generic semi-positivity of the sheaf of differentials and the pseudo-effectivity of the second Chern class via positive characteristic. We derive abundance from non-vanishing after replacing the threefold by a special divisorially log terminal pair. The abundance for $v = 1$ is obtained by moving the boundary infinitesimally, whilst that for $v = 2$ is by a certain pseudo-effectivity involving the second Chern class.

Birational minimal models have many features in common. They are connected by flops and have the same Betti and Hodge numbers. In dimension three, they turn out to have the same analytic singularities. This is a consequence of the explicit description of a threefold flop by involution due to Kollár.

By Birkar, Cascini, Hacon and McKernan, a variety of general type admits only finitely many birational maps to minimal models. This fails in general but one can still expect the finiteness of the number of minimal models ignoring the structure of the birational map. This is a part of Kawamata and Morrison's cone conjecture for Calabi–Yau fibrations. We explain Kawamata's work on the conjecture for threefold fibrations with non-trivial base.

It is desirable to control the behaviour of the l-th pluricanonical map in terms of l. In dimension three, there exists a uniform number for l such that the l-th map is birational to the Iitaka fibration. We find this number explicitly in the case of general type, as well as an explicit number for which the l-th pluricanonical divisor has a global section.

10.1 Non-Vanishing

Let X be a terminal projective threefold such that K_X is nef. By Theorem 6.3.3, X is never uniruled. In this and the next section, we shall prove the semi-ampleness of K_X according to the numerical Kodaira dimension $\nu = \nu(X)$ in Definition 1.7.6. Thanks to Theorem 1.7.12, the semi-ampleness is reduced to the abundance $\kappa = \nu$ with the Kodaira dimension $\kappa = \kappa(X)$ in Definition 1.7.3. The abundance is known when ν equals the dimension from the base-point free theorem, and when ν equals zero from Theorem 1.7.13. We need to investigate the cases $\nu = 1$ and 2. The purpose of this section is to prove that $\kappa \geq 0$, namely the non-vanishing of $H^0(\mathscr{O}_X(lK_X))$ for some positive integer l.

Theorem 10.1.1 *Let X be a terminal projective threefold. If K_X is nef, then $\kappa(X) \geq 0$.*

We take the *irregularity* $q = h^1(\mathscr{O}_X)$. When X is smooth, it equals $h^0(\Omega_X)$ via the Hodge symmetry in Theorem 6.3.14. If q is positive, then one can make use of the Albanese map of X explained below. A reference is [458, section 9].

By a *complex torus*, we mean a compact Kähler manifold V/Γ which is the quotient of the complex vector space $V = \mathbf{C}^n$ by a lattice $\Gamma \simeq \mathbf{Z}^{2n}$ spanning V as a real vector space of dimension $2n$. A complex torus is an abelian variety if it is projective.

Lemma 10.1.2 *Let T be a complex torus and let Z be an integral closed analytic subspace of T. Then $H^0(\omega_{Z'}) \neq 0$ for any resolution Z' of Z.*

Proof Write $T = V/\Gamma$ with $V = \mathbf{C}^n$ and $\Gamma \simeq \mathbf{Z}^{2n}$. Let m denote the dimension of Z. After translation, the origin o of T is a smooth point of Z. Then there exists a basis $\{z_1, \ldots, z_n\}$ of V such that $\bigwedge^m \Omega_Z$ is generated at o by $\theta = dz_1 \wedge \cdots \wedge dz_m$. The pull-back to Z' of the global m-form θ on T is a non-zero element in $H^0(\omega_{Z'})$. $\qquad\square$

Let X be a compact Kähler manifold. The map

$$H^0(X, \Omega_X) \times H_1(X, \mathbf{Z}) \to \mathbf{C}, \qquad (\omega, [\gamma]) \mapsto \int_\gamma \omega$$

defines the quotient $T = H^0(X, \Omega_X)^\vee / H_1(X, \mathbf{Z})$ as a complex torus. It becomes an abelian variety if X is projective [458, proposition 9.15]. For a fixed base point $o \in X$, the integral along a path from o to a point in X defines a morphism $\alpha: X \to T$. It satisfies the universal property that every morphism $\alpha': X \to T'$ to a complex torus factors through α uniquely as $\alpha' = t \circ \alpha$ for a composite $t: T \to T'$ of a group homomorphism and a translation.

Definition 10.1.3   The morphism $\alpha: X \to T = H^0(X, \Omega_X)^\vee / H_1(X, \mathbf{Z})$ is called the *Albanese map* of X.

Proof of Theorem 10.1.1 *for* $q \geq 1$ Recall that Kodaira dimension is a birational invariant. Let X be a smooth projective threefold with $q = h^1(\mathcal{O}_X) \geq 1$. Take the Albanese map $\alpha: X \to T$, where T is an abelian variety of dimension q. It follows from the universal property of the Albanese map that the image $Z = \alpha(X)$ is not a point. Let $m \geq 1$ be the dimension of Z. We take smooth birational models X' and Z' of X and Z for which α induces a morphism $\alpha': X' \to Z'$. By Lemma 10.1.2, Z' has Kodaira dimension $\kappa(Z') \geq 0$.

If $m = 3$, then α' is generically finite and the inequality $\kappa(X') \geq \kappa(Z')$ follows from the ramification formula in Theorem 2.2.20. Hence $\kappa(X) = \kappa(X') \geq 0$. We assume that $m = 1$ or 2. It suffices to prove that if $\kappa(X') = -\infty$, then X' is uniruled. We may assume that α' is the composite of a contraction $X' \to Y$ to a smooth variety and a generically finite morphism $Y \to Z'$, for which $\kappa(Y) \geq \kappa(Z') \geq 0$. We apply the assertion $\kappa(X') \geq \kappa(F) + \kappa(Y)$ of the Iitaka conjecture $C_{3,m}$ in Theorem 1.7.18 where F is the very general fibre of $X' \to Y$. Remark that $C_{3,1}$ was first obtained by Viehweg [461]. This inequality gives the implication from $\kappa(X') = -\infty$ to $\kappa(F) = -\infty$. Thus F is a ruled surface if $m = 1$ and a rational curve if $m = 2$. In both cases, X' is uniruled by Theorem 6.3.3. Note that $K_{X'}|_F = K_F$.   □

The non-vanishing in the case of irregularity zero was proved by Miyaoka [327]. The main tool is the generic semi-positivity of the sheaf Ω_X of differentials. In order to formulate this, we prepare the notion of stability for torsion-free sheaves, in which we work over an algebraically closed field k in general. The book [194] is a good reference. See also [325, section 2].

Let X be a normal projective variety of dimension n over k. An $(n-1)$-tuple $H = (H_1, \dots, H_{n-1})$ of ample divisors on X will be called a *polarised system* on X. By a *general* curve C in mH, where m is a positive integer, we assume that mH_i is very ample for all i and mean that C is the intersection $\bigcap_{i=1}^{n-1} A_i$ of general members A_i of $|mH_i|$. Let $\mathscr{E}$ be a torsion-free coherent sheaf on X. Its rank is denoted by $\mathrm{rk}\, \mathscr{E}$. The restriction $\mathscr{E}|_C$ to a general curve C in mH is a locally free sheaf on C. We make the natural definition of the *degree* $\deg_H \mathscr{E}$ of $\mathscr{E}$ along H by $\deg_H \mathscr{E} = m^{-(n-1)} \deg \mathscr{E}|_C$, where the degree of $\mathscr{E}|_C$ is that of the determinant of $\mathcal{E}|_C$. We define the *slope* $\mu_H(\mathcal{E})$ of $\mathcal{E}$ with respect to H as

$$\mu_H(\mathcal{E}) = \frac{\deg_H \mathcal{E}}{\operatorname{rk} \mathcal{E}}.$$

Definition 10.1.4 We say that a torsion-free coherent sheaf $\mathcal{E}$ is H-*semistable* (resp. H-*stable*) if $\mu_H(\mathcal{F}) \leq \mu_H(\mathcal{E})$ (resp. $\mu_H(\mathcal{F}) < \mu_H(\mathcal{E})$) for any non-zero subsheaf $\mathcal{F}$ of $\mathcal{E}$ with smaller rank $\operatorname{rk} \mathcal{F} < \operatorname{rk} \mathcal{E}$. An H-semistable sheaf is said to be H-*polystable* if it is a direct sum of H-stable sheaves, all of which have the same slope by definition. When $H = (H, \ldots, H)$, the H-stability is referred to as stability with respect to H. We omit reference to H when it is clear and thus write $\mu(\mathcal{E}) = \mu_H(\mathcal{E})$.

Remark 10.1.5 The stability in the above definition is also called *slope stability* or *Mumford–Takemoto stability*. There is another notion of stability, called *Gieseker stability*. It is defined in terms of the reduced Hilbert polynomial instead of the slope. These notions coincide on curves.

Thanks to the following filtrations, a (semi)stable sheaf is considered to be a building block of a torsion-free sheaf. See [194, theorem 1.6.7] for example.

Definition 10.1.6 (i) Let $\mathcal{E}$ be a torsion-free coherent sheaf. There exists a unique filtration $0 = \mathcal{E}_0 \subset \mathcal{E}_1 \subset \cdots \subset \mathcal{E}_l = \mathcal{E}$ by subsheaves such that every quotient $\mathcal{F}_i = \mathcal{E}_i/\mathcal{E}_{i-1}$ is torsion-free and H-semistable and such that $\mu_H(\mathcal{F}_1) > \mu_H(\mathcal{F}_2) > \cdots > \mu_H(\mathcal{F}_l)$. It is called the *Harder–Narasimhan filtration* of $\mathcal{E}$. We define $\mu_{H,\min}(\mathcal{E}) = \mu_H(\mathcal{F}_l)$ and $\mu_{H,\max}(\mathcal{E}) = \mu_H(\mathcal{F}_1)$.

(ii) Let $\mathcal{F}$ be an H-semistable sheaf. There exists a filtration $0 = \mathcal{F}_0 \subset \mathcal{F}_1 \subset \cdots \subset \mathcal{F}_m = \mathcal{F}$ by subsheaves such that every quotient $\mathcal{G}_i = \mathcal{F}_i/\mathcal{F}_{i-1}$ is torsion-free and H-stable with $\mu_H(\mathcal{G}_i) = \mu_H(\mathcal{F})$. It is called a *Jordan–Hölder filtration* of $\mathcal{F}$. The direct sum $\bigoplus_i \mathcal{G}_i$ is unique up to isomorphism.

For torsion-free coherent sheaves $\mathcal{E}$ and $\mathcal{F}$, if $\mu_{H,\min}(\mathcal{E})$ is greater than $\mu_{H,\max}(\mathcal{F})$, then $\operatorname{Hom}(\mathcal{E}, \mathcal{F})$ is zero. In like manner if $\mathcal{E}$ is H-stable, then $\operatorname{Hom}(\mathcal{E}, \mathcal{E}) \simeq k$. If $\mathcal{E}$ is H-(semi)stable, then so is the dual $\mathcal{E}^\vee$. Further in characteristic zero, tensor products preserve semistability and polystability as below. See [194, theorems 3.1.4, 3.2.11].

Theorem 10.1.7 *Let X be a normal projective variety in characteristic zero endowed with a polarised system* H. *If $\mathcal{E}$ and $\mathcal{F}$ are H-semistable (or H-polystable) sheaves on X, then so is the quotient of $\mathcal{E} \otimes \mathcal{F}$ by torsion.*

A torsion-free sheaf $\mathcal{E}$ is (semi)stable if so is the restriction $\mathcal{E}|_C$ to a general curve C in mH. Actually, the converse is also true.

Theorem 10.1.8 (Mehta–Ramanathan [313], [314]) *Let X be a normal projective variety in any characteristic with a polarised system* H. *Let $\mathcal{E}$ be an* H-*semistable* (*resp.* H-*stable*) *sheaf on X. Then for any sufficiently large and divisible integer m, the restriction $\mathcal{E}|_C$ to a general curve C in mH is semistable* (*resp. stable*) *on C.*

Finally the (semi)stability is an open condition in a flat family. Apply the argument in [194, proposition 2.3.1].

Miyaoka's *generic semi-positivity* is stated as follows.

Theorem 10.1.9 (Miyaoka [324, corollary 8.6]) *Let X be a normal projective complex variety with a polarised system* H. *If X is not uniruled, then for any sufficiently large and divisible m, every locally free quotient of $\Omega_X|_C$ on a general curve C in mH has non-negative degree.*

We take the dual $\mathcal{T}_X = \Omega_X^\vee$, which is reflexive [179, corollary 1.2]. The theorem is equivalent to the statement that if $\mu_{\max}(\mathcal{T}_X)$ is positive, then X is uniruled. We shall prove it along the simplified argument by Shepherd-Barron [416]. We make a reduction to positive characteristic.

Henceforth we assume that $\mu_{\max}(\mathcal{T}_X) > 0$. Let $\mathcal{F}$ be the first piece of the Harder–Narasimhan filtration of $\mathcal{T}_X$. To be precise, $\mathcal{F}$ is a semistable subsheaf of $\mathcal{T}_X$ with torsion-free quotient $\mathcal{G} = \mathcal{T}_X/\mathcal{F}$ such that $\mu(\mathcal{F}) = \mu_{\max}(\mathcal{T}_X)$ and such that $\mu(\mathcal{F}) > \mu_{\max}(\mathcal{G})$ or $\mathcal{G} = 0$. It fits into the exact sequence

$$0 \to \mathcal{F} \to \mathcal{T}_X \to \mathcal{G} \to 0.$$

Let C be a general curve in mH for sufficiently large and divisible m. By Theorem 10.1.8, the restriction $\mathcal{F}|_C$ is a semistable locally free sheaf on C of positive degree.

Lemma 10.1.10 $\operatorname{Hom}(\wedge^2 \mathcal{F}, \mathcal{G}) = 0.$

Proof By Theorem 10.1.7, the quotient $\mathcal{Q}$ of $\wedge^2 \mathcal{F}$ by torsion is semistable. It has slope $\mu(\mathcal{Q}) = 2\mu(\mathcal{F}) > \mu(\mathcal{F}) > \mu_{\max}(\mathcal{G})$ or $\mathcal{G} = 0$. Hence $\operatorname{Hom}(\mathcal{Q}, \mathcal{G}) = 0$, from which the lemma follows. □

Let n denote the dimension of X. Fix an ample Cartier divisor H on X. We seek a rational curve D through a general point in X such that

$$(H \cdot D) \leq \frac{2n(H \cdot C)}{\deg \mathcal{F}|_C}.$$

The uniruledness of X is derived from the existence of such D by using the Hilbert scheme parametrising curves in X of bounded degree.

We reduce the existence of D to that in positive characteristic. The variety X is realised as the base change of a scheme X_R of finite type over a subring R of $\mathbf{C}$ which is finitely generated over $\mathbf{Z}$. For a closed point t in $\operatorname{Spec} R$, we write the geometric fibre $X_{\bar{t}} = X_R \times_{\operatorname{Spec} R} \bar{t}$ at $\bar{t} = \operatorname{Spec} \overline{k(t)}$, where $\overline{k(t)}$ is the algebraic closure of the residue field $k(t)$ at t. It is an algebraic scheme over $\overline{k(t)}$ of positive characteristic. One can extend to X_R every necessary object on X such as H, H, C or $\mathscr{F}$, enlarging R if necessary. For example, H is extended to H_R on X_R so that $\mathsf{H} = \mathsf{H}_R \times_{\operatorname{Spec} R} \operatorname{Spec} \mathbf{C}$. We express the fibre at $\bar{t}$ of the extended object by adding the subscript $\bar{t}$, such as $\mathsf{H}_{\bar{t}}$, $H_{\bar{t}}$, $C_{\bar{t}}$ or $\mathscr{F}_{\bar{t}}$.

When t is general, the geometric fibre $X_{\bar{t}}$ inherits the assumptions X has. For example, $\mu(\mathscr{G}|_C) = \mu(\mathscr{G}_{\bar{t}}|_{C_{\bar{t}}})$. Hence taking the Hilbert scheme of $X_R/\operatorname{Spec} R$, one has only to prove the corresponding existence of $D_{\bar{t}}$ on a general geometric fibre $X_{\bar{t}}$. Namely for a general geometric point $\bar{t}$, it suffices to find a rational curve $D_{\bar{t}}$ through a general point in $X_{\bar{t}}$ with $(H_{\bar{t}} \cdot D_{\bar{t}}) \leq 2n(H_{\bar{t}} \cdot C_{\bar{t}})/\deg \mathscr{F}_{\bar{t}}|_{C_{\bar{t}}}$. By Lemma 10.1.10, $\operatorname{Hom}(\bigwedge^2 \mathscr{F}_{\bar{t}}, \mathscr{G}_{\bar{t}})$ is zero for general $\bar{t}$. To sum up, Theorem 10.1.9 is reduced to the next proposition.

Proposition 10.1.11 *For a quintuple (n, g, μ, μ', d) of integers n, g and rational numbers μ, μ', d with $d > 0$, there exists an integer $p_0 = p_0(n, g, \mu, \mu', d)$ which satisfies the following property.*

Let X be a normal projective variety of dimension n over an algebraically closed field k of characteristic p. Let H be a polarised system on X and let H be an ample Cartier divisor on X. Let $\mathscr{F}$ be a subsheaf of $\mathscr{T}_X = \Omega_X^\vee$ with torsion-free quotient $\mathscr{G} = \mathscr{T}_X/\mathscr{F}$ such that $\operatorname{Hom}(\bigwedge^2 \mathscr{F}, \mathscr{G}) = 0$. Let C be a general curve in $m\mathsf{H}$ for some m. We write g for the genus of C and write $\mu = \mu(\mathscr{G}|_C)$ and $\mu' = \mu_{\max}(\mathscr{G}|_C)$. Suppose that $\mathscr{F}|_C$ is semistable with $\deg \mathscr{F}|_C > d$. If $p \geq p_0$, then for any point x in C, there exists a rational curve D in X through x such that $(H \cdot D) < 2n(H \cdot C)/d$.

We shall work over k of positive characteristic p until the end of the proof of the proposition. Our strategy is to make the quotient of X by $\mathscr{F}$ by showing that $\mathscr{F}$ is a foliation.

Definition 10.1.12 (Ekedahl [121], Miyaoka [324]) Let X be a normal variety over k. By a *foliation* on X, we mean a subsheaf $\mathscr{F}$ of $\mathscr{T}_X = \Omega_X^\vee$ with torsion-free quotient $\mathscr{T}_X/\mathscr{F}$ such that $\mathscr{F}$ is closed under Lie bracket and p-th power.

We shall use the observation that $\mathscr{F}$ is a foliation if $\operatorname{Hom}(\bigwedge^2 \mathscr{F}, \mathscr{T}_X/\mathscr{F})$ and $\operatorname{Hom}(F^* \mathscr{F}, \mathscr{T}_X/\mathscr{F})$ are zero, where $F \colon X \to X$ is the absolute Frobenius morphism. The next correspondence is found in [121]. The reader may consult the exposition [329, I lecture III].

Theorem 10.1.13   *For a normal variety X over k, there exists a one-to-one correspondence between foliations $\mathscr{F}$ on X and factorisations $X \to Y \to X^{(1)}$ of the geometric Frobenius morphism $X \to X^{(1)}$ by a normal variety Y. Given $\mathscr{F}$, the quotient Y is defined by the subalgebra $\mathcal{O}_Y$ of $\mathcal{O}_X$ which consists of functions annihilated by all derivations in $\mathscr{F}$. Conversely for $\rho \colon X \to Y$, $\mathscr{F}$ is defined as the kernel of $d\rho \colon \mathscr{T}_X \to \rho^* \mathscr{T}_Y$. If $\mathscr{F}$ is of rank r, then ρ is of degree p^r.*

Proof of Proposition 10.1.11   Take rational numbers d_1 and d_2 so that $d < d_1 < d_2 < \lfloor d \rfloor + 1$. Then the integer $\deg \mathscr{F}|_C$ is greater than d_2.

We shall verify that $\mathscr{F}$ is a foliation. By assumption, $\mathrm{Hom}(\wedge^2 \mathscr{F}, \mathscr{G}) = 0$. It suffices to check $\mathrm{Hom}(F^* \mathscr{F}, \mathscr{G}) = 0$ for the absolute Frobenius morphism $F \colon X \to X$. The pull-back $F^* \mathscr{F}|_C$ has slope $\mu(F^* \mathscr{F}|_C) = p\mu(\mathscr{F}|_C)$. It is no longer semistable in general, but Shepherd-Barron [417, corollary 2] proved the estimate

$$\mu_{\max}(F^* \mathscr{F}|_C) - \mu_{\min}(F^* \mathscr{F}|_C) \le (\mathrm{rk}\, \mathscr{F} - 1)(2g - 2).$$

Hence as far as $p \ge p_1 = d^{-1}n(c + \mu')$ for $c = \max\{(n - 1)(2g - 2), 0\}$, one obtains

$$\mu_{\min}(F^* \mathscr{F}|_C) \ge p\mu(\mathscr{F}|_C) - c > \frac{pd}{n} - c \ge \mu' = \mu_{\max}(\mathscr{G}|_C).$$

Thus $\mathrm{Hom}(F^* \mathscr{F}|_C, \mathscr{G}|_C) = 0$ for a general curve C in mH. This implies that $\mathrm{Hom}(F^* \mathscr{F}, \mathscr{G}) = 0$.

Take the quotient $\rho \colon X \to Y$ by the foliation $\mathscr{F}$ as in Theorem 10.1.13. The geometric Frobenius morphism $F_X \colon X \to X^{(1)}$ factors through ρ as $F_X = \sigma \circ \rho$ for $\sigma \colon Y \to X^{(1)}$. One has an ample Cartier divisor $G = \sigma^* H^{(1)}$ on Y such that $\rho^* G = pH$. By construction, $\mathscr{G}$ is isomorphic to the image of $d\rho \colon \mathscr{T}_X \to \rho^* \mathscr{T}_Y$. It fits into the exact sequence

$$0 \to \mathscr{G} \to \rho^* \mathscr{T}_Y \to F_X^* \mathscr{F}^{(1)} \to 0$$

as in [121, corollary 3.4]. Consequently $(-K_Y \cdot \rho_* C)$ equals the sum of degrees of the restrictions to C of $\mathscr{G}$ and $F_X^* \mathscr{F}^{(1)}$. Thus

$$(-K_Y \cdot \rho_* C) = \mathrm{rk}\, \mathscr{G} \cdot \mu(\mathscr{G}|_C) + p \deg \mathscr{F}|_C > -n|\mu| + pd_2.$$

By Theorem 5.4.7, the local dimension δ of the Hom scheme $\mathrm{Hom}(C, Y)$ at $[\rho|_C]$ is greater than $-n|\mu| + pd_2 + n(1 - g)$. Hence as far as $p \ge p_1$ and $p \ge p_2 = (d_2 - d_1)^{-1}n(|\mu| + g - 1)$, one has the inequality

$$\delta > pd_1 \ge \left\lfloor \frac{pd_1}{n} \right\rfloor n.$$

For any point x in C, the bend and break in Theorem 6.3.4 finds a rational curve D_Y in Y through $\rho(x)$ such that $(G \cdot D_Y) \leq 2(G \cdot \rho_* C)/\lfloor pd_1/n \rfloor$. If $p \geq p_1, p_2$ and $p \geq p_3 = (d_1 - d)^{-1}n$, then $pd/n < \lfloor pd_1/n \rfloor$ and

$$(G \cdot D_Y) < \frac{2n}{pd}(G \cdot \rho_* C) = \frac{2n}{d}(H \cdot C).$$

The reduced locus $D = \rho^{-1}(D_Y)$ is a rational curve through x which satisfies $(H \cdot D) = (G \cdot \rho_* D)/p \leq (G \cdot D_Y) < 2n(H \cdot C)/d$. Therefore any $p_0 \geq p_1, p_2, p_3$ has the required property. $\qquad \square$

On a smooth projective complex surface, a semistable vector bundle $\mathscr{E}$ with respect to an ample divisor satisfies *Bogomolov's inequality*

$$(\mathrm{rk}\,\mathscr{E} - 1)c_1^2(\mathscr{E}) \leq 2\,\mathrm{rk}\,\mathscr{E} \cdot c_2(\mathscr{E})$$

proved in [50]. Generalising this to higher dimensions, Miyaoka derived from Theorem 10.1.9 the pseudo-effectivity of the second Chern class on a minimal variety. Langer [285] studied a generalisation in positive characteristic.

Theorem 10.1.14 (Miyaoka [325]) *Let X be a terminal projective complex variety of dimension n such that K_X is nef. Let $H_1, \ldots, H_{n-2}$ be nef divisors on X and let $\mu \colon X' \to X$ be a strong resolution. Then*

$$0 \leq \mu^* H_1 \cdots \mu^* H_{n-2} \cdot c_1^2(X') \leq 3\mu^* H_1 \cdots \mu^* H_{n-2} \cdot c_2(X').$$

We commence the proof of Theorem 10.1.1 in the case of irregularity zero. Let X be a terminal projective complex threefold such that K_X is nef. Let $\nu = \nu(X)$ denote the numerical Kodaira dimension of X. As explained at the beginning of the section, the abundance is known for $\nu = 0$ and 3. We shall deal with the cases $\nu = 1$ and 2.

There is nothing to prove if $H^0(\mathscr{O}_X(K_X)) \neq 0$. We may and shall assume that $h^3(\mathscr{O}_X) = h^0(\mathscr{O}_X(K_X)) = 0$ as well as $q = h^1(\mathscr{O}_X) = 0$. Then $\chi(\mathscr{O}_X) = 1 + h^2(\mathscr{O}_X) \geq 1$. We fix an integer $r \geq 2$ such that rK_X is Cartier. The Riemann–Roch formula provides

$$\chi(\mathscr{O}_X(rK_X)) = \frac{r(r-1)(2r-1)}{12}K_X^3 + \frac{r}{12}K_X c_2(X) + \chi(\mathscr{O}_X),$$

in which $K_X c_2(X)$ is defined as $\mu^* K_X \cdot c_2(X')$ for a strong resolution $\mu \colon X' \to X$. Since $K_X c_2(X) \geq 0$ by Theorem 10.1.14, one has

$$h^0(\mathscr{O}_X(rK_X)) + h^2(\mathscr{O}_X(rK_X)) \geq \chi(\mathscr{O}_X(rK_X)) \geq \chi(\mathscr{O}_X) \geq 1.$$

We take a sufficiently very ample effective divisor H on X generally, which is a smooth surface. The case $\nu = 2$ is easy.

Proof of Theorem 10.1.1 *for q* = 0, *v* = 2   By the observation above, it suffices to derive the vanishing of $H^2(\mathscr{O}_X(rK_X))$. Since $H^i(\mathscr{O}_X(rK_X + H)) = 0$ for $i \geq 1$ by Serre vanishing, the exact sequence

$$0 \to \mathscr{O}_X(rK_X) \to \mathscr{O}_X(rK_X + H) \to \mathscr{O}_H(K_H + (r-1)K_X|_H) \to 0$$

yields the isomorphism $H^1(\mathscr{O}_H(K_H + (r-1)K_X|_H)) \simeq H^2(\mathscr{O}_X(rK_X))$. The condition $v = 2$ implies that $K_X|_H$ is nef and big. Thus $H^1(\mathscr{O}_H(K_H + (r-1)K_X|_H)) = 0$ by Kawamata–Viehweg vanishing, which completes the proof for $v = 2$. □

Henceforth we assume that $v = 1$ besides $h^1(\mathscr{O}_X) = h^3(\mathscr{O}_X) = 0$. Let U denote the smooth locus in X.

We consider the *étale fundamental group* $\pi_1^{\text{ét}}(U)$ of U. As described in [317, section I.5], it is the profinite completion of the usual fundamental group $\pi_1(U)$ as an analytic space, where the *profinite completion* of a group G is the inverse limit $\varprojlim_N G/N$ by all normal subgroups $N \lhd G$ of finite index. Whilst $\pi_1(U)$ comes from covering spaces of U, $\pi_1^{\text{ét}}(U)$ comes from étale coverings of U. Every étale cover $U' \to U$ uniquely extends to a finite morphism $X' \to X$ from a normal variety X', which is again terminal with $K_{X'}$ nef. It has $v(X) = v(X')$, and also $\kappa(X) = \kappa(X')$ [458, theorem 5.13]. In fact, the implication from $\kappa(X') \geq 0$ to $\kappa(X) \geq 0$, which we need, follows from the observation that the norm of a pluricanonical form on X' in the canonical ring descends to a form on X. Thus one can replace X by X' for the theorem.

We divide the proof according to whether $\pi_1^{\text{ét}}(U)$ is finite or infinite.

Proof of Theorem 10.1.1 *for q* = 0, *v* = 1, $\pi_1^{\text{ét}}(U)$ *finite*   Once $\pi_1^{\text{ét}}(U)$ is finite, one can assume that $\pi_1^{\text{ét}}(U) = 1$ after the replacement of X mentioned above. The pair $H \subset X$ satisfies the *effective Lefschetz condition* Leff(H, X) in Definition 6.4.5 as in [161, X exemple 2.2] and hence Leff(H, U) also holds by definition. Then $\pi_1^{\text{ét}}(H) \simeq \pi_1^{\text{ét}}(U) = 1$ by [161, X theorem 3.10].

By the argument preceding the proof for $v = 2$, we may assume that

$$\text{Ext}^1(\mathscr{O}_X(rK_X), \mathscr{O}_X(K_X)) \simeq H^1(\mathscr{O}_X((1-r)K_X)) = H^2(\mathscr{O}_X(rK_X))^\vee \neq 0,$$

in which Serre duality in Theorem 1.1.12 is used. Note that $\mathscr{O}_X(rK_X)$ is invertible. Then there exists a non-split exact sequence

$$0 \to \mathscr{O}_X(K_X) \to \mathscr{E} \to \mathscr{O}_X(rK_X) \to 0.$$

The sheaf $\mathscr{E}$ is reflexive and the restriction $\mathscr{E}_H = \mathscr{E}|_H$ to H is locally free.

Step 1   We shall see that $\mathscr{E}$ is unstable with respect to H. If $\mathscr{E}$ were stable, then by Theorem 10.1.8, $\mathscr{E}_H$ would be a stable vector bundle on H, possibly after replacing H by the general member of the linear system of its multiple.

Let $\mathscr{F}_H = \mathscr{H}om(\mathscr{E}_H, \mathscr{E}_H) = \mathscr{E}_H \otimes \mathscr{E}_H^\vee$. Then $H^0(\mathscr{F}_H) \simeq \mathbf{C}$, and by Theorem 10.1.7, $\mathscr{F}_H$ is polystable.

One computes the total Chern class $c(\mathscr{F}_H)$ as

$$c(\mathscr{F}_H) = c(\mathscr{E}_H \otimes \mathscr{O}_H(-K_X|_H)) \cdot c(\mathscr{E}_H \otimes \mathscr{O}_H(-rK_X|_H))$$
$$= c(\mathscr{O}_H((r-1)K_X|_H)) \cdot c(\mathscr{O}_H(-(r-1)K_X|_H)) = 1$$

using $K_X^2 H = 0$ from $\nu = 1$. Donaldson [107, p.5] proved that every polystable vector bundle on a smooth complex surface with trivial Chern class is induced from a unitary representation of the fundamental group. Let $\pi_1(H) \to \mathrm{U}(4)$ be the associated representation of $\mathscr{F}_H$, where $\mathrm{U}(n)$ stands for the unitary group. Its image Γ is a finitely generated subgroup of $U(4)$, which is residually finite by a result of Mal'cev [299]. Because $\pi_1(H)$ has profinite completion $\pi_1^{\text{ét}}(H) = 1$, it follows that $\Gamma = 1$ and hence $\mathscr{F}_H \simeq \mathscr{O}_H^{\oplus 4}$. This contradicts $H^0(\mathscr{F}_H) \simeq \mathbf{C}$.

Step 2 Thus $\mathscr{E}$ is unstable. This means the existence of an exact sequence

$$0 \to \mathscr{L} \to \mathscr{E} \to \mathscr{M} \to 0$$

with torsion-free sheaves $\mathscr{L}$ and $\mathscr{M}$ on X such that $\mu(\mathscr{L}) \geq \mu(\mathscr{E})$. Take Weil divisors L and M on X such that $\mathscr{L}^{\vee\vee} \simeq \mathscr{O}_X(L)$ and $\mathscr{M}^{\vee\vee} \simeq \mathscr{O}_X(M)$. Let $l = L|_H$, $m = M|_H$, $h = H|_H$ and $k = K_X|_H$. Then $\mu(\mathscr{L}) = lh$, $\mu(\mathscr{M}) = mh$ and $\mu(\mathscr{E}) = (r+1)kh/2$. Note that $kh > 0$ by $\nu = 1$. One has

$$l + m \sim (r+1)k, \qquad mh \leq \frac{r+1}{2}kh \leq lh.$$

In particular, $\mu(\mathscr{L}) > \mu(\mathscr{O}_X(K_X))$ and hence $\mathscr{L}$ is not contained in $\mathscr{O}_X(K_X)$ as a subsheaf of $\mathscr{E}$. Thus the natural map $\mathscr{L} \to \mathscr{E}/\mathscr{O}_X(K_X) \simeq \mathscr{O}_X(rK_X)$ is non-zero and

$$l \lesssim rk, \qquad k \lesssim m, \qquad l - m \lesssim (r-1)k,$$

where $A \lesssim B$ stands for $A \leq B$ up to linear equivalence.

Whilst the induced sequence $0 \to \mathscr{O}_H(k) \to \mathscr{E}_H \to \mathscr{O}_H(rk) \to 0$ is exact, only the sequence $0 \to \mathscr{O}_H(l) \to \mathscr{E}_H \to \mathscr{O}_H(m)$ is exact and the last map $\mathscr{E}_H \to \mathscr{O}_H(m)$ has cokernel of finite length. As a result,

$$\chi(\mathscr{O}_H(k)) + \chi(\mathscr{O}_H(rk)) = \chi(\mathscr{E}_H) \leq \chi(\mathscr{O}_H(l)) + \chi(\mathscr{O}_H(m)),$$

which is equivalent to the inequality

$$k(k - K_H) + rk(rk - K_H) \leq l(l - K_H) + m(m - K_H).$$

From this with $l + m \sim (r+1)k$ and $k^2 = 0$, one obtains $l^2 + m^2 \geq 0$. Hence $(l - m)^2 = 2(l^2 + m^2) - (l + m)^2 \geq 0$. If $(l - m)^2$ were positive, then $\chi(\mathscr{O}_H(t(l-m)))$ grows quadratically with $t \geq 0$, and so does $h^0(\mathscr{O}_H(t(l-m)))$

as $(l - m)h \geq 0$. This contradicts the inequality $l - m \lesssim (r - 1)k$ for the nef but not big divisor k. Thus $(l - m)^2$ is zero.

Step 3 We have seen that $(l - m)^2 = k^2 = 0$ and $(l - m)h \geq 0$, $kh > 0$, $h^2 > 0$. Further $(l - m)k \leq 0$ from $l - m \lesssim (r - 1)k$. By the Hodge index theorem, Theorem 9.1.5, the 3×3 matrix of intersection numbers of $l - m$, k, h has non-negative determinant, that is,

$$2((l - m)k) \cdot ((l - m)h) \cdot (kh) - ((l - m)k)^2 \cdot (h^2) \geq 0.$$

This gives $(l - m)k = 0$, and the same theorem shows the numerical dependence of $l - m$ and k. Together with $l + m \sim (r + 1)k$, one can write $l \equiv ck$ with a rational number $c \leq r$.

From the assumption $H^1(\mathscr{O}_X) = 0$ with $H^2(\mathscr{O}_X(-H)) = H^1(\mathscr{O}_X(K_X + H))^\vee = 0$, one has $H^1(\mathscr{O}_H) = 0$ and hence the injection $\mathrm{Pic}\, H \hookrightarrow H^2(H, \mathbf{Z})$. Thus $l \sim_{\mathbf{Q}} ck$. As in the proof of Lemma 6.4.4, the Grothendieck–Lefschetz theory provides the injection $\mathrm{Cl}\, X \simeq \mathrm{Pic}\, U \hookrightarrow \mathrm{Pic}\, H$. Therefore $cK_X \sim_{\mathbf{Q}} L \lesssim rK_X$. If $c < r$, then $K_X \gtrsim 0$, that is, $\kappa(X) \geq 0$. If $c = r$, then $L \sim_{\mathbf{Q}} rK_X$ and further $L \sim rK_X$ by $\pi_1^{\text{ét}}(U) = 1$. In this case, the injection $\mathscr{O}_X(rK_X) \simeq \mathscr{L}^{\vee\vee} \hookrightarrow \mathscr{E}$ would yield a splitting of $0 \to \mathscr{O}_X(K_X) \to \mathscr{E} \to \mathscr{O}_X(rK_X) \to 0$ because $\mathscr{L} \to \mathscr{E}/\mathscr{O}_X(K_X)$ is non-zero. This is absurd. □

Proof of Theorem 10.1.1 *for* $q = 0$, $\nu = 1$, $\pi_1^{\text{ét}}(U)$ *infinite* When $\pi_1^{\text{ét}}(U)$ is infinite, there exists an infinite sequence $\cdots \to X_2 \to X_1 \to X_0 = X$ of terminal threefolds such that every $X_{i+1} \to X_i$ is non-trivial and finite and the restriction over U is étale. Recall that by Theorem 2.4.17, a singular point $x \in X$ of index s has an analytic neighbourhood V with $\pi_1(V \setminus x) \simeq \mathbf{Z}_s$. Hence there exists an index i such that $X_j \to X_i$ is étale over the whole X_i for all $j \geq i$. Replace X by X_i. One can still assume that $h^1(\mathscr{O}_X) = h^3(\mathscr{O}_X) = 0$.

After this replacement, all $X_i \to X$ are étale. By Theorem 9.2.3, $\chi(\mathscr{O}_{X_i}) = d_i\chi(\mathscr{O}_X)$ for the degree d_i of $X_i \to X$. Replacing X by X_i with $d_i \geq 4$, we may assume that $\chi(\mathscr{O}_X) \geq 4$, by which $h^2(\mathscr{O}_X) = \chi(\mathscr{O}_X) - 1 \geq 3$.

Fix a resolution $\mu \colon Y \to X$, in which $h^0(\Omega_Y^2) = h^2(\mathscr{O}_Y) = h^2(\mathscr{O}_X) \geq 3$. There exists an injection $H^0(\mathscr{O}_Y)^{\oplus 3} \hookrightarrow H^0(\Omega_Y^2)$ and it defines a homomorphism $\alpha \colon \mathscr{O}_Y^{\oplus 3} \to \Omega_Y^2$ of positive rank e. Let $\mathscr{E}$ be the maximal subsheaf of Ω_Y^2 containing the image $\alpha(\mathscr{O}_Y^{\oplus 3})$ such that $\mathscr{E}/\alpha(\mathscr{O}_Y^{\oplus 3})$ is torsion. For the double dual $\mathscr{L}$ of $\bigwedge^e \mathscr{E}$, α induces an injection $\bigwedge^e H^0(\mathscr{O}_Y)^{\oplus 3} \hookrightarrow H^0(\mathscr{L})$ and hence $\mathscr{L} \simeq \mathscr{O}_Y(D)$ for some effective divisor D on Y. If $e = 3$, then $\mathscr{E} = \Omega_Y^2$ and $\mathscr{L} = \omega_Y^{\otimes 2}$, for which $H^0(\mathscr{O}_X(2K_X)) \simeq H^0(\omega_Y^{\otimes 2}) = H^0(\mathscr{O}_Y(D)) \neq 0$.

Suppose that $e = 1$ or 2. Then by the injection $\mathbf{C}^3 \simeq \bigwedge^e H^0(\mathscr{O}_Y)^{\oplus 3} \hookrightarrow H^0(\mathscr{O}_Y(D))$, $|D|$ has dimension at least two. In particular, there exists a mobile divisor $0 < M \leq D$. Note that $\mu_*M \neq 0$.

The sheaf $\mathscr{E}^\vee \otimes \omega_Y$ is a quotient of $(\Omega_Y^2)^\vee \otimes \omega_Y \simeq \Omega_Y$. By Theorem 10.1.9, its determinant $\mathscr{O}_Y(eK_Y - D)$ has non-negative degree along $A \cdot B$ for any ample **Q**-divisors A and B on Y, that is, $(eK_Y - D)AB \geq 0$. Approaching A and B to μ^*K_X and μ^*H in $N^1(Y)$, one has $(eK_Y - D)\mu^*K_X \cdot \mu^*H \geq 0$. Thus

$$0 \leq (\mu_*M)K_XH \leq (\mu_*D)K_XH \leq e(\mu_*K_Y)K_XH = eK_X^2H = 0,$$

where $K_X^2 \equiv 0$. Hence $mk = 0$ for $m = \mu_*M|_H \neq 0$ and $k = K_X|_H$.

Since M is mobile, $m^2 \geq 0$. For $h = H|_H$, $kh > 0$ but $k^2 = 0$. The Hodge index theorem provides $m^2 = 0$ and $k \equiv cm$ with a positive rational number c. Now one deduces that $K_X \sim_{\mathbf{Q}} c\mu_*M \geq 0$ by the same argument as in the last paragraph of the proof in the case of finite $\pi_1^{\text{ét}}(U)$. □

10.2 Abundance

In this section, we prove the abundance $\kappa = \nu$ for a terminal threefold X with K_X nef, where $\kappa = \kappa(X)$ and $\nu = \nu(X)$. This was completed for $\nu = 1$ by Miyaoka [326] and for $\nu = 2$ by Kawamata [238].

Theorem 10.2.1 *Let X be a terminal projective threefold. If K_X is nef, then $\kappa(X) = \nu(X)$ and K_X is semi-ample.*

By **Q**-factorialisation in Proposition 1.5.17 or Corollary 5.1.11, we may and shall assume that X is a *minimal* threefold, namely a **Q**-factorial terminal projective threefold with K_X nef. By Theorem 1.7.12, the semi-ampleness of K_X is reduced to the *abundance* $\kappa = \nu$. The abundance is known for $\nu = 0$ and for $\nu = \dim X$ as explained at the beginning of the preceding section. We keep studying the cases $\nu = 1$ and 2. By Theorem 10.1.1, we have already obtained an effective **Q**-divisor $D \sim_{\mathbf{Q}} K_X$. We shall first construct by the MMP a special dlt model (X', B') with $\kappa = \kappa(X', K_{X'} + B')$ and $\nu = \nu(X', K_{X'} + B')$, and then derive the abundance for (X', B') from that for the restriction to a prime component of B'. Thus our argument uses the abundance for lc surfaces as well as the MMP for dlt threefold pairs.

Theorem 10.2.2 (T. Fujita [138]) *Let (S, Δ) be an lc pair on a projective surface S. If $K_S + \Delta$ is nef, then it is semi-ample.*

We shall construct the dlt model by which X will be replaced.

Proposition 10.2.3 *Let X_0 be a minimal threefold. Let $D_0 \sim_{\mathbf{Q}} K_{X_0}$ be an effective **Q**-divisor and let B_0 denote the support of D_0. Then there exist a **Q**-factorial dlt projective pair (X, B) with B reduced and an effective **Q**-divisor $D \sim_{\mathbf{Q}} K_X + B$ such that*

- $X_0 \setminus B_0$ *is isomorphic to* $X \setminus B$,
- $K_X + B$ *is nef,*
- B *equals the support of* D *and*
- $\kappa(X_0) = \kappa(X, K_X + B)$ *and* $\nu(X_0) = \nu(X, K_X + B)$

and such that for each prime component S *of* B, *which is normal, and for a Cartier divisor* $L = l(K_X + B)$ *with a positive integer* l,

- *if* $\nu(X_0) = 1$, *then* $\nu(S, L|_S) = 0$ *and* $S \cap (B - S) = \emptyset$ *and*
- *if* $\nu(X_0) = 2$, *then* $\nu(S, L|_S) = 1$ *and* $L(B - S)S = 0$.

Proof The surface S is Cohen–Macaulay by Proposition 2.2.23 and satisfies R_1 by the classification of dlt surface singularities in Theorem 4.4.14. Hence S is normal.

Step 1 Since X_0 has only isolated singularities, there exists a projective morphism $\mu \colon X_1 \to X_0$ which is isomorphic outside B_0 and which is a log resolution of (X_0, B_0) about B_0. Then $B_1 = \mu^{-1}(B_0)$ is an snc divisor and it supports $\mu^* D_0$. The **Q**-divisor $E = K_{X_1/X_0} + B_1$ is effective and has support B_1. Let $D_1 = \mu^* D_0 + E \sim_{\mathbf{Q}} K_{X_1} + B_1$, which still has support B_1.

Run the $(K_{X_1} + B_1)$-MMP to produce a **Q**-factorial dlt pair (X, B) with $K_X + B$ nef, together with the strict transform $D \sim_{\mathbf{Q}} K_X + B$ of D_1. The induced birational map $X_1 \dashrightarrow X$ is isomorphic outside the support B_1 of D_1 because every new contraction in the MMP is negative with respect to $D_1 \sim_{\mathbf{Q}} K_{X_1} + B_1$. Hence there exists a projective morphism $\mu_1 \colon Y \to X_1$ isomorphic outside B_1 such that μ_1 is a log resolution of (X_1, B_1) about B_1 and such that it induces a morphism $\pi \colon Y \to X$. Let $B_Y = \mu_1^{-1}(B_1)$. Then B is the divisorial part of the image $\pi(B_Y)$. Note that $D_Y = (\mu \circ \mu_1)^* D_0$ is nef and has support B_Y.

Step 2 We claim that $B = \pi(B_Y)$. Suppose that $B \neq \pi(B_Y)$ and consider an irreducible component Z of the closure of $\pi(B_Y) \setminus B$. If Z is a point, then we take the general hyperplane section T_Y of Y and let $T = \pi(T_Y)$. If Z is a curve, then we take the general hyperplane section T of X and let $T_Y = \pi^* T$. The surface T_Y is smooth and the restriction $\pi_T \colon T_Y \to T$ is not isomorphic at $Z|_T$. Define the **Q**-divisor G on T_Y to be the part of $D_Y|_{T_Y}$ contained in $\pi_T^{-1}(Z|_T)$. It is a connected component of $D_Y|_{T_Y}$ and would have self-intersection number $G^2 = (D_Y|_{T_Y} \cdot G) \geq 0$, which contradicts Grauert's criterion in Theorem 4.4.21.

The equality $B = \pi(B_Y)$ implies not only the isomorphism $X_0 \setminus B_0 \simeq Y \setminus B_Y \simeq X \setminus B$ but also the coincidence of the supports of the nef **Q**-divisors $D_Y \sim_{\mathbf{Q}} (\mu \circ \mu_1)^* K_{X_0}$ and $\pi^* D \sim_{\mathbf{Q}} \pi^*(K_X + B)$, which are B_Y. By Lemma 10.2.4 below, they have the same Iitaka and numerical Iitaka dimension.

Step 3 Thus the model (X, B) possesses all the required properties but the property on a prime component S of B. To achieve this property, we perform the $(K_X + B - S)$-MMP crepantly with respect to $K_X + B$. We shall explain this only in the case $\nu(X_0) = 1$. The other case $\nu(X_0) = 2$ is similar.

Assume that $\nu(X_0) = 1$ and hence $\nu(X, K_X+B) = 1$. Then $L = l(K_X+B) \sim_{\mathbf{Q}} lD$ satisfies $LD \equiv 0$. Since L is nef, this implies that $LS \equiv 0$ for every prime component S of the support B of D. Thus $\nu(S, L|_S) = 0$.

Take the general hyperplane section H of X. By the $\mathbf{Q}$-factoriality of X, as far as S meets another component S' of B, there exists a curve C in $S'|_H$ which meets $S|_H$. Hence $(S \cdot C) > 0$, whilst $(L \cdot C) = 0$ by $\nu(S', L|_{S'}) = 0$. It follows that $((K_X + B) \cdot C) = 0$ and $((K_X + B - S) \cdot C) = -(S \cdot C) < 0$, from which there exists a $(K_X + B - S)$-negative extremal ray of $\overline{\mathrm{NE}}(X)$ perpendicular to $K_X + B$. The associated contraction is a step of the $(K_X + B - S)$-MMP. The MMP directed in this manner eventually attains a new model on which $S \cap (B - S) = \emptyset$. Since this MMP is crepant with respect to $K_X + B$, the new model inherits all the other properties. □

Lemma 10.2.4 *Let D_1 and D_2 be nef effective $\mathbf{Q}$-divisors on a normal complete variety X. If they have the same support, then $\kappa(X, D_1) = \kappa(X, D_2)$ and $\nu(X, D_1) = \nu(X, D_2)$.*

Proof Write $\kappa_i = \kappa(X, D_i)$ and $\nu_i = \nu(X, D_i)$. It suffices to show that $\kappa_1 \le \kappa_2$ and $\nu_1 \le \nu_2$. Multiplying D_2, we may assume that $D_1 \le D_2$. Then the inequality $\kappa_1 \le \kappa_2$ is obvious. By Chow's lemma, we may assume that X is projective. For an ample divisor H on X, one has $0 < D_1^{\nu_1} H^{n-\nu_1} \le D_1^{\nu_1-1} D_2 H^{n-\nu_1} \le \cdots \le D_2^{\nu_1} H^{n-\nu_1}$ where n is the dimension of X. Thus $\nu_1 \le \nu_2$. □

We proceed to the proof of Theorem 10.2.1 in the case when $\nu(X) = 1$. Assume that $\nu(X) = 1$. Instead of X, we work on the dlt model (X, B) with $\nu(X, L) = 1$ constructed by Proposition 10.2.3, where $L = l(K_X + B)$ is a nef Cartier divisor with a positive integer l. Multiplying l, we may and shall assume the existence of a member $D \in |L|$ with support B. In our case $\nu = 1$, every prime component S of B satisfies $\nu(S, L|_S) = 0$ and $S \cap (B - S) = \emptyset$. In particular, (X, B) is plt. Our goal is the equality $\kappa(X, L) = 1$.

We fix S, which is normal, henceforth. By adjunction, (S, B_S) is klt for the different B_S in Definition 1.4.24 given by $K_S + B_S = (K_X + B)|_S = (K_X + S)|_S$. By Theorem 10.2.2, $L|_S \sim l(K_S + B_S)$ is semi-ample with $\kappa(S, L|_S) = \nu(S, L|_S) = 0$. Hence $L|_S$ is a torsion in $\mathrm{Pic}\, S$. Replacing l, we may assume that $L|_S \sim 0$.

By Theorem 2.3.10, we fix an open analytic neighbourhood U in X about S with $B|_U = S$ such that $H^i(U, \mathbf{Z}) \simeq H^i(S, \mathbf{Z})$ for $i = 1, 2$. For the coefficient c of S in D, one has $cS \sim l(K_U + S) = L|_U$ on U. Hence cS is Cartier on U and $cS|_S \sim 0$, that is, $\mathscr{O}_U(cS) \otimes \mathscr{O}_S \simeq \mathscr{O}_S$.

Lemma 10.2.5 *Let U be an analytic space and let S be a reduced compact analytic subspace of U. Suppose that $H^i(U, \mathbf{Z}) \simeq H^i(S, \mathbf{Z})$ for $i = 1, 2$.*

(i) *The kernel of the group homomorphism $\operatorname{Pic} U \to \operatorname{Pic} S$ is divisible and torsion-free.*

(ii) *Suppose that U is normal. Let D be a divisor on U such that lD is Cartier and $lD|_S \sim 0$ for some positive integer l. Then there exists a finite surjective morphism $\pi \colon \bar{U} \to U$ from a normal analytic space, étale in codimension one, such that $\bar{D} = \pi^* D$ is a Cartier divisor on $\bar{U}$ and $\bar{D}|_{\bar{S}} \sim 0$ on $\bar{S} = \pi^{-1}(S)$.*

Proof The commutative diagram

$$
\begin{array}{ccccccc}
H^1(U, \mathbf{Z}) & \longrightarrow & H^1(\mathscr{O}_U) & \longrightarrow & \operatorname{Pic} U & \longrightarrow & H^2(U, \mathbf{Z}) \\
\downarrow{\scriptstyle \wr} & & \downarrow & & \downarrow & & \downarrow{\scriptstyle \wr} \\
0 \longrightarrow H^1(S, \mathbf{Z}) & \longrightarrow & H^1(\mathscr{O}_S) & \longrightarrow & \operatorname{Pic} S & \longrightarrow & H^2(S, \mathbf{Z})
\end{array}
$$

induces an isomorphism from the kernel of $H^1(\mathscr{O}_U) \to H^1(\mathscr{O}_S)$ to the kernel K of $\operatorname{Pic} U \to \operatorname{Pic} S$. Hence K is a complex vector space, which implies (i).

For (ii), we choose l minimally. By (i), the element lD in K is expressed as $lD \sim lL$ by some Cartier divisor L on U such that $L|_S \sim 0$. Take the cyclic cover $\pi \colon \bar{U} \to U$ associated with $D - L$ as in Definition 2.2.17, for which $\pi^*(D - L)$ is principal by Lemma 2.2.18. Then $\bar{D} = \pi^* D \sim \pi^* L$ is a Cartier divisor and $\bar{D}|_{\bar{S}} \sim \pi_S^*(L|_S) \sim 0$ for $\pi_S \colon \bar{S} \to S$. $\qquad\square$

In our situation, where $cS \sim l(K_U + S) = L|_U$ on U, we write $c = gc'$ and $l = gl'$ with $g = \gcd(c, l)$ and choose integers a and b such that $ac' + bl' = 1$. Let $T = a(K_U + S) + bS$. Then

$$
g(K_U + S - c'T) \sim 0, \quad g(S - l'T) \sim 0 \quad \text{on } U, \qquad gc'l'T|_S \sim 0.
$$

Applying Lemma 10.2.5(ii) to T and to the pull-backs of $K_U + S - c'T$ and $S - l'T$ successively, we obtain a finite Galois morphism $\pi \colon \bar{U} \to U$ from a normal analytic surface $\bar{U}$, étale in codimension one, such that for $\bar{S} = \pi^* S$ and $\bar{T} = \pi^* T$,

- $K_{\bar{U}} + \bar{S} \sim c'\bar{T}$ and $\bar{S} \sim l'\bar{T}$,
- $\bar{S}$ and $\bar{T}$ are Cartier and $\bar{T}|_{\bar{S}} \sim 0$,
- $\bar{U}$ is Gorenstein and canonical and $(\bar{U}, \bar{S})$ is lc by Corollary 2.2.21 and
- $\bar{S}$ is connected and Gorenstein and $\omega_{\bar{S}} \simeq \mathscr{O}_{\bar{S}}$ by Proposition 2.2.23.

The next step is crucial, in which we move S infinitesimally. For a positive integer i, we let $i\bar{S}$ denote the analytic subspace of $\bar{U}$ defined by the ideal sheaf $\mathcal{O}_{\bar{U}}(-i\bar{S})$ in $\mathcal{O}_{\bar{U}}$, which is a covering of iS with the scheme structure. Let $A_i = \operatorname{Spec} \mathbf{C}[\varepsilon]/(\varepsilon^i)$. We set the natural embeddings $s_i \colon i\bar{S} \hookrightarrow (i+1)\bar{S}$ and $a_i \colon A_i \hookrightarrow A_{i+1}$.

Lemma 10.2.6 *There exists a collection of flat proper morphisms $\xi_i \colon i\bar{S} \to A_i$ for $i \geq 1$ compatible with s_i and a_i in the sense that $\xi_{i+1} \circ s_i = a_i \circ \xi_i$ for all i. Every such collection satisfies the properties for all i that $R^p \xi_{i*} \mathcal{O}_{i\bar{S}}$ is free for all p, $\omega_{i\bar{S}} \simeq \mathcal{O}_{i\bar{S}}$ and $\mathcal{O}_{\bar{U}}(\bar{S}) \otimes \mathcal{O}_{i\bar{S}} \simeq \mathcal{O}_{i\bar{S}}$.*

Proof We shall construct ξ_i by induction on i from the structure morphism $\xi_1 \colon \bar{S} \to A_1 = \operatorname{Spec} \mathbf{C}$. Assuming that the flat proper morphism ξ_i has been obtained, we first show that ξ_i has the required properties, and then construct ξ_{i+1} so that $\xi_{i+1} \circ s_i = a_i \circ \xi_i$.

It follows from the log canonicity of $(\bar{U}, \bar{S})$ that the Cartier divisor $\bar{S}$ is semi-log canonical and has *Du Bois singularities*. The reader should consult [269, chapters 5, 6] for these notions and assertions. For such $\bar{S}$, the map $H^p(\bar{S}, \mathbf{C}) \to H^p(\bar{S}, \mathcal{O}_{\bar{S}})$ is surjective for all p. In fact the class of Du Bois singularities is formulated by this surjectivity. Using this, one derives the surjection $H^p(i\bar{S}, \mathcal{O}_{i\bar{S}}) \twoheadrightarrow H^p(\bar{S}, \mathcal{O}_{\bar{S}})$ from the diagram

$$
\begin{array}{ccc}
H^p(i\bar{S}, \mathbf{C}) & \longrightarrow & H^p(i\bar{S}, \mathcal{O}_{i\bar{S}}) \\
\downarrow & & \downarrow \\
H^p(\bar{S}, \mathbf{C}) & \twoheadrightarrow & H^p(\bar{S}, \mathcal{O}_{\bar{S}}) \quad .
\end{array}
$$

The freedom of $R^p \xi_{i*} \mathcal{O}_{i\bar{S}}$ is then a consequence of cohomology and base change, Theorem 1.1.5.

To show that $\omega_{i\bar{S}} \simeq \mathcal{O}_{i\bar{S}}$, we need *Grothendieck duality* [177] for proper morphisms. This provides an isomorphism

$$
\mathbf{R}\xi_{i*}\omega_{i\bar{S}}^{\bullet} \simeq \mathbf{R}\operatorname{Hom}_{A_i}(\mathbf{R}\xi_{i*}\mathcal{O}_{i\bar{S}}, \omega_{A_i}^{\bullet})
$$

in the derived category, where $\omega_{i\bar{S}}^{\bullet}$ and $\omega_{A_i}^{\bullet}$ are the dualising complexes. One has the descriptions $\omega_{i\bar{S}}^{\bullet} = \omega_{i\bar{S}}[2]$ and $\omega_{A_i}^{\bullet} = \mathcal{O}_{A_i}$ since $i\bar{S}$ and A_i are Gorenstein. Hence $\mathbf{R}\xi_{i*}\omega_{i\bar{S}}[2] \simeq \mathbf{R}\operatorname{Hom}_{A_i}(\mathbf{R}\xi_{i*}\mathcal{O}_{i\bar{S}}, \mathcal{O}_{A_i})$.

Because every $R^p \xi_{i*} \mathcal{O}_{i\bar{S}}$ is free, the complex which represents $\mathbf{R}\xi_{i*}\mathcal{O}_{i\bar{S}}$ admits a quasi-isomorphism from the complex $\{R^p \xi_{i*} \mathcal{O}_{i\bar{S}}\}_p$ endowed with zero maps $R^p \xi_{i*} \mathcal{O}_{i\bar{S}} \to R^{p+1} \xi_{i*} \mathcal{O}_{i\bar{S}}$. Thus the above duality yields an isomorphism

$$
R^p \xi_{i*} \omega_{i\bar{S}} \simeq \operatorname{Hom}_{A_i}(R^{2-p} \xi_{i*} \mathcal{O}_{i\bar{S}}, \mathcal{O}_{A_i}).
$$

In particular, $R^p \xi_{i*} \omega_{i\bar{S}}$ is free for all p. By cohomology and base change, the free A_i-module $\xi_{i*} \omega_{i\bar{S}}$ has the isomorphism $\xi_{i*} \omega_{i\bar{S}} \otimes_{A_i} \mathbf{C} \simeq H^0(\omega_{i\bar{S}}|_{\bar{S}})$. Thus the isomorphism $\mathcal{O}_{\bar{S}} \simeq \mathcal{O}_{\bar{U}}(K_{\bar{U}} + i\bar{S}) \otimes \mathcal{O}_{\bar{S}} = \omega_{i\bar{S}}|_{\bar{S}}$ is lifted to an isomorphism $\mathcal{O}_{i\bar{S}} \simeq \omega_{i\bar{S}}$.

Recall that $\omega_{i\bar{S}} = \mathcal{O}_{\bar{U}}(K_{\bar{U}} + i\bar{S}) \otimes \mathcal{O}_{i\bar{S}} \simeq \mathcal{O}_{\bar{U}}((c' + (i-1)l')\bar{T}) \otimes \mathcal{O}_{i\bar{S}}$ with $c' + (i-1)l' \neq 0$. Hence $\mathcal{O}_{\bar{U}}(\bar{T}) \otimes \mathcal{O}_{i\bar{S}}$ is a torsion in the kernel of $\mathrm{Pic}\, i\bar{S} \to \mathrm{Pic}\, \bar{S}$, and it is zero by Lemma 10.2.5(i). Thus $\mathcal{O}_{\bar{U}}(\bar{S}) \otimes \mathcal{O}_{i\bar{S}} \simeq \mathcal{O}_{i\bar{S}}$ for $\bar{S} \sim l'\bar{T}$.

We shall complete the induction by constructing ξ_{i+1}. Let $\mathscr{L}$ be the kernel of the natural surjection $\mathcal{O}_{(i+1)\bar{S}} \twoheadrightarrow \mathcal{O}_{\bar{S}}$ fitting into

$$0 \to \mathscr{L} \to \mathcal{O}_{(i+1)\bar{S}} \to \mathcal{O}_{\bar{S}} \to 0.$$

Then $\mathscr{L} = \mathcal{O}_{\bar{U}}(-\bar{S})/\mathcal{O}_{\bar{U}}(-(i+1)\bar{S}) \simeq \mathcal{O}_{\bar{U}}(-\bar{S}) \otimes \mathcal{O}_{i\bar{S}} \simeq \mathcal{O}_{i\bar{S}}$, and the image of the composite $\mathscr{L} \hookrightarrow \mathcal{O}_{(i+1)\bar{S}} \twoheadrightarrow \mathcal{O}_{i\bar{S}}$ is generated by $\xi_i^* \varepsilon$. Take an element $e \in H^0(\mathscr{L})$ such that $s_i^* e = \xi_i^* \varepsilon$. Then $\mathscr{L} = \mathcal{O}_{i\bar{S}} \cdot e$ as an $\mathcal{O}_{(i+1)\bar{S}}$-module. From the exact sequence

$$0 \to (\mathbf{C}[\varepsilon]/(\varepsilon^i)) \cdot e \to H^0(\mathcal{O}_{(i+1)\bar{S}}) \to \mathbf{C} \to 0,$$

one obtains an isomorphism $\mathbf{C}[\varepsilon]/(\varepsilon^{i+1}) \simeq H^0(\mathcal{O}_{(i+1)\bar{S}})$ of $\mathbf{C}$-algebras which sends ε to e. It defines a flat proper morphism $\xi_{i+1} \colon (i+1)\bar{S} \to A_{i+1}$, which satisfies the compatibility $\xi_{i+1} \circ s_i = a_i \circ \xi_i$ since $s_i^* \xi_{i+1}^* \varepsilon = s_i^* e = \xi_i^* \varepsilon$. $\qquad\square$

Proof of Theorem 10.2.1 *for* $\nu = 1$ Let G be the Galois group associated with the covering π and write d for the order of G, which equals the degree of π. Consider $\xi_i \colon i\bar{S} \to A_i$ in Lemma 10.2.6. For a generator u of the $\mathcal{O}_{id\bar{S}}$-module $\mathcal{O}_{\bar{U}}(\bar{S}) \otimes \mathcal{O}_{id\bar{S}}$, which is isomorphic to $\mathcal{O}_{id\bar{S}}$, the norm $\prod_{\sigma \in G} \sigma^* u$ to the power of i is an G-invariant generator of $\mathcal{O}_{\bar{U}}(id\bar{S}) \otimes \mathcal{O}_{id\bar{S}}$. Sending 1 to this generator, one has an G-equivariant isomorphism $\mathcal{O}_{id\bar{S}} \simeq \mathcal{O}_{\bar{U}}(id\bar{S}) \otimes \mathcal{O}_{id\bar{S}}$ of $\mathcal{O}_{id\bar{S}}$-modules.

Via $\xi_{id*}\mathcal{O}_{id\bar{S}} = \mathcal{O}_{A_{id}}$ from Lemma 10.2.6, there exists an isomorphism

$$\mathbf{C}[\varepsilon]/(\varepsilon^{id}) \simeq H^0(\mathcal{O}_{id\bar{S}}) \simeq H^0(\mathcal{O}_{\bar{U}}(id\bar{S}) \otimes \mathcal{O}_{id\bar{S}})$$

of vector spaces. We make an G-action on $\mathbf{C}[\varepsilon]/(\varepsilon^{id})$ so that this isomorphism is G-equivariant. Then $h = \prod_{\sigma \in G} \sigma^* \varepsilon$ is G-invariant, and $1, h, \ldots, h^{i-1}$ are linearly independent. Hence the G-invariant part of $H^0(\mathcal{O}_{\bar{U}}(id\bar{S}) \otimes \mathcal{O}_{id\bar{S}})$ is of dimension at least i. This means that $h^0(\mathcal{O}_X(idS) \otimes \mathcal{O}_{idS}) \geq i$ as far as the divisor idS on X is Cartier. For any such i, one has the estimate $h^0(\mathcal{O}_X(idS)) \geq 1 + i - h^1(\mathcal{O}_X)$ from the exact sequence

$$0 \to H^0(\mathcal{O}_X) \to H^0(\mathcal{O}_X(idS)) \to H^0(\mathcal{O}_X(idS) \otimes \mathcal{O}_{idS}) \to H^1(\mathcal{O}_X).$$

Thus $\kappa(X, S) \geq 1$, and $\kappa(X, L) \geq 1$ by $L \sim D \geq S$. $\qquad\square$

We turn to the case $\nu(X) = 2$ in Theorem 10.2.1. The assertion is reduced to the positivity $\kappa(X) \geq 1$. Indeed if $\kappa(X) \geq 1$, then by Theorem 1.7.15, X has a good minimal model X'. The abundance $\kappa(X) = \nu(X)$ on X follows from that on X' because of the birational invariance of κ and ν, where the invariance of ν is derived from Corollary 1.5.7.

We work on the dlt model (X, B) constructed by Proposition 10.2.3. We have a nef Cartier divisor $L = l(K_X + B)$ with $\nu(X, L) = 2$ and an effective divisor $D \sim L$ with support B such that $\nu(S, L|_S) = 1$ and $L(B - S)S = 0$ for any prime component S of B. We want $\kappa(X, L) \geq 1$.

By adjunction, S admits an lc pair (S, B_S) by the different B_S given by $K_S + B_S = (K_X + B)|_S$. Note that lB_S is integral with $lK_S + lB_S \sim L|_S$. By the log abundance for surfaces in Theorem 10.2.2, a multiple of $L|_S$ defines a contraction $\varphi_S \colon S \to C_S$ to a curve. Using Miyaoka's generic semi-positivity and analysing the structure of φ_S, Kawamata performed the following computation of the second Chern class. See [238, lemma 3.6] for the proof.

Lemma 10.2.7 *One has $\mu^* L \cdot (K_Y^2 + c_2(Y)) \geq 0$ for a resolution $\mu \colon Y \to X$, and the equality holds only if the general fibre of $\varphi_S \colon S \to C_S$ is an elliptic curve for any component S of B.*

Proof of Theorem 10.2.1 for $\nu = 2$ Recall that X has rational singularities by Theorem 1.4.20. One has $L^3 = 0$ from $\nu(X, L) = 2$, and $L^2 K_X = L^2(l^{-1}L - B) = 0$ from $L^2 S = 0$ for any component S of B. Thus on a resolution $\mu \colon Y \to X$, the Riemann–Roch formula computes $\chi(\mathcal{O}_X(K_X + iL)) = -\chi(\mathcal{O}_X(-iL)) = -\chi(\mathcal{O}_Y(-i\mu^* L))$ as

$$\chi(\mathcal{O}_X(K_X + iL)) = \frac{i}{12}\mu^* L(K_Y^2 + c_2(Y)) - \chi(\mathcal{O}_Y).$$

Take the general hyperplane section H of X, which is lt as so is X. We consider $i \geq 1$. Then $H^p(\mathcal{O}_X(K_X + H + iL)) = 0$ for $p \geq 1$ by Kawamata–Viehweg vanishing. Since $L|_H$ is nef and big by $\nu(X, L) = 2$, the restriction $(K_X + H + iL)|_H = K_H + iL|_H$ has the vanishing $H^p(\mathcal{O}_H((K_X + H + iL)|_H)) = 0$ for $p \geq 1$. Hence $H^p(\mathcal{O}_X(K_X + iL)) = 0$ for $p \geq 2$, and the expression of $\chi(\mathcal{O}_X(K_X + iL))$ for $i \geq 1$ becomes

$$h^0(\mathcal{O}_X(K_X + iL)) = h^1(\mathcal{O}_X(K_X + iL)) + \frac{i}{12}\mu^* L(K_Y^2 + c_2(Y)) - \chi(\mathcal{O}_Y).$$

As $(il + 1)(K_X + B) \geq K_X + iL$, the Iitaka dimension $\kappa(X, L) = \kappa(X, K_X + B)$ is positive if $\mu^* L(K_Y^2 + c_2(Y)) > 0$. Thus by Lemma 10.2.7, one has only to care about the case $\mu^* L(K_Y^2 + c_2(Y)) = 0$, in which the general fibre of $\varphi_S \colon S \to C_S$ defined by a multiple of $L|_S$ is an elliptic curve for any S.

Suppose that $\mu^* L(K_Y^2 + c_2(Y)) = 0$ and fix S. Observe that $\kappa(X, L)$ is positive if either $h^3(\mathcal{O}_X(-S - iL))$, which equals $h^3(\mathcal{H}om(\mathcal{O}_X(K_X + S + iL), \omega_X)) =$

$h^0(\mathscr{O}_X(K_X + S + iL))$, or $h^2(\mathscr{O}_X(-iL)) = h^1(\mathscr{O}_X(K_X + iL))$ grows at least linearly with $i \geq 0$. From the exact sequence

$$H^2(\mathscr{O}_X(-iL)) \to H^2(\mathscr{O}_S(-iL|_S)) \to H^3(\mathscr{O}_X(-S - iL)),$$

it suffices to show the linear growth of $h^0(\mathscr{O}_S(K_S + iL|_S)) = h^2(\mathscr{O}_S(-iL|_S))$.

The direct image $\varphi_{S*}\omega_S$ is invertible since it is torsion-free of rank one. We write $\varphi_{S*}\omega_S \simeq \mathscr{O}_{C_S}(P)$ with a divisor P on C_S and write $dL|_S \sim \varphi_S^* A$ with a positive integer d and an ample divisor A on C_S. Then $H^0(\mathscr{O}_S(K_S + idL|_S)) = H^0(\mathscr{O}_{C_S}(P + iA))$ by the projection formula, from which the linear growth of $h^0(\mathscr{O}_S(K_S + iL|_S))$ follows. $\square$

10.3 Birational Minimal Models

By definition, a *minimal model* is a **Q**-factorial terminal projective variety such that the canonical divisor is nef. As is the case with Mori fibre spaces, it is fundamental to understand a birational map $X \dashrightarrow X'$ of minimal models. The structure is relatively simple compared with a link of Mori fibre spaces. By Corollary 1.5.7, the map $X \dashrightarrow X'$ is small and crepant. We further prove that it is decomposed into *flops* in Definition 5.1.6.

Theorem 10.3.1 (Kawamata [246], Birkar [44, corollary 3.3]) *Let $(X_i/S, \Delta_i)$ for $i = 1, 2$ be **Q**-factorial klt pairs projective over a quasi-projective variety with $K_{X_i} + \Delta_i$ relatively nef. Let $f : X_1 \dashrightarrow X_2$ be a small birational map over S with $\Delta_2 = f_*\Delta_1$. Then f is expressed as a finite sequence of elementary log flips of a klt pair $(X_1/S, \Delta_1 + B_1)$ with some effective B_1 such that every log flip is crepant with respect to $K_{X_1} + \Delta_1$.*

Proof This was first proved by Kawamata when the boundary Δ_i is a **Q**-divisor. He used the MMP with big boundary and the boundedness of length of an extremal ray. Then Birkar generalised this to the case of boundary with real coefficients by showing the rationality of Shokurov's polytope as in Remark 1.5.9. For simplicity, we shall demonstrate the theorem in the substantial case when Δ_i is a **Q**-divisor. Then there exists a positive integer r such that $r(K_{X_1} + \Delta_1)$ is integral and Cartier.

We shall write down the proof in the case $\Delta_i = 0$, because the verbatim replacement of K_{X_i} by $K_{X_i} + \Delta_i$ gives the proof for Δ_i with rational coefficients. Hence we assume that Δ_i is zero, where rK_{X_1} is Cartier. We further assume that $S = \operatorname{Spec} \mathbf{C}$ since the relativisation is also straightforward. Let n be the dimension of X and let $e = 1/(2nr + 1) < 1$.

Fix an ample effective **Q**-divisor G_2 on X_2 such that $(X_1, G_1 = f_*^{-1}G_2)$ is klt and then take an ample effective **Q**-divisor H_1 on X_1 such that $(X_1, eG_1 + H_1)$

is klt and such that $K_{X_1} + eG_1 + H_1$ is nef. Since G_1 is big, one can run the $(K_{X_1} + eG_1)$-MMP with scaling of H_1 by Corollary 1.5.13 to produce a log minimal model (Y, eG_Y) of (X_1, eG_1). Note that $K_{X_1} + eG_1$ is big as K_{X_1} is nef. By the base-point free theorem, $K_Y + eG_Y$ is semi-ample and its multiple defines a birational contraction $Y \to X_3$. This gives the log canonical model of (X_1, eG_1).

Since $K_{X_2} + eG_2$ is ample, the negativity lemma shows that the small map $f \colon X_1 \dashrightarrow X_2$ is $(K_{X_1} + eG_1)$-non-positive. Thus by Proposition 1.5.4, f is also the log canonical model of (X_1, eG_1). Hence $X_2 \simeq X_3$ by Lemma 1.5.3. In particular, $X_1 \dashrightarrow Y \to X_3 \simeq X_2$ is small and every step in the MMP from X_1 to Y is a log flip. Further, the small contraction $Y \to X_3$ is an isomorphism because X_2 is $\mathbf{Q}$-factorial. One has only to prove that every log flip in this MMP is crepant.

Consider the initial log flip $f_1 \colon X_1 \dashrightarrow X_1'$. Let R be the extremal ray of $\overline{\mathrm{NE}}(X_1)$ operated on by f_1. By construction, R is $(K_{X_1} + eG_1)$-negative and perpendicular to $K_{X_1} + eG_1 + t_1 H_1$ for some positive t_1. Since K_{X_1} is nef, the former condition implies that R is $(K_{X_1} + G_1)$-negative. In particular, f_1 is also a step in the $(K_{X_1} + G_1)$-MMP. By the boundedness of the length of R in Theorem 1.4.10, there exists a rational curve C belonging to R such that $0 < -((K_{X_1} + G_1) \cdot C) \le 2n$. Then

$$(K_{X_1} \cdot C) = \frac{1}{1-e}((K_{X_1} + eG_1) \cdot C) - \frac{e}{1-e}((K_{X_1} + G_1) \cdot C) < \frac{2ne}{1-e} = \frac{1}{r}.$$

Hence $(K_{X_1} \cdot C) \in r^{-1} \mathbf{Z}_{\ge 0}$ equals zero, which means that f_1 is crepant. Moreover, $(X_1', G_1' = f_{1*} G_1)$ is klt and $r K_{X_1'}$ is Cartier from the exact sequence in Theorem 1.4.7. Thus one can continue the same argument for the $(K_{X_1'} + eG_1')$-MMP with scaling of $t_1 f_{1*} H_1$, and the crepancy follows. $\square$

Corollary 10.3.2 *Every birational map of minimal models is a composite of finitely many flops.*

The Atiyah flop in Example 1.3.18 is the simplest example of a threefold flop. We shall describe threefold flops explicitly in the analytic category using involution. For this, we formulate the notion of an analytic threefold flop analogously to an extremal neighbourhood in Definition 4.4.6. Compare it with Definition 5.1.6.

Definition 10.3.3 The analytic germ $\pi \colon C = \pi^{-1}(q) \subset X \to q \in Y$ of a projective bimeromorphic morphism of terminal threefolds is called an *analytic terminal threefold flopping contraction* if π is isomorphic outside the locus C of dimension one. Choose a π-ample divisor $-D$ on X. Then for $D_Y = \pi_* D$, the

$\mathscr{O}_Y$-algebra $\mathscr{R}(Y, D_Y) = \bigoplus_{i \in \mathbf{N}} \mathscr{O}_Y(iD_Y)$ is finitely generated by the analytic version [236, theorem 4.1] of Proposition 5.1.10. The *flop*, or *D-flop*, of π is the analytic germ $\pi^+ \colon C^+ = (\pi^+)^{-1}(q) \subset X^+ = \mathrm{Proj}_Y \mathscr{R}(Y, D_Y) \to q \in Y$. The flop π^+ will turn out to be independent of D, even if C is reducible, from the assumption that X is a terminal threefold.

Theorem 10.3.4 (Kollár [262, theorem 2.4]) *Let* $\pi \colon C \subset X \to q \in Y$ *be an analytic terminal threefold flopping contraction. Then the flop* $\pi^+ \colon C^+ \subset X^+ \to q \in Y$ *of* π *is described in one of the following ways, where r is the index of* $q \in Y$. *In particular,* π^+ *as the D-flop is independent of the choice of D. The set of singularities of X coincides with that of* X^+. *In other words, analytic terminal threefold flops preserve the set of analytic singularities.*

(i) *The germ* $q \in Y$ *is isomorphic to*

$$o \in (x_1^2 + f(x_2, x_3, x_4) = 0) \subset \mathfrak{D}^4/\mathbf{Z}_r(a_1, a_2, a_3, a_4).$$

Then $\pi^+ = \iota \circ \pi$ *for the involution* $\iota \colon Y \to Y$ *given by* $\iota^*(x_1, x_2, x_3, x_4) = (-x_1, x_2, x_3, x_4)$.

(ii) *The germ* $q \in Y$ *is isomorphic to*

$$o \in (x_1 x_2 - f(x_3, x_4) = 0) \subset \mathfrak{D}^4/\mathbf{Z}_r(1, -1, 0, a).$$

The index-one cover $\bar{Y} = (x_1 x_2 - f(x_3, x_4) = 0) \subset \mathfrak{D}^4$ *admits an involution* $\bar{\iota} \colon \bar{Y} \to \bar{Y}$ *by* $\iota^*(x_1, x_2, x_3, x_4) = (x_2, x_1, x_3, x_4)$. *The action on* $\bar{Y}$ *by the group* $\bar{\iota} \mathbf{Z}_r \bar{\iota}^{-1}$ *extends to an action on* $X \times_Y \bar{Y}$ *regularly. Take the quotient* $X^+ = (X \times_Y \bar{Y})/\bar{\iota} \mathbf{Z}_r \bar{\iota}^{-1} \to Y^+ = \bar{Y}/\bar{\iota} \mathbf{Z}_r \bar{\iota}^{-1}$, *for which* Y^+ *is isomorphic to* $o \in (x_1 x_2 - f = 0) \subset \mathfrak{D}^4/\mathbf{Z}_r(-1, 1, 0, a)$. *Then* π^+ *is* $X^+ \to Y^+ \simeq Y$, *where the isomorphism* $Y^+ \simeq Y$ *is given by the interchange of* x_1 *and* x_2.

The analytic germ $q \in Y$ is described as in (i) or (ii) by the classification of terminal threefold singularities in Theorem 2.4.8. The first case is easy.

Proof of Theorem 10.3.4(i)   The involution ι defines the quotient $\varphi \colon Y \to Z = Y/\mathbf{Z}_2 \simeq \mathfrak{D}^3/\mathbf{Z}_r(a_2, a_3, a_4)$. Take a π-ample divisor $-D$ on X. Let $D_Y = \pi_* D$. Then $D_Y + \iota(D_Y)$ is the pull-back of the **Q**-Cartier divisor $\varphi_* D_Y$. Hence $D_Y \sim_{\mathbf{Q}} -\iota(D_Y)$ and the D-flop π^+ of π is given as $\mathrm{Proj}_Y \mathscr{R}(Y, -\iota(D_Y)) \to Y$, which is $\iota \circ \pi$. $\square$

In the second case, we describe the divisor class group $\mathrm{Cl}\, Y \simeq \mathrm{Pic}(Y \setminus q)$ of the analytic germ $q \in Y$ explicitly.

Lemma 10.3.5   *Let* $o \in X = (x_1 x_2 - f(x_3, x_4) = 0) \subset \mathfrak{D}^4/\mathbf{Z}_r(1, -1, 0, a)$ *be the germ of a terminal threefold analytic singularity of index r. Take the* *irreducible decomposition* $f = \prod_{i=1}^{l} f_i$ *in* $\mathbf{C}\{x_3, x_4^r\}$ *and define the Weil divisor* $D_i = (x_1 = f_i = 0)$ *on* X. *Then*

$$\operatorname{Cl} X = \mathbf{Z}_r[K_X] \oplus \left(\bigoplus_{i=1}^{l} \mathbf{Z}[D_i] \right) \Big/ \mathbf{Z}[D_1 + \cdots + D_l].$$

Proof We follow [263, proposition 2.2.7]. No multiple factor appears in $\prod_i f_i$ since $o \in X$ is an isolated singularity. The divisor $\sum_{i=1}^{l} D_i$ is the principal divisor $(x_1)_X$. Hence $[\sum_{i=1}^{l} D_i] = 0$ in $\operatorname{Cl} X$. The lemma is equivalent to the expression $\operatorname{Cl} X = \mathbf{Z}_r[K_X] \oplus \bigoplus_{i=1}^{l-1} \mathbf{Z}[D_i]$.

Step 1 First we treat the case $r = l = 1$, in which X is given in $\mathfrak{D}^4$ by $x_1 x_2 - f$ with irreducible $f \in \mathbf{C}\{x_3, x_4\}$. By a result of Flenner quoted as Theorem 10.3.6, it suffices to show the vanishing of $H^2(X \setminus o, \mathbf{Z})$.

As argued in the proof of Theorem 2.3.19, the punctured germ $X \setminus o$ is homotopic to $X \cap S_\varepsilon^7$ for small positive ε, where $S_\varepsilon^{2n-1} = \{z \in \mathfrak{D}^n \mid ||z|| = \varepsilon\}$. On the other hand, the irreducibility of f implies that $C \setminus o \simeq \mathfrak{D}^1 \setminus o$ for the germ $o \in C = (f = 0) \subset \mathfrak{D}^2$. Hence $C \cap S_\varepsilon^3$ is homeomorphic to the circle S^1. Then it follows from a result of Giblin [142] that $X \cap S_\varepsilon^7$ is homeomorphic to the sphere S^5. Thus $H^2(X \setminus o, \mathbf{Z}) \simeq H^2(S^5, \mathbf{Z}) = 0$.

Step 2 Next we treat the case $r = 1$ generally, in which $f = \prod_{i=1}^{l} f_i$. Define the divisor $E_i = (x_1 - \prod_{j=1}^{i} f_j = x_2 - \prod_{j=i+1}^{l} f_j = 0)$. Writing for brevity $g = \prod_{j=1}^{i} f_j$ and $h = \prod_{j=i+1}^{l} f_j$ with $f = gh$, one has

$$(x_1 - g)\mathcal{O}_X = (x_1 - g, h(x_1 - g))\mathcal{O}_X$$
$$= (x_1 - g, x_1(x_2 - h))\mathcal{O}_X = (x_1 - g, x_2 - h)\mathcal{O}_X \cap (x_1, g)\mathcal{O}_X.$$

Hence $(x_1 - g)_X = E_i + \sum_{j=1}^{i} D_j$ and $E_i \sim - \sum_{j=1}^{i} D_j$. It suffices to prove that $\operatorname{Cl} X$ is freely generated by $E_1, \ldots, E_{l-1}$.

Assume that $l \geq 2$ and let X_1 be the blow-up of X along D_1. It is covered by the two affine charts

- $U = (x_1 x_2' - f_1 = 0) \subset \mathfrak{D}^4$ with coordinates x_1, x_2', x_3, x_4 for $x_2 = x_2' \prod_{j \geq 2} f_j$, which is factorial by Step 1, and
- $V = (x_1' x_2 - \prod_{j \geq 2} f_j = 0) \subset \mathfrak{D}^4$ with coordinates x_1', x_2, x_3, x_4 for $x_1 = x_1' f_1$.

The exceptional locus in X_1 is a smooth rational curve. Let E_i^1 be the strict transform in X_1 of E_i. Whilst E_1^1 is away from the origins of U and V, E_i^1 for $i \geq 2$ passes through the origin v of V, at which $E_i^1 = (x_1' - \prod_{j=2}^{i} f_j = x_2 - \prod_{j=i+1}^{l} f_j) = 0$. Thus as far as $l \geq 3$, one can blow up the germ $v \in X_1$ along the locus $(x_1' = f_2 = 0)$ in the same manner to make $X_2 \to X_1$. Iterating this operation $l - 1$ times, one finally obtains a factorial threefold $X' = X_{l-1}$.

The exceptional locus C of $X' \to X$ is the union of smooth rational curves $C_1, \ldots, C_{l-1}$, where C_i is the strict transform of the exceptional locus of $X_i \to X_{i-1}$ with $X_0 = X$. One has

$$\operatorname{Pic} X' \simeq H^2(X', \mathbf{Z}) \simeq H^2(C, \mathbf{Z}) = \mathbf{Z}^{\oplus(l-1)},$$

where the first isomorphism is from the rationality of the singularity $o \in X$ and the second is from Theorem 2.3.10. The strict transform E_i' of E_i intersects C_i with $(E_i' \cdot C_i) = 1$ but does not intersect any other C_j. Therefore $E_1', \ldots, E_{l-1}'$ generate $\operatorname{Pic} X'$ freely and the assertion follows via $\operatorname{Cl} X \simeq \operatorname{Pic} X'$.

Step 3 Finally we shall reduce the lemma to the case of index one. Take the index-one cover $\bar{X} = (x_1 x_2 - f = 0) \subset \mathfrak{D}^4$ of X. Each f_i is decomposed into irreducible but not necessarily invariant factors f_{ij} in $\mathbf{C}\{x_3, x_4\}$ as $f_i = \prod_j f_{ij}$. Let $\bar{D}_{ij}$ be the divisor $(x_1 = f_{ij} = 0)$ on $\bar{X}$. We have known that $\operatorname{Cl} \bar{X} = (\bigoplus_{i,j} \mathbf{Z}[\bar{D}_{ij}])/\mathbf{Z}[\sum_{i,j} \bar{D}_{ij}]$ by Step 2, from which $(\operatorname{Cl} \bar{X})^{Z_r} = (\bigoplus_i \mathbf{Z}[D_i])/\mathbf{Z}[\sum_i D_i]$.

Now it suffices to show that the kernel of the natural surjection $\operatorname{Cl} X \twoheadrightarrow (\operatorname{Cl} \bar{X})^{Z_r}$ is generated by K_X. If a divisor D belongs to the kernel, then D is defined by a semi-invariant function δ in $\mathscr{O}_{\bar{X}}$. By the definition of the index-one cover, there exists an integer l such that lK_X is defined by a semi-invariant function of the same weight as δ. Then $D \sim lK_X$. $\qquad\square$

Theorem 10.3.6 (Flenner [124, Satz 6.1]) *Let $x \in X$ be the germ of a normal analytic space and let Σ be the singular locus in X. Suppose that $R^i \mu_* \mathscr{O}_{X'} = 0$ for $i = 1, 2$ for a resolution $\mu \colon X' \to X$. Then $\operatorname{Cl} X \simeq \varinjlim_U H^2(U \setminus \Sigma, \mathbf{Z})$, where the direct limit is taken over all open neighbourhoods U in X at x.*

Proof of Theorem 10.3.4(ii) As in Lemma 10.3.5, we take the irreducible decomposition $f = \prod_{i=1}^l f_i$ in $\mathbf{C}\{x_3, x_4^r\}$ and define the divisor $D_i = (x_1 = f_i = 0)$ on Y. Take a π-ample divisor $-D$ on X and let $D_Y = \pi_* D$. By Lemma 10.3.5, after reordering D_i, there exists a strictly increasing sequence $0 = n(0) < n(1) < \cdots < n(s) < n(s+1) = l$ of indices such that D_Y is up to torsion linearly equivalent to $e_1 S_1 + \cdots + e_s S_s$ with $e_i \in \mathbf{Z}_{>0}$ and with $S_i = D_1 + \cdots + D_{n(i)}$. Multiplying D, we may assume that $D_Y \sim e_1 S_1 + \cdots + e_s S_s$.

Using the decomposition

$$f = \prod_{i=1}^{s+1} g_i, \qquad g_i = \prod_{j=n(i-1)+1}^{n(i)} f_j,$$

we construct a tower $\bar{Y}_s \to \cdots \to \bar{Y}$ of blow-ups over the index-one cover $\bar{Y}$ of Y as in Step 2 of the proof of Lemma 10.3.5. For example, $\bar{Y}_1$ is the blow-up of $\bar{Y}$

along the locus $(x_1 = g_1 = 0)$. It has the germ $y_1 \in \bar{Y}_1$ given by $x_1' x_2 - \prod_{i=2}^{s+1} g_i$, and $\bar{Y}_2$ is the blow-up of $\bar{Y}_1$ along the locus $(x_1' = g_2 = 0)$.

Let $\bar{D}_Y$, $\bar{D}_i$ and $\bar{S}_i$ be the Weil divisors on $\bar{Y}$ which coincide with the pullbacks of D_Y, D_i and S_i respectively on $Y \setminus q$. The divisor $\bar{D}_i$ is given as $(x_1 = f_i = 0)$, and as observed in the proof of Lemma 10.3.5, $-\bar{S}_i$ is linearly equivalent to the divisor $\bar{E}_i = (x_1 - \prod_{j=1}^{i} g_j = x_2 - \prod_{j=i+1}^{s+1} g_j = 0)$. Hence by the construction of $\bar{Y}_s$, the strict transform in $\bar{Y}_s$ of $-\bar{D}_Y \sim e_1 \bar{E}_1 + \cdots + e_s \bar{E}_s$ is ample over $\bar{Y}$. This shows the isomorphism $\bar{Y}_s \simeq X \times_Y \bar{Y}$ over $\bar{Y}$. The action of $\mathbf{Z}_r$ on $\bar{Y}$ extends to $\bar{Y}_s$ biregularly, by which $X \simeq \bar{Y}_s/\mathbf{Z}_r$.

The involution $\bar{\iota} \colon \bar{Y} \to \bar{Y}$ interchanging x_1 and x_2 is lifted to that of $\bar{Y}_s$. It yields a commutative diagram

$$
\begin{array}{ccc}
X \simeq \bar{Y}_s/\mathbf{Z}_r & \dashrightarrow & X^+ = \bar{Y}_s/\bar{\iota}\,\mathbf{Z}_r\bar{\iota}^{-1} \\
\downarrow & & \downarrow \\
Y = \bar{Y}/\mathbf{Z}_r & \xrightarrow{\ \sim\ } & Y^+ = \bar{Y}/\bar{\iota}\,\mathbf{Z}_r\bar{\iota}^{-1}
\end{array}
$$

of quotients, where the isomorphism $Y \simeq Y^+$ is given by the interchange of x_1 and x_2. The divisor $\bar{\iota}(\bar{D}_i)$ is $(x_2 = f_i = 0)$, and $\bar{D}_i + \bar{\iota}(\bar{D}_i) = (f_i)_{\bar{X}}$. Hence $-\bar{\iota}(\bar{D}_Y) \sim \bar{D}_Y$ and $X^+ \simeq \mathrm{Proj}_{Y^+} \mathscr{R}(Y^+, D_Y^+)$ for the image D_Y^+ in Y^+ of D_Y. Thus $X^+ \to Y^+ \simeq Y$ is the D-flop π^+ of π.

By the above explicit description, the set of singularities of X consists of $o \in (x_1 x_2 - g_i = 0) \subset \mathfrak{D}^4/\mathbf{Z}_r(1, -1, 0, a)$ for $1 \le i \le s + 1$ and that of X^+ consists of $o \in (x_1 x_2 - g_i = 0) \subset \mathfrak{D}^4/\mathbf{Z}_r(-1, 1, 0, a)$ for $1 \le i \le s + 1$. As a set of singularities, they are the same. □

In higher dimensions, flops no longer preserve the set of singularities as observed by Matsuki [306, section III.2]. Morrison and Stevens [349, theorem 2.4(ii)] proved that in dimension four, an isolated terminal cyclic quotient singularity of index one is of type $\frac{1}{r}(1, -1, a, -a)$ with a coprime to r. For any such a and r, Matsushita [311] constructed a toric fourfold with a quotient singularity of type $\frac{1}{r}(1, -1, a, -a)$ which is connected to a smooth fourfold by a sequence of flops.

We mention the birational invariance of Betti and Hodge numbers of smooth minimal models. For a smooth projective complex variety X, the *Betti number* $b_i(X)$ is the rank of $H_i(X, \mathbf{Z})$ as in Remark 2.5.5 and the *Hodge number* $h^{p,q}(X)$ is the dimension of $H^{p,q}(X) = H^q(X, \Omega_X^p)$ as introduced right after Theorem 6.3.14.

Theorem 10.3.7 *Let X_1 and X_2 be smooth minimal models which are birational. Then they have the same Betti and Hodge numbers, that is, $b_i(X_1) = b_i(X_2)$ and $h^{p,q}(X_1) = h^{p,q}(X_2)$.*

Batyrev [32] first proved the birational invariance of Betti numbers of Calabi–Yau manifolds by means of p-adic integration and the Weil conjecture. Extending this approach, Wang [469] obtained the invariance for smooth minimal models. At the same time, it turned out that the *motivic integration* invented by Kontsevich [279] and developed by Denef–Loeser [104] can be used to prove Theorem 10.3.7. This idea was realised by Batyrev [31] in the notion of stringy Hodge numbers. The birational invariance of Hodge numbers was stated by Veys as a corollary to [459, theorem 2.7]. Ito [213] and Wang [470] reproved it by the original p-adic method.

10.4 Nef and Mobile Cones

In the preceding section, we have seen that birational minimal models are connected by flops. It is natural to ask how many minimal models a given variety has. For this question, we clarify how to count the number of minimal models.

Definition 10.4.1 Let $(X/S, \Delta)$ be an lc pair projective over a variety. As in Definition 1.5.5, a log minimal model $(X'/S, \Delta')$ of $(X/S, \Delta)$ is equipped with a $(K_X + \Delta)$-negative birational contraction map $f \colon X \dashrightarrow X'/S$ with $\Delta' = f_* \Delta$, where X' is **Q**-factorial and $K_{X'} + \Delta'$ is relatively nef. We call $(X'/S, \Delta')$ endowed with f a *marked log minimal model* of (X, Δ) over S with *marking* f. It is called a *marked (relative) minimal model* if X is terminal and Δ is zero.

The first way of counting is to take the cardinality of the set of isomorphism classes of marked log minimal models (X', f). The second is to look at the set of isomorphism classes of log minimal models X' by ignoring the marking $f \colon X \dashrightarrow X'$. If $(X/S, \Delta)$ itself is a log minimal model, then by Corollary 1.5.7, every marking f is small and crepant with respect to $K_X + \Delta$. In this case, the latter set is considered to be the quotient of the former by the group of birational self-maps of $(X/S, \Delta)$.

In the case of general type, we have a satisfactory answer from Theorem 1.5.8(ii).

Theorem 10.4.2 *Let $(X/S, \Delta)$ be a klt pair projective over a variety such that $K_X + \Delta$ is relatively big. Then the number of marked log minimal models of $(X/S, \Delta)$ is finite.*

This number is bounded in terms of volume. For a smooth threefold without boundary, it is topologically bounded.

Theorem 10.4.3 (Martinelli–Schreieder–Tasin [304]) *For positive integers n, r and a positive rational number v, there exists a number $N(n,r,v)$ with the following property. If (X, Δ) is a klt projective pair of dimension n such that $r\Delta$ is integral and such that $K_X + \Delta$ is big with volume $\mathrm{vol}(K_X + \Delta) = v$, then the number of marked log minimal models of (X, Δ) is at most $N(n,r,v)$.*

Together with the work of Cascini and Tasin [69], if X is a smooth projective threefold of general type, then the number of marked minimal models of X is bounded in terms of the second and third Betti numbers of X.

Without the condition of general type, there may exist infinitely many marked minimal models as will be explained in Examples 10.4.11 and 10.4.12. One can still expect the finiteness of minimal models ignoring marking. As an approach to this, we shall explain the *Kawamata–Morrison conjecture* on the effective nef and mobile cones for Calabi–Yau fibrations. It has the origin in mirror symmetry. In the following definition, neither fundamental group nor irregularity is assumed to be trivial.

Definition 10.4.4 A *log Calabi–Yau fibration* $(X, \Delta) \to S$ is a contraction from a **Q**-factorial klt pair (X, Δ) such that $K_X + \Delta$ is relatively numerically trivial. It is called a *Calabi–Yau fibration* if X is terminal and Δ is zero. A *log Calabi–Yau pair* means a log Calabi–Yau fibration over $S = \mathrm{Spec}\,\mathbf{C}$. A *Calabi–Yau variety* is a Calabi–Yau fibration over $S = \mathrm{Spec}\,\mathbf{C}$.

Remark 10.4.5 Let $\pi\colon (X, \Delta) \to S$ be a log Calabi–Yau fibration. It follows from the abundance in Theorems 1.7.12 and 1.7.13 that $K_X + \Delta$ is relatively **R**-linearly trivial. If S is projective, then by Theorem 1.4.28, it admits a klt pair (S, Δ_S) such that $K_X + \Delta \sim_{\mathbf{R}} \pi^*(K_S + \Delta_S)$.

Definition 10.4.6 Let (X, Δ) and (Y, Γ) be pairs proper over a variety S. A *pseudo-isomorphism* $(X/S, \Delta) \dashrightarrow (Y/S, \Gamma)$ of pairs means a small birational map $X \dashrightarrow Y$ over S which transforms Δ to Γ. It is an *isomorphism* of pairs if $X \dashrightarrow Y$ is an isomorphism. A *pseudo-automorphism* (resp. *automorphism*) of $(X/S, \Delta)$ is a pseudo-isomorphism (resp. isomorphism) from $(X/S, \Delta)$ to itself. We let $\mathrm{PAut}(X/S, \Delta)$ (resp. $\mathrm{Aut}(X/S, \Delta)$) denote the group of pseudo-automorphisms (resp. automorphisms) of $(X/S, \Delta)$.

For example by Corollary 1.5.7, the birational map of two log minimal models of a given pair is a pseudo-isomorphism.

Definition 10.4.7 Let $\pi\colon X \to S$ be a proper morphism from a normal variety to a variety. We defined the ample cone $A(X/S)$, the nef cone $\mathrm{Nef}(X/S)$ and the pseudo-effective cone $P(X/S)$ in Section 1.2 and the mobile cone $\mathrm{Mb}(X/S)$

in Definition 8.4.7. We also defined the big cone $B(X/S)$ when π is projective. We write $\overline{\mathrm{Mb}}(X/S)$ for the closure of $\mathrm{Mb}(X/S)$ in $N^1(X/S)$.

The *effective cone* $B^e(X/S)$ is the convex cone in $N^1(X/S)$ spanned by the classes of Cartier divisors D on X with $\pi_*\mathscr{O}_X(D) \neq 0$, that is, *relatively effective* divisors introduced in Remark 1.2.16. Its closure is the pseudo-effective cone $P(X/S)$. Its interior is the big cone $B(X/S)$ when π is projective. The *effective nef cone* $\mathrm{Nef}^e(X/S)$ and the *effective mobile cone* $\mathrm{Mb}^e(X/S)$ are defined as

$$\mathrm{Nef}^e(X/S) = \mathrm{Nef}(X/S) \cap B^e(X/S),$$
$$\mathrm{Mb}^e(X/S) = \overline{\mathrm{Mb}}(X/S) \cap B^e(X/S).$$

A pseudo-isomorphism $f: X \dashrightarrow Y/S$ of $\mathbf{Q}$-factorial normal varieties projective over S induces an isomorphism $f_*: N^1(X/S) \to N^1(Y/S)$ of real vector spaces. This yields an isomorphism $\mathrm{Mb}^e(X/S) \simeq \mathrm{Mb}^e(Y/S)$. By Lemma 3.1.2, f is an isomorphism if and only if $A(X/S) \cap f_*^{-1}A(Y/S) \neq \emptyset$. If f is an isomorphism, then it induces an isomorphism $\mathrm{Nef}^e(X/S) \simeq \mathrm{Nef}^e(Y/S)$.

We state the Kawamata–Morrison cone conjecture in the logarithmic form following Totaro [445, section 8]. We say that an extremal face F of $\mathrm{Nef}^e(X/S)$ *corresponds to* a contraction $\varphi: X \to Y$ over S if the relative interior of F is the pull-back $\varphi^*A(Y/S)$ of the ample cone. A *rational polyhedral cone* in $N^1(X/S)$ means a convex cone spanned by a finite number of Cartier divisors.

Conjecture 10.4.8 (Kawamata–Morrison cone conjecture [243], [347], [348]) *Let $(X, \Delta) \to S$ be a log Calabi–Yau fibration.*

(i) *The group $\mathrm{Aut}(X/S, \Delta)$ has finitely many orbits in the set of faces of $\mathrm{Nef}^e(X/S)$ corresponding to contractions. Moreover, there exists a rational polyhedral cone Π which is a fundamental domain for the action of $\mathrm{Aut}(X/S, \Delta)$ on $\mathrm{Nef}^e(X/S)$ in the sense that*

$$\mathrm{Nef}^e(X/S) = \bigcup_{\sigma \in \mathrm{Aut}(X/S, \Delta)} \sigma_*\Pi$$

where the interiors of Π and $\sigma_*\Pi$ are disjoint unless $\sigma = 1$.

(ii) *The group $\mathrm{PAut}(X/S, \Delta)$ has finitely many orbits in the set of chambers $f_*^{-1}\mathrm{Nef}^e(X'/S)$ in $\mathrm{Mb}^e(X/S)$ corresponding to marked log minimal models $f: (X/S, \Delta) \dashrightarrow (X'/S, \Delta')$. Equivalently, the number of log minimal models of $(X/S, \Delta)$ is finite. Moreover, there exists a rational polyhedral cone Σ which is a fundamental domain for the action of $\mathrm{PAut}(X/S, \Delta)$ on $\mathrm{Mb}^e(X/S)$.*

Remark 10.4.9 The conjecture with the abundance implies the finiteness of the number of log minimal models of a klt pair without marking. Indeed if a klt log minimal model (X, Δ) over S has relatively semi-ample $K_X + \Delta$, then its ample model T/S makes a log Calabi–Yau fibration $(X, \Delta) \to T$. Every marked log minimal model $(X, \Delta) \dashrightarrow (Y, \Gamma)$ of (X, Δ) over S is over T and it is a pseudo-isomorphism by Corollary 1.5.7. Thus Conjecture 10.4.8(ii) implies the finiteness of the number of $(Y/S, \Gamma)$ without marking.

Proposition 10.4.10 *In the item* (i) *in Conjecture* 10.4.8, *the first statement follows from the second. The same is the case with the item* (ii) *once Theorem* 1.5.8(i) *holds without the assumption* $A_\Delta \le \Delta$.

Proof Suppose the existence of a rational polyhedral fundamental domain Π of $\mathrm{Nef}^e(X/S)$ in (i). Let F be a face of $\mathrm{Nef}^e(X/S)$ which corresponds to a contraction $\varphi \colon X \to Y/S$. Up to automorphism of $(X/S, \Delta)$, we can assume that the relative interior of F meets Π. Then $F \cap \Pi$ is a face of Π, and any divisor D belonging to the relative interior of $F \cap \Pi$ defines φ in the sense that a multiple of D is the pull-back of a relatively ample divisor on Y. Thus the orbit of F by $\mathrm{Aut}(X/S, \Delta)$ is identified by one of the finitely many faces of Π.

Next suppose the existence of Σ in (ii). Let $(X'/S, \Delta')$ be a log minimal model of $(X/S, \Delta)$, which is accompanied by a pseudo-isomorphism $f \colon (X/S, \Delta) \dashrightarrow (X'/S, \Delta')$ by Corollary 1.5.7. Take a relatively ample divisor A' on X'. Up to pseudo-automorphism of $(X/S, \Delta)$, we can assume that $A = f_*^{-1} A'$ belongs to Σ. We want the finiteness of the number of such marked log minimal models f, for which we may assume that S is quasi-projective. Take effective **Q**-divisors $B_1, \ldots, B_r$ spanning Σ such that $(X, \Delta + B)$ is klt for all $B \in \mathscr{E}$ with

$$\mathscr{E} = \left\{ \sum_{i=1}^{r} b_i B_i \;\middle|\; b_i \in \mathbf{R}_{\ge 0}, \; \sum_{i=1}^{r} b_i = 1 \right\}.$$

Then $A \equiv_S aB$ for some $a \in \mathbf{Q}_{>0}$ and $B \in \mathscr{E}$, and X' is the ample model of $K_X + \Delta + B \equiv_S a^{-1}A$ over S. The finiteness follows at once if one can relax the relative bigness of boundary in Theorem 1.5.8(i) for X/S. □

We exhibit two examples from [243, example 3.8], which have infinitely many marked minimal models.

Example 10.4.11 This has already appeared in [399, example 6.8]. Consider the analytic family

$$X^\circ = (x_2^2 = (x_1^2 - t_1)((x_1 - 1)^2 - t_2)) \subset \mathbf{A}^2 \times \mathfrak{D}^2 \to o \in \mathfrak{D}^2$$

of one-dimensional affine schemes over the base $\mathfrak{D}^2$ with coordinates t_1, t_2. It is compactified to a projective morphism $\pi \colon X \to S = \mathfrak{D}^2$ such that the central fibre $C = X \times_S o$ consists of two smooth rational curves C_1 and C_2 intersecting transversally at two points. The fibre C is of type I_2 in *Kodaira's table* of singular elliptic fibres [258, theorem 6.2]. In fact, π is a miniversal deformation of C from Example 4.4.3 and the second statement of Theorem 4.4.5, which holds without the vanishing of $H^1(\mathscr{O}_W)$ when V is a point. The space X/S is a Calabi–Yau fibration in the analytic category. The t_i-axis l_i for $i = 1, 2$ corresponds to the smoothing of one of the two nodes of C. Outside $l_1 \cup l_2$, every fibre of π is an elliptic curve.

The relative Picard number of X/S is two and $\overline{\mathrm{NE}}(X/S)$ is spanned by $[C_1]$ and $[C_2]$. For each i, the flopping contraction of C_i over S exists similarly to Lemma 5.1.4. Let $f_i \colon X \dashrightarrow X_i$ be the corresponding flop. By Theorem 10.3.4, $\pi_i \colon X_i \to S$ is again a miniversal deformation of $X_i \times_S o$ isomorphic to C. Hence π and π_i are isomorphic over S after the marking f_i is ignored.

The real vector space $N^1(X/S)$ has coordinates y_1, y_2 defined by $y_i(z) = (z \cdot C_i)$ for $z \in N^1(X/S)$. The effective nef cone $\mathrm{Nef}^e(X/S)$ coincides with the nef cone $\mathrm{Nef}(X/S)$ spanned by $(1, 0)$ and $(0, 1)$. We shall see that the chamber $f_{1*}^{-1} \mathrm{Nef}(X/S)$ is spanned by $(0, 1)$ and $(-1, 2)$. Take Cartier divisors H_1 and H_2 on X which represent $(1, 0)$ and $(0, 1)$ respectively. Let C_1^+ be the flopped curve in X_1 and let C_2^+ be the strict transform in X_1 of C_2. Let H_i^+ be the strict transform in X_1 of H_i. Then $(H_1^+ \cdot C_1^+) < 0$, $(H_2^+ \cdot C_j^+) = (H_2 \cdot C_j)$ and $(H_1^+ \cdot C_1^+) + (H_1^+ \cdot C_2^+) = 1$. Thus the matrix $((H_i^+ \cdot C_j^+))_{ij}$ is expressed as $\begin{pmatrix} -n & n+1 \\ 0 & 1 \end{pmatrix}$ with some $n \in \mathbf{Z}_{>0}$. It must have determinant ± 1. Hence $n = 1$. By symmetry, $f_{2*}^{-1} \mathrm{Nef}(X/S)$ is spanned by $(2, -1)$ and $(1, 0)$.

In this manner, the chambers Π_n spanned by $(n+1, -n)$ and $(n, -n+1)$ for $n \in \mathbf{Z}$ are transforms of the nef cone by the reciprocal composites $f_1 f_2 f_1 \cdots$, $f_2 f_1 f_2 \cdots$.

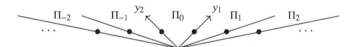

In particular, X/S has infinitely many marked minimal models, all of which are isomorphic without marking.

Example 10.4.12 Let $A = P_1 \times P_2 \times P_3$ with $P_1 = P_2 = \mathbf{P}^1$ and $P_3 = \mathbf{P}^2$. Let X be the general hypersurface in A of tridegree $(2, 2, 3)$, that is, $\mathscr{O}_A(X) \simeq \mathscr{O}_A(2, 2, 3)$ where $\mathscr{O}_A(a_1, a_2, a_3) = \bigotimes_{i=1}^3 p_i^* \mathscr{O}_{P_i}(a_i)$ with the projection p_i to P_i. It is a Calabi–Yau threefold and $\mathrm{Pic}\, X \simeq \mathrm{Pic}\, A \simeq \mathbf{Z}^{\oplus 3}$ by the Grothendieck–Lefschetz theorem [161, XII corollaire 4.9]. The projection $p_{i3} \colon X \to P_i \times P_3$ for $i = 1, 2$ is generically finite of degree two. Its Stein factorisation $q_i \colon X \to Y_i$ is a flopping contraction. In fact, q_i contracts $54 = (\mathscr{O}_{P_i \times P_3}(2, 3))^3$ rational curves. The involution of Y_i over $P_i \times P_3$ yields the flop $f_i \colon X \dashrightarrow X$ of q_i. Like the preceding example, the reciprocal composites $f_1 f_2 f_1 \cdots$, $f_2 f_1 f_2 \cdots$ are distinct marked minimal models of X.

If $X \to S$ is birational, then $B(X/S) = N^1(X/S)$ and $\mathrm{PAut}(X/S) = 1$. In this case, the cone conjecture is a corollary to Theorem 1.5.8(ii).

Theorem 10.4.13 *Let $(X, \Delta) \to S$ be a log Calabi–Yau fibration. For any rational polyhedral cone C inside $B(X/S) \cup 0$, the cone $\mathrm{Nef}^e(X/S) \cap C$ is rational polyhedral and the number of marked log minimal models $f : (X/S, \Delta) \dashrightarrow (X'/S, \Delta')$ with $f_*^{-1} \mathrm{Nef}^e(X'/S) \cap C \setminus 0 \neq \emptyset$ is finite.*

Proof Notice that $\mathrm{Nef}^e(X/S) \cap C = \mathrm{Nef}(X/S) \cap C$. We may assume that S is quasi-projective from the expression

$$\mathrm{Nef}(X/S) \cap C = \left(\bigcap_i (\iota_i^*)^{-1} (\mathrm{Nef}(X_i/S_i) \cap \iota_i^* C) \right) \cap C$$

for a covering of S by quasi-projective open subsets S_i with $\iota_i : X_i = X \times_S S_i \hookrightarrow X$. One deduces from Theorem 1.5.8(ii) or Remark 1.5.9 that the cone $\mathrm{Nef}(X/S) \cap C$ is rational polyhedral, by working on a polytope $\mathscr{E}$ spanning C such that $(X, \Delta + B)$ is klt for all $B \in \mathscr{E}$ as in the proof of Proposition 10.4.10 for (ii). Let $f : (X/S, \Delta) \dashrightarrow (X'/S, \Delta')$ be a marked log minimal model such that the chamber $f_*^{-1} \mathrm{Nef}^e(X'/S)$ intersects $C \setminus 0$. It is also a marked log minimal model of $(X/S, \Delta + B)$ for $B \in f_*^{-1} \mathrm{Nef}^e(X'/S) \cap \mathscr{E} \neq \emptyset$. The number of such f is finite again by Theorem 1.5.8(ii). □

Corollary 10.4.14 *Conjecture 10.4.8 holds when $X \to S$ is birational.*

Proof By Theorem 10.4.13, $\mathrm{Nef}^e(X/S) = \mathrm{Nef}(X/S)$ itself is rational polyhedral and $\mathrm{Mb}^e(X/S) = \overline{\mathrm{Mb}}(X/S)$ is a finite union of rational polyhedral chambers $f_*^{-1} \mathrm{Nef}(X'/S)$. □

We are particularly interested in a Calabi–Yau fibration $X \to S$. By Corollary 1.5.7, every birational self-map of such X/S is a pseudo-automorphism and hence $\mathrm{PAut}(X/S)$ coincides with the *birational automorphism group*

$$\mathrm{Bir}(X/S) = \{X \dashrightarrow X \mid \text{a birational map over } S\}$$

over S. The main theorem in this section asserts the first statements of both items in the cone conjecture for threefold Calabi–Yau fibrations over the base of positive dimension.

Theorem 10.4.15 (Kawamata [243]) *Let $X \to S$ be a Calabi–Yau fibration from a threefold to a variety of positive dimension. Then the first statements of (i) and (ii) in Conjecture 10.4.8 hold for X/S. Namely,*

(i) *$\mathrm{Aut}(X/S)$ has finitely many orbits in the set of faces of $\mathrm{Nef}^e(X/S)$ corresponding to contractions and*

(ii) *the number of relative minimal models of X/S is finite.*

Since we know the abundance for minimal threefolds in Theorem 10.2.1, we obtain the following finiteness from Remark 10.4.9.

Corollary 10.4.16 *The number of minimal models of a terminal projective threefold of positive Kodaira dimension is finite.*

In dimension two, the assertions (i) and (ii) in Conjecture 10.4.8 are the same because every pseudo-isomorphism is an isomorphism. As pointed out by Kawamata [243, theorem 2.1], the conjecture for Calabi–Yau surfaces was essentially proved by Sterk [433] and Namikawa [371]. Extending it logarithmically, Totaro [446] settled the conjecture in dimension two.

Theorem 10.4.17 *Conjecture* 10.4.8 *holds when X is a surface.*

For example, the proof for a projective K3 surface X is performed as follows. See also [193, section 8.4] or [447]. The *positive cone* $\mathscr{C}$ of X is by definition the connected component of $\{z \in N^1(X) \mid z^2 > 0\}$ which contains the class of an ample divisor. Let $O(\mathrm{Pic}\,X)$ be the *orthogonal group* of the free abelian group Pic X with respect to the intersection pairing, and let O^+ be the subgroup of $O(\mathrm{Pic}\,X)$ of index two which preserves $\mathscr{C}$. The subgroup G of Aut X acting trivially on $H^0(\Omega_X^2)$ is of finite index and regarded as a subgroup of O^+. The *Weyl group* W is the subgroup of O^+ generated by reflections in $z \in \mathrm{Pic}\,X$ with $z^2 = -2$. By the *Torelli theorem* of Piatetski-Shapiro and Shafarevich [382], O^+ contains the semi-product $G \ltimes W$ as a subgroup of finite index.

By the general theory of lattices, O^+ acts on the *effective* positive cone $\mathscr{C}^e$, which is the convex hull of integral points in the closure of $\mathscr{C}$, with a rational polyhedral fundamental domain Π. On the other hand, W acts on $\mathscr{C}^e$ with a fundamental domain $\mathrm{Nef}^e(X)$. Hence one can choose Π inside $\mathrm{Nef}^e(X)$, by which Π is a fundamental domain for the action of Aut X on $\mathrm{Nef}^e(X)$ up to finite index.

We remark a complete generalisation to abelian varieties.

Theorem 10.4.18 (Prendergast-Smith [384]) *Conjecture* 10.4.8 *holds for an abelian variety* $X \to \mathrm{Spec}\,\mathbf{C}$.

We return to Theorem 10.4.15. The case when S is a threefold is done by Corollary 10.4.14. We shall provide the entire proof in the case when S is a surface. The other case when S is a curve is similar though it requires properties of surfaces such as Theorem 10.4.17.

Lemma 10.4.19 *Let $\pi\colon X \to S$ be a Calabi–Yau fibration. Then there exists a contraction $\tau\colon X \to T/S$ with T relatively projective and birational to S which is maximal in the sense that any contraction $\tau_1\colon X \to T_1/S$ with T_1/S projective and birational factors through τ as $\tau_1 = t \circ \tau$ for a morphism $t\colon T \to T_1$.*

Proof A contraction $X \to T/S$ with T/S projective and birational is determined by a divisor H on X which is the pull-back of a relatively ample divisor on T. If two such $X \to T_i/S$ for $i = 1, 2$ are given by H_i, then the sum $H_1 + H_2$ is π-semi-ample and defines a contraction $X \to T$ over T_1 and T_2. Hence the maximal T exists by the decrease $\rho(X/T) < \rho(X/T_i)$, unless $T = T_i$, in the relative Picard number. $\qquad\square$

Definition 10.4.20 For a Calabi–Yau fibration $\pi\colon X \to S$, we call the unique maximal Calabi–Yau fibration $\tau\colon X \to T$ in Lemma 10.4.19 the *maximal base modification* of π.

If a Calabi–Yau fibration X/S is from a threefold to a surface, then the maximal base modification has further properties as below. By Remark 10.4.5 and Theorem 4.4.14, the surface S has only quotient singularities. Note that by the MMP, the restriction to each germ $s \in S$ can be compactified to a Calabi–Yau fibration over a projective surface.

Lemma 10.4.21 *Let $\pi\colon X \to S$ be a Calabi–Yau fibration from a threefold to a surface and let $\tau\colon X \to T$ be the maximal base modification. Then there exists a relative minimal model $\pi_m\colon X_m \to S$ of π such that all fibres of the maximal base modification $\tau_m\colon X_m \to T_m$ of π_m are of dimension one. For any such π_m, the induced birational map $T_m \dashrightarrow T$ is a morphism.*

Proof We shall prove the existence of π_m by induction on $\rho(X/T)$. Suppose that there exists a fibre $\tau^{-1}(t)$ of dimension two with $t \in T$. Let C be a general curve in T through t and let D be the unique prime divisor on X mapped onto C, which is τ-mobile. Run the $K_X + \varepsilon D \equiv_T \varepsilon D$-MMP over T with small positive ε to produce a Calabi–Yau fibration $\tau'\colon X' \to T$ such that the strict transform D' in X' of D is τ'-nef. Every step in the MMP is a D-flop as D is τ-mobile.

By Theorem 1.7.12 for $(X'/T, \varepsilon D')$, the divisor D' is τ'-semi-ample. Its multiple defines a contraction $X' \to T_1'/T$ to a surface T_1'. Then $T_1' \neq T$. For, D' is never numerically trivial over T since $\emptyset \neq (\tau')^{-1}(t) \cap D' \neq (\tau')^{-1}(t)$. The maximal base modification $X' \to T'$ of $X' \to T \to S$ has a morphism $T' \to T_1'$ with $\rho(X'/T') \leq \rho(X'/T_1') < \rho(X'/T) = \rho(X/T)$. Thus a model π_m exists by the inductive assumption after replacement of X/S by X'/S.

Working locally on S, we shall prove that $T_m \to S \dashrightarrow T$ is a morphism. Let G be the general hyperplane section of T and define the prime divisor $M = \tau^* G$ on X. Let G_m be the strict transform in T_m of G and let M_m be that in X_m of M. Then M_m is a mobile prime divisor on X_m with $\tau_m(M_m) = G_m$. Recall that T_m has quotient singularities and in particular it is $\mathbf{Q}$-factorial. By the general choice of G, the pull-back $\tau_m^* G_m$ coincides with M_m at the generic point of

G_m. It follows that $M_m = \tau_m^* G_m$ globally because τ_m has no two-dimensional fibres. Here G_m is mobile and hence it is nef on the surface T_m.

Now we have two log minimal models X and X_m of the pair $(X/S, \varepsilon M)$ with small positive ε, and T is the ample model. Thus the target T_m of τ_m with $M_m = \tau_m^* G_m$ admits a morphism $T_m \to T$.   □

Let $\pi\colon X \to S$ be a Calabi–Yau fibration as in Theorem 10.4.15, with S assumed to be a surface. Every divisor D that belongs to $\mathrm{Nef}^e(X/S)$ is π-semi-ample by Theorem 1.7.12 for $(X/S, \varepsilon D)$ with small positive ε. Let $\tau\colon X \to T$ be the maximal base modification of π. By Lemma 10.4.21, we construct a relative minimal model $\pi_m\colon X_m \to S$ with maximal base modification $\tau_m\colon X_m \to T_m$, for which a natural morphism $T_m \to T$ exists. Let T be the set of normal surfaces $T_1 \neq S$ between T_m and S. Precisely,

$$\mathsf{T} = \{T_1 \mid T_m \to T_1 \text{ a contraction over } S,\ T_1 \neq S\},$$

which is a finite set. Define the subset

$$\mathsf{T}_X = \{T_1 \in \mathsf{T} \mid T \to T_1 \text{ a morphism}\} \subset \mathsf{T}.$$

Note that $T \in \mathsf{T}_X$ unless $T = S$.

Let $I = \{(f, \varphi')\}$ be the set of pairs of a marked minimal model $f\colon X \dashrightarrow X'/S$ and a contraction $\varphi'\colon X' \to Y'$ over some $T_1' \in \mathsf{T}_{X'}$. The target Y' may be X'. We define

$$J(X/S) = \mathrm{Mb}^e(X/S) \setminus \bigcup_{(f,\varphi') \in I} f_*^{-1}(\varphi')^* A(Y'/S),$$

where $A(Y'/S)$ is the ample cone. By the MMP, any divisor in $J(X/S) \setminus 0$ is transformed to a relatively semi-ample divisor on some minimal model X'/S of π. It follows from the definition of $J(X/S)$ that its multiple defines a birational contraction from X'. Thus $J(X/S) \setminus 0$ is contained in the big cone $B(X/S)$.

For $d \in \mathbf{R}$, we define the affine subspace

$$\Pi_d = \{z \in N^1(X/S) \mid \deg z = d\} \subset N^1(X/S)$$

of codimension one using the degree $\deg z = (z \cdot f)$ restricted to the general fibre f of $\pi\colon X \to S$. We consider the section

$$J(X/S)_1 = J(X/S) \cap \Pi_1.$$

Lemma 10.4.22 $J(X/S)_1$ *is a closed subset of* Π_1.

Proof   The section $\mathrm{Mb}(X/S)^e \cap \Pi_1$ is contained in $B(X/S)$. Hence by Theorem 10.4.13, it is a locally finite union of local rational polytopes of form $f_*^{-1} \mathrm{Nef}^e(X'/S) \cap \Pi_1$ with $f\colon X \dashrightarrow X'/S$. It thus suffices to check that for two faces F_i of $\text{Nef}^e(X'/S) \cap \Pi_1$ corresponding to contractions $\varphi'_i \colon X' \to Y'_i$ for $i = 1, 2$, if $F_2 \subset F_1$ and $(f, \varphi'_2) \in I$, then $(f, \varphi'_1) \in I$. This follows from the existence of a morphism $g \colon Y'_1 \to Y'_2$ such that $\varphi'_2 = g \circ \varphi'_1$. □

Let $V(X/S)$ be the vector subspace of $N^1(X/S)$ generated by all prime divisors P with $\deg P = 0$. It is contained in Π_0 and generated by the finite set V which consists of

- the pull-backs $\tau^* C$ of curves C in T exceptional over S,
- prime divisors P on X such that $\tau(P)$ is a curve with $P \neq \tau^{-1}\tau(P)$ and
- prime divisors P on X such that $\tau(P)$ is a point.

We take the quotient vector space $W(X/S) = N^1(X/S)/V(T/S)$ and the affine subspace

$$W(X/S)_d = \{z \in W(X/S) \mid \deg z = d\} \subset W(X/S).$$

The group $\text{Bir}(X/S)$ acts on $W(X/S)$ and $W(X/S)_d$ naturally. It also acts on $J(X/S)$ and $J(X/S)_1$. We shall need the following.

Lemma 10.4.23 *The projection $J(X/S)_1 \to W(X/S)_1$ is a proper map.*

Lemma 10.4.24 *Let G be the image of the natural representation $\text{Bir}(X/S) \to GL(N^1(X/S), \mathbf{Z})$. Then there exists a subgroup H of G of finite index acting trivially on $V(X/S)$ and properly discontinuously on $W(X/S)_1$ so that the quotient $W(X/S)_1/H$ is a real torus.*

Corollary 10.4.25 *The quotient $J(X/S)_1/H$ by the group H in Lemma 10.4.24 is compact.*

First we complete the proof of the theorem by admitting this.

Proof of Theorem 10.4.15 for a surface S We shall prove the theorem by induction on $\rho(X/S)$ using Corollary 10.4.25. It is obvious if $\rho(X/S) = 1$. We shall assume that $\rho(X/S) \geq 2$.

First we prove (i). By induction on $\rho(X/S)$, we may assume that for every $T_1 \in \mathsf{T}_X$, the number of faces corresponding to contractions over T_1 is finite up to automorphism of X/S. We need to prove the finiteness up to automorphism of X/S of the number of faces F of $\text{Nef}^e(X/S)$ corresponding to contractions $\varphi \colon X \to Y/S$ not over any $T_1 \in \mathsf{T}_X$. Such φ is necessarily birational and F has relative interior inside $J(X/S) \subset B(X/S) \cup 0$. Since $\text{Nef}^e(X/S) \cap B(X/S)$ is locally rational polyhedral by Theorem 10.4.13, we have only to count faces F of codimension one or equivalently extremal contractions φ.

Fix a point x in $N^1(X/S)$. For faces F and F' of $\text{Nef}^e(X/S)$ of codimenison one and for $f \in \text{Bir}(X/S)$, if the relative interiors of F and $f_* F'$ contain x, then by Lemma 3.1.2, F and f_*F' coincides about x and f sends the interior of $\mathrm{Nef}^e(X/S)$ to the interior or exterior of $\mathrm{Nef}^e(X/S)$. If f sends the interior to the interior, then f is an automorphism. If f and another f' send the interior to the exterior, then $f' \circ f^{-1}$ is an automorphism. Thus the number of faces of codimenison one which have relative interior meeting the orbit of x by $\mathrm{Bir}(X/S)$ is at most two up to automorphism. Because every face we need to take into account has relative interior inside $J(X/S)$, the finiteness is derived from Corollary 10.4.25.

Next we prove (ii). By Proposition 1.5.6, one can replace π by the relative minimal model π_m in Lemma 10.4.21, by which $\mathsf{T} = \mathsf{T}_X$. Then by induction on $\rho(X/S)$, we may assume that for every $T_1 \in \mathsf{T}$, the number of relative minimal models of X over T_1 is finite. We need to show the finiteness of the number of orbits by $\mathrm{Bir}(X/S)$ in the set of chambers $\mathscr{C} = f_*^{-1} \mathrm{Nef}^e(X'/S)$ in $\mathrm{Mb}^e(X/S)$ for marked minimal models $f \colon X/S \dashrightarrow X'/S$ such that $X' \to S$ itself is its maximal base modification. These $\mathscr{C}$ are contained in $J(X/S)$ and the local finiteness for chambers at each point in $J(X/S)_1$ is known from Theorem 10.4.13. The global finiteness for orbits of $\mathscr{C}$ by $\mathrm{Bir}(X/S)$ is derived from Corollary 10.4.25.   □

Proof of Lemma 10.4.23   Set $p \colon J(X/S)_1 \to W(X/S)_1$. By Lemma 10.4.22, the statement means that the inverse image $p^{-1}(Z)$ of a bounded subset Z of $W(X/S)_1$ is again bounded. Since this property is independent of $\pi \colon X \to S$ amongst its relative minimal models, we may replace π by π_m in Lemma 10.4.21 and thus assume that all fibres of the maximal base modification $\tau \colon X \to T$ are of dimension one. Recall that T is **Q**-factorial.

Let $\{C_i\}_{i \in I_1}$ be the set of curves in T exceptional over S. We fix a basis $\{y_i\}_{i \in I_1 \sqcup I_2 \sqcup I_3}$ of $N^1(X/S)$ in such a manner that

- $y_i = [\tau^*C_i]$ for $i \in I_1$,
- $y_i = [P_i]$ with a prime divisor P_i such that $\deg P_i = 0$ and $P_i \neq \tau^{-1}\tau(P_i)$ for $i \in I_2$ and
- $\deg y_i = 1$ for $i \in I_3$,

where the degree has been defined as $\deg z = (z \cdot f)$ by the general fibre f of π. Then $\{y_i\}_{i \in I_1 \sqcup I_2}$ and $\{y_i\}_{i \in I_3}$ are bases of $V(X/S)$ and $W(X/S)$ respectively. For a point $z \in p^{-1}(Z)$, we write $z = \sum_i r_i y_i$ with $r_i \in \mathbf{R}$. Since $p(z)$ belongs to the bounded set Z, there exists a constant M such that $|r_i| \leq M$ for all $i \in I_3$.

For $i \in I_2$, we consider the fibre $\tau^{-1}(p_i)$ at a general point p_i in $\tau(P_i)$. There exist one-cycles l_i^+ and l_i^- on X supported in $\tau^{-1}(p_i)$ such that $(y_i \cdot l_i^+) > 0$, $(y_i \cdot l_i^-) < 0$ and $(y_j \cdot l_i^{\pm}) = 0$ for all $j \in I_1 \sqcup I_2 \setminus \{i\}$. One can make $l_i^{\pm}$ effective by adding a multiple of the one-cycle defined by $p_i \times_T X$. This implies that $(z \cdot l_i^{\pm}) \geq 0$ since z belongs to $\mathrm{Mb}^e(X/S)$. Then for both l_i^+ and l_i^-,

$$-r_i(y_i \cdot l_i^{\pm}) = -(z \cdot l_i^{\pm}) + \sum_{j \in I_3} r_j(y_j \cdot l_i^{\pm}) \leq M \sum_{j \in I_3} |(y_j \cdot l_i^{\pm})|.$$

Hence $|r_i|$ is also bounded for $i \in I_2$. Replacing M, we may assume that $|r_i| \leq M$ for all $i \in I_2 \sqcup I_3$.

For $i \in I_1$, Theorem 4.4.21 gives a morphism $t_i \colon T \to T_i$ in the analytic category which contracts all C_j for $j \neq i$. We take the pull-back D_i to T of a general curve in S through the image of C_i. Then $t_i^* t_i(C_i) \leq D_i$, where the pull-back by t_i is defined as in Definition 4.4.16. A general effective one-cycle γ_i^+ on X with $\tau_* \gamma_i^+$ proportional to $D_i - t_i^* t_i(C_i)$ has $(y_i \cdot \gamma_i^+) = (C_i \cdot \tau_* \gamma_i^+) > 0$, $(y_j \cdot \gamma_i^+) = 0$ for $j \in I_1 \setminus \{i\}$ and $(z \cdot \gamma_i^+) \geq 0$. Likewise a general effective one-cycle γ_i^- on X with $\tau_* \gamma_i^-$ proportional to $t_i^* t_i(C_i)$ has $(y_i \cdot \gamma_i^-) < 0$, $(y_j \cdot \gamma_i^-) = 0$ for $j \in I_1 \setminus \{i\}$ and $(z \cdot \gamma_i^-) \geq 0$. Then

$$-r_i(y_i \cdot \gamma_i^{\pm}) = -(z \cdot \gamma_i^{\pm}) + \sum_{j \in I_2 \sqcup I_3} r_j(y_j \cdot \gamma_i^{\pm}) \leq M \sum_{j \in I_2 \sqcup I_3} |(y_j \cdot \gamma_i^{\pm})|$$

and hence $|r_i|$ is also bounded for $i \in I_1$. Thus $p^{-1}(Z)$ is bounded. $\qquad \square$

Proof of Lemma 10.4.24   Let $\eta = \mathrm{Spec}\, K$ denote the generic point of S for the function field K of S and take the generic fibre $X_\eta = X \times_S \eta$. Then $W(X/S) \simeq N^1(X_\eta/\eta)$. Let d be the minimum of degrees of prime divisors P on X such that $\pi(P) = S$. Equivalently, there exists a point $P_\eta = P \times_S \eta$ in X_η of degree d. Take a finite Galois extension L of K which contains the function field of P_η. Then $X_L = X_\eta \times_\eta \mathrm{Spec}\, L$ is an elliptic curve over L with the fixed L-point $\mathrm{Spec}\, L \to P_\eta \times_\eta \mathrm{Spec}\, L \to X_L$.

The finitely generated abelian group $\Gamma_L = (\mathrm{Pic}^0 X_L)(L)$ of divisors on X_L of degree zero modulo linear equivalence acts on X_L by translation. It is a basic property of an elliptic curve [429, III theorem 10.1] that Γ_L is a subgroup of $\mathrm{Aut}(X_L/\mathrm{Spec}\, L)$ of finite index. The invariant part $\Gamma = \Gamma_L \cap \mathrm{Aut}(X_\eta/\eta)$ by the action of the Galois group of L/K is also of finite index in $\mathrm{Aut}(X_\eta/\eta) = \mathrm{Bir}(X/S)$. The group Γ acts on the finite set V and hence it has a subgroup Γ_1 of finite index which acts trivially on V and hence trivially on $V(X/S)$. Then the image H_1 in $\mathrm{GL}(N^1(X/S), \mathbf{Z})$ of Γ_1 is isomorphic to its image in $\mathrm{GL}(W(X/S), \mathbf{Z})$.

Since H_1 is finitely generated, it has a unique free subgroup H of finite index. By the inclusion $H \subset \mathrm{GL}(W(X/S), \mathbf{Z})$, the group H acts on $W(X/S)_d$ properly discontinuously. It remains to see that $W(X/S)_d/H$ is a real torus. This means that the lattice of the orbit by H of $[P] \in W(X/S)_d$ is of maximal rank.

The lattice in $W(X/S)_d$ of Weil divisors on X is expressed as $[P] + \Lambda$ with $\Lambda = \{[Q] \in W(X/S)_0 \mid Q \text{ integral}\}$. For each $[P+Q] \in W(X/S)_d$, the divisor $Q_\eta = Q \times_S \eta$ on X_η of degree zero defines a translation belonging to Γ which sends $[P] \in W(X/S)_d$ to $[P + dQ]$. Hence the orbit of $[P]$ by Γ contains the lattice $[P] + d\Lambda$ of maximal rank in $W(X/S)_d$. Since H is of finite index in the image in $\mathrm{GL}(W(X/S), \mathbf{Z})$ of Γ, the orbit of $[P]$ by H is also of maximal rank. $\qquad\square$

As pointed out with the example below by Totaro [445], Conjecture 10.4.8 no longer holds for log canonical Calabi–Yau pairs.

Example 10.4.26 Let X be the blow-up of $\mathbf{P}^2$ at nine very general points and let Δ be the strict transform of the unique cubic through the nine points. Then (X, Δ) is lc with $K_X + \Delta \sim 0$. The automorphism group of X is trivial [143, proposition 8]. However, the effective nef cone $\mathrm{Nef}^e(X)$ is not rational polyhedral since X contains infinitely many (-1)-curves as shown by Nagata [366, theorem 4a].

10.5 Pluricanonical Maps

Let X be a smooth projective variety of non-negative Kodaira dimension. By Theorem 1.5.16, the *canonical ring* $R = \bigoplus_{i \in \mathbf{N}} H^0(\mathscr{O}_X(iK_X))$ is finitely generated and the rational map $X \dashrightarrow \mathrm{Proj}\, R$ is birational to the Iitaka fibration associated with K_X. It is realised as the pluricanonical map φ_{lK_X} for a sufficiently large and divisible integer l.

Notation 10.5.1 Let X be a normal projective variety. Let l be a positive integer. We use the notation $P_l = P_l(X) = h^0(\mathscr{O}_X(lK_X))$ of the *l-th plurigenus* of X. We usually write $p_g = p_g(X)$ for P_1 and call it the *geometric genus* of X. In this section, the *l-th pluricanonical map* $\varphi_l = \varphi_{lK_X} : X \dashrightarrow Z_l$ means the rational map defined by the linear system $|lK_X|$ to the image Z_l in $\mathbf{P}H^0(\mathscr{O}_X(lK_X))$, in which $P_l \geq 1$ is assumed.

In explicit birational geometry, it is desirable to control the behaviour of φ_l in terms of l. The next conjecture is fundamental.

Conjecture 10.5.2 *Given a positive integer n, there exists a positive integer $l = l(n)$ depending only on n such that if X is a smooth projective variety of dimension n and of non-negative Kodaira dimension, then φ_l is birational to the Iitaka fibration associated with K_X.*

We have an affirmative answer in the case of general type.

Theorem 10.5.3 (Hacon–McKernan [170], Takayama [438], see also [450]) *Conjecture* 10.5.2 *holds for varieties of general type. To be precise, there exists l depending only on n such that if X is a smooth projective variety of dimension n and of general type, then φ_l is a birational map.*

The conjecture has also been settled for threefolds, according to the Kodaira dimension $\kappa = \kappa(X)$. Besides the case $\kappa = 3$ of general type, the cases $\kappa = 0$, 1 and 2 in dimension three are done by Kawamata [235], Fujino–Mori [134, corollary 6.2] and Viehweg–Zhang [465] respectively.

Theorem 10.5.4 *Conjecture* 10.5.2 *holds for n = 3.*

The purpose of this section is to find an explicit number l for which the pluricanonical map φ_l of a threefold of general type has a good property.

Theorem 10.5.5 (J. A. Chen–M. Chen [78], [79], [80], [83]) *Let X be a terminal projective threefold of general type. Then $P_{12} \geq 1$, $P_{24} \geq 2$ and φ_l is birational for all $l \geq 57$.*

The worst known example is the following.

Example 10.5.6 Let $X = X_{46} \subset \mathbf{P}(4, 5, 6, 7, 23)$ be a general hypersurface of degree 46, which is a member of the family of no 23 in the list [195, 15.1] of Fletcher. It has $P_2 = P_3 = 0$, $P_4 = \cdots = P_9 = 1$, $P_{10} = 2$, and φ_l is birational if and only if $l = 23$ or $l \geq 27$.

One may replace X by its minimal model because the space $H^0(\mathscr{O}_X(lK_X))$ remains the same. If a minimal threefold of general type is Gorenstein, then P_2 is at least four as will be seen in Corollary 10.5.12, and φ_5 is birational [81]. The latter is best possible as so is it for a smooth surface S of general type. For such S, φ_4 is not birational if and only if $(K_S^2, p_g) = (1, 2)$ and φ_3 is not if and only if $(K_S^2, p_g) = (1, 2)$ or $(2, 3)$ [51], [52], [70], [323].

Theorem 10.5.7 (Bombieri [51]) *Let S be a smooth projective surface of general type. Then φ_l is birational for all $l \geq 5$.*

This pertains to the *Fujita conjecture* below. Reider [404] settled it in dimension two, and the statement on the freedom has been proved up to dimension five [114], [242], [475].

Conjecture 10.5.8 (T. Fujita [139]) *Let X be a smooth projective variety of dimension n and let A be an ample divisor on X. Then $K_X + lA$ is free for all $l \geq n + 1$ and very ample for all $l \geq n + 2$.*

The assertion on P_l in Theorem 10.5.5 is proved by a detailed computation of P_l by means of the singular Riemann–Roch formula. We extract the essential ideas by demonstrating the following weaker statement.

Theorem 10.5.9   *If X is a terminal projective threefold of general type, then P_8, P_{10} or P_{12} is positive.*

We defined the *basket* of a terminal threefold in Definition 2.5.22 as the set of fictitious singularities, which are cyclic quotient singularities of type $\frac{1}{r_\iota}(1, -1, b_\iota)$. It is considered to be the set of pairs (r_ι, b_ι) allowing multiplicities. This leads us to the following notation. We arrange r and b as (r, b) to keep consistency with previous chapters, rather than the notation (b, r) in the papers by J. A. Chen and M. Chen.

Notation 10.5.10   In this section, we let

$$S = \{(r, b) \in \mathbf{N}_{>0}^2 \mid r, b \text{ coprime, } b \le r/2\}.$$

Associating $(r, b) \in S$ with the fraction b/r, one may consider S to be the set of irreducible fractions $0 < b/r \le 1/2$. The *basket* $B = \{(r_\iota, b_\iota)\}_{\iota \in I}$ means a finite collection of pairs $(r_\iota, b_\iota) \in S$ allowing multiplicities. We use the expression $B = \{n_j \times (r_j, b_j)\}_{j \in J}$ in which (r_j, b_j) appears n_j times. For example, $\{n \times (r, b)\}$ stands for the collection $\{(r, b), \ldots, (r, b)\}$ of n pairs (r, b). We define the integers

$$\sigma(B) = \sum_{\iota \in I} b_\iota, \qquad \Delta_l(B) = \sum_{\iota \in I} \left(\frac{\overline{lb_\iota}(r_\iota - \overline{lb_\iota})}{2r_\iota} - \frac{lb_\iota(r_\iota - lb_\iota)}{2r_\iota} \right)$$

for $l \in \mathbf{Z}$, where $\bar{a}$ denotes the residue of a modulo r_ι as in Notation 2.2.14. Note that $\Delta_l(B) \ge 0$ and $\Delta_0(B) = \Delta_1(B) = \Delta_2(B) = 0$.

Proposition 10.5.11   *Let X be a terminal projective threefold with basket B of fictitious singularities. Let $\chi_l = \chi(\mathscr{O}_X(lK_X))$ for $l \in \mathbf{Z}$. Then*

$$\chi_{l+1} - \chi_l = l^2 \chi_2 + (3l^2 - 2)\chi_0 - \frac{l(l-1)}{2} \sigma(B) + \Delta_l(B).$$

Proof   This is a direct consequence of the singular Riemann–Roch formula with the duality $\chi_l = -\chi_{1-l}$. Express χ_l by Theorem 2.5.23 and take the difference $\chi_{l+1} - \chi_l$ with the aid of Lemma 2.6.9(i). Then

$$\chi_{l+1} - \chi_l = \frac{l^2}{2} K_X^3 + \frac{1}{12} K_X c_2(X) - \sum_{\iota \in I} \frac{r_\iota^2 - 1}{12 r_\iota} + B(l)$$

where $B(l) = \sum_{\iota \in I} \overline{lb_\iota}(r_\iota - \overline{lb_\iota})/2r_\iota$. Substituting zero for l and using $\chi_1 = -\chi_0$, one obtains

$$-2\chi_0 = \frac{1}{12}K_X c_2(X) - \sum_{\iota \in I} \frac{r_\iota^2 - 1}{12r_\iota}$$

and hence $\chi_{l+1} - \chi_l = l^2 K_X^3/2 - 2\chi_0 + B(l)$. Substituting one for l, one has

$$\chi_2 = \frac{1}{2}K_X^3 - 3\chi_0 + B(1).$$

Eliminating the term K_X^3 and using the relation $B(l) - l^2 B(1) = -2^{-1}l(l - 1)\sigma(B) + \Delta_l(B)$, one attains the desired expression of $\chi_{l+1} - \chi_l$. □

Corollary 10.5.12 *Let X be a terminal projective threefold of general type. Then $P_2 \geq 1 - 3\chi(\mathscr{O}_X)$. If X has a Gorenstein minimal model, then $\chi(\mathscr{O}_X) < 0$ and hence $P_2 \geq 4$.*

Proof We may assume that X is minimal. By Kawamata–Viehweg vanishing, $P_l = \chi_l$ for $l \geq 2$. The first assertion follows from the expression $\chi_2 = K_X^3/2 - 3\chi_0 + B(1)$. If X is Gorenstein, then the expression of χ_0 becomes $\chi_0 = -K_X c_2(X)/24$. Miyaoka's pseudo-effectivity of c_2 in Theorem 10.1.14 provides $0 < K_X^3 \leq 3K_X c_2(X)$. Thus $\chi_0 < 0$. □

For the estimate of $\Delta_l(B)$, we introduce the notion of packing of a basket.

Definition 10.5.13 Let (r, b), (p_1, q_1) and (p_2, q_2) be elements in S such that $q_1/p_1 \leq b/r \leq q_2/p_2$ and $p_1 q_2 - p_2 q_1 = 1$. The collection

$$B' = \{(rq_2 - bp_2) \times (p_1, q_1), \ (bp_1 - rq_1) \times (p_2, q_2)\}$$

is called an *unpacking* of the pair (r, b), and conversely (r, b) is called a *packing* of B'. Notice the equality $(r, b) = (rq_2 - bp_2)(p_1, q_1) + (bp_1 - rq_1)(p_2, q_2)$ as a vector. More generally if a basket B' is obtained from a basket B by finitely many replacements of an element by an unpacking of it, then B' is called an *unpacking* of B and B is called a *packing* of B'. This defines a partial order $\leqslant$ in the set of baskets in such a manner that $B \leqslant B'$ if and only if B is a packing of B'.

Example 10.5.14 $\{(7, 3)\} \leqslant \{(2, 1), (5, 2)\} \leqslant \{2 \times (2, 1), (3, 1)\}$.

Lemma 10.5.15 *If $B \leqslant B'$, then $\sigma(B) = \sigma(B')$ and $\Delta_l(B) \leq \Delta_l(B')$.*

Proof It suffices to check the assertions for $B = \{(r, b)\}$ and its unpacking $B' = \{(rq_2 - bp_2) \times (p_1, q_1), \ (bp_1 - rq_1) \times (p_2, q_2)\}$ in the above definition. The equality on σ is obvious. For a pair (r, b) of positive integers, we let

$$f_l(r, b) = 2lb\left\lfloor \frac{lb}{r} \right\rfloor - \left\lfloor \frac{lb}{r} \right\rfloor\left(\left\lfloor \frac{lb}{r} \right\rfloor + 1\right)r.$$

Then $2\Delta_l(\{(r,b)\}) = f_l(r,b)$ for $(r,b) \in S$. Hence the inequality on Δ_l is reduced to the convexity $f_l(p_1 + p_2, q_1 + q_2) \leq f_l(p_1,q_1) + f_l(p_2,q_2)$ for any pairs (p_1,q_1), (p_2,q_2) of positive integers.

Write $m_i = \lfloor lq_i/p_i \rfloor$ and $n = \lfloor l(q_1+q_2)/(p_1+p_2) \rfloor$. Without loss of generality, we may assume that $m_1 \leq n \leq m_2$. Then $f_l(p_1,q_1) + f_l(p_2,q_2) - f_l(p_1 + p_2, q_1 + q_2) = g_1 + g_2$ for

$$g_i = (m_i - n)(2lq_i - (m_i + n + 1)p_i).$$

One has $g_1 \geq (n - m_1)(n - m_1 - 1)p_1 \geq 0$ from $lq_1 < (m_1 + 1)p_1$. In like manner, $g_2 \geq (m_2 - n)(m_2 - n - 1)p_2 \geq 0$ from $m_2 p_2 \leq lq_2$. These show the convexity of f_l. □

Starting with $S_2 = \{(n,1) \mid n \in \mathbf{N}_{\geq 2}\}$, we define a sequence $\{S_l\}_{l \geq 2}$ of subsets of S by the recurrence relation $S_l = S_{l-1} \cup \{(l,a) \mid (l,a) \in S\}$. For instance, $S_2 = S_3 = S_4$ and $S_5 = S_4 \cup \{(5,2)\}$. Then $p_1 q_2 - p_2 q_1 = 1$ for any *adjacent* elements (p_1,q_1), (p_2,q_2) in S_l defined by the two conditions $q_1/p_1 < q_2/p_2$ and $\{(r,b) \in S_l \mid q_1/p_1 < b/r < q_2/p_2\} = \emptyset$. Indeed for adjacent (p_1,q_1), $(p_2,q_2) \in S_{l-1}$ with $p_1 q_2 - p_2 q_1 = 1$ and for $q_1/p_1 < a/l < q_2/p_2$, the equalities $p_1 a - lq_1 = lq_2 - p_2 a = 1$ follow from the observation that $(a - q_i)/(l - p_i)$ lies outside the open interval $(q_1/p_1, q_2/p_2)$.

Definition 10.5.16 Let B be a basket and fix an integer l at least two. For each $(r,b) \in B$, we take adjacent elements (p_1,q_1), (p_2,q_2) in S_l such that $q_1/p_1 \leq b/r \leq q_2/p_2$ and replace (r,b) by its unpacking $\{(rq_2 - bp_2) \times (p_1,q_1), (bp_1 - rq_1) \times (p_2,q_2)\}$. This makes an unpacking B_l of B every member of which belongs to S_l. We call B_l the unpacking at *level l* of B.

Example 10.5.17 The unpacking B_l at level l of the basket $B = \{(7,3)\}$ is $\{2 \times (2,1), (3,1)\}$ for $l = 2,3,4$, $\{(2,1), (5,2)\}$ for $l = 5,6$ and remains $\{(7,3)\}$ for $l \geq 7$.

Lemma 10.5.18 (i) $B \leqslant \cdots \leqslant B_{l+1} \leqslant B_l \leqslant \cdots \leqslant B_2$.
(ii) $\Delta_l(B) = \cdots = \Delta_l(B_{l+1}) = \Delta_l(B_l)$.

Proof By definition, B_l coincides with the unpacking $(B_m)_l$ at level l of B_m as far as $2 \leq l \leq m$, which implies the first assertion. In the proof of Lemma 10.5.15, if q_1/p_1 and q_2/p_2 belong to the same closed interval $[n/l, (n+1)/l]$ with $n = \lfloor l(q_1 + q_2)/(p_1 + p_2) \rfloor$, then one computes g_i to be zero. The second assertion is derived from this observation. □

Proposition 10.5.19 *Notation and assumptions as in Proposition* 10.5.11. *Then* $\chi_0 + 10\chi_2 + 4\chi_3 + \chi_7 + \chi_{11} + \chi_{13} \leq 2\chi_5 + 3\chi_6 + \chi_8 + \chi_{10} + \chi_{12}$.

Proof Let $\delta_l = \chi_{l+1} - \chi_l$. The proposition is equivalent to the non-negativity of the integer $\alpha = -\delta_{12} - \delta_{10} + \delta_7 + 3\delta_5 + 5\delta_4 + 5\delta_3 + \delta_2 - 9\chi_2 - \chi_0$. One computes α by Proposition 10.5.11 as

$$\alpha = 14\sigma - \Delta_{12} - \Delta_{10} + \Delta_7 + 3\Delta_5 + 5\Delta_4 + 5\Delta_3,$$

writing $\sigma = \sigma(B)$ and $\Delta_l = \Delta_l(B)$ for simplicity. Recall that $\Delta_2 = 0$.

We express the unpacking at level l of B as $B_l = \{n_{r,b}^l \times (r,b)\}_{(r,b)\in S_l}$ with $n_{r,b}^l \in \mathbf{N}$. By Lemmata 10.5.15 and 10.5.18, $\beta = 14\sigma - \Delta_{12} - \Delta_{10} + \Delta_7$ is at least $\beta(B_7) = 14\sigma(B_7) - \Delta_{12}(B_7) - \Delta_{10}(B_7) + \Delta_7(B_7)$. By direct computation, the number $c_{r,b}(l) = \Delta_l(\{(r,b)\})$ for $(r,b) \in S_7$ and $l = 12, 10, 7$ is as follows.

l	$c_{2,1}$	$c_{7,3}$	$c_{5,2}$	$c_{3,1}$	$c_{7,2}$	$c_{4,1}$	$c_{5,1}$	$c_{6,1}$	$c_{7,1}$	$c_{8,1}$	$c_{9,1}$	$c_{10,1}$	$c_{11,1}$
12	30	75	46	18	30	12	9	6	5	4	3	2	1
10	20	50	30	12	19	8	5	4	3	2	1		
7	9	21	13	5	7	3	2	1					

One has $\sigma(B_7) = \sum_{(r,b)\in S_7} b n_{r,b}^7$ and $\Delta_l(B_7) = \sum_{(r,b)\in S_7} c_{r,b}(l) n_{r,b}^7$. It follows from $\{(7,3)\} \preccurlyeq \{(2,1),(5,2)\}$ and $\{(7,2)\} \preccurlyeq \{(3,1),(4,1)\}$ that $n_{2,1}^5 = n_{2,1}^7 + n_{7,3}^7$, $n_{5,2}^5 = n_{7,3}^7 + n_{5,2}^7$, $n_{3,1}^5 = n_{3,1}^7 + n_{7,2}^7$ and $n_{4,1}^5 = n_{7,2}^7 + n_{4,1}^7$. Hence

$$\beta \geq \beta(B_7) \geq -27n_{2,1}^7 - 62n_{7,3}^7 - 35n_{5,2}^7 - 11n_{3,1}^7 - 14n_{7,2}^7 - 3n_{4,1}^7$$
$$= -27n_{2,1}^5 - 35n_{5,2}^5 - 11n_{3,1}^5 - 3n_{4,1}^5.$$

Together with $\Delta_5 = \Delta_5(B_5) = 4n_{2,1}^5 + 5n_{5,2}^5 + 2n_{3,1}^5 + n_{4,1}^5$, one obtains

$$\beta + 3\Delta_5 \geq -15n_{2,1}^5 - 20n_{5,2}^5 - 5n_{3,1}^5 = -15n_{2,1}^2 - 5n_{3,1}^2.$$

Finally with $\Delta_4 = 2n_{2,1}^2 + n_{3,1}^2$ and $\Delta_3 = n_{2,1}^2$, one attains the non-negativity of $\alpha = \beta + 3\Delta_5 + 5\Delta_4 + 5\Delta_3$. □

Proof of Theorem 10.5.9 We may assume that X is minimal. Keep the notation $\chi_l = \chi(\mathscr{O}_X(lK_X))$. Recall that $P_l = \chi_l$ for $l \geq 2$. If P_8, P_{10} and P_{12} were all zero, then $P_5 = P_6 = 0$ and thus $\chi(\mathscr{O}_X) = \chi_0 \leq 0$ by Proposition 10.5.19. Then $P_2 \geq 1$ by Corollary 10.5.12, which contradicts $P_8 = 0$. □

We finally demonstrate the following theorem of Kollár on which the assertion on φ_l in Theorem 10.5.5 is founded. M. Chen [82] improved it by showing the birationality of φ_m for all $m \geq 5l + 6$.

Theorem 10.5.20 (Kollár [260, corollary 4.8]) *Let X be a terminal projective threefold of general type. If $P_l \geq 2$, then φ_{11l+5} is birational.*

Proof By resolution, we may assume that X is smooth. Consider the rational map $X \dashrightarrow \mathbf{P}^1$ defined by a one-dimensional linear subsystem Λ of $|lK_X|$.

Resolve its indeterminacy by a resolution $\mu \colon X' \to X$, in which the mobile part Λ' of the pull-back $\mu^*\Lambda$ defines a morphism $\pi \colon X' \to \mathbf{P}^1$. Here the general member of Λ' is the strict transform of the general member of the mobile part of Λ and one can write $\mu^*\Lambda = \Lambda' + E$ with an effective divisor E. One has the inclusion $\pi^* \mathscr{O}_{\mathbf{P}^1}(1) = \mathscr{O}_{X'}(l\mu^*K_X - E) \subset \mathscr{O}_{X'}(lK_{X'})$. Replacing X by X', we have a morphism $\pi \colon X \to \mathbf{P}^1$ with $\pi^* \mathscr{O}_{\mathbf{P}^1}(1) \subset \mathscr{O}_X(lK_X)$. Let $\mathscr{L} = \pi^* \mathscr{O}_{\mathbf{P}^1}(1) \otimes \mathscr{O}_X(5K_{X/\mathbf{P}^1})$ with $K_{X/\mathbf{P}^1} = K_X - \pi^* K_{\mathbf{P}^1}$. Then $\mathscr{L} \simeq \pi^* \mathscr{O}_{\mathbf{P}^1}(11) \otimes \mathscr{O}_X(5K_X)$ and it is embedded into $\mathscr{O}_X((11l + 5)K_X)$ by $\pi^* \mathscr{O}_{\mathbf{P}^1}(1) \subset \mathscr{O}_X(lK_X)$.

The locally free sheaf $\pi_* \mathscr{O}_X(5K_{X/\mathbf{P}^1})$ is nef by [462, Satz V]. Compare this with Theorem 9.2.10. Hence with Theorem 1.3.5, the direct image $\pi_* \mathscr{L} = \mathscr{O}_{\mathbf{P}^1}(1) \otimes \pi_* \mathscr{O}_X(5K_{X/\mathbf{P}^1})$ is a direct sum of ample invertible sheaves. It follows that the map $H^0(\mathscr{L}) \to \pi_* \mathscr{L} \otimes (k(p_1) \oplus k(p_2))$ is surjective for any two points p_1 and p_2 in $\mathbf{P}^1$, where $k(p_i) \simeq \mathbf{C}$ is the residue field at p_i. If p_1 and p_2 are taken generally, then $\pi_* \mathscr{L} \otimes k(p_i) \simeq H^0(\mathscr{O}_{F_i}(5K_{F_i}))$ for the fibre $F_i = X \times_{\mathbf{P}^1} p_i$ and one has the surjection

$$H^0(\mathscr{L}) \twoheadrightarrow H^0(\mathscr{O}_{F_1}(5K_{F_1})) \oplus H^0(\mathscr{O}_{F_2}(5K_{F_2})).$$

By the easy addition, Theorem 1.7.4, the general fibre F_i is a disjoint union of smooth surfaces of general type. Thus by Theorem 10.5.7, the linear system $|5K_{F_i}|$ defines a birational map. Together with the above surjection, $\mathscr{L}$ defines a birational map from X to the image in $\mathbf{P}H^0(\mathscr{L})$. From the inclusion $\mathscr{L} \subset \mathscr{O}_X((11l + 5)K_X)$, one concludes that φ_{11l+5} is also birational. $\qquad\square$

References

[1] H. Abban, M. Fedorchuk and I. Krylov. Stability of fibrations over one-dimensional bases. Duke Math. J. **171** (2022), 2461–2518.

[2] M. Abe and M. Furushima. On non-normal del Pezzo surfaces. Math. Nachr. **260** (2003), 3–13.

[3] H. Ahmadinezhad. Singular del Pezzo fibrations and birational rigidity. *Automorphisms in birational and affine geometry*, 3–15. Springer Proc. Math. Stat. **79**, Springer, 2014.

[4] H. Ahmadinezhad. On pliability of del Pezzo fibrations and Cox rings. J. Reine Angew. Math. **723** (2017), 101–125.

[5] H. Ahmadinezhad and F. Zucconi. Circle of Sarkisov links on a Fano 3-fold. Proc. Edinb. Math. Soc. (2) **60** (2017), 1–16.

[6] V. A. Alekseev. Rationality conditions for three-dimensional varieties with sheaf of del Pezzo surfaces of degree 4. Mat. Zametki **41** (1987), 724–730; translation in Math. Notes **41** (1987), 408–411.

[7] V. Alexeev. Two two-dimensional terminations. Duke Math. J. **69** (1993), 527–545.

[8] V. Alexeev. Boundedness and K^2 for log surfaces. Int. J. Math. **5** (1994), 779–810.

[9] S. Altınok, G. Brown and M. Reid. Fano 3-folds, K3 surfaces and graded rings. *Topology and geometry: Commemorating SISTAG*, 25–53. Contemp. Math. **314**, Am. Math. Soc., 2002.

[10] A. Altman and S. Kleiman. *Introduction to Grothendieck duality theory*. Lecture Notes in Mathematics **146**, Springer-Verlag, 1970.

[11] F. Ambro. Ladders on Fano varieties. J. Math. Sci. (N.Y.) **94** (1999), 1126–1135.

[12] F. Ambro. On minimal log discrepancies. Math. Res. Lett. **6** (1999), 573–580.

[13] F. Ambro. The moduli b-divisor of an lc-trivial fibration. Compos. Math. **141** (2005), 385–403.

[14] T. Ando. On extremal rays of the higher dimensional varieties. Invent. Math. **81** (1985), 347–357.

[15] M. Andreatta and J. A. Wiśniewski. On contractions of smooth varieties. J. Algebraic Geom. **7** (1998), 253–312.

[16] M. Artin. Some numerical criteria for contractability of curves on algebraic surfaces. Am. J. Math. **84** (1962), 485–496.

[17] M. Artin. On isolated rational singularities of surfaces. Am. J. Math. **88** (1966), 129–136.

[18] M. Artin. On the solutions of analytic equations. Invent. Math. **5** (1968), 277–291.

[19] M. Artin. Algebraic approximation of structures over complete local rings. Publ. Math. Inst. Hautes Études Sci. **36** (1969), 23–58.

[20] M. Artin. The implicit function theorem in algebraic geometry. *Algebraic geometry (Int. Colloq., Bombay, 1968)*, 13–34. Oxford University Press, 1969.

[21] M. Artin. Algebraization of formal moduli: II. Existence of modifications. Ann. Math. (2) **91** (1970), 88–135.

[22] M. Artin and D. Mumford. Some elementary examples of unirational varieties which are not rational. Proc. Lond. Math. Soc. (3) **25** (1972), 75–95.

[23] I. Arzhantsev, U. Derenthal, J. Hausen and A. Laface. *Cox rings*. Cambridge Studies in Advanced Mathematics **144**, Cambridge University Press, 2015.

[24] M. F. Atiyah. On analytic surfaces with double points. Proc. R. Soc. Lond. Ser. A **247** (1958), 237–244.

[25] M. F. Atiyah and R. Bott. A Lefschetz fixed point formula for elliptic complexes: II. Applications. Ann. Math. (2) **88** (1968), 451–491.

[26] M. F. Atiyah and G. B. Segal. The index of elliptic operators: II. Ann. Math. (2) **87** (1968), 531–545.

[27] M. F. Atiyah and I. M. Singer. The index of elliptic operators: I, III. Ann. Math. (2) **87** (1968), 484–530, 546–604.

[28] L. Bădescu. *Algebraic surfaces*. Translated from the 1981 Romanian original by V. Maşek and revised by the author. Universitext, Springer-Verlag, 2001.

[29] C. Bănică and O. Stănăşilă. *Algebraic methods in the global theory of complex spaces*. Editura Academiei; John Wiley & Sons, 1976.

[30] W. P. Barth, K. Hulek, C. A. M. Peters and A. Van de Ven. *Compact complex surfaces*. Second edition. Ergebnisse der Mathematik und ihrer Grenzgebiete (3) **4**, Springer-Verlag, 2004.

[31] V. V. Batyrev. Stringy Hodge numbers of varieties with Gorenstein canonical singularities. *Integrable systems and algebraic geometry*, 1–32. World Scientific Publishing, 1998.

[32] V. V. Batyrev. Birational Calabi–Yau *n*-folds have equal Betti numbers. *New trends in algebraic geometry*, 1–11. London Math. Soc. Lecture Note Ser. **264**, Cambridge University Press, 1999.

[33] A. Beauville. Prym varieties and the Schottky problem. Invent. Math. **41** (1977), 149–196.

[34] A. Beauville. Variétés de Prym et jacobiennes intermédiaires. Ann. Sci. Éc. Norm. Supér. (4) **10** (1977), 309–391.

[35] A. Beauville. *Complex algebraic surfaces*. Translated from the 1978 French original by R. Barlow, with assistance from N. I. Shepherd-Barron and M. Reid. Second edition. London Mathematical Society Student Texts **34**, Cambridge University Press, 1996.

[36] A. Beauville. The Lüroth problem. *Rationality problems in algebraic geometry*, 1–27. Lecture Notes in Math. **2172**, Springer, 2016.

[37] A. Beauville, J.-L. Colliot-Thélène, J.-J. Sansuc and P. Swinnerton-Dyer. Variétés stablement rationnelles non rationnelles. Ann. Math. (2) **121** (1985), 283–318.

[38] M. Beltrametti. On the Chow group and the intermediate Jacobian of a conic bundle. Ann. Mat. Pura Appl. (4) **141** (1985), 331–351.

[39] R. J. Berman. K-polystability of **Q**-Fano varieties admitting Kähler–Einstein metrics. Invent. Math. **203** (2016), 973–1025.

[40] R. J. Berman, S. Boucksom and M. Jonsson. A variational approach to the Yau–Tian–Donaldson conjecture. J. Am. Math. Soc. **34** (2021), 605–652.

[41] P. Berthelot, A. Grothendieck and L. Illusie. *Théorie des intersections et théorème de Riemann–Roch (SGA 6)*. Avec la collaboration de D. Ferrand, J. P. Jouanolou, O. Jussila, S. Kleiman, M. Raynaud et J. P. Serre. Lecture Notes in Mathematics **225**, Springer-Verlag, 1971.

[42] J. Bingener. *Lokale Modulräume in der analytischen Geometrie*. Band 1, 2. With the cooperation of S. Kosarew. Aspects of Mathematics **D2**, **D3**, Friedr. Vieweg & Sohn, 1987.

[43] C. Birkar. On existence of log minimal models. Compos. Math. **146** (2010), 919–928.

[44] C. Birkar. On existence of log minimal models II. J. Reine Angew. Math. **658** (2011), 99–113.

[45] C. Birkar. Existence of log canonical flips and a special LMMP. Publ. Math. Inst. Hautes Études Sci. **115** (2012), 325–368.

[46] C. Birkar. Anti-pluricanonical systems on Fano varieties. Ann. Math. (2) **190** (2019), 345–463.

[47] C. Birkar. Singularities of linear systems and boundedness of Fano varieties. Ann. Math. (2) **193** (2021), 347–405.

[48] C. Birkar, P. Cascini, C. D. Hacon and J. McKernan. Existence of minimal models for varieties of log general type. J. Am. Math. Soc. **23** (2010), 405–468.

[49] H. Blum, Y. Liu and C. Xu. Openness of K-semistability for Fano varieties. Duke Math. J. **171** (2022), 2753–2797.

[50] F. A. Bogomolov. Holomorphic tensors and vector bundles on projective varieties. Izv. Akad. Nauk SSSR Ser. Mat. **42** (1978), 1227–1287; translation in Math. USSR-Izv. **13** (1979), 499–555.

[51] E. Bombieri. Canonical models of surfaces of general type. Publ. Math. Inst. Hautes Études Sci. **42** (1973), 171–219.

[52] E. Bombieri and F. Catanese. The tricanonical map of a surface with $K^2 = 2$, $p_g = 0$. *C. P. Ramanujam – a tribute*, 279–290. Tata Inst. Fund. Res. Stud. Math. **8**, Springer, 1978.

[53] A. Borel and J.-P. Serre. Le théorème de Riemann–Roch. Bull. Soc. Math. France **86** (1958), 97–136.

[54] A. A. Borisov and L. A. Borisov. Singular toric Fano varieties. Mat. Sb. **183** (1992), 134–141; translation in Sb. Math. **75** (1993), 277–283.

[55] S. Boucksom, J.-P. Demailly, M. Păun and T. Peternell. The pseudo-effective cone of a compact Kähler manifold and varieties of negative Kodaira dimension. J. Algebraic Geom. **22** (2013), 201–248.

[56] N. Bourbaki. *Éléments de mathématique. Algèbre. Chapitre 9*. Reprint of the 1959 original. Springer-Verlag, 2007.

[57] E. Brieskorn. Über die Auflösung gewisser Singularitäten von holomorphen Abbildungen. Math. Ann. **166** (1966), 76–102.

[58] E. Brieskorn. Rationale Singularitäten komplexer Flächen. Invent. Math. **4** (1967/68), 336–358.

[59] E. Brieskorn. Die Auflösung der rationalen Singularitäten holomorpher Abbildungen. Math. Ann. **178** (1968), 255–270.

[60] J. W. Bruce and C. T. C. Wall. On the classification of cubic surfaces. J. Lond. Math. Soc. (2) **19** (1979), 245–256.

[61] W. Bruns and J. Herzog. *Cohen–Macaulay rings*. Revised edition. Cambridge Studies in Advanced Mathematics **39**, Cambridge University Press, 1998.

[62] D. A. Buchsbaum and D. Eisenbud. Algebra structures for finite free resolutions, and some structure theorems for ideals of codimension 3. Am. J. Math. **99** (1977), 447–485.

[63] L. Burch. On ideals of finite homological dimension in local rings. Proc. Camb. Philos. Soc. **64** (1968), 941–948.

[64] F. Campana. Connexité rationnelle des variétés de Fano. Ann. Sci. Éc. Norm. Supér. (4) **25** (1992), 539–545.

[65] F. Campana, V. Koziarz and M. Păun. Numerical character of the effectivity of adjoint line bundles. Ann. Inst. Fourier (Grenoble) **62** (2012), 107–119.

[66] H. Cartan. Quotient d'un espace analytique par un groupe d'automorphismes. *Algebraic geometry and topology*, 90–102. Princeton University Press, 1957.

[67] H. Cartan. Quotients of complex analytic spaces. *Contributions to function theory (Int. Colloq., Bombay, 1960)*, 1–15. Tata Inst. Fund. Res., 1960.

[68] R. Carter, G. Segal and I. Macdonald. *Lectures on Lie groups and Lie algebras*. London Mathematical Society Student Texts **32**, Cambridge University Press, 1995.

[69] P. Cascini and L. Tasin. On the Chern numbers of a smooth threefold. Trans. Am. Math. Soc. **370** (2018), 7923–7958.

[70] F. Catanese. Pluricanonical mappings of surfaces with $K^2 = 1, 2$, $q = p_g = 0$. *Algebraic surfaces*, 247–266. C.I.M.E. Summer Sch. **76**, Springer, 2010.

[71] I. Cheltsov. Nonrational nodal quartic threefolds. Pacific J. Math. **226** (2006), 65–81.

[72] I. Cheltsov. Log canonical thresholds of del Pezzo surfaces. Geom. Funct. Anal. **18** (2008), 1118–1144.

[73] I. A. Cheltsov. Extremal metrics on two Fano manifolds. Mat. Sb. **200** (2009), 97–136; translation in Sb. Math. **200** (2009), 95–132.

[74] I. Cheltsov. Factorial threefold hypersurfaces. J. Algebraic Geom. **19** (2010), 781–791.

[75] I. Cheltsov and M. Grinenko. Birational rigidity is not an open property. Bull. Korean Math. Soc. **54** (2017), 1485–1526.

[76] I. Cheltsov and J. Park. Birationally rigid Fano threefold hypersurfaces. Mem. Am. Math. Soc. **246**, no 1167, (2017).

[77] J.-J. Chen. On threefold canonical thresholds. Adv. Math. **404** (2022), 108447, 36pp.

[78] J. A. Chen and M. Chen. Explicit birational geometry of threefolds of general type, I. Ann. Sci. Éc. Norm. Supér. (4) **43** (2010), 365–394.

[79] J. A. Chen and M. Chen. Explicit birational geometry of 3-folds of general type, II. J. Differ. Geom. **86** (2010), 237–271.

[80] J. A. Chen and M. Chen. Explicit birational geometry of 3-folds and 4-folds of general type, III. Compos. Math. **151** (2015), 1041–1082.

[81] J. A. Chen, M. Chen and D.-Q. Zhang. The 5-canonical system on 3-folds of general type. J. Reine Angew. Math. **603** (2007), 165–181.

[82] M. Chen. On the **Q**-divisor method and its application. J. Pure Appl. Algebra **191** (2004), 143–156.

[83] M. Chen. On minimal 3-folds of general type with maximal pluricanonical section index. Asian J. Math. **22** (2018), 257–268.

[84] X. Chen, S. Donaldson and S. Sun. Kähler–Einstein metrics on Fano manifolds. III: Limits as cone angle approaches 2π and completion of the main proof. J. Am. Math. Soc. **28** (2015), 235–278.

[85] S.-S. Chern. Characteristic classes of Hermitian manifolds. Ann. Math. (2) **47** (1946), 85–121.

[86] C. Chevalley. Invariants of finite groups generated by reflections. Am. J. Math. **77** (1955), 778–782.

[87] C. H. Clemens. Double solids. Adv. Math. **47** (1983), 107–230.

[88] C. H. Clemens and P. A. Griffiths. The intermediate Jacobian of the cubic three-fold. Ann. Math. (2) **95** (1972), 281–356.

[89] J.-L. Colliot-Thélène. Arithmétique des variétés rationnelles et problèmes birationnels. *Proceedings of the International Congress of Mathematicians (Berkeley, 1986). Vol. 1*, 641–653. Am. Math. Soc., 1987.

[90] J.-L. Colliot-Thélène and A. Pirutka. Hypersurfaces quartiques de dimension 3: Non-rationalité stable. Ann. Sci. Éc. Norm. Supér. (4) **49** (2016), 371–397.

[91] A. Corti. Factoring birational maps of threefolds after Sarkisov. J. Algebraic Geom. **4** (1995), 223–254.

[92] A. Corti. Del Pezzo surfaces over Dedekind schemes. Ann. Math. (2) **144** (1996), 641–653.

[93] A. Corti. Singularities of linear systems and 3-fold birational geometry. *Explicit birational geometry of 3-folds*, 259–312. London Math. Soc. Lecture Note Ser. **281**, Cambridge University Press, 2000.

[94] A. Corti, ed. *Flips for 3-folds and 4-folds*. Oxford Lecture Series in Mathematics and its Applications **35**, Oxford University Press, 2007.

[95] A. Corti and M. Mella. Birational geometry of terminal quartic 3-folds, I. Am. J. Math. **126** (2004), 739–761.

[96] A. Corti, A. Pukhlikov and M. Reid. Fano 3-fold hypersurfaces. *Explicit birational geometry of 3-folds*, 175–258. London Math. Soc. Lecture Note Ser. **281**, Cambridge University Press, 2000.

[97] A. Corti and M. Reid, eds. *Explicit birational geometry of 3-folds*. London Mathematical Society Lecture Note Series **281**, Cambridge University Press, 2000.

[98] S. Cutkosky. Elementary contractions of Gorenstein threefolds. Math. Ann. **280** (1988), 521–525.

[99] S. D. Cutkosky and V. Srinivas. On a problem of Zariski on dimensions of linear systems. Ann. Math. (2) **137** (1993), 531–559.

[100] V. I. Danilov. Birational geometry of toric 3-folds. Izv. Akad. Nauk SSSR Ser. Mat. **46** (1982), 971–982; translation in Math. USSR-Izv. **21** (1983), 269–280.

[101] O. Debarre. *Higher-dimensional algebraic geometry*. Universitext, Springer-Verlag, 2001.

[102] T. de Fernex. Birationally rigid hypersurfaces. Invent. Math. **192** (2013), 533–566; erratum ibid. **203** (2016), 675–680.

[103] T. de Fernex, L. Ein and M. Mustaţă. Multiplicities and log canonical threshold. J. Algebraic Geom. **13** (2004), 603–615.

[104] J. Denef and F. Loeser. Germs of arcs on singular algebraic varieties and motivic integration. Invent. Math. **135** (1999), 201–232.

[105] A. Dimca. Betti numbers of hypersurfaces and defects of linear systems. Duke Math. J. **60** (1990), 285–298.

[106] I. Dolgachev. Weighted projective varieties. *Group actions and vector fields*, 34–71. Lecture Notes in Math. **956**, Springer, 1982.

[107] S. K. Donaldson. Anti self-dual Yang–Mills connections over complex algebraic surfaces and stable vector bundles. Proc. Lond. Math. Soc. (3) **50** (1985), 1–26.

[108] S. K. Donaldson. Scalar curvature and stability of toric varieties. J. Differ. Geom. **62** (2002), 289–349.

[109] A. Douady. Le problème des modules pour les sous-espaces analytiques compacts d'un espace analytique donné. Ann. Inst. Fourier (Grenoble) **16** (1966), 1–95.

[110] T. Ducat. Mori extractions from singular curves in a smooth 3-fold. PhD thesis, University of Warwick, 2015.

[111] T. Ducat. Divisorial extractions from singular curves in smooth 3-folds. Int. J. Math. **27** (2016), 1650005, 23pp.

[112] T. Ducat. Constructing **Q**-Fano 3-folds à la Prokhorov & Reid. Bull. Lond. Math. Soc. **50** (2018), 420–434.

[113] A. H. Durfee. Fifteen characterizations of rational double points and simple critical points. Enseign. Math. **25** (1979), 131–163.

[114] L. Ein and R. Lazarsfeld. Global generation of pluricanonical and adjoint linear series on smooth projective threefolds. J. Am. Math. Soc. **6** (1993), 875–903.

[115] L. Ein, R. Lazarsfeld, M. Mustaţă, M. Nakamaye and M. Popa. Restricted volumes and base loci of linear series. Am. J. Math. **131** (2009), 607–651.

[116] L. Ein and M. Mustaţă. Inversion of adjunction for local complete intersection varieties. Am. J. Math. **126** (2004), 1355–1365.

[117] L. Ein, M. Mustaţă and T. Yasuda. Jet schemes, log discrepancies and inversion of adjunction. Invent. Math. **153** (2003), 519–535.

[118] D. Eisenbud. Homological algebra on a complete intersection, with an application to group representations. Trans. Am. Math. Soc. **260** (1980), 35–64.

[119] D. Eisenbud. *Commutative algebra. With a view toward algebraic geometry*. Graduate Texts in Mathematics **150**, Springer-Verlag, 1995.

[120] D. Eisenbud and J. Harris. 3264 *and all that. A second course in algebraic geometry*. Cambridge University Press, 2016.

[121] T. Ekedahl. Foliations and inseparable morphisms. *Algebraic geometry, Bowdoin, 1985*, 139–149. Proc. Sympos. Pure Math. **46**, Part 2, Am. Math. Soc., 1987.

[122] S. Endraß. On the divisor class group of double solids. Manuscripta Math. **99** (1999), 341–358.

[123] G. Fischer. *Complex analytic geometry*. Lecture Notes in Mathematics **538**, Springer-Verlag, 1976.

[124] H. Flenner. Divisorenklassengruppen quasihomogener Singularitäten. J. Reine Angew. Math. **328** (1981), 128–160.

[125] P. Francia. Some remarks on minimal models I. Compos. Math. **40** (1980), 301–313.

[126] A. Fujiki. Deformation of uniruled manifolds. Publ. Res. Inst. Math. Sci. **17** (1981), 687–702.

[127] O. Fujino. Termination of 4-fold canonical flips. Publ. Res. Inst. Math. Sci. **40** (2004), 231–237; addendum ibid. **41** (2005), 251–257.

[128] O. Fujino. On the Kleiman–Mori cone. Proc. Jpn. Acad. Ser. A Math. Sci. **81** (2005), 80–84.

[129] O. Fujino. Fundamental theorems for the log minimal model program. Publ. Res. Inst. Math. Sci. **47** (2011), 727–789.

[130] O. Fujino. On Kawamata's theorem. *Classification of algebraic varieties*, 305–315. EMS Ser. Congr. Rep., Eur. Math. Soc., 2011.

[131] O. Fujino. *Foundations of the minimal model program.* MSJ Memoirs **35**, Mathematical Society of Japan, 2017.

[132] O. Fujino. Semipositivity theorems for moduli problems. Ann. Math. (2) **187** (2018), 639–665.

[133] O. Fujino and Y. Gongyo. On canonical bundle formulas and subadjunctions. Michigan Math. J. **61** (2012), 255–264.

[134] O. Fujino and S. Mori. A canonical bundle formula. J. Differ. Geom. **56** (2000), 167–188.

[135] O. Fujino and H. Sato. Introduction to the toric Mori theory. Michigan Math. J. **52** (2004), 649–665.

[136] K. Fujita. A valuative criterion for uniform K-stability of **Q**-Fano varieties. J. Reine Angew. Math. **751** (2019), 309–338.

[137] T. Fujita. On the structure of polarized manifolds with total deficiency one, I, II, III. J. Math. Soc. Jpn. **32** (1980), 709–725, **33** (1981), 415–434, **36** (1984), 75–89.

[138] T. Fujita. Fractionally logarithmic canonical rings of algebraic surfaces. J. Fac. Sci. Univ. Tokyo Sect. IA Math. **30** (1984), 685–696.

[139] T. Fujita. Problems. *Birational geometry of algebraic varieties. Open problems*, 42–45. The 23rd International Symposium, Division of Mathematics, Taniguchi Foundation, 1988.

[140] W. Fulton. *Introduction to toric varieties.* Annals of Mathematics Studies **131**, Princeton University Press, 1993.

[141] W. Fulton. *Intersection theory.* Second edition. Ergebnisse der Mathematik und ihrer Grenzgebiete (3) **2**, Springer-Verlag, 1998.

[142] P. J. Giblin. On the singularities of two related hypersurfaces. J. Lond. Math. Soc. (2) **7** (1973), 367–375.

[143] M. H. Gizatullin. Rational *G*-surfaces. Izv. Akad. Nauk SSSR Ser. Mat. **44** (1980), 110–144; translation in Math. USSR-Izv. **16** (1981), 103–134.

[144] Y. Gongyo. On the minimal model theory for dlt pairs of numerical log Kodaira dimension zero. Math. Res. Lett. **18** (2011), 991–1000.

[145] M. Goresky and R. MacPherson. *Stratified Morse theory.* Ergebnisse der Mathematik und ihrer Grenzgebiete (3) **14**, Springer-Verlag, 1988.

[146] S. Goto, K. Nishida and K.-i. Watanabe. Non-Cohen–Macaulay symbolic blow-ups for space monomial curves and counterexamples to Cowsik's question. Proc. Am. Math. Soc. **120** (1994), 383–392.

[147] T. Graber, J. Harris and J. Starr. Families of rationally connected varieties. J. Am. Math. Soc. **16** (2003), 57–67.

[148] H. Grauert. Über Modifikationen und exzeptionelle analytische Mengen. Math. Ann. **146** (1962), 331–368.

[149] H. Grauert. Über die Deformation isolierter Singularitäten analytischer Mengen. Invent. Math. **15** (1972), 171–198.

[150] H. Grauert. Der Satz von Kuranishi für kompakte komplexe Räume. Invent. Math. **25** (1974), 107–142.

[151] H. Grauert and R. Remmert. *Coherent analytic sheaves*. Grundlehren der mathematischen Wissenschaften **265**, Springer-Verlag, 1984.

[152] H. Grauert and O. Riemenschneider. Verschwindungssätze für analytische Kohomologiegruppen auf komplexen Räumen. Invent. Math. **11** (1970), 263–292.

[153] G.-M. Greuel, C. Lossen and E. Shustin. *Introduction to singularities and deformations*. Springer Monographs in Mathematics, Springer-Verlag, 2007.

[154] P. Griffiths and J. Harris. *Principles of algebraic geometry*. Reprint of the 1978 original. Wiley Classics Library, John Wiley & Sons, Inc., 1994.

[155] M. M. Grinenko. Birational properties of pencils of del Pezzo surfaces of degrees 1 and 2. Mat. Sb. **191** (2000), 17–38; translation in Sb. Math. **191** (2000), 633–653.

[156] M. M. Grinenko. On fibrations into del Pezzo surfaces. Mat. Zametki **69** (2001), 550–565; translation in Math. Notes **69** (2001), 499–513.

[157] M. M. Grinenko. On fibrewise modifications of fibrings into del Pezzo surfaces of degree 2. Uspekhi Mat. Nauk **56** (2001), 145–146; translation in Russian Math. Surveys **56** (2001), 753–754.

[158] M. M. Grinenko. Birational properties of pencils of del Pezzo surfaces of degrees 1 and 2. II. Mat. Sb. **194** (2003), 31–60; translation in Sb. Math. **194** (2003), 669–696.

[159] A. Grothendieck. Sur la classification des fibrés holomorphes sur la sphère de Riemann. Am. J. Math. **79** (1957), 121–138.

[160] A. Grothendieck. Eléments de géométrie algébrique. Publ. Math. Inst. Hautes Études Sci. **4** (1960), **8** (1961), **11** (1961), **17** (1963), **20** (1964), **24** (1965), **28** (1966), **32** (1967).

[161] A. Grothendieck. *Cohomologie locale des faisceaux cohérents et théorèmes de Lefschetz locaux et globaux (SGA 2)*. Augmenté d'un exposé par M. Raynaud. Advanced Studies in Pure Mathematics **2**, North-Holland Publishing Co.; Masson & Cie, Éditeur, 1968.

[162] A. Grothendieck. Le groupe de Brauer I, II, III. *Dix exposés sur la cohomologie des schémas*, 46–66, 67–87, 88–188. Adv. Stud. Pure Math. **3**, North-Holland, 1968.

[163] A. Grothendieck. *Revêtements étales et groupe fondamental (SGA 1)*. Augmenté de deux exposés de M. Raynaud. Lecture Notes in Mathematics **224**, Springer-Verlag, 1971.

[164] A. Grothendieck and J. A. Dieudonné. *Eléments de géométrie algébrique I*. Grundlehren der mathematischen Wissenschaften **166**, Springer-Verlag, 1971.

[165] N. P. Gushel'. On Fano varieties of genus 6. Izv. Akad. Nauk SSSR Ser. Mat. **46** (1982), 1159–1174; translation in Math. USSR-Izv. **21** (1983), 445–459.

[166] N. P. Gushel'. Fano varieties of genus 8. Uspekhi Mat. Nauk **38** (1983), 163–164; translation in Russian Math. Surveys **38** (1983), 192–193.

[167] P. Hacking and Y. Prokhorov. Smoothable del Pezzo surfaces with quotient singularities. Compos. Math. **146** (2010), 169–192.

[168] P. Hacking, J. Tevelev and G. Urzúa. Flipping surfaces. J. Algebraic Geom. **26** (2017), 279–345.

[169] C. D. Hacon and S. J. Kovács. *Classification of higher dimensional algebraic varieties*. Oberwolfach Seminars **41**, Birkhäuser Verlag, 2010.

[170] C. D. Hacon and J. McKernan. Boundedness of pluricanonical maps of varieties of general type. Invent. Math. **166** (2006), 1–25.

[171] C. D. Hacon and J. McKernan. Extension theorems and the existence of flips. *Flips for 3-folds and 4-folds*, 76–110. Oxford Lecture Ser. Math. Appl. **35**, Oxford University Press, 2007.

[172] C. D. Hacon and J. McKernan. Existence of minimal models for varieties of log general type II. J. Am. Math. Soc. **23** (2010), 469–490.

[173] C. D. Hacon and J. McKernan. The Sarkisov program. J. Algebraic Geom. **22** (2013), 389–405.

[174] C. D. Hacon, J. McKernan and C. Xu. ACC for log canonical thresholds. Ann. Math. (2) **180** (2014), 523–571.

[175] C. D. Hacon and C. Xu. Existence of log canonical closures. Invent. Math. **192** (2013), 161–195.

[176] C. D. Hacon and C. Xu. Boundedness of log Calabi–Yau pairs of Fano type. Math. Res. Lett. **22** (2015), 1699–1716.

[177] R. Hartshorne. *Residues and duality*. With an appendix by P. Deligne. Lecture Notes in Mathematics **20**, Springer-Verlag, 1966.

[178] R. Hartshorne. *Algebraic geometry*. Graduate Texts in Mathematics **52**, Springer-Verlag, 1977.

[179] R. Hartshorne. Stable reflexive sheaves. Math. Ann. **254** (1980), 121–176.

[180] B. Hassett, A. Pirutka and Y. Tschinkel. Stable rationality of quadric surface bundles over surfaces. Acta Math. **220** (2018), 341–365.

[181] T. Hayakawa. Blowing ups of 3-dimensional terminal singularities. Publ. Res. Inst. Math. Sci. **35** (1999), 515–570.

[182] T. Hayakawa. Blowing ups of 3-dimensional terminal singularities, II. Publ. Res. Inst. Math. Sci. **36** (2000), 423–456.

[183] T. Hayakawa. Divisorial contractions to 3-dimensional terminal singularities with discrepancy one. J. Math. Soc. Jpn. **57** (2005), 651–668.

[184] T. Hayakawa and K. Takeuchi. On canonical singularities of dimension three. Jpn. J. Math. **13** (1987), 1–46.

[185] F. Hidaka and K.-i. Watanabe. Normal Gorenstein surfaces with ample anticanonical divisor. Tokyo J. Math. **4** (1981), 319–330.

[186] H. Hironaka. On the theory of birational blowing-up. PhD thesis, Harvard University, 1960.

[187] H. Hironaka. Resolution of singularities of an algebraic variety over a field of characteristic zero: I, II. Ann. Math. (2) **79** (1964), 109–203, 205–326.

462   *References*

[188] H. Hironaka and H. Rossi. On the equivalence of imbeddings of exceptional complex spaces. Math. Ann. **156** (1964), 313–333.

[189] F. Hirzebruch. *Topological methods in algebraic geometry*. Translated from the German and Appendix One by R. L. E. Schwarzenberger. Appendix Two by A. Borel. Reprint of the 1978 edition. Classics in Mathematics, Springer-Verlag, 1995.

[190] H. Holmann. Quotienten komplexer Räume. Math. Ann. **142** (1961), 407–440.

[191] H. Holmann. Komplexe Räume mit komplexen Transformationsgruppen. Math. Ann. **150** (1963), 327–360.

[192] H. P. Hudson. *Cremona transformations in plane and space*. Cambridge University Press, 1927.

[193] D. Huybrechts. *Lectures on K3 surfaces*. Cambridge Studies in Advanced Mathematics **158**, Cambridge University Press, 2016.

[194] D. Huybrechts and M. Lehn. *The geometry of moduli spaces of sheaves*. Second edition. Cambridge Mathematical Library, Cambridge University Press, 2010.

[195] A. R. Iano-Fletcher. Working with weighted complete intersections. *Explicit birational geometry of 3-folds*, 101–173. London Math. Soc. Lecture Note Ser. **281**, Cambridge University Press, 2000.

[196] S. Iitaka. *Algebraic geometry. An introduction to birational geometry of algebraic varieties*. Graduate Texts in Mathematics **76**, Springer-Verlag, 1982.

[197] M.-N. Ishida and N. Iwashita. Canonical cyclic quotient singularities of dimension three. *Complex analytic singularities*, 135–151. Adv. Stud. Pure Math. **8**, North-Holland, 1987.

[198] S. Ishii. *Introduction to singularities*. Second edition. Springer, 2018.

[199] V. A. Iskovskih. Rational surfaces with a pencil of rational curves. Mat. Sb. **74(116)** (1967), 608–638; translation in Math. USSR-Sb. **3** (1967), 563–587.

[200] V. A. Iskovskih. Rational surfaces with a pencil of rational curves and with positive square of the canonical class. Mat. Sb. **83(125)** (1970), 90–119; translation in Math. USSR-Sb. **12** (1970), 91–117.

[201] V. A. Iskovskih. Birational properties of a surface of degree 4 in $\mathbf{P}_k^4$. Mat. Sb. **88(130)** (1972), 31–37; translation in Math. USSR-Sb. **17** (1972), 30–36.

[202] V. A. Iskovskih. Fano 3-folds. I. Izv. Akad. Nauk SSSR Ser. Mat. **41** (1977), 516–562; translation in Math. USSR-Izv. **11** (1977), 485–527.

[203] V. A. Iskovskih. Fano 3-folds. II. Izv. Akad. Nauk SSSR Ser. Mat. **42** (1978), 506–549; translation in Math. USSR-Izv. **12** (1978), 469–506.

[204] V. A. Iskovskih. Anticanonical models of three-dimensional algebraic varieties. *Current problems in mathematics. Vol. 12*, 59–157. VINITI, 1979; translation in J. Soviet Math. **13** (1980), 745–814.

[205] V. A. Iskovskih. Minimal models of rational surfaces over arbitrary fields. Izv. Akad. Nauk SSSR Ser. Mat. **43** (1979), 19–43; translation in Math. USSR-Izv. **14** (1980), 17–39.

[206] V. A. Iskovskikh. On the rationality problem for conic bundles. Duke Math. J. **54** (1987), 271–294.

[207] V. A. Iskovskikh. Double projection from a line on Fano threefolds of the first kind. Mat. Sb. **180** (1989), 260–278; translation in Math. USSR-Sb. **66** (1990), 265–284.

[208] V. A. Iskovskikh. On the rationality problem for conic bundles. Mat. Sb. **182** (1991), 114–121; translation in Math. USSR-Sb. **72** (1992), 105–111.

[209] V. A. Iskovskikh. A rationality criterion for conic bundles. Mat. Sb. **187** (1996), 75–92; translation in Sb. Math. **187** (1996), 1021–1038.

[210] V. A. Iskovskikh. Factorization of birational maps of rational surfaces from the viewpoint of Mori theory. Uspekhi Mat. Nauk **51** (1996), 3–72; translation in Russian Math. Surveys **51** (1996), 585–652.

[211] V. A. Iskovskih and J. I. Manin. Three-dimensional quartics and counterexamples to the Lüroth problem. Mat. Sb. **86(128)** (1971), 140–166; translation in Math. USSR-Sb. **15** (1971), 141–166.

[212] V. A. Iskovskikh and Y. G. Prokhorov. Fano varieties. *Algebraic geometry V*, 1–247. Encyclopaedia Math. Sci. **47**, Springer, 1999.

[213] T. Ito. Birational smooth minimal models have equal Hodge numbers in all dimensions. *Calabi–Yau varieties and mirror symmetry*, 183–194. Fields Inst. Commun. **38**, Am. Math. Soc., 2003.

[214] Y. Ito and I. Nakamura. Hilbert schemes and simple singularities. *New trends in algebraic geometry*, 151–233. London Math. Soc. Lecture Note Ser. **264**, Cambridge University Press, 1999.

[215] K. Iwasawa. *Lectures on p-adic L-functions*. Annals of Mathematics Studies **74**, Princeton University Press; University of Tokyo Press, 1972.

[216] D. B. Jaffe. Local geometry of smooth curves passing through rational double points. Math. Ann. **294** (1992), 645–660.

[217] P. Jahnke and I. Radloff. Gorenstein Fano threefolds with base points in the anticanonical system. Compos. Math. **142** (2006), 422–432.

[218] H. Kajiura, K. Saito and A. Takahashi. Matrix factorization and representations of quivers II: Type ADE case. Adv. Math. **211** (2007), 327–362.

[219] A.-S. Kaloghiros. Relations in the Sarkisov program. Compos. Math. **149** (2013), 1685–1709.

[220] B. Kaup. *Äquivalenzrelationen auf allgemeinen komplexen Räumen*. Schr. Math. Inst. Univ. Münster **39**, 1968.

[221] M. Kawakita. Divisorial contractions in dimension three which contract divisors to smooth points. Invent. Math. **145** (2001), 105–119.

[222] M. Kawakita. Divisorial contractions in dimension three which contract divisors to compound A_1 points. Compos. Math. **133** (2002), 95–116.

[223] M. Kawakita. General elephants of three-fold divisorial contractions. J. Am. Math. Soc. **16** (2003), 331–362.

[224] M. Kawakita. Three-fold divisorial contractions to singularities of higher indices. Duke Math. J. **130** (2005), 57–126.

[225] M. Kawakita. Inversion of adjunction on log canonicity. Invent. Math. **167** (2007), 129–133.

[226] M. Kawakita. Towards boundedness of minimal log discrepancies by the Riemann–Roch theorem. Am. J. Math. **133** (2011), 1299–1311.

[227] M. Kawakita. Supplement to classification of threefold divisorial contractions. Nagoya Math. J. **206** (2012), 67–73.

[228] M. Kawakita. The index of a threefold canonical singularity. Am. J. Math. **137** (2015), 271–280.

[229] M. Kawakita. Divisors computing the minimal log discrepancy on a smooth surface. Math. Proc. Camb. Philos. Soc. **163** (2017), 187–192.

[230] Y. Kawamata. A generalization of Kodaira–Ramanujam's vanishing theorem. Math. Ann. **261** (1982), 43–46.

[231] Y. Kawamata. Kodaira dimension of algebraic fiber spaces over curves. Invent. Math. **66** (1982), 57–71.

[232] Y. Kawamata. The cone of curves of algebraic varieties. Ann. Math. (2) **119** (1984), 603–633.

[233] Y. Kawamata. Minimal models and the Kodaira dimension of algebraic fiber spaces. J. Reine Angew. Math. **363** (1985), 1–46.

[234] Y. Kawamata. Pluricanonical systems on minimal algebraic varieties. Invent. Math. **79** (1985), 567–588.

[235] Y. Kawamata. On the plurigenera of minimal algebraic 3-folds with $K \equiv 0$. Math. Ann. **275** (1986), 539–546.

[236] Y. Kawamata. Crepant blowing-up of 3-dimensional canonical singularities and its application to degenerations of surfaces. Ann. Math. (2) **127** (1988), 93–163.

[237] Y. Kawamata. On the length of an extremal rational curve. Invent. Math. **105** (1991), 609–611.

[238] Y. Kawamata. Abundance theorem for minimal threefolds. Invent. Math. **108** (1992), 229–246.

[239] Y. Kawamata. Boundedness of **Q**-Fano threefolds. *Proceedings of the International Conference on Algebra*, 439–445. Contemp. Math. **131**, Part 3, Am. Math. Soc., 1992.

[240] Y. Kawamata. Termination of log flips for algebraic 3-folds. Int. J. Math. **3** (1992), 653–659.

[241] Y. Kawamata. Divisorial contractions to 3-dimensional terminal quotient singularities. *Higher dimensional complex varieties*, 241–246. De Gruyter, 1996.

[242] Y. Kawamata. On Fujita's freeness conjecture for 3-folds and 4-folds. Math. Ann. **308** (1997), 491–505.

[243] Y. Kawamata. On the cone of divisors of Calabi–Yau fiber spaces. Int. J. Math. **8** (1997), 665–687.

[244] Y. Kawamata. Subadjunction of log canonical divisors, II. Am. J. Math. **120** (1998), 893–899.

[245] Y. Kawamata. On effective non-vanishing and base-point-freeness. Asian J. Math. **4** (2000), 173–181.

[246] Y. Kawamata. Flops connect minimal models. Publ. Res. Inst. Math. Sci. **44** (2008), 419–423.

[247] Y. Kawamata. On the abundance theorem in the case of numerical Kodaira dimension zero. Am. J. Math. **135** (2013), 115–124.

[248] Y. Kawamata. *Higher dimensional algebraic varieties*. Iwanami Shoten, 2014.

[249] Y. Kawamata, K. Matsuda and K. Matsuki. Introduction to the minimal model problem. *Algebraic geometry, Sendai, 1985*, 283–360. Adv. Stud. Pure Math. **10**, North-Holland, 1987.

[250] S. Keel, K. Matsuki and J. McKernan. Log abundance theorem for threefolds. Duke Math. J. **75** (1994), 99–119; corrections ibid. **122** (2004), 625–630.

[251] G. Kempf, F. Knudsen, D. Mumford and B. Saint-Donat. *Toroidal embeddings I*. Lecture Notes in Mathematics **339**, Springer-Verlag, 1973.

[252] R. Kiehl. Äquivalenzrelationen in analytischen Räumen. Math. Z. **105** (1968), 1–20.

[253] I.-K. Kim, T. Okada and J. Won. K-stability of birationally superrigid Fano 3-fold weighted hypersurfaces. arXiv:2011.07512.

[254] S. L. Kleiman. Toward a numerical theory of ampleness. Ann. Math. (2) **84** (1966), 293–344.

[255] F. Klein. *Vorlesungen über das Ikosaeder und die Auflösung der Gleichungen vom fünften Grade*. Reprint of the 1884 original. Birkhäuser Verlag; B. G. Teubner, 1993.

[256] T. Kobayashi. *Lie Group and Representation Theory*. To appear from Springer.

[257] K. Kodaira. On a differential-geometric method in the theory of analytic stacks. Proc. Nat. Acad. Sci. USA **39** (1953), 1268–1273.

[258] K. Kodaira. On compact analytic surfaces, I, II, III. Ann. Math. (2) **71** (1960), 111–152, **77** (1963), 563–626, **78** (1963), 1–40.

[259] J. Kollár. The cone theorem: Note to a paper of Y. Kawamata [232]. Ann. Math. (2) **120** (1984), 1–5.

[260] J. Kollár. Higher direct images of dualizing sheaves I. Ann. Math. (2) **123** (1986), 11–42.

[261] J. Kollár. Subadditivity of the Kodaira dimension: Fibers of general type. *Algebraic geometry, Sendai, 1985*, 361–398. Adv. Stud. Pure Math. **10**, North-Holland, 1987.

[262] J. Kollár. Flops. Nagoya Math. J. **113** (1989), 15–36.

[263] J. Kollár. Flips, flops, minimal models, etc. *Surveys in differential geometry*, 113–199. Lehigh Univ., 1991.

[264] J. Kollár, ed. *Flips and abundance for algebraic threefolds*. Astérisque **211**, Société Mathématique de France, 1992.

[265] J. Kollár. Nonrational hypersurfaces. J. Am. Math. Soc. **8** (1995), 241–249.

[266] J. Kollár. *Rational curves on algebraic varieties*. Ergebnisse der Mathematik und ihrer Grenzgebiete (3) **32**, Springer-Verlag, 1996.

[267] J. Kollár. Polynomials with integral coefficients, equivalent to a given polynomial. Electron. Res. Announc. Am. Math. Soc. **3** (1997), 17–27.

[268] J. Kollár. Real algebraic threefolds, III. Conic bundles. J. Math. Sci. (N.Y.) **94** (1999), 996–1020.

[269] J. Kollár. *Singularities of the minimal model program*. With the collaboration of S. Kovács. Cambridge Tracts in Mathematics **200**, Cambridge University Press, 2013.

[270] J. Kollár and S. J. Kovács. Log canonical singularities are Du Bois. J. Am. Math. Soc. **23** (2010), 791–813.

[271] J. Kollár and T. Matsusaka. Riemann–Roch type inequalities. Am. J. Math. **105** (1983), 229–252.

[272] J. Kollár, Y. Miyaoka and S. Mori. Rational curves on Fano varieties. *Classification of irregular varieties*, 100–105. Lecture Notes in Math. **1515**, Springer, 1992.

[273] J. Kollár, Y. Miyaoka and S. Mori. Rationally connected varieties. J. Algebraic Geom. **1** (1992), 429–448.

[274] J. Kollár, Y. Miyaoka and S. Mori. Rational connectedness and boundedness of Fano manifolds. J. Differ. Geom. **36** (1992), 765–779.

[275] J. Kollár, Y. Miyaoka, S. Mori and H. Takagi. Boundedness of canonical **Q**-Fano 3-folds. Proc. Jpn. Acad. Ser. A Math. Sci. **76** (2000), 73–77.

[276] J. Kollár and S. Mori. Classification of three-dimensional flips. J. Am. Math. Soc. **5** (1992), 533–703; errata by S. Mori ibid. **20** (2007), 269–271.

[277] J. Kollár and S. Mori. *Birational geometry of algebraic varieties*. With the collaboration of C. H. Clemens and A. Corti. Translated from the 1998 Japanese original. Cambridge Tracts in Mathematics **134**, Cambridge University Press, 1998.

[278] J. Kollár and N. I. Shepherd-Barron. Threefolds and deformations of surface singularities. Invent. Math. **91** (1988), 299–338.

[279] M. Kontsevich. Lecture at Orsay, 7 December 1995.

[280] M. Kontsevich and Y. Tschinkel. Specialization of birational types. Invent. Math. **217** (2019), 415–432.

[281] A. R. Kustin and M. Miller. Constructing big Gorenstein ideals from small ones. J. Algebra **85** (1983), 303–322.

[282] C.-J. Lai. Varieties fibered by good minimal models. Math. Ann. **350** (2011), 533–547.

[283] S. Lang. *Algebra*. Revised third edition. Graduate Texts in Mathematics **211**, Springer-Verlag, 2002.

[284] C. Lange. When is the underlying space of an orbifold a manifold? Trans. Am. Math. Soc. **372** (2019), 2799–2828.

[285] A. Langer. Semistable sheaves in positive characteristic. Ann. Math. (2) **159** (2004), 251–276; addendum ibid. **160** (2004), 1211–1213.

[286] H. B. Laufer. Taut two-dimensional singularities. Math. Ann. **205** (1973), 131–164.

[287] H. B. Laufer. On minimally elliptic singularities. Am. J. Math. **99** (1977), 1257–1295.

[288] R. Lazarsfeld. *Positivity in algebraic geometry I*. Ergebnisse der Mathematik und ihrer Grenzgebiete (3) **48**, Springer-Verlag, 2004.

[289] R. Lazarsfeld and M. Mustaţă. Convex bodies associated to linear series. Ann. Sci. Éc. Norm. Supér. (4) **42** (2009), 783–835.

[290] M. N. Levine. Deformations of uni-ruled varieties. Duke Math. J. **48** (1981), 467–473.

[291] M. Levine. Some examples from the deformation theory of ruled varieties. Am. J. Math. **103** (1981), 997–1020.

[292] C. Li. K-semistability is equivariant volume minimization. Duke Math. J. **166** (2017), 3147–3218.

[293] C. Li. *G*-uniform stability and Kähler–Einstein metrics on Fano varieties. Invent. Math. **227** (2022), 661–744.

[294] J. Lin. Birational unboundedness of **Q**-Fano threefolds. Int. Math. Res. Not. IMRN **2003** (2003), 301–312.

[295] J. Lipman. Rational singularities, with applications to algebraic surfaces and unique factorization. Publ. Math. Inst. Hautes Études Sci. **36** (1969), 195–279.

[296] Y. Liu, C. Xu and Z. Zhuang. Finite generation for valuations computing stability thresholds and applications to K-stability. Ann. Math. (2) **196** (2022), 507–566.

[297] S. Lojasiewicz. Triangulation of semi-analytic sets. Ann. Sc. Norm. Super. Pisa Cl. Sci. (3) **18** (1964), 449–474.

[298] D. Maclagan. Antichains of monomial ideals are finite. Proc. Am. Math. Soc. **129** (2001), 1609–1615.

[299] A. I. Mal'cev. On the faithful representation of infinite groups by matrices. Mat. Sb. **8(50)** (1940), 405–422; translation in Am. Math. Soc. Transl. (2) **45** (1965), 1–18.

[300] J. I. Manin. Rational surfaces over perfect fields. Publ. Math. Inst. Hautes Études Sci. **30** (1966), 55–113.

[301] J. I. Manin. Rational surfaces over perfect fields. II. Mat. Sb. **72(114)** (1967), 161–192; translation in Math. USSR-Sb. **1** (1967), 141–168.

[302] Y. I. Manin. *Cubic forms. Algebra, geometry, arithmetic.* Translated from the Russian by M. Hazewinkel. Second edition. North-Holland Mathematical Library **4**, North-Holland Publishing Co., 1986.

[303] D. Markushevich. Minimal discrepancy for a terminal cDV singularity is 1. J. Math. Sci. Univ. Tokyo **3** (1996), 445–456.

[304] D. Martinelli, S. Schreieder and L. Tasin. On the number and boundedness of log minimal models of general type. Ann. Sci. Éc. Norm. Supér. (4) **53** (2020), 1183–1207.

[305] L. Masiewicki. Universal properties of Prym varieties with an application to algebraic curves of genus five. Trans. Am. Math. Soc. **222** (1976), 221–240.

[306] K. Matsuki. Weyl groups and birational transformations among minimal models. Mem. Am. Math. Soc. **116**, no 557, (1995).

[307] K. Matsuki. *Introduction to the Mori program.* Universitext, Springer-Verlag, 2002.

[308] H. Matsumura. *Commutative ring theory.* Translated from the Japanese by M. Reid. Paperback edition. Cambridge Studies in Advanced Mathematics **8**, Cambridge University Press, 1989.

[309] T. Matsusaka. Algebraic deformations of polarized varieties. Nagoya Math. J. **31** (1968), 185–245; corrections ibid. **33** (1968), 137, **36** (1968), 119.

[310] T. Matsusaka. Polarized varieties with a given Hilbert polynomial. Am. J. Math. **94** (1972), 1027–1077.

[311] D. Matsushita. On smooth 4-fold flops. Saitama Math. J. **15** (1997), 47–54.

[312] J. McKay. Graphs, singularities, and finite groups. *The Santa Cruz Conference on Finite Groups*, 183–186. Proc. Sympos. Pure Math. **37**, Am. Math. Soc., 1980.

[313] V. B. Mehta and A. Ramanathan. Semistable sheaves on projective varieties and their restriction to curves. Math. Ann. **258** (1981/82), 213–224.

[314] V. B. Mehta and A. Ramanathan. Restriction of stable sheaves and representations of the fundamental group. Invent. Math. **77** (1984), 163–172.

[315] M. Mella. Birational geometry of quartic 3-folds II: The importance of being $\mathbb{Q}$-factorial. Math. Ann. **330** (2004), 107–126.

[316] J. C. Migliore. *Introduction to liaison theory and deficiency modules.* Progress in Mathematics **165**, Birkhäuser Boston, Inc., 1998.

[317] J. S. Milne. *Étale cohomology.* Princeton Mathematical Series **33**, Princeton University Press, 1980.

[318] J. S. Milne. Jacobian varieties. *Arithmetic geometry*, 167–212. Springer, 1986.

[319] J. Milnor. *Morse theory.* Annals of Mathematics Studies **51**, Princeton University Press, 1963.

[320] J. Milnor. *Singular points of complex hypersurfaces.* Annals of Mathematics Studies **61**, Princeton University Press; University of Tokyo Press, 1968.

[321] M. Miyanishi. Algebraic methods in the theory of algebraic threefolds – surrounding the works of Iskovskikh, Mori and Sarkisov. *Algebraic varieties and analytic varieties*, 69–99. Adv. Stud. Pure Math. **1**, North-Holland, 1983.

[322] M. Miyanishi and D. Q. Zhang. Gorenstein log del Pezzo surfaces of rank one. J. Algebra **118** (1988), 63–84.

[323] Y. Miyaoka. Tricanonical maps of numerical Godeaux surfaces. Invent. Math. **34** (1976), 99–111.

[324] Y. Miyaoka. Deformations of a morphism along a foliation and applications. *Algebraic geometry, Bowdoin, 1985*, 245–268. Proc. Sympos. Pure Math. **46**, Part 1, Am. Math. Soc., 1987.

[325] Y. Miyaoka. The Chern classes and Kodaira dimension of a minimal variety. *Algebraic geometry, Sendai, 1985*, 449–476. Adv. Stud. Pure Math. **10**, North-Holland, 1987.

[326] Y. Miyaoka. Abundance conjecture for 3-folds: Case $v = 1$. Compos. Math. **68** (1988), 203–220.

[327] Y. Miyaoka. On the Kodaira dimension of minimal threefolds. Math. Ann. **281** (1988), 325–332.

[328] Y. Miyaoka and S. Mori. A numerical criterion for uniruledness. Ann. Math. (2) **124** (1986), 65–69.

[329] Y. Miyaoka and T. Peternell. *Geometry of higher dimensional algebraic varieties.* DMV Seminar **26**, Birkhäuser Verlag, 1997.

[330] B. G. Moĭšezon. Algebraic homology classes on algebraic varieties. Izv. Akad. Nauk SSSR Ser. Mat. **31** (1967), 225–268; translation in Math. USSR-Izv. **1** (1967), 209–251.

[331] S. Mori. Projective manifolds with ample tangent bundles. Ann. Math. (2) **110** (1979), 593–606.

[332] S. Mori. Threefolds whose canonical bundles are not numerically effective. Ann. Math. (2) **116** (1982), 133–176.

[333] S. Mori. On 3-dimensional terminal singularities. Nagoya Math. J. **98** (1985), 43–66.

[334] S. Mori. Classification of higher-dimensional varieties. *Algebraic geometry, Bowdoin, 1985*, 269–331. Proc. Sympos. Pure Math. **46**, Part 1, Am. Math. Soc., 1987.

[335] S. Mori. Flip theorem and the existence of minimal models for 3-folds. J. Am. Math. Soc. **1** (1988), 117–253.

[336] S. Mori. On semistable extremal neighborhoods. *Higher dimensional birational geometry*, 157–184. Adv. Stud. Pure Math. **35**, Math. Soc. Japan, 2002.

[337] S. Mori and S. Mukai. Classification of Fano 3-folds with $B_2 \geq 2$. Manuscripta Math. **36** (1981), 147–162; erratum ibid. **110** (2003), 407.

[338] S. Mori and Y. Prokhorov. On **Q**-conic bundles. Publ. Res. Inst. Math. Sci. **44** (2008), 315–369.

[339] S. Mori and Y. Prokhorov. On **Q**-conic bundles, II. Publ. Res. Inst. Math. Sci. **44** (2008), 955–971.

[340] S. Mori and Y. Prokhorov. On **Q**-conic bundles, III. Publ. Res. Inst. Math. Sci. **45** (2009), 787–810.

[341] S. Mori and Y. G. Prokhorov. Multiple fibers of del Pezzo fibrations. Tr. Mat. Inst. Steklova **264** (2009), 137–151; reprinted in Proc. Steklov Inst. Math. **264** (2009), 131–145.

[342] S. Mori and Y. Prokhorov. Threefold extremal contractions of type (IA). Kyoto J. Math. **51** (2011), 393–438.

[343] S. Mori and Y. Prokhorov. 3-fold extremal contractions of types (IC) and (IIB). Proc. Edinb. Math. Soc. (2) **57** (2014), 231–252.

[344] S. Mori and Y. G. Prokhorov. Threefold extremal contractions of type (IIA). I. Izv. Ross. Akad. Nauk Ser. Mat. **80** (2016), 77–102; reprinted in Izv. Math. **80** (2016), 884–909.

[345] S. Mori and Y. Prokhorov. Threefold extremal contractions of type (IIA). Part II. *Geometry and physics. Vol. II*, 623–652. Oxford University Press, 2018.

[346] D. R. Morrison. Canonical quotient singularities in dimension three. Proc. Am. Math. Soc. **93** (1985), 393–396.

[347] D. R. Morrison. Compactifications of moduli spaces inspired by mirror symmetry. *Journées de géométrie algébrique d'Orsay*. Astérisque **218** (1993), 243–271.

[348] D. R. Morrison. Beyond the Kähler cone. *Proceedings of the Hirzebruch 65 Conference on Algebraic Geometry*, 361–376. Israel Math. Conf. Proc. **9**, Bar-Ilan Univ., 1996.

[349] D. R. Morrison and G. Stevens. Terminal quotient singularities in dimensions three and four. Proc. Am. Math. Soc. **90** (1984), 15–20.

[350] S. Mukai. On the moduli space of bundles on K3 surfaces, I. *Vector bundles on algebraic varieties*, 341–413. Tata Inst. Fund. Res. Stud. Math. **11**, Tata Inst. Fund. Res., 1987.

[351] S. Mukai. Curves, K3 surfaces and Fano 3-folds of genus ≤ 10. *Algebraic geometry and commutative algebra. Vol. I*, 357–377. Kinokuniya, 1988.

[352] S. Mukai. Biregular classification of Fano 3-folds and Fano manifolds of coindex 3. Proc. Nat. Acad. Sci. USA **86** (1989), 3000–3002.

[353] S. Mukai. Fano 3-folds. *Complex projective geometry*, 255–263. London Math. Soc. Lecture Note Ser. **179**, Cambridge University Press, 1992.

[354] S. Mukai. Curves and Grassmannians. *Algebraic geometry and related topics*, 19–40. Conf. Proc. Lecture Notes Algebraic Geom. I, Int. Press, 1993.

[355] S. Mukai. Curves and symmetric spaces, I. Am. J. Math. **117** (1995), 1627–1644.

[356] S. Mukai. New developments in the theory of Fano threefolds: Vector bundle method and moduli problems. Sugaku Expositions **15** (2002), 125–150.

[357] S. Mukai. *An introduction to invariants and moduli*. Translated from the 1998 and 2000 Japanese editions by W. M. Oxbury. Cambridge Studies in Advanced Mathematics **81**, Cambridge University Press, 2003.

[358] S. Mukai. Curves and symmetric spaces, II. Ann. Math. (2) **172** (2010), 1539–1558.

[359] S. Mukai and H. Umemura. Minimal rational threefolds. *Algebraic geometry*, 490–518. Lecture Notes in Math. **1016**, Springer, 1983.

[360] D. Mumford. The topology of normal singularities of an algebraic surface and a criterion for simplicity. Publ. Math. Inst. Hautes Études Sci. **9** (1961), 5–22.

[361] D. Mumford. *Abelian varieties*. Tata Institute of Fundamental Research Studies in Mathematics **5**, Oxford University Press, 1970.

[362] D. Mumford. Theta characteristics of an algebraic curve. Ann. Sci. Éc. Norm. Supér. (4) **4** (1971), 181–192.

[363] D. Mumford. Prym varieties I. *Contributions to analysis*, 325–350. Academic Press, 1974.

[364] D. Mumford, J. Fogarty and F. Kirwan. *Geometric invariant theory*. Third edition. Ergebnisse der Mathematik und ihrer Grenzgebiete (2) **34**, Springer-Verlag, 1994.

[365] M. Mustaţă and Y. Nakamura. A boundedness conjecture for minimal log discrepancies on a fixed germ. *Local and global methods in algebraic geometry*, 287–306. Contemp. Math. **712**, Am. Math. Soc., 2018.

[366] M. Nagata. On rational surfaces, II. Mem. Coll. Sci. Univ. Kyoto Ser. A Math. **33** (1960/61), 271–293.

[367] N. Nakayama. Invariance of the plurigenera of algebraic varieties under minimal model conjectures. Topology **25** (1986), 237–251.

[368] N. Nakayama. The lower semi-continuity of the plurigenera of complex varieties. *Algebraic geometry, Sendai, 1985*, 551–590. Adv. Stud. Pure Math. **10**, North-Holland, 1987.

[369] N. Nakayama. *Zariski-decomposition and abundance*. MSJ Memoirs **14**, Mathematical Society of Japan, 2004.

[370] Y. Namikawa. Smoothing Fano 3-folds. J. Algebraic Geom. **6** (1997), 307–324.

[371] Y. Namikawa. Periods of Enriques surfaces. Math. Ann. **270** (1985), 201–222.

[372] J. Nicaise and E. Shinder. The motivic nearby fiber and degeneration of stable rationality. Invent. Math. **217** (2019), 377–413.

[373] T. Oda. *Convex bodies and algebraic geometry*. Translated from the Japanese. Ergebnisse der Mathematik und ihrer Grenzgebiete (3) **15**, Springer-Verlag, 1988.

[374] Y. Odaka and Y. Sano. Alpha invariant and K-stability of **Q**-Fano varieties. Adv. Math. **229** (2012), 2818–2834.

[375] T. Okada. On the birational unboundedness of higher dimensional **Q**-Fano varieties. Math. Ann. **345** (2009), 195–212.

[376] T. Okada. On birational rigidity of singular del Pezzo fibrations of degree 1. J. Lond. Math. Soc. (2) **102** (2020), 1–21.

[377] C. Okonek, M. Schneider and H. Spindler. *Vector bundles on complex projective spaces*. Corrected reprint of the 1988 edition. With an appendix by S. I. Gelfand. Modern Birkhäuser Classics, Birkhäuser/Springer Basel AG, 2011.

[378] I. Pan. Une remarque sur la génération du groupe de Cremona. Bol. Soc. Brasil. Mat. (N.S.) **30** (1999), 95–98.

[379] I. A. Panin. Rationality of bundles of conics with degenerate curve of degree five and even theta-characteristic. Zap. Nauchn. Sem. Leningrad. Otdel. Mat. Inst. Steklov. (LOMI) **103** (1980), 100–105; translation in J. Soviet Math. **24** (1984), 449–452.

[380] S. A. Papadakis and M. Reid. Kustin–Miller unprojection without complexes. J. Algebraic Geom. **13** (2004), 563–577.

[381] J. Park. Birational maps of del Pezzo fibrations. J. Reine Angew. Math. **538** (2001), 213–221.

[382] I. I. Pjateckiĭ-Šapiro and I. R. Šafarevič. A Torelli theorem for algebraic surfaces of type K3. Izv. Akad. Nauk SSSR Ser. Mat. **35** (1971), 530–572; translation in Math. USSR-Izv. **5** (1971), 547–588.

[383] G. Pourcin. Théorème de Douady au-dessus de *S*. Ann. Sc. Norm. Super. Pisa Cl. Sci. (3) **23** (1969), 451–459.

[384] A. Prendergast-Smith. The cone conjecture for abelian varieties. J. Math. Sci. Univ. Tokyo **19** (2012), 243–261.

[385] Y. G. Prokhorov. On the existence of complements of the canonical divisor for Mori conic bundles. Mat. Sb. **188** (1997), 99–120; translation in Sb. Math. **188** (1997), 1665–1685.

[386] Y. G. Prokhorov. On the degree of Fano threefolds with canonical Gorenstein singularities. Mat. Sb. **196** (2005), 81–122; translation in Sb. Math. **196** (2005), 77–114.

[387] Y. G. Prokhorov. The degree of **Q**-Fano threefolds. Mat. Sb. **198** (2007), 153–174; translation in Sb. Math. **198** (2007), 1683–1702.

[388] Y. Prokhorov. **Q**-Fano threefolds of large Fano index, I. Doc. Math. **15** (2010), 843–872.

[389] Y. Prokhorov. A note on degenerations of del Pezzo surfaces. Ann. Inst. Fourier (Grenoble) **65** (2015), 369–388.

[390] Y. Prokhorov. Log-canonical degenerations of del Pezzo surfaces in **Q**-Gorenstein families. Kyoto J. Math. **59** (2019), 1041–1073.

[391] Y. Prokhorov and M. Reid. On **Q**-Fano 3-folds of Fano index 2. *Minimal models and extremal rays (Kyoto, 2011)*, 397–420. Adv. Stud. Pure Math. **70**, Math. Soc. Japan, 2016.

[392] A. V. Pukhlikov. Birational automorphisms of algebraic threefolds with a pencil of Del Pezzo surfaces. Izv. Ross. Akad. Nauk Ser. Mat. **62** (1998), 123–164; translation in Izv. Math. **62** (1998), 115–155.

[393] A. V. Pukhlikov. Birational automorphisms of Fano hypersurfaces. Invent. Math. **134** (1998), 401–426.

[394] A. V. Pukhlikov. Fiber-wise birational correspondences. Mat. Zametki **68** (2000), 120–130; translation in Math. Notes **68** (2000), 103–112.

[395] D. Rees. On a problem of Zariski. Illinois J. Math. **2** (1958), 145–149.

[396] M. Reid. Canonical 3-folds. *Journées de géometrie algébrique d'Angers, Juillet 1979*, 273–310. Sijthoff & Noordhoff, 1980.

[397] M. Reid. Lines on Fano 3-folds according to Shokurov. Institut Mittag-Leffler report, no 11, 1980.

[398] M. Reid. Decomposition of toric morphisms. *Arithmetic and geometry. Vol. II*, 395–418. Progr. Math. **36**, Birkhäuser Boston, 1983.

[399] M. Reid. Minimal models of canonical 3-folds. *Algebraic varieties and analytic varieties*, 131–180. Adv. Stud. Pure Math. **1**, North-Holland, 1983.

[400] M. Reid. Projective morphisms according to Kawamata. Warwick preprint, 1983.

[401] M. Reid. Young person's guide to canonical singularities. *Algebraic geometry, Bowdoin, 1985*, 345–414. Proc. Sympos. Pure Math. **46**, Part 1, Am. Math. Soc., 1987.

[402] M. Reid. Nonnormal del Pezzo surfaces. Publ. Res. Inst. Math. Sci. **30** (1994), 695–727.

[403] M. Reid. Chapters on algebraic surfaces. *Complex algebraic geometry*, 3–159. IAS/Park City Math. Ser. **3**, Am. Math. Soc., 1997.

[404] I. Reider. Vector bundles of rank 2 and linear systems on algebraic surfaces. Ann. Math. (2) **127** (1988), 309–316.

[405] B. Saint-Donat. Sur les équations définissant une courbe algébrique. C. R. Acad. Sci. Paris Sér. A **274** (1972), 324–327, 487–489.

[406] B. Saint-Donat. On Petri's analysis of the linear system of quadrics through a canonical curve. Math. Ann. **206** (1973), 157–175.

[407] B. Saint-Donat. Projective models of K-3 surfaces. Am. J. Math. **96** (1974), 602–639.

[408] T. Sano. Deforming elephants of **Q**-Fano 3-folds. J. Lond. Math. Soc. (2) **95** (2017), 23–51.

[409] V. G. Sarkisov. Birational automorphisms of conic bundles. Izv. Akad. Nauk SSSR Ser. Mat. **44** (1980), 918–945; translation in Math. USSR-Izv. **17** (1981), 177–202.

[410] V. G. Sarkisov. On conic bundle structures. Izv. Akad. Nauk SSSR Ser. Mat. **46** (1982), 371–408; translation in Math. USSR-Izv. **20** (1983), 355–390.

[411] V. G. Sarkisov. Birational maps of standard **Q**-Fano fiberings. Kurchatov Institute of Atomic Energy preprint, 1989.

[412] S. Schreieder. Stably irrational hypersurfaces of small slopes. J. Am. Math. Soc. **32** (2019), 1171–1199.

[413] H. Seifert and W. Threlfall. *A textbook of topology*. Translated from the 1934 German edition by M. A. Goldman. With Topology of 3-dimensional fibered spaces by H. Seifert, translated by W. Heil. Pure and Applied Mathematics **89**, Academic Press, Inc., 1980.

[414] J.-P. Serre. Géométrie algébrique et géométrie analytique. Ann. Inst. Fourier (Grenoble) **6** (1956), 1–42.

[415] G. C. Shephard and J. A. Todd. Finite unitary reflection groups. Canad. J. Math. **6** (1954), 274–304.

[416] N. I. Shepherd-Barron. Miyaoka's theorems on the generic seminegativity of T_X and on the Kodaira dimension of minimal regular threefolds. *Flips and abundance for algebraic threefolds*, 103–114. Astérisque **211**, Soc. Math. France, 1992.

[417] N. I. Shepherd-Barron. Semi-stability and reduction mod p. Topology **37** (1998), 659–664.

[418] N. I. Shepherd-Barron. Stably rational irrational varieties. *The Fano Conference*, 693–700. Univ. Torino, 2004.

[419] K.-H. Shin. 3-dimensional Fano varieties with canonical singularities. Tokyo J. Math. **12** (1989), 375–385.

[420] V. V. Šokurov. Smoothness of the general anticanonical divisor on a Fano 3-fold. Izv. Akad. Nauk SSSR Ser. Mat. **43** (1979), 430–441; translation in Math. USSR-Izv. **14** (1980), 395–405.

[421] V. V. Šokurov. The existence of a straight line on Fano 3-folds. Izv. Akad. Nauk SSSR Ser. Mat. **43** (1979), 922–964; translation in Math. USSR-Izv. **15** (1980), 173–209.

[422] V. V. Shokurov. Prym varieties: Theory and applications. Izv. Akad. Nauk SSSR Ser. Mat. **47** (1983), 785–855; translation in Math. USSR-Izv. **23** (1984), 83–147.

[423] V. V. Shokurov. The nonvanishing theorem. Izv. Akad. Nauk SSSR Ser. Mat. **49** (1985), 635–651; translation in Math. USSR-Izv. **26** (1986), 591–604.

[424] V. V. Shokurov. 3-fold log flips. With an appendix by Y. Kawamata. Izv. Ross. Akad. Nauk Ser. Mat. **56** (1992), 105–203; translation in Izv. Math. **40** (1993), 95–202.

[425] V. V. Shokurov. 3-fold log models. J. Math. Sci. (N.Y.) **81** (1996), 2667–2699.

[426] V. V. Shokurov. Prelimiting flips. Tr. Mat. Inst. Steklova **240** (2003), 82–219; reprinted in Proc. Steklov Inst. Math. **240** (2003), 75–213.

[427] V. V. Shokurov. Letters of a bi-rationalist V: Mld's and termination of log flips. Tr. Mat. Inst. Steklova **246** (2004), 328–351; translation in Proc. Steklov Inst. Math. **246** (2004), 315–336.

[428] K. A. Shramov. On the rationality of non-singular threefolds with a pencil of Del Pezzo surfaces of degree 4. Mat. Sb. **197** (2006), 133–144; translation in Sb. Math. **197** (2006), 127–137.

[429] J. H. Silverman. *The arithmetic of elliptic curves*. Second edition. Graduate Texts in Mathematics **106**, Springer, 2009.

[430] E. H. Spanier. *Algebraic topology*. Corrected reprint of the 1966 original. Springer-Verlag, 1995.

[431] T. A. Springer. *Invariant theory*. Lecture Notes in Mathematics **585**, Springer-Verlag, 1977.

[432] D. A. Stepanov. Smooth three-dimensional canonical thresholds. Mat. Zametki **90** (2011), 285–299; translation in Math. Notes **90** (2011), 265–278.

[433] H. Sterk. Finiteness results for algebraic K3 surfaces. Math. Z. **189** (1985), 507–513.

[434] C. Stibitz and Z. Zhuang. K-stability of birationally superrigid Fano varieties. Compos. Math. **155** (2019), 1845–1852.

[435] K. Suzuki. On Fano indices of $\mathbf{Q}$-Fano 3-folds. Manuscripta Math. **114** (2004), 229–246.

[436] H. P. F. Swinnerton-Dyer. Rational points on del Pezzo surfaces of degree 5. *Algebraic geometry, Oslo 1970*, 287–290. Wolters-Noordhoff, 1972.

[437] E. Szabó. Divisorial log terminal singularities. J. Math. Sci. Univ. Tokyo **1** (1994), 631–639.

[438] S. Takayama. Pluricanonical systems on algebraic varieties of general type. Invent. Math. **165** (2006), 551–587.

[439] K. Takeuchi. Some birational maps of Fano 3-folds. Compos. Math. **71** (1989), 265–283.

[440] B. R. Tennison. *Sheaf theory*. London Mathematical Society Lecture Note Series **20**, Cambridge University Press, 1975.

[441] G. Tian. On Kähler–Einstein metrics on certain Kähler manifolds with $C_1(M) > 0$. Invent. Math. **89** (1987), 225–246.

[442] G. Tian. Kähler–Einstein metrics with positive scalar curvature. Invent. Math. **130** (1997), 1–37.

[443] G. Tian. K-stability and Kähler–Einstein metrics. Comm. Pure Appl. Math. **68** (2015), 1085–1156; corrigendum ibid. **68** (2015), 2082–2083.

[444] B. Totaro. Towards a Schubert calculus for complex reflection groups. Math. Proc. Camb. Philos. Soc. **134** (2003), 83–93.

[445] B. Totaro. Hilbert's 14th problem over finite fields and a conjecture on the cone of curves. Compos. Math. **144** (2008), 1176–1198.

[446] B. Totaro. The cone conjecture for Calabi–Yau pairs in dimension 2. Duke Math. J. **154** (2010), 241–263.

[447] B. Totaro. Algebraic surfaces and hyperbolic geometry. *Current developments in algebraic geometry*, 405–426. Math. Sci. Res. Inst. Publ. **59**, Cambridge University Press, 2012.

[448] B. Totaro. Hypersurfaces that are not stably rational. J. Am. Math. Soc. **29** (2016), 883–891.

[449] J.-C. Tougeron. Idéaux de fonctions différentiables I. Ann. Inst. Fourier (Grenoble) **18** (1968), 177–240.

[450] H. Tsuji. Pluricanonical systems of projective varieties of general type I. Osaka J. Math. **43** (2006), 967–995.

[451] A. N. Tyurin. On intersections of quadrics. Uspekhi Mat. Nauk **30** (1975), 51–99; translation in Russian Math. Surveys **30** (1975), 51–105.

[452] G. N. Tyurina. Resolution of singularities of plane deformations of double rational points. Funktsional. Anal. i Prilozhen. **4** (1970), 77–83; translation in Funct. Anal. Appl. **4** (1970), 68–73.

[453] N. Tziolas. Terminal 3-fold divisorial contractions of a surface to a curve I. Compos. Math. **139** (2003), 239–261.

[454] N. Tziolas. Families of D-minimal models and applications to 3-fold divisorial contractions. Proc. Lond. Math. Soc. (3) **90** (2005), 345–370; corrigendum ibid. **93** (2006), 82–84.

[455] N. Tziolas. Three dimensional divisorial extremal neighborhoods. Math. Ann. **333** (2005), 315–354.

[456] N. Tziolas. $\mathbf{Q}$-Gorenstein deformations of nonnormal surfaces. Am. J. Math. **131** (2009), 171–193.

[457] N. Tziolas. Three-fold divisorial extremal neighborhoods over cE_7 and cE_6 compound DuVal singularities. Int. J. Math. **21** (2010), 1–23.

[458] K. Ueno. *Classification theory of algebraic varieties and compact complex spaces*. Notes written in collaboration with P. Cherenack. Lecture Notes in Mathematics **439**, Springer-Verlag, 1975.

[459] W. Veys. Zeta functions and 'Kontsevich invariants' on singular varieties. Canad. J. Math. **53** (2001), 834–865.

[460] E. Viehweg. Canonical divisors and the additivity of the Kodaira dimension for morphisms of relative dimension one. Compos. Math. **35** (1977), 197–223; correction ibid. **35** (1977), 336.

[461] E. Viehweg. Klassifikationstheorie algebraischer Varietäten der Dimension drei. Compos. Math. **41** (1980), 361–400.

[462] E. Viehweg. Die Additivität der Kodaira Dimension für projektive Faserräume über Varietäten des allgemeinen Typs. J. Reine Angew. Math. **330** (1982), 132–142.

[463] E. Viehweg. Vanishing theorems. J. Reine Angew. Math. **335** (1982), 1–8.

[464] E. Viehweg. Weak positivity and the additivity of the Kodaira dimension for certain fibre spaces. *Algebraic varieties and analytic varieties*, 329–353. Adv. Stud. Pure Math. **1**, North-Holland, 1983.

[465] E. Viehweg and D.-Q. Zhang. Effective Iitaka fibrations. J. Algebraic Geom. **18** (2009), 711–730.

[466] C. Voisin. *Hodge theory and complex algebraic geometry I*. Translated from the French by L. Schneps. Cambridge Studies in Advanced Mathematics **76**, Cambridge University Press, 2002.

[467] C. Voisin. *Hodge theory and complex algebraic geometry II*. Translated from the French by L. Schneps. Cambridge Studies in Advanced Mathematics **77**, Cambridge University Press, 2003.

[468] C. Voisin. Unirational threefolds with no universal codimension 2 cycle. Invent. Math. **201** (2015), 207–237.

[469] C.-L. Wang. On the topology of birational minimal models. J. Differ. Geom. **50** (1998), 129–146.

[470] C.-L. Wang. Cohomology theory in birational geometry. J. Differ. Geom. **60** (2002), 345–354.

[471] J. Werner. Kleine Auflösungen spezieller dreidimensionaler Varietäten. Dissertation, Rheinische Friedrich-Wilhelms-Universität, 1987.

[472] G. K. White. Lattice tetrahedra, Canad. J. Math. **16** (1964), 389–396.

[473] C. Xu. A minimizing valuation is quasi-monomial. Ann. Math. (2) **191** (2020), 1003–1030.

[474] Y. Yamamoto. Divisorial contractions to cDV points with discrepancy greater than 1. Kyoto J. Math. **58** (2018), 529–567.

[475] F. Ye and Z. Zhu. On Fujita's freeness conjecture in dimension 5. Adv. Math. **371** (2020), 107210, 56pp.

[476] A. A. Zagorskii. Three-dimensional conical fibrations. Mat. Zametki **21** (1977), 745–758; translation in Math. Notes **21** (1977), 420–427.

[477] Q. Zhang. Rational connectedness of log **Q**-Fano varieties. J. Reine Angew. Math. **590** (2006), 131–142.

[478] Z. Zhuang. Birational superrigidity is not a locally closed property. Selecta Math. (N.S.) **26** (2020), paper no 11, 20pp.

Notation

476

Index